Graduate Texts in Mathematics

Volume 296

Graduate Texts in Mathematics bridge the gap between passive study and creative understanding, offering graduate-level introductions to advanced topics in mathematics. The volumes are carefully written as teaching aids and highlight characteristic features of the theory. Although these books are frequently used as textbooks in graduate courses, they are also suitable for individual study.

Mihran Papikian

Drinfeld Modules

 Springer

Mihran Papikian
Department of Mathematics
Pennsylvania State University
University Park, PA, USA

This work was supported in part by the Simons Foundation.

ISSN 0072-5285   ISSN 2197-5612   (electronic)
Graduate Texts in Mathematics
ISBN 978-3-031-19709-3   ISBN 978-3-031-19707-9   (eBook)
https://doi.org/10.1007/978-3-031-19707-9

Mathematics Subject Classification: Primary: 11G09, Secondary: 11R58

This Springer imprint is published by the registered company Springer Nature Switzerland AG
The registered company address is: Gewerbestrasse 11, 6330 Cham, Switzerland

This book is dedicated to my children, Anahit and Sevan

Preface

This is an introductory textbook on Drinfeld modules for beginning graduate students. This book has emerged from notes for courses on this topic that the author taught at Stanford University in 2006, Göttingen University in summer of 2012, and the Pennsylvania State University in 2018.

Drinfeld modules are certain function field analogues of elliptic curves.[1] In fact, due to this analogy, in his seminal paper [Dri74], Vladimir Drinfeld introduced these objects under the name of "elliptic modules." Drinfeld's motivation was to construct function field analogues of classical modular curves classifying elliptic curves with some additional data and use these to relate automorphic forms and Galois representations, in line with the program envisioned by Langlands. The theory of Drinfeld modules and their generalizations, called shtukas, has since led to a successful resolution of the Langlands correspondence over function fields, cf. [Dri80, LRS93, Laf02, Laf18], and it continues to play a central role in number theory because of its applications to many other important problems, such as the Birch and Swinnerton-Dyer conjecture, cf. [YZ17].

This book does not discuss any of the applications of Drinfeld modules and their modular varieties to the Langlands program. Our goal is much more modest: to develop the theory of Drinfeld modules in parallel with the theory of elliptic curves assuming as little background material as possible. For most of the book, we only assume that the reader is familiar with graduate-level abstract algebra as presented, for example, in the book by Dummit and Foote [DF04]; we review some of the algebra results important for our purposes in Chap. 1. In addition, in some proofs we use facts from algebraic number theory, which can be found, for example, in the book by Marcus [Mar18].

---

[1] More precisely, Drinfeld modules of rank 1 are similar to the multiplicative group of a field, Drinfeld modules of rank 2 are similar to elliptic curves, and Drinfeld modules of rank ≥ 3, although again similar to elliptic curves or abelian varieties, do not have a proper classical analogue.

A major inspiration for the overall structure of this book and the choice of specific topics was the book [Sil86] "The Arithmetic of Elliptic Curves" by Joseph Silverman. By now, Silverman's book has become a classic, and essentially all Ph.D. students working in algebraic number theory read this book during the years of their study. We have tried to write a book on Drinfeld modules intended for the same audience. Also, while learning this subject, the author benefited immensely from the papers by Ernst-Ulrich Gekeler, and the style of exposition in Gekeler's papers had a big influence on the exposition in this book.

Because Drinfeld modules are essentially linear algebra objects, one can present their theory without any appeal to algebraic geometry, which is impossible to do for elliptic curves.[2] Moreover, to make our treatment as simple as possible, we consider only Drinfeld modules for the polynomial ring $\mathbb{F}_q[T]$, as opposed to "ring of integers" of a general function field.[3] This has several advantages: The first is that a Drinfeld $\mathbb{F}_q[T]$-module can be uniquely defined and studied using a single equation, akin to the Weierstrass equation for an elliptic curve. The second is that $\mathbb{F}_q[T]$ is a Euclidean domain, which simplifies some of the technicalities of the theory. The third is that the $\mathbb{F}_q[T]$ case is the most studied in literature, so there are more results available for this case. And finally, the main ideas of the theory are already present in this special case, so the loss of generality is not substantial.

Most of the foundational results in this book are due to Drinfeld and had appeared in some form in [Dri74] or [Dri77], although the proofs of these results in *loc. cit.* are quite terse and other authors had filled in the details later. Those results that appear in the book without a specific attribution can be assumed to be due to Drinfeld; for all other nontrivial results, observations, and examples, we have included specific references.

There are several books in the literature that include a discussion of Drinfeld modules, but they are all either aimed at a more advanced audience or Drinfeld modules play a secondary role in them; see [Gek86, Gos96, Lau96, GvdPRVG97, Ros02, Tha04, VS06, Leh09]. The closest books that have similar aims as the present textbook are perhaps the following two: first, the book by David Goss [Gos96], which is about Drinfeld modules for general function rings, their generalizations (Anderson modules, shtukas), and positive characteristic L-series; second, the proceedings volume [GvdPRVG97], which contains 12 expository articles by various authors providing an introduction to Drinfeld modules and their modular varieties.[4] We should mention that neither of these books has exercises.

---

[2] To understand the motivation for some of the methods and terminology used in the theory of Drinfeld modules, some knowledge of elliptic curves is helpful, so from time to time we recall facts about elliptic curves as side remarks, but these are not essential and can be safely skipped.

[3] In Appendix A, we briefly discuss the theory of Drinfeld modules for general function rings.

[4] There is also an earlier expository paper [DH87] on Drinfeld modules and their modular varieties based on lectures by Deligne from 1975.

Before briefly discussing the contents of the book, we add a caveat that the theory of Drinfeld modules has become a vast subject and many important topics were left out from this book. We hope that the book will prompt the readers to explore further some of the more advanced topics on their own.

In Chap. 1, we review some of the terminology and facts from abstract algebra that are used in the book.

In Chap. 2, we review the basic theory of non-Archimedean fields and non-Archimedean analysis. This material is essential for the analytic theory of Drinfeld modules discussed in Chaps. 5 and 6.

In Chap. 3, we start with a discussion of the ring of twisted polynomials, which is fundamental in the theory of Drinfeld modules. We then define Drinfeld modules and study their homomorphisms and torsion submodules. This chapter discusses those properties of Drinfeld modules which are valid over all base fields.

In Chap. 4, we study Drinfeld modules over finite fields. The theory of these modules is closely related to the theory of central division algebras. In this chapter, we develop an analogue of the Honda-Tate theory for Drinfeld modules and prove an analogue of the celebrated Hasse-Weil bound. The Frobenius endomorphism plays a key role in this chapter.

In Chap. 5, we develop the theory of analytic uniformization of Drinfeld modules over the analogue of complex numbers. This theory gives a bijection between the sets of Drinfeld modules and certain lattices. In this chapter, we also discuss the work of Carlitz [Car35] on special values of zeta-functions, which implicitly contains the theory of Drinfeld modules of rank 1, but preceded Drinfeld's paper [Dri74] by about 40 years.

In Chap. 6, we study Drinfeld modules over local fields. Here an important concept is the concept of reduction of a Drinfeld module. We examine the relationship between the reduction properties of a given Drinfeld module, the action of the Galois group on its torsion points, and the existence of a uniformization similar to the analytic uniformization. We then prove that the rational torsion submodule of a Drinfeld module over a local field is finite.

In Chap. 7, we study Drinfeld modules over global fields. We show that the torsion submodules of Drinfeld modules give rise to interesting Galois representations and then examine the analogues of the famous Mordell-Weil theorem and the class number formula in the context of Drinfeld modules.

In Appendix A, we define Drinfeld modules for general function rings and briefly discuss how the main results in the book transfer to this more general setting.

Each section of the book is followed by a series of exercises of various difficulty, which serve both to cement the understanding of the content of the section and to introduce additional results. Some of the exercises are routine and require no more

than a straightforward calculation, the others are more challenging,[5] but we believe
that all the exercises are doable by an average student who understands the material
in the corresponding section. Appendix B contains references, hints, and solutions
for selected exercises.

University Park, PA, USA   Mihran Papikian

---

[5] Some of these exercises are based on results that have appeared in the literature.

Acknowledgements

This book owes a big intellectual debt to Vladimir Drinfeld and Leonard Carlitz. In my personal journey to Function Field Arithmetic, many other mathematicians have played important roles – I am especially grateful to Ara Alexanian, Brian Conrad, Ernst-Ulrich Gekeler, Barry Mazur, and Douglas Ulmer. I am also grateful to my colleagues at Penn State, especially Dale Brownawell, Winnie Li, and Yuri Zarhin, for advice and encouragement over the years.

Many thanks go to Florian Breuer, Alina Cojocaru, Sjoerd de Vries, Kevin Ho, Valentijn Karemaker, Jeff Katen, as well as the anonymous referees, for helpful comments on the preliminary versions of this book. I also thank Ute McCrory at Springer for her assistance during the review process.

The final preparation of this book was done while I was visiting the Max Planck Institute for Mathematics in Bonn. I thank the institute for excellent working conditions and financial support.

Finally, I dedicate this book to my wonderful children, Anahit and Sevan, who made writing the book so much harder.

Contents

Notation and Conventions

As a generally accepted standard notation, we use the symbols

$$\mathbb{Z}, \quad \mathbb{Q}, \quad \mathbb{R}, \quad \mathbb{C}, \quad \mathbb{Z}/n\mathbb{Z}$$

to denote the integers, rational numbers, real numbers, complex numbers, and the ring of integers modulo $n \geq 1$, respectively. For a prime number p, we interchangeably denote $\mathbb{Z}/p\mathbb{Z}$ by $\mathbb{F}_p$, especially if we want to stress that $\mathbb{Z}/p\mathbb{Z}$ is a field. We denote by $\mathbb{R}_{>0}$ the (strictly) positive real numbers, and by $\mathbb{R}_{\geq 0}$ the non-negative real numbers. Similarly, we denote

$$\mathbb{Z}_{>0} = \{1, 2, \dots\}, \quad \text{and} \quad \mathbb{Z}_{\geq 0} = \{0, 1, 2, \dots\}.$$

The symbols

$$\sideset{}{'}\sum_{s \in S} f(s) \quad \text{and} \quad \sideset{}{'}\prod_{s \in S} f(s)$$

denote the sum and the product of the values of a function f over the nonzero elements of a given set S. If S is a finite set, then $\#S$ denotes the number of elements in S. Given two sets S and S', the notations $S \subset S'$ and $S \subseteq S'$ have the same meaning of S being a subset of S'; to indicate a strict inclusion we write $S \subsetneq S'$.

A ring will always be assumed to have a *multiplicative identity* (aka *unity*). In some algebra textbooks, e.g., [DF04], this is not part of the definition. Thus, a *ring* is a set R with two operations $+$ and $\times$ such that R is an abelian group with respect to $+$, while multiplication is associative and there is an element $1 \in R$ such that $1 \cdot a = a \cdot 1 = a$ for all $a \in R$. Moreover, the two operations are related by the distributivity law $a(b + c) = ab + ac$ and $(a + b)c = ac + bc$ for all $a, b, c \in R$. It is trivial to prove that multiplicative identity of a ring is unique. The zero element (the identity for $+$) will be denoted by 0, so $0 + a = a + 0 = a$ for all $a \in R$, and the additive inverse of a will be denoted by $-a$. If the multiplication is commutative, the ring is called *commutative*. Every module M over a ring R is assumed *unital*, that is, $1 \in R$ acts as the identity operator of M.

A *subring* of a ring R is a subset $R' \subset R$ which is a ring under the same operations $+$ and $\times$ as R and shares the same multiplicative identity. We note that our definition of a subring is more restrictive than the one in [DF04]. For example, with our definition $\mathbb{Z}$ has no subrings other than itself since 1 additively generates $\mathbb{Z}$. On the other hand, the subset $n\mathbb{Z} \subset \mathbb{Z}$ is closed under addition and multiplication of $\mathbb{Z}$, hence $n\mathbb{Z}$ is a subring of $\mathbb{Z}$ according to the definition in [DF04, p. 228].[6]

We will assume that a homomorphism of rings $\varphi\colon R \to S$ necessarily maps the multiplicative identity of R to the multiplicative identity of S. Thus, given two rings R and S, we define a *homomorphism* $\varphi\colon R \to S$ to be a map such that $\varphi(a + b) = \varphi(a) + \varphi(b)$ and $\varphi(ab) = \varphi(a)\varphi(b)$ for all $a, b \in R$, and $\varphi(1) = 1$. A homomorphism $\varphi\colon R \to R$ is called an *endomorphism* of R. Again, compared to the definition in [DF04, p. 239], the condition $\varphi(1) = 1$ is extra.[7] It guarantees that the image of φ is a subring according to our definition, and prevents the trivial map, $\varphi(a) = 0$ for all $a \in R$, being a ring homomorphism, unless $S = \{0\}$ is the zero ring.

Example N.1 The map $\varphi\colon \mathbb{Z} \to R$ defined by

$$\varphi(n) = \begin{cases} 1 + 1 + \cdots + 1 \ (n \text{ times}) & \text{if } n > 0 \\ 0 & \text{if } n = 0 \\ -1 - 1 + \cdots - 1 \ (-n \text{ times}) & \text{if } n < 0 \end{cases}$$

is a ring homomorphism. The kernel of φ is $n\mathbb{Z}$ for some $n \in \mathbb{Z}_{\geq 0}$; this number n is called the *characteristic* of R. It is easy to check that if R has no *zero divisors* (i.e., nonzero $a, b \in R$ such that $a \cdot b = 0$) then its characteristic is either 0 or a positive prime number. In particular, the characteristic of a field K is either 0, in which case $\mathbb{Q} \subset K$, or it is $p > 0$, in which case $\mathbb{F}_p \subset K$.

Example N.2 Let R be a commutative ring of positive characteristic p, where p is prime. Consider the map

$$\mathrm{Fr}_p \colon R \to R$$

$$a \mapsto a^p.$$

Obviously $\mathrm{Fr}_p(1) = 1^p = 1$ and

$$\mathrm{Fr}_p(ab) = (ab)^p = a^p b^p = \mathrm{Fr}_p(a)\,\mathrm{Fr}_p(b).$$

---

[6] Also note that $R = \{0, 3\} \subset \mathbb{Z}/6\mathbb{Z}$ is a ring with multiplicative identity with respect to the addition and multiplication on $\mathbb{Z}/6\mathbb{Z}$, but R is not a subring of $\mathbb{Z}/6\mathbb{Z}$.

[7] Note that $\varphi\colon \mathbb{Z}/6\mathbb{Z} \to \mathbb{Z}/6\mathbb{Z}$, $\varphi(a) = 3a$, satisfies $\varphi(a+b) = \varphi(a)+\varphi(b)$ and $\varphi(ab) = \varphi(a)\varphi(b)$ for all $a, b \in \mathbb{Z}/6\mathbb{Z}$, but φ is not an endomorphism.

On the other hand, the binomial coefficient $\binom{p}{i}$ is divisible by p for $1 \leq i \leq p-1$, so for any $a \in R$ we have $\binom{p}{i}a = 0$. This implies that

$$\mathrm{Fr}_p(a+b) = (a+b)^p = \sum_{i=0}^{p} \binom{p}{i} a^i b^{p-i} = a^p + b^p = \mathrm{Fr}_p(a) + \mathrm{Fr}_p(b). \qquad \text{(N.1)}$$

Therefore, Fr_p is an endomorphism of R, called the *Frobenius endomorphism* of R. The Frobenius endomorphism plays a prominent role in this book.

More generally, if R is not necessarily commutative but has prime characteristic p, then for any $a, b \in R$ which commute with each other, i.e., $ab = ba$, the calculation (N.1) gives $(a + b)^p = a^p + b^p$. Since a^{p^m} and b^{p^m} also commute with each other for all $m \geq 0$, we get by induction

$$(a + b)^{p^n} = a^{p^n} + b^{p^n} \quad \text{for all } n \geq 0.$$

A *unit* $u \in R$ of a ring is an element which has a left and right multiplicative inverse, i.e., there is $v \in R$ such that $uv = vu = 1$; the set of units of R, denoted $R^\times$, is a group with respect to the multiplication on R. A ring in which every nonzero element is a unit is called a *division ring* (or *skew-field*). A commutative division ring is a *field*. Note that if $\varphi \colon R \to S$ is a homomorphism of rings, then φ restricts to a homomorphism of groups $\varphi \colon R^\times \to S^\times$.[8] An *integral domain* is a nonzero commutative ring without zero divisors.

Given a commutative ring R and $a, b \in R$ with $b \neq 0$, we say that b *divides* a, written $b \mid a$, if $a = bc$ for some $c \in R$. A *greatest common divisor* of $a_1, a_2, \ldots, a_s \in R$ is an element $d \in R$ such that d divides all a_i's, and if $d' \in R$ is some other element that divides all a_i's then $d' \mid d$. The elements $a_1, a_2, \ldots, a_s$ are said to be *coprime* (or *relatively prime*) if 1 is a greatest common divisor for them. An *irreducible element* (resp. *prime element*) of an integral domain is a nonzero element a which is not a unit and cannot be decomposed $a = bc$ as a product of two non-unit elements b and c (resp. if $a \mid bc$ then $a \mid b$ or $a \mid c$).

Remarks N.3

(1) Let R be the subring of $\mathbb{Q}[x]$ consisting of those polynomials $a_0 + a_2 x^2 + \cdots + a_n x^n$ whose coefficient of x is 0. We leave it as an exercise for the reader to show the following:

- x^5 and x^6 have no greatest common divisor as elements of R.
- x^2 is irreducible in R but not prime.

---

[8] This would not always be true if we did not insist on $\varphi(1) = 1$.

(2) Let R be an integral domain. If d and d' are greatest common divisors of nonzero elements $a_1, a_2, \ldots, a_s \in R$, then $d = bd'$ and $d' = cd$ for some $b, c \in R$. This implies that $d(bc - 1) = 0$, so $b, c \in R^\times$. Thus, two greatest common divisors of the same set of elements in R differ by a unit.

(3) If $a_1, \ldots, a_s$ are nonzero elements in the commutative ring R such that the ideal generated by them in R,

$$(a_1, a_2, \ldots, a_s) = Ra_1 + \cdots Ra_s,$$

is a principal ideal (d), then d is a greatest common divisor of $a_1, \ldots, a_s$. (Indeed, d divides a_i's, and if d' divides all a_i's, then $(d) = (a_1, \ldots, a_s) \subseteq (d')$, so $d' \mid d$.) In particular, if R is a *principal ideal domain* (PID), i.e., every ideal of R is generated by one element, then a greatest common divisor of $a_1, a_2, \ldots, a_s \in R$ always exists. But this condition is not a necessary condition, e.g., in $\mathbb{Z}[x]$ the elements 2 and x are coprime but $(2, x) \neq (1)$.

(4) Note that $b \mid a$ if and only if we have the inclusion of ideals $(a) \subseteq (b)$.

(5) In an integral domain, every prime element is irreducible,[9] but, as we observed in (1), the converse is not true.[10] The converse is true for *unique factorization domains* (UFDs), i.e., integral domains in which every nonzero non-unit decomposes into a product of irreducible elements, uniquely up to order and units. A PID is a UFD (see [DF04, p. 287]), but the converse is not always true, e.g., $\mathbb{Z}[x]$ is a UFD but not a PID.

(6) When working with polynomials in one variable over a field, one often chooses the monic generator of $(f_1, \ldots, f_s)$ as their greatest common divisor, similar to choosing the greatest common divisor of integers being positive. We will denote the monic generator of $(f_1, \ldots, f_s)$ by $\gcd(f_1, \ldots, f_s)$ and call it *the* greatest common divisor of $f_1, \ldots, f_s$.

For a ring R and a subset $S \subseteq R$, let $\mathrm{Cent}_R(S)$ be the *centralizer of S*, which is the subring of R consisting of those elements that commute with all elements of S. The *center* of R is

$$Z(R) = \mathrm{Cent}_R(R).$$

Let R be a commutative ring. An *R-algebra* (or an *algebra over R*) is a ring A together with a ring homomorphism $f : R \to A$ such that the subring $f(R)$ of A is contained in $Z(A)$. If A is an R-algebra then it has a natural R-module structure defined by $r \cdot a = f(r)a$, where $f(r)a$ is just the multiplication in the ring A. If A and B are R-algebras, an *R-algebra homomorphism* is a ring homomorphism $\varphi : A \to B$ such that $\varphi(r \cdot a) = r \cdot \varphi(a)$ for all $r \in R$ and $a \in A$.

---

[9] Suppose p is prime and $p = ab$. By definition, $p \mid a$ or $p \mid b$. Assume without loss of generality that $p \mid a$. Then $a = pc$, which implies that $p(1 - cb) = 0$, so $cb = 1$, i.e., $b \in R^\times$.

[10] A more famous example is the ring $\mathbb{Z}[\sqrt{-5}]$; in this ring the element 3 is irreducible but not prime.

Example N.4 Let R be a ring and let n be a positive integer. Let $\mathrm{Mat}_n(R)$ be the set of all $n \times n$ *matrices with entries from* R. The element (a_{ij}) of $\mathrm{Mat}_n(R)$ is an $n \times n$ square array of elements of R whose entry in row i and column j is $a_{ij} \in R$. The set $\mathrm{Mat}_n(R)$ becomes a ring if one defines addition by $(a_{ij}) + (b_{ij}) = (a_{ij} + b_{ij})$, i.e., (i, j)-th entry of $(a_{ij}) + (b_{ij})$ is $a_{ij} + b_{ij}$, and multiplication by $(a_{ij})(b_{ij}) = (c_{ij})$, where $c_{ij} = \sum_{k=1}^{n} a_{ik} b_{kj}$ (the order of elements is important if R is non-commutative). A *diagonal matrix* is a matrix of the form

$$\mathrm{diag}(a_1, a_2, \ldots, a_n) = \begin{pmatrix} a_1 & 0 & \cdots & 0 \\ 0 & a_2 & \cdots & 0 \\ \vdots & & \ddots & \vdots \\ 0 & 0 & \cdots & a_n \end{pmatrix}.$$

Note that the multiplicative identity of $\mathrm{Mat}_n(R)$ is $I_n := \mathrm{diag}(1, 1, \ldots, 1)$, and

$$Z(\mathrm{Mat}_n(R)) = Z(R) \cdot I_n.$$

If R is commutative, then $\mathrm{Mat}_n(R)$ is an R-algebra with $f : R \to \mathrm{Mat}_n(R)$ defined by $f(r) = \mathrm{diag}(r, r, \ldots, r)$. The group of units $\mathrm{GL}_n(R) := \mathrm{Mat}_n(R)^{\times}$ is called the *general linear group over* R.

Chapter 1
Algebraic Preliminaries

In this chapter, we review some facts from abstract algebra. Our main reference will be [DF04]. Another source that we recommend is Keith Conrad's online expository notes on algebra and number theory [Con].[1]

Since we assume that the reader is familiar with basic abstract algebra, we put the emphasis on the concepts that are particularly important in this book, such as the ring of polynomials, modules over this ring, algebraic and inseparable field extensions, finite fields, and central simple algebras.

1.1 Polynomials

The reader has certainly encountered the ring of polynomials before, but due to the central role it plays in this book (as well as in algebra in general), we recall its basic notions and properties in this section.

Given a commutative ring R, let $R[\![x]\!]$ be the *ring of formal power series* in the indeterminate x with coefficients in R. The elements of $R[\![x]\!]$ are the formal sums

$$f(x) = a_0 + a_1 x + a_2 x^2 + \cdots ,$$

where $a_i \in R$ are the *coefficients* of $f(x)$. It is assumed that $x^0 = 1$ and $0 x^i = 0$ for any $i \geq 0$. The addition and multiplication on $R[\![x]\!]$ are defined by the formulas

$$\sum_{n=0}^{\infty} a_n x^n + \sum_{n=0}^{\infty} b_n x^n = \sum_{n=0}^{\infty} (a_n + b_n) x^n, \qquad (1.1.1)$$

---

[1] Unfortunately, for now, Conrad's notes exist only electronically, with a separate file for each topic, and the notes are revised occasionally, which makes making precise references to them difficult.

© The Author(s), under exclusive license to Springer Nature Switzerland AG 2023
M. Papikian, *Drinfeld Modules*, Graduate Texts in Mathematics 296,
https://doi.org/10.1007/978-3-031-19707-9_1

$$\left(\sum_{n=0}^{\infty} a_n x^n\right)\left(\sum_{n=0}^{\infty} b_n x^n\right) = \sum_{n=0}^{\infty}\left(\sum_{m=0}^{n} a_m b_{n-m}\right) x^n. \tag{1.1.2}$$

The zero power series is the power series all of whose coefficients are 0. The multiplication on $R[[x]]$ is obviously commutative, so $R[[x]]$ is a commutative ring with unity $1 = 1 + 0x + 0x^2 + \cdots$. Moreover, it is easy to check that $R[[x]]$ is an integral domain if and only if R is an integral domain.

A power series with only finitely many nonzero coefficients is a called a *polynomial*. A polynomial is usually written as a finite formal sum

$$a_n x^n + a_{n-1} x^{n-1} + \cdots + a_1 x + a_0.$$

The set of polynomials $R[x]$ is a subring of $R[[x]]$. The *degree* of a nonzero polynomial $f(x) = \sum_{i \geq 0} a_i x^i$, denoted $\deg(f)$, is the largest index n such that $a_n \neq 0$; the *leading term* of $f(x)$ is $a_n x^n$; the *leading coefficient* of $f(x)$ is a_n; the *constant term* of $f(x)$ is a_0. The polynomial $f(x)$ of degree n is *monic* if $a_n = 1$. Sometimes, to simplify the notation, we write f for $f(x)$. The polynomial f is *constant* if $f = 0$ or $\deg(f) = 0$; the polynomial f is *linear* if $\deg(f) = 1$. One identifies R with the subring of $R[x]$ consisting of constant polynomials. We formally put

$$\deg(0) = -\infty,$$

so that the degree of 0 is less than the degree of any other polynomial. It is clear that if R is an integral domain, then $R[x]$ is also an integral domain and

$$\deg(fg) = \deg(f) + \deg(g) \qquad \text{for all} \quad f, g \in R[x].$$

Lemma 1.1.1

(1) A power series $f(x) = \sum_{n \geq 0} a_n x^n$ is a unit in $R[[x]]$ if and only if $a_0 \in R^\times$.
(2) Assume R is an integral domain. A polynomial $f(x) = \sum_{n \geq 0} a_n x^n$ is a unit in $R[x]$ if and only if $f(x) = a_0 \in R^\times$.

In particular, a polynomial might be a unit in $R[[x]]$, but its inverse usually will not be a polynomial. As an easy example, the inverse of $1 - x$ is $1 + x + x^2 + \cdots$. Note that (2) might be false if R is not an integral domain, e.g., $(1 + 2x)^2 = 1$ in $\mathbb{Z}/4\mathbb{Z}[x]$.

Proof of Lemma 1.1.1

(1) We want to find $g(x) \in R[[x]]$ such that $f(x)g(x) = 1$. Put $g(x) = \sum_{n \geq 0} b_n x^n$, where b_i's are to be determined. We need

$$a_0 b_0 = 1,$$

$$a_0 b_n + \sum_{i=1}^{n} a_i b_{n-i} = 0, \quad \text{for all } n > 0.$$

We can recursively solve the previous equations for $b_0, b_1, b_2 \ldots$ if and only if a_0 is invertible in R. Moreover, the solution is unique:

$$b_0 = a_0^{-1}, \tag{1.1.3}$$

$$b_n = -a_0^{-1}(b_0 a_n + b_1 a_{n-1} + \cdots + b_{n-1} a_1), \quad \text{for all } n > 0. \tag{1.1.4}$$

(2) This is clear since $\deg(fg) = \deg(f) + \deg(g)$ implies that if $\deg(f) > 0$, then $\deg(fg) > 0$ for any nonzero $g(x) \in R[x]$, whereas $\deg(1) = 0$. □

Evaluating a power series at $c \in R$ generally makes no sense[2] since we will be adding infinitely many elements of R. On the other hand, evaluating a polynomial at $c \in R$ is always meaningful, and the map

$$R[x] \longrightarrow R$$

$$f(x) = \sum_{i=0}^{n} a_i x^i \longmapsto f(c) = \sum_{i=0}^{n} a_i c^i$$

is a ring homomorphism, called the *evaluation at c*. We say that c is a *root* (or *zero*) of $f(x)$ if $f(c) = 0$. The kernel of the evaluation at c consists of polynomials for which c is a root.

The following theorem is a prototype of several theorems that we will prove later for some other rings.

Theorem 1.1.2 (Division Algorithm for Polynomials) *Given two polynomials* $f(x), g(x) \in R[x]$ *such that the leading coefficient of* $g(x)$ *is a unit in R, there exist unique* $h(x), r(x) \in R[x]$ *such that*

$$f(x) = h(x)g(x) + r(x) \quad \text{and} \quad \deg(r) < \deg(g).$$

Proof First, we prove the existence using induction on the degree of f. Let n be the degree of f, and let m be the degree of g. Let a be the leading coefficient of f, and let b be the leading coefficient of g. If $n < m$, then we take $h = 0$ and $r = f$. Now suppose $n \geq m$. Since b is a unit by assumption, we can consider $f_1 := f - \frac{a}{b} x^{n-m} g \in R[x]$. Then $\deg(f_1) < n$, so by the induction assumption, there exist h_1 and r_1 with $\deg(r_1) < m$ such that $f_1 = h_1 g + r_1$. Now we can take $h = \frac{a}{b} x^{n-m} + h_1$ and $r = r_1$.

---

[2] A notable exception is the case when R is complete with respect to a metric and there is an analytic notion of convergence with respect to that metric, e.g., $\sum_{n \geq 0} x^n / n!$ can be evaluated at an arbitrary $c \in \mathbb{C}$. We will return to this in Chap. 2.

To prove the uniqueness, suppose $f = hg + r = h'g + r'$. In this case, $(h - h')g = r' - r$ and $\deg(r' - r) < m$. Since b is a unit, we have $\deg((h - h')g) \geq m$ unless $h = h'$. Hence, $h = h'$, and this implies that $r = r'$. $\qquad\qquad\qquad\qquad\square$

For the rest of this section, we assume that the coefficient ring is a field K. In that case, Theorem 1.1.2 applies to every nonzero polynomial in $K[x]$.

Corollary 1.1.3 *An ideal $0 \neq I \lhd K[x]$ is generated by a polynomial $g \in I$ having the minimal degree among the nonzero polynomials in I. In particular, $K[x]$ is a PID.*

Proof Suppose $0 \neq g \in I$ has minimal degree among the polynomials in I. If $f \in I$, then applying Theorem 1.1.2 one finds $h \in K[x]$ such that $\deg(f - gh) < \deg(g)$. But $f - gh \in I$, so $f = gh$. $\qquad\qquad\qquad\qquad\qquad\qquad\qquad\qquad\square$

Example 1.1.4 Theorem 1.1.2 gives an efficient algorithm for computing the greatest common divisor of two polynomials $f, g \in K[x]$. Instead of recalling the details, we give one explicit example from which the idea of the algorithm should be clear.

Let $f = x^5 + 4x^4 - 5x^3 + 3x - 3$ and $g = x^3 - 4x^2 + 7x - 4$ be polynomials in $\mathbb{Q}[x]$. Then

$$f = h_1 g + r_1, \qquad h_1 = x^2 + 8x + 20, \qquad r_1 = 28x^2 - 105x + 77;$$

$$g = h_2 r_1 + r_2, \qquad h_2 = \frac{1}{28}x - \frac{1}{112}, \qquad r_2 = \frac{53}{16}(x - 1);$$

$$r_1 = h_3 r_2 + r_3, \qquad h_3 = \frac{448}{53}x - \frac{1232}{53}, \qquad r_3 = 0.$$

Thus r_2 divides r_1; thus r_2 divides g (from the second equation); thus r_2 divides f (from the first equation). Conversely, if r divides f and g, then r divides r_1 (from the first equation), and thus r divides r_2 (from the second equation). This implies that $(f, g) = (r_2) = (x - 1)$. Moreover,

$$x - 1 = \frac{16}{53}(g - h_2 r_1) = \frac{16}{53}(g - h_2(f - h_1 g)) = \frac{16}{53}(1 + h_1)g - \frac{16}{53}h_2 f$$

is an explicit expression of $\gcd(f, g)$ as a $\mathbb{Q}[x]$-linear combination of f and g.

Definition 1.1.5 Assume R is an integral domain. A polynomial $f(x) \in R[x]$ is *irreducible* if it is an irreducible element of the ring $R[x]$. If $f(x)$ is not irreducible, then it is called *reducible*. Note that when R is a field, $f(x)$ is irreducible if and only if it is non-constant and cannot be decomposed as a product $f = g \cdot h$ with $\deg(g) < \deg(f)$ and $\deg(h) < \deg(f)$.

The fact that $K[x]$ is a PID implies that $K[x]$ is a UFD. Hence, every nonzero polynomial $f \in K[x]$ decomposes into a product $f = cP_1 \cdot P_2 \cdots P_m$, where $c \in K^\times$ and $P_1, \ldots, P_m$ are irreducible monic polynomials (not necessarily distinct), and this decomposition is unique, up to permutation of the factors. The fact that $K[x]$ is a PID also implies that nonzero prime ideals of $K[x]$ are maximal (see [DF04, p. 280]), and the maximal ideals in $K[x]$ are the ideals (P) generated by irreducible polynomials P (see [DF04, p. 284]).

Generally, it is hard to determine whether a given polynomial $f(x) \in K[x]$ is irreducible or not. In certain situations, the so-called Gauss' Lemma simplifies this problem considerably.

Theorem 1.1.6 (Gauss' Lemma) *Let R be a UFD with field of fractions K, and let $f(x) \in R[x]$ be a non-constant polynomial. Suppose $f(x) = g(x)h(x)$ for some non-constant polynomials $g(x)$ and $h(x)$ in $K[x]$. Then there are nonzero elements $a, b \in K$ such that $a \cdot g(x)$ and $b \cdot h(x)$ both lie in $R[x]$, and*

$$f(x) = (a \cdot g(x))(b \cdot h(x))$$

is a factorization in $R[x]$.

Proof See [DF04, p. 303]. □

Corollary 1.1.7 *Let R be a UFD with field of fractions K and let $f(x) \in R[x]$ be a non-constant polynomial. Suppose $f(x) = g(x)h(x)$ for some non-constant polynomials $g(x)$ and $h(x)$ in $K[x]$. If $f(x)$ is monic, then one can assume that $g(x)$ and $h(x)$ are also monic and have coefficients in R.*

Example 1.1.8 Consider the polynomial $f(x) = x^4 - 10x^2 + 1 \in \mathbb{Q}[x]$. We claim that it is irreducible. Suppose to the contrary that $f(x) = g(x)h(x)$ with $\deg(g) = 1$ or 2. By Corollary 1.1.7, we may assume that $g(x), h(x)$ have integer coefficients and are monic. If $g(x) = x - a$, then a is a root of $f(x)$ and divides the constant term of $f(x)$. On the other hand, $f(\pm 1) \neq 0$, so $g(x)$ cannot be a linear polynomial. Now assume $\deg(g) = 2$, and write $g(x) = x^2 + ax + b$ and $h(x) = x^2 + cx + d$. Then

$$bd = 1$$

$$bc + ad = 0$$

$$b + d + ac = -10.$$

It is easy to check that this system of equations has no solutions in integers. We reach a contradiction, so $f(x)$ is irreducible.

Let $g(x) \in K[x]$ be a fixed nonzero polynomial. For any polynomial $f(x) \in K[x]$, Theorem 1.1.2 guarantees the existence of a unique polynomial $r(x) \in K[x]$, called the *residue of $f(x)$ modulo $g(x)$*, such that $\deg(r) < \deg(g)$ and $f = gh + r$.

The map $\varphi_g : K[x] \to K[x]/(g)$, $f \mapsto r$, is a ring homomorphism. Note that if $g(x) = x - c$, then $K[x]/(g) \cong K$, and φ_g is the evaluation-at-c map.

Now suppose $P(x) \in K[x]$ is irreducible. As we mentioned, $(P) \lhd K[x]$ is a maximal ideal, so $L = K[x]/(P)$ is a field. It is clear that φ_P restricted to K is an injection $K \hookrightarrow L$. Hence, L is a *field extension* of K, that is, L is a field containing an isomorphic copy of K. We can consider any polynomial $f(x) \in K[x]$ as a polynomial in $L[x]$, by identifying K with its image in L under φ_P. If we denote $\alpha = \varphi_P(x)$, then φ_P can be described as the evaluation map $f(x) \mapsto f(\alpha)$. Since the residue of $P(x)$ modulo $P(x)$ is 0, we see that $P(\alpha) = 0$. Thus, there exists a field L containing an isomorphic copy of K in which $P(x)$ has a root. Iterating this construction, one shows that for any non-constant $f(x) \in K[x]$, there is a field extension L of K such that $f(x)$ factors completely into linear factors (or *splits completely*) in $L[x]$. An extension L/K is called a *splitting field* of $f(x)$ if $f(x)$ splits completely in $L[x]$ and $f(x)$ does not split completely over any proper subfield of L containing K. An important fact in field theory is that any two splitting fields of $f(x)$ are isomorphic, so a splitting field is unique up to isomorphism; see [DF04, p. 542].

Let $f(x)$ be a non-constant polynomial with leading coefficient a_n. Over the splitting field L for $f(x)$, we have a factorization

$$f(x) = a_n(x - \alpha_1)^{m_1}(x - \alpha_2)^{m_2} \cdots (x - \alpha_k)^{m_k},$$

where $\alpha_1, \alpha_2, \ldots, \alpha_k$ are distinct elements of the splitting field and $m_i \geq 1$ for all i. Since a linear polynomial is irreducible, the fact that $L[x]$ is a UFD implies that the above factorization is unique. Hence, $f(x)$ has exactly n roots in L, counted with multiplicities.[3]

For any given non-constant polynomials $f(x), g(x) \in K[x]$, we can find a field extension L/K over which $f(x)$ and $g(x)$ split completely (for example, we can consider $g(x)$ over the splitting field K' of $f(x)$ and take L to be the splitting field of $g(x)$ as a polynomial in $K'[x]$). Over such a field, $f(x)$ and $g(x)$ are relatively prime if and only if they do not have common roots. On the other hand, from the Division Algorithm, it is easy to see that $\gcd(f, g)$ does not change under field extensions of K, i.e., $\gcd(f, g)$ is the same whether we consider f and g as polynomials in $K[x]$ or $L[x]$ for any extension $K \subset L$. Thus, $f(x)$ and $g(x)$ are relatively prime if and only if they do not have common roots over any extension of K.

If $f(x)$ is divisible by $(x - \alpha)^m$ over its splitting field but not by $(x - \alpha)^{m+1}$, then α is said to be a root of *multiplicity m*. A *multiple root* (or a *repeated root*) is a root with multiplicity $m \geq 2$; a *simple root* is a root with multiplicity $m = 1$. A polynomial $f(x) \in K[x]$ is called *separable* if it has no multiple roots in its splitting field; otherwise, it is called *inseparable*.

---

[3] Consider $f(x) = x^2 - 1$ in $\mathbb{Z}/8\mathbb{Z}[x]$. Then 1, 3, 5, 7 are roots of $f(x)$, so $f(x)$ has 4 roots even though it has degree 2. This, of course, does not contradict the claim, as $\mathbb{Z}/8\mathbb{Z}$ is not a field. Also note that in $\mathbb{Z}/6\mathbb{Z}[x]$, we have the factorization $x = (3x + 4)(4x + 3)$. Since neither of these factors is a unit, x is reducible in $\mathbb{Z}/6\mathbb{Z}[x]$.

If the roots of the polynomial $f(x) = a_n x^n + a_{n-1} x^{n-1} + \cdots + a_1 x + a_0$ are $\alpha_1, \ldots, \alpha_n$, then the *discriminant of* $f(x)$ is

$$\text{disc}(f) = a_n^{2n-2} \prod_{i<j} (\alpha_i - \alpha_j)^2 \tag{1.1.5}$$

$$= (-1)^{n(n-1)/2} \cdot a_n^{2n-2} \cdot \prod_{i \neq j} (\alpha_i - \alpha_j).$$

Note that $\text{disc}(f) = 0$ if and only if f is inseparable. Also note that $\text{disc}(f)$ is symmetric in the roots of $f(x)$, so the fundamental theorem on symmetric functions [DF04, p. 608] implies that there is a polynomial $G(x_0, \ldots, x_n) \in K[x_0, \ldots, x_n]$ such that $\text{disc}(f) = G(a_0, \ldots, a_n)$, i.e., $\text{disc}(f)$ can be expressed as a polynomial in the coefficients of $f(x)$. In particular, $\text{disc}(f) \in K$. For example (see [DF04, p. 612]), we have

if $f(x) = ax^2 + bx + c$, then $\text{disc}(f) = b^2 - 4ac$;

if $f(x) = ax^3 + bx^2 + cx + d$, then

$$\text{disc}(f) = b^2 c^2 - 4ac^3 - 4b^3 d - 27a^2 d^2 + 18abcd.$$

The polynomials $G(x_0, \ldots, x_n)$ become very complicated as n increases.

Definition 1.1.9 Given a commutative ring R, the *derivative* of

$$f(x) = a_n x^n + a_{n-1} x^{n-1} + \cdots + a_1 x + a_0 \in R[x],$$

denoted $f'(x)$, is the polynomial

$$f'(x) = na_n x^{n-1} + (n-1)a_{n-1} x^{n-2} + \cdots + a_1. \tag{1.1.6}$$

It is easy to verify that for all $a, b \in R$ and $f(x), g(x) \in R[x]$, the formal derivative satisfies the usual identities:

- $(a \cdot f(x) + b \cdot g(x))' = a \cdot f'(x) + b \cdot g'(x)$.
- $(f(x) \cdot g(x))' = f'(x) \cdot g(x) + f(x) \cdot g'(x)$.
- $(f(g(x)))' = f'(g(x)) \cdot g'(x)$.

We leave the proof of the following lemma as an exercise:

Lemma 1.1.10 $f(x) \in K[x]$ *is separable if and only if* $f(x)$ *and* $f'(x)$ *are coprime in* $K[x]$.

This is a very useful lemma since, combined with the Division Algorithm, it provides an efficient method for checking whether $f(x)$ is separable without factoring $f(x)$. In fact, in certain important cases, we do not even need the Division Algorithm.

Example 1.1.11 Let n be a positive integer. Suppose that either the characteristic of K is 0 or n is relatively prime to the characteristic of K. Then $x^n - a \in K[x]$ is separable for any $a \in K^\times$ because $(x^n - a)' = nx^{n-1}$ is obviously coprime to $x^n - a$.

Example 1.1.12 Let K be a field of positive characteristic p, and let $a \in K$. The polynomial $x^p - a$ has only one root in its splitting field L. Indeed, if $\alpha \in L$ is such that $\alpha^p = a$, then $x^p - a = (x - \alpha)^p$. Moreover, if $a \in K$ is not equal to the p-th power of another element of K, then $x^p - a$ is irreducible over K. Indeed, any nontrivial proper monic factor of $x^p - a$ is $(x - \alpha)^m$, where $1 \leq m \leq p - 1$. The coefficient of x^{m-1} in $(x - \alpha)^m$ is $-m\alpha$, so if $x^p - a$ has a nontrivial proper factor $(x - \alpha)^m$ in $K[x]$, then $-m\alpha \in K$. In that case, $m \in \mathbb{F}_p^\times \subset K^\times$, so $\alpha \in K$, which means that $a = \alpha^p$, a contradiction.

Now consider $f(x) = x^{p^n} - x \in K[x]$. This polynomial is separable since $f'(x) = -1$. More generally, if $g(x) \in K[x]$ is any polynomial and $a \in K^\times$, then $f(x) = g(x^{p^n}) + ax$ is separable since $f'(x) = a$.

Corollary 1.1.13 *We have:*

(1) *An irreducible polynomial $f(x) \in K[x]$ is inseparable if and only if $f'(x) = 0$.*
(2) *If K has characteristic 0, then every irreducible polynomial in $K[x]$ is separable.*
(3) *If K has characteristic p, then an irreducible polynomial $f(x)$ is separable if and only if $f(x) \neq g(x^p)$ for all $g(x) \in K[x]$.*

Proof

(1) By Lemma 1.1.10, $f(x)$ is separable if and only if it is relatively prime to $f'(x)$. If $f(x)$ is irreducible, then it is not relatively prime to $f'(x)$ if and only if $f(x)$ divides $f'(x)$. But $f'(x)$ has degree strictly less than $f(x)$, so $f(x)$ divides $f'(x)$ if and only if $f'(x) = 0$.
(2) In characteristic 0, $\deg(f') = \deg(f) - 1$, so $f'(x)$ is never the zero polynomial.
(3) In characteristic p, we have $f'(x) = 0$ if and only if $f(x) = g(x^p)$ for some $g(x) \in K[x]$, as easily follows by looking at the formula (1.1.6) for $f'(x)$. □

Lemma 1.1.14 *Let K be a field of characteristic p. Let $f(x) \in K[x]$ be an irreducible polynomial. There is a unique separable irreducible polynomial $g(x) \in K[x]$ and a unique integer $k \geq 0$ such that $f(x) = g(x^{p^k})$.*

Proof If $f(x)$ is not separable, then by Corollary 1.1.13 there is a polynomial $f_1(x) \in K[x]$ such that $f(x) = f_1(x^p)$. If $f_1(x)$ is reducible, say $f_1(x) = g_1(x)g_2(x)$, then $f(x) = g_1(x^p)g_2(x^p)$ is a decomposition of $f(x)$, which contradicts the irreducibility of $f(x)$. Hence, $f_1(x)$ is irreducible. If the polynomial $f_1(x)$ is not separable, then by the same argument $f_1(x) = f_2(x^p)$, with $f_2(x)$ irreducible, and $f(x) = f_2(x^{p^2})$. Continuing in this manner, we will eventually obtain a uniquely defined power p^k such that $f(x) = g(x^{p^k})$, with $g(x)$ irreducible and separable. □

Definition 1.1.15 We will denote the polynomial $g(x)$ in the previous lemma by $f_{\text{sep}}(x)$, so

$$f(x) = f_{\text{sep}}(x^{p^k}).$$

The degree of $f_{\text{sep}}(x)$ is called the *separable degree* of $f(x)$, and p^k is called the *inseparable degree* of $f(x)$. If $\deg(f_{\text{sep}}) = 1$, then we say that $f(x)$ is *purely inseparable*.

Exercises

1.1.1 Let K be a field. Let $K(\!(x)\!)$ be the set of *formal Laurent series*

$$\sum_{n=N}^{\infty} a_n x^n, \qquad a_n \in K, \qquad N \in \mathbb{Z},$$

that is, each element of $K(\!(x)\!)$ is a formal power series in x with coefficients in K such that finitely many terms are allowed to have negative powers of x:

(a) Check that addition (1.1.1) and multiplication (1.1.2) on $K[\![x]\!]$ extend to $K(\!(x)\!)$, and make $K(\!(x)\!)$ into a commutative ring containing $K[\![x]\!]$ as a subring.
(b) Prove that $K(\!(x)\!)$ is a field. In fact, $K(\!(x)\!)$ is the fraction field of $K[\![x]\!]$.

1.1.2 Let K be a field and let $f(x) \in K[x]$. Prove that $\alpha \in K$ is a root of $f(x)$ if and only if $x - \alpha$ divides $f(x)$, and deduce that the number of roots of $f(x)$ in K is at most $\deg(f)$.

1.1.3 Prove the claim of Lemma 1.1.10.

1.1.4 Let K be a field, and let $f(x), g(x) \in K[x]$. Let L be a field extension of K. Prove that $f(x) \mid g(x)$ in $K[x]$ if and only if $f(x) \mid g(x)$ in $L[x]$.

1.1.5 Assume K is a field of positive characteristic p and $f(x) \in K[x]$ is irreducible. Prove that $f(x^p)$ is either irreducible in $K[x]$ or $f(x^p) = g(x)^p$ for some irreducible $g(x) \in K[x]$. Give examples that show that both possibilities occur if the Frobenius $\text{Fr}_p \colon K \to K$ is not surjective.

1.1.6 Let R be a commutative ring, let $f(x) \in R[x]$, and let $a \in R$. Show that

$$f(x) = f(a) + f'(x)(x - a) + g(x)(x - a)^2$$

for a unique $g(x) \in R[x]$.

1.2 Modules over Polynomial Rings

The main theorem of this section gives a description of the structure of finitely generated $K[x]$-modules, where K is a field. The theorem says that such modules are isomorphic to a direct sum of finitely many cyclic modules. The same statement is true more generally for modules over principal ideal domains (cf. [DF04, Ch. 12]), but we will use it only for $K[x]$ and $\mathbb{Z}$, where a simpler proof is available (in fact, the same proof can be easily adapted to any Euclidean domain).

We start by recalling some relevant terminology from the theory of modules.

Let R be a commutative ring and $M_1, \ldots, M_r$ be a finite collection of R-modules. The *direct product* of these modules is the R-module

$$M_1 \times \cdots \times M_r = \{(m_1, \ldots, m_r) \mid m_i \in M_i, 1 \le i \le r\}$$

with the addition and the action of R defined componentwise:

$$(m_1, \ldots, m_r) + (m'_1, \ldots, m'_r) = (m_1 + m'_1, \ldots, m_r + m'_r),$$

$$a(m_1, \ldots, m_r) = (am_1, \ldots, am_r), \qquad a \in R.$$

The direct product of $M_1, \ldots, M_r$ is also sometimes referred to as the *direct sum* of $M_1, \ldots, M_r$ and denoted

$$M_1 \oplus \cdots \oplus M_r.$$

An R-module M is said to be *free of (finite) rank r* if M is isomorphic to a direct product of r copies of R:

$$M \cong \overbrace{R \times R \times \cdots \times R}^{r} =: R^r.$$

In the notation involving direct sums, the same module is denoted by $R^{\oplus r}$.

Lemma 1.2.1 *A free R-module of finite rank is uniquely determined by its rank, i.e., $R^n \cong R^m$ if and only if $n = m$.*

Proof Let $\varphi \colon R^n \xrightarrow{\sim} R^m$ be an isomorphism, and let I be an ideal of R. Then $R^n/IR^n \cong R^m/\varphi(IR^n)$. Because φ is an R-module homomorphism, we have $\varphi(IR^n) = IR^m$. Therefore,

$$R^n/IR^n \cong (R/I)^n \cong (R/I)^m \cong R^m/IR^m.$$

When I is a maximal ideal, this reduces the statement of the lemma to the basic fact from linear algebra that two finite dimensional vector spaces over the same field are isomorphic if and only if they have the same dimension. $\square$

An R-module M is said to be *finitely generated* if there exist elements $y_1, \ldots, y_r \in M$ such that every $y \in M$ can be written as an R-linear combination of these elements:

$$y = a_1 y_1 + \cdots + a_r y_r, \qquad a_1, \ldots, a_r \in R.$$

The following fact is not hard to prove: An R-module M is free of rank r if and only if there exist elements $y_1, \ldots, y_r \in M$ such that every $y \in M$ can be written **uniquely** as an R-linear combination of these elements. In that case, the collection $\{y_1, \ldots, y_r\}$ is called an *R-basis* of M. More generally, an R-module M is said to be *free* if there is a subset Y of elements of M such that every nonzero $y \in M$ can be written uniquely as a finite R-linear combination $a_1 y_1 + \cdots + a_s y_s$ for some nonzero elements $y_1, \ldots, y_s \in Y, s \in \mathbb{Z}_{>0}$.

An R-module M is called *cyclic* if M is generated by one element, i.e., $M = Ry$ for some $y \in M$. Note that a cyclic R-module generated by y is free if and only if the *annihilator* of y

$$\mathrm{Ann}_R(y) := \{r \in R \mid ry = 0\}$$

is the zero ideal of R.

An element y of an R-module M is called a *torsion element* if $ry = 0$ for some nonzero $r \in R$. The set of torsion elements is denoted

$$M_{\mathrm{tor}} = \{y \in M \mid ry = 0 \text{ for some } 0 \neq r \in R\}.$$

An R-module M is said to be a *torsion module* if $M = M_{\mathrm{tor}}$; it is *torsion-free* if $M_{\mathrm{tor}} = 0$.

From now on the ring that we will be working with will be $A := K[x]$.

Proposition 1.2.2 *Let N be a free A-module of rank n. Let $M \subseteq N$ be an A-submodule. Then M is a free A-module of rank $m \leq n$.*

Proof If $M = 0$, then $m = 0$, and there is nothing to prove. So assume $M \neq 0$. After fixing an A-basis $\{e_1, \ldots, e_n\}$ of N, one obtains an isomorphism $N \xrightarrow{\sim} A^n$, which maps $y = a_1 e_1 + \cdots + a_n e_n$ to $(a_1, \ldots, a_n)$, where $a_1, \ldots, a_n \in A$. We call $a_1, \ldots, a_n$ the coordinates of y. From now on, we identify N with A^n and consider M as being a submodule of A^n.

Let I be the set of elements in A that occur as the first coordinate of an element in M. Since M is an A-module, I is an ideal of A. Because A is a PID, $I = (a)$ is generated by some $a \in A$. Fix some $y' = (a, a_2, \ldots, a_n) \in M$, and let $M' = Ay'$. Let M_0 be the submodule of M consisting of elements whose first coordinate is 0. We claim that $M \cong M' \oplus M_0$. Indeed, if $y = (b_1, \ldots, b_n) \in M$, then b_1 is divisible by a, i.e., $b_1 = ab$ for some $b \in A$, so $y - by' \in M_0$. Therefore, $y = y_1 + y_2$ with $y_1 \in M'$ and $y_2 \in M_0$. If $y = y_1' + y_2'$ is a second representation of y in this form, then $y_1 - y_1' = y_2' - y_2 \in M' \cap M_0$. On the other hand, it is obvious that

$M' \cap M_0 = 0$. Thus, each $y \in M$ can be uniquely represented as a sum of elements of M' and M_0. This implies that $M \cong M' \oplus M_0$, with the isomorphism given by $y = y_1 + y_2 \mapsto (y_1, y_2)$.

Obviously, M' is a free A-module of rank 1. If $n = 1$, then $M = M'$ is also free of rank 1. Now assume that $n \geq 2$, and we proved the claim for A-modules of rank $\leq n - 1$. Clearly, $M_0 \subseteq N_0 \cong R^{n-1}$, where N_0 is the submodule of N consisting of elements whose first coordinate is 0. By induction, M_0 is free of rank $\leq n - 1$. Then $M \cong M' \oplus M_0$ is free of rank $\leq 1 + (n - 1) = n$.   □

Let N be a free A-module of rank n, and let $M \subseteq N$ be a submodule. As we just proved, M is also free of rank $m \leq n$. Choose an A-basis $\{e_1, \ldots, e_n\}$ of N, an A-basis $\{y_1, \ldots, y_m\}$ of M, and then expand each y_i in terms of e_j's:

$$y_i = a_{i,1}e_1 + \cdots + a_{i,n}e_n.$$

Let

$$S = \begin{pmatrix} a_{1,1} & \cdots & a_{1,n} \\ a_{2,1} & \cdots & a_{2,n} \\ \vdots & \ddots & \vdots \\ a_{m,1} & \cdots & a_{m,n} \end{pmatrix}. \tag{1.2.1}$$

An *elementary row operation* on S is one of the following:

(i) Interchange two rows of S.
(ii) For $a \in A$ and $i \neq j$, add a times row j to row i.

The elementary column operations are defined analogously. ·

The elementary row operations correspond to changing the basis of M. More precisely, interchanging the i-th row with the j-th row in S corresponds to the expansion in terms of $\{e_1, \ldots, e_n\}$ of the elements of a basis of M obtained by interchanging y_i with y_j in $\{y_1, \ldots, y_m\}$; adding a times row j to row i corresponds to the expansion of the elements of the basis

$$\{y_1, \ldots, y_{i-1}, y_i + ay_j, \ldots, y_j, \ldots, y_m\}$$

of M in terms of $\{e_1, \ldots, e_n\}$. (It is clear that the set above is still an A-basis of M since $y_i = (y_i + ay_j) - ay_j$.) Similarly, the elementary column operations on S correspond to changing the basis of N. For example, if we add a times the j-th column to the i-th column, then the resulting matrix corresponds to the expansion of $\{y_1, \ldots, y_m\}$ in terms of $\{e_1, \ldots, e_i, \ldots, e_j - ae_i, \ldots, e_n\}$. Thus, the elementary row and column operations only change our chosen bases of M and N, but not the modules themselves.

Next, we show that using the elementary row and column operations, we can reduce S to a diagonal matrix with an additional property. Let $\delta(S)$ be the minimum of the degrees of nonzero entries of S. We apply the following algorithm to S.

Step 1 After interchanging some rows and columns, i.e., performing the elementary row and column operation (i), we can assume that $\deg(a_{1,1}) = \delta(S)$. Now using (ii), we can transform S into a matrix S' whose $(i, 1)$-th entry is the residue of $a_{i,1}$ modulo $a_{1,1}$ for all $2 \leq i \leq s$, and similarly the $(1, i)$-th entry is the residue of $a_{1,i}$ modulo $a_{1,1}$ for all $2 \leq i \leq s$. From Theorem 1.1.2, it is clear that $\delta(S') < \delta(S)$ unless all $a_{i,1}$ and $a_{1,i}$ are zero for $i \neq 1$. Since $\delta(S)$ is a natural number, repeating this process a finite number of times, we will eventually obtain a matrix whose first column and first row consist of zeros except for the $(1, 1)$-entry,

$$S'' = \begin{pmatrix} b_{1,1} & 0 & \cdots & 0 \\ 0 & b_{2,2} & \cdots & b_{2,n} \\ \vdots & \vdots & \ddots & \vdots \\ 0 & b_{m,2} & \cdots & b_{m,n} \end{pmatrix},$$

and $\delta(S'') = \deg(b_{1,1})$.

Step 2 Suppose $b_{1,1}$ does not divide one of the entries in S'', e.g., $b_{1,1} \nmid b_{i,j}$. We add the i-th row to the first row and then subtract an appropriate multiple of the first column from the j-th column to replace the $(1, j)$-th entry by its residue modulo $b_{1,1}$. This again produces a matrix with strictly smaller δ than $\delta(S'')$. Now we apply Step 1 to the resulting matrix.

Since each step of the above process makes δ smaller, eventually we will arrive at a matrix S'', where $b_{1,1}$ divides all other entries of the matrix. In that case, set

$$S''' = \begin{pmatrix} b_{2,2} & \cdots & b_{2,n} \\ \vdots & \ddots & \vdots \\ b_{m,2} & \cdots & b_{m,n} \end{pmatrix},$$

and apply the algorithm to this matrix. Note that a matrix obtained from S''' by any sequence of elementary row and column operations retains the property that all its entries are divisible by $b_{1,1}$. Thus, working with smaller-and-smaller matrices, we eventually will get an $m \times n$ diagonal matrix

$$\begin{pmatrix} c_{1,1} & 0 & \cdots & & \cdots & 0 \\ 0 & c_{2,2} & \cdots & & \cdots & 0 \\ \vdots & \vdots & \ddots & & & \vdots \\ 0 & 0 & \cdots & c_{m,m} & \cdots & 0 \end{pmatrix},$$

with the additional property that

$$c_{1,1} \mid c_{2,2} \mid \cdots \mid c_{m,m},$$

i.e., $c_{1,1}$ divides $c_{2,2}$, $c_{2,2}$ divides $c_{3,3}$, and so on.

As we have observed, each step of the above algorithm corresponds to choosing a new A-basis of M or N. Therefore, we have proved the following:

Proposition 1.2.3 *Let N be a free A-module of rank n, and let $M \subseteq N$ be a free submodule of rank m. There is an A-basis $\{e_1, e_2, \ldots, e_n\}$ of N such that $\{a_1 e_1, \ldots, a_m e_m\}$ is an A-basis of M for some nonzero $a_1, a_2, \ldots, a_m \in A$ satisfying $a_1 \mid a_2 \mid \cdots \mid a_m$.*

Theorem 1.2.4 *A finitely generated A-module M is isomorphic to a direct sum*

$$M \cong A^{\oplus r} \oplus A/(a_1) \oplus \cdots \oplus A/(a_m),$$

where $r \geq 0$ is an integer, and either $m = 0$ or $a_1 \mid a_2 \mid \cdots \mid a_m$ all have positive degrees.

*The integer r is called the **rank** of M, and the elements $a_1, \ldots, a_m$, chosen to be monic, are called the **invariant factors** of M.*

Two finitely generated A-modules are isomorphic if and only if they have the same rank and the same list of invariant factors.

Proof Suppose M is generated by n elements $y_1, \ldots, y_n$. Let $N = A^{\oplus n}$, and define a map

$$\pi : N \to M,$$

$$\pi(a_1, \ldots, a_n) = a_1 y_1 + \cdots + a_n y_n.$$

It is trivial to check that this is a well-defined surjective homomorphism of A-modules. Let $N' = \ker(\pi)$. Then, by Proposition 1.2.2, $N' \subseteq N$ is a free A-module of rank $m \leq n$. By Proposition 1.2.3, we can choose a basis $\{e_1, \ldots, e_n\}$ of N such that $\{a_1 e_1, \ldots, a_m e_m\}$ is a basis of N', and moreover, $a_1 \mid a_2 \mid \cdots \mid a_m$. So we can assume

$$N' = a_1 A \oplus \cdots \oplus a_m A \subseteq A \oplus \cdots \oplus A = N.$$

For this specific submodule of N, it is clear that

$$M \cong N/N' \cong A/(a_1) \oplus \cdots \oplus A/(a_m) \oplus A^{\oplus r}, \qquad \text{where} \quad r = n - m.$$

We can omit from the right-hand side those $A/(a_i)$ for which $(a_i) = A$.

From the above isomorphism, we see that M_{tor} is a submodule of M and $M = M_{\text{tor}} \oplus M_{\text{free}}$, where $M_{\text{free}} \cong A^{\oplus r}$ is a free submodule.[4] Since M_{tor} is uniquely determined by M, the module $M_{\text{free}} \cong M/M_{\text{tor}}$ is also uniquely determined by M. On the other hand, by Lemma 1.2.1, a free module is uniquely determined by its rank. Hence, r is an isomorphism invariant of M.

---

[4] In general, M_{tor} need not be a submodule of M, and even if it is a submodule, there might not exist another submodule $M' \subset M$ such that $M \cong M' \oplus M_{\text{tor}}$; see Exercise 1.2.1.

Next, note that $M/M_{\text{free}} \cong M_{\text{tor}}$, so it is enough to show that isomorphic finitely generated torsion A-modules have the same invariant factors. Let $\text{Ann}_A(M) = \{a \in A \mid ay = 0 \text{ for all } y \in M\}$. If $M \cong A/(a_1) \oplus \cdots \oplus A/(a_m)$ with $a_1 \mid a_2 \mid \cdots \mid a_m$, then $\text{Ann}_A(M) = (a_m)$. Since isomorphic modules have the same annihilators, the highest degree invariant factor of M is an isomorphism invariant. On the other hand, the quotient of $A/(a_1) \oplus \cdots \oplus A/(a_m)$ by its last direct summand $A/(a_m)$ is isomorphic to $A/(a_1) \oplus \cdots \oplus A/(a_{m-1})$. Now an obvious iteration of this argument eventually establishes that $a_1, \ldots, a_m$ are isomorphism invariants of M. □

Definition 1.2.5 For a finitely generated A-module M of rank 0, define $\chi(M)$ to be the product of its invariant factors. In a different terminology, the ideal generated by $\chi(M)$ is the *fitting ideal* of M; see [DF04, p. 671]. If we replace A by $\mathbb{Z}$, then $\chi(M)$ is the order of the finite abelian group M.

Lemma 1.2.6 *Let N be a free A-module of rank n, and let $M \subseteq N$ be an A-submodule of the same rank. Let S be the $n \times n$-matrix (1.2.1). Then we have an equality of ideals of A*

$$(\chi(N/M)) = (\det(S)).$$

Proof The claim is obvious for diagonal matrices. In general, this follows from the fact that elementary row and column operations do not change $(\det(S))$. □

Remark 1.2.7 Let F be the fraction field of A, and let V be an F-vector space of finite dimension. An A-*lattice* in V is a free A-submodule Λ of V of rank $r = \dim_F V$ that spans V as an F-vector space. Let Λ_1 and Λ_2 be two A-lattices in V. Then $\Lambda_3 = \Lambda_1 \cap \Lambda_2$ is also an A-lattice. One can extend $\chi(N/M)$ from Lemma 1.2.6 to an arbitrary pair of lattices by defining

$$\chi(\Lambda_1, \Lambda_2) = \chi(\Lambda_1/\Lambda_3) \cdot \chi(\Lambda_2/\Lambda_3)^{-1}.$$

Moreover, Lemma 1.2.6 itself extends to this setting; see Exercise 1.2.10.

Theorem 1.2.8 (Chinese Remainder Theorem) *Let R be a PID. Let $a_1, \ldots, a_n \in R$ be elements such that a_i and a_j are coprime for all $1 \leq i < j \leq n$. Then*

$$R/(a_1 \cdots a_n) \cong R/(a_1) \times \cdots \times R/(a_n).$$

Proof See [DF04, p. 265]. □

Given $0 \neq a \in A$ with prime decomposition $a = P_1^{t_1} \cdots P_k^{t_k}$, Theorem 1.2.8 implies that

$$A/(a) \cong A/(P_1^{t_1}) \oplus \cdots \oplus A/(P_k^{t_k}).$$

Applying this to each a_i in Theorem 1.2.4 allows us to write each of the direct summands $A/(a_i)$ as a direct sum of cyclic modules whose annihilators are the prime power divisors of a_i,

$$M \cong A^{\oplus r} \oplus A/(P_1^{t_1}) \oplus \cdots \oplus A/(P_h^{t_h}),$$

where $P_1^{t_1}, \ldots, P_h^{t_h}$ are positive powers of (not necessarily distinct) monic irreducibles in A. The polynomials $P_1^{t_1}, \ldots, P_h^{t_h}$ are the *elementary divisors* of M. It is possible to recover the invariant factors from the elementary divisors, so the elementary divisors themselves are isomorphism invariants of M. Indeed, choose among the elementary divisors of M the highest powers of distinct irreducibles, and let a be their product. For example, if the elementary divisors are P, P^3, Q^3, Q^5, R, where P, Q, R are distinct, then $a = P^3 Q^5 R$. Then by the Chinese Remainder Theorem, $M \cong M' \oplus A/(a)$. The elementary divisors of M' are among the elementary divisors of M. Repeating the same process for M', we obtain $M \cong M'' \oplus A/(a') \oplus A/(a)$. (In the previous example, $a' = P Q^3$ and $M \cong A/(P Q^3) \oplus A/(P^3 Q^5 R)$.) It is easy to see that $a' \mid a$. Continuing this process, we eventually obtain

$$M \cong A^{\oplus r} \oplus A/(a_1) \oplus \cdots \oplus A/(a_m)$$

for nonzero monic polynomials $a_1, \ldots, a_m$ of positive degrees that satisfy the divisibility relations $a_1 \mid a_2 \mid \cdots \mid a_m$. Because of the uniqueness property in Theorem 1.2.4, these must be the invariant factors.

The proof of Theorem 1.2.4 can be easily adapted to the setting of finitely generated $\mathbb{Z}$-modules, i.e., to finitely generated abelian groups, where it is known as the Fundamental Theorem of Finitely Generated Abelian Groups. An interesting application of this theorem is the following useful fact:

Theorem 1.2.9 *Let K be a field, and let $G \subset K^\times$ be a finite subgroup of the group of nonzero elements of K. Then G is a cyclic group.*

Proof Theorem 1.2.4 for $A = \mathbb{Z}$ gives an isomorphism $G \cong \mathbb{Z}/(a_1) \oplus \cdots \mathbb{Z}/(a_m)$ where $a_1 \mid \cdots \mid a_m$. We claim that $m = 1$, so $G \cong \mathbb{Z}/(a_1)$ is cyclic. To show this, put $a = a_m$ and consider the polynomial $x^a - 1 \in K[x]$. Every element of G has order dividing a, so $g^a = 1$ for all $g \in G$. This implies that g, when considered as an element of K, is a root of the polynomial $x^a - 1 \in K[x]$. By Exercise 1.1.2, the polynomial $x^a - 1$ has at most a distinct roots, so $\#G \leq a_m$. On the other hand, $\#G = a_1 \cdots a_m$. Thus, $m = 1$. $\square$

Exercises

In the exercises, A denotes the ring of polynomials $K[x]$, where K is a field.

1.2.1 Let R be a commutative ring and M be an R-module.

(a) Prove that if R is an integral domain, then M_{tor} is a submodule of M (called the *torsion* submodule of M).

(b) Let $R = \mathbb{Z}/pq\mathbb{Z}$, where $p \neq q$ are primes. Let $M = R$. Determine M_{tor} and show that it is not a submodule. Do the same for $R = \mathbb{Z} \times \mathbb{Z}$.

(c) Let $R = \mathbb{Z}/p^n\mathbb{Z}$, where p is prime and $n \geq 2$. Let $M = R$. Determine M_{tor}, and show that it is a submodule, but there does not exist another submodule $M' \subseteq M$ such that $M \cong M_{\text{tor}} \oplus M'$.

1.2.2 Prove that the fraction field F of A is not free as a module over A.

1.2.3 An R-module M is called *divisible* if for every $0 \neq m \in M$ and $0 \neq r \in R$, there is $m' \in M$ such that $rm' = m$ (that is, $rM = M$). Prove the following statements:

(a) A divisible R-module cannot be finitely generated.
(b) If F is the fraction field of A, then F/A is a divisible torsion A-module.

1.2.4 Prove that a_1 in Proposition 1.2.3 is equal to the greatest common divisor of the nonzero entries of the matrix (1.2.1).

1.2.5 Prove that every ideal of the ring $A \times A \times \cdots \times A$, the direct product of finitely many copies of A, is principal. What if the number of copies of A is infinite? Note that there are two possible versions of A^∞: the *direct product* $\prod_{n=1}^{\infty} A$, whose elements are $\prod a_n$ with $a_n \in A$ arbitrary (and operations defined componentwise), and the *direct sum*, $\bigoplus_{n=1}^{\infty} A$, whose elements are $\prod a_n$ with all but finitely many a_n equal to zero.

1.2.6 Prove the following:

(a) Let $n \geq 2$. If the greatest common divisor of $a_1, a_2, \ldots, a_n \in A$ is 1, then there is an $n \times n$-matrix in $GL_n(A)$ in which $a_1, a_2, \ldots, a_n$ appear as the elements of the first row.
(b) Let M be a finitely generated A-module with a set of generators $x_1, \ldots, x_n$. If $a_1, \ldots, a_n \in A$ are elements whose greatest common divisor is 1, then $a_1 x_1 + \cdots + a_n x_n$ can be chosen as an element of a set of n generators of M.

1.2.7 Let V be a finite dimensional vector space over K. Let $T: V \to V$ be a linear transformation. Consider V as an A-module, where x acts as T.

(a) Prove that V is a finitely generated torsion A-module. Hence,

$$V \cong A/(a_1) \oplus \cdots \oplus A/(a_s), \quad \text{with} \quad a_1 \mid \cdots \mid a_s.$$

(b) Prove that $\chi(V)$ is the characteristic polynomial of T, whereas a_s is the minimal polynomial of T. In particular, the minimal polynomial of T divides the characteristic polynomial of T, which is the Cayley–Hamilton theorem.
(c) Conclude that the characteristic and minimal polynomials of T have the same irreducible factors in $K[x]$, apart from multiplicities.

1.2.8 Let M be a free A-module of rank 3 generated by e_1, e_2, e_3. Let $N \subset M$ be the submodule generated by $xe_1 + x^2 e_2 + x^2 e_3$ and $x^2 e_1 - xe_2 + x^2 e_3$. Determine the rank and the invariant factors of M/N.

1.2.9 Let $M_1 \subseteq M_2 \subseteq M_3$ be A-modules. Prove that

$$\chi(M_3/M_1) = \chi(M_3/M_2) \cdot \chi(M_2/M_1).$$

1.2.10 Let the notation be as in Remark 1.2.7.

(a) Show that $\Lambda_1 \cap \Lambda_2$ and $\Lambda_1 + \Lambda_2$ are A-lattices in V, where

$$\Lambda_1 + \Lambda_2 = \{\lambda_1 + \lambda_2 \mid \lambda_1 \in \Lambda_1, \lambda_2 \in \Lambda_2\}.$$

(b) Let $S \in \mathrm{GL}_n(F)$ be a matrix that maps a basis of Λ_1 onto a basis of Λ_2, i.e., $S\Lambda_1 = \Lambda_2$. Show that we have an equality $(\chi(\Lambda_1, \Lambda_2)) = (\det(S))$ of fractional A-ideals in F.

(c) Let $\langle \cdot, \cdot \rangle$ be a bilinear non-degenerate pairing on V. Let Λ be an A-lattice in V. The *dual lattice* (with respect to the pairing $\langle \cdot, \cdot \rangle$) is the A-module

$$\Lambda^* := \{v \in V \mid \langle v, \lambda \rangle \in A \text{ for all } \lambda \in \Lambda\}.$$

Prove that Λ^* is an A-lattice in V.

(d) Let $\{e_1, \dots, e_n\}$ be an A-basis of Λ. Show that

$$(\chi(\Lambda^*, \Lambda)) = (\det(\langle e_i, e_j \rangle)).$$

This is the *discriminant* of the lattice Λ with respect to the pairing $\langle \cdot, \cdot \rangle$.

(e) Assume the pairing $\langle \cdot, \cdot \rangle$ is symmetric, i.e., $\langle v_1, v_2 \rangle = \langle v_2, v_1 \rangle$ for all $v_1, v_2 \in V$. Prove that if Λ_1 and Λ_2 are two A-lattices in V, then

$$(\Lambda_1 + \Lambda_2)^* = \Lambda_1^* \cap \Lambda_2^*.$$

1.3 Algebraic Extensions

In this section, we recall some definitions and facts from Field Theory.

Let K be a field, and let L/K a field extension. The field L can be considered as a vector space over K, with the action of $\alpha \in K$ on $v \in L$ being the multiplication $\alpha \cdot v$ in L. The *degree* of the extension L/K, denoted $[L : K]$, is the dimension of L as a vector space over K. The extension L/K is said to be *finite* if $[L : K]$ is finite. For a tower of finite extensions $K \subset L \subset M$, we have

$$[M : K] = [K : L][L : K].$$

An element $\alpha \in L$ is *algebraic* over K if it is a root of some nonzero polynomial $f(x) \in K[x]$; the elements that are not algebraic over K are called *transcendental*. The extension L/K is *algebraic* if every element of L is algebraic over K. It is easy to show that if $[L : K]$ is finite, then L is algebraic over K. (The set $\{1, \alpha, \alpha^2, \ldots, \alpha^n\}$ must be linearly dependent for any $\alpha \in L$ if $n \geq [L : K]$, and a linear relationship gives the required polynomial.) The field L is called an *algebraic closure* of K if L is algebraic over K and every polynomial $f(x) \in K[x]$ splits completely over L. Due to the existence of the splitting field of a single polynomial, it is intuitively clear that an algebraic closure of K exists and is unique up to isomorphism, as it is essentially the union of the splitting fields of all the polynomials in $K[x]$, but the actual details of the proof are somewhat messy; we refer to [DF04, p. 543] for these details. The algebraic closure of K will be denoted by $\overline{K}$. Usually, $\overline{K}$ has infinite degree over K, although a notable exception is $\mathbb{C}/\mathbb{R}$, which has degree 2. A field is said to be *algebraically closed* if it is its own algebraic closure. For example, it is not hard to show that $\overline{K}$ is algebraically closed; see [DF04, p, 543].

Let F and K be subfields of L. The *composite of F and K*, denoted FK, is the intersection of all the subfields of L that contain both F and K. Let $\alpha \in L$ be algebraic over K. Let $K(\alpha)$ be the intersection of all the subfields of L containing α and K. The set of polynomials $\{f(x) \in K[x] \mid f(\alpha) = 0\}$ is an ideal of $K[x]$ and hence is generated by a unique monic polynomial $m_{\alpha,K}(x) \in K[x]$, called the *minimal polynomial* of α over K. Clearly, $m_{\alpha,K}(x)$ divides any other polynomial in $K[x]$ having α as a root. This implies that $m_{\alpha,K}(x)$ is irreducible: if $m_{\alpha,K}(x) = f_1(x)f_2(x)$ in $K[x]$, then $0 = m_{\alpha,K}(\alpha) = f_1(\alpha)f_2(\alpha)$ implies that α is a root of either f_1 or f_2, so $m_{\alpha,K}$ divides either f_1 or f_2. Because $m_{\alpha,K}(x)$ is irreducible, we have $K(\alpha) \cong K[x]/(m_{\alpha,K}(x))$, so

$$\deg m_{\alpha,K}(x) = [K(\alpha) : K].$$

Two elements $\alpha, \beta \in L$, algebraic over K, are said to be *conjugate* if $m_{\alpha,K}(x) = m_{\beta,K}(x)$.

An algebraic element α over K is called *separable* (respectively, *inseparable*) if its minimal polynomial is separable (respectively, inseparable). An algebraic extension L/K is called *separable* if every element of L is separable over K. When L/K is not separable, it is called *inseparable*. An algebraic extension L/K is *normal* if it is the splitting field of some (possibly infinite) set of polynomials in $K[x]$. An algebraic extension L/K is *Galois* if it is normal and separable.

By Corollary 1.5.6, the splitting field of a separable polynomial in $K[x]$ is a separable extension of K. This implies that a finite extension L/K is Galois if and only if it is the splitting field of a separable polynomial in $K[x]$. One can also show that a finite extension L/K is normal if and only if every irreducible polynomial in $K[x]$ with a root in L splits completely in $L[x]$; see [DF04, p. 650]. Therefore, if an irreducible polynomial $f(x) \in K[x]$ has a root in a Galois extension L/K, then $f(x)$ is separable and splits completely in $L[x]$.

Next, we recall the basics of Galois theory. Let L/K be a field extension. A K-*automorphism* of L is a field automorphism $\sigma : L \to L$ that fixes the elements of K, i.e., $\sigma(c) = c$ for all $c \in K$. The set of K-automorphisms of L is a group under composition and is denoted $\text{Aut}(L/K)$. Note that for any $f(x) \in K[x], \alpha \in L$, and

$\sigma \in \mathrm{Aut}(L/K)$, we have $\sigma(f(\alpha)) = f(\sigma(\alpha))$. In particular, any K-automorphism of L permutes the roots of $f(x)$ in L.

If L/K is finite, then $\mathrm{Aut}(L/K)$ is also finite. Moreover, one can show that $\#\mathrm{Aut}(L/K)$ always divides $[L : K]$. More importantly, we have the following alternative characterization of Galois extensions (see [DF04, p. 562]), which sometimes is given as the definition of Galois extensions:

Theorem 1.3.1 *A finite extension L/K is Galois if and only if $\#\mathrm{Aut}(L/K) = [L : K]$.*

If L/K is Galois, then $\mathrm{Aut}(L/K)$ is called the *Galois group* of L/K and is denoted $\mathrm{Gal}(L/K)$. An extension L/K is called *abelian* (respectively, *cyclic*) if L/K is Galois and $\mathrm{Gal}(L/K)$ is an abelian (respectively, cyclic) group.

By Corollary 1.5.6, the set of elements of $\overline{K}$ that are separable over K forms a subfield, denoted K^{sep}, called the *separable closure of K*. The extension K^{sep}/K is Galois, but it is usually infinite. The Galois group $\mathrm{Gal}(K^{\mathrm{sep}}/K)$ is the *absolute Galois group of K*. We will frequently denote this group by

$$G_K = \mathrm{Gal}(K^{\mathrm{sep}}/K).$$

Remark 1.3.2 Generally, the absolute Galois groups are very complicated infinite groups. Especially, $G_{\mathbb{Q}}$ is a central object of study in number theory. Many basic questions about $G_{\mathbb{Q}}$ remain open, for example, it is expected, but yet unproven, that every finite group is isomorphic to the Galois group of some finite Galois extension of $\mathbb{Q}$.

The Galois group of a separable polynomial $f(x) \in K[x]$ is the Galois group of the splitting field of $f(x)$ over K. Because K-automorphisms permute the roots of an irreducible $f(x) \in K[x]$, the Galois group of an irreducible separable $f(x)$ of degree n is a subgroup of the symmetric group S_n. Determining the Galois group of a given polynomial is generally hard, but there are methods that work well in certain situations (see [DF04, p. 640]).

If L/K is a Galois extension, then for any intermediate field $K \subset E \subset L$ the extension L/E is Galois,[5] so we get a subgroup $\mathrm{Gal}(L/E) \subset \mathrm{Gal}(L/K)$. In the other direction, given a subgroup $H \subset \mathrm{Gal}(L/K)$, we get a subfield

$$L^H := \{\alpha \in L \mid \sigma(\alpha) = \alpha \text{ for all } \sigma \in H\},$$

which lies between K and L.

Theorem 1.3.3 (Fundamental Theorem of Galois Theory) *Let L/K be a finite Galois extension with Galois group G. Then the maps*

$$E \longmapsto \mathrm{Gal}(L/E), \qquad H \longmapsto L^H,$$

---

[5] But note that if L/E is normal and E/K is normal, then L/K is not necessarily normal; for example, $\mathbb{Q} \subset \mathbb{Q}(\sqrt{2}) \subset \mathbb{Q}(\sqrt[4]{2})$. Hence, a tower of Galois extensions is not necessarily Galois.

are inverse bijections between the set of subgroups of G and the set of intermediate fields between L and K. Moreover, L/L^H is always Galois with $\mathrm{Gal}(L/L^H) = H$, and L^H/K is Galois if and only if H is normal, in which case $\mathrm{Gal}(L^H/K) \cong G/H$.

Now let L/K be a field extension of possibly infinite degree. It is easy to show that L is Galois over K if and only if it is a union of finite Galois extensions of K. Unfortunately, the bijection of the fundamental theorem of Galois theory

$$H \subset G \quad \longleftrightarrow \quad L^H \subset L$$

does not quite work for infinite extensions. For example, $\mathrm{Gal}(\overline{\mathbb{Q}}/\mathbb{Q})$ has normal subgroups of index 2^n for any $n \geq 1$ whose fixed field is $\mathbb{Q}$; see [Mil12a, Prop. 7.25]. (Such subgroups are not easy to construct.) To remedy this, one introduces a topology on G by taking as a base for the closed sets the subgroups of G that are the fixing subgroups of finite extensions of K in L, together with all the left and right cosets of these subgroups.

Theorem 1.3.4 (Fundamental Theorem of Infinite Galois Theory) *Let L/K be a Galois extension with Galois group G. Then the maps*

$$E \longmapsto \mathrm{Gal}(L/E), \quad H \longmapsto L^H,$$

are inverse bijections between the set of closed subgroups of G and the set of intermediate fields between L and K. Moreover:

(1) For an arbitrary subgroup H of G, the extension L/L^H is Galois and $\mathrm{Gal}(L/L^H)$ is the closure of H.

(2) A closed subgroup H of G is open if and only if L^H has finite degree over K, in which case $[G : H] = [L^H : K]$.

(3) A closed subgroup H of G is normal if and only if L^H is Galois over K, in which case $\mathrm{Gal}(L^H/K) \cong G/H$.

Proof See [Mil12a, Chapter 7]. $\qquad\qquad\qquad\qquad\qquad\qquad\qquad\qquad\qquad\qquad$ □

Remark 1.3.5 Let G be a topological group, i.e., the set G has a group structure and a topology such that $G \times G \to G$, $(g, h) \mapsto gh$, and $G \to G$, $g \mapsto g^{-1}$, are both continuous. For any subgroup H of G, the coset gH of H is open or closed if H is open or closed. As the complement of H in G is a union of such cosets, this shows that

$$H \text{ is open} \quad \Longrightarrow \quad H \text{ is closed;}$$

$$H \text{ is closed of finite index} \quad \Longrightarrow \quad H \text{ is open.}$$

One can give an alternative description of the topology of the Galois group of infinite extensions, which relies on the concept of inverse limit.

Definition 1.3.6 Assume $(I, \preceq)$ is a partially ordered set such that for any $i, j \in I$ there is $k \in I$ such that $i, j \preceq k$. Let $(G_i)_{i \in I}$ be a family of finite groups equipped with a family of homomorphisms $\varphi_{i,j} : G_j \to G_i$ for all $i \preceq j$ such that:

(a) $\varphi_{i,i}$ is the identity on G_i for all $i \in I$.
(b) $\varphi_{i,k} = \varphi_{i,j} \circ \varphi_{j,k}$ for all $i \preceq j \preceq k$.

The *inverse limit* of $((G_i)_{i \in I}, (\varphi_{i,j})_{i,j \in I})$ is defined by

$$\varprojlim_{i \in I} G_i := \left\{ \vec{g} \in \prod_{i \in I} G_i \;\middle|\; g_i = \varphi_{i,j}(g_j) \text{ for all } i \preceq j \right\}.$$

The inverse limit is equipped with the relative product topology, where each G_i has the discrete topology.[6,7]

Now let I be the set of finite Galois extensions of K contained in L, with the inclusion of fields being the ordering. Note that for any two finite Galois extensions E_1 and E_2 of K, their composite $E_1 E_2$ is a finite Galois extension that contains both E_1 and E_2. To $E \in I$, we associate the group $\text{Gal}(E/K)$. If $E_1 \subseteq E_2$, then there is a natural quotient homomorphism $\text{Gal}(E_2/K) \to \text{Gal}(E_1/K)$ given by Theorem 1.3.3. For this family of homomorphisms, we have

$$\text{Gal}(L/K) = \varprojlim_{E \in I} \text{Gal}(E/K).$$

Moreover, the topology on $\text{Gal}(L/K)$ defined earlier agrees with the topology defined on the inverse limit.

Returning to finite extensions L/K, we consider the problem of choosing a basis of L over K with special properties. We will discuss two such bases, one that can be generated by the powers of a single element, and the other that consists of the Galois conjugates of a single element.

Definition 1.3.7 Given a finite extension L/K, an element $\alpha \in L$ is called *primitive* if $L = K(\alpha)$. In other words, $1, \alpha, \ldots, \alpha^{n-1}$ is a basis of L over K, where $n = [L : K]$, or equivalently, $\deg(m_{\alpha,K}) = n$.

There is a nice characterization of those extensions that have primitive elements (see [DF04, p. 594]):

Theorem 1.3.8 (Primitive Element Theorem) *A finite extension L/K has a primitive element if and only if there are finitely many intermediate fields between K and L.*

---

[6] Recall that in product topology the open sets in $\prod_{i \in I} G_i$ are the unions of sets of the form $\prod_{i \in I} U_i$, where $U_i \subseteq G_i$ and $U_i \neq G_i$ for only finitely many i.
[7] With respect to this topology, $\varprojlim G_i$ is compact and totally disconnected.

Corollary 1.3.9 *Any finite separable extension L/K contains a primitive element.*

Proof One can enlarge L to a Galois extension L' of K; for example, one could take L' to be the splitting field of the minimal polynomials of a finite set of generators of L over K (such L'/K is normal by definition, and it is separable by Corollary 1.5.6). Hence, we may assume that L/K itself is Galois. On the other hand, by the fundamental theorem of Galois theory, a finite Galois extension has finitely many intermediate fields since these fields are in bijection with the subgroups of $\mathrm{Gal}(L/K)$. □

Remark 1.3.10 For a finite separable extension L/K, the intersection in K^{sep} of all the Galois extensions of K containing L is itself Galois and is called the *Galois closure* of L over K.

Example 1.3.11 The assumption that L/K is separable in Corollary 1.3.9 is important. Let F be an algebraically closed field of characteristic p. Let $K = F(T, U)$, where T and U are indeterminates, and $L = K(T^{1/p}, U^{1/p})$. We have $[L : K] = p^2$. Moreover, the subfields

$$K(T^{1/p} + cU^{1/p}), \quad c \in F,$$

all have degree p over K since $(T^{1/p} + cU^{1/p})^p = T + c^p U \in K$. If any two of these subfields were equal, say $K' = K(T^{1/p} + cU^{1/p}) = K(T^{1/p} + c'U^{1/p})$ for $c \neq c'$, then

$$(T^{1/p} + cU^{1/p}) - (T^{1/p} + c'U^{1/p}) = (c - c')U^{1/p} \in K',$$

which implies that $U^{1/p} \in K'$, and therefore also $(T^{1/p} + cU^{1/p}) - cU^{1/p} = T^{1/p} \in K'$. We get $K' = L$, which is impossible by degree considerations. Thus, all the subfields $K(T^{1/p} + cU^{1/p})$ of L are distinct, so L has no primitive element by Theorem 1.3.8 (since an algebraically closed field is infinite[8]).

Proposition 1.3.12 *Let $n \geq 1$ be an integer, and let K be a field of characteristic not dividing n. Assume K contains the n-th roots of unity. Then the extension $K(\sqrt[n]{a})$ for any $a \in K^\times$ is cyclic over K of degree dividing n. Conversely, any cyclic extension of degree n is of the form $K(\sqrt[n]{a})$ for some $a \in K^\times$.*

Proof We only sketch the argument; see [DF04, p. 625–626] for the details.

The set μ_n of roots of $x^n - 1$ forms a subgroup of $\overline{K}^\times$, so μ_n is cyclic of order n by Theorem 1.2.9. The polynomial $x^n - a$ is separable for $a \neq 0$ (see Example 1.1.11). If $\alpha = \sqrt[n]{a}$ is a fixed root of $x^n - a$, then the roots of $x^n - a$ are $\zeta\alpha$, $\zeta \in \mu_n$. By assumption, $\mu_n \subset K$, so $x^n - a$ splits completely over $K(\alpha)$. Thus, $K(\alpha)/K$ is a Galois extension. Let $\sigma \in \mathrm{Gal}(K(\alpha)/K)$ and suppose $\sigma\alpha = \zeta_\sigma\alpha$ for some $\zeta_\sigma \in \mu_n$.

---

[8] If F is a finite field of characteristic p, then $x^n - 1$ cannot split completely over F if $n > \#F$ and $p \nmid n$.

Then

$$\sigma(\zeta\alpha) = \zeta\sigma(\alpha) = \zeta_\sigma(\zeta\alpha),$$

so ζ_σ does not depend on the choice of a root of $x^n - a$ and uniquely determines the action of σ on $K(\alpha)$. Thus,

$$\text{Gal}(K(\alpha)/K) \longrightarrow \mu_n, \qquad \sigma \longmapsto \zeta_\sigma,$$

is an injective homomorphism. We conclude that $\text{Gal}(K(\alpha)/K)$ is isomorphic to a subgroup of μ_n, so it is cyclic.

Conversely, suppose L/K is a Galois extension with $\text{Gal}(L/K) \cong \mathbb{Z}/n\mathbb{Z}$. Let σ be a generator of $\text{Gal}(L/K)$, and let ζ a generator of μ_n. Using the fact that the K-automorphisms of L are linearly independent over K, one shows that there is $\alpha \in L$ such that

$$\beta = \alpha + \zeta\sigma(\alpha) + \zeta^2\sigma^2(\alpha) + \cdots + \zeta^{n-1}\sigma^{n-1}(\alpha)$$

is nonzero. A simple calculation shows that $\sigma(\beta) = \zeta^{-1}\beta$. Iterating this calculation gives $\sigma^i(\beta) = \zeta^{-i}\beta$ for all $0 \leq i \leq n-1$, which shows that there are n distinct elements in the orbit of β under $\text{Gal}(L/K)$. Hence, the minimal polynomial $m_{\beta,K}(x)$ of β has degree n and $L = K(\beta)$. Finally, $\sigma(\beta^n) = (\zeta^{-1})^n\beta^n = \beta^n$, so β^n is fixed by $\text{Gal}(L/K)$. This implies that $a := \beta^n \in K$, i.e., β is a root of $x^n - a \in K[x]$.   □

Now suppose that L/K is a finite Galois extension with Galois group $G = \{\sigma_i \mid 1 \leq i \leq n\}, n = [L : K]$. If $\alpha_1, \ldots, \alpha_t$ is the orbit of $\alpha \in L$ under the action of G, then

$$m_{\alpha,K}(x) = \prod_{i=1}^{t}(x - \alpha_i).$$

(To see this, note that the coefficients of $\prod_{i=1}^{t}(x - \alpha_i)$ are fixed by G, so lie in K, and if this polynomial were reducible in $K[x]$, then its roots could not lie in a single orbit.) Hence, α is primitive if and only if $\sigma_1(\alpha), \sigma_2(\alpha), \ldots, \sigma_n(\alpha)$ are distinct. A stronger condition than this is that $\sigma_1(\alpha), \sigma_2(\alpha), \ldots, \sigma_n(\alpha)$ are linearly independent over K, so form a basis of L over K. Such a basis is called a *normal basis* for L/K.

Theorem 1.3.13 (Normal Basis Theorem) *Any finite Galois extension L/K has a normal basis.*

Proof We will prove this assuming $G = \text{Gal}(L/K)$ is cyclic. In that case, the proof is relatively easy and is a nice application of the theory of finitely generated modules over $K[x]$. The proof of the general case is much more complicated (see [Lan02, Thm. 13.1]).

Let σ be a generator of $G = \{1, \sigma, \ldots, \sigma^{n-1}\}$. We consider σ as a K-linear transformation of L, and give L a $K[x]$-module structure with x acting as σ. Then (cf. Exercise 1.2.7)

$$L \cong K[x]/(a_1) \oplus \cdots \oplus K[x]/(a_s), \quad a_1 \mid a_2 \mid \cdots \mid a_s,$$

is a direct sum of cyclic modules, where a_s is the minimal polynomial of σ and $a_1 \cdots a_s$ is the characteristic polynomial of σ. Let $a_s(x) = x^m + c_{m-1}x^{m-1} + \cdots + c_1 x + c_0 \in K[x]$. If $m < n$, then $a_s(\sigma) = 0$ implies that $\{1, \sigma, \ldots, \sigma^{n-1}\}$ are linearly dependent over K, which contradicts one of the key preliminary theorems of Galois theory, the so-called Linear Independence of Characters Theorem (see [DF04, p. 569]). Hence, $m \geq n$. On the other hand, $\sigma^n = 1$, so $a_s(x) \mid (x^n - 1)$. We conclude that $a_s(x) = x^n - 1$. Since $[L : K] = n$, the characteristic polynomial of σ has degree n. Thus, $L \cong K[x]/(x^n - 1)$ is a cyclic $K[x]$-module (we stress that this is an isomorphism of $K[x]$-modules, not rings). This means that there is an element $\alpha \in L$ such that $\alpha, \sigma(\alpha), \ldots, \sigma^{n-1}(\alpha)$ is a basis of L over K. □

Exercises

1.3.1 Let k be a field, and let F and K be finite extensions of k contained in a common field L. Let FK be the composite of F and K in L. Prove that the following are equivalent:

(a) The homomorphism of k-algebras $F \otimes_k K \to FK$ induced by $(a, b) \mapsto ab$ is an isomorphism.
(b) Any k-basis of F remains linearly independent over K.
(c) If $\{v_i\}$ and $\{w_j\}$ are k-bases of F and K respectively, then $\{v_i w_j\}$ is a basis of FK.
(d) $[FK : k] = [F : k] \cdot [K : k]$.

The fields F and K are said to be *linearly disjoint over k* if they satisfy these equivalent conditions. Conclude that if $[F : k]$ and $[K : k]$ are coprime, then F and K are linearly disjoint over k.

1.3.2 Let D_{2n} be the dihedral group with $2n$ elements, where $n \geq 3$. Suppose L/K is a Galois extension with $\mathrm{Gal}(L/K) \cong D_{2n}$. Prove that L is the splitting field of an irreducible polynomial of degree n over K.

1.3.3 Let $f(x)$ and $g(x)$ be nonzero relatively prime polynomials in $K[x]$ that are not both constant. Let $h(x) = f(x)/g(x)$ and $E = K(h(x))$. Prove the following statements:

(a) $h(x)$ is not algebraic over K.
(b) x is a root of the polynomial $p(y) = h(x)g(y) - f(y) \in E[y]$.
(c) $p(y)$ is irreducible in $E[y]$.
(d) $[K(x) : K(f(x)/g(x))] = \max(\deg(f), \deg(g))$.

1.3.4

(a) Determine the subgroup of $\mathrm{Aut}(K(x)/K)$ generated by

$$\sigma: f(x) \longmapsto f(1/x) \quad \text{and} \quad \tau: f(x) \longmapsto f(1-x).$$

(b) Prove that the subgroup of $\mathrm{Aut}(K(x)/K)$ generated by

$$\sigma: f(x) \longmapsto f(-1/x) \quad \text{and} \quad \tau: f(x) \longmapsto f(1+x)$$

is isomorphic to $\mathrm{SL}_2(\mathbb{Z})$ if the characteristic of K is zero. What happens if the characteristic of K is positive?

(c) Show that the group $\mathrm{Aut}(K(x)/K)$ consists of all linear fractional transformations

$$f(x) \longmapsto f((ax+b)/(cx+d)),$$

where $\begin{pmatrix} a & b \\ c & d \end{pmatrix} \in \mathrm{GL}_2(K)$.

1.3.5 Let $L = K(\alpha)$ be an algebraic field extension. Show that

$$\# \mathrm{Hom}_K(L, \overline{K}) = \#\{\beta \in \overline{K} \mid m_{\alpha, K}(\beta) = 0\} \le [L : K],$$

with an equality if and only if α is separable.

1.4   Trace and Norm

Let K be a field, let n be a positive integer, and let D be an n-dimensional algebra over K, i.e., D is a K-algebra which has dimension n as a vector space over K. Each $\alpha \in D$ determines a K-linear transformation $T_\alpha: t \mapsto \alpha t$ on D. It is clear that for any $\alpha, \beta \in D$ and $r, s \in K$, we have

$$T_{r\alpha+s\beta} = rT_\alpha + sT_\beta \quad \text{and} \quad T_{\alpha\beta} = T_\alpha T_\beta.$$

Moreover, since D has a unity element, $T_\alpha = 0$ if and only if $\alpha = 0$. Therefore, the map

$$D \longrightarrow \mathrm{End}_K(D) \approx \mathrm{Mat}_n(K)$$

$$\alpha \longmapsto T_\alpha$$

is an injective homomorphism of K-algebras. Note that if $\alpha \in K \subset D$, then $T_\alpha = \alpha I_n$.

Definition 1.4.1 Fix a basis of D over K, so that each T_α, $\alpha \in D$, is represented by an $n \times n$-matrix. The *norm* $\mathrm{Nr}_{D/K} : D \to K$ and the *trace* $\mathrm{Tr}_{D/K} : D \to K$ are the maps defined by

$$\mathrm{Nr}_{D/K}(\alpha) = \det(T_\alpha) \quad \text{and} \quad \mathrm{Tr}_{D/K}(\alpha) = \mathrm{Tr}(T_\alpha).$$

The *characteristic polynomial* of α is

$$\chi_{\alpha, D/K}(x) = \det(x I_n - T_\alpha)$$
$$= x^n - \mathrm{Tr}_{D/K}(\alpha) x^{n-1} + \cdots + (-1)^n \mathrm{Nr}_{D/K}(\alpha).$$

The *minimal polynomial* of α relative to K is the monic generator $m_{\alpha, D/K}(x)$ of the ideal

$$(m_{\alpha, D/K}(x)) = \{ f(x) \in K[x] \mid f(T_\alpha) = 0 \};$$

cf. Exercise 1.2.7. The characteristic and the minimal polynomial of α do not depend on the choice of a basis of D over K.

The properties of the determinant imply that $\mathrm{Nr}_{D/K}$ is a homomorphism $D^\times \to K^\times$ of multiplicative groups, i.e.,

$$\mathrm{Nr}_{D/K}(\alpha\beta) = \mathrm{Nr}_{D/K}(\alpha) \cdot \mathrm{Nr}_{D/K}(\beta) \quad \text{for all } \alpha, \beta \in D^\times. \tag{1.4.1}$$

Furthermore, if $\alpha \in K$, then $\mathrm{Nr}_{D/K}(\alpha) = \alpha^n$. Similarly, the properties of the trace of a matrix imply that $\mathrm{Tr}_{D/K}$ is a K-linear transformation $D \to K$ of vector spaces, i.e.,

$$\mathrm{Tr}_{D/K}(\alpha + \beta) = \mathrm{Tr}_{D/K}(\alpha) + \mathrm{Tr}_{D/K}(\beta) \quad \text{for all } \alpha, \beta \in D,$$

and, if $\alpha \in K$, then $\mathrm{Tr}_{D/K}(\alpha\beta) = \alpha \, \mathrm{Tr}_{D/K}(\beta)$.

Remark 1.4.2 Neither $\mathrm{Nr}_{D/K}$ nor $\mathrm{Tr}_{D/K}$ are necessarily surjective. Let $D = L$ be a field extension of K. As we will see, the surjectivity of the trace $\mathrm{Tr}_{L/K} : L \to K$ is equivalent to the separability of L/K (see Theorem 1.5.5). The surjectivity of the norm $\mathrm{Nr}_{L/K} : L^\times \to K^\times$ is more subtle and depends on the type of fields involved. The norm is surjective for extensions of finite fields (cf. Proposition 1.6.8), but it is not surjective for extensions of local or global fields (cf. Exercise 1.6.7). In fact, the norm plays a central role in the statements of the main theorems of class field theory; essentially, the abelian extensions of local and global fields are classified by the images of norms of such extensions; cf. [Mil13].

Theorem 1.4.3 *If D is a division algebra, then*

$$\chi_{\alpha, D/K}(x) = m_{\alpha, D/K}(x)^{[D:K(\alpha)]},$$

where $[D : K(\alpha)]$ is the dimension of D as a vector space over $K(\alpha)$. Moreover,

$$m_{\alpha,D/K}(x) = m_{\alpha,K}(x)$$

is the minimal polynomial of α in the sense of field theory of Sect. 1.3.

Proof If D is a division algebra, then $K(\alpha)/K$ is a field extension. The minimal polynomial $m_{\alpha,K}(x)$ of α annihilates T_α since $m_{\alpha,K}(T_\alpha) = m_{\alpha,K}(\alpha) = 0$. Hence, by definition, $m_{\alpha,D/K}(x)$ divides $m_{\alpha,K}(x)$. But $m_{\alpha,K}(x)$ is irreducible, so $m_{\alpha,D/K}(x) = m_{\alpha,K}(x)$ is irreducible. Now Exercise 1.2.7 implies that $\chi_{\alpha,D/K}(x)$ is equal to a power of $m_{\alpha,D/K}(x)$, and comparing the dimensions, we see that that power is $[D : K(\alpha)]$. $\qquad\qquad\square$

Corollary 1.4.4 *If D is an n-dimensional division algebra over K, then*

$$\mathrm{Nr}_{D/K}(\alpha) = (-1)^n \cdot m_{\alpha,K}(0)^{[D:K(\alpha)]}$$

$$= \mathrm{Nr}_{K(\alpha)/K}(\alpha)^{[D:K(\alpha)]}.$$

Proof This follows from $\chi_{\alpha,D/K}(0) = (-1)^n \cdot \mathrm{Nr}_{D/K}(\alpha)$ and Theorem 1.4.3. $\qquad\square$

For the rest of this section, we consider only field extensions of K.

Lemma 1.4.5 *Let $K \subset L \subset M$ be a tower of finite field extensions. For any $\alpha \in M$, we have*

$$\mathrm{Nr}_{M/K}(\alpha) = \mathrm{Nr}_{L/K}(\mathrm{Nr}_{M/L}(\alpha))$$

$$\mathrm{Tr}_{M/K}(\alpha) = \mathrm{Tr}_{L/K}(\mathrm{Tr}_{M/L}(\alpha)).$$

Proof Let $m = [M : L]$ and $n = [L : K]$. Choose a basis $\{\beta_1, \ldots, \beta_m\}$ of M over L. The L-linear transformation of M given by multiplication by $\alpha \in M$ with respect to this basis corresponds to an $m \times m$-matrix $\tilde{T}_\alpha = (\tilde{a}_{ij})$ over L. Now choose a basis $\{\gamma_1, \ldots, \gamma_n\}$ of L over K. The K-linear transformation of L given by multiplication by $\tilde{a}_{ij}$ with respect to this basis corresponds to an $n \times n$-matrix $\tilde{A}_{ij}$ over K. The K-linear transformation of M given by multiplication by α with respect to the basis $\{\beta_i \gamma_j\}$, $1 \le i \le m$, $1 \le j \le n$, corresponds to the $mn \times mn$-matrix $T_\alpha = (\tilde{A}_{ij})$; this can be checked by expanding $\alpha(\beta_i \gamma_j)$, which is a straightforward but tedious calculation.

We can choose the basis $\{\beta_1, \ldots, \beta_m\}$ so that $\tilde{T}_\alpha$ is upper-triangular. Then

$$\mathrm{Nr}_{M/L}(\alpha) = \det \tilde{T}_\alpha = \prod_{i=1}^{m} \tilde{a}_{ii}.$$

Hence,

$$\mathrm{Nr}_{L/K}(\mathrm{Nr}_{M/L}(\alpha)) = \mathrm{Nr}_{L/K}\left(\prod_{i=1}^{m}\tilde{a}_{ii}\right)$$

$$= \prod_{i=1}^{m}\mathrm{Nr}_{L/K}(\tilde{a}_{ii})$$

$$= \prod_{i=1}^{m}\det(\tilde{A}_{ii})$$

$$= \det T_{\alpha} = \mathrm{Nr}_{M/K}(\alpha).$$

The claim about the trace can be proved by a similar argument. □

Definition 1.4.6 The *discriminant* of an n-tuple $\alpha_1, \dots, \alpha_n \in L$ is

$$\mathrm{disc}_{L/K}(\alpha_1, \dots, \alpha_n) = \det(\mathrm{Tr}_{L/K}(\alpha_i \alpha_j))_{1 \le i, j \le n}.$$

The discriminant of an irreducible polynomial can be computed using the trace or norm as follows:

Theorem 1.4.7 Let $f(x) \in K[x]$ be a monic irreducible polynomial of degree n and $\alpha \in \overline{K}$ a root of $f(x)$. Then

$$\mathrm{disc}(f) = \mathrm{disc}_{K(\alpha)/K}(1, \alpha, \dots, \alpha^{n-1}),$$

$$\mathrm{disc}(f) = (-1)^{n(n-1)/2}\, \mathrm{Nr}_{K(\alpha)/K}(f'(\alpha)).$$

Proof Let $f(x) = (x - \alpha_1) \cdots (x - \alpha_n)$ be the decomposition of $f(x)$ into linear factors over $\overline{K}$. Then $\mathrm{Tr}_{K(\alpha)/K}(\alpha) = \alpha_1 + \cdots + \alpha_n$, so by Exercise 1.4.2,

$$\mathrm{Tr}_{K(\alpha)/K}(\alpha^i \alpha^j) = \alpha_1^i \alpha_1^j + \cdots + \alpha_n^i \alpha_n^j \quad \text{for all} \quad 0 \le i, j \le n - 1.$$

Using this observation, it is not hard to show that

$$(\mathrm{Tr}_{K(\alpha)/K}(\alpha^i \alpha^j))_{0 \le i, j \le n-1} = V^{\top} V,$$

where

$$V = \begin{pmatrix} 1 & \alpha_1 & \cdots & \alpha_1^{n-1} \\ 1 & \alpha_2 & \cdots & \alpha_2^{n-1} \\ \vdots & \vdots & \ddots & \vdots \\ 1 & \alpha_n & \cdots & \alpha_n^{n-1} \end{pmatrix}$$

is the Vandermonde matrix and $V^\top$ denotes its transpose. As is well-known,

$$\det V = \prod_{i>j}(\alpha_i - \alpha_j).$$

Hence,

$$\mathrm{disc}_{K(\alpha)/K}(1, \alpha, \dots, \alpha^{n-1}) = \det(V^\top) \cdot \det(V)$$

$$= \prod_{i>j}(\alpha_i - \alpha_j)^2$$

$$= \mathrm{disc}(f).$$

To prove the second formula, note that since $\mathrm{Nr}_{K(\alpha)/K}(\alpha) = \prod_{i=1}^{n}\alpha_i$, Exercise 1.4.2 implies that

$$\mathrm{Nr}_{K(\alpha)/K}(f'(\alpha)) = \prod_{i=1}^{n} f'(\alpha_i).$$

On the other hand, from the product rule for the derivatives, we have

$$f'(\alpha_i) = \prod_{\substack{1 \le j \le n \\ j \ne i}}(\alpha_j - \alpha_i).$$

Thus,

$$\mathrm{Nr}_{K(\alpha)/K}(f'(\alpha)) = \prod_{i=1}^{n} \prod_{\substack{1 \le j \le n \\ j \ne i}}(\alpha_j - \alpha_i)$$

$$= \prod_{i \ne j}(\alpha_i - \alpha_j)$$

$$= (-1)^{n(n-1)/2} \mathrm{disc}(f).$$

$\square$

Now suppose L/K is a finite Galois extension with Galois group $G = \mathrm{Gal}(L/K)$. Then we can express the characteristic polynomials, traces, and norms in terms of G.

Theorem 1.4.8 *Let L/K be a finite Galois extension, with Galois group $G = \mathrm{Gal}(L/K)$. For $\alpha \in L$, we have*

$$\chi_{\alpha, L/K}(x) = \prod_{\sigma \in G}(x - \sigma(\alpha)).$$

Proof Let $H = \mathrm{Gal}(L/K(\alpha))$. Note that $\sigma \in H$ if and only if $\sigma(\alpha) = \alpha$. Let $\sigma_1, \ldots, \sigma_s$ be representatives of distinct left cosets of H in G. Then the orbit $G\alpha$ consists of $\{\sigma_1(\alpha), \ldots, \sigma_s(\alpha)\}$. If we put

$$f(x) = \prod_{i=1}^{s}(x - \sigma_i(\alpha)),$$

then $\prod_{\sigma \in G}(x - \sigma(\alpha)) = f(x)^{\#H} = f(x)^{[L:K(\alpha)]}$. Using Theorem 1.4.3, it is enough to show that $f(x) = m_{\alpha, K}(x)$.

Any $\sigma \in G$ permutes the elements of the orbit $G\alpha$. Hence,

$$\sigma\left(\prod_{i=1}^{s}(x - \sigma_i(\alpha))\right) = \prod_{i=1}^{s}(x - \sigma\sigma_i(\alpha)) = \prod_{i=1}^{s}(x - \sigma_i(\alpha)) = f(x).$$

This implies that the coefficients of $f(x)$ are fixed by σ. By Galois theory, $L^G = K$, so the fact that the coefficients of $f(x)$ are fixed by all elements of G implies that $f(x) \in K[x]$. Since we can assume $\sigma_1 = 1$, we have $f(\alpha) = 0$. By the definition of the minimal polynomial, $m_{\alpha, K}(x)$ must divide $f(x)$. On the other hand,

$$\deg(f(x)) = [G : H] = [K(\alpha) : K] = \deg(m_{\alpha, K}(x)).$$

Hence, $f(x) = m_{\alpha, K}(x)$, as both are monic. □

Corollary 1.4.9 *Let L/K be a finite Galois extension with Galois group $G = \mathrm{Gal}(L/K)$. For $\alpha \in L$, we have*

$$\mathrm{Tr}_{L/K}(\alpha) = \sum_{\sigma \in G}\sigma(\alpha), \qquad \mathrm{Nr}_{L/K}(\alpha) = \prod_{\sigma \in G}\sigma(\alpha).$$

Example 1.4.10 Let K be a field of characteristic not equal to 2. Assume $x^2 - c \in K[x]$ is irreducible, and let $L = K(\gamma)$ be the splitting field of this quadratic polynomial, where $\gamma^2 = c$. Then $\mathrm{Gal}(L/K) = \{1, \sigma\}$ and $\sigma(\gamma) = -\gamma$. Given an element $\alpha = a + b\gamma$, we have

$$\mathrm{Tr}_{L/K}(\alpha) = (a + b\gamma) + (a - b\gamma) = 2a,$$

$$\mathrm{Nr}_{L/K}(\alpha) = (a + b\gamma)(a - b\gamma) = a^2 - cb^2,$$

$$\chi_{\alpha, L/K}(x) = x^2 - 2ax + (a^2 - cb^2).$$

Exercises

1.4.1 Show that $f(x) = x^3 - 3$ is irreducible over $K = \mathbb{Q}(\sqrt[3]{2})$. Trying to do this by brute force methods is quite messy. Instead, use the following strategy:

(i) If $f(x)$ is reducible over K, then $\sqrt[3]{3} \in K$. Write

$$\sqrt[3]{3} = a + b\sqrt[3]{2} + c(\sqrt[3]{2})^2, \quad a, b, c \in \mathbb{Q}.$$

  Compute $\mathrm{Tr}_{K/\mathbb{Q}}$ of both sides to deduce that $a = 0$.

(ii) Multiply both sides of the above equation by $\sqrt[3]{2}$, and again compute the trace $\mathrm{Tr}_{K/\mathbb{Q}}$ to deduce that $c = 0$.

(iii) Show that $\sqrt[3]{3} = b\sqrt[3]{2}$ is not possible.

1.4.2 Let L/K be a finite extension of degree n and $\alpha \in L$. Let

$$\chi_{\alpha,L/K}(x) = \prod_{i=1}^{n}(x - \beta_i)$$

be the decomposition of the characteristic polynomial of α into linear factors over $\overline{K}$. Show that for any $g(x) \in K[x]$,

$$\chi_{g(\alpha),L/K}(x) = \prod_{i=1}^{n}(x - g(\beta_i)).$$

1.4.3 Let L/K be an extension of degree n and M be an $n \times n$-matrix over K. Show that if $\alpha_i, \beta_j \in L$, $1 \le i, j \le n$, are such that $(\beta_1, \ldots, \beta_n) = (\alpha_1, \ldots, \alpha_n)M$, then

$$\mathrm{disc}_{L/K}(\beta_1, \ldots, \beta_n) = \det(M)^2 \cdot \mathrm{disc}_{L/K}(\alpha_1, \ldots, \alpha_n).$$

1.5   Inseparable Extensions

In this section, we study inseparable extensions of a field K. Note that for K to have such extensions it must have positive characteristic.

We start with a useful result about tensor products, which will be used in this section and also later in the book.

Proposition 1.5.1 *Let R be a subring of a commutative ring S. For an ideal $I \lhd R[x]$, we have*

$$(R[x]/I) \otimes_R S \cong S[x]/I\,S[x].$$

Proof First, note that $R[x] \times S \to S[x]$, $(a, s) \mapsto as$, is R-bilinear and surjective since sx^i is the image of (x^i, s), $i \geq 0$. Hence, $R[x] \otimes_R S \to S[x]$ induced by this map is surjective. We can construct an inverse because we have a well-defined R-linear map $S[x] \to R[x] \otimes_R S$, $\sum_{i=0}^n s_i x^i \mapsto \sum_{i=0}^n x^i \otimes s_i$. Thus, $R[x] \otimes_R S \cong S[x]$. Moreover, it is clear that the image of $I \otimes_R S$ under this isomorphism is $IS[x] = \{\sum_{i=0}^n a_i s_i \mid a_i \in I, s_i \in S[x], n \geq 0\}$.

Next, recall that $(R[x]/I) \otimes_R S$ is generated as an R-module by the elements $\alpha \otimes s$, $\alpha \in R[x]/I$, and $s \in S$. Let $a \in R[x]$ be an element that maps to α under the quotient map $R[x] \to R[x]/I$. Then $a \otimes s \in R[x] \otimes_R S$ maps to $\alpha \otimes s$. Therefore, the image of $R[x] \otimes_R S \to (R[x]/I) \otimes_R S$ contains a set of generators for $(R[x]/I) \otimes_R S$, and so it is equal to it. The kernel of this map is $I \otimes_R S$. Thus, $(R[x] \otimes_R S)/(I \otimes_R S) \cong (R[x]/I) \otimes_R S$. Combining this with the previous paragraph, we obtain the desired isomorphism. □

Lemma 1.5.2 *Assume $f(x) \in K[x]$ is irreducible. Let L/K be a finite separable extension. Then over L the polynomial $f(x)$ decomposes into a product of **distinct** irreducible polynomials, even if $f(x)$ is not separable.*

Proof Let $f = f_1^{s_1} f_2^{s_2} \cdots f_n^{s_n}$ be the decomposition of f over L into a product of powers of distinct irreducibles. We have

$$L \otimes_K (K[x]/(f(x))) \cong L[x]/(f(x)) \qquad \text{(Proposition 1.5.1)} \qquad (1.5.1)$$

$$\cong \prod_{i=1}^n L[x]/(f_i(x)^{s_i}) \qquad \text{(Theorem 1.2.8)}.$$

If some $s_i \geq 2$, then $L[x]/(f_i(x)^{s_i})$ contains nilpotent elements, e.g., the image of $f_i(x)$. On the other hand, because L/K is separable, $L \cong K[x]/(g(x))$ for some separable polynomial $g(x)$. Let $K' = K[x]/(f(x))$. Over K', g decomposes into a product of distinct irreducibles $g = g_1 \cdots g_m$ (because g is separable). Hence, again using Theorem 1.2.8 and Proposition 1.5.1, we have

$$L \otimes_K (K[x]/(f(x))) \cong K[x]/(g(x)) \otimes_K K' \cong \prod_{i=1}^m K'[x]/(g_i(x)).$$

Each $K'[x]/(g_i(x))$ is a field extension of K', so $L \otimes_K (K[x]/(f(x)))$ is isomorphic to a direct product of fields. Such a ring has no nilpotents, so all s_i must be equal to 1. □

Corollary 1.5.3 *Let $f(x) \in K[x]$ be a monic irreducible polynomial, and let L/K be an arbitrary field extension. Assume that $f(x)$ factors in $L[x]$ into a product of distinct monic irreducible polynomials $f(x) = P_1(x) \cdots P_r(x)$. Then*

$$K[x]/(f(x)) \otimes_K L \cong \prod_{i=1}^r L[x]/(P_i(x)),$$

where each $L[x]/(P_i(x))$ is a finite field extension of L. The assumption is satisfied if $f(x)$ is a separable polynomial, or if L/K is a separable extension.

Example 1.5.4 The assumption of the corollary is necessary. Let $K = k(T)$, where k is a field of characteristic p and T is an indeterminate. The polynomial $f(x) = x^p - T \in K[x]$ is irreducible. Let $K' = K[x]/(f(x)) = k(T^{1/p})$. Over K', we have $f(x) = (x - T^{1/p})^p$. Thus,

$$K' \otimes_K K' \cong K'[x]/((x - T^{1/p})^p).$$

Obviously, $x - T^{1/p}$ is a nilpotent element in $K'[x]/((x - T^{1/p})^p)$, so $K' \otimes_K K'$ is not isomorphic to a direct product of fields.

Theorem 1.5.5 *Let L/K be a finite extension. The following are equivalent:*

(1) L/K is separable.
(2) $L \otimes_K \overline{K}$ is reduced, i.e., has no nonzero nilpotent elements.
(3) $\mathrm{Tr}_{L/K} : L \to K$ is surjective.
(4) The bilinear trace form $L \times L \to K$, $(\alpha, \beta) \mapsto \mathrm{Tr}_{L/K}(\alpha\beta)$, is non-degenerate.

Proof Let $\alpha \in L$. We have $K(\alpha) \cong K[x]/(m_{\alpha,K}(x))$. Let $m_{\alpha,K}(x) = \prod_{i=1}^s (x - \alpha_i)^{t_i}$ be the decomposition of $m_{\alpha,K}(x)$ over $\overline{K}$, where $\alpha_1, \alpha_2, \ldots, \alpha_s$ are its distinct roots. Then, as in (1.5.1),

$$K(\alpha) \otimes_K \overline{K} \cong \overline{K}[x]/(m_{\alpha,K}(x)) \cong \prod_{i=1}^s \overline{K}[x]/\big((x - \alpha_i)^{t_i}\big).$$

Hence, $K(\alpha) \otimes_K \overline{K}$ is reduced if and only if $t_i = 1$ for all $1 \leq i \leq s$, i.e., if and only if α is separable over K.

(1) $\Leftrightarrow$ (2): Suppose L/K is separable. By the Primitive Element Theorem (see Corollary 1.3.9), there exists $\alpha \in L$ such that $L = K(\alpha)$. By the previous paragraph, $L \otimes_K \overline{K}$ is reduced, since α is separable. Conversely, suppose L/K is inseparable. Let $\alpha \in L$ be an element that is inseparable over K. Since $K(\alpha) \otimes_K \overline{K}$ is not reduced, it is enough to show that the injection $K(\alpha) \to L$ remains an injection after tensoring with $\overline{K}$. It is even enough to do this by considering the fields $K(\alpha) \subseteq L$ as vector spaces over K. But as vector spaces, we can decompose $L = K(\alpha) \oplus W$ for some K-subspace W of L. Since tensor product commutes with direct sums, we get $L \otimes_K \overline{K} = (K(\alpha) \otimes_K \overline{K}) \oplus (W \otimes_K \overline{K})$. This implies $K(\alpha) \otimes_K \overline{K} \subseteq L \otimes_K \overline{K}$.

(1) $\Leftrightarrow$ (3): Since $\mathrm{Tr}_{L/K}$ is K-linear, it is either surjective or is the zero map. If L/K is separable, then by the Primitive Element Theorem $L = K(\alpha)$ for some $\alpha \in L$, and we need to show that $\mathrm{Tr}_{K(\alpha)/K}$ is nonzero. On the other hand, if L/K is inseparable, then there is $\alpha \in L$, which is inseparable over K. Since $\mathrm{Tr}_{L/K} = \mathrm{Tr}_{K(\alpha)/K}(\mathrm{Tr}_{L/K(\alpha)})$, to show that $\mathrm{Tr}_{L/K}$ is the zero map, it is enough to show that $\mathrm{Tr}_{K(\alpha)/K}$ is the zero map. In either case, we can assume $L = K(\alpha)$.

Observe that the trace of the K-linear transformation induced by multiplication by some $\beta \in L$ on L does not change under extension of scalars, i.e., the trace of the $\overline{K}$-linear transformation induced by multiplication by $\beta \otimes 1$ on $L \otimes_K \overline{K}$ is equal to $\mathrm{Tr}_{L/K}(\beta)$. Therefore, it is enough to show that $\mathrm{Tr}_{L \otimes \overline{K}/K \otimes \overline{K}}$ is nonzero if and only if L/K is separable. By (1.5.1), $K(\alpha) \otimes_K \overline{K} \cong \prod_{i=1}^{s} \overline{K}[x]/((x - \alpha_i)^{t_i})$. For a direct sum of finite dimensional $\overline{K}$-vector spaces $V = V_1 \oplus V_2$ and a linear transformation S preserving this decomposition, we have $\mathrm{Tr}_{V/\overline{K}}(S) = \mathrm{Tr}_{V_1/\overline{K}}(S_1) + \mathrm{Tr}_{V_2/\overline{K}}(S_2)$, where S_1 and S_2 are the linear transformations induced by S on V_1 and V_2, respectively. Hence,

$$\mathrm{Tr}_{\overline{K}(\alpha)/\overline{K}} = \sum_{i=1}^{s} \mathrm{Tr}_{\overline{K}[x]/((x-\alpha_i)^{t_i}))/\overline{K}} .$$

If α is separable, then $t_i = 1$ for all i, so $\overline{K}[x]/((x - \alpha_i)^{t_i}) \cong \overline{K}$. The trace of the idempotent $(1, 0, \ldots, 0) \in \overline{K}^{\oplus s}$ is 1, so $\mathrm{Tr}_{\overline{K}(\alpha)/\overline{K}}$ is not the zero map. Now suppose α is inseparable. Then, by Corollary 1.1.13, the characteristic p of K divides all the t_i's. Since $\overline{K}[x]/((x - \alpha_i)^{t_i}) \cong \overline{K}[y]/(y^{t_i})$, it is enough to show that the trace on $A = \overline{K}[y]/(y^{pt})$ is identically zero. Any element of A can be written as $z = \sum_{j=0}^{pt-1} a_j y^j$ with $a_j \in \overline{K}$. By the linearity of the trace, we have

$$\mathrm{Tr}_{A/\overline{K}}(z) = \sum_{j=0}^{pt-1} a_j \, \mathrm{Tr}_{A/\overline{K}}(y^j).$$

Since y is nilpotent, $\mathrm{Tr}_{A/\overline{K}}(y^j) = 0$ for all $j \geq 1$; cf. [DF04, p. 502]. On the other hand, $\mathrm{Tr}_{A/\overline{K}}(1) = pt = 0$. Thus, $\mathrm{Tr}_{A/\overline{K}}(z) = 0$.

(3) $\Leftrightarrow$ (4): This is easy using our earlier observation that $\mathrm{Tr}_{L/K}$ is surjective if and only if it is not the zero map. Indeed, if the bilinear form is non-degenerate, then taking $\beta = 1$, there is some $\alpha \in L$ such that $\mathrm{Tr}_{L/K}(\alpha) \neq 0$. On the other hand, if $\mathrm{Tr}_{L/K}(\alpha) \neq 0$ and β is an arbitrary nonzero element, then $(\alpha/\beta, \beta)$ maps to a nonzero element of K, so the bilinear form is non-degenerate. □

Corollary 1.5.6 *If $\alpha_1, \ldots, \alpha_n \in \overline{K}$ are separable over K, then the extension $K(\alpha_1, \ldots, \alpha_n)/K$ is separable.*

Proof Denote $L_0 = K$ and $L_i = K(\alpha_1, \ldots, \alpha_i)$, so that $L_0 \subseteq L_1 \subseteq \cdots \subseteq L_n$. From the proof of Theorem 1.5.5, we see that $\mathrm{Tr}_{L_i/L_{i-1}}$ is surjective. Hence,

$$\mathrm{Tr}_{L/K} = \mathrm{Tr}_{L_1/L_0} \circ \mathrm{Tr}_{L_2/L_1} \circ \cdots \circ \mathrm{Tr}_{L_n/L_{n-1}}$$

is also surjective, so L/K is separable. □

The previous corollary implies that the set of separable elements of a finite extension L/K forms a subfield L_s, called the *maximal separable subextension* of L/K (or the *separable closure* of K in L). A finite extension L/K is *purely inseparable* if $L_s = K$. Note that L/L_s is purely inseparable, since if $\alpha \in L$ is separable over L_s, then the extension $L_s(\alpha)$ is separable over K, so $\alpha \in L_s$. The degree $[L : K]_s := [L_s : K]$ is the *separable degree* of L/K, and $[L : K]_i := [L : L_s]$ is the *inseparable degree*. Note that

$$[L : K] = [L : K]_s \cdot [L : K]_i.$$

The next lemma implies that $[L : K]_i$ is equal to a power of the characteristic of K.

Lemma 1.5.7 *A finite extension L/K in characteristic p is purely inseparable if and only if each $\alpha \in L$ has minimal polynomial of the form $x^{p^k} - a \in K[x]$ for some $k \geq 0$.*

Proof Suppose $L_s = K$. Write the minimal polynomial $f(x)$ of $\alpha \in L$ over K as $f(x) = f_{\text{sep}}(x^{p^k})$; see Definition 1.1.15. The element α^{p^k} is a root of f_{sep} and hence is separable over K; hence, $K(\alpha^{p^k})$ is separable over K; hence, $\alpha^{p^k} \in K$; hence, $\deg(f_{\text{sep}}) = 1$; hence, $f(x) = x^{p^k} - a$ for some $a \in K$. The converse is clear.   □

Example 1.5.8 Assume K has positive characteristic p. Let $f(x) = f_{\text{sep}}(x^{p^k})$ be an irreducible polynomial in $K[x]$. Let α be a root of $f(x)$, $L' = K(\alpha^{p^k})$, and $L = K(\alpha)$. We have the tower of extensions $K \subset L' \subset L$. Note that L/L' is purely inseparable since any $\beta \in L$ can be written as $\beta = a_0 + a_1\alpha + \cdots + a_n\alpha^n$ for some $a_0, \ldots, a_n \in L'$, so $\beta^{p^k} = a_0^{p^k} + a_1^{p^k}\alpha^{p^k} + \cdots + a_n^{p^k}(\alpha^{p^k})^n \in L'$, so β is the root of the purely inseparable polynomial $x^{p^k} - \beta^{p^k} \in L'[x]$. On the other hand, L'/K is separable since α^{p^k} is a root of the separable polynomial $f_{\text{sep}}(x) \in K[x]$. We conclude that $L_s = K(\alpha^{p^k})$ and

$$[L : K]_s = \deg(f_{\text{sep}}),$$

$$[L : K]_i = p^k.$$

Theorem 1.5.9 *Let $K \subset L \subset F$ be a tower of finite extensions. Then*

$$[F : K]_s = [F : L]_s \cdot [L : K]_s,$$

$$[F : K]_i = [F : L]_i \cdot [L : K]_i.$$

Proof By Exercise 1.5.6 (see also Exercise 1.3.5), $[L : K]_s$ is equal to the number of distinct embeddings of L into $\overline{K}$. Once we fix an embedding of $\sigma : L \to \overline{K}$, the separable degree $[F : L]_s$ becomes equal to the number of distinct embeddings of F into $\overline{K}$ extending σ. Let $\sigma_1, \ldots, \sigma_n$ be the distinct embeddings of L into $\overline{K}$, and for each i, let $\tau_{i,1}, \ldots, \tau_{i,m}$ be the distinct extensions of σ_i to F. The set of embeddings $\{\tau_{i,j}\}$ contains precisely $[F : L]_s[L : K]_s$ elements. On the other hand, any embedding of F into $\overline{K}$ must be one of the $\tau_{i,j}$, and thus $[F : K]_s = [F : L]_s[L : K]_s$.

The multiplicativity of inseparable degrees in towers follows from the multiplicativity of separable degrees, and the relation $[F : K] = [F : L][L : K]$. □

Proposition 1.5.10 *Suppose L/K is a finite separable extension, and let F/K be a finite purely inseparable extension. Then L and F are linearly disjoint over K.*

Proof We will prove that $L \otimes_K F$ is a field (cf. Exercise 1.3.1). Using the Primitive Element Theorem, we can write $L = K(\alpha)$. Let $f(x)$ be the minimal polynomial of α over K. Note that $LF = F(\alpha)$ and $L \otimes_K F \cong F[x]/(f(x))$. Hence, it is enough to show that $f(x)$ remains irreducible over F. The polynomial $f(x)$ is separable, so all of its roots are separable over K. Suppose $f(x) = g(x)h(x)$ decomposes over F into a product of monic polynomials of positive degrees. Since the roots of $g(x)$ and $h(x)$ are among the roots of $f(x)$, the coefficients of $g(x)$ and $h(x)$, being symmetric combinations of the roots, are separable over K. But F/K is purely inseparable, which implies that $g(x)$ and $h(x)$ have coefficients in K. This contradicts the irreducibility of $f(x)$ over K. □

For the rest of this section, we discuss certain complications that arise only in inseparable extensions, so we assume from now on that K has characteristic $p > 0$. Let L/K be a finite extension. One possible problem with L/K is that L might not be generated by a single element over K, i.e., might not have a primitive element; see Example 1.3.11. Another problem is that there might not exist a purely inseparable extension L'/K such that $L = L_s L'$, i.e., the purely inseparable part of L/K cannot be moved to the bottom field, as the next example demonstrates.

Example 1.5.11 Let k be an algebraically closed field of characteristic p. Let $K = k(T, U)$, where T and U are indeterminates. Let $g(x) = x^p + Tx + U \in k[T, U, x]$ and $f(x) = g(x^p)$. By considering $f(x)$ as a polynomial in $k[T, x][U]$ and applying Theorem 1.1.6, we conclude that $f(x)$ is irreducible over K. Let α be a root of $f(x)$ and $L = K(\alpha)$. Note that $g'(x) = T$, so $g(x)$ is separable. As in Example 1.5.8, we conclude that $L_s = K(\alpha^p)$ and $[L : K]_s = [L : K]_i = p$. Now suppose $K \subset F \subset L$, $[F : K] = p$, and F/K is purely inseparable. If L/F is inseparable, then L/K would be purely inseparable, which it is not. Hence, L/F is separable. Let $h(x) \in F[x]$ be the minimal polynomial of α over F. Then $h(x)$ is separable, has degree p, and divides $f(x)$. Let $f(x) = (x - \alpha_1)^p \cdots (x - \alpha_p)^p$ be the decomposition of $f(x)$ over $\overline{K}$. The stated properties of $h(x)$ leave only one possibility, namely $h(x) = (x - \alpha_1) \cdots (x - \alpha_p)$. Hence, $f(x) = h(x)^p$. On the other hand, $f(x) = x^{p^2} + Tx^p + U = (x^p + T^{1/p}x + U^{1/p})^p$. Therefore, $h(x) = x^p + T^{1/p}x + U^{1/p}$. This implies that $T^{1/p}, U^{1/p} \in F$. Since $[k(T^{1/p}, U^{1/p}) : k(T, U)] = p^2$, we get $[F : K] \geq p^2$, which is a contradiction.

In this book, we will mostly be dealing with $\mathbb{F}_p(T)$ and its finite extensions. We show that the problems indicated earlier are not present for such fields.

Definition 1.5.12 Let K be a field of characteristic $p > 0$, and let $n \geq 1$. Let

$$K^{p^n} = \{\alpha^{p^n} \mid \alpha \in K\}.^9$$

It is easy to check that K^{p^n} is a subfield of K. The field K is *perfect* if $K^p = K$.

Remark 1.5.13 It is not hard to check that K is perfect if and only if every algebraic extension of K is separable[10]; cf. Exercise 1.5.3. Note that any finite field or algebraically closed field is perfect since $K \to K^p$, $\alpha \mapsto \alpha^p$, is a bijection. Also note that for a (not necessarily algebraic) field extension $K(\alpha)$ of K, we have $K(\alpha)^p = K^p(\alpha^p)$.

Let

$$K^{1/p^n} = \{\alpha^{1/p^n} \in \overline{K} \mid \alpha \in K\},$$

where α^{1/p^n} denotes the unique root of the polynomial $x^{p^n} - \alpha$ in $\overline{K}$. The extension K^{1/p^n} is a purely inseparable algebraic extension of K of possibly infinite degree and $(K^{1/p^n})^{p^n} = K$. The *perfect closure* of K is

$$K^{p^{-\infty}} = \left\{\alpha \in \overline{K} \mid \alpha^{p^n} \in K \text{ for some } n \geq 0\right\}.$$

It is not hard to check that $K^{p^{-\infty}}$ is a perfect field, and moreover, $K^{p^{-\infty}}$ is the smallest perfect subfield of $\overline{K}$ containing K; see Exercise 1.5.4.

Lemma 1.5.14 *Let k be a perfect field of characteristic p, and let K be a finite extension of $k(T)$, where T is an indeterminate. Then $[K : K^{p^n}] = p^n$.*

Proof Note that $K \to K^{p^n}$, $\alpha \mapsto \alpha^{p^n}$, is an isomorphism of fields, which maps $k(T)$ isomorphically onto $(k(T))^{p^n}$. Since k is perfect, $(k(T))^{p^n} = k(T^{p^n})$. Hence, $[K : k(T)] = [K^{p^n} : k(T^{p^n})]$. On the other hand,

$$[K : k(T^{p^n})] = [K : K^{p^n}] \cdot [K^{p^n} : k(T^{p^n})]$$

and

$$[K : k(T^{p^n})] = [K : k(T)] \cdot [k(T) : k(T^{p^n})].$$

This implies $[K : K^{p^n}] = [k(T) : k(T^{p^n})] = p^n$.   □

---

[9] Sometimes the notation K^m is used for a vector space of dimension m over K, or for the direct product of m copies of K. Here K^{p^n} has a different meaning.

[10] This equivalent definition applies to any field, not necessarily of positive characteristic, so a field of characteristic 0 is perfect.

To motivate the next lemma, we first consider an example:

Example 1.5.15 Let k be a perfect field of characteristic p and $K = k(T)$. Write $z \in K$ as a ratio of polynomials

$$z = (a_0 + a_1 T + \cdots + a_n T^n)/(b_0 + b_1 T + \cdots + b_m T^m),$$

where $a_i, b_j \in k$. The polynomial $f(x) = x^{p^s} - z \in K[x]$ is purely inseparable. Since k is perfect, we can find unique $c_i, d_j \in k$ such that $(c_i)^{p^s} = a_i$ and $(d_j)^{p^s} = b_j$. The unique root of $f(x)$ in $\overline{K}$ is

$$z^{1/p^s} = \frac{c_0 + c_1 T^{1/p^s} + \cdots + c_n (T^{1/p^s})^n}{d_0 + d_1 T^{1/p^s} + \cdots + d_m (T^{1/p^s})^m}.$$

Hence, $z^{1/p^s} \in k(T^{1/p^s})$ and $K(z^{1/p^s}) = K^{1/p^s}$.

Lemma 1.5.16 *Let k be a perfect field of characteristic p, and let K a finite extension of $k(T)$. Let L/K be a finite purely inseparable extension of degree p^n. Then $L = K^{1/p^n}$.*

Proof By Lemma 1.5.7, L is obtained by adjoining to K the roots of finitely many irreducible polynomials of the form $x^{p^{t_i}} - \alpha_i \in K[x]$. Let t be the maximal t_i. Then $t \le n$. On the other hand, $L^{p^t} \subseteq K$. Hence,

$$p^n = [L : K] \le [L : L^{p^t}] = p^t,$$

where the last equality follows from Lemma 1.5.14. This implies $n \le t$; hence, $n = t$ and $K = L^{p^n}$. □

Theorem 1.5.17 *Let k be a perfect field of characteristic p, and let K be a finite extension of $k(T)$. Let L/K be a finite extension with $[L : K]_i = p^n$. Then $L = K^{1/p^n} L_s$, i.e., L is the composition of the separable extension L_s/K and the purely inseparable extension $K^{1/p^n}/K$.*

Proof Since L/L_s is purely inseparable, $L = L_s^{1/p^n}$ by Lemma 1.5.16. On the other hand, we obviously have $K^{1/p^n} \subset L_s^{1/p^n} = L$. Thus, the purely inseparable extension K^{1/p^n} of K is a subfield of L and so is $K^{1/p^n} L_s$. Now

$$[K^{1/p^n} L_s : K] = [K^{1/p^n} : K] \cdot [L_s : K] \qquad \text{(Proposition 1.5.10)}$$
$$= p^n \cdot [L : K]_s \qquad \text{(Lemma 1.5.14)}$$
$$= [L : K]_i \cdot [L : K]_s$$
$$= [L : K].$$

This implies $L = K^{1/p^n} L_s$. □

Remark 1.5.18 If K is an arbitrary field of characteristic p, but L/K is a finite *normal* extension, then the fixed field F of $\text{Aut}(L/K)$ is purely inseparable over K and $L = L_s F$; see Exercise 1.5.5. Thus, the purely inseparable portion of the extension L/K can be "moved to the bottom."

Theorem 1.5.19 *Let k be a perfect field of characteristic p, and let K be a finite extension of $k(T)$. If L/K is a finite extension, then $L = K(\alpha)$ for some $\alpha \in L$.*

Proof By Theorem 1.3.8, we need to show that L/K has finitely many intermediate subfields. Let $K \subseteq F \subseteq L$ be a tower of fields. Let $[F : K]_i = p^m$ and $[L : K]_i = p^n$. It is clear that $m \le n$, so $K^{1/p^m} \subseteq K^{1/p^n}$. It is also clear that $F_s \subseteq L_s$. By Theorem 1.5.17, $F = F_s K^{1/p^m}$. Hence, F is a composition of subfields of K^{1/p^n} and L_s. The extension $K^{1/p^n}/K$ has only finitely many intermediate fields, namely K^{1/p^t}, $0 \le t \le n$. On the other hand, the Fundamental Theorem of Galois theory implies that L_s/K also has only finitely many intermediate fields. Therefore, there are only finitely many possibilities for F. $\square$

Definition 1.5.20 Let K be a field of characteristic p. An *Artin–Schreier polynomial* is a polynomial of the form $f(x) = x^{p^n} - x - a \in K[x]$. The splitting field of an Artin–Schreier polynomial is called an *Artin–Schreier extension* of K. Note that $f(x)$ is separable, since $f'(x) = -1$.

Theorem 1.5.21 *Let K be a field of characteristic p.*

(1) Let $f(x) = x^p - x - a \in K[x]$ be an Artin–Schreier polynomial and α be a root of $f(x)$. If $\alpha \notin K$, then $K(\alpha)$ is the splitting field of $f(x)$ and $\text{Gal}(K(\alpha)/K) \cong \mathbb{Z}/p\mathbb{Z}$.
(2) Any cyclic extension of K of degree p is an Artin–Schreier extension.

Proof

(1) Let α be a root of $f(x)$. For any $i \in \mathbb{F}_p$, we have

$$(\alpha + i)^p - (\alpha + i) - a = \alpha^p + i^p - \alpha - i - a = \alpha^p - \alpha - a = 0,$$

so the roots of $f(x)$ are $\alpha + i$, $i \in \mathbb{F}_p$. If one of the roots lies in K, then all the roots lie in K. Assume no root lies in K. We claim that in that case $f(x)$ is irreducible over K. Suppose $f(x)$ is reducible:

$$f(x) = g(x)h(x), \qquad g(x), h(x) \in K[x].$$

The roots of $g(x)$ are $\alpha + j$, where j runs over a proper subset $S \subset \mathbb{F}_p$. Let d, $1 \le d \le p - 1$, be the degree of $g(x)$. The coefficient of $-x^{d-1}$ in $g(x)$ is $\sum_{j \in S}(\alpha + j) = d\alpha + t$ for some $t \in \mathbb{F}_p$. Since this coefficient is in K and $d \ne 0$, we get that $\alpha \in K$, a contradiction. Therefore, $f(x)$ is irreducible, and $[K(\alpha) : K] = p$. As we observed, $f(x)$ is separable and splits over $K(\alpha)$, so $K(\alpha)/K$ is a Galois extension with a Galois group of prime order p. Thus, $\text{Gal}(K(\alpha)/K) \cong \mathbb{Z}/p\mathbb{Z}$.

(2) Suppose L/K is a cyclic extension of degree p. Since L/K is separable, there is $\theta \in L$ such that $\mathrm{Tr}_{L/K}(\theta) \neq 0$. Fix a generator σ of $\mathrm{Gal}(L/K)$, and let

$$\alpha = -\frac{1}{\mathrm{Tr}_{L/K}(\theta)}[\sigma(\theta) + 2\sigma^2(\theta) + \cdots + (p-1)\sigma^{p-1}(\theta)].$$

It is straightforward to check that $\alpha - \sigma(\alpha) = -1$. Hence, $\sigma(\alpha) = \alpha + 1$, which implies that $\alpha \notin K$ and $K(\alpha) = L$ (since $[K(\alpha) : K]$ divides $p = [L : K]$). Let $a = \alpha^p - \alpha$. Then

$$\sigma(a) = (\alpha + 1)^p - (\alpha + 1) = \alpha^p - \alpha = a,$$

which implies that a is fixed by $\mathrm{Gal}(L/K)$. Thus, $a \in K$, and α is a root of the Artin–Schreier polynomial $f(x) = x^p - x - a \in K[x]$.

$\square$

We conclude this section with a theorem of number theoretic nature, which we will use later in the book.

Definition 1.5.22 Given a subring R of a commutative ring S, an element $a \in S$ is *integral over R* if it is a root of a monic polynomial in $R[x]$. The ring S is *integral over R* if all its elements are. The *integral closure* of R in S is the subset of S consisting of elements that are integral over R. It is not hard to show that $s \in S$ is integral over R if and only if $R[s]$ is a finitely generated R-module, where $R[s]$ is the ring of all R-linear combinations of powers of s; see [DF04, p. 692]. This then implies that the integral closure of R in S is a subring of S and that integrality is transitive, i.e., if a commutative ring S' is integral over S and S is integral over R, then S' is integral over R; see [DF04, p. 692]. The ring R is *integrally closed in S* if the integral closure of R in S is R itself.

Theorem 1.5.23 *Let k be a perfect field of characteristic p, and let K be a finite extension of $k(T)$. Let B be the integral closure of $A = k[T]$ in K. Then B is a free A-module of rank $[K : k(T)]$.*

Proof If we assume that $K/k(T)$ is a separable extension, then the standard proof that the ring of integers in a number field $\mathcal{K}$ is a free $\mathbb{Z}$-module of rank $[\mathcal{K} : \mathbb{Q}]$ works also in our situation; see [Mar18, p. 22] or [DF04, p. 696]. The problem with that argument for inseparable extensions is that it relies on the discriminant of K or rather the non-degeneracy of the trace map $\mathrm{Tr}_{K/k(T)}$, which is actually degenerate in the inseparable case by Theorem 1.5.5.

Instead, using Theorem 1.5.17, we decompose $K/k(T)$ into a tower of extensions $k(T) \subseteq K' \subseteq K$, where $K' = k(T^{1/p^n})$ is purely inseparable over $k(T)$ and K/K' is separable. It is easy to check that the integral closure B' of $k[T]$ in K' is $k[T^{1/p^n}]$, which is a free A-module with basis $1, u, u^2 \ldots, u^{p^n-1}$, where $u := T^{1/p^n}$. By the remark at the beginning of the proof, B is a free $k[u]$-module of rank $[K : K']$. Hence, B is a free A-module of rank $p^n \cdot [K : K'] = [K : k(T)]$, as claimed.

$\square$

Remark 1.5.24 The issue of separability is quite subtle for the conclusion of Theorem 1.5.23. In Exercise 11 on page 205 of [BS66], one finds an example of a PID A with fraction field F and an inseparable finite extension K/F such that the integral closure of A in K is not finitely generated as an A-module.

Exercises

1.5.1 Let L/K be a finite extension. Let $\alpha_1, \dots, \alpha_n \in L$. Prove that $\text{disc}_{L/K}(\alpha_1, \dots, \alpha_n) = 0$ if and only if either L/K is inseparable or $\alpha_1, \dots, \alpha_n$ are linearly dependent over K.

1.5.2 Let K be a field, and let D be a finite dimensional commutative K-algebra. Assume every element of D is separable over K. Prove the following:

(a) $D \cong L_1 \times L_2 \times \cdots \times L_m$, where each L_i is a finite separable extension of K.
(b) If K is infinite, then $D \cong K[x]/(f(x))$ for some separable $f(x) \in K[x]$.

1.5.3 Let K be a field of characteristic p. Prove that the following are equivalent:

(i) K is perfect, i.e., $K^p = K$.
(ii) Every irreducible polynomial in $K[x]$ is separable.
(iii) Every finite extension of K is separable.

1.5.4 Let K be a field of characteristic p. Prove that:

(a) $K^{p^{-\infty}}$ is a perfect field.
(b) $\overline{K}/K^{p^{-\infty}}$ is a separable extension.
(c) If L is a perfect field such that $K \subseteq L \subseteq \overline{K}$, then $K^{p^{-\infty}} \subseteq L$.
(d) If L/K is a purely inseparable extension, then $L \subseteq K^{p^{-\infty}}$.

1.5.5 Let L/K be a normal extension.

(a) Prove that L_s is normal over K.
(b) Let F be the fixed field of $\text{Aut}(L/K)$. Prove that F/K is purely inseparable and $L = L_s F$.

1.5.6 Let L/K be a finite extension. Let $s = [L : K]_s$ and $p^k = [L : K]_i$. Show that there are exactly s distinct embeddings $\sigma_1, \dots, \sigma_s : L \to \overline{K}$ and that

$$\text{Nr}_{L/K}(\alpha) = \left(\prod_{i=1}^{s} \sigma_i(\alpha)\right)^{p^k}.$$

1.5.7 Let $K = \mathbb{F}_q(T)$ and $f(x) = x^q - x - T \in K[x]$, where $\mathbb{F}_q$ is the finite field with $q = p^n$ elements. Prove that $f(x)$ is irreducible, and determine its Galois group over K.

1.5.8 In the setup of Theorem 1.5.23, assume $K/k(T)$ is purely inseparable of degree p^n. Let $\mathfrak{p}$ be a maximal ideal of A. Prove that $\mathfrak{p}B = \mathfrak{P}^{p^n}$ for some maximal ideal $\mathfrak{P} \lhd B$.

1.5.9 Prove that a UFD is integrally closed in its fraction field.

1.6 Finite Fields

In this section, we briefly review some parts of the theory of finite fields, i.e., fields with finitely many elements. The general theory of fields discussed earlier easily implies some of the fundamental facts about finite fields: for example, the existence and uniqueness of splitting fields of polynomials implies the existence and uniqueness of fields of given prime powers. Nevertheless, the theory of finite fields is vast due to its diverse applications in combinatorics, coding theory, and cryptography. The reader interested in more in-depth discussion of finite fields, as well as their applications, might consult the textbook by Lidl and Niederreiter [LN94].

Assume k is a finite field. The characteristic of k has to be positive, so k contains $\mathbb{F}_p$ as a subfield for some prime p. When we consider k as a finite dimensional vector space over $\mathbb{F}_p$, k is isomorphic with the space of n-tuples $(\alpha_1, \dots, \alpha_n)$, where $\alpha_1, \dots, \alpha_n \in \mathbb{F}_p$ and $n = \dim_{\mathbb{F}_p} k$. Therefore, since we have p choices of each α_i, we get $\#k = p^n$, i.e., the order of k is a power of p. Next, since $k^\times$ is a group of order $p^n - 1$, using Lagrange's theorem from group theory, we deduce that for every element $\alpha \in k$, we have $\alpha^{p^n} = \alpha$ (see [DF04, p. 90]). Thus, k contains the splitting field of $f(x) = x^{p^n} - x \in \mathbb{F}_p[x]$, as the number of distinct roots of $f(x)$ is at most $p^n = \#k$. On the other hand, since $f'(x) = -1$, $f(x)$ has no multiple roots, so the splitting field of $f(x)$ must have at least p^n elements. We conclude that k is the splitting field of $x^{p^n} - x$. Thus, k is uniquely determined by its order, up to isomorphism.

Conversely, to show the existence of a finite field of order p^n for any $n \geq 1$, let k be the splitting field of $f(x) = x^{p^n} - x \in \mathbb{F}_p[x]$. For any two roots α, β of $f(x)$, we have

$$(\alpha + \beta)^{p^n} = \alpha^{p^n} + \beta^{p^n} \qquad \text{(by Example N.2)}$$
$$= \alpha + \beta,$$
$$(\alpha^{-1})^{p^n} = (\alpha^{p^n})^{-1} = \alpha^{-1}, \qquad \text{(if } \alpha \neq 0),$$
$$(\alpha\beta)^{p^n} = \alpha^{p^n}\beta^{p^n} = \alpha\beta.$$

Thus, the roots of $f(x)$ form a subfield $\mathbb{F}$ of k, and since k is the splitting field of f by assumption, we must have $k = \mathbb{F}$. Since the roots of $f(x)$ are distinct, $\#\mathbb{F} = p^n$. We arrive at the following:

Theorem 1.6.1 *For each prime power $q = p^n$, there is a finite field of order q, unique up to isomorphism.*

We denote the field of order q by $\mathbb{F}_q$. If instead of $\mathbb{F}_p$ we start with an arbitrary $\mathbb{F}_q$, then we can obtain the field $\mathbb{F}_{q^n}$ as the splitting field of $x^{q^n} - x$ considered as a polynomial in $\mathbb{F}_q[x]$.

Theorem 1.6.2 $\mathbb{F}_{q^n}$ *is the unique field extension of $\mathbb{F}_q$ of degree n in $\overline{\mathbb{F}}_q$. The extension $\mathbb{F}_{q^n}/\mathbb{F}_q$ is Galois with a cyclic Galois group $\mathrm{Gal}(\mathbb{F}_{q^n}/\mathbb{F}_q) \cong \mathbb{Z}/n\mathbb{Z}$ generated by the automorphism*

$$\tau : \mathbb{F}_{q^n} \to \mathbb{F}_{q^n}$$

$$\alpha \mapsto \alpha^q.$$

*The automorphism τ is called the **Frobenius automorphism** of $\mathbb{F}_{q^n}$.*

Proof By counting the number of elements of $\mathbb{F}_{q^n}$ as a vector space over $\mathbb{F}_q$, we get

$$[\mathbb{F}_{q^n} : \mathbb{F}_q] = n.$$

Moreover, as we have discussed above, any extension of $\mathbb{F}_q$ of degree n is the splitting field of $x^{q^n} - x \in \mathbb{F}_q[x]$ and thus coincides with $\mathbb{F}_{q^n}$. This also implies that $\mathbb{F}_{q^n}/\mathbb{F}_q$ is Galois since $x^{q^n} - x$ is a separable polynomial.

To show that $\mathrm{Gal}(\mathbb{F}_{q^n}/\mathbb{F}_q)$ is cyclic, we first show that τ is an automorphism. By Example N.2, $\tau : \mathbb{F}_{q^n} \to \mathbb{F}_{q^n}$ is a ring homomorphism. Moreover, since $\mathbb{F}_{q^n}$ has no nilpotent elements, τ is injective. But $\mathbb{F}_{q^n}$ is a finite set, so τ must also be surjective. Hence, τ is an automorphism of $\mathbb{F}_{q^n}$. Next, note that τ fixes $\mathbb{F}_q$ since every $\alpha \in \mathbb{F}_q$ is a root of $x^q - x$. Thus, $\tau \in \mathrm{Aut}(\mathbb{F}_{q^n}/\mathbb{F}_q)$. We claim that the cyclic subgroup of $\mathrm{Aut}(\mathbb{F}_{q^n}/\mathbb{F}_q)$ generated by τ has order n, so that $\mathrm{Aut}(\mathbb{F}_{q^n}/\mathbb{F}_q) = \langle \tau \rangle \cong \mathbb{Z}/n\mathbb{Z}$. Indeed, if $\tau^m = 1$ for some $m < n$, then $\alpha^{q^m} = \alpha$ for all $\alpha \in \mathbb{F}_{q^n}$. But this implies that $x^{q^m} - x$ has at least q^n distinct roots, in contradiction with the fact that a polynomial over a field cannot have more roots than its degree. $\square$

Remark 1.6.3 From the previous proof, it is clear that p-th power map $\alpha \mapsto \alpha^p$ is an automorphism of $\mathbb{F}_q$; in fact, it is the Frobenius of $\mathbb{F}_q$ over $\mathbb{F}_p$. Thus, $(\mathbb{F}_q)^p = \mathbb{F}_q$, so any finite field is perfect.

Corollary 1.6.4 *Let $n \geq 1$ be an integer. There is an irreducible polynomial in $\mathbb{F}_q[x]$ of degree n.*

Proof Later in this section, we will prove that the number of irreducible polynomials of degree n is quite large—it is approximately q^n/n.

To prove the weaker statement that there is at least one irreducible polynomial of degree n, note that by Corollary 1.3.9 there is $\alpha \in \mathbb{F}_{q^n}$ such that $\mathbb{F}_{q^n} = \mathbb{F}_q(\alpha)$ (since $\mathbb{F}_{q^n}/\mathbb{F}_q$ is separable). The minimal polynomial $m_{\alpha, \mathbb{F}_q}(x)$ is irreducible and has degree n. □

Corollary 1.6.5 *The trace and norm from $\mathbb{F}_{q^n}$ to $\mathbb{F}_q$ are the following maps:*

$$\mathrm{Tr}_{\mathbb{F}_{q^n}/\mathbb{F}_q}(\alpha) = \sum_{i=0}^{n-1} \alpha^{q^i},$$

$$\mathrm{Nr}_{\mathbb{F}_{q^n}/\mathbb{F}_q}(\alpha) = \prod_{i=0}^{n-1} \alpha^{q^i} = \alpha^{(q^n-1)/(q-1)}.$$

Proof This follows from Corollary 1.4.9 and Theorem 1.6.2. □

Theorem 1.6.6 *Let $f(x) \in \mathbb{F}_q[x]$ be an irreducible polynomial of degree n. Let $\alpha \in \overline{\mathbb{F}}_q$ be a root of $f(x)$. Then $f(x)$ has n distinct roots that are $\alpha, \alpha^q, \ldots, \alpha^{q^{n-1}}$.*

Proof The extension $\mathbb{F}_q(\alpha)/\mathbb{F}_q$ has degree n since $f(x)$ is irreducible. By Theorem 1.6.2, $\mathbb{F}_q(\alpha)/\mathbb{F}_q$ is Galois with $\mathrm{Gal}(\mathbb{F}_q(\alpha)/\mathbb{F}_q) \cong \mathbb{Z}/n\mathbb{Z}$. In particular, $f(x)$ is separable. The Galois group permutes the roots of $f(x)$. If $\tau^i(\alpha) = \tau^j(\alpha)$ for some $0 \le i < j \le n-1$, then τ^{j-i} fixes $\mathbb{F}_q(\alpha)$, which contradicts Theorem 1.3.3. Thus, $\alpha, \tau(\alpha) = \alpha^q, \ldots, \tau^{n-1}(\alpha) = \alpha^{q^{n-1}}$ are the distinct roots of $f(x)$. □

The following important result about the structure of the multiplicative group $\mathbb{F}_q^{\times}$ follows from Theorem 1.2.9.

Theorem 1.6.7 $\mathbb{F}_q^{\times}$ *is a cyclic group, i.e., there is an element $\alpha \in \mathbb{F}_q$ such that*

$$\mathbb{F}_q^{\times} = \{\alpha, \alpha^2, \ldots, \alpha^{q-1} = 1\}.$$

Proposition 1.6.8 *Both $\mathrm{Tr}_{\mathbb{F}_{q^n}/\mathbb{F}_q}$ and $\mathrm{Nr}_{\mathbb{F}_{q^n}/\mathbb{F}_q}$ are surjective.*

Proof The surjectivity of the trace follows from Theorem 1.5.5 because $\mathbb{F}_{q^n}/\mathbb{F}_q$ is separable, but this can also be seen directly as follows. Since $\mathrm{Tr}_{\mathbb{F}_{q^n}/\mathbb{F}_q} : \mathbb{F}_{q^n} \to \mathbb{F}_q$ is an $\mathbb{F}_q$-linear map, its image is at most 1 dimensional, and hence, the kernel must have dimension either $n-1$ or n. If the dimension of the kernel is n, then this means that $\mathrm{Tr}_{\mathbb{F}_{q^n}/\mathbb{F}_q}$ is 0 on all of $\mathbb{F}_{q^n}$. But by Corollary 1.6.5, this implies that the polynomial $x + x^q + \cdots + x^{q^{n-1}}$ has q^n roots, which is larger than its degree. Thus, the image of $\mathrm{Tr}_{\mathbb{F}_{q^n}/\mathbb{F}_q}$ is $\mathbb{F}_q$.

For the surjectivity of $\mathrm{Nr}_{\mathbb{F}_{q^n}/\mathbb{F}_q}$, we again use Corollary 1.6.5. According to that corollary, $\mathrm{Nr}_{\mathbb{F}_{q^n}/\mathbb{F}_q}$ is the homomorphism of cyclic groups $\mathbb{F}_{q^n}^{\times} \to \mathbb{F}_q^{\times}$ given by $\alpha \mapsto \alpha^{(q^n-1)/(q-1)}$. Thus, if we choose a generator θ of $\mathbb{F}_{q^n}^{\times}$, then the image of $\mathrm{Nr}_{\mathbb{F}_{q^n}/\mathbb{F}_q}$ is the subgroup of $\mathbb{F}_{q^n}^{\times}$ generated by $\theta^{(q^n-1)/(q-1)}$. The order of $\theta^{(q^n-1)/(q-1)}$ is $q-1$, so $\mathrm{Nr}_{\mathbb{F}_{q^n}/\mathbb{F}_q}$ is surjective. □

Remarks 1.6.9

(1) The finite field $\mathbb{F}_p$ for prime p is easy to describe and work with since we can simply take the numbers $\{0, 1, \ldots, p-1\}$ to represent the elements of $\mathbb{F}_p$ and perform multiplication and addition modulo p. How can one work explicitly with $\mathbb{F}_{p^n}$? First of all, $\mathbb{F}_{p^n}$ is *not* isomorphic to $\mathbb{Z}/p^n\mathbb{Z}$, so we cannot simply do arithmetic modulo p^n. Instead, choose a monic irreducible polynomial $f(x) = a_0 + a_1 x + \cdots + x^n$ of degree n. The field $\mathbb{F}_{p^n}$ is isomorphic to $\mathbb{F}_p[x]/(f)$, so $\mathbb{F}_{p^n}$ can be identified with the set of polynomials in $\mathbb{F}_p[x]$ of degree $\leq n - 1$, with addition and multiplication done modulo f.

As an explicit example, let $q = 3$ and $f(x) = x^2 + 2x + 2$. Then $\mathbb{F}_9$ is the set

$$0, \quad 1, \quad 2, \quad \alpha, \quad 1 + \alpha, \quad 2 + \alpha, \quad 2\alpha, \quad 1 + 2\alpha, \quad 2 + 2\alpha$$

with the obvious addition and multiplication, except anytime we get α^2, we must replace it with $\alpha + 1$, e.g.,

$$(1 + \alpha)(2 + 2\alpha) = 2 + \alpha + 2\alpha^2 = 2 + \alpha + 2(1 + \alpha) = 1.$$

Note that $g(x) = x^2 + 1$ is also irreducible over $\mathbb{F}_3$. Since $\mathbb{F}_9 \cong \mathbb{F}_3[x]/(g)$, the polynomial $g(x)$ splits in $\mathbb{F}_9$. In fact, the previous calculation shows that $1 + \alpha$ and $2(\alpha + 1)$ are the roots of $g(x)$ in $\mathbb{F}_9$. This is also in agreement with Theorem 1.6.6 since

$$(1 + \alpha)^3 = 1 + \alpha^3 = 1 + \alpha(1 + \alpha) = 1 + \alpha + \alpha^2 = 1 + \alpha + 1 + \alpha = 2(1 + \alpha).$$

Finally, note that α is a generator of $\mathbb{F}_9^\times$:

$$\alpha, \quad \alpha^2 = 1 + \alpha, \quad \alpha^3 = 1 + 2\alpha, \quad \alpha^4 = 2,$$
$$\alpha^5 = 2\alpha, \quad \alpha^6 = 2 + 2\alpha, \quad \alpha^7 = 2 + \alpha, \quad \alpha^8 = 1.$$

It is not true in general that if we represent $\mathbb{F}_{q^n}$ as $\mathbb{F}_q[x]/(f(x))$ for an irreducible polynomial of degree n, then a root of $f(x)$ generates $\mathbb{F}_{q^n}^\times$. For example, if we take $\mathbb{F}_9 = \mathbb{F}_3[x]/(x^2 + 1)$, then $x^4 \equiv 1 \pmod{x^2 + 1}$, so a root of $x^2 + 1$ in $\mathbb{F}_9^\times$ generates a subgroup of order 4.

General finite fields also naturally arise as quotients of general number rings by prime ideals, similar to how finite fields of prime order arise as quotients of $\mathbb{Z}$ by prime ideals. For example, $\mathbb{Z}[i]/(3) \cong \mathbb{F}_9$, where $i^2 = -1$ is the imaginary unit.

(2) We point out that finite fields have several rather special properties:

  (a) $\mathbb{F}_q$ has a unique field extension of degree n in $\overline{\mathbb{F}}_q$ for every $n \in \mathbb{Z}_{>0}$.
  (b) Any finite extension of $\mathbb{F}_q$ is Galois.
  (c) $\mathbb{F}_q^\times$ is a cyclic group.
  (d) Every finite integral domain is a finite field, i.e., every nonzero element has a multiplicative inverse. (This is an easy exercise.)
  (e) Every finite division ring is a finite field, i.e., is commutative. (See Exercise 1.7.4.)

These properties are false for most other fields. For example, take the field of rational numbers $\mathbb{Q}$. All $\mathbb{Q}(\sqrt{p})$ for different primes p are mutually non-isomorphic quadratic extensions of $\mathbb{Q}$, and most cubic extensions of $\mathbb{Q}$ are not Galois since the splitting field of a typical irreducible cubic in $\mathbb{Q}[x]$ has degree 6 over $\mathbb{Q}$. Moreover, $\mathbb{Q}^\times$ is not a cyclic group (it is not even finitely generated), and $\mathbb{Z}$ is not a field. Finally, the division quaternion algebras over $\mathbb{Q}$ are not commutative; see Example 1.7.3.

(3) The subfields of $\overline{\mathbb{F}}_q$ containing $\mathbb{F}_q$ are indexed by $m \in \mathbb{Z}_{>0}$, where m corresponds to $\mathbb{F}_{q^m}$. With the partial ordering $\preceq$ on $\mathbb{Z}_{>0}$ given by divisibility, i.e., $m \preceq n$ if $m \mid n$, we have

$$G_{\mathbb{F}_q} = \varprojlim_m \mathrm{Gal}(\mathbb{F}_{q^m}/\mathbb{F}_q) = \varprojlim_m \mathbb{Z}/m\mathbb{Z},$$

where the transition homomorphisms $\mathbb{Z}/n\mathbb{Z} \to \mathbb{Z}/m\mathbb{Z}$, $m \mid n$, are the natural surjective homomorphisms mapping 1 to 1. We will show in Chap. 2 that

$$\varprojlim_m \mathbb{Z}/m\mathbb{Z} \cong \prod_\ell \mathbb{Z}_\ell,$$

where the product on the right-hand side is over all prime numbers and $\mathbb{Z}_\ell$ denotes the group of ℓ-adic integers. The group $\prod_\ell \mathbb{Z}_\ell$ is usually denoted by $\widehat{\mathbb{Z}}$. Thus, we have an isomorphism $G_{\mathbb{F}_q} \cong \widehat{\mathbb{Z}}$ sending the Frobenius automorphism to 1. The subgroup $\langle 1 \rangle = \mathbb{Z}$ is dense in $\widehat{\mathbb{Z}}$.

Theorem 1.6.10 *Let $\mathcal{P}_d$ denote the set of monic irreducible polynomials in $\mathbb{F}_q[x]$ of degree d. In $\mathbb{F}_q[x]$, the polynomial $x^{q^n} - x$ decomposes as*

$$x^{q^n} - x = \prod_{d \mid n} \prod_{f(x) \in \mathcal{P}_d} f(x).$$

Proof Let $m_{\alpha, \mathbb{F}_q}(x) \in \mathbb{F}_q[x]$ be the minimal polynomial of $\alpha \in \mathbb{F}_{q^n}$ over $\mathbb{F}_q$. Since α is a root of $x^{q^n} - x$, the minimal polynomial $m_{\alpha, \mathbb{F}_q}(x)$ divides $x^{q^n} - x$. If $d = \deg(m_{\alpha, \mathbb{F}_q})$, then $\mathbb{F}_q(\alpha) \cong \mathbb{F}_{q^d} \subset \mathbb{F}_{q^n}$. We have

$$n = [\mathbb{F}_{q^n} : \mathbb{F}_q] = [\mathbb{F}_{q^n} : \mathbb{F}_{q^d}] \cdot [\mathbb{F}_{q^d} : \mathbb{F}_q] = [\mathbb{F}_{q^n} : \mathbb{F}_{q^d}] \cdot d,$$

so d divides n. We see that every root of $G(x) = x^{q^n} - x$ is a root of $H(x) = \prod_{d|n} \prod_{f(x) \in \mathcal{P}_d} f(x)$. This implies that $G(x)$ divides $H(x)$, as both polynomials have no multiple roots ($H(x)$ has no multiple roots since any two irreducible polynomials over $\mathbb{F}_q$ are coprime). Conversely, if α is a root of H, then α is a root of some irreducible f of degree d. Since $\mathbb{F}_q[x]/(f) = \mathbb{F}_q(\alpha) \cong \mathbb{F}_{q^d}$, and d divides n, we see that $\alpha \in \mathbb{F}_{q^n}$ (here we use the uniqueness of the extension of $\mathbb{F}_q$ of degree d). Thus, $G(\alpha) = 0$. This implies that $H(x)$ divides $G(x)$, so these polynomials must be equal as they are both monic.   □

Example 1.6.11 Let $q = 2$ and $n = 4$. In this case, the formula in Theorem 1.6.10 corresponds to the decomposition $x^{16} - x$ into irreducibles in $\mathbb{F}_2[x]$:

$$x^{16} - x = x(x+1)(x^2+x+1)(x^4+x+1)(x^4+x^3+1)(x^4+x^3+x^2+x+1).$$

The proof of Theorem 1.6.10 shows that the subfields of $\mathbb{F}_{q^n}/\mathbb{F}_q$ are in natural bijection with the positive divisors of n. For example, the subfields of $\mathbb{F}_{q^{20}}$ are represented by the lattice:

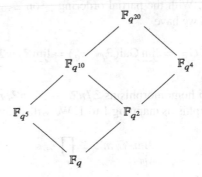

Let $N_q(d)$ be the number of monic irreducible polynomials of degree d in $\mathbb{F}_q[x]$. Computing the degrees of both sides of the equality in Theorem 1.6.10, we get

$$q^n = \sum_{d|n} N_q(d) \cdot d. \tag{1.6.1}$$

Using the Möbius Inversion Formula, one can obtain from this a formula for $N_q(n)$. First, recall that the Möbius function μ on $\mathbb{Z}_{>0}$ can be recursively defined by $\mu(1) = 1$, and

$$\mu(n) = -\sum_{\substack{d|n \\ d<n}} \mu(d), \quad \text{if } n > 1. \tag{1.6.2}$$

A more common (equivalent) definition of the Möbius function is

$$
\mu(n) = \begin{cases} 1 & \text{if } n = 1, \\ (-1)^k & \text{if } n \text{ is a product of } k \text{ distinct primes,} \\ 0 & \text{if } n \text{ is divisible by a square of a prime.} \end{cases}
$$

The Möbius Inversion Formula is an analogue of the inclusion–exclusion principle, in which divisibility of integers replaces inclusion of sets.

Lemma 1.6.12 (Möbius Inversion Formula) *Let h and H be two functions from $\mathbb{Z}_{>0}$ into an additively written abelian group. If*

$$
H(n) = \sum_{d \mid n} h(d) \quad \text{for all } n \in \mathbb{Z}_{>0},
$$

then

$$
h(n) = \sum_{d \mid n} \mu(n/d) H(d).
$$

Proof We have

$$
\sum_{d \mid n} \mu(n/d) H(d) = \sum_{d \mid n} \mu(d) H(n/d) = \sum_{d \mid n} \mu(d) \sum_{c \mid \frac{n}{d}} h(c)
$$

$$
= \sum_{c \mid n} \sum_{d \mid \frac{n}{c}} \mu(d) h(c) = \sum_{c \mid n} h(c) \sum_{d \mid \frac{n}{c}} \mu(d) = h(n).
$$

□

Using (1.6.1) and Lemma 1.6.12, we obtain

$$
N_q(n) = \frac{1}{n} \sum_{d \mid n} \mu(d) q^{n/d} = \frac{1}{n} \left(q^n + \sum_{\substack{d \mid n \\ d \geq 2}} \mu(d) q^{n/d} \right). \tag{1.6.3}
$$

The rough estimate $\#\{d \; : \; d \mid n, d \geq 2\} < q^{n/2}$ implies that $N_q(n) > 0$, so there are irreducible polynomials of degree n for any $n \geq 1$. In fact, it is known that the number of divisors of n is $o(n^\varepsilon)$ for all $\varepsilon > 0$ (cf. [Apo76, p. 296]), so we get[11]

$$N_q(n) = \frac{q^n}{n} + O\left(\frac{q^{n/2}}{n}\right). \tag{1.6.4}$$

Remark 1.6.13 The asymptotic formula (1.6.4) is a stronger analogue of the famous Prime Number Theorem for $\mathbb{F}_q[x]$. Indeed, the number of monic polynomials of degree n is q^n, and the formula implies that $N_q(n) \sim q^n / \log_q(q^n)$, which is similar to the asymptotic formula $\pi(n) \sim n/\log(n)$ for the number of primes less than n. The analogue of (1.6.4) for $\pi(n)$ is expected to be true, but this is essentially equivalent to the Riemann Hypothesis.

Example 1.6.14 Using (1.6.3), we list the first few values of $N_q(n)$:

$$N_q(1) = q, \quad N_q(2) = (q^2 - q)/2, \quad N_q(3) = (q^3 - q)/3,$$
$$N_q(4) = (q^4 - q^2 - q)/4.$$

For example, if $q = 3$, then $N(2) = 3$, and the three quadratic irreducibles in $\mathbb{F}_3[x]$ are

$$x^2 + 1, \quad x^2 + x + 2, \quad x^2 + 2x + 2.$$

Exercises

1.6.1 Show that there are exactly four commutative rings of cardinality p^2.

1.6.2 Suppose $f(x) \in \mathbb{F}_q[x]$ is a polynomial of degree 5. Show that $f(x)$ is irreducible if and only if $f(x)$ is relatively prime to $x^{q^2} - x$.

1.6.3 Let $n \geq 0$ be a fixed integer. Show that

$$\sum_{c \in \mathbb{F}_q^\times} c^n = \begin{cases} 0 & \text{if } n \not\equiv 0 \pmod{q-1}, \\ -1 & \text{if } n \equiv 0 \pmod{q-1}. \end{cases}$$

1.6.4 Let $f(x) \in \mathbb{F}_q[x]$ be an irreducible polynomial of degree n. Prove that over $\mathbb{F}_{q^m}$ the polynomial $f(x)$ decomposes into a product of $\gcd(n, m)$ distinct irreducible polynomials, each of degree $n/\gcd(n, m)$.

---

[11] Recall that, given two functions $f, g \colon \mathbb{R}_{>0} \to \mathbb{R}_{>0}$, the notation $f(x) = o(g(x))$ means that $f(x) \leq M \cdot g(x)$ for any $M > 0$ as $x \to \infty$, and $f(x) = O(g(x))$ means that $f(x) \leq M \cdot g(x)$ for some $M > 0$ as $x \to \infty$.

1.6.5 Prove that the intersection $\mathbb{F}_{q^n} \cap \mathbb{F}_{q^m}$ inside $\overline{\mathbb{F}}_q$ is $\mathbb{F}_{q^d}$, where d is the greatest common divisor of n and m.

1.6.6 The following two statements are special cases of the so-called Hilbert Theorem 90, but they are much easier to prove for finite fields compared to general fields:

(a) For $\alpha \in \mathbb{F}_{q^n}$, $\mathrm{Tr}_{\mathbb{F}_{q^n}/\mathbb{F}_q}(\alpha) = 0$ if and only if there exists $\beta \in \mathbb{F}_{q^n}$ with $\alpha = \beta^q - \beta$.
(b) For $\alpha \in \mathbb{F}_{q^n}$, $\mathrm{Nr}_{\mathbb{F}_{q^n}/\mathbb{F}_q}(\alpha) = 1$ if and only if there exists $\beta \in \mathbb{F}_{q^n}$ with $\alpha = \beta^q/\beta$.

1.6.7 Let $K = \mathbb{F}_q(T)$ and $L = \mathbb{F}_{q^n}(T)$. Consider L as a finite Galois extension of K. Prove that $\mathrm{Nr}_{L/K}$ is not surjective.

1.6.8

(a) Let $H(n)$ and $h(n)$ be two functions from $\mathbb{Z}_{>0}$ into a multiplicatively written abelian group G. Prove that

$$II(n) = \prod_{d|n} h(d) \qquad \text{for all } n \subset \mathbb{Z}_{>0}$$

if and only if

$$h(n) = \prod_{d|n} H(d)^{\mu(n/d)}.$$

(b) Prove that

$$\prod_{d|n}(x^{q^d} - x)^{\mu(n/d)}$$

is equal to the product of all monic irreducible polynomials in $\mathbb{F}_q[x]$ of degree n.

1.6.9 Let $\alpha \in \mathbb{F}_p$ be an arbitrary nonzero element. Prove that $P(x) = x^p - x - \alpha \in \mathbb{F}_p[x]$ is irreducible.

1.6.10 Let ℓ and p be primes. Determine the number and the degrees of irreducible polynomials in the decomposition of $x^{\ell-1} + x^{\ell-2} + \cdots + x + 1$ in $\mathbb{F}_p[x]$. When is this polynomial irreducible?

1.6.11 Prove that the order of $\mathrm{GL}_n(\mathbb{F}_q)$ is

$$(q^n - 1)(q^n - q) \cdots (q^n - q^{n-1}).$$

1.7   Central Simple Algebras

In this section, we introduce some terminology and facts from the theory of central simple algebras. The theory of central simple algebras is a vast area of algebra, with important applications in number theory, but in this book we will need only a few facts from this theory. We will omit most of the proofs; although these proofs do not involve much beyond basic abstract algebra, they lie somewhat outside of the scope of this book. We direct the interested reader to [Rei03], especially Sections 7 and 9 in *loc. cit.*, for more details.

Definition 1.7.1 Let K be a field. A K-algebra is said to be *simple* if it contains no nontrivial proper two-sided ideals. A *central simple K-algebra* is a simple K-algebra whose center is K. A central simple K-algebra is *finite dimensional* if it is finite dimensional as a vector space over K.

In this section, we assume that all K-algebras under consideration are finite dimensional.

Example 1.7.2 Obviously, a division K-algebra is simple. More generally, let D be a division K-algebra, and let n be a positive integer. We claim that the K-algebra $A = \mathrm{Mat}_n(D)$ is simple. Indeed, denote by $e_{i,j}$ the $n \times n$ matrix that has 1 in its (i, j)-th entry and zeros elsewhere. Clearly, $\{e_{i,j} \mid 1 \leq i, j \leq n\}$ is a D-basis of A. Now let $\mathcal{I}$ be a two-sided nonzero ideal in A. Then there is an element $M \in \mathcal{I}$ that has a nonzero entry, say its (i, j)-th entry is $a_{i,j} \neq 0$. Multiplying M by $a_{i,j}^{-1} I_n$ on the left, we may assume that $a_{i,j} = 1$. Then $e_{i,i} M e_{j,j} = e_{i,j}$. Since $e_{k,i} e_{i,j} e_{j,t} = e_{k,t}$, we get $\mathcal{I} = A \mathcal{I} A = A$. Similarly, it is easy to check that a matrix $S \in A$ commutes with every other matrix in A if and only if $S = \mathrm{diag}(\alpha, \alpha, \ldots, \alpha)$ for some $\alpha \in Z(D)$, where $Z(D)$ denotes the center of D. Thus, A is central if and only if D is central.

Example 1.7.3 Assume the characteristic of K is not 2. Given $a, b \in K^\times$, let $\Delta(a, b)$ be the 4-dimensional algebra over K defined by generators and relations:

$$\Delta(a, b) = K + Ki + Kj + Kij,$$

$$i^2 = a, \quad j^2 = b, \quad ij = -ji.$$

We leave it to the reader the (easy) verification that $\Delta(a, b)$ is central and simple. Given

$$\alpha = x_1 + x_2 i + x_3 j + x_4 ij \in \Delta(a, b), \quad \text{let}$$

$$\bar{\alpha} = x_1 - x_2 i - x_3 j - x_4 ij.$$

A straightforward calculation shows that

$$\alpha\bar{\alpha} = x_1^2 - ax_2^2 - bx_3^2 + abx_4^2 \in K.$$

Hence, if $\alpha\bar{\alpha} \neq 0$, then $\bar{\alpha}/(\alpha\bar{\alpha})$ is the inverse of α. This implies that $\Delta(a, b)$ is a division algebra if and only if the quadratic equation

$$x_1^2 - ax_2^2 - bx_3^2 + abx_4^2 = 0 \tag{1.7.1}$$

has only the trivial solution $x_1 = x_2 = x_3 = x_4 = 0$ in K. Next, note that if a and b are squares in K, then

$$i \mapsto \begin{pmatrix} \sqrt{a} & 0 \\ 0 & -\sqrt{a} \end{pmatrix}, \qquad j \mapsto \begin{pmatrix} 0 & \sqrt{b} \\ \sqrt{b} & 0 \end{pmatrix}$$

defines an isomorphism $\Delta(a, b) \cong \mathrm{Mat}_2(K)$. In fact, one can show that $\Delta(a, b)$ is either a division algebra or $\Delta(a, b) \cong \mathrm{Mat}_2(K)$; see Exercise 1.7.1. Finally, note that for a field extension L/K, we have $\Delta(a, b) \otimes_K L = L + Li + Lj + Lij$, subject to the same relations. Thus, there is a finite extension L/K such that $\Delta(a, b) \otimes_K L \cong \mathrm{Mat}_2(L)$.

Remark 1.7.4 A central simple K-algebra of dimension 4 is called a *quaternion algebra*. A famous example of a quaternion algebra is the Hamilton quaternions $\Delta(-1, -1)$ over $\mathbb{R}$. It is not hard to show that any quaternion algebra over a field of characteristic not equal to 2 is isomorphic to $\Delta(a, b)$ for some $a, b \in K^\times$. If the characteristic of K is 2, then a quaternion algebra is isomorphic to the K-algebra $\Delta(a, b) = K + Ki + Kj + Kij$ defined by $i^2 + i = a$, $j^2 = b$, and $ij = j(1 + i)$ for some $a, b \in K^\times$; see [Vig80, Sec. 1].

Example 1.7.5 Let L/K be a cyclic extension of degree n. Let σ be a generator of $\mathrm{Gal}(L/K)$. Let $a \in K^\times$ be any element. Consider the associative algebra

$$A = (L/K, \sigma, a) = \bigoplus_{i=0}^{n-1} Lu^i, \qquad u \cdot x = \sigma(x) \cdot u, \qquad u^n = a, \qquad x \in L.$$

Such an algebra A is called a *cyclic algebra*. It is not difficult to show that A is a central simple algebra, and $A \otimes_K L \cong \mathrm{Mat}_n(L)$; see [Rei03, §30]. Note that if $b \in K^\times$ is a not a square, then $\Delta(a, b) \cong (K(\sqrt{b})/K, \sigma, a)$, so quaternion algebras are cyclic.

It turns out that, in a sense, Example 1.7.2 captures all possible simple K-algebras, and Example 1.7.5 captures all possible central simple K-algebras for global fields. Recall that a *global field* is a finite extension of $\mathbb{F}_q(T)$ or $\mathbb{Q}$.

Theorem 1.7.6 (Wedderburn) *Any simple K-algebra A is isomorphic to $\mathrm{Mat}_n(D)$ for some $n \geq 1$ and some division K-algebra D. Moreover, A determines n uniquely and determines D up to isomorphism.*

Proof See Theorem 7.4 in [Rei03]. □

Theorem 1.7.7 *If K is a global field, then every central simple K-algebra is cyclic.*

Proof See Theorem 32.20 in [Rei03]. We should mention that, compared to the other theorems in this section, this is a deeper result.   □

Theorem 1.7.8 *Let A be a central simple K-algebra, and let L/K be a field extension. Then $A \otimes_K L$ is a central simple L-algebra.*

Proof See Corollary 7.8 in [Rei03].   □

Definition 1.7.9 Let A be a central simple K-algebra. We say that an extension L of K *splits* A, or is a *splitting field* for A, if

$$A \otimes_K L \cong \mathrm{Mat}_n(L) \quad \text{for some } n \geq 1.$$

It is easy to see that $\overline{K}$ splits any central simple K-algebra A. Indeed, by Theorems 1.7.6 and 1.7.8, $\overline{K} \otimes_K A \cong \mathrm{Mat}_n(\Delta)$ for some division $\overline{K}$-algebra Δ. But every element of a finite dimensional $\overline{K}$-algebra is algebraic over $\overline{K}$, so lies in $\overline{K}$. Hence, $\Delta = \overline{K}$. This implies that the dimension of a central simple K-algebra A is necessarily a square of a natural number since

$$\dim_K A = \dim_{\overline{K}} A \otimes_K \overline{K} = \dim_{\overline{K}} \mathrm{Mat}_n(\overline{K}) = n^2.$$

For example, there are no non-commutative division $\mathbb{R}$-algebras of dimension 3.

It is important in certain problems that splitting fields exist with special properties:

Theorem 1.7.10 *Let D be a central division algebra over K.*

*(1) Every maximal subfield L of D contains K and is a splitting field for D; a subfield is called **maximal** if it is not properly contained in another subfield of D.*

(2) If L is a maximal subfield of D and $[L : K] = n$, then $[D : K] = n^2$. In particular, the degree $[L : K]$ of a maximal subfield L of D does not depend on L.

Proof See Theorem 7.15 in [Rei03].   □

Theorem 1.7.11 (Jacobson–Noether) *Let D be a non-commutative central division algebra over K. Then there is an element in $D \setminus K$ that is separable over K.*

Proof The statement is obvious if the characteristic of K is 0, so we assume that the characteristic of K is $p > 0$. Assume contrary to the claim that every element of $D \setminus K$ is inseparable over K. Then, by Lemma 1.5.7, for every $a \in D \setminus K$, there is an integer $n \geq 0$ such that $a^{p^n} \in K$. It follows that there is an element $a \in D \setminus K$ such that $a^p \in K$.

Consider the map $\delta \colon D \to D$ defined by $\delta(x) = xa - ax$. Clearly, $\delta = f - g$, where $f, g \colon D \to D$ are the K-linear maps defined by $f(x) = ax$ and $g(x) = xa$.

Denote by δ^m the composition of δ with itself m-times. Since $f \circ g = f(g(x)) = axa = g \circ f$, we have

$$\delta^m = \sum_{i=1}^{n} (-1)^{n-i} \binom{n}{i} f^i \circ g^{n-i}.$$

Hence, for all $x \in D$, we have

$$\delta^p(x) = f^p(x) - g^p(x) \qquad \left(\text{because } \binom{p}{i} = 0 \text{ for all } 1 \le i \le p-1\right)$$

$$= a^p x - x a^p$$

$$= 0 \qquad\qquad\qquad\qquad\qquad\qquad\qquad\qquad (\text{because } a^p \in K).$$

This implies that $\delta^n = 0$ for $n \ge p$. On the other hand, since D is central and $a \notin K$, there is $b \in D$ such that $\delta(b) \ne 0$. Thus, there exists an integer $m \ge 1$ such that $c := \delta^m(b) \ne 0$ but $\delta(c) = 0$, i.e., c commutes with a. Let $d = c^{-1} a \delta^{m-1}(b)$. Now

$$ad - da = ac^{-1} a \delta^{m-1}(b) - c^{-1} a \delta^{m-1}(b) a$$

$$= ac^{-1} \left(a \delta^{m-1}(b) - \delta^{m-1}(b) a \right)$$

$$= ac^{-1} \delta^m(b) = ac^{-1} c = a.$$

Hence, $d = 1 + a^{-1} da$. By our assumption, there is an integer $s \ge 1$ such that $d^{p^s} \in K$. Then

$$d^{p^s} = 1 + a^{-1} d^{p^s} a = 1 + d^{p^s},$$

which leads to a contradiction. $\qquad\qquad\qquad\qquad\qquad\qquad\qquad\qquad\qquad\qquad$ □

Corollary 1.7.12 *Let A be a central simple K-algebra of dimension n^2. Then $A \otimes K^{\mathrm{sep}} \cong \mathrm{Mat}_n(K^{\mathrm{sep}})$.*

Proof By Theorem 1.7.6, $A \otimes_K K^{\mathrm{sep}} \cong \mathrm{Mat}_n(\Delta)$ for some central division algebra Δ over K^{sep}. If $[\Delta : K^{\mathrm{sep}}] > 1$, then Δ is necessarily non-commutative. In that case, by Theorem 1.7.11, there is $a \in \Delta \setminus K^{\mathrm{sep}}$, which is separable over K^{sep}, and this is a contradiction. Hence, we must have $\Delta = K^{\mathrm{sep}}$. $\qquad\qquad$ □

Given a K-algebra A, we have seen in Sect. 1.4 how to assign to each $\alpha \in A$ a characteristic polynomial, norm, and trace, using the K-linear transformation of A

arising from the left multiplication by α. In case A is a central simple algebra, there are more refined versions of these concepts defined as follows:

Definition 1.7.13 Given a central simple K-algebra A, choose an extension field L of K that splits A, i.e., there is an isomorphism of L-algebras

$$\iota \colon L \otimes_K A \cong \mathrm{Mat}_n(L), \quad \text{where} \quad [A : K] = n^2.$$

The *reduced characteristic polynomial* of $\alpha \in A$ is

$$\widetilde{\chi}_{\alpha, A/K}(x) = \det(x I_n - \iota(1 \otimes \alpha)).$$

The *reduced norm* and the *reduced trace* of $\alpha \in A$ are

$$\mathrm{nr}_{A/K}(\alpha) = \det(\iota(1 \otimes \alpha)) \quad \text{and} \quad \mathrm{tr}_{A/K}(\alpha) = \mathrm{Tr}(\iota(1 \otimes \alpha)).$$

Note that

$$\widetilde{\chi}_{\alpha, A/K}(x) = x^n - \mathrm{tr}_{A/K}(\alpha) x^{n-1} + \cdots + (-1)^n \, \mathrm{nr}_{A/K}(\alpha).$$

A priori, Definition 1.7.13 depends on the choice of L, but we have the following fact:

Theorem 1.7.14 *Let A be a central simple K-algebra of dimension n^2. For each $\alpha \in A$, $\widetilde{\chi}_{\alpha, A/K}(x)$ lies in $K[x]$ and is independent of the choice of the splitting field L of A used to define $\widetilde{\chi}$. Hence, $\mathrm{nr}_{A/K}(\alpha)$ and $\mathrm{tr}_{A/K}(\alpha)$ also do not depend on the choice of L and lie in K. Moreover,*

$$\chi_{\alpha, A/K}(x) = \widetilde{\chi}_{\alpha, A/K}(x)^n.$$

Hence,

$$\mathrm{Nr}_{A/K}(\alpha) = \mathrm{nr}_{A/K}(\alpha)^n \quad \text{and} \quad \mathrm{Tr}_{A/K}(\alpha) = n \cdot \mathrm{tr}_{A/K}(\alpha).$$

Proof See Theorems 9.3 and 9.5 in [Rei03].   □

Example 1.7.15 Let $A = \Delta(a, b)$ be the quaternion algebra of Example 1.7.3. Then $L = K(\sqrt{a}, \sqrt{b})$ is a splitting field of A. We define $\iota \colon L \otimes_K A \xrightarrow{\sim} \mathrm{Mat}_2(L)$ by

$$\alpha = x_1 + x_2 i + x_3 j + x_4 i j \longmapsto \begin{pmatrix} x_1 + x_2\sqrt{a} & x_3\sqrt{b} + x_4\sqrt{ab} \\ x_3\sqrt{b} - x_4\sqrt{ab} & x_1 - x_2\sqrt{a} \end{pmatrix}.$$

Hence,

$$\mathrm{nr}_{A/K}(\alpha) = x_1^2 - ax_2^2 - bx_3^2 + abx_4^2 = \alpha\bar{\alpha},$$

$$\mathrm{tr}_{A/K}(\alpha) = 2x_1 = \alpha + \bar{\alpha},$$

$$\widetilde{\chi}_{\alpha,A}(x) = x^2 - (\alpha + \bar{\alpha})x + \alpha\bar{\alpha} = (x - \alpha)(x - \bar{\alpha}).$$

Theorem 1.7.16 *Let A be a central simple K-algebra, and let L be a maximal subfield of A. For any $\alpha \in L$, we have*

$$\widetilde{\chi}_{\alpha,A/K}(x) = \chi_{\alpha,L/K}(x) = m_{\alpha,K}(x)^{[L:K(\alpha)]}.$$

In particular, $\mathrm{nr}_{A/K}(\alpha) = \mathrm{Nr}_{L/K}(\alpha)$ for all $\alpha \in L$.

Proof First, observe that the proof of Theorem 1.4.3 applies in this setting because $K(\alpha)/K$ is a field extension. Hence,

$$\chi_{\alpha,A/K}(x) = m_{\alpha,K}(x)^{[A:K(\alpha)]}. \tag{1.7.2}$$

Next, we have

$$[A : K(\alpha)] = [A : L] \cdot [L : K(\alpha)] = n \cdot [L : K(\alpha)], \quad \text{(Theorem 1.7.10),}$$

$$\chi_{\alpha,A/K}(x) = \widetilde{\chi}_{\alpha,A/K}(x)^n, \quad \text{(Theorem 1.7.14),}$$

$$\chi_{\alpha,L/K}(x) = m_{\alpha,K}(x)^{[L:K(\alpha)]}, \quad \text{(Theorem 1.4.3).}$$

Combining these with (1.7.2), we get

$$\widetilde{\chi}_{\alpha,A/K}(x) = m_{\alpha,K}(x)^{[L:K(\alpha)]} = \chi_{\alpha,L/K}(x).$$

The final claim of the theorem follows from

$$\mathrm{nr}_{A/K}(\alpha) = (-1)^n \widetilde{\chi}_{\alpha,A/K}(0) = (-1)^n \chi_{\alpha,L/K}(0) = \mathrm{Nr}_{L/K}(\alpha).$$

$\square$

Corollary 1.7.17 *Let D be a division algebra over K. Let K' be the center of D, so that D is a central division algebra over K'. Define*

$$N : D \longrightarrow K$$

$$\alpha \longmapsto \mathrm{Nr}_{K'/K}(\mathrm{nr}_{D/K'}(\alpha)).$$

A maximal subfield L of D contains K', and N restricted to L agrees with the field norm $\mathrm{Nr}_{L/K}$.

Proof A maximal subfield of D as a K-algebra is also a maximal subfield of D as a K'-algebra. Therefore, by Theorem 1.7.10, L is an extension of K'. Now, using Lemma 1.4.5, it is enough to show that the restriction of $\mathrm{nr}_{D/K'}$ to L agrees with the field norm $\mathrm{Nr}_{L/K'}$; equivalently, it is enough to prove the corollary assuming that D is central over K. But this latter claim is a special case of Theorem 1.7.16. □

Definition 1.7.18 Let A be an n-dimensional K-algebra. A *norm form* on A is a function

$$N : A \to K$$

such that:

(i) $N(\alpha) = \alpha^k$ for all $\alpha \in K$ and a fixed integer $k \geq 0$.
(ii) $N(ab) = N(a)N(b)$ for all $a, b \in A$.
(iii) N is a polynomial function, i.e., given a basis $e_1, \ldots, e_n$ of A over K, there is a polynomial $f(x_1, \ldots, x_n) \in K[x_1, \ldots, x_n]$ such that whenever $a = \sum_{i=1}^{n} \alpha_i e_i$ then

$$N(a) = f(\alpha_1, \ldots, \alpha_n).$$

Example 1.7.19 The norm on a finite dimensional K-algebra defined in Sect. 1.4 is a norm form. The reduced norm on a central simple algebra is also a norm form, as follows from the well-known properties of the determinant function. Finally, the map $N : D \to K$ in Corollary 1.7.17 is a norm form.

Theorem 1.7.20 *Assume K is infinite. Let A be a simple K-algebra with center K', and let N be a norm form on A. Then K' is a field, and there is an integer $m \geq 0$ such that*

$$N = \left(\mathrm{Nr}_{K'/K} \circ \mathrm{nr}_{A/K'} \right)^m.$$

Proof The center of a division algebra is a commutative division algebra, so it is a field. Now by Wedderburn's Theorem 1.7.6, we have $A \cong \mathrm{Mat}_n(D)$ for some integer n and a division algebra D. Thus, $K' := Z(A) \cong Z(D)$ is a field.

Next, we consider the claim about the norm form in three special cases:

(1) When $A = K'$ is field, the claim is that any norm form on K' is a power of the field norm $\mathrm{Nr}_{K'/K}$. This is a result of Flanders [Fla53], [Fla55]. The proof of this result is not easy but does not require anything beyond basic abstract algebra.
(2) When $A \cong \mathrm{Mat}_n(K)$, the norm form N restrict to a group homomorphism

$$N : \mathrm{GL}_n(K) \to K^\times.$$

Since $K^\times$ is abelian, ker N contains the commutator subgroup of $\mathrm{GL}_n(K)$. It is not that difficult to show that the commutator subgroup of $\mathrm{GL}_n(K)$ is $\mathrm{SL}_n(K)$ (here we implicitly use the assumption that K is infinite); cf. [Lan02, p. 541]. Since $\mathrm{SL}_n(K)$ is the kernel of the determinant restricted to $\mathrm{GL}_n(K)$, we get that $N|_{\mathrm{GL}_n(K)} = \varphi \circ \det$ for some homomorphism $\varphi\colon K^\times \to K^\times$. Because N and det are polynomial functions, φ is given by a polynomial in $K[x]$. It is easy to check that the only polynomials that induce endomorphisms of $K^\times$ are x^m, $m \geq 0$. Thus, $N|_{\mathrm{GL}_n(K)} = \det^m$. Since N is a polynomial function, we must have $N = \det^m$ also on $\mathrm{Mat}_n(K)$.

(3) Let K' be a finite extension of K, and let $A = \mathrm{Mat}_n(K')$ considered as a K-algebra. A norm form $N\colon A \to K$ can be considered as a norm form $A \to K'$. Hence, by (2), $N = N' \circ \det^t$ is the composition of $\det^t$ for some $t \geq 0$ with a norm form $N'\colon K' \to K$. By (1), $N' = (\mathrm{Nr}_{K'/K})^s$ for some $s \geq 0$, so $N = (\mathrm{Nr}_{K'/K} \circ \det)^{ts}$.

Now let A be a general simple K-algebra. Let L/K be the maximal separable subextension of K'/K, and let $\sigma_j\colon L \to K^{\mathrm{sep}}$, $1 \leq j \leq h = [L : K]$, be the various embeddings of L into K^{sep} over K. Let K_j^{sep} be the field K^{sep} considered as an L-algebra through σ_j. Then

$$K' \otimes_K K^{\mathrm{sep}} \cong K' \otimes_L (L \otimes_K K^{\mathrm{sep}}) \cong K' \otimes_L \prod_{j=1}^h K_j^{\mathrm{sep}} \cong \prod_{j=1}^h K' \otimes_L K_j^{\mathrm{sep}}.$$

From Proposition 1.5.10, one concludes that $K_j' := K' \otimes_L K_j^{\mathrm{sep}}$ is a purely inseparable extension of K_j^{sep}. We have

$$A \otimes_K K^{\mathrm{sep}} \cong A \otimes_{K'} (K' \otimes_K K^{\mathrm{sep}}) \cong \prod_{j=1}^h A \otimes_{K'} K_j'. \tag{1.7.3}$$

Since K_j' is separably closed, from Corollary 1.7.12, we get $A \otimes_{K'} K_j' \cong \mathrm{Mat}_n(K_j')$, where n^2 is the dimension of A over K'. A norm form N on A extends to a norm form on $A \otimes_K K^{\mathrm{sep}}$ over K^{sep} and defines a norm form N_j on $A \otimes_{K'} K_j'$ by restriction to the j-th component of the above decomposition. We have

$$N(\alpha_1, \ldots, \alpha_h) = \prod_{j=1}^h N_j(\alpha_j).$$

By the special case (3) discussed earlier, $N_j = (\mathrm{Nr}_{K_j/K^{\mathrm{sep}}} \circ \det)^{m_j}$. Since N is defined over K, for any automorphism σ of K^{sep} over K, we have

$$N(\sigma(\alpha_1, \ldots, \alpha_h)) = \sigma N(\alpha_1, \ldots, \alpha_h) = \prod_{j=1}^h \sigma N_j(\alpha_j).$$

Considering σ as a permutation of the integers from 1 to h, write $\sigma(j)$ for the image of j under this permutation. Then $\prod_{j=1}^{h} \sigma N_j(\alpha_j) = \prod_{j=1}^{h} N_{\sigma(j)} (\sigma \alpha_{\sigma(j)})$. Substituting, we see that $m_j = m_{\sigma(j)}$ for all j. Since the Galois group of K^{sep} over K acts transitively on the embeddings of L into K^{sep}, we must have all the m_j equal to some $m \geq 0$. Thus,

$$N = \prod_{j=1}^{h} \left(\mathrm{Nr}_{K_j/K^{\mathrm{sep}}} \circ \det \right)^m = \mathrm{Nr}_{L/K} \circ \left(\mathrm{Nr}_{K'/L} \circ \mathrm{nr}_{A/K'} \right)^m = \left(\mathrm{Nr}_{K'/K} \circ \mathrm{nr}_{A/K'} \right)^m.$$

□

Remark 1.7.21 Theorem 1.7.20 appears as a lemma on page 179 in [Mum70], under the assumption that K'/K is separable. Our proof is modeled on Mumford's proof.

The final two theorems of this section are consequences of the so-called Double Centralizer Theorem [Rei03, §7b] and the Noether–Skolem Theorem [Rei03, §7d], respectively. We state them in a slightly specialized form that we will actually need.

Theorem 1.7.22 *Let D be a central division algebra over K. Let K'/K be a finite extension of K in D, i.e., $K \subset K' \subset D$. The centralizer of K' in D,*

$$\mathrm{Cent}_D(K') = \{\alpha \in D \mid \alpha\beta = \beta\alpha \text{ for all } \beta \in K'\},$$

is a central division algebra over K', and

$$[D : K] = [\mathrm{Cent}_D(K') : K] \cdot [K' : K].$$

Theorem 1.7.23 *Let Δ be a central division algebra over K. Let D and D' be two division subalgebras of Δ. If $D \cong D'$ as K-algebras, then there exists $0 \neq u \in \Delta$ such that $D' = uDu^{-1}$.*

Exercises

1.7.1 Let the notation and assumptions be those of Example 1.7.3.

(a) Without appealing to the theorems of this section, show that if $\Delta(a, b)$ is not a division algebra, then it is isomorphic to $\mathrm{Mat}_2(K)$.

(b) Let $L = K(i)$. Construct an explicit isomorphism $L \otimes \Delta(a, b) \xrightarrow{\sim} \mathrm{Mat}_2(L)$.

(c) Show that (1.7.1) always has nontrivial solutions over a finite field, so if K is a finite field, then $\Delta(a, b) \cong \mathrm{Mat}_2(K)$.

1.7.2 Let $(L/K, \sigma, a)$ be the cyclic algebra from Example 1.7.5. Show that:

(a) $(L/K, \sigma, 1) \cong \mathrm{Mat}_n(K)$.

(b) If $b = (\mathrm{Nr}_{L/K}(c))a$ for some $c \in L^\times$, then $(L/K, \sigma, a) \cong (L/K, \sigma, b)$.

1.7.3 Let D be a division algebra over K, and let L be a subfield of D. Prove that L is maximal if and only if $\operatorname{Cent}_D(L) = L$.

1.7.4 Let D be a finite dimensional division algebra over a finite field.

(a) Show that all the maximal subfields of D are isomorphic to each other.
(b) Let L be a maximal subfield of D. Show that

$$D^\times = \bigcup_{\alpha \in D^\times} \alpha L^\times \alpha^{-1}.$$

(c) Show that the distinct subgroups $\alpha L^\times \alpha^{-1}$, $\alpha \in D^\times$, are in bijection with the left cosets of $L^\times$ in $D^\times$. Hence, there are $[D^\times : L^\times]$ sets $\{\alpha L^\times \alpha^{-1}\}$ occurring in the above union, each of cardinality $\#L^\times$.
(d) Conclude that $D = L$, so D is necessarily commutative; this fact, i.e., that every finite division algebra is a field, is known as *Wedderburn's Little Theorem*.

1.7.5 Prove that every central division K-algebra contains a maximal subfield that is separable over K.

1.7.6 Let $K \subset L \subset M$ be a tower of field extensions of finite degrees. Use Flanders' result mentioned in the proof of Theorem 1.7.20 to show that (cf. Lemma 1.4.5)

$$\operatorname{Nr}_{M/K} = \operatorname{Nr}_{L/K} \circ \operatorname{Nr}_{M/L}.$$

Chapter 2
Non-Archimedean Fields

Non-Archimedean valued fields, and the analytic tools available over such fields, will be extensively used in Chaps. 5 and 6. The present chapter contains the necessary preliminaries from the theory of non-Archimedean fields. We start by defining the notion of a non-Archimedean absolute value and then study the completions of fields with respect to such absolute values. In Sect. 2.3, we prove that the non-Archimedean absolute value on a complete field K extends uniquely to a finite extension L of K. In Sects. 2.4 and 2.5, we discuss Hensel's Lemma and the Newton polygon method—two extremely versatile tools available over complete non-Archimedean fields. Hensel's lemma gives successive approximations of the roots of a polynomial, whereas the Newton polygon of a polynomial gives information about the absolute values of its roots. In Sect. 2.6, we introduce the notion of ramification and show that any finite extension L/K of local fields is composed of an unramified extension followed by a totally ramified extension. Then we discuss the inertia subgroup of the Galois group of an extension of a local field, which is closely related to ramification. In Sect. 2.7, we explore some basic notions of analysis in the setting of complete non-Archimedean fields, such as the radius of convergence of a power series, the Weierstrass factorization theorem, and the existence and distribution of zeros of entire functions. In the final section, Sect. 2.8, we discuss the valuations and completions of global function fields, which are essentially the only valuations that are relevant for our purposes. Here the important difference from the case of local fields is that a given valuation might have several non-equivalent extensions in L/K. We also discuss the Galois groups of extensions of global function fields, and the Frobenius elements of these groups.

Non-Archimedean methods have become a natural and indispensable tool in number theory, and there are many excellent introductory textbooks on this topic, such as [Ser79, Kob84, Cas86, Gou20] to name a few, where the reader will find much more in-depth discussions of the topics of this chapter, as well as interesting applications of the theory.

© The Author(s), under exclusive license to Springer Nature Switzerland AG 2023
M. Papikian, *Drinfeld Modules*, Graduate Texts in Mathematics 296,
https://doi.org/10.1007/978-3-031-19707-9_2

2.1 Valuations

Definition 2.1.1 Let D be a division ring. A *(non-Archimedean) valuation* on D is a map

$$v: D \to \mathbb{R} \cup \{+\infty\}$$

that satisfies the following three conditions:

  (i) $v(a) = +\infty$ if and only if $a = 0$.
  (ii) $v(ab) = v(a) + v(b)$ for all $a, b \in D$.
  (iii) $v(a + b) \geq \min(v(a), v(b))$ for all $a, b \in D$.

Remark 2.1.2 We point out some easy consequences of this definition:

- Since $v(1) = v(1 \cdot 1) = v(1) + v(1)$, we have $v(1) = 0$.
- Given a nonzero $a \in D$ and $n \geq 0$, using (ii), we get $v(a^n) = n \cdot v(a)$. On the other hand, since $0 = v(1) = v(aa^{-1}) = v(a) + v(a^{-1})$, we must have $v(a^{-1}) = -v(a)$. Hence,

$$v(a^n) = n \cdot v(a) \quad \text{for all } a \in D \text{ and } n \in \mathbb{Z}.$$

- Suppose $\zeta \in D^\times$ has finite order, i.e., $\zeta^n = 1$ for some $n \geq 1$. Then $n \cdot v(\zeta) = 0$, which implies $v(\zeta) = 0$. In particular, $v(-a) = v(-1) + v(a) = v(a)$ for all $a \in D$.
- (iii) can be sharpened to

    (iii)$'v(a + b) \geq \min(v(a), v(b))$ with equality if $v(a) \neq v(b)$.

  To see this, assume $v(a) > v(b)$. Then

$$v(b) = v(a + b - a)$$
$$\geq \min(v(a + b), v(a))$$
$$= v(a + b) \qquad\qquad (\text{otherwise } v(b) \geq v(a))$$
$$\geq \min(v(a), v(b)) = v(b).$$

- A valuation v on D defines a homomorphism $v: D^\times \to \mathbb{R}$ from the multiplicative group of D into the additive group of real numbers. Its image $v(D^\times) = \{v(a) \mid a \in D^\times\}$ is a subgroup of $\mathbb{R}$, called the *value group* of v.

Example 2.1.3 The map $v: D \to \mathbb{R} \cup \{+\infty\}$ defined by

$$v(a) = \begin{cases} 0 & \text{if } a \neq 0, \\ +\infty & \text{if } a = 0 \end{cases}$$

is a valuation on D, called the *trivial valuation*. (Note that if D is a finite field, then, because $D^\times$ consists of roots of 1, the trivial valuation is the only possible valuation on D.)

We will mostly work with valuations on fields, so from now on we assume that $D = K$ is a field.

Example 2.1.4 Let $K = \mathbb{Q}$, and let p be a prime number. Any nonzero $r \in \mathbb{Q}$ can be written as

$$r = p^n \cdot \frac{s}{t},$$

where $n, s, t \in \mathbb{Z}$ and $p \nmid s$, $p \nmid t$. By the unique factorization in $\mathbb{Z}$, the number n depends only on r. Put

$$\mathrm{ord}_p(r) = n.$$

Then ord_p is a valuation on $\mathbb{Q}$. Indeed, (i) and (ii) are obvious. To see (iii), let $r' = p^{n'} s'/t'$. We may assume $n \geq n'$. Then

$$r + r' = p^{n'} \frac{p^{n-n'} st' + ts'}{tt'}.$$

Clearly, $p \nmid tt'$, so $\mathrm{ord}_p(r + r') = n' + \mathrm{ord}_p(p^{n-n'} st' + ts')$. Since $\mathrm{ord}_p(p^{n-n'} st' + ts') \geq 0$, we get

$$\mathrm{ord}_p(r + r') \geq n' = \min(\mathrm{ord}_p(r), \mathrm{ord}_p(r')),$$

which proves (iii). Note that if $n > n'$, then p divides $p^{n-n'} st'$, which implies that p does not divide $p^{n-n'} st' + ts'$, as otherwise p would divide ts'. Thus, if $n > n'$, then $\mathrm{ord}_p(r + r') = \mathrm{ord}_p(r')$.

Example 2.1.5 Let k be a field, and let $K = k(x)$ be the fraction field of the polynomial ring $k[x]$. Let $P(x)$ be a monic irreducible polynomial in $k[x]$. Any nonzero $f(x) \in k[x]$ can be uniquely decomposed as a product $f(x) = P(x)^n \cdot g(x)$, where $n \geq 0$ and $g(x)$ is relatively prime to $P(x)$. Put

$$\mathrm{ord}_P(f) = n,$$

and extend this map to K by $\mathrm{ord}_P(f/g) = \mathrm{ord}_P(f) - \mathrm{ord}_P(g)$. This is the analogue of ord_p on $\mathbb{Q}$. As in the previous example, it is easy to check that ord_P is a valuation on K.

Example 2.1.6 Let again $K = k(x)$. For a polynomial $f(x) \in k[x]$, put

$$\mathrm{ord}_\infty(f) = -\deg(f),$$

and extend this map to K by $\mathrm{ord}_\infty(f/g) = \mathrm{ord}_\infty(f) - \mathrm{ord}_\infty(g)$. Again, it is easy to check that ord_∞ is a valuation on K. In fact, this example is a special case of the previous example: after replacing x by $y = x^{-1}$, we have $\mathrm{ord}_\infty = \mathrm{ord}_y$ on $K = k(x) = k(y)$. For example, $\mathrm{ord}_\infty(x) = \mathrm{ord}_y(y^{-1}) = -1$.

Definition 2.1.7 We define two valuations v_1, v_2 on K as being *equivalent* if $v_1 = c \cdot v_2$ for some positive real number c. An equivalence class of nontrivial valuations on K is called a *place* of K.

The next theorem is the analogue for $k(x)$ of a famous result of Ostrowski for $\mathbb{Q}$.

Theorem 2.1.8 *Let v be a nontrivial valuation on $k(x)$, whose restriction to k is trivial. If v is non-negative on the polynomial ring $k[x] \subset k(x)$, then there is a monic irreducible polynomial $P(x)$ such that v is equivalent to ord_P. If v takes a negative value on $k[x]$, then v is equivalent to ord_∞.*

Proof First, assume v is non-negative on $k[x]$. Then there is a monic polynomial $P(x) \in k[x]$ of minimal degree such that $v(P) > 0$. We claim that P is irreducible. Indeed, if $P = f \cdot g$, then $v(P) = v(f) + v(g) > 0$, so $v(f) > 0$ or $v(g) > 0$. This would contradict the choice of P. After scaling by a positive real number, we may assume $v(P) = 1$. Suppose $f(x) \in k[x]$ is not divisible by $P(x)$. By the division algorithm, there exist nonzero polynomials $g(x)$ and $r(x)$ such that $f = Pg + r$ and $0 \le \deg(r) < \deg(P)$. By the choice of P, we have $v(r) = 0$. Hence,

$$0 = v(f - Pg) \ge \min(v(f), v(P) + v(g)) \ge \min(v(f), v(P)).$$

Since $v(f) \ge 0$ and $v(P) > 0$, we conclude that $v(f) = 0 = \mathrm{ord}_P(f)$. On the other hand, if $f = P^n g$ with g relatively prime to P, then

$$v(f) = v(P^n) + v(g) = v(P^n) = nv(P) = n = \mathrm{ord}_P(f).$$

Thus, $v = \mathrm{ord}_P$ on $k[x]$, and this implies that $v = \mathrm{ord}_P$ on K.

Now assume v takes some negative values on $k[x]$. As before, there exists a monic polynomial $P(x)$ of minimal degree such that $v(P) < 0$. Write $P(x) = x^n + a_{n-1}x^{n-1} + \cdots + a_0 = x^n + f(x)$, where $n \ge 1$ and $v(f) \ge 0$. Since $0 > v(P) \ge \min(v(x^n), v(f))$, we must have $v(x^n) = nv(x) < 0$. Therefore, $v(x) < 0$. After scaling by a positive real number, we may assume $v(x) = -1$. Denote $y = 1/x$, and consider the subring $k[y] \subset K$. Since $k[y]$ is a polynomial ring and $v(y) = 1$, by our earlier argument, we get $v = \mathrm{ord}_y$ on $k[y]$. Thus, $v = \mathrm{ord}_y$ also on $k(y) = K$. Finally, we write $f(x) \in k[x]$ of degree n as

$$f(x) = y^{-n}(a_n + a_{n-1}y + \cdots + a_0 y^n).$$

Since $a_n \ne 0$, we get

$$v(f) = \mathrm{ord}_y(f) = -n + \mathrm{ord}_y(a_n + a_{n-1}y + \cdots + a_0 y^n) = -n + 0 = \mathrm{ord}_\infty(f).$$

This implies that $v = \mathrm{ord}_\infty$ on $k[x]$, and therefore, $v = \mathrm{ord}_\infty$ on K. $\qquad\square$

Let v be a nontrivial valuation on K. The *valuation ring* (or the *ring of integers*) of v is

$$R_v = \{a \in K \mid v(a) \geq 0\}.$$

We also denote

$$M_v = \{a \in K \mid v(a) > 0\}.$$

It is easy to check that R_v is a subring of K. Note that a nonzero $u \in R_v$ is a unit if and only if its multiplicative inverse u^{-1} in K lies in R_v. Since $v(u^{-1}) = -v(u)$, we conclude that

$$R_v^\times = \{a \in K \mid v(a) = 0\}.$$

It is also easy to check that M_v is an ideal. Since any element of R_v with valuation 0 is a unit, any proper ideal of R_v is contained in M_v. Hence, M_v is the unique maximal ideal of R_v. The quotient field

$$k_v = R_v/M_v$$

is called the *residue field* at v.

Example 2.1.9 The valuation ring of $v = \mathrm{ord}_P$ on $k(x)$ is

$$R_v = \{f(x)/g(x) \in k(x) \mid P(x) \text{ does not divide } g(x)\},$$

and its maximal ideal $M_v = P(x) \cdot R_v$ is the ideal generated by $P(x)$.

Lemma 2.1.10 *Two nontrivial valuations v_1 and v_2 on a field K are equivalent if and only if $M_{v_1} = M_{v_2}$.*

Proof If v_1 and v_2 equivalent, then obviously $M_{v_1} = M_{v_2}$.

Now assume $M_{v_1} = M_{v_2}$. Let $0 \neq a \in M_{v_1}$. Then $c = v_1(a)/v_2(a)$ is a positive real number. We claim that $v_1(b) = c \cdot v_2(b)$ for all $b \in K$. If this claim is false, then there exists b such that $c \cdot v_2(b) < v_1(b)$ (after possibly replacing b with b^{-1}). Choose a rational number n/m, $m > 0$, such that

$$c \cdot v_2(b) < \frac{n}{m} c \cdot v_2(a) = \frac{n}{m} v_1(a) < v_1(b).$$

Then

$$c \cdot v_2(b^m) < c \cdot v_2(a^n) = v_1(a^n) < v_1(b^m)$$

and

$$c \cdot v_2(b^m) - c \cdot v_2(a^n) < 0 < v_1(b^m) - v_1(a^n).$$

Thus,

$$v_2(b^m/a^n) < 0 \quad \text{and} \quad 0 < v_1(b^m/a^n),$$

which contradicts $M_{v_1} = M_{v_2}$.   □

A valuation v is said to be *discrete* if its value group $v(K^\times)$ is a nontrivial discrete subgroup of $\mathbb{R}$ (discrete with respect to the usual topology on $\mathbb{R}$). Note that an additive subgroup G of $\mathbb{R}$ is discrete if and only if there is $\varepsilon > 0$ such that the only $G \cap (-\varepsilon, \varepsilon) = \{0\}$.[1]

Suppose $G \subset \mathbb{R}$ is a discrete subgroup, and let $c \in G$ be its smallest positive element. We claim that $G' = c^{-1}G = \mathbb{Z}$. Indeed, clearly $1 \in G'$ is the smallest positive element, so $\mathbb{Z} \subset G'$. If there is some $g \in G'$, which is not an integer, then for an appropriate $n \in \mathbb{Z}$, we have $0 < g - n < 1$, which leads to a contradiction since $g - n \in G'$. Thus, a discrete valuation is equivalent to a valuation such that $v(K^\times) = \mathbb{Z}$. A discrete valuation on K such that $v(K^\times) = \mathbb{Z}$ is said to be *normalized*.

An integral domain R is called a *discrete valuation ring* (DVR) if R is the valuation ring of a discrete valuation v on the field of fractions of R.

Lemma 2.1.11 *Suppose R is a DVR with respect to the valuation v. Let $\pi \in R$ be any element with minimal positive valuation; such an element is called a* **uniformizer**.

(1) *Any nonzero $a \in R$ can be uniquely decomposed as $a = \pi^n u$, where $u \in R^\times$ and $n \geq 0$.*
(2) *Every nonzero ideal of R is a principal ideal generated by π^n for some $n \geq 0$.*
(3) *$M = (\pi)$ is the unique nonzero prime ideal of R.*

Proof After scaling v by a positive constant, we may assume $v(\pi) = 1$; by the above argument, such valuation is normalized.

(1) Suppose $a \in R$ has valuation $v(a) = n$. Then $u = a/\pi^n \in K$ has valuation 0, so u is a unit in R, and we have $a = \pi^n u$.
(2) Let $I \lhd R$ be a nonzero ideal. Let $n := \min\{v(a) > 0 \mid a \in I\}$. Let $a \in I$ be an element with valuation n. Then $a = \pi^n u$, where $u \in R^\times$. Hence, $u^{-1}a = \pi^n \in I$. This implies that $(\pi^n) \subseteq I$. Now let $b \in I$ be an arbitrary element with $v(b) = m \geq n$. We can write $b = \pi^m u' = \pi^n(\pi^{m-n} u')$, where $u' \in R^\times$. Since $\pi^{m-n} u' \in R$, we see that $b \in (\pi^n)$. Thus, $I = (\pi^n)$.
(3) This is now obvious.   □

---

[1] This condition is obviously necessary. Conversely, if G is not discrete, then there is a Cauchy sequence $g_1, g_2, \ldots$ with $g_n \in G$. In that case, the sequence $\{g_m - g_{m+1}\}$ consists of elements of G converging to 0.

Exercises

2.1.1 A *valuation ring* is an integral domain R with fraction field K such that for every $\alpha \in K$, either $\alpha \in R$ or $\alpha^{-1} \in R$. Prove that every valuation ring is integrally closed.

2.1.2 Show that if R is a PID with a unique maximal ideal $\mathcal{M} \neq 0$, then R is a DVR.

2.1.3 Let $v_1, \ldots, v_n$ be nontrivial valuations on K. Suppose that no two are equivalent. Show that there is $a \in K$ with $v_1(a) < 0$ and $v_i(a) > 0$ for $i = 2, \ldots, n$.

2.2 Completions

A nontrivial valuation on a field K can be used to make K into a metric space. To do this, we introduce the concept of "absolute value" on K that is equivalent to the concept of valuation. Valuations are more natural from the arithmetic perspective since in many cases they measure the divisibility by prime elements, as in the examples of Sect. 2.1. On the other hand, absolute values are more natural if we want to consider K as a metric space. One can define a more general notion of absolute value on a field that includes the usual absolute values on $\mathbb{R}$ and $\mathbb{C}$. We are primarily interested in fields of positive characteristic, and such fields admit only non-Archimedean absolute values, if any; see Exercise 2.2.9.

Definition 2.2.1 Let D be a division ring. A *non-Archimedean absolute value* on D is a function $|\cdot| : D \to \mathbb{R}_{\geq 0}$ such that for all $a, b \in D$ we have:

(i) $|a| = 0$ if and only if $a = 0$.
(ii) $|ab| = |a| \cdot |b|$.
(iii) $|a + b| \leq \max\{|a|, |b|\}$ with equality if $|a| \neq |b|$.

The last property is called the *strong triangle inequality*.

  We define two absolute values $|\cdot|$ and $|\cdot|'$ on D to be *equivalent* and write $|\cdot| \sim |\cdot|'$, if $|a|' = |a|^s$ for all $a \in D$ and a fixed positive constant s.

  Although the concept of completion makes sense for general division rings, we will be mostly working with absolute values on fields, so from now on we assume that $D = K$ is a field. Given a non-Archimedean absolute value $|\cdot|$ on K and a fixed positive constant $c > 1$, the function $v(a) := -\log_c |a|$ is a valuation on K. Conversely, given a valuation v on K, the function $|a| := c^{-v(a)}$ is a non-Archimedean absolute value. Under these constructions, the set of equivalence classes of valuations on K is in natural bijection with the set of equivalence classes of non-Archimedean absolute values on K. Note that the trivial valuation corresponds to the *trivial absolute value*, defined as

$$|a| = \begin{cases} 1 & \text{if } a \neq 0, \\ 0 & \text{if } a = 0. \end{cases}$$

Also note that the valuation ring of v, its maximal ideal, and the units can now be defined as

$$R_v = \{a \in K : |a| \leq 1\},$$
$$M_v = \{a \in K : |a| < 1\},$$
$$R_v^{\times} = \{a \in K : |a| = 1\},$$

where $|\cdot|$ is an absolute value associated to v.

From now on, we fix a nontrivial non-Archimedean absolute value $|\cdot|$ on K. Defining the distance between two elements $a, b \in K$ to be $|a - b|$ makes K into a metric space.[2] Working with a non-Archimedean metric requires some psychological adjustments. For example, every point of the disk $\mathbb{D}^{\circ}(a, 1) = \{x : |x - a| < 1\}$ has an equal right to be regarded as its center, since for a fixed $b \in \mathbb{D}^{\circ}(a, 1)$ and any $x \in K$ the strong triangle inequality implies

$$|x - b| = |(x - a) + (a - b)| < 1 \quad \Longleftrightarrow \quad x \in \mathbb{D}^{\circ}(a, 1).$$

Note that the next lemma implies that the "unit circle" is open in K. (See also Exercises 2.2.1 and 2.2.7.)

Lemma 2.2.2 *The sets R_v, M_v, $R_v^{\times}$ are both open and closed in K.*

Proof For $a_0 \in R_v$, the open disk $\{a \in K : |a - a_0| < 1\}$ around a_0 is contained in R_v, because if $|a| > 1$, then the strong triangle inequality gives $|a - a_0| = |a| > 1$. Thus, R_v is open. Now consider the complement $U = \{a \in K : |a| > 1\}$ of R_v in K. For $a_0 \in U$, the open disk $\{a \in K : |a - a_0| < 1\}$ around a_0 is contained in U, because if $|a - a_0| < 1$, then the strong triangle inequality gives $|a| = |a_0 - (a_0 - a)| = |a_0| > 1$. Thus, R_v is closed. Similar argument shows that M_v is both open and closed. Finally, $R_v^{\times}$ is the complement of M_v in R_v, so it is also open and closed.   □

We will show that K can be "completed" with respect to $|\cdot|$ in the same way as the real numbers are constructed from $\mathbb{Q}$ by completing with respect to the usual absolute value.

First, we recall some standard definitions from analysis: A sequence $\{a_n\}_{n \geq 1}$ in K is *Cauchy* if

$$\lim_{n,m \to \infty} |a_n - a_m| = 0;$$

---

[2] It is easy to show that $|\cdot| \sim |\cdot|'$ if and only if $|\cdot|$, and $|\cdot|'$ define the same topology on K, i.e., if and only if K has the same open sets with respect to $|\cdot|$ and $|\cdot|'$.

it is *convergent* if there exists $a \in K$ such that

$$\lim_{n \to \infty} |a_n - a| = 0.$$

It is clear that every convergent sequence is Cauchy. The field K is *complete* with respect to $|\cdot|$ if every Cauchy sequence is convergent. We note that the completeness is a property of the metric topology; it does not depend on the particular absolute value in the equivalence class of $|\cdot|$.

Example 2.2.3 Let $K = k(x)$, $v = \mathrm{ord}_x$, and $|a| = c^{-\mathrm{ord}_x(a)}$ be the corresponding absolute value for some fixed $c > 1$. Consider the sequence $a_0 = 1$, $a_n = 1 + x + \cdots + x^n$ for $n \geq 1$. Then

$$|a_m - a_n| = c^{-m} \quad (m < n).$$

Hence, $\{a_n\}_{n \geq 0}$ is Cauchy. Now note that

$$\left| \frac{1}{1-x} - a_n \right| = \left| \frac{x^{n+1}}{1-x} \right| = c^{-(n+1)} \to 0,$$

so $\{a_n\}_{n \geq 0}$ converges to $1/(1-x)$ in x-adic sense. On the other hand, there are naturally occurring Cauchy sequences in $k(x)$ that are not convergent. Consider the polynomial $f(y) = y^2 - y + x$ in $K[y]$. This polynomial has no roots in K, but one can construct a sequence $\{a_n\}_{n \geq 1}$ in $k[x]$ such that $|f(a_n)| \leq c^{-n}$ and $a_{n+1} = a_n + b_n x^n$ for some $b_n \in k$. To do so, put $a_1 = 1$, and assume inductively that we have found $a_2, \ldots, a_n$ with the required properties. Note that $a_m \equiv 1 \pmod{x}$ for $1 \leq m \leq n$. To continue the induction, we need to find $b_n \in k$ such that $f(a_{n+1}) \equiv 0 \pmod{x^{n+1}}$. By expanding

$$f(a_{n+1}) = (a_n^2 - a_n + x) + b_n x^n (2a_n - 1) + b_n^2 x^{2n}$$

$$= t x^n + b_n x^n (2a_n - 1) + b_n^2 x^{2n}, \quad t \in k[x],$$

we see that we need to find $b_n \in k$ such that $t + b_n(2a_n - 1) \equiv 0 \pmod{x}$. Equivalently, we need to find $b_n \in k$ such that $b_n \equiv -t \pmod{x}$, which obviously can be done since the elements of k represent all the cosets of (x) in $k[x]$. The sequence $\{a_n\}_{n \geq 1}$ converges to a root of $f(x)$ that is not in K. (In essence, Hensel's lemma, which we will discuss later in this chapter, is a generalization of this type of approximation of a root of a polynomial.)

Definition 2.2.4 Let L be a field containing K. We say that an absolute value $\|\cdot\|$ on L *extends* the absolute value $|\cdot|$ on K if $\|a\| = |a|$ for all $a \in K$.

We say that a field $\widehat{K}$ with absolute value $\|\cdot\|$ is a *completion* of K with respect to $|\cdot|$ if:

(i) $\widehat{K}$ is an extension of K.
(ii) $\|\cdot\|$ extends $|\cdot|$.

(iii) $\widehat{K}$ is complete with respect to $\|\cdot\|$.

(iv) K is dense in $\widehat{K}$ with respect to the topology induced by $\|\cdot\|$.

Theorem 2.2.5 *A completion of K exists, and any two completions are isomorphic.*

Proof The proof consists of several routine but tedious verifications.

Let $\mathscr{C}$ be the set of all Cauchy sequences in K. Declare two Cauchy sequences as being equivalent $\{a_n\} \sim \{b_n\}$ if $\lim_{n\to\infty} |a_n - b_n| = 0$. Let $\widehat{K}$ be $\mathscr{C}$ modulo the equivalence relation. Define addition and multiplication on $\widehat{K}$ by

$$\{a_n\} + \{b_n\} = \{a_n + b_n\},$$

$$\{a_n\} \cdot \{b_n\} = \{a_n \cdot b_n\}.$$

These operations are well-defined (i.e., the resulting sequences are Cauchy and, up to equivalence, do not depend on the choice of representatives of the equivalence classes of $\{a_n\}$ and $\{b_n\}$) and make $\widehat{K}$ into a commutative ring. If $\{a_n\} \not\sim \{0\}$, then $a_n \neq 0$ for $n \gg 0$, i.e., for all large enough n, so $\{a_n^{-1}\}_{n \gg 0}$ is the multiplicative inverse of $\{a_n\}$. Thus, $\widehat{K}$ is a field. It contains a subfield isomorphic to K as the set of equivalence classes of sequences $\{a_n\}$ with $a_n = a$ for some fixed $a \in K$ and all $n \geq 1$. Define $\|\cdot\| : \widehat{K} \to \mathbb{R}_{\geq 0}$ by $\|\{a_n\}\| = \lim_{n\to\infty} |a_n|$. It is easy to check that $\|\cdot\|$ is an absolute value on $\widehat{K}$ that extends the absolute value $|\cdot|$ on K.

Let $x = \{a_n\}_{n \geq 1}$ be an element of $\widehat{K}$. For a given $n \geq 1$, let $\{b_{n,m}\}_{m \geq 1}$ be the element of $K \subset \widehat{K}$ such that $b_{n,m} = a_n$ for all $m \geq 1$. Then, using the fact that $\{a_n\}_{n \geq 1}$ is Cauchy, one checks that the sequence $\{\{b_{n,m}\}_{m \geq 1}\}_{n \geq 1}$ converges to x, so K is dense in $\widehat{K}$.

Suppose $\{x_n\}_{n \geq 1}$ is a Cauchy sequence in $\widehat{K}$. Since K is dense in $\widehat{K}$, there exists $a_n \in K$ such that $\|x_n - a_n\| < 1/n$ for all $n \geq 1$. The sequence $y = \{a_n\}_{n \geq 1}$ is Cauchy in K, so $y \in \widehat{K}$. One easily checks that $x_n \to y$, so $\widehat{K}$ is complete.

Finally, let $(\widehat{K}_1, \|\cdot\|_1)$ and $(\widehat{K}_2, \|\cdot\|_2)$ be two completions of $(K, |\cdot|)$. Then there is an isomorphism $\phi : K \to K$ such that $\|\phi(a)\|_2 = \|a\|_1$ for all $a \in K$. Since K is dense in $\widehat{K}_1$ and $\widehat{K}_2$, ϕ uniquely extends to an embedding $\varphi : \widehat{K}_1 \to \widehat{K}_2$ such that $\|\varphi(a)\|_2 = \|a\|_1$ for all $a \in \widehat{K}_1$. Since K is dense in $\varphi(\widehat{K}_1)$ and $\widehat{K}_2$, we must have $\varphi(\widehat{K}_1) = \widehat{K}_2$. $\square$

Hardly anybody works with $\mathbb{R}$ as the set of equivalence classes of Cauchy sequences in $\mathbb{Q}$. Fortunately, completions of many non-Archimedean fields that will be important to us have simple explicit descriptions, similar to the decimal expansion of real numbers.

Example 2.2.6 The completion of $k(x)$ with respect to an absolute value associated with ord_x is especially easy to construct; this completion turns out to be the field $k((x))$ of formal Laurent series. Given a nonzero element $\sum_{n=N}^{\infty} a_n x^n \in k((x))$ with $N \in \mathbb{Z}$ and $a_N \neq 0$, define

$$v_x \left(\sum_{n=N}^{\infty} a_n x^n \right) = N. \tag{2.2.1}$$

Then v_x is a discrete valuation on $k((x))$ that extends ord_x on $k(x) \subset k((x))$. Fix $c > 1$, and define an absolute value on $k((x))$ by $\|a\| = c^{-v_x(a)}$, $0 \neq a \in k((x))$. This absolute value extends $|a| = c^{-\mathrm{ord}_x(a)}$ on $k(x)$. To show that $k((x))$ is the completion of $k(x)$ with respect to $|\cdot|$, first note that any power series $a = \sum_{n \geq 0} a_n x^n$ is the limit of the sequence of polynomials $\alpha_m = \sum_{n=0}^m a_n x^n$, $m \geq 1$. Hence, $k[x]$ is dense in $k[[x]]$, and consequently, $k(x)$ is dense in $k((x))$. It remains to show that $k((x))$ is complete. Suppose $\{\alpha_m\}_{m \geq 1}$ is a Cauchy sequence in $k((x))$. By Exercise 2.2.8, for any $N > 0$, there is $m(N)$ such that $\mathrm{ord}_x(\alpha_{m+1} - \alpha_m) > N$ for all $m > m(N)$. This means that for any n the coefficient of x^n in α_m is the same for all large enough m. Denote this common coefficient by a_n. Now it is clear that $\alpha_m \to \sum a_n x^n \in k((x))$, so $\{\alpha_m\}_{m \geq 1}$ is a convergent sequence.

The field of formal Laurent series is actually a prototypical example of many complete fields.

Example 2.2.7 Let $\widehat{K}$ be a field complete with respect to a nontrivial non-Archimedean absolute value $|\cdot|$. Let $\widehat{R}$ be the ring of integers of $\widehat{K}$, and let $\widehat{M}$ be the maximal ideal of $\widehat{R}$. Assume the valuation v associated with $|\cdot|$ is discrete. Furthermore, assume there is a subfield k of $\widehat{R}$ that maps isomorphically onto the residue field $\widehat{R}/\widehat{M}$. Fix a generator π of $\widehat{M}$ (see Lemma 2.1.11). We claim that π is transcendental over k, $\widehat{R} \cong k[[\pi]]$, and $\widehat{K} \cong k((\pi))$.

Since $k \cap \widehat{M} = \{0\}$, we have $|a| = 1$ for all $a \in k^\times$. If π is algebraic over k, then $\pi^n + a_{n-1}\pi^{n-1} + \cdots + a_1\pi + a_0 = 0$ for some $a_0, a_1, \ldots, a_{n-1} \in k$ and $a_0 \neq 0$. Because $0 < |\pi| < 1$, $|a_i| \leq 1$, and $|a_0| = 1$, the strong triangle inequality implies that

$$\left| \pi^n + a_{n-1}\pi^{n-1} + \cdots + a_1\pi + a_0 \right| = 1,$$

which is a contradiction. Hence, π is transcendental over k.

By Exercise 2.2.8, any series $\sum_{n \geq 0} a_n \pi^n$, $a_n \in k$, converges in $\widehat{R}$, so $k[[\pi]] \subseteq \widehat{R}$ is a subring. It remains to show that this inclusion is an equality. Let $\beta \in \widehat{R}$ be an arbitrary element. There is a unique $a_0 \in k$ and a unique $\beta_1 \in \widehat{M}$ such that $\beta = a_0 + \beta_1$. Now $\beta_1/\pi \in \widehat{R}$; hence, there is a unique $a_1 \in k$ and a unique $\beta_2 \in \widehat{M}$ such that $\beta_1/\pi = a_1 + \beta_2$, or equivalently $\beta_1 = a_1\pi + \beta_2\pi$. Continuing this process, i.e., considering β_2/π and so on, gives a unique expansion $\beta = \sum_{n \geq 0} a_n \pi^n$.

Now normalize v so that $v(\pi) = 1$. Let $0 \neq \beta = \sum_{n \geq N} a_n \pi^n \in \widehat{R}$ with $a_N \neq 0$. Then $\beta/\pi^N = a_N + \pi\gamma$ with $v(a_N) = 0$ and $\gamma \in \widehat{R}$. By the strong triangle inequality, we get $v(\beta/\pi^N) = 0$, so $v(\beta) = N$. Thus, once we identify $\widehat{K}$ with $k((\pi))$, the normalized valuation v becomes v_π from (2.2.1).

Remark 2.2.8 The construction of Example 2.2.7 can be generalized to the case when k is not necessarily isomorphic to a subfield of $\widehat{K}$ as follows. Assume again that the valuation on $\widehat{K}$ is discrete. Fix a uniformizer π, and let $S \subset \widehat{R}$ be a complete set of distinct coset representatives for the residue field $\widehat{R}/\widehat{M}$. For each sequence $\{a_n\}_{n \geq N}$ of elements of S, consider the series $\sum_{n \geq N} a_n \pi^n$. Such a series converges in $\widehat{K}$ by Exercise 2.2.8. One can use the argument in Example 2.2.7 to show that every element $\alpha \in \widehat{K}$ is uniquely represented by such a series. For example, for the completion $\mathbb{Q}_p$ of $\mathbb{Q}$ with respect to the valuation ord_p, one can take $S = \{0, 1, \ldots, p-1\}$ and $\pi = p$, so every element of $\mathbb{Q}_p$ is represented by a series $\sum_{n \geq N} a_n p^n$ for uniquely determined $a_0, a_1, \ldots \in S$ and $N \in \mathbb{Z}$. The difference from Example 2.2.7 is that S in $\mathbb{Q}_p$ is no longer closed under addition and multiplication. Thus, when performing arithmetic operations on $\mathbb{Q}_p$, with its elements represented by series $\sum_{n \geq N} a_n p^n$, one has to do digit carry-overs, e.g., for $p = 3$, we have

$$(2 + 2p + p^2)(p + 2p^2 + p^3 + \cdots)$$
$$= (2p + 4p^2 + 2p^3 + \cdots) + (2p^2 + 4p^3 + \cdots) + (p^3 + \cdots)$$
$$= (2p + p^2 + 0p^3 + \cdots) + (2p^2 + p^3 + \cdots) + (p^3 + \cdots)$$
$$= 2p + 3p^2 + 2p^3 + \cdots$$
$$= 2p + 0p^2 + 0p^3 + \cdots$$

Theorem 2.2.9 *Let K be a field equipped with a nontrivial discrete valuation v. Let $\widehat{K}$ be the completion of K with respect to v. If K and $k = R_v/M_v$ have the same characteristic, then $\widehat{K} \cong k((\pi))$, where π is a uniformizer of R_v.*

Proof We will prove this under the assumption that k is a finite field. For a proof of the theorem under the weaker assumption that k is perfect, see [Ser79, Sec. II.4].

Let $\widehat{R}$ be the valuation ring of $\widehat{K}$, and let $\widehat{M}$ be the maximal ideal of $\widehat{R}$. Thanks to Example 2.2.7, it is enough to prove the following three facts:

  (i) The value groups of K and $\widehat{K}$ are the same, i.e., $v(K^\times) = v(\widehat{K}^\times)$. Hence, the valuation on $\widehat{K}$ is discrete.
 (ii) The residue field $\widehat{R}/\widehat{M}$ is isomorphic to k.
(iii) $\widehat{R}$ contains a subfield isomorphic to k.

The fact that $v(K^\times) = v(\widehat{K}^\times)$ easily follows from the fact that K is dense in $\widehat{K}$ and v is discrete (see Exercise 2.2.11).

Now we prove that for any $n \geq 1$, the map that sends a coset $a + M_v^n$ to the coset $a + \widehat{M}^n$, $a \in R_v$ gives an isomorphism

$$R_v/M_v^n \xrightarrow{\sim} \widehat{R}/\widehat{M}^n. \tag{2.2.2}$$

Suppose $a_1, a_2 \in R_v$ are such that $a_1 + \widehat{M}^n = a_2 + \widehat{M}^n$. Then $a_1 - a_2 \in \widehat{M}^n$, so $v(a_1 - a_2) \geq n$. This implies that $a_1 - a_2 \in M_v^n$. Hence, the map is injective. On the other hand, since K is dense in $\widehat{K}$, for any $a' \in \widehat{R}$, there is an element $a \in K$ such that $v(a' - a) \geq n$. Then $a \in R_v$ and $a' + \widehat{M}^n = a + \widehat{M}^n$, which implies that the above map is also surjective.

To prove (iii), let p be the characteristic of k and q be its order. By assumption, $\mathbb{F}_p$ is a subfield of $\widehat{K}$. Since k is the splitting field of $x^q - x \in \mathbb{F}_p[x]$, Hensel's Lemma (see Theorem 2.4.2)[3] implies that $x^q - x$ splits completely in $\widehat{K}$. As in Sect. 1.6, one shows that the roots of $x^q - x$ in $\widehat{K}$ are distinct and form a subfield isomorphic to k. Since the only valuation on a finite field is the trivial valuation, $|a| = 1$ for all $a \in k^\times \subset K^\times$. Thus $k \subset \widehat{R}$.   □

Remark 2.2.10 If K is equipped with a discrete valuation v, has the same characteristic as its residue field k, but is not complete, then generally it is false that K contains a subfield isomorphic to k. For example, we can take $K = \mathbb{F}_p(T)$ and $v = \mathrm{ord}_P$ the valuation corresponding to an irreducible polynomial P in $\mathbb{F}_p[T]$ of degree $d \geq 2$. Then $k \cong \mathbb{F}_{p^d}$, which is not a subfield of K.

Remark 2.2.11 Let the notation and assumptions be as Theorem 2.2.9 and its proof. We have $\widehat{R} \cong k[\![\pi]\!]$. For any $m \geq n$, we have the surjective homomorphism

$$f_{n,m} : \widehat{R}/\widehat{\mathcal{M}}^m \longrightarrow \widehat{R}/\widehat{\mathcal{M}}^n, \quad a + \widehat{\mathcal{M}}^m \longmapsto a + \widehat{\mathcal{M}}^n.$$

It is clear that $f_{s,n} \circ f_{n,m} = f_{s,m}$ for all $m \geq n \geq s$. Thus, we can consider the *inverse limit*

$$\varprojlim_n \widehat{R}/\widehat{\mathcal{M}}^n$$

with respect to these maps. By definition, $\varprojlim(\widehat{R}/\widehat{\mathcal{M}}^n)$ is the subring of $\prod_{n=1}^\infty \widehat{R}/\widehat{\mathcal{M}}^n$ consisting of those tuples $(b_1, b_2, \dots)$ such that $f_{n,m}(b_m) = b_n$, i.e., $b_m \equiv b_n \pmod{\pi^n}$. In particular, $b_{n+1} = b_n + a_{n+1}\pi^{n+1}$ for some $a_{n+1} \in k$. Now it is easy to see that the homomorphism

$$\widehat{R} \longrightarrow \varprojlim_n \widehat{R}/\widehat{\mathcal{M}}^n,$$

$$\sum_{n \geq 0} a_n \pi^n \longmapsto (a_0, a_0 + a_1\pi, a_0 + a_1\pi + a_2\pi^2, \dots)$$

is an isomorphism. Thus, $\widehat{R} \cong \varprojlim \widehat{R}/\widehat{\mathcal{M}}^n$. On the other hand, we proved that $R_v/\mathcal{M}_v^n \xrightarrow{\sim} \widehat{R}/\widehat{\mathcal{M}}^n$. Thus

$$\widehat{R} \cong \varprojlim_n R_v/\mathcal{M}_v^n,$$

where the transition maps in the inverse limit are defied similarly to the transition maps $f_{n,m}$. This gives an algebraic interpretation to the completion of R_v.

---

[3] Although Hensel's Lemma will be proved in a later section, its proof does not use the isomorphism $\widehat{K} \cong k(\!(\pi)\!)$, so the argument here is not circular.

Example 2.2.12 Let $P \in k[x]$ be an irreducible polynomial and $v = \mathrm{ord}_P$ be the associated valuation on $K = k(x)$. Recall that the valuation ring of v is

$$R_v = \{f(x)/g(x) \in k(x) \: : \: P(x) \text{ does not divide } g(x)\}.$$

Let $f/g \in R_v$ be a nonzero element and $n \geq 1$. Since g and P^n are coprime in $k[x]$, there exists a polynomial $h \in k[x]$ such that $gh - 1$ is divisible by P^n. Let $r \in k[x]$ be the residue of fh modulo P^n, i.e., $fh - r$ is divisible by P^n and $\deg(r) < \deg(P^n)$. Now

$$\frac{f}{g} - r = \frac{f - rg}{g} \equiv \frac{g(fh - r)}{g} \equiv 0 \; (\mathrm{mod} \; P^n).$$

Hence, modulo $\mathcal{M}_v^n = (P^n)$, every element of R_v is congruent to some polynomial of degree less than $\deg(P^n)$. Since such polynomials are not congruent to each other modulo (P^n), we conclude that $R_v/\mathcal{M}_v^n \cong k[x]/(P^n)$. Therefore, from the previous remark, we conclude that the completion $\widehat{R}$ of $k[x]$ with respect to ord_P is isomorphic to the inverse limit

$$\widehat{R} \cong \varprojlim_n k[x]/(P^n).$$

Definition 2.2.13 A field that is complete with respect to a nontrivial discrete valuation and has finite residue field is called a *local field*. If K is a local field, v is the normalized valuation on K, and k is its residue field, then the absolute value on K defined by $|a| = (\#k)^{-v(a)}$ is said to be *normalized*.

What we essentially proved in Theorem 2.2.9 is that a local field of positive characteristic is isomorphic to $k(\!(\pi)\!)$, where k is its residue field. One can show that a local field of characteristic zero is a finite extension of $\mathbb{Q}_p$ for some prime p; see [Ser79].

We conclude this section with an alternative (topological) characterization of local fields.

Lemma 2.2.14 *Let $\widehat{K}$ be a field complete with respect to a nontrivial non-Archimedean absolute value $|\cdot|$. Then $\widehat{K}$ is a local field if and only if $\widehat{R}$ is compact.*

Let $r \in |\widehat{K}^\times|$, and pick $\alpha \in \widehat{K}^\times$ with $|\alpha| = r$. Because $\{x \in \widehat{K} \: : \: |x| \leq r\} = \alpha \widehat{R}$, all discs in $\widehat{K}$ are homeomorphic to $\widehat{R}$. Therefore, the statement of the lemma can be reformulated as saying that $\widehat{K}$ is a local field if and only if $\widehat{K}$ is locally compact.

Proof Recall that a subset X of a metric space is compact (respectively, sequentially compact) if any cover of X by open sets contains a finite subcover (respectively, every infinite sequence of elements of X contains a subsequence converging to a point in X). According to the Heine–Borel theorem, compact and sequentially compact are equivalent notions.

First, we prove that if $\widehat{K}$ is not a local field, then $\widehat{R}$ is not compact. Suppose the residue field k is infinite. Let $S \subset \widehat{R}$ be a set of representatives of distinct cosets of $\widehat{M}$ in $\widehat{R}$. Since $|\alpha - \beta| = 1$ for any distinct $\alpha, \beta \in S$, no subsequence of the infinite set S can be Cauchy. Thus, $\widehat{R}$ is not compact. Suppose the valuation v is not discrete. Then there is a sequence $\{x_n\}_{n \geq 1}$ in $\widehat{R}$ such that $|x_i| < |x_{i+1}|$ for all $i \geq 1$. In that case, by the strong triangle inequality, for any $m \neq n$, we have

$$|x_m - x_n| = \max(|x_n|, |x_m|) > |x_1|.$$

Thus, no subsequence of $\{x_n\}_{n \geq 1}$ is Cauchy.

Now we prove that if $\widehat{K}$ is a local field, then $\widehat{R}$ is compact. Assume k is finite and $\widehat{M} = (\pi)$ is principal; cf. Lemma 2.1.11. Let $\{x_n\}_{n \geq 1}$ be a sequence in $\widehat{R}$. Let S be a set of representatives of the distinct cosets of $\widehat{M}$ in $\widehat{R}$. Write a typical element $x_n = \sum_{i \geq 0} a_{n,i} \pi^i$ in terms of an infinite series with all $a_{n,i} \in S$. Since S is finite, there are infinitely many x_n with same $a_{n,0}$; denote this common value by b_0. Among the elements of this latter subsequence, there are infinitely many elements with the same $a_{n,1}$; denote this common value by b_1. We can continue this argument with $a_{n,2}$ and so on. Now it should be clear that there is a subsequence of $\{x_n\}_{n \geq 1}$ converging to $b = \sum_{i \geq 0} b_i \pi^i$. $\qquad\square$

Exercises

Let $|\cdot|$ be a nontrivial non-Archimedean absolute value on K.

2.2.1 Suppose $K = \mathbb{F}_q(\!(\pi)\!)$ and $|\pi| = q^{-1}$. For $\alpha \in K$ and $r > 0$, denote

$$D_K(\alpha, r) = \{\beta \in K \ : \ |\beta - \alpha| \leq r\}.$$

Prove that the unit circle $S^1 := \{\beta \in K \ : \ |\beta| = 1\}$ is a disjoint union of closed discs

$$S^1 = \bigsqcup_{\alpha \in \mathbb{F}_q^\times} D_K(\alpha, 1/q).$$

2.2.2 Prove that for $\alpha_1, \ldots, \alpha_n \in K$, we have

$$|\alpha_n - \alpha_1| \leq \max_{1 \leq i \leq n-1} |\alpha_{i+1} - \alpha_i|.$$

2.2.3 Prove that two absolute values $|\cdot|, |\cdot|'$ on K are equivalent if and only if the following holds:

A sequence is Cauchy with respect to $|\cdot|$ if and only if it is Cauchy with respect to $|\cdot|'$.

2.2.4 Given a polynomial $f(x) = \sum_{i=0}^{n} a_n x^n \in K[x]$, define

$$\|f\| = \max_{0 \le i \le n} |a_i|.$$

Extend $\|\cdot\| : K[x] \to \mathbb{R}$ to the field of rational functions $K(x)$ by $\|f/g\| = \|f\| / \|g\|$, where $f(x), g(x) \in K[x]$. Prove that $\|\cdot\| : K(x) \to \mathbb{R}$ is a non-Archimedean absolute value that coincides with $|\cdot|$ on K.

2.2.5 This exercise is a version of Gauss' Lemma (Theorem 1.1.6).

Let $R = \{a \in K : |a| \le 1\}$ be the ring of integers of K. Suppose $f \in R[x]$ is a product $f = g \cdot h$ of two non-constant elements of $g, h \in K[x]$.

(a) Let $\|\cdot\|$ be the absolute value on $K[x]$ constructed in Exercise 2.2.4. Show that by multiplying g by an appropriate $a \in K$, we can assume that $f = g \cdot h$ and $\|g\| = 1$. Conclude that in that case we also have $\|h\| \le 1$.
(b) Prove that f is a product of two non-constant elements of $R[x]$.

2.2.6 This exercise is a version of the Chinese Remainder Theorem.

Let $|\cdot|_1, \ldots, |\cdot|_n$ be pairwise inequivalent nontrivial absolute values on K. Let $b_1, \ldots, b_n \in K$ be given arbitrarily. Prove that for any $\varepsilon > 0$, there is an $a \in K$ such that $|a - b_i|_i < \varepsilon$ for all $1 \le i \le n$.

2.2.7 Suppose the valuation associated to $|\cdot|$ is discrete. Prove that K is totally disconnected. (A topological space X is called *disconnected* if it can be represented as a union of two disjoint nonempty open subsets, and X is *totally disconnected* if the only connected subsets of X are the empty set and the points of X.)

2.2.8 Prove the following statements:

(a) A sequence $\{a_n\}_{n\ge 1}$ in $(K, |\cdot|)$ is Cauchy if and only if $\lim_{n\to\infty} |a_{n+1} - a_n| = 0$.
(b) A series $\sum_{n=1}^{\infty} a_n$ in $(K, |\cdot|)$ converges in $\widehat{K}$ if and only if $\lim_{n\to\infty} |a_n| = 0$.
(c) Give examples that show that (a) and (b) are false for $\mathbb{R}$ with its usual absolute value.

2.2.9 Let K be a field and $|\cdot| : K \to \mathbb{R}_{\ge 0}$ be an absolute value on K, not necessarily non-Archimedean. This means that $|\cdot|$ satisfies:

(i) $|x| = 0$ if and only if $x = 0$.
(ii) $|xy| = |x| \cdot |y|$ for all $x, y \in K$.
(iii) $|x + y| \le |x| + |y|$ for all $x, y \in K$.

Note that (iii) here is a weaker condition than the corresponding condition for non-Archimedean absolute values. Prove that if there is a constant c such that $|n| \le c$ for all $n \in \mathbb{Z}$, then $|\cdot|$ is non-Archimedean, where we consider $n = n \cdot 1$ as an element of K. Deduce that an absolute value on a field of positive characteristic is necessarily non-Archimedean.

2.2.10 Let P_1 and P_2 be two monic irreducibles in $\mathbb{F}_q[T]$ of the same degree. Let K_1 and K_2 be the completions of $\mathbb{F}_q(T)$ with respect to ord_{P_1} and ord_{P_2}, respectively. Are K_1 and K_2 isomorphic as fields? Are K_1 and K_2 isomorphic as $\mathbb{F}_q(T)$-algebras?

2.2.11 Prove that the value groups of K and its completion $\widehat{K}$ are the same, i.e., $v(K^\times) = v(\widehat{K}^\times)$, even if $v(K^\times)$ is not discrete in $\mathbb{R}$. (This is of course not true for the completion of $\mathbb{Q}$ with respect to the usual Archimedean absolute value.)

2.2.12 Let K be a local field of positive characteristic p. Let k be its residue field, and let π be a uniformizer.

(a) Suppose $p \neq 2$. Prove that $K^\times/(K^\times)^2 \cong \mathbb{Z}/2\mathbb{Z} \times \mathbb{Z}/2\mathbb{Z}$, and $1, \pi, c, c\pi$ can be taken as representatives of the cosets, where $c \in k^\times$ is any non-square.
(b) Suppose $p = 2$. Prove that $K^\times/(K^\times)^2$ is an infinite group of exponent 2.

2.2.13 Assume k is a perfect field of characteristic p and $K = k(\!(T)\!)$. Show that every finite purely inseparable extension of K is isomorphic to $k(\!(T^{1/p^n})\!)$ for some $n \geq 1$ (cf. Example 1.5.15).

2.3 Extensions of Valuations

Let K be a field equipped with a non-Archimedean absolute value $|\cdot|$. Let L/K be a field extension. Recall from Definition 2.2.4 that an absolute value $\|\cdot\|$ on L *extends the absolute value* $|\cdot|$ on K if $\|a\| = |a|$ for all $a \in K$. In Sect. 2.2, we were interested in the completion $\widehat{K}$ of K and the extension of $|\cdot|$ to $\widehat{K}$. The extension $\widehat{K}/K$ is usually transcendental. In this section, we are interested in extending $|\cdot|$ to an algebraic extension L of K. In general, $|\cdot|$ can have several non-equivalent extensions to L. On the other hand, when K is complete, there is a unique extension of $|\cdot|$ to L. This is the main result of this section. We will prove the existence and uniqueness separately, as the proofs have very different flavors.

In this section, K is a field complete with respect to a nontrivial non-Archimedean absolute value $|\cdot|$. First, we prove the existence of an extension of the absolute value. In fact, the proof works in the more general setting of finite dimensional division algebras over K. This more general version has an important application in the theory of division algebras, so we will prove the theorem in that setting.

Proposition 2.3.1 *Let D be an n-dimensional division algebra over K. The map $\|\cdot\| : D \to \mathbb{R}_{\geq 0}$ defined by*

$$\|a\| = \left| \mathrm{Nr}_{D/K}(a) \right|^{1/n} = \left| \mathrm{Nr}_{K(a)/K}(a) \right|^{1/[K(a):K]}$$

is an absolute value on D extending the absolute value on K.

Proof Recall that $\mathrm{Nr}_{D/K}(a)$ is defined as the determinant of the K-linear transformation on D induced by multiplication by a. Since this linear transformation is obviously invertible for $a \neq 0$ and the determinant is multiplicative on matrices, $\|\cdot\| = \left|\mathrm{Nr}_{D/K}(\cdot)\right|^{1/n}$ satisfies (i) and (ii) of Definition 2.2.1. It remains to prove the strong triangle inequality $\|a + b\| \leq \max\{\|a\|, \|b\|\}$ for arbitrary $a, b \in D$. To accomplish this, we can assume without loss of generality that $\|a\| \geq \|b\|$. Then

$$\|a + b\| = \|a\| \cdot \|1 + b/a\|,$$

where $\|b/a\| \leq 1$. Now to prove $\|a + b\| \leq \max\{\|a\|, \|b\|\} = \|a\|$, it is enough to prove that $\|1 + b/a\| \leq 1$. Thus, assuming $\|b\| \leq 1$, it is enough to prove that

$$\|1 + b\| \leq 1. \tag{2.3.1}$$

To prove (2.3.1), let $m_{b,K}(x) = x^t + a_{t-1}x^{t-1} + \cdots + a_1 x + a_0$ be the minimal polynomial of b over K. By Corollary 1.4.4, $\mathrm{Nr}_{D/K}(b) = (-1)^n a_0^{n/t}$. Hence, $\|b\| \leq 1$ implies that $|a_0| \leq 1$. On the other hand, a corollary of Hensel's Lemma (see Corollary 2.4.6) implies that

$$\max\{|1|, |a_{t-1}|, \ldots, |a_0|\} = \max\{|1|, |a_0|\}.$$

The minimal polynomial of $1 + b$ is $m_{b,K}(x - 1)$, so

$$\mathrm{Nr}_{D/K}(1 + b) = (-1)^n m_{b,K}(-1)^{n/t}.$$

Combining these facts, we get

$$\|1 + b\| = \left|\mathrm{Nr}_{D/K}(1 + b)\right|^{1/n}$$

$$= \left|(-1)^t + a_{t-1}(-1)^{t-1} + \cdots + a_0\right|^{1/t}$$

$$\leq \max\{1, |a_{t-1}|, \ldots, |a_0|\}^{1/t}$$

$$= \max\{|1|, |a_0|\}^{1/t}$$

$$= 1.$$

Finally, by Corollary 1.4.4, $\mathrm{Nr}_{D/K}(a) = \mathrm{Nr}_{K(a)/K}(a)^{n/[K(a):K]}$, which implies that

$$\|a\| = \left|\mathrm{Nr}_{K(a)/K}(a)\right|^{1/[K(a):K]}.$$

$\square$

To prove the uniqueness of the extension of the absolute value to D, we will prove a more general statement applicable to norms on vector spaces over K.

Definition 2.3.2 Let V be a vector space over K. A *norm* on V is a function $\|\cdot\| : V \to \mathbb{R}_{\geq 0}$ such that:

(i) $\|v\| = 0$ if and only if $v = 0$.
(ii) $\|a \cdot v\| = |a| \cdot \|v\|$ for all $a \in K$ and $v \in V$.
(iii) $\|v + u\| \leq \|v\| + \|u\|$ for all $v, u \in V$.

A vector space over K equipped with a norm is called a *normed vector space*. We define the distance between $v, u \in V$ to be $\|v - u\|$. This induces a topology in V.

Example 2.3.3 Assume V is a finite dimensional vector space over K. Fix a basis $e_1, \ldots, e_n$ of V. Let $v = a_1 e_1 + \cdots + a_n e_n$ be the expansion of a vector $v \in V$ with respect to this basis. Define

$$\|v\|_{\sup} = \max(|a_1|, \ldots, |a_n|).$$

It is obvious that (i) and (ii) of the definition of norm hold for $\|\cdot\|_{\sup}$. As for (iii), we have the stronger inequality

$$\|v - u\|_{\sup} = \max(|a_1 - b_1|, \ldots, |a_n - b_n|)$$

$$\leq \max(\max(|a_1|, |b_1|), \ldots, \max(|a_n|, |b_n|))$$

$$= \max(\max(|a_1|, \ldots, |a_n|), \max(|b_1|, \ldots, |b_n|))$$

$$= \max(\|v\|_{\sup}, \|u\|_{\sup}),$$

where $u = b_1 e_1 + \cdots + b_n e_n$ is the expansion of u. Hence, $\|\cdot\|_{\sup}$ is a norm on V, called *sup-norm*.

Let $\{v_m\}_{m \geq 1}$ be a Cauchy sequence in V with respect to $\|\cdot\|_{\sup}$, i.e., $\|v_m - v_s\|_{\sup} \to 0$ as $m, s \to \infty$. Let $v_m = \sum_{i=1}^n a_{m,i} e_i$ be the expansion of v_m with respect to our fixed basis. For all $1 \leq i \leq n$, we have

$$|a_{m,i} - a_{s,i}| \leq \|v_m - v_s\|_{\sup}.$$

It follows that for each $1 \leq i \leq n$ the sequence $\{a_{m,i}\}_{m \geq 1}$ is Cauchy, hence converges to some $a_i \in K$. Therefore, $\{v_m\}_{m \geq 1}$ converges to $v = a_1 e_1 + \cdots + a_n e_n$, so V is complete with respect to $\|\cdot\|_{\sup}$.

The next result essentially says that all norms on V are equivalent.

Proposition 2.3.4 *Let V be a finite dimensional vector space over K. Let $\|\cdot\|$ be a norm on V. Then there are positive constants c_1 and c_2 such that*

$$c_1 \|v\|_{\sup} \leq \|v\| \leq c_2 \|v\|_{\sup} \qquad \text{for all } v \in V,$$

where $\|v\|_{\sup}$ is the norm defined in Example 2.3.3 with respect to a fixed basis $e_1, \ldots, e_n$ of V. Moreover, V is complete with respect to $\|\cdot\|$.

Proof One of the inequalities is easy to prove. Expand $v = a_1 e_1 + \cdots + a_n e_n$. Then

$$\|v\| \le |a_1|\,\|e_1\| + \cdots + |a_n|\,\|e_n\| \le c_2 \|v\|_{\sup},$$

where $c_2 = \|e_1\| + \cdots + \|e_n\| > 0$.

To prove the other inequality, we use induction on $n = \dim(V)$. When $n = 1$, the claim is clear. Assume $n \ge 2$. It is enough to show that there is a positive constant c_1 such that $c_1 \le \|v\|$ for all $v \in V$ with $\|v\|_{\sup} = 1$ (since any nonzero vector can be scaled so that $\|v\|_{\sup} = 1$). Suppose this is not the case. Then there is a sequence of vectors $\{v_i\}_{i \ge 1}$ such that $\|v_i\|_{\sup} = 1$ but $\|v_i\| \to 0$ as $i \to \infty$. After possibly permuting the basis elements $e_1, \ldots, e_n$ and taking a subsequence of $\{v_i\}_{i \ge 1}$, we can assume that $\|v_i\|_{\sup} = |a_{i,n}| = 1$, where $v_i = a_{i,1} e_1 + \cdots + a_{i,n} e_n$. Now replacing v_i by $a_{i,n}^{-1} v_i$, we can assume that $v_i = u_i + e_n$, where u_i is in the subspace W spanned by $e_1, \ldots, e_{n-1}$. Overall, if there is no positive c_1 such that $c_1 \|v\|_{\sup} \le \|v\|$ for all $v \in V$, then we can find a sequence $\{u_i\}_{i \ge 1}$ in W such that $\|u_i + e_n\| \to 0$ as $i \to \infty$. Note that property (iii) in Definition 2.3.2 implies

$$\|u_i - u_j\| \le \|u_i + e_n\| + \|u_j + e_n\| \to 0 \quad \text{as } i, j \to \infty.$$

Since W has dimension $n - 1$, by the inductive hypothesis, it is complete under $\|\cdot\|$. Hence, there is $u \in W$ such that

$$\|u_i - u\| \to 0 \quad \text{as} \quad i \to \infty.$$

Now

$$\|u + e_n\| = \lim_{i \to \infty} \|(u - u_i) + (u_i + e_n)\|$$

$$\le \lim_{i \to \infty} (\|u - u_i\| + \|u_i + e_n\|)$$

$$= \lim_{i \to \infty} \|u_i + e_n\|$$

$$= 0.$$

Since $u \in W$, it is not a multiple of e_n. Therefore, $u + e_n \ne 0$ but $\|u + e_n\| = 0$. This contradicts (i) of Definition 2.3.2. Thus, there is $c_1 > 0$ such that $c_1 \|v\|_{\sup} \le \|v\|$ for all $v \in V$.

To prove that V is complete with respect to $\|\cdot\|$, consider a Cauchy sequence $\{v_i\}_{i \ge 1}$ with respect to $\|\cdot\|$. Since $c_1 \|v_i - v_j\|_{\sup} \le \|v_i - v_j\|$, this sequence is Cauchy also with respect to $\|\cdot\|_{\sup}$. By Example 2.3.3, there is $v \in V$ such that $\|v - v_i\|_{\sup} \to 0$ as $i \to \infty$. Now

$$\|v - v_i\| \le c_2 \|v - v_i\|_{\sup} \to 0 \quad \text{as} \quad i \to 0,$$

so $\{v_i\}_{i \ge 1}$ converges to v also with respect to $\|\cdot\|$. $\square$

Theorem 2.3.5 *Let D be a finite dimensional division K-algebra. There is a unique absolute value $\|\cdot\|$ on D extending the absolute value $|\cdot|$ on K. Moreover, D is complete with respect to this absolute value.*

Proof The fact that the absolute value on K extends to D was proved in Proposition 2.3.1.

An absolute value on D extending the absolute value on K can be considered as a norm on the finite dimensional K-vector space D. Suppose $\|\cdot\|_1$ and $\|\cdot\|_2$ are two distinct extensions of $|\cdot|$ on K. Then there is $a \in D^\times$ such that $\|a\|_1 < \|a\|_2$. In that case, for any fixed $c, c' > 0$, we can find a sufficiently large m such that $c' \cdot \|a^m\|_1 < c \cdot \|a^m\|_2$. On the other hand, applying Proposition 2.3.4 to $\|\cdot\|_1$ and $\|\cdot\|_2$, we conclude that there are positive constants c, c' such that for all $b \in D$, we have

$$c \cdot \|b\|_2 \leq \|b\|_{\sup} \leq c' \cdot \|b\|_1 .$$

In particular, $c \cdot \|a^m\|_2 \leq c' \cdot \|a^m\|_1$, which leads to a contradiction. Finally, the completeness of D with respect to the extended absolute value follows from Proposition 2.3.4.   □

Let R be the valuation ring of K. Theorem 2.3.5 allows us to define an analogue of R for D by setting

$$R_D = \{a \in D \ : \ \left|\mathrm{Nr}_{D/K}(a)\right| \leq 1\} \tag{2.3.2}$$
$$= \{a \in D \ : \ \|a\| \leq 1\}.$$

Corollary 2.3.6 *R_D is a subring of D containing R, called the **valuation ring** of D.*

Theorem 2.3.7 *The absolute value $|\cdot|$ on K can be uniquely extended to an algebraic closure $\overline{K}$ of K.*

Proof For $a \in \overline{K}$, we choose a finite extension L/K such that $a \in L$ and then define $\|a\| = |\mathrm{Nr}_{L/K}(a)|^{1/[L:K]}$. Because the extension of $|\cdot|$ to $K(a)$ is unique, $\|a\|$ does not depend on the extension L. This gives us a well-defined map $\|\cdot\| : \overline{K} \to \mathbb{R}_{\geq 0}$. To check that $\|\cdot\|$ satisfies (ii) and (iii) of Definition 2.2.1, one can restrict $\|\cdot\|$ to the finite extension $L = K(a, b)$ of K, where we have shown that $|\mathrm{Nr}_{L/K}(\cdot)|^{1/[L:K]}$ has these properties. The uniqueness of the extension of $|\cdot|$ to $\overline{K}$ follows from the uniqueness of the restriction of $\|\cdot\|$ to each finite extension of K.   □

We record two useful facts about the absolute values of conjugate elements. Recall that two elements $\overline{K}$ are said to be conjugate if they have the same minimal polynomial over K.

Lemma 2.3.8 *The elements of $\overline{K}$, which are conjugate over K, have the same absolute value.*

Proof Assume $a, b \in \overline{K}$ are conjugate, and let $m(x)$ be their common minimal polynomial over K. If L/K is a finite extension such that $a \in L$, then by Corollary 1.4.4

$$\mathrm{Nr}_{L/K}(a) = (-1)^{[L:K]} m(0)^{[L:K(a)]}.$$

Let $L = K(a, b)$. Since $[K(a) : K] = [K(b) : K] = \deg m(x)$, we get

$$\|a\| = \left|\mathrm{Nr}_{L/K}(a)\right|^{1/[L:K]} = \left|\mathrm{Nr}_{L/K}(b)\right|^{1/[L:K]} = \|b\|.$$

$\square$

Proposition 2.3.9 (Krasner's Lemma) *Let $a, b \in \overline{K}$. Assume a is separable over $K(b)$. If $|a - b| < |a - a'|$ for all the conjugates $a' \neq a$ of a over $K(b)$, then $a \in K(b)$.*

Proof Let L be the splitting field of the minimal polynomial of a over $K(b)$. By assumption, $L/K(b)$ is a Galois extension. It is enough to show that $\sigma(a) = a$ for all $\sigma \in \mathrm{Gal}(L/K(b))$. By Lemma 2.3.8, $|a - b| = |\sigma(a) - b|$ for all $\sigma \in \mathrm{Gal}(L/K(b))$. On the other hand,

$$|a - \sigma(a)| = |(a - b) + (b - \sigma(a))|$$
$$\leq \max(|a - b|, |\sigma(a) - b|)$$
$$= |a - b|.$$

This contradicts the assumption of the proposition, unless $a = \sigma(a)$. $\square$

Unfortunately, the algebraic closure $\overline{K}$ of a local field is not complete with respect to the extended absolute value. To show this, we construct an element in the completion of $\overline{K}$ that is transcendental over K.

Lemma 2.3.10 *If K is a local field, then $\overline{K}$ is not complete with respect to the extension of the absolute value on K.*

Proof Let p be the characteristic of the residue field of K. Choose a prime $\ell \neq p$, and for each $n \geq 0$, let ω_n be a primitive ℓ^n-th root of 1 in $\overline{K}$ (i.e., a generator of the cyclic group of ℓ^n-th roots of 1, cf. Lemma 1.2.9). Consider the sequence $\{a_n\}_{n \geq 0}$ in $\overline{K}$ defined, recursively, by $a_0 = \omega_0 = 1$ and $a_n = a_{n-1} + \omega_n \pi^n$ for $n \geq 1$, where π is a uniformizer of K. This sequence is Cauchy, since

$$|a_n - a_{n-1}| = |\omega_n \pi^n| = |\pi|^n \to 0 \quad \text{as} \quad n \to \infty$$

(see Exercise 2.2.8). Denote $K_n = K(a_0, a_1, a_2, \ldots, a_n) = K(\omega_n)$. The degrees of cyclotomic extensions $K(\omega_n)$ over K go to infinity with n since otherwise all ω_n are contained in some finite extension L of K. In that case, reducing $x^{\ell^n} - 1$ modulo

$\mathcal{M}_L$, one concludes that all these polynomials split completely over the residue field of L, which is false since the residue field is finite.

Now suppose $\{a_n\}_{n\geq 0}$ converges to $a \in \overline{K}$. Then a belongs to some finite extension of K. We will show that this leads to a contradiction. Note that a can be represented as a series $a = \sum_{i\geq 0} \omega_i \pi^i$. The coefficients $\omega_0, \omega_1, \ldots$ of this series generate larger and larger extensions of K, but the whole sum might, in principle, be in some finite extension. We need to show that that cannot happen.

For an element $\sigma \in \mathrm{Gal}(K^{\mathrm{sep}}/K)$, we have

$$a_n - \sigma(a_n) = \sum_{i=0}^{n} (\omega_i - \sigma(\omega_i))\pi^i.$$

If $a_n \neq \sigma(a_n)$, then there is some $0 \leq i \leq n$ such that $\omega_i \neq \sigma(\omega_i)$. Let i be the minimal such index. Then $|a_n - \sigma(a_n)| = |\pi|^i \geq |\pi|^n$. On the other hand, $|a_n - a| = |\pi|^{n+1}$. Thus, $|a_n - a| < |a_n - a'_n|$ for all the conjugates $a'_n \neq a_n$ of a_n. By Krasner's Lemma, $a_n \in K(a)$ for all $n \geq 0$. Hence, $K(a)$ cannot be a finite extension of K. $\qquad\square$

Although $\overline{K}$ is not complete, it is a field equipped with a non-Archimedean absolute value, so we can form its completion $\widehat{\overline{K}}$, which from now on we will denote by $\mathbb{C}_K$.

Proposition 2.3.11 *The field $\mathbb{C}_K$ is algebraically closed.*

Proof We only give a sketch of the proof. Let $f(x) \in \mathbb{C}_K[x]$ be an irreducible polynomial. One can approximate the coefficients of f by elements from $\overline{K}$ to obtain a sequence of polynomials $\{f_n\}$ of the same degree as f that converge to f coefficientwise. Then the roots of $f(x)$ are approximated by the roots of $f_n(x)$. The roots of $f_n(x)$ form Cauchy sequences in $\overline{K}$, so they converge in $\mathbb{C}_K$. This implies that the roots of $f(x)$ are in $\mathbb{C}_K$, so $f(x)$ is linear. $\qquad\square$

Remarks 2.3.12

(1) As we saw in Sect. 2.2, if the valuation associated with $|\cdot|$ is discrete and K has positive characteristic, then K has a simple description as the field of Laurent series $k(\!(\pi)\!)$. This is no longer the case for $\overline{K}$ and $\mathbb{C}_K$. Note that the valuation on $\overline{K}$ and $\mathbb{C}_K$ is no longer discrete; in fact, if $v(K^\times) = \mathbb{Z}$, then $v(\overline{K}^\times) = v(\mathbb{C}_K^\times) = \mathbb{Q}$. (This follows from Exercise 2.2.11 and the discussion of ramifications in Sect. 2.6.) Nevertheless, the algebraic closure of $\overline{\mathbb{F}}_p(\!(T)\!)$ can be described in terms of certain generalized power series; see [Ked17]. An example of such a series is

$$T^{-1/p} + T^{-1/p^2} + T^{-1/p^3} + \cdots,$$

which is a root of $x^p - x - T^{-1}$.

(2) The proof of Lemma 2.3.10 also shows that K^{sep} is not complete, but one can prove that the completions of $\overline{K}$ and K^{sep} are the same; see Exercise 2.3.6. This is a theorem of Ax [Ax70]. Of course, this result is nontrivial only when K has positive characteristic, so that K^{sep} is strictly contained in $\overline{K}$. Note that the action of the absolute Galois group G_K on K^{sep} extends by continuity to an action on $\widehat{K^{\text{sep}}}$; cf. Lemma 2.3.8. Since this latter field is $\mathbb{C}_K$, the group G_K naturally acts on $\mathbb{C}_K$. Ax also proved in [Ax70] that the fixed field $\mathbb{C}_K^{G_K}$ under this action is $\widehat{K^{p^{-\infty}}}$, the completion of the perfect closure of K. In particular, $\mathbb{C}_K^{G_K} = K$ if the characteristic of K is zero,[4] but $\mathbb{C}_K^{G_K}$ is strictly larger than K if K has positive characteristic.

Exercises

2.3.1 Let K be a local field with normalized valuation v and associated absolute value $|\cdot|$. Denote the extension of $|\cdot|$ to $\mathbb{C}_K$ by the same symbol. For $z \in \mathbb{C}_K$, define the "imaginary" part of z as

$$\Im(z) = \inf_{x \in K} |z - x|,$$

which replaces the distance of a complex number from the real line in the setting of complex numbers. Prove the following properties of $\Im(z)$:

(a) $\Im(z) = \min_{x \in K} |z - x|$.
(b) $\Im(z) = 0$ if and only if $z \in K$.
(c) If $v(z) \notin \mathbb{Z}$, then $|z| = \Im(z)$.
(d) If $|z| = 1$, then $\Im(z) = 1$ if and only if the image of z in the residue field of $\mathbb{C}_K$ is not in the residue field of K.

2.3.2 Give an example of an infinite dimensional vector space over a complete non-Archimedean field K with norms that induce different topologies.

2.3.3 Let K be a complete non-Archimedean field with absolute value $|\cdot|$. For a polynomial $f(x) = \sum_{i=0}^{n} a_n x^n \in K[x]$ define as in Exercise 2.2.4

$$\|f\| = \max(|a_0|, |a_1|, \dots, |a_n|).$$

Suppose $f(x)$ is a monic irreducible separable polynomial in $K[x]$. Show that any monic polynomial $g(x) \in K[x]$ such that $\|f - g\|$ is sufficiently small is also irreducible. Moreover, for any root α of $f(x)$, there is a root β of $g(x)$ such that $K(\alpha) = K(\beta)$.

---

[4] This was originally proved by Tate.

2.3.4 Let K be a field equipped with a nontrivial non-Archimedean absolute value $|\cdot|$. For a polynomial $f(x) = \sum_{i=0}^{n} a_n x^n \in K[x]$, define

$$\|f\| = |a_0| + |a_1| + \cdots + |a_n|.$$

Prove the following:

(a) $\|\cdot\|$ is a norm on the K-vector space $K[x]$.
(b) If $f(x)$ is monic, then for any root $\alpha \in \overline{K}$ of $f(x)$, we have $|\alpha| < \|f\|$.

2.3.5 Let $K = \mathbb{F}_q((T))$ and $L = \overline{\mathbb{F}}_q K$ be the infinite extension of K obtained by adjoining all the elements in $\overline{K}$ that are algebraic over $\mathbb{F}_q$. Prove that L is not complete with respect to the extension of the absolute value on K and show that $\widehat{L} = \overline{\mathbb{F}}_q((T))$.

2.3.6 Let K be a local field of characteristic p. Assume $\alpha \in \overline{K}$ is not separable over K, so that $\alpha^q = a \in K^{\text{sep}}$ for some $q = p^n$. Choose a sequence $\{b_i\}_{i \geq 1}$ of nonzero elements of K converging to 0. Let β_i be a root of $x^q - b_i x - a$. Show the following:

(a) If $|b_i|$ is small enough, then $|\beta_i|^q = |a|$.
(b) $(\beta_i - \alpha)^q = b_i \beta_i \to 0$ as $i \to \infty$.
(c) All $\beta_i \in K^{\text{sep}}$, and $\beta_i \to \alpha$ as $i \to \infty$.
(d) $\widehat{K^{\text{sep}}} = \mathbb{C}_K$.

2.4 Hensel's Lemma

Let K be a field complete with respect to a nontrivial non-Archimedean absolute value. Denote by R the ring of integers of K, by $\mathcal{M}$ the maximal ideal of R, and by $k = R/\mathcal{M}$ the residue field.

Hensel's Lemma essentially says that the decomposition of a polynomial over R, and in particular the existence of roots over K, can be detected by examining the decomposition of the reduction of the polynomial modulo $\mathcal{M}$. The idea behind Hensel's Lemma and its proof is the successive approximation of a solution of a polynomial equation.[5] Note that we have already used Hensel's Lemma to prove two important results; namely, the existence of an extension of the absolute value on K to its finite extensions, and the explicit description of K as $k((\pi))$ in the case when K has positive characteristic and its valuation is discrete. (The proof of Hensel's Lemma does not use either of these results, so our reasoning will not be circular.)

We start with an example.

Example 2.4.1 Let $F = \mathbb{F}_q(T)$ and $\widehat{F} = \mathbb{F}_q((T))$ be the completion of F with respect to ord_T. Consider the Artin–Schreier polynomial

$$f(X) = X^q - X - T \in \widehat{F}[X].$$

---

[5] This is an idea that goes back to Isaac Newton; cf. [Gou20, p. 91].

Substituting

$$\alpha = T + T^q + T^{q^2} + \cdots$$

into $f(X)$, we get $f(\alpha) = 0$. Hence, $\alpha \in \widehat{F}$ is a root of $f(X)$. As in the proof of Theorem 1.5.21, the roots of $f(X)$ are $\alpha + \beta$, $\beta \in \mathbb{F}_q$. It is clear that $\alpha \notin F$, so $f(X)$ has no roots in F. The clear advantage of passing to the completion in this example is that $f(x)$ splits completely over $\widehat{F}$, although it is irreducible over F. The machinery of Hensel's Lemma applied to $f(X)$ produces its roots via successive approximations $\beta + \sum_{i=0}^{n} T^{q^n}$, $n \geq 0$, in F. Since the roots of $f(X)$ are not in F, for Hensel's Lemma to apply, the assumption that the base field K is complete is important, although it can be weakened.[6]

Given a polynomial $f(x) \in R[x]$, denote by $\bar{f}(x) \in k[x]$ the polynomial obtained by reducing the coefficients of $f(x)$ modulo $\mathcal{M}$.

Theorem 2.4.2 (Hensel's Lemma) *Let $f(x) \in R[x]$ be a polynomial such that $\bar{f}(x) \in k[x]$ is nonzero. Suppose:*

(i) $\bar{f}(x) = g_0(x) \cdot h_0(x)$, where $g_0(x), h_0(x) \in k[x]$.
(ii) $g_0(x)$ is monic.
(iii) $g_0(x)$ and $h_0(x)$ are relatively prime in $k[x]$.

Then $f(x)$ factors as

$$f(x) = g(x) \cdot h(x)$$

for uniquely determined $g(x), h(x) \in R[x]$ such that $\bar{g}(x) = g_0(x)$, $\bar{h}(x) = h_0(x)$, and $g(x)$ is monic.

For the proof, we need two preliminary lemmas. Note that in these lemmas we do not need K to be complete with respect to the given non-Archimedean absolute value $|\cdot|$.

Lemma 2.4.3 (Nakayama's Lemma) *Let B be a finitely generated R-module and*

$$\mathcal{M}B = \left\{ \sum a_i b_i \mid a_i \in \mathcal{M}, b_i \in B \right\}.$$

If $\mathcal{M}B = B$, then $B = 0$.

Proof Assume $B \neq 0$, and let $e_1, \ldots, e_n \in B$ be a minimal set of generators of B as an R-module. If $\mathcal{M}B = B$, then we can write

$$e_1 = a_1 e_1 + \cdots + a_n e_n, \quad a_i \in \mathcal{M},$$

---

[6] For Hensel's Lemma to apply, instead of working over the completion $\widehat{K}$ of a given field K equipped with a nontrivial non-Archimedean absolute value, it is enough to pass to the smaller field $\widehat{K} \cap K^{\mathrm{sep}}$, called the *Henselization of K*. The ability to work over the Henselization, rather than completion, is important in some arithmetic problems, although it is not essential for our purposes.

or equivalently, $(1 - a_1)e_1 = a_2 e_2 + \cdots + a_n e_n$. Since $|a_1| < 1$, we have $|1 - a_1| = 1$. Therefore, $1 - a_1$ is a unit. This implies that $e_1 = (1 - a_1)^{-1}(a_2 e_2 + \cdots + a_n e_n)$, and so B can be generated by $\{e_2, \ldots, e_n\}$, which contradicts our choice of $\{e_1, \ldots, e_n\}$. □

Lemma 2.4.4 *Suppose $g(x), h(x) \in R[x]$ are such that $g(x)$ is monic and $\bar{g}(x), \bar{h}(x) \in k[x]$ are relatively prime. Let $r(x) \in R[x]$ be a nonzero polynomial with $\deg(r) < \deg(gh)$. Then there exist $u(x), w(x) \in R[x]$ with $\deg(u) < \deg(h)$, $\deg(w) < \deg(g)$ such that*

$$g(x)u(x) + h(x)w(x) = r(x).$$

Proof Let $B = R[x]/(g, h)$. Since g is monic, it is easy to check that B is a finitely generated R-module. Since $(\bar{g}, \bar{h}) = k[x]$, we have $(g, h) + M[x] = R[x]$. But this implies that $MB = B$. By Nakayama's lemma, $B = 0$. Hence, there are $u, w \in R[x]$ such that $gu + hw = r$. Since g is monic, we can use the division algorithm in $R[x]$ to write $w = gq + s$ with $\deg(s) < \deg(g)$. Now $g(u + hq) + hs = r$ and $\deg(s) < \deg(g)$. It remains to show that $\deg(u + hq) < \deg(h)$. Suppose this is not the case. Then $\deg(g(u + hq)) \geq \deg(gh)$. On the other hand, $g(u + hq) = r - hs$, so

$$\deg(g(u + hq)) = \deg(r - hs) \leq \max(\deg(r), \deg(hs)).$$

Since $\deg(hs) < \deg(gh)$ and $\deg(r) < \deg(gh)$, we get $\max(\deg(r), \deg(hs)) < \deg(gh)$. Thus, $\deg(g(u + hq)) < \deg(gh)$, which contradicts our earlier inequality. □

Now we are ready to prove the main theorem.

Proof of Theorem 2.4.2 We construct two sequences of polynomials $g_n(x)$, $h_n(x) \in R[x]$, $n \geq 1$, such that:

(1) $g_n(x)$ is monic, and the leading coefficient of $h_n(x)$ is the leading coefficient of $f(x)$.
(2) $\bar{g}_n = g_0, \bar{h}_n = h_0$.
(3) $\deg(g_n h_n) = \deg(f)$.
(4) $f - g_n h_n \in M^n[x]$.

We use induction for our construction. Choose $g_1(x), h_1(x) \in R[x]$ satisfying (1), (2), (3). By the assumption of the theorem, we have $\bar{f} = \bar{g}_1 \bar{h}_1$. Hence, $f - g_1 h_1 \in M[x]$, so (4) is also satisfied. Assume the polynomials $g_m(x), h_m(x)$ satisfying (1)–(4) exist for $1 \leq m \leq n - 1$. Then we can write $f - g_{n-1} h_{n-1}$ as a finite sum

$$f - g_{n-1} h_{n-1} = \sum_i c_i r_i, \quad \text{with } c_i \in M^{n-1}, r_i \in R[x], \text{ and } \deg(r_i) < \deg(f).$$

Since $\bar{g}_{n-1} = g_0$ and $\bar{h}_{n-1} = h_0$ are relatively prime, by Lemma 2.4.4, there exist $u_i, w_i \in R[x]$ such that $g_{n-1}u_i + h_{n-1}w_i = r_i$ and $\deg(u_i) < \deg(h_{n-1})$, $\deg(w_i) < \deg(g_{n-1})$. Set

$$g_n = g_{n-1} + \sum_i c_i w_i, \qquad h_n = h_{n-1} + \sum_i c_i u_i.$$

It is clear that (1)–(3) are satisfied for g_n, h_n. On the other hand,

$$g_n h_n = g_{n-1}h_{n-1} + \sum_i c_i w_i h_{n-1} + \sum_i c_i u_i g_{n-1} + \sum_{i,j} c_i c_j u_i w_j$$

$$= g_{n-1}h_{n-1} + \sum_i c_i r_i + \sum_{i,j} c_i c_j w_i u_j$$

$$= f + \sum_{i,j} c_i c_j w_i u_j.$$

Note that $\sum_{i,j} c_i c_j w_i u_j \in \mathcal{M}^{2(n-1)}[x]$. As $n \geq 2$, we have $2(n-1) \geq n$, so the previous calculation shows that $f - g_n h_n \in \mathcal{M}^n[x]$, i.e., (4) is satisfied.

Now write $g_n(x) = \sum_{i=0}^t a_i^{(n)} x^i$ with all $a_i^{(n)} \in R$. Since by construction $g_n - g_{n-1} \in \mathcal{M}^{n-1}[x]$, for each $0 \leq i \leq t$, the sequence $\left\{a_i^{(n)}\right\}_{n=1}^{\infty}$ is Cauchy. As R is complete, the limit $\lim_{n\to\infty} a_i^{(n)} = a_i$ exists in R. We put $g(x) = \sum_{i=0}^t a_i x^i$. Similarly, we define $h(x) = \lim_{n\to\infty} h_n(x) \in R[x]$. Then $f(x) = g(x)h(x)$, $g(x)$ is monic, $\bar{g}(x) = g_0(x)$, and $\bar{h}(x) = h_0(x)$.

To prove the uniqueness of g and h, suppose $f(x) = G(x)H(x)$, with $G(x)$ monic, $\bar{G}(x) = g_0(x)$, and $\bar{H}(x) = h_0(x)$. Applying Lemma 2.4.4 to g, H, and $r = 1$, we can write $1 = gu + Hw$ with $w, u \in R[x]$. Then

$$G = gGu + GHw = gGu + ghw.$$

This implies that g divides G. As both are monic and have the same degree, they must be equal. Now $gH = gh$ implies $h = H$. $\square$

Corollary 2.4.5 Let $f(x) \in R[x]$. Assume $\bar{f}(x)$ has a simple root α_0, i.e., $\bar{f}(\alpha_0) = 0$ and $\bar{f}'(\alpha_0) \neq 0$. Then there exists a unique root $\alpha \in R$ of $f(x)$ with $\alpha \equiv \alpha_0 \pmod{\mathcal{M}}$.

In some textbooks, this consequence of Theorem 2.4.2 is given as "Hensel's Lemma."

Proof By assumption, $\bar{f}(x) = (x - \alpha_0) \cdot \bar{h}(x)$ and $\bar{h}(\alpha_0) \neq 0$. Hence, $x - \alpha_0$ and $\bar{h}(x)$ are relatively prime in $k[x]$. Theorem 2.4.2 implies that $f(x)$ splits in $R[x]$ as $f(x) = (x - \alpha) \cdot h(x)$ and $\alpha \equiv \alpha_0 \pmod{\mathcal{M}}$. Obviously, $\alpha \in R$ is a root of $f(x)$. $\square$

Corollary 2.4.6 *Let* $f(x) = a_n x^n + a_{n-1} x^{n-1} + \cdots + a_1 x + a_0 \in K[x]$ *be an irreducible polynomial. Then*

$$\max\{|a_n|, |a_{n-1}|, \ldots, |a_0|\} = \max\{|a_n|, |a_0|\}.$$

Proof The claim is trivial for $n \le 1$. Hence, assume $n \ge 2$. If the claim is false, then there exists a positive integer $1 \le m \le n - 1$ such that

$$|a_m| > \max\{|a_n|, |a_0|\}.$$

Choose the largest such m. The polynomial $a_m^{-1} f(x)$ has coefficients in $R[x]$, and its reduction $g_0(x)$ in $k[x]$ is monic of degree m. Put $h_0(x) = 1$. By Theorem 2.4.4, $a_m^{-1} f(x)$ factors in $R[x]$ as $a_m^{-1} f(x) = g(x) h(x)$ with $\deg(g) = m$. This contradicts the assumption that $f(x)$ is irreducible in $K[x]$. $\qquad\qquad\square$

Exercises

2.4.1 Let $f(x) = x^n + a_{n-1} x^{n-1} + \cdots + a_1 x + a_0 \in R[x]$ be an irreducible polynomial. Prove the following statements:

(a) $f(x)$ modulo $\mathcal{M}$ is a positive integer power e of an irreducible polynomial in $k[x]$.
(b) If K is a local field and $L = K[x]/(f(x))$, then $e \ge e(L/K)$, where $e(L/K)$ is the ramification index (see Definition 2.6.1). Give an explicit example where this inequality is strict.

2.4.2 Let $f(x) \in R[x]$. Suppose $\alpha_0 \in R$ is such that

$$|f(\alpha_0)| < |f'(\alpha_0)|^2.$$

For $n \ge 0$, define

$$\alpha_{n+1} := \alpha_n - \frac{f(\alpha_n)}{f'(\alpha_n)}.$$

(a) Prove that the sequence $\{\alpha_n\}_{n \ge 0}$ is well-defined and converges to a unique root $\alpha \in R$ of $f(x)$.
(b) Show that (a) implies Corollary 2.4.5.
(c) Let $f(x) = x^2 - 17 \in \mathbb{Z}_2[x]$. Show that $f(x)$ has two simple roots in $\mathbb{Z}_2$, but Corollary 2.4.5 does not apply in this case.

2.4.3 Let $f(x_1, \ldots, x_n) \in R[x_1, \ldots, x_n]$ be a polynomial in n variables. Let

$$(\nabla f)(x_1, \ldots, x_n) = \left(\frac{df}{dx_1}, \ldots, \frac{df}{dx_n} \right)$$

be its gradient, which is the vector of partial derivatives of f. Suppose some $\mathbf{a} = (a_1, \ldots, a_n) \in R^n$ satisfies

$$|f(\mathbf{a})| < \|(\nabla f)(\mathbf{a})\|_{\text{sup}}^2 .$$

Prove that there is $\mathbf{b} \in R^n$ such that $f(\mathbf{b}) = 0$ and $\|\mathbf{b} - \mathbf{a}\|_{\text{sup}} < \|(\nabla f)(\mathbf{a})\|_{\text{sup}}$. In particular, if

$$|f(\mathbf{a})| < 1 \quad \text{and} \quad \|(\nabla f)(\mathbf{a})\|_{\text{sup}} = 1,$$

then there is $\mathbf{b} \in R^n$ such that $f(\mathbf{b}) = 0$ and $b_i \equiv a_i \pmod{\mathcal{M}}$ for all $i = 1, \ldots, n$.

2.4.4 Let D be a finite dimensional division algebra over K. Prove that an element $a \in D$ is integral over R if and only if $a \in R_D$, where R_D was defined in (2.3.2). Hence, R_D is the integral closure of R in D.

2.4.5 Let R be a commutative ring with a unique maximal ideal $\mathcal{M}$. Let B be a finitely generated R-module.

(a) Prove that the statement of Lemma 2.4.3 holds for R, i.e., if $\mathcal{M}B = B$, then $B = 0$.
(b) Let $A \subseteq B$ be a submodule of B. Prove that $A + \mathcal{M}B = B$ implies $A = B$.
(c) Suppose the images of $e_1, \ldots, e_n \in B$ generate $B/\mathcal{M}B$ as an $R/\mathcal{M}$-module. Prove that the elements $e_1, \ldots, e_n$ generate B as an R-module.

2.5 Newton Polygon

Let K be a field equipped with a nontrivial valuation v, and let $R \subset K$ be the valuation ring of v. We assume that K is complete with respect to an absolute value associated with v, so that v uniquely extends to $\overline{K}$. The Newton polygon method is a method that efficiently determines the valuations of the roots of a given polynomial $f(x) \in K[x]$ from the valuations of its coefficients. In special cases, it is possible to deduce from this a factorization of $f(x)$, or the existence of a rational root (i.e., a root in K), although the Newton polygon does not determine the roots even approximately, as Hensel's Lemma does.

Definition 2.5.1 Let

$$f(x) = a_0 + a_1 x + \cdots + a_n x^n \in K[x]$$

be a polynomial such that $a_0 a_n \neq 0$. The *Newton polygon* of $f(x)$, denoted NP(f), is the lower convex hull of the set of points

$$P_i = (i, v(a_i)), \quad 0 \le i \le n, \quad a_i \neq 0.$$

More explicitly, take the vertical line through $P_0 = (0, v(a_0))$, and rotate it counterclockwise until it hits any of the points P_i—the first segment of the Newton polygon is the line segment $P_0 P_{i_1}$, where P_{i_1} is the point furthest from P_0 on the rotated vertical line. Repeat the process with a vertical line through P_{i_1} and so on. The resulting polygon NP(f) has the following properties:

- NP(f) starts at P_0 and ends at P_n.
- Each of the line segments of NP(f) starts at some P_i and ends at some P_j with $i < j$.
- No point P_i lies below NP(f).
- Any line joining two points of NP(f) has no points lying below the polygon.
- The slopes of the line segments of NP(f) strictly increase as one moves along the x-axis.

See Fig. 2.1 for an example of a Newton polygon.

Theorem 2.5.2 *Suppose a line segment of the Newton polygon of $f(x)$ joins P_i with P_j, $i < j$. Then $f(x)$ has exactly $(j - i)$ roots α in $\overline{K}$ such that $v(\alpha) = -\frac{v(a_j) - v(a_i)}{j - i}$, i.e., the valuation of these roots is the negative of the slope of the given line segment, and the number of roots with this valuation is the horizontal length of the line segment.*

Proof It is convenient to replace $f(x)$ by $a_0^{-1} f(x)$, so that the constant term of the resulting polynomial is 1. This does not change the roots. Moreover, the Newton

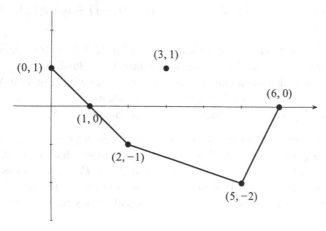

Fig. 2.1 The Newton polygon of $x^6 - \frac{1}{\pi^2} x^5 + \pi x^3 + \frac{1}{\pi} x^2 - x + \pi$, $\quad v(\pi) = 1$

polygon of $a_0^{-1} f(x)$ is the Newton polygon of $f(x)$ shifted vertically by $-v(a_0)$. In particular, the horizontal lengths and the slopes of the line segments of both polygons are the same. Therefore, it is enough to prove the theorem assuming $a_0 = 1$. Let $\lambda_1, \ldots, \lambda_n \in \overline{K}$ be the inverse roots of $f(x)$, so that

$$f(x) = \prod_{i=1}^{n} (1 - \lambda_i x). \tag{2.5.1}$$

The coefficients of $f(x)$ are given by the elementary symmetric functions of λ_i's:

$$a_s = (-1)^s \sum_{1 \leq i_1 < i_2 < \cdots < i_s \leq n} \lambda_{i_1} \lambda_{i_2} \cdots \lambda_{i_s}. \tag{2.5.2}$$

Reindex the λ_i's so that $v(\lambda_1) = v(\lambda_2) = \cdots = v(\lambda_m) < v(\lambda_i), i > m$. Observe the following:

- For $1 \leq s \leq m$, there is a summand in (2.5.2) with valuation $s \cdot v(\lambda_1)$, for example, $\lambda_1 \lambda_2 \cdots \lambda_s$, and every other summand has valuation $\geq s \cdot v(\lambda_1)$.
- For $s = m$, there is a unique summand in (2.5.2) with valuation $m \cdot v(\lambda_1)$, namely $\lambda_1 \lambda_2 \cdots \lambda_m$, and every other summand has valuation $> m \cdot v(\lambda_1)$.
- For $s > m$, every summand in (2.5.2) has valuation $> s \cdot v(\lambda_1)$.

Recall that $v(a + b) \geq \min\{v(a), v(b)\}$, with equality if $v(a) \neq v(b)$. Combining this with the above observations, one concludes that $v(a_s) \geq s \cdot v(\lambda_1)$ if $s < m$, $v(a_m) = m \cdot v(\lambda_1)$, and $v(a_s) > s \cdot v(\lambda_1)$ if $s > m$. Sketching the points $(i, v(a_i))$ on the plane, it is easy to see that the first line segment of $\mathrm{NP}(f)$ is the line joining $P_0 = (0, 0)$ to $P_m = (m, m \cdot v(\lambda_1))$, and any other line segment of $\mathrm{NP}(f)$ has larger slope. This agrees with the statement of the theorem.

Next, reindex $\{\lambda_{m+1}, \ldots, \lambda_n\}$ so that $v(\lambda_{m+1}) = v(\lambda_{m+2}) = \cdots = v(\lambda_{m+r}) < v(\lambda_i)$ for all $i > m + r$. Put $N_s = (s - m) \cdot v(\lambda_{m+1}) + m \cdot v(\lambda_1)$. Observe the following:

- For $m + 1 \leq s \leq m + r$, there is a summand in (2.5.2) with valuation N_s, for example, $\lambda_1 \lambda_2 \cdots \lambda_s$, and every other summand has valuation $\geq N_s$.
- For $s = m + r$, there is a unique summand in (2.5.2) with valuation N_s, namely $\lambda_1 \lambda_2 \cdots \lambda_{m+r}$, and every other summand has valuation $> N_s$.
- For $s > m + r$, every summand in (2.5.2) has valuation $> N_s$.

As before, this implies that $v(a_s) \geq N_s$ for $m+1 \leq s \leq m+r-1$, $v(a_{m+r}) = N_{m+r}$, and $v(a_s) > N_s$ for $s > m + r$. From this, it is easy to see that the second segment of the Newton polygon of $f(x)$ is the line joining P_m to P_{m+r}, which has horizontal length r and slope $v(\lambda_{m+1})$. This agrees with the claim of the theorem.

This argument can be iterated, which eventually leads to the statement of the theorem. $\square$

Remark 2.5.3 The assumption that K is complete is not really used in the proof of Theorem 2.5.2; it is used implicitly to have well-defined valuations of the roots of f. Hence, Theorem 2.5.2 is valid without the completeness assumption if all the roots of f are in K.

Proposition 2.5.4 *Let $f(x) = \prod_{i=1}^{n}(1 - \lambda_i x)$ be the polynomial in (2.5.1). Then*

$$g(x) = \prod_{v(\lambda_i)=v_0} (1 - \lambda_i x)$$

has coefficients in K.

Proof Let λ be one of the λ_i's with $v(\lambda_i) = v_0$. Let $m_{\lambda^{-1},K}(x)$ be the minimal polynomial of $1/\lambda$ over K. Because $f(1/\lambda) = 0$, the polynomial $m_{\lambda^{-1},K}(x)$ divides $f(x)$. By Lemma 2.3.8, conjugate elements of $\overline{K}$ have the same valuation, so $v(1/\lambda) = v(1/\lambda')$ for all the roots λ' of $m_{\lambda^{-1},K}(x)$. This implies that the polynomial $m_{\lambda^{-1},K}(x)$ divides $g(x)$. If η is a root of $g(x)/m_{\lambda^{-1},K}(x)$, then a similar argument shows that $m_{\eta^{-1},K}(x)$ divides $g(x)/m_{\lambda^{-1},K}(x)$. Continuing in this manner, we find that $g(x)$ is a product of polynomials with coefficients in K.   □

Corollary 2.5.5 *We have:*

(1) *If the Newton polygon of $f(x)$ has more than one line segment, then $f(x)$ is reducible.*

(2) *If the Newton polygon of $f(x)$ has n line segments, then all the roots of $f(x)$ are in K.*

Corollary 2.5.6 (Eisenstein Irreducibility Criterion) *Let*

$$f(x) = a_0 + a_1 x + \cdots + a_{n-1} x^{n-1} + a_n x^n \in K[x]$$

be a polynomial such that $v(a_0) = 1$, $v(a_n) = 0$, and $v(a_i) \geq 1$ for all $i \neq 0, n$. Then $f(x)$ is irreducible.

A polynomial satisfying the above assumption is called an *Eisenstein polynomial*.

Proof The Newton polygon of f has one line segment joining $(0, 1)$ to $(n, 0)$. Hence, all the roots of $f(x)$ have valuation $1/n$. To conclude that $f(x)$ is irreducible, first observe that by Proposition 2.3.1 every root of an irreducible polynomial of degree d over K has valuation in $\frac{1}{d}\mathbb{Z}$. Therefore, if $f(x)$ is reducible, then the valuations of its roots are in $\frac{1}{d}\mathbb{Z}$ for various $d < n$. This contradicts our earlier conclusion.   □

Remark 2.5.7 More generally, the argument in the previous proof shows that if $NP(f)$ consists of a single line segment whose slope is m/n, and m is coprime to n, then f is irreducible over K.

Alternative Proof Suppose $f(x)$ is reducible $f(x) = g(x)h(x)$. By Gauss' Lemma (see Theorem 1.1.6 and Exercise 2.2.5), we can assume that $g(x) = b_0 + b_1 x + $

Fig. 2.2 Newton polygon of
$x^q - x - T$ over $\mathbb{F}_q((T))$

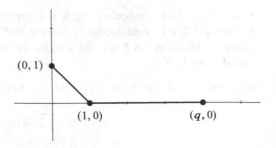

... $+ b_r x^r$ and $h(x) = c_0 + c_1 x + \cdots + c_s x^s$ are in $R[x]$, and $r, s \geq 1$. Reducing the coefficients of both sides of $f(x) = g(x)h(x)$ modulo $\mathcal{M}$, we obtain an equality $\bar{a}_n x^n = \bar{g}(x)\bar{h}(x)$, where the bar denotes the homomorphism from R onto the residue field k and also the induced map from $R[x]$ to $k[x]$. This implies that $\bar{g}(x) = \bar{b}_r x^r$ and $\bar{h}(x) = \bar{c}_s x^s$. In particular, $v(b_0) \geq 1$ and $v(c_0) \geq 1$. Therefore, $v(a_0) = v(b_0) + v(c_0) \geq 2$, contrary to our assumption. $\qquad\square$

Example 2.5.8 Let $K = \mathbb{F}_q((T))$, with valuation v normalized by $v(T) = 1$. Consider the Artin–Schreier polynomial $f(x) = x^q - x - T$. Its Newton polygon has two line segments, one of which has length one; see Fig. 2.2. Hence, $f(x)$ has a unique root α of valuation 1, and $q - 1$ roots of valuation 0. By Proposition 2.5.4, α must be in K. In fact, from Example 2.4.1, we know that $f(x)$ splits over K since the roots of $f(x)$ are $\alpha + \beta$, $\beta \in \mathbb{F}_q$, where $\alpha = \sum_{n \geq 0}^{\infty} T^{q^n} \in K$. Note that α is not in $\mathbb{F}_q(T)$. In fact, $f(x)$ is irreducible over $\mathbb{F}_q(T)$ because, by Example 2.5.9, it is irreducible over the completion $\mathbb{F}_q((1/T))$ of $\mathbb{F}_q(T)$ with respect to ord_∞. This shows that Proposition 2.5.4 is false without the assumption that K is complete. The reader should determine where exactly in the proof of that proposition we use the completeness assumption.

Example 2.5.9 With the setup of the previous example, consider the Artin–Schreier polynomial $g(x) = x^q - x - 1/T$. Its Newton polygon has one line segment joining $(0, -1)$ to $(q, 0)$; see Fig. 2.3. Hence, all the roots of $g(x)$ have valuation $-1/q$. By Remark 2.5.7, this implies that $g(x)$ is irreducible over K. But $g(x)$ splits completely over the field $K(\alpha)$ obtained from K by adjoining one of the roots of $g(x)$, since, as before, the roots of $g(x)$ are $\alpha + \beta$, $\beta \in \mathbb{F}_q$. Let $\sigma \in \mathrm{Gal}(K(\alpha)/K)$.

Fig. 2.3 Newton polygon of
$x^q - x - 1/T$ over $\mathbb{F}_q((T))$

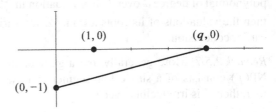

Then $\sigma(\alpha) = \alpha + \beta_\sigma$ for some $\beta_\sigma \in \mathbb{F}_q$. Moreover,

$$\sigma(\alpha + \gamma) = \sigma(\alpha) + \sigma(\gamma) = \alpha + \beta_\sigma + \gamma = (\alpha + \gamma) + \beta_\sigma, \quad \forall \gamma \in \mathbb{F}_q,$$
$$\sigma\tau(\alpha) = \sigma(\alpha + \beta_\tau) = \sigma(\alpha) + \beta_\tau = \alpha + \beta_\sigma + \beta_\tau, \quad \forall \sigma, \tau \in \mathrm{Gal}(K(\alpha)/K).$$

Therefore, $\mathrm{Gal}(K(\alpha)/K) \to \mathbb{F}_q$, $\sigma \mapsto \beta_\sigma$, is an injective group homomorphism, where we consider $\mathbb{F}_q$ as an additive group. On the other hand, $\#\mathrm{Gal}(K(\alpha)/K) = [K(\alpha) : K] = q$. Thus, $\mathrm{Gal}(K(\alpha)/K) \cong \mathbb{F}_q$. We conclude that the Galois group of $g(x)$ over K is isomorphic to

$$\mathbb{Z}/p\mathbb{Z} \times \cdots \times \mathbb{Z}/p\mathbb{Z} \quad (n \text{ times}), \quad \text{where } q = p^n.$$

Exercises

2.5.1 Let

$$f(x) = x^3 + x^2 - Tx + T^3.$$

Prove that this polynomial is irreducible over $\mathbb{F}_5(T)$ but splits completely over $\mathbb{F}_5(\!(T)\!)$.

2.5.2 Give a different proof of Corollary 2.4.6 using Newton polygons.

2.5.3 Let $a, b \in K^\times$. Describe the relationship between the Newton polygons of $af(bx)$ and $f(x)$.

2.5.4 Let A_+ denote the set of monic polynomials in $\mathbb{F}_q[T]$. Describe the Newton polygon of

$$f(x) = \prod_{\substack{a \in A_+ \\ \deg(a) \leq n}} (x - a)$$

over $K = \mathbb{F}_q(\!(T)\!)$.

2.5.5 For a polynomial $f(x) \in K[x]$ of degree n, let $\mathrm{NP}_f(z)$ be the real valued function on $[0, n]$ whose graph is the Newton polygon of f. Suppose $f(x), g(x) \in K[x]$ are polynomials of degrees n and m, respectively, such that $v(\alpha) \geq v(\beta)$ whenever α is a root of $f(x)$ and β is a root of $g(x)$. Show that

$$\mathrm{NP}_{fg}(z) = \begin{cases} \mathrm{NP}_f(z) + \mathrm{NP}_g(0) & \text{if } z \in [0, n] \\ \mathrm{NP}_g(z - n) + \mathrm{NP}_f(n) & \text{if } z \in [n, n + m]. \end{cases}$$

2.5.6 Let $f(x) \in K[x]$ be a polynomial of degree n, and let $f(x) = \prod_{i=1}^{n}(x - \alpha_i)$ be its decomposition over $\overline{K}$. Assume the roots of f are indexed so that

$$v(\alpha_1) \leq v(\alpha_2) \leq \cdots \leq v(\alpha_n).$$

Prove that for $z \in [m-1, m]$ with $1 \leq m \leq n$, we have

$$\mathrm{NP}_f(z) = \sum_{i=m+1}^{n} v(\alpha_i) + (m - z)v(\alpha_m).$$

2.6 Ramification and Inertia Group

Let K be a local field, and let L/K be a finite extension of degree n. To distinguish the rings of integers of K and L, denote them by R_K and R_L, respectively. Furthermore, denote by $\mathcal{M}_K$ and $\mathcal{M}_L$ the maximal ideals of R_K and R_L, respectively, and let $k = R_K/\mathcal{M}_K$, $l = R_L/\mathcal{M}_L$ be the residue fields. To slightly simplify the discussion in this section, we assume that K has positive characteristic p (in later sections of this book, we work only with such local fields), although most of the results are also true for local fields of characteristic 0. Fix a uniformizer π_K of K and a uniformizer π_L of L. Then, by Theorem 2.2.9, we have $K \cong k((\pi_K))$ and $L \cong l((\pi_L))$. Also, recall that by Lemma 2.1.11 we have $\mathcal{M}_K = (\pi_K)$ and $\mathcal{M}_L = (\pi_L)$.

Let v_K be the normalized valuation on K, i.e., $v_K(\pi_K) = 1$. By Proposition 2.3.1, the unique extension of v_K to L is given by

$$v_L(\alpha) = \frac{1}{n} v_K(\mathrm{Nr}_{L/K}(\alpha)) \quad \text{for all } \alpha \in L.$$

Therefore, the value group $v_L(L^\times)$ is a subgroup of $\frac{1}{n}\mathbb{Z}$ that contains $\mathbb{Z}$. Thus,

$$v_L(L^\times) = \frac{1}{e}\mathbb{Z}, \quad \text{for some } e \mid n.$$

Definition 2.6.1 The number $e = e(L/K) \geq 1$ is called the *ramification index* of L/K. The extension L/K is said to be *unramified* if $e = 1$, *ramified* if $e > 1$, and *totally ramified* if $e = n$. The extension L/K is said to be *tamely ramified* if $p \nmid e$, and *wildly ramified* otherwise.

It is clear that $\mathcal{M}_L \cap R_K = \mathcal{M}_K$ as both sides are the elements of K with strictly positive valuations. Hence, the map $R_K \to R_L/\mathcal{M}_L$ factors through $R_K/\mathcal{M}_K \to R_L/\mathcal{M}_L$. This implies that the residue field l is a field extension of k.

Definition 2.6.2 The *residue degree* (also called *inertial degree*) is

$$f(L/K) = [l : k].$$

Next, we show that the residue degree and the ramification index are related through the formula $n = e(L/K) \cdot f(L/K)$, so L/K is totally ramified if and only if $f(L/K) = 1$, and L/K is unramified if and only if $f(L/K) = n$.

Theorem 2.6.3 *With notation and assumptions as above, we have:*

(1) R_L is the integral closure of R_K in L.
(2) $M_K R_L = M_L^e$.
(3) $[L : K] = e(L/K) \cdot f(L/K)$.

Proof

(1) Suppose $\alpha \in L$ is integral over R_K. Let $m(x)$ be the minimal polynomial of α over K. Then $m(x)$ is monic with coefficients in R_K. In particular, $\mathrm{Nr}_{L/K}(\alpha) = (-1)^n m(0)^{[L:K(\alpha)]} \in R_K$, so

$$v_L(\alpha) = \frac{1}{n} v_K(\mathrm{Nr}_{L/K}(\alpha)) \geq 0.$$

This means that $\alpha \in R_L$. Conversely, suppose that $\alpha \in R_L$, i.e., $v_L(\alpha) \geq 0$. By definition of the extension of v_K, the constant term of $m(x)$ is in R_K. But then, Corollary 2.4.6 implies that $m(x) \in R_K[x]$, so α is integral over R_K.

(2) By definition, $v_L(\pi_L) > 0$ generates $v_L(L^\times)$; hence, $v_L(\pi_L) = 1/e$. On the other hand, $v_L(\pi_K) = v_K(\pi_K) = 1$. Therefore, $u = \pi_L^e/\pi_K$ is a unit in R_L, as $v_L(u) = 0$. This implies that π_L^e and π_K generate the same ideal in R_L, so $M_L^e = (\pi_L^e) = \pi_K R_L = M_K R_L$.

(3) Let $e = e(L/K)$ and $f = f(L/K)$. Let $\{\lambda_1, \ldots, \lambda_f\}$ be a basis of l as a k-vector space. We claim that the elements

$$\lambda_j \pi_L^i, \qquad 1 \leq j \leq f, \quad 0 \leq i \leq e - 1, \tag{2.6.1}$$

form a basis of L as a K-vector space, which would imply $[L : K] = e \cdot f$. Suppose

$$\sum_{i=0}^{e-1} \sum_{j=1}^{f} a_{ij} \lambda_j \pi_L^i = \sum_{i=0}^{e-1} s_i \pi_L^i = 0, \quad \text{for some } a_{ij} \in K,$$

where $s_i = \sum_{j=1}^{f} a_{ij} \lambda_j$. Note that $s_i \in l(\!(\pi_K)\!)$, so the valuation of each nonzero s_i is an integer. Assume that not all $a_{ij} = 0$. Then in the sum $\sum_{i=0}^{e-1} s_i \pi_L^i$, two of the summands must have the same valuation, say

$$v_L(s_i \pi_L^i) = v_L(s_j \pi_L^j), \qquad i \neq j.$$

But this implies that

$$\frac{i}{e} - \frac{j}{e} = v_L(\pi_L^i) - v_L(\pi_L^j) = v_L(s_j) - v_L(s_i) \in \mathbb{Z},$$

which is a contradiction.   □

Remark 2.6.4 One can show that R_L as a free R_K-module of rank $e(L/K) \cdot f(L/K)$ and the elements in (2.6.1) form a basis of R_L over R_K; see Exercise 2.6.7.

Lemma 2.6.5 *Let $K \subseteq L \subseteq M$ be a chain of finite field extensions. Then*

$$f(M/K) = f(M/L) \cdot f(L/K),$$

$$e(M/K) = e(M/L) \cdot e(L/K).$$

Proof Let m be the residue field of M. Then we have the chain of extensions $k \subseteq l \subseteq m$. The claim $f(M/K) = f(M/L) \cdot f(L/K)$ is equivalent to $[m : k] = [m : l] \cdot [l : k]$, which is an easy fact from the theory of fields. The second claim $e(M/K) = e(M/L) \cdot e(L/K)$ follows from the first claim and Theorem 2.6.3. (It can also be easily proved directly from the definition of the ramification index.)   □

The unramified and totally ramified extensions have a special significance, since, as we will show next, for any finite extension L/K, there is a uniquely determined intermediate field $K \subseteq L' \subseteq L$ such that L'/K is unramified and L/L' is totally ramified.

Let L'/K be an unramified subextension of L/K. Let l' be the residue field of L'. By assumption, $[L' : K] = [l' : k]$. Because L' has positive characteristic, l' is a subfield of L, so $l'K \subseteq L'$. Let $q = \#l'$. By Hensel's Lemma, the decomposition of $x^q - x \in K[x]$ into irreducibles over a finite extension K' of K is the same as the decomposition of $x^q - x \in k[x]$ over the residue field of K'. This implies that $[l'K : K] = [l' : k]$. Therefore, $[l'K : K] = [L' : K]$. Since $l'K \subseteq L'$, we conclude that $L' = l'K$. Conversely, given a subfield $k \subset l' \subset l$, the field $l'K \cong l'(\!(\pi_K)\!)$ is unramified over K. Finally, we have

$$[lK : K] = [l : k] = f(lK/K) = f(L/K)$$

and $e(lK/K) = 1$, which implies that $f(L/lK) = 1$ and $e(L/lK) = e(L/K)$. Thus, we have proved the following:

Proposition 2.6.6 *The map $l' \mapsto l'K$ gives a bijection between the fields l' with $k \subseteq l' \subseteq l$ and the fields L' with $K \subseteq L' \subseteq L$ that are unramified over K. In particular, the field $lK \cong l(\!(\pi_K)\!)$ is the maximal unramified extension of K contained in L, and L/lK is totally ramified.*

The extension $K^{\mathrm{nr}} := \bar{k}K$ is called the *maximal unramified extension of K*. It is an infinite Galois extension of K with $\mathrm{Gal}(K^{\mathrm{nr}}/K) \cong \mathrm{Gal}(\bar{k}/k)$. By Exercise 2.3.5, K^{nr} is not complete, and its completion is isomorphic to $\bar{k}(\!(\pi_K)\!)$.

Now we consider more carefully the totally ramified extensions of K. There is an explicit description of such extensions, although it is not as simple as the description of the unramified extensions:

Proposition 2.6.7 *With our previous notation and assumptions, we have:*

(1) L/K is totally ramified if and only if L is obtained from K by adjoining a root α of an Eisenstein polynomial of degree n. Moreover, α is a uniformizer of L, and $\mathrm{Nr}_{L/K}(\alpha)$ is a uniformizer of K.

(2) L/K is totally tamely ramified if and only if L is obtained by adjoining to K a root of a polynomial of the form $x^n - \beta \pi_K$, where $\beta \in k^\times$ and $p \nmid n$.

Proof

(1) Let $f(x) \in R_K[x]$ be an Eisenstein polynomial of degree n. Let α be a root of $f(x)$. By Corollary 2.5.6 and its proof, $f(x)$ is irreducible, and $v_L(\alpha) = 1/n$. Therefore, $[K(\alpha) : K] = n$ and $n \mid e(K(\alpha)/K)$. Hence, $e(K(\alpha)/K) = n$, so $K(\alpha)/K$ is totally ramified. Moreover, $v_L(\alpha)$ generates the value group $v_L(L^\times) = \frac{1}{n}\mathbb{Z}$, so α is a uniformizer. Since $(-1)^n \mathrm{Nr}_{L/K}(\alpha)$ is the constant term of $f(x)$, we have $v_K(\mathrm{Nr}_{L/K}(\alpha)) = 1$. Hence, $\mathrm{Nr}_{L/K}(\alpha)$ is a uniformizer of K.

Conversely, suppose L/K is totally ramified. Since $v(L^\times)$ is generated by $v_L(\pi_L)$, we must have $v_L(\pi_L) = 1/n$. Let $f(x) = a_0 + a_1 x + \cdots + a_{d-1}x^{d-1} + x^d$ be the minimal polynomial of π_L over K. Note that π_L is integral over R_K, so $f(x) \in R_K[x]$. Since $f(x)$ is irreducible, $\mathrm{NP}(f)$ consists of one line segment joining $(0, v_K(a_0))$ to $(d, 0)$; see Corollary 2.5.5. But for such f, as follows from Theorem 2.5.2, all its roots have valuation $v_K(a_0)/d$. Thus, $v_K(a_0)/d = 1/n$. Since $d \mid n$ and $v_K(a_0)$ is an integer, we must have $v_K(a_0) = 1$ and $d = n$. Now it is easy to see that for $\mathrm{NP}(f)$ to have only one line segment, the valuations $v_K(a_i)$ must be strictly positive for all $1 \le i \le d - 1$. This means that $f(x)$ is an Eisenstein polynomial.

(2) It is clear that adjoining to K a root of $x^n - \beta \pi_K$ produces a totally ramified extension of K of degree n, which will be tame if $p \nmid n$. Conversely, suppose L/K is totally tamely ramified of degree n. Since the residue field of L is k, we can expand $\pi_L^n/\pi_K \in R_L$ as $\sum_{i \ge 0} \beta_i \pi_L^i$ with $\beta_i \in k$. We must have $\beta_0 \ne 0$ because the valuation of π_L^n/π_K is zero. Let $u := \beta_0^{-1} \sum_{i \ge 0} \beta_i \pi_L^i \in R_L^\times$, and consider $f(x) = x^n - u \in R_L[x]$. The reduction of $f(x)$ modulo M_L is $x^n - 1 \in k[x]$, which has no multiple roots (because $p \nmid n$), and 1 is a k-rational root. By Hensel's Lemma, $f(x)$ has a root w in R_L. Then $(\pi_L/w)^n = \beta_0 \pi_K$ and $L = K(\pi_L/w)$.   □

Remarks 2.6.8

(1) Assume K has positive characteristic p, and let n be a positive integer not divisible by p. Let $f(x) = x^n - \beta \pi_K$, $\beta \in k^\times$, be a polynomial as in the previous proposition. Let α be a root of $f(x)$, and let L be the splitting field of $f(x)$ over K. Let K_n be the splitting field of $x^n - 1$ over K. Then L is Galois over K and contains K_n as a subfield. Over K_n, the polynomial $f(x)$ is still

Eisenstein because π_K is a uniformizer of K_n. Therefore, L is a cyclic totally tamely ramified extension of K_n of degree n; cf. Proposition 1.3.12. On the other hand, if $K_n \neq K$, then $K(\alpha)/K$ is not Galois,[7] and L/K is not abelian.[8]

(2) Note that, in contrast to the situation with unramified extensions, there is no "largest" totally ramified extension of K in $\overline{K}$, since the composite of two totally ramified extensions need not be totally ramified. For example, let ζ be primitive n-th root of 1; the extension $K(\zeta\alpha)/K$ is a totally ramified extension contained in L, but $L = K(\alpha, \zeta\alpha)$ is not totally ramified.

Example 2.6.9 For each $p^n, n \geq 1$, there is a unique purely inseparable extension of K, namely the splitting field of $x^{p^n} - \pi_K$ (see the proof of Theorem 2.8.5). Clearly, this extension is totally wildly ramified.

Now assume $p = 3$, and let L be the splitting field of $f(x) = x^3 + \pi_K x + \pi_K$. This cubic is separable and irreducible over K since it is Eisenstein. The discriminant of $f(x)$ is equal to $-\pi_K^3$, which is not a square in K. Therefore, by [DF04, p. 612], $L = K(\alpha, \sqrt{-\pi_K})$, where α is a root of $f(x)$, and $\mathrm{Gal}(L/K) \cong S_3$ is the permutation group on three elements. Because $e(K(\sqrt{-\pi_K})/K) = 2$ and $e(K(\alpha)/K) = 3$, we have $e(L/K) \geq 6$. Hence, $e(L/K) = 6$, and L/K is totally ramified. We have the chain of field extensions

$$K \subset K(\sqrt{-\pi_K}) \subset L,$$

where $K(\sqrt{-\pi_K})/K$ is tamely ramified and $e(L/K(\sqrt{-\pi_K})) = 3$ is a power of p. As we explain next, in general, totally ramified extensions can be decomposed into similar chains of subextensions.

Assume L/K is a Galois extension, and let $G = \mathrm{Gal}(L/K)$ be its Galois group. By Lemma 2.3.8, for any $\sigma \in G$ and $a \in L$, we have $|\sigma a| = |a|$. Therefore, σ induces an automorphism of R_L and $\mathcal{M}_L^i$, $i \geq 1$; thus, it induces an automorphism of $R_L/\mathcal{M}_L^i$. We define

$$G_i = \left\{ \sigma \in G \mid \sigma \text{ acts trivially on } R_L/\mathcal{M}_L^{i+1} \right\}, \quad i \geq 0. \tag{2.6.2}$$

Since G_i is the kernel of the homomorphism $G \to \mathrm{Aut}\left(R_L/\mathcal{M}_L^{i+1}\right)$, it is a normal subgroup of G called the *i-th ramification group*. Some of these groups also have special names:

- G_0 is the *inertia subgroup* of G.
- G_1 is the *wild inertia subgroup* of G.
- G_0/G_1 is the *tame inertia group*.

---

[7] If $K(\alpha)/K$ is Galois, then $K(\alpha) = L$, so $n = [K(\alpha) : K] = [L : K_n][K_n : K] = n[K_n : K]$, which implies $K_n = K$.

[8] If L/K is abelian, then every intermediate field $K \subset K' \subset L$ is Galois over K. On the other hand, $K(\alpha)$ is such a field.

Proposition 2.6.10 *With notation and assumptions as above, let* $L_i := L^{G_i}, i \geq 0$. *Then:*

(1) L_0 is the maximal unramified extension of K contained in L. In particular, L/K is unramified if and only if $G_0 = 1$.

(2) L_1 is the maximal tamely ramified extension of K contained in L. In particular, L/K is tamely ramified if and only if $G_1 = 1$.

(3) $G/G_0 \cong \mathrm{Gal}(l/k)$.

(4) G_0/G_1 is isomorphic to a subgroup of the multiplicative group $l^\times$.

(5) For $i \geq 1$, G_i/G_{i+1} is isomorphic to a subgroup of the additive group $(l, +)$.

(6) $G_i = 1$ for all sufficiently large i.

Proof If we identify $K = k(\!(\pi_K)\!)$, then the maximal unramified extension of K in L is $l(\!(\pi_K)\!)$. From this description, it is obvious that $\sigma \in G$ fixes $l(\!(\pi_K)\!)$ if and only if σ fixes $l = R_L/M_L$ and thus belongs to G_0.

To prove the other claims, after passing to the maximal unramified subextension lK of L, we can assume that L/K is totally ramified and $G = G_0$. By Proposition 2.6.7, $L = K(\pi)$, where π is a root of an Eisenstein polynomial $f(x) \in K[x]$. Furthermore, π is a uniformizer of L and $l = k$, so $L = k(\!(\pi)\!)$ and $R_L = k[\![\pi]\!]$. Once we identify R_L with $k[\![\pi]\!]$, it becomes clear that $\sigma \in G$ belongs to $G_i, i \geq 1$, if and only if $\sigma\pi \equiv \pi \pmod{\pi^{i+1}}$.

First, we consider the case $i = 1$. Since $\sigma\pi$ is also a uniformizer of L, we have $\sigma\pi = u_\sigma\pi$ for some unit $u_\sigma \in R_L^\times$. Now write $u_\sigma = a_\sigma + \pi u'$, where $a_\sigma \in k^\times$ and $u' \in R_L$. Then $\sigma \in G_1$ if and only if $a_\sigma = 1$. For $\sigma, \tau \in G$, we have

$$(\tau\sigma)\pi = \tau(\sigma\pi) = \tau(u_\sigma\pi) = \tau(u_\sigma)\tau(\pi) = \tau(u_\sigma)u_\tau\pi.$$

Thus, $u_{\tau\sigma} = \tau(u_\sigma)u_\tau$. This implies that $a_{\tau\sigma} = \tau(a_\sigma)a_\tau$. On the other hand, as $a_\sigma \in k^\times$, we have $\tau(a_\sigma) = a_\sigma$. Thus, $a_{\tau\sigma} = a_\sigma a_\tau$. We conclude that $G \to k^\times$, $\sigma \mapsto a_\sigma$, is a homomorphism whose kernel is G_1. Therefore, G/G_1 is isomorphic to a subgroup of $k^\times$. (Keep in mind that we have reduced to the case where $l = k$ and $G = G_0$, so we did not prove a stronger claim than (4).)

Now suppose $\sigma \in G_i, i \geq 1$. Then

$$\sigma\pi = \pi + u_\sigma\pi^{i+1}, \qquad u_\sigma \in R_L.$$

Write $u_\sigma = a_\sigma + \pi u'$, where $a_\sigma \in k$ and $u' \in R_L$. It is easy to see that $\sigma \in G_{i+1}$ if and only if $a_\sigma = 0$. On the other hand, modulo π^{i+2}, we have the congruences

$$(\tau\sigma)\pi \equiv \pi + a_{\tau\sigma}\pi^{i+1}$$

$$\equiv \tau(\pi) + a_\sigma(\tau\pi)^{i+1}$$

$$\equiv \pi + (a_\tau + a_\sigma)\pi^{i+1},$$

for all $\tau, \sigma \in G_i$, which implies that $a_{\tau\sigma} = a_\tau + a_\sigma$. Hence, $\sigma \mapsto a_\sigma$ gives a homomorphism $G_i \to k$ whose kernel is G_{i+1}. This proves (5).

To prove (6), let $\sigma \in G$ and $\sigma\pi = u_\sigma\pi$ with $u_\sigma \in R_L^\times$. Note that $\sigma \neq 1$ if and only if $u_\sigma \neq 1$. On the other hand, if $u_\sigma \neq 1$, then for all sufficiently large i, we have $u_\sigma \not\equiv 1 \pmod{\pi^{i+1}}$. This implies that for any $\sigma \neq 1$ and all sufficiently large i, we have $\sigma \notin G_i$. Since G is a finite group, $G_i = \{1\}$ for all sufficiently large i. $\qquad\square$

Corollary 2.6.11 *With notation and assumptions as before, we have:*

(1) $\#G_0 = e(L/K)$.
(2) G_1 *is the unique Sylow p-subgroup of G_0.*
(3) L/L_1 *is a totally ramified extension whose Galois group is a p-group.*
(4) L_1/L_0 *is a totally ramified cyclic extension whose Galois group is isomorphic to a subgroup of $l^\times$. In particular, $L_1 = L_0(\sqrt[n]{a})$ for some uniformizer a of L_0 and $n = [G_0 : G_1]$.*

Proof By Proposition 2.6.10, $[G : G_0] = \#G/\#G_0 = f(L/K)$. On the other hand, by Theorem 2.6.3, $\#G = [L : K] = e(L/K) \cdot f(L/K)$. Thus, $\#G_0 = e(L/K)$. This proves (1).

By Proposition 2.6.10, L/L_0 is totally ramified, G_0/G_1 is isomorphic to a subgroup of $l^\times$, while G_1 has a composition series $G_1 \rhd G_2 \rhd G_3 \rhd \cdots \rhd \{1\}$ such that each G_i/G_{i+1} is isomorphic to a subgroup of l. Note that l, as an additive group, is an elementary p-group, i.e., is isomorphic to a direct product of $\mathbb{Z}/p\mathbb{Z}$'s. Hence, G_1 is a p-group. Since the order of G_0/G_1 is not divisible by p, G_1 is a Sylow p-subgroup of G_0. On the other hand, G_1 is normal in G_0, so by Sylow's theorem G_1 is the unique Sylow p-subgroup of G_0. This implies (2) and (3), while Proposition 1.3.12 implies (4). $\qquad\square$

Now passing to infinite extensions of K, let

$$K^{\mathrm{nr}} := \bar{k}K \quad \text{and} \quad K^{\mathrm{t}} := \bigcup_{p \nmid n} K^{\mathrm{nr}}(\pi_K^{1/n}).$$

Then we have a tower of infinite field extensions

$$K \subset K^{\mathrm{nr}} \subset K^{\mathrm{t}} \subset K^{\mathrm{sep}} \subset \overline{K}.$$

The subgroup $I_K := \mathrm{Gal}(K^{\mathrm{sep}}/K^{\mathrm{nr}})$ of $G_K = \mathrm{Gal}(K^{\mathrm{sep}}/K)$ is called the *inertia group*. The normal subgroup $I_K^{\mathrm{w}} := \mathrm{Gal}(K^{\mathrm{sep}}/K^{\mathrm{t}})$ of I_K is the *wild inertia group*, while the quotient group $I_K^{\mathrm{t}} := \mathrm{Gal}(K^{\mathrm{t}}/K^{\mathrm{nr}})$ is the *tame inertia group*:

$$0 \longrightarrow I_K^{\mathrm{w}} \longrightarrow I_K \longrightarrow I_K^{\mathrm{t}} \longrightarrow 0.$$

Using the isomorphism $G/G_0 \cong \mathrm{Gal}(\ell/k)$ of Proposition 2.6.10, one deduces that there is also the exact sequence

$$0 \longrightarrow I_K \longrightarrow G_K \longrightarrow G_k \longrightarrow 0.$$

Remark 2.6.12 By Remark 1.6.9,

$$\mathrm{Gal}(K^{\mathrm{nr}}/K) \cong G_k \cong \varprojlim \mathbb{Z}/n\mathbb{Z} \cong \prod_\ell \mathbb{Z}_\ell.$$

It is also not hard to show that

$$\mathrm{Gal}(K^{\mathrm{t}}/K^{\mathrm{nr}}) \cong \varprojlim_{p \nmid n} \mathbb{Z}/n\mathbb{Z} \cong \prod_{\ell \neq p} \mathbb{Z}_\ell,$$

where the second product is over all primes ℓ not equal to p. In particular, K^{nr}/K and $K^{\mathrm{t}}/K^{\mathrm{nr}}$ are abelian extensions, and, in fact, each is "topologically cyclic," i.e., contains a cyclic dense subgroup. For $\mathrm{Gal}(K^{\mathrm{nr}}/K)$, the Frobenius automorphism $\mathrm{Fr}_k : x \mapsto x^q$ of $\bar{k}$ is a canonical topological generator. In contrast to the tame inertia group, the wild inertia group I_K^{w} is a complicated non-abelian group. It is also worth pointing out that the extension K^{t}/K is not abelian: one can show that if σ is a topological generator of $\mathrm{Gal}(K^{\mathrm{t}}/K^{\mathrm{nr}})$, then $\mathrm{Fr}_k \circ \mathrm{Fr}_k^{-1} = \sigma^q$, and $\mathrm{Gal}(K^{\mathrm{t}}/K)$ is topologically generated by σ and Fr_k subject to the given single relation.

Exercises

2.6.1 Assume K'/K is unramified and K''/K is totally ramified. Prove that K' and K'' are linearly disjoint over K, so that $[K'K'' : K] = [K' : K] \cdot [K'' : K]$.

2.6.2 Assume K has characteristic 2. Let L be the splitting field of $x^4 + \pi_K x + \pi_K$. Determine $\mathrm{Gal}(L/K)$ and $e(L/K)$.

2.6.3 Prove that there are exactly two non-isomorphic cubic extensions of $\mathbb{Q}_2$.

2.6.4 Let $K = \mathbb{F}_p((T))$. Prove the following:

(a) K has only finitely many extensions of degree n, up to isomorphism, if $p \nmid n$.
(b) K has infinitely many non-isomorphic extensions of degree p.

2.6.5 Let L/K be a finite Galois extension. Let $I_{L/K} \subseteq \mathrm{Gal}(L/K)$ be the image of I_K in $\mathrm{Gal}(L/K)$ under the quotient map $G_K \to \mathrm{Gal}(L/K)$, and let $I_{L/K}^{\mathrm{w}}$ be the image of I_K^{w}. In the notation of (2.6.2), prove that $I_{L/K} = G_0$ and $I_{L/K}^{\mathrm{w}} = G_1$.

2.6.6 Suppose L/K is a finite abelian extension, i.e., $G = \mathrm{Gal}(L/K)$ is an abelian group. Prove that in this case G_0/G_1 is isomorphic to a subgroup of $k^\times$.

2.6.7 Prove that R_L is a free R_K-module of rank n, and in fact, the elements in (2.6.1) form a basis of R_L over R_K.

2.6.8 Let $\bar{\alpha}$ be a primitive element of l over k, i.e., $l = k[\bar{\alpha}]$, and $\bar{f}(x) \in k[x]$ be the minimal polynomial of $\bar{\alpha}$ over k. Let $f(x) \in R_K[x]$ be a monic polynomial such that $f \equiv \bar{f} \pmod{\pi_K}$. Prove the following statements:

(a) If $\alpha \in R_L$ is such that $\alpha \equiv \bar{\alpha} \pmod{\pi_L}$, then $\alpha^i \pi_L^j$, $0 \le i \le f(L/K) - 1$ and $0 \le j \le e(L/K) - 1$, are linearly independent over K.
(b) There exists $\alpha \in R_L$ such that $\alpha \equiv \bar{\alpha} \pmod{\pi_L}$ and $v_L(f(\alpha)) = 1$.
(c) $R_L = R_K[\alpha]$ for some $\alpha \in R_L$. In particular, R_L is generated by a single element as an R_K-algebra. (This is generally false for the integral closure of $A = \mathbb{F}_q[T]$ in finite extensions of $\mathbb{F}_q(T)$.)

2.7 Power Series

Let K be a field complete with respect to a nontrivial non-Archimedean absolute value $|\cdot|$. As in earlier sections, we denote by R the ring of integers in K, $\mathcal{M}$ the maximal ideal of R, and $\mathbb{C}_K$ the completion of an algebraic closure of K. We fix $q > 1$ and denote $v(\cdot) = -\log_q |\cdot|$; then $v \colon K^\times \to \mathbb{R}$ is a valuation.

2.7.1 Convergence

We use formal power series to define functions on K. More precisely, given a power series $f(x) = \sum_{n \ge 0} a_n x^n \in K[[x]]$, we think of it as defining a function whose domain is the set of $\alpha \in K$ for which the series $f(\alpha) = \sum_{n \ge 0} a_n \alpha^n$ converges.

Lemma 2.7.1 *An infinite series $\sum_{n \ge 0} a_n$ over K converges if and only if $|a_n| \to 0$. Moreover, if $|a_n| \to 0$, then*

$$\left| \sum_{n \ge 0} a_n \right| \le \max_{n \ge 0} |a_n|.$$

Proof For $m \ge 0$, denote $S_m = \sum_{n=0}^m a_n$. Since K is complete, the series $\sum_{n \ge 0} a_n$ converges if and only if the sequence of partial sums $\{S_m\}_{m \ge 0}$ is Cauchy in K. Thus, if $\sum_{n \ge 0} a_n$ converges, then $|S_m - S_{m-1}| = |a_m| \to 0$ as m increases. Conversely, if $|a_m| \to 0$, then for $n \ge m$, we have

$$|S_n - S_m| \le \max(|a_{m+1}|, \dots, |a_n|) \longrightarrow 0,$$

as $n, m \to \infty$, so $\sum_{n \ge 0} a_n$ converges.

To prove the second claim, assume $|a_n| \to 0$ and denote $M = \max_{n \geq 0} |a_n|$. Then, by the strong triangle inequality, we have

$$|S_m| \leq \max(|a_0|, \ldots, |a_m|) \leq M.$$

Hence, $|\sum_{n \geq 0} a_n| = \lim |S_m| \leq M$. $\qquad\qquad\qquad\qquad\qquad\qquad\qquad\qquad$ □

Remark 2.7.2 Note that the claim of Lemma 2.7.1 is false over $\mathbb{R}$ since $\sum_{n \geq 1} 1/n$ diverges.

Lemma 2.7.3 *Assume $\sum_{n \geq 0} a_n$ and $\sum_{m \geq 0} b_m$ are convergent series over K. For $t \geq 0$, let*

$$c_t = \sum_{n+m=t} a_n b_m.$$

Then the series $\sum_{t \geq 0} c_t$ is convergent and

$$\left(\sum_{n \geq 0} a_n \right) \left(\sum_{m \geq 0} b_m \right) = \sum_{t \geq 0} c_t.$$

Proof For $N \geq 0$, let

$$s_1(N) := \sum_{n=0}^{N} a_n, \quad \varepsilon_1(N) := \sum_{n > N} a_n$$

und

$$s_2(N) := \sum_{m=0}^{N} b_m, \quad \varepsilon_2(N) := \sum_{m > N} b_m.$$

Then

$$\left(\sum_{n \geq 0} a_n \right) \left(\sum_{m \geq 0} b_m \right) = s_1(N)s_2(N) + s_1(N)\varepsilon_2(N) + s_2(N)\varepsilon_1(N) + \varepsilon_1(N)\varepsilon_2(N).$$

Since the absolute values of $s_1(N)$ and $s_2(N)$ remain bounded as $N \to \infty$, whereas $\varepsilon_1(N)$, $\varepsilon_2(N) \to 0$, for any $\varepsilon > 0$, we can find N such that

$$\left| \left(\sum_{n \geq 0} a_n \right) \left(\sum_{m \geq 0} b_m \right) - s_1(N)s_2(N) \right| < \varepsilon.$$

On the other hand, using the fact that $|a_n| \to 0$ and $|b_m| \to 0$, it is not hard to show that for any $\varepsilon > 0$, we can find N' such that

$$\left| \sum_{t \geq 0} c_t - s_1(N') s_2(N') \right| < \varepsilon.$$

By choosing the larger of N and N', we can assume that both estimates hold for the same N. Then

$$\left| \left(\sum_{n \geq 0} a_n \right) \left(\sum_{m \geq 0} b_m \right) - \sum_{t \geq 0} c_t \right|$$

$$= \left| \left(\left(\sum_{n \geq 0} a_n \right) \left(\sum_{m \geq 0} b_m \right) - s_1(N) s_2(N) \right) + \left(s_1(N) s_2(N) - \sum_{t \geq 0} c_t \right) \right| < \varepsilon,$$

which implies the claim of the lemma.   □

We leave as exercises a few more similar technical problems, such as the rearrangement of series and the summation of double series (see Exercises 2.7.1 and 2.7.2).

Corollary 2.7.4 *If $f(x), g(x) \in K[\![x]\!]$ both converge at $\alpha \in K$, then their product $h(x) = f(x)g(x)$ in $K[\![x]\!]$ also converges at α and $f(\alpha)g(\alpha) = h(\alpha)$.*

Definition 2.7.5 The *radius of convergence* of $f(x) = \sum_{n \geq 0} a_n x^n \in K[\![x]\!]$, denoted $\rho(f)$, is

$$\frac{1}{\rho(f)} = \overline{\lim_{n \to \infty}} \sqrt[n]{|a_n|},$$

where $1/\infty = 0$, $1/0 = \infty$, and $\overline{\lim}_{n \to \infty} \sqrt[n]{|a_n|}$ (the limit supremum) is the largest limit point of the sequence $\{\sqrt[n]{|a_n|}\}$ in $\mathbb{R}_{\geq 0} \cup \{\infty\}$. Note that the radius of convergence can be equivalently defined by the formula

$$\log_q \rho(f) = \lim_{n \to \infty} \frac{v(a_n)}{n}.$$

The power series $f(x)$ is called *entire* if $\rho(f) = \infty$.

Lemma 2.7.6 *The power series $f(x) = \sum_{n \geq 0} a_n x^n \in K[\![x]\!]$ converges at $\alpha \in K$ if $|\alpha| < \rho(f)$ and diverges at α if $|\alpha| > \rho(f)$. Thus, $f(x)$ defines a function*

$$\{z \in K : |z| < \rho(f)\} \longrightarrow K, \quad z \longmapsto f(z),$$

on the open disk of radius $\rho(f)$.

Proof By Lemma 2.7.1, the series $f(\alpha) = \sum_{n \geq 0} a_n \alpha^n$ converges if and only if

$$|a_n \alpha^n| = (\sqrt[n]{|a_n|} \cdot |\alpha|)^n \to 0.$$

If $|\alpha| > \rho(f)$, then $1/|\alpha| < 1/\rho(f) = \overline{\lim}_{n \to \infty} \sqrt[n]{|a_n|}$. Hence, $1/|\alpha| < \sqrt[n]{|a_n|}$ infinitely often, so $|a_n \alpha^n| > 1$ infinitely often. Hence, $f(\alpha)$ does not converge.

Assume now $|\alpha| < \rho(f)$. Then $1/|\alpha| > \overline{\lim}_{n \to \infty} \sqrt[n]{|a_n|}$. This implies that there is $0 < \varepsilon < 1$ such that $1 - \varepsilon > \sqrt[n]{|a_n|} \cdot |\alpha|$ for all $n \gg 0$. Hence, $0 \leq |a_n \alpha^n| < (1 - \varepsilon)^n \to 0$, so $f(\alpha)$ converges. $\qquad\square$

Example 2.7.7 For $f(x) = \sum_{n \geq 0} x^n$, we have $\rho(f) = 1$. Note that $f(\alpha)$ does not converge for any $\alpha \in K$ with $|\alpha| = 1$. Now fix $0 \neq \pi \in M$ and consider $g(x) = \sum_{n \geq 0} \pi^n x^{n^2}$. Since $0 < |\pi| < 1$, we have

$$1/\rho(g) = \lim_{n \to \infty} |\pi|^{n/n^2} = 1.$$

Thus, $\rho(g) = 1$, but $g(\alpha)$ converges for all α with $|\alpha| \leq 1$, including those that have absolute value 1. More generally, a power series $f(x)$ converges at α with $|\alpha| = \rho(f)$ if and only if $\lim_{n \to \infty} |a_n| \rho(f)^n = 0$.[9]

Lemma 2.7.8 *Let $u(x) \in R[\![x]\!]^\times$. Then $u(\alpha)$ converges and is nonzero for all $\alpha \in \mathbb{C}_K$ with $|\alpha| < 1$.*

Proof Since the coefficients of $u(x)$ are in R, we have $\rho(f) \geq 1$, so $u(\alpha)$ converges for all $\alpha \in \mathbb{C}_K$ with $|\alpha| < 1$. The same is true for its inverse $u(x)^{-1}$ in $R[\![x]\!]$. On the other hand, by Corollary 2.7.4, we have $u(\alpha)u(\alpha)^{-1} = 1$, so $u(\alpha) \neq 0$. $\qquad\square$

2.7.2 Weierstrass Preparation Theorem

The key technical result of this section is the non-Archimedean version of the so-called Weierstrass Preparation Theorem from classical complex analysis.

Definition 2.7.9 A polynomial $\sum_{n=0}^{d} a_n x^n \in R[x]$ is called *distinguished* if $a_0, \ldots, a_{d-1} \in M$ and $a_d = 1$.

Theorem 2.7.10 *Let $f = \sum_{n \geq 0} a_n x^n \in R[\![x]\!]$ be such that $a_0, \ldots, a_{d-1} \in M$ and $a_d \in R^\times$. Then there is a unit $u(x) \in R[\![x]\!]^\times$ and a distinguished polynomial $g(x) \in R[x]$ of degree d such that $f(x) = u(x)g(x)$. Moreover, $u(x)$ and $g(x)$ with these properties are uniquely determined by $f(x)$.*

---

[9] This is a non-Archimedean phenomenon: the series $\sum_{n \geq 1} \frac{(-1)^n}{n} x^n \in \mathbb{R}[\![x]\!]$ converges for $|x| < 1$ and $x = 1$ but diverges for $x = -1$.

Proof First, we prove the uniqueness. Suppose $u_1 g_1 = u_2 g_2$, where u_1 and u_2 are units, and g_1 and g_2 are distinguished polynomials of the same degree. Let $\alpha \in \mathbb{C}_K$ be a root of g_1. The slopes of the Newton polygon of g_1 are all strictly negative (because it is distinguished). This implies $|\alpha| < 1$. By Lemma 2.7.8, $u_2(\alpha) \neq 0$. Thus $0 = u_1(\alpha)g_1(\alpha) = u_2(\alpha)g_2(\alpha)$ implies $g_2(\alpha) = 0$. Canceling $(x-\alpha)$ from the decompositions of $g_1(x)$ and $g_2(x)$ into linear factors over $\mathbb{C}_K$, and using induction on the degrees of g_1 and g_2, one concludes that g_1 and g_2 have the same roots in $\mathbb{C}_K$ and thus differ by a constant multiple. As they are both monic, we must have $g_1 = g_2 = g$. Now $(u_1 - u_2) \cdot g = 0$ implies $u_1 = u_2$, because $R[\![x]\!]$ is an integral domain.

To prove the existence, we will inductively construct elements

$$b_1(i), \ldots, b_d(i) \in R, \qquad i \geq 1,$$

and units

$$u_i(x) \in R[\![x]\!], \quad u_i(x) = 1 + \text{higher degree terms}, \qquad i \geq 1,$$

such that

$$b_m(1) \equiv a_m \ (\text{mod } \mathcal{M}), \qquad\qquad m = 1, \ldots, d,$$
(2.7.1)

$$b_m(i+1) \equiv b_m(i) \ (\text{mod } \mathcal{M}^i), \qquad i \geq 1, \quad m = 1, \ldots, d,$$
(2.7.2)

$$u_{i+1} \equiv u_i \ (\text{mod } \mathcal{M}^i), \qquad i \geq 1, \quad \text{(congruence of coefficients)}$$
(2.7.3)

$$u_i f \equiv a_0 + \sum_{m=1}^{d} b_m(i)x^m \ (\text{mod } \mathcal{M}^i), \quad i \geq 1, \quad \text{(congruence of coefficients)}.$$
(2.7.4)

Assume for the moment that such elements have been constructed. Because of the congruence conditions (2.7.2), $\{b_m(i)\}_{i \geq 1}$ form a convergent sequence in R, and we define $b_m = \lim_{i \to \infty} b_m(i)$ for $m = 1, \ldots, d$. Similarly, due to (2.7.3), the coefficients of u_i's form convergent sequences, and we define $\tilde{u} = \lim_{i \to \infty} u_i$. Note that the constant term of $\tilde{u}$ is 1, so it is a unit in $R[\![x]\!]$. Also, $b_d \in R$ is a unit because $b_d \equiv a_d \ (\text{mod } \mathcal{M})$ is nonzero. If we let $\tilde{g} := a_0 + b_1 x + \cdots + b_d x^d$, then $\tilde{u} f = \tilde{g}$. Now put $g = b_d^{-1}\tilde{g}$ and $u = (b_d^{-1}\tilde{u})^{-1}$. Then g is distinguished of degree d, $u \in R[\![x]\!]$ is a unit, and $f = u \cdot g$.

We now proceed to our inductive construction.

Put $b_1(1) = b_2(1) = \cdots = b_{d-1}(1) = 0$, $b_d(1) = a_d$, and $u_1 = a_d \left(\sum_{m \geq 0} a_{m+d}x^m\right)^{-1}$. Note that $\sum_{m \geq 0} a_{m+d}x^m$ is invertible (because a_d is a unit), and the constant term of u_1 is 1. Then

$$u_1 f \equiv a_d \left(\sum_{m \geq 0} a_{m+d} x^m \right)^{-1} \left(\sum_{m \geq 0} a_{m+d} x^m \right) x^d$$

$$\equiv a_d x^d$$

$$\equiv a_0 + \sum_{i=1}^{d} b_i(1) x^i \pmod{\mathcal{M}}.$$

Assume we have constructed $b_1(n), b_2(n), \ldots, b_d(n)$ and u_n such that the required congruences hold. Let $b_m(n+1) = b_m(n) + c_m$, $1 \leq m \leq d$, and $u_{n+1} = u_n + w$, where we need $c_1, \ldots, c_d \in \mathcal{M}^n$ and $w \in \mathcal{M}^n[\![x]\!]$ such that

$$u_{n+1} f \equiv a_0 + \sum_{m=1}^{d} b_m(n+1) x^m \pmod{\mathcal{M}^{n+1}}.$$

We can write

$$u_n f = a_0 + \sum_{m=1}^{d} b_m(n) x^m + h,$$

with $h = \sum_{i \geq 1} h_i x^i \in \mathcal{M}^n[\![x]\!]$. (Note that the constant term of h is zero because $u_n f = a_0 +$ higher degree terms.) Now, we have

$$u_{n+1} f = u_n f + w f$$

$$= a_0 + \sum_{m=1}^{d} b_m(n) x^m + (wf + h)$$

$$= a_0 + \sum_{m=1}^{d} b_m(n+1) x^m + \left(wf + h - \sum_{m=1}^{d} c_m x^m \right).$$

We need to choose $w \in \mathcal{M}^n[\![x]\!]$ and $c_1, \ldots, c_m \in \mathcal{M}^n$ such that

$$wf + h - \sum_{m=1}^{d} c_m x^m \equiv 0 \pmod{\mathcal{M}^{n+1}}.$$

Put $c_m = h_m$ for $1 \leq m \leq d$. By assumption, $w \sum_{i=0}^{d-1} a_i x^i \in \mathcal{M}^{n+1}[\![x]\!]$, so we need

$$w \cdot \left(\sum_{m \geq 0} a_{m+d} x^m \right) x^d \equiv - \left(\sum_{m \geq 0} h_{m+d} x^m \right) x^d \ (\mathrm{mod}\ \mathcal{M}^{n+1}).$$

Since $\sum_{m \geq 0} a_{m+d} x^m$ is invertible in $R[\![x]\!]$ (as $a_d \in R^\times$), to make the above congruence true, we may take

$$w = - \left(\sum_{m \geq 0} h_{m+d} x^m \right) \left(\sum_{m \geq 0} a_{m+d} x^m \right)^{-1}.$$

Note that $w \in \mathcal{M}^n[\![x]\!]$, since $\sum_{m \geq 0} h_{m+d} x^m \in \mathcal{M}^n[\![x]\!]$. This completes the inductive step and the proof of the theorem. $\square$

2.7.3 Weierstrass Factorization Theorem

The Weierstrass factorization theorem from complex analysis asserts that every entire function can be represented as a (possibly infinite) product involving its zeros, which may be viewed as an extension of the fundamental theorem of algebra. Moreover, every sequence tending to infinity has an associated entire function with zeros at precisely the points of that sequence. Here we prove the non-Archimedean analogues of these results.

Definition 2.7.11 We say that $\alpha \in \mathbb{C}_K$ is a *zero with multiplicity* $m \geq 1$ of $f(x) \in \mathbb{C}_K[\![x]\!]$ if there exists $g(x) \in \mathbb{C}_K[\![x]\!]$ such that $f(x) = (x - \alpha)^m \cdot g(x)$, $g(x)$ converges at α, and $g(\alpha) \neq 0$.

Given $r \in \mathbb{R}_{\geq 0}$, denote

$$\mathbb{D}(r) = \{ \alpha \in \mathbb{C}_K \ : \ |\alpha| \leq r \},$$
$$\mathbb{D}^\circ(r) = \{ \alpha \in \mathbb{C}_K \ : \ |\alpha| < r \}.$$

Proposition 2.7.12 *Suppose* $f(x) \in K[\![x]\!]$ *is entire and not a constant. Then:*

(1) $f(x)$ has at least one zero in $\mathbb{C}_K$.
(2) The zeros of $f(x)$ are algebraic over K.
(3) The number of zeros of $f(x)$ in $\mathbb{D}(r)$ is finite for any $r \geq 0$.
(4) $f : \mathbb{C}_K \to \mathbb{C}_K$, $\alpha \mapsto f(\alpha)$, is surjective.

Proof Let $f(x) = \sum_{n\geq 0} a_n x^n$, and let

$$d = \min\{m \geq 0 \;:\; |a_n| \leq |a_m| \text{ for all } n \geq 0\}.$$

Note that $f_c(x) := f(cx) = \sum_{n\geq 0} a_n c^n x^n$ is entire for any $c \in K$. After replacing f by f_c for an appropriate c, we may assume that $d \geq 1$ and $r < 1$. Now replacing $f(x)$ by $a_d^{-1} f(x)$, we may assume that $f(x)$ satisfies the assumptions of Theorem 2.7.10. Hence, $f(x) = u(x)g(x)$, where $g(x) \in R[x]$ is a distinguished polynomial of degree d and $u(x) \in R[\![x]\!]^{\times}$ is a unit. By Lemma 2.7.8, $u(x)$ has no zeros in $\mathbb{D}^{\circ}(1)$. Hence, the zeros of $f(x)$ in $\mathbb{D}^{\circ}(1)$ are the zeros of the polynomial $g(x)$. Since the zeros of $g(x)$ are algebraic over K and finite in number, we get (1), (2), and (3).

To prove (4), let $\beta \in \mathbb{C}_K$. Applying (1) to $f(x) - \beta$, which is still entire, one concludes that there is $\alpha \in \mathbb{C}_K$ such that $f(\alpha) = \beta$. $\qquad\square$

Definition 2.7.13 Let V be a K-vector space equipped with a norm $\|\cdot\|$; see Definition 2.3.2. A subset $S \subset V$ is *weakly discrete* if S is discrete with respect to the topology on V, i.e., every point of S has a neighborhood that contains no other points of the set. A subset $S \subset V$ is *discrete* if $\{v \in S \;:\; \|v\| < r\}$ is a finite set for any positive r. We say that a subset $S \subset \mathbb{C}_K$ is weakly discrete (respectively, discrete), if it is weakly discrete (respectively, discrete) when we consider $\mathbb{C}_K$ as a vector space over K with the norm being the absolute value on $\mathbb{C}_K$.

Remark 2.7.14 A discrete set is weakly discrete, but the converse is false: if $K = \mathbb{F}_q(\!(T)\!)$, then $S = \overline{\mathbb{F}}_q$ is weakly discrete in $\mathbb{C}_K$ since $|\alpha - \beta| = 1$ for all $\alpha \neq \beta \in \overline{\mathbb{F}}_q$, but S is not discrete since $S \subset \mathbb{D}(1)$.

Let $f(x) \in \mathbb{C}_K[\![x]\!]$ be an entire function, and let α be a zero of $f(x)$. Then we can write $f(x) = (x - \alpha)^{c_\alpha} g(x)$, where $g(\alpha) \neq 0$ and $c_\alpha \geq 1$ is the multiplicity of α. Let

$$Z(f) = \{(\alpha, c_\alpha) \mid f(\alpha) = 0\} \subset \mathbb{C}_K \times \mathbb{Z}_{>0}$$

be the set of zeros of $f(x)$ in $\mathbb{C}_K$, counted with multiplicities. Proposition 2.7.12 implies that the projection of $Z(f)$ onto $\mathbb{C}_K$ is discrete. We will show that $Z(f)$ uniquely determines $f(x)$, up to a constant multiple, and moreover, $f(x)$ has a product decomposition over its zeros similar to the decomposition of polynomials.[10] This allows a complete classification of entire functions over $\mathbb{C}_K$.

Let $S \subset \mathbb{C}_K$ be a set. We assign to each $\alpha \in S$ some positive integer c_α, which we call the multiplicity of α. Let

$$Z = \{(\alpha, c_\alpha) \mid \alpha \in S, c_\alpha \in \mathbb{Z}_{>0}\} \subset \mathbb{C}_K \times \mathbb{Z}_{>0}.$$

---

[10] This is a special property of entire functions over non-Archimedean fields since both $\exp(x) = \sum_{n\geq 0} x^n/n!$ and 1 have empty sets of zeros over $\mathbb{C}$.

We treat Z as a multiset in $\mathbb{C}_K$, i.e., Z is the set of elements in S but each $\alpha \in S$ occurs c_α times in Z. We say that Z is discrete if $S \subset \mathbb{C}_K$ is discrete. Assume Z is discrete in $\mathbb{C}_K$ and define

$$f_Z(x) = x^{c_0} \prod_{\alpha \in Z}' \left(1 - \frac{x}{\alpha}\right) = x^{c_0} \prod_{\alpha \in S}' \left(1 - \frac{x}{\alpha}\right)^{c_\alpha},$$

where c_0 is the multiplicity of 0 if $0 \in S$, and $c_0 = 0$ if $0 \notin S$. We expand $f_Z(x)$ into a power series

$$f_Z(x) = x^{c_0} \sum_{n \geq 0} (-1)^n G_n x^n,$$

where

$$G_n = \sum_{\substack{\alpha_1,\ldots,\alpha_n \in Z \\ \text{all nonzero}}} \frac{1}{\alpha_1 \cdots \alpha_n}.$$

It is clear that Z is countable because we can enumerate its elements by enumerating the elements contained in discs with increasing integer radii. We enumerate the elements of $Z = \{\beta_1, \beta_2, \ldots\}$ so that $|\beta_1| \leq |\beta_2| \leq \ldots$. Since S is discrete and the multiplicities c_α are finite, $|\beta_{i_1} \cdots \beta_{i_n}| \to \infty$ as $\max\{i_1, \ldots, i_n\} \to \infty$. Hence, the series G_n converge, and we can consider f_Z as a power series over $\mathbb{C}_K$. (The order in which we add the elements in the series G_n does not affect its value by Exercise 2.7.1.)

Proposition 2.7.15 *With above notation and assumptions, $f_Z(x)$ is an entire function such that $Z(f_Z) = Z$.*

Proof Using Lemma 2.7.1, we see that

$$|G_n| \leq (|\beta_1| \cdots |\beta_n|)^{-1}.$$

Let $z \in \mathbb{C}_K$ and $0 < \varepsilon < 1$. Since S is discrete, there is an integer $N(\varepsilon)$ such that $|z|/|\beta_i| < \varepsilon$ for all $i \geq N(\varepsilon)$. Hence, for $n \geq N(\varepsilon)$, we have $|G_n z^n| \leq c \cdot \varepsilon^{n-N(\varepsilon)}$, where c is a positive constant that does not depend on n. This makes it obvious that $|G_n z^n| \to 0$, so $f_Z(x)$ converges at z. Thus, $f_Z(x)$ is entire.

To prove that $Z(f_Z) = Z$, it is enough to show that $Z \cap (\mathbb{D}(r) \times \mathbb{Z}_{>0})$ is the set of zeros of f_Z in $\mathbb{D}(r)$, counted with multiplicities, for any $r \in |\mathbb{C}_K^\times|$. For this last statement, it is enough to show that

$$h(x) = \prod_{\substack{\alpha \in S \\ |\alpha| > r}} \left(1 - \frac{x}{\alpha}\right)^{c_\alpha}$$

has no zeros in $\mathbb{D}(r)$. But this follows by applying Lemma 1.1.1 and Lemma 2.7.8 to $u(x) = h(\beta x)$ for any $\beta \in \mathbb{C}_K^\times$ with $|\beta| = r$. $\qquad \square$

Theorem 2.7.16 *Assume $f(x) \in \mathbb{C}_K[\![x]\!]$ is entire. Then there is $\beta \in \mathbb{C}_K^\times$ such that*

$$f(x) = \beta \cdot x^{c_0} \prod_{\alpha \in Z(f)}{}' \left(1 - \frac{x}{\alpha}\right),$$

where c_0 is the multiplicity of 0 in $Z(f)$. Thus, up to a nonzero constant, $f(x)$ is uniquely determined by its multiset of zeros $Z(f)$.

Proof We proved that $Z := Z(f)$ is discrete, so by Proposition 2.7.15 the function f_Z is entire and Z is its multiset of zeros. We will show that $f = \beta \cdot f_Z$ for some $\beta \in \mathbb{C}_K^\times$.

Since $f(x) = x^{c_0}\tilde{f}(x)$, where $\tilde{f}(x)$ is entire and $0 \notin Z(\tilde{f})$, by considering $\tilde{f}(x)$ instead of $f(x)$, we may assume that $0 \notin Z$, so that the constant term of $f(x)$ is nonzero. In that case, $f(x)$ has an inverse $f(x)^{-1}$ in the formal power series ring $\mathbb{C}_K[\![x]\!]$ (see Lemma 1.1.1). Consider $h(x) = f_Z(x)f(x)^{-1}$ as a formal power series in $\mathbb{C}_K[\![x]\!]$. We claim that $h(x)$ is entire and has no zeros. Assuming this, $h(x) = \beta$ is a constant by Proposition 2.7.12, so $f(x) = \beta \cdot f_Z(x)$.

Let $r > 0$ be any real number. It is enough to show that $h(x)$ converges on $\mathbb{D}(r)$ and has no zeros in this disk. Choose $c \in \mathbb{C}_K^\times$ with $|c| > r$; after replacing $h(x)$ by $h(cx)$, we may assume that $r < 1$ (note that $f_c(x) = f(cx)$ is entire and $f_{Z(f_c)}(x) = f_Z(cx)$). Furthermore, since the coefficients of the series $f(x)$ tend to 0, after replacing $f(x)$ by $c'f(x)$ for an appropriate $c' \in \mathbb{C}_K^\times$, we may assume that all the coefficients of $f(x)$ have absolute values ≤ 1. Now, using Theorem 2.7.10, we decompose $f(x) = g(x)u(x)$ into a product of a polynomial $g(x)$ and an entire function $u(x) = 1 + \sum_{n\geq 1} u_n x^n$ with $|u_n| \leq 1$ for all n and no zeros in $\mathbb{D}(r)$ (see Lemma 2.7.8). Hence, the zeros of $f(x)$ in $\mathbb{D}(r)$ are the zeros of $g(x)$. From the formulas (1.1.3) for the coefficients of $u(x)^{-1}$, it is clear that $u(x)^{-1} = 1 + \sum_{n\geq 1} w_n x^n$ has coefficients that also satisfy $|w_n| \leq 1$ for all $n \geq 1$. But then $u(x)^{-1}$ converges on $\mathbb{D}(r)$ and has no zeros in this disk. By construction, $f_Z(x)/g(x)$ is entire and has no zeros in $\mathbb{D}(r)$. Finally, we conclude that $h(x) = (f_Z(x)/g(x))u(x)^{-1}$ converges and has no zeros in $\mathbb{D}(r)$. $\qquad \square$

2.7.4 Newton Polygon of Power Series

The Newton polygon of a polynomial is an extremely useful tool for analyzing the absolute values of its roots. We will show that similar information can be extracted from the Newton polygon of a general power series; the proof of this reduces to the case of polynomials with the help of Theorem 2.7.10.

Let $f(x) = \sum_{n \geq 0} a_n x^n \in K[\![x]\!]$ be a power series with $a_0 \neq 0$. Similar to the case of polynomials, the *Newton polygon* of f, denoted NP(f), is defined as the lower convex hull in $\mathbb{R}^2$ of the set of points $\{(n, v(a_n)) \mid n \geq 0, a_n \neq 0\}$. But unlike the case of polynomials, several distinct possibilities arise due to the fact that NP(f) is the convex hull of infinitely many points:

(1) NP(f) has infinitely many segments of finite length. For example, $f(x) = \sum_{n \geq 0} \pi^{n^2} x^n$ is such a series if $v(\pi) > 0$.
(2) NP(f) has finitely many segments of finite length and one last segment of infinite length that passes through infinitely many points $(n, v(a_n))$. For example, the Newton polygon of $f(x) = \sum_{n \geq 0} \pi^n x^n$ consists of a single half-line starting at $(0, 0)$ with slope $v(\pi)$, and all the points $(n, v(a_n))$ are on this line.
(3) NP(f) has finitely many segments of finite length and one last segment of infinite length that passes through finitely many points $(n, v(a_n))$. For example, the Newton polygon of $f(x) = 1 + \sum_{n \geq 1} \pi x^n$ consists of a single horizontal half-line starting at $(0, 0)$, and only the point $(0, 0)$ is on this line. Note that if we rotate the horizontal half-line even slightly counterclockwise, then all but finitely many of the points $(n, v(\pi))$ will lie below this rotated line.
(4) A degenerate case of (3), when we cannot rotate the vertical line through $(0, v(a_0))$ counterclockwise without leaving infinitely many of the points $(n, v(a_n))$ below that line. This happens for example for $f(x) = 1 + \sum_{n \geq 1} \pi^{-n^2} x^n$. In fact, it is easy to check that this happens if and only if $\rho(f) = 0$.

Proposition 2.7.17 *Assume $f(x) = 1 + \sum_{n \geq 1} a_n x^n \in K[\![x]\!]$ is not a polynomial and $\rho(f) \neq 0$. Then the sequence of the slopes of NP(f) is increasing, and its limit is $\log_q \rho(f)$.*

Proof It is not hard to see by drawing a picture of a convex polygon, infinite in the positive direction of the x-axis, that the limit of the slopes of NP(f) is equal to $\varliminf_{n \to \infty} v(a_n)/n$ (cf. Fig. 2.4). On the other hand, by the definition of $\rho(f)$, this limit infimum is $\log_q \rho(f)$.   □

Proposition 2.7.18 *Let $u(x) \in R[\![x]\!]^{\times}$ be a unit and $g(x) \in R[x]$ be a distinguished polynomial of degree d. The Newton polygon of $f(x) = u(x)g(x)$ is obtained by translating NP(u) by d to the right so that the first vertex of NP(u) coincides with the last vertex $(d, 0)$ of NP(g); cf. Fig. 2.5.*

Proof Let $f(x) = \sum_{n \geq 0} f_n x^n$, $u(x) = \sum_{n \geq 0} u_n x^n$, and $g(x) = \sum_{n=0}^{d} g_n x^n$. Let NP$_f(z)$ denote the piecewise linear function on $[0, +\infty)$ defined by the Newton polygon of f, and similarly for g and u. For $0 \leq n \leq d$, we have

$$f_n = g_n u_0 + g_{n-1} u_1 + \cdots + g_0 u_n.$$

Fig. 2.4 The dashed lines have slopes converging to $\underline{\lim}_{n \to \infty} v(a_n)/n$

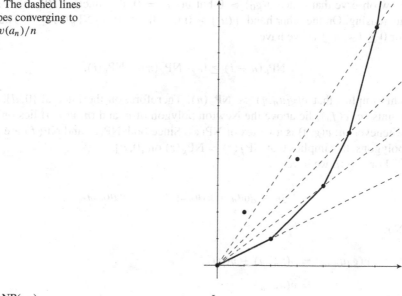

Fig. 2.5 NP(gu)

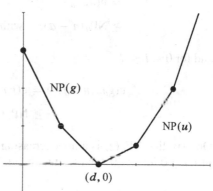

NP(g)

NP(u)

$(d, 0)$

Now

$$v(g_n u_0) = v(g_n) + v(u_0)$$
$$= v(g_n)$$
$$\geq \text{NP}_g(n), \quad \text{with equality if } \text{NP}_g(n) = v(g_n),$$

and for $0 \leq i < n$

$$v(g_i u_{n-i}) = v(g_i) + v(u_{n-i})$$
$$\geq \text{NP}_g(i) + \text{NP}_u(n - i).$$

Now observe that, since $v(g_0) > 0$ but $v(g_d) = 0$, the function $\mathrm{NP}_g(z)$ is strictly decreasing. On the other hand, $v(u_i) \geq 0$ for all $i \geq 0$, so $\mathrm{NP}_u(z) \geq 0$. Therefore, for $0 \leq i < n \leq d$, we have

$$\mathrm{NP}_u(n - i) \geq 0 > \mathrm{NP}_g(n) - \mathrm{NP}_g(i).$$

This implies that $v(g_i u_{n-i}) > \mathrm{NP}_g(n)$. Therefore, on the interval $[0, d]$, all the points $(n, v(f_n))$ lie above the Newton polygon of g, and $(n, v(f_n))$ lies on $\mathrm{NP}(g)$ whenever $(n, v(g_n))$ is a vertex of $\mathrm{NP}(g)$. Since both $\mathrm{NP}(g)$ and $\mathrm{NP}(f)$ are convex polygons, this implies that $\mathrm{NP}_f(z) = \mathrm{NP}_g(z)$ on $[0, d]$.

For $n \geq d$, we have

$$f_n = g_0 u_n + g_1 u_{n-1} + \cdots + g_d u_{n-d}.$$

Now

$$v(g_d u_{n-d}) = v(u_{n-d}) + v(g_d)$$

$$= v(u_{n-d})$$

$$\geq \mathrm{NP}_u(n - d), \quad \text{with equality if } \mathrm{NP}_u(n - d) = v(u_{n-d}),$$

and for $0 < i \leq d$

$$v(g_{d-i} u_{n-d+i}) = v(g_{d-i}) + v(u_{n-d+i})$$

$$\geq \mathrm{NP}_g(d - i) + \mathrm{NP}_u(n - d + i).$$

Observe that $\mathrm{NP}_u(z)$ is a nondecreasing function on $[0, +\infty)$ since $v(u_0) = 0$ and $v(u_i) \geq 0$ for $i > 0$. On the other hand, $\mathrm{NP}_g(z) > 0$ for $x \in [0, d)$. Therefore,

$$\mathrm{NP}_u(n - d + i) - \mathrm{NP}_u(n - d) \geq 0 > -\mathrm{NP}_g(d - i).$$

This implies that

$$v(g_{d-i} u_{n-d+i}) > \mathrm{NP}_u(n - d).$$

Therefore, on $[d, +\infty)$, all the points $(n, v(f_n))$ lie above $\mathrm{NP}_u(z-d)$, and $(n, v(f_n))$ lies on $\mathrm{NP}_u(z - d)$ whenever $(n - d, v(u_{n-d}))$ is a vertex of $\mathrm{NP}(u)$. As in the earlier case, since both $\mathrm{NP}(u)$ and $\mathrm{NP}(f)$ are convex polygons, this implies that $\mathrm{NP}_f(z) = \mathrm{NP}_u(z - d)$ on $[d, +\infty)$. □

Theorem 2.7.19 Let $f(x) = \sum_{n \geq 0} a_n x^n \in \mathbb{C}_K[\![x]\!]$ be a power series such that $a_0 \neq 0$ and $\rho(f) \neq 0$. If a segment of $\mathrm{NP}(f)$ has finite horizontal length N and slope λ, then $f(x)$ has precisely N zeros, counted with multiplicities, with valuation $-\lambda$.

Proof First, we make an observation about the Newton polygon of $f_c(x) := f(cx)$ for $c \in \mathbb{C}_K^\times$. It is easy to check that the vertices of $NP(f_c)$ and $NP(f)$ have the same x-coordinates; thus the line segments of these polygons are in natural bijection. Moreover, the finite segments that correspond to each other have the same horizontal length, but the slope of a segment of $NP(f_c)$ is equal to the slope of the corresponding segment of $NP(f)$ plus $v(c)$. Hence, it is enough to prove the theorem for f_c for any nonzero c.

If $NP(f)$ has a segment of finite horizontal length N and slope λ, then by Proposition 2.7.17 $f(\alpha)$ converges for all $\alpha \in \mathbb{C}_K$ with $|\alpha| \le q^\lambda$. This implies that if $c \ne 0$ and $|c| \le q^\lambda$, then $|a_n c^n| \to 0$. Thus, after replacing $f(x)$ by $f(cx)$, we may assume that $v(a_n) \to +\infty$. Now, multiplying $f(x)$ by an appropriate constant, $c' \cdot f(x)$, we may assume that all the coefficients of $f(x)$ have non-negative valuations, and at least one of its coefficients has valuation 0 (multiplication by c' only shifts $NP(f)$ vertically without changing its segments or their slopes; it also does not change the valuations of zeros). This puts us in the setting of Theorem 2.7.10, so we can factor $f(x) = g(x)u(x)$, with $g(x)$ a distinguished polynomial of degree d (=smallest index i such that $v(a_i) = 0$) and $u(x)$ a unit.

If the slope of the segment in question is negative, then by Proposition 2.7.18 this segment corresponds to a segment of $NP(g)$ of the same horizontal length and slope. The claim of the theorem then follows from Theorem 2.5.2 since, by Lemma 2.7.8, $u(\alpha) \ne 0$ if $|\alpha| < 1$.

Now assume the slope in question is non-negative. In that case, using Proposition 2.7.18 again, we may assume $f(x) = u(x)$. Then the first vertex of $NP(f)$ is $(0, 0)$, and the slopes of $NP(f)$ are strictly increasing $\lambda_1 < \lambda_2 < \dots$ (the number of finite length segments of $NP(f)$ can be finite or infinite, but by assumption there is at least one finite segment). Assume $\lambda = \lambda_n$, and let a_m be the coefficient of $f(x)$ such that the end vertex of our segment is $(m, v(a_m))$. We can choose $c \in \mathbb{C}_K$ such that $\lambda_n - v(c) < 0 < \lambda_{n+1} - v(c)$. Now it is not hard to check that $\frac{c^m}{a_m} f(x/c) = \sum_{i \ge 0} b_i x^i$ is such that $|b_i| < 1$ for $0 \le i \le m - 1$, $b_m = 1$, and $|b_i| \le 1$ for $i \ge m + 1$. Moreover, the slope of our segment becomes negative, so we end up in the previous case.   □

2.7.5 Formal Substitutions

For now, let R be a commutative ring, and let $R[\![x]\!]$ be the ring of formal power series with coefficients in R. For the same reason that we cannot generally evaluate $f(x) = \sum_{n \ge 0} a_n x^n \in R[\![x]\!]$ at $a \in R$, we cannot substitute $g(x) \in R[\![x]\!]$ into $f(x)$ if $g(x)$ has a nonzero constant term. However, if the constant term of $g(x)$ is zero, then

$$f(g(x)) = \sum_{n \ge 0} a_n g(x)^n$$

is defined because for any $m \geq 0$ the first m coefficients of $f(g(x))$ coincide with the first m coefficients of $\sum_{n=0}^{m} a_n g(x)^n$. Denote this substitution in $R[\![x]\!]$ by $f \circ g$.

Lemma 2.7.20 *Let $f \in R[\![x]\!]$, and let $g, h \in x R[\![x]\!]$. Then*

$$(f \circ g) \circ h = f \circ (g \circ h).$$

Proof Note that $(x^n \circ g) \circ h = g^n \circ h$. On the other hand, in general,

$$(g_1 g_2) \circ h = (g_1 \circ h)(g_2 \circ h).$$

Hence, $(x^n \circ g) \circ h = g^n \circ h = (g \circ h)^n$. This implies that for $f = \sum_{n \geq 0} a_n x^n$, we have

$$(f \circ g) \circ h = \sum_{n \geq 0} a_n (g \circ h)^n = f \circ (g \circ h).$$

$\square$

According to the previous lemma, substitution in $x R[\![x]\!]$ is an associative binary operation. Moreover,[11]

$$(f + g) \circ h = f \circ h + g \circ h \qquad \text{for all} \qquad f, g, h \in x R[\![x]\!].$$

Obviously, x is the identity for $\circ$. The next lemma is the analogue of Lemma 1.1.1.

Lemma 2.7.21 *Let $f = \sum_{n \geq 1} a_n x^n \in x R[\![x]\!]$. There exists $g \in x R[\![x]\!]$ such that $f \circ g = x$ if and only if $a_1 \in R^\times$, in which case g is unique and also has the property that $g \circ f = x$. In particular, if R is a field, then the elements of $x R[\![x]\!]$ with nonzero first coefficient form a non-abelian group under substitution.*

Proof Put $g = \sum_{n \geq 1} b_n x^n$, where the b_i's are indeterminates. Formally expanding $f \circ g = \sum_{n \geq 1} c_n x^n$, it is easy to see that $c_1 = a_1 b_1$ and, for $n \geq 1$,

$$c_n = a_1 b_n + (\text{polynomial in } a_2, \ldots, a_n \text{ and } b_1, \ldots, b_{n-1}).$$

We want to have $c_1 = 1$ and $c_n = 0$ for $n \geq 2$. One can recursively solve the previous equations for $b_1, b_2, \ldots$ if and only if a_1 is invertible in R. Moreover, the solution is unique. Next, note that $b_1 = a_1^{-1}$ is invertible, and so there is a unique $h \in x R[\![x]\!]$ such that $g \circ h = x$. To show that $h = f$, we compute

$$h = x \circ h = (f \circ g) \circ h = f \circ (g \circ h) = f \circ x = f.$$

$\square$

---

[11] But note that $\circ$ is not commutative, and $f \circ (g + h)$ is not always equal to $f \circ g + f \circ h$.

Now we return to our initial non-Archimedean setting and consider the convergence of substitutions in $K[[x]]$. Suppose we are given two formal power series $f(x)$ and $g(x)$ with $g(0) = 0$, so that $(f \circ g)(x)$ is a well-defined formal power series. Suppose $\alpha \in K$ is such that $g(\alpha)$ converges, and let $\beta = g(\alpha)$ be its limit; in addition, suppose $f(x)$ converges at β. A basic question is whether $(f \circ g)(x)$ converges at α and, if it does, whether $(f \circ g)(\alpha) = f(g(\alpha))$. This turns out to be a subtle question, and the answer is generally negative as the next example demonstrates.

Example 2.7.22 Let $f(x) = \sum_{n \geq 1} a_n x^n \in K[[x]]$, and assume $a_1 = 1$, so that by Lemma 2.7.21 there exists $g(x) \in x K[[x]]$ such that $(f \circ g)(x) = x$. Assume there exists $0 \neq \alpha \in Z(g)$. Then we have $g(\alpha) = 0$ and $f(0) = 0$, but $(f \circ g)(\alpha) = \alpha \neq 0$, so that $(f \circ g)(\alpha) \neq g(f(\alpha))$.

Lemma 2.7.23 *Suppose $f(x) \in K[x]$ is a polynomial and $g(x) \in x K[[x]]$. Then $(f \circ g)(x)$ converges on $\mathbb{D}^\circ(\rho(g))$ and $(f \circ g)(\alpha) = f(g(\alpha))$ for all $\alpha \in \mathbb{D}^\circ(\rho(g))$.*

Proof If $f(x) = a_0 + a_1 x + \cdots + a_n x^n$, then $(f \circ g)(x) = a_0 + a_1 g(x) + \cdots + a_n g(x)^n$ is a sum of finitely many terms. Hence, it is enough to prove the lemma for $f(x) = x^n$. But in this case

$$(f \circ g)(x) = g(x) \cdots g(x) \qquad (n \text{ times}),$$

so the claim follows from Corollary 2.7.4.   □

Theorem 2.7.24 *Let $f(x) = \sum_{n \geq 0} a_n x^n$ and $g(x) = \sum_{n \geq 1} b_n x^n$ be formal power series in $K[[x]]$. If*

$$|\alpha| < \rho(g) \quad \text{and} \quad \max\{|b_n \alpha^n| : n \geq 0\} < \rho(f),$$

then $(f \circ g)(\alpha)$ and $f(g(\alpha))$ converge, and $(f \circ g)(\alpha) = f(g(\alpha))$.

Proof The full proof, which is not very difficult, can be found in [Rob00, VI.1.5]. We only indicate the main idea. Define the polynomials

$$f_N(x) = \sum_{n=0}^{N} a_n x^n.$$

By Lemma 2.7.23, we have $(f_N \circ g)(\alpha) = f_N(g(\alpha))$. Since

$$|g(\alpha)| \leq \max\{|b_n \alpha^n|\} < \rho(f),$$

we have $f_N(g(\alpha)) \to f(g(\alpha))$ as $N \to \infty$. Finally, using the assumption $\max\{|b_n \alpha^n|\} < \rho(f)$, one shows that $(f_N \circ g)(\alpha) \longrightarrow (f \circ g)(\alpha)$ as $N \to \infty$.   □

Corollary 2.7.25 *Let $f(x) = \sum_{n\geq 0} a_n x^n$ and $g(x) = \sum_{n\geq 1} b_n x^n$ be formal power series in $K[\![x]\!]$. Assume one of these power series is entire and the other is a polynomial. Then $(f \circ g)(\alpha) = f(g(\alpha))$ for all $\alpha \in \mathbb{C}_K$.*

Proof Note that $\rho(f) = \rho(g) = +\infty$, so the claim follows from Lemma 2.7.23 or Theorem 2.7.24, depending on which of these power series is a polynomial. □

Exercises

2.7.1 Let $\sum_{n\geq 0} a_n$ be a convergent series in K. Let $\{a_n'\}$ be a rearrangement of $\{a_n\}$. Show that $\sum_{n\geq 0} a_n'$ converges and is equal to $\sum_{n\geq 0} a_n$.

2.7.2 Let $a_{m,n} \in K$ for $m, n = 0, 1, 2, \ldots$ Suppose that for every $\varepsilon > 0$, there exists N such that $|a_{m,n}| < \varepsilon$ whenever $\max(m, n) \geq N$. Show that the series

$$\sum_{n\geq 0} \left(\sum_{m\geq 0} a_{m,n} \right) \quad \text{and} \quad \sum_{m\geq 0} \left(\sum_{n\geq 0} a_{m,n} \right)$$

both converge, and the corresponding limits are equal.

2.7.3 Let $f(x) \in K[\![x]\!]$ be a function convergent on $\mathbb{D}(r)$ for some $r > 0$. Assume $f(0) = 0$. Prove that f maps $\mathbb{D}(r)$ onto $\mathbb{D}(r')$ for some $r' > 0$.

2.7.4 Let K be a *local field*, and let V be a finite dimensional K-vector space with norm $\|\cdot\|$. Show that a weakly discrete set $S \subset V$ is discrete.

2.7.5 Let $f(x) = \sum_{n\geq 0} a_n x^n \in \mathbb{C}_K[\![x]\!]$. Suppose

$$d = \max\{m \geq 0 \mid v(a_n) \geq v(a_m) \text{ for all } n \geq 0\}$$

is a well-defined integer. Use Theorem 2.7.19 to show that the number of zeros of $f(x)$ in $\mathbb{D}(1)$ is d, counted with multiplicities.

2.7.6 Let

$$\mathbb{T} = \left\{ \sum_{n\geq 0} a_n x^n \in K[\![x]\!] \;\middle|\; \lim_{n\to\infty} |a_n| = 0 \right\}.$$

Prove that $f(x) \in K[\![x]\!]$ lies in $\mathbb{T}$ if and only if $f(x)$ converges on $\mathbb{D}(1)$. Conclude that $\mathbb{T}$ is a subalgebra of $K[\![x]\!]$, called the *Tate algebra*.

2.7.7 The *norm* of $f(x) = \sum_{n\geq 0} a_n x^n \in \mathbb{T}$ is

$$\|f\| = \max_{n\geq 0} |a_n|.$$

Note that by assumption $|a_n| \to 0$, so the maximum exists. Prove the following statements:

(a) $\|f\| \geq 0$ with equality if and only if $f = 0$.
(b) $\|f + g\| \leq \max(\|f\|, \|g\|)$.
(c) $\|a \cdot f\| = |a| \cdot \|f\|$ for all $a \in K$.
(d) $\|f \cdot g\| = \|f\| \cdot \|g\|$.
(e) If $f_1, f_2, \ldots, f_m, \cdots \in \mathbb{T}$ is a sequence such that $\|f_n\| \to 0$, then $\sum_{n \geq 0} f_n \in \mathbb{T}$.

The above properties mean that $(\mathbb{T}, \|\cdot\|)$ is a *Banach algebra*. Also, prove that

(f) $\mathbb{T}$ is the completion of the polynomial ring $K[x]$ with respect to the norm $\|\cdot\|$ defined in Exercise 2.2.4.
(g) For $f(x) \in \mathbb{T}$, we have the "maximum modulus principle"

$$\|f\| = \max_{\alpha \in \mathbb{D}(1)} |f(\alpha)|.$$

2.8 Extensions of Valuations of Global Fields

Let $\mathbb{F}_q$ be a finite field of positive characteristic p, let $A = \mathbb{F}_q[T]$ be the ring of polynomials in T, and let $F = \mathbb{F}_q(T)$ be the fraction field of A. Theorem 2.1.8 gives a complete classification of the valuations on F. In this section, we are interested in describing the extensions of these valuations to finite extensions of F. This problem is closely related to the problem of splitting of primes in extensions of global fields, and the following terminology comes from that setup:

Definition 2.8.1 Let K be a field, and let L/K be a finite extension. If the valuation w of L extends the valuation v on K, then we say that the place corresponding to w *divides* (or *lies over*) the place of K corresponding to v. Sometimes, with abuse of notation, we will denote by w the place corresponding to the valuation w and write $w \mid v$ if w divides v. We say that v *splits* in L if there are at least two distinct places of L dividing v.

Recall that a *Dedekind domain* is an integral domain R with the following properties:

- R is integrally closed in its field of fractions.
- Every ideal of R is finitely generated.
- Every nonzero prime ideal of R is maximal.

Theorem 2.8.2 *Let* $F = \mathbb{F}_q(T)$, *and let* K/F *be a finite extension. The integral closure of* $A = \mathbb{F}_q[T]$ *in* K, *denoted* B, *is a Dedekind domain.*

Proof This is a special case of a basic fact about integral closures of Dedekind domains, usually proved in a course on algebraic number theory under the assumption that K/F is separable. We need to verify that B has the properties in the definition of a Dedekind domain:

- The fact that B is integrally closed in K follows from the transitivity of integrality; cf. Definition 1.5.22.
- By Theorem 1.5.23, B is a free A-module of rank $[K : F]$. Since an ideal $I \lhd B$ is an A-submodule, I is also finitely generated over A and thus especially finitely generated over B.
- Let $\mathfrak{P} \lhd B$ be a nonzero prime ideal. For nonzero $b \in \mathfrak{P}$, the norm $\mathrm{Nr}_{K/F}(b)$ is nonzero and lies in $\mathfrak{p} = \mathfrak{P} \cap A$. Hence, $\mathfrak{p}$ is a nonzero prime ideal, so a maximal ideal in A. Now $B/\mathfrak{P}$ is an integral extension of the field $A/\mathfrak{p}$; hence, it is a field. This implies that $\mathfrak{P}$ is maximal.

$\square$

Remark 2.8.3 Let A be an arbitrary Dedekind domain, let F be its fraction field, let K/F be a finite extension, and let B be the integral closure of A in K. Then B is always a Dedekind domain, even if it is not finitely generated over A; cf. Remark 1.5.24. This follows from the Krull–Akizuki Theorem; see [Mat89, Thm. 11.7].

A property that distinguishes Dedekind domains among all other integral domains is the fact that every nonzero proper ideal factors uniquely into a product of prime ideals; see [DF04, p. 765]. Let K and B be as in the previous theorem. For a given nonzero prime ideal $\mathfrak{P} \lhd B$, we can define a valuation $\mathrm{ord}_{\mathfrak{P}}$ on K by a construction very similar to the case of F. Namely, given $a \in B$, we decompose the ideal (a) into a product $\mathfrak{P}^n I$, where $n \geq 0$ and $\mathfrak{P} + I = B$, then define $\mathrm{ord}_{\mathfrak{P}}(a) = n$, and extend this from B to K by $\mathrm{ord}_{\mathfrak{P}}(a/b) = \mathrm{ord}_{\mathfrak{P}}(a) - \mathrm{ord}_{\mathfrak{P}}(b)$.

For simplicity, we call nonzero prime ideals of Dedekind domains *primes*. Note that this slightly conflicts with the common terminology used for integral domains, where a prime is an element that generates a prime ideal [DF04, p. 284].

Now let $\mathfrak{p} = (P)$ be the prime ideal generated by a monic irreducible polynomial $P \in A$. Let $\mathfrak{p}B = PB$ be the ideal generated by P as an element of B. By the prime decomposition theorem in Dedekind domains, we have

$$\mathfrak{p}B = \mathfrak{P}_1^{e_1} \cdots \mathfrak{P}_r^{e_r}, \tag{2.8.1}$$

where $\mathfrak{P}_1, \ldots, \mathfrak{P}_r$ are distinct primes of B and e_i are positive integers. The primes $\mathfrak{P}_1, \ldots, \mathfrak{P}_r$ are said to *lie over* $\mathfrak{p}$, and $\mathfrak{p}$ is said to *lie under* each $\mathfrak{P}_i$. It is easy to show that $\mathrm{ord}_{\mathfrak{P}}$ extends ord_P if and only if $\mathfrak{P}$ lies over $\mathfrak{p}$. Thus, the number of inequivalent extensions of ord_P to K (i.e., places of K over ord_P) is at least the number of distinct primes in the prime decomposition of $\mathfrak{p}$ in B. On the other hand, suppose w on K extends ord_P. Then $\mathcal{M}_w = \{a \in B \mid w(a) > 0\}$ is a maximal ideal of B which divides $\mathfrak{p}$. Hence, w is equivalent to $\mathrm{ord}_{\mathfrak{P}}$ for one of the primes over $\mathfrak{p}$. The extensions of ord_∞ can be put into a similar context by considering F as the fraction field of the polynomial ring $\mathbb{F}_q[1/T]$, similar to the proof of Theorem 2.1.8. Finally, if we have a nontrivial valuation w on K, then its restriction to F is a nontrivial valuation v; hence, w extends v. We have proved the following:

Theorem 2.8.4 *Let $F = \mathbb{F}_q(T)$ and K/F be a finite field extension. The number of inequivalent extensions of ord_P to K is the number of distinct primes in the prime decomposition of P in the integral closure of A in K. Similarly, the number of inequivalent extensions of ord_∞ to K is the number of distinct primes in the prime decomposition of $1/T$ in the integral closure of $\mathbb{F}_q[1/T]$ in K. A nontrivial valuation on K that is trivial on $\mathbb{F}_q$ extends either ord_P for some prime $P \in A$ or ord_∞.*

Theorem 2.8.5 *Let v be a nontrivial valuation on $F = \mathbb{F}_q(T)$. Denote by F_v the completion of F with respect to the absolute value associated to v. If K/F is a finite extension, then*

$$K \otimes_F F_v \cong \prod_{w|v} K_w,$$

where the product is over the valuations of K extending the valuation v.

Proof By Theorem 1.5.19, there is a primitive element $\alpha \in K$ such that $K = F(\alpha)$. Let $f(x) \in F[x]$ be the minimal polynomial of α, so that $K \cong F[x]/(f(x))$.

Let $f = g_1 g_2 \cdots g_s$ be the decomposition of f into irreducible monic polynomials in $F_v[x]$. We claim that all g_i are distinct, even if f is inseparable. The coefficients of g_i are polynomial expressions in the roots of f, so they are algebraic over F. Therefore, f decomposes into the product $g_1 g_2 \cdots g_s$ over some finite algebraic extension L of F that can be embedded into F_v. Using Lemma 1.5.2, it is enough to show that L is separable over F. Suppose on the contrary that L is inseparable over F. Then, by Theorem 1.5.17, L contains $T^{1/p}$. On the other hand, by Theorem 2.2.9, either $F_v \cong k((P))$ or $F_v \cong \mathbb{F}_q((T^{-1}))$, depending on whether v is equivalent to ord_P or ord_∞, where $k = A/(P)$. We can express $P^{1/p}$ as a polynomial in $T^{1/p}$; hence, if $T^{1/p} \in F_v$, then $F_v^p = k((P))^p = k((P^p))$ contains P, which is false. A similar argument applies to $\mathbb{F}_q((T^{-1}))$.

Now, by Corollary 1.5.3,

$$K \otimes_F F_v \cong (F[x]/(f(x))) \otimes_F F_v$$

$$\cong F_v[x]/(f(x))$$

$$= F_v[x]/(g_1(x)g_2(x) \cdots g_s(x))$$

$$\cong \prod_{i=1}^{s} F_v[x]/(g_i(x)),$$

where in the last isomorphism we use the Chinese Remainder Theorem 1.2.8 and the fact that the irreducible polynomials $g_1, \ldots, g_s$ are distinct. Note that $\mathcal{K}_i :=$

$F_v[x]/(g_i(x))$ is a finite extension of F_v. According to Theorem 2.3.5, the valuation v extends uniquely to $\mathcal{K}_i$, and $\mathcal{K}_i$ is complete with respect to that extended absolute value. On the other hand, by its definition, $\mathcal{K}_i$ is generated over F_v by a root of f; hence, it must be isomorphic to a completion of K at a place w over v. Conversely, let w be a place of K over v, and let $g(x)$ be the minimal polynomial of $\alpha \in K_w$ over F_v, so that $K_w \cong F_v[x]/(g(x))$. Then g is irreducible and divides f since $f(\alpha) = 0$. Therefore, g is one of the g_i's. This gives a bijection between the fields $F_v[x]/(g_i(x))$, $1 \leq i \leq s$, and the completions K_w, $w \mid v$. Therefore, $K \otimes_F F_v \cong \prod_{w|v} K_w$. □

Corollary 2.8.6 *Let v be a nontrivial valuation on $F = \mathbb{F}_q(T)$, and let K/F be a finite extension. Then $K \otimes_F F_v$ is a field if and only if there is a unique place of K over v.*

We record a useful fact that was established in the proof of Theorem 2.8.5 for F_v. This is generally true for the completions of function fields and can be proved by a similar argument (we leave this as an exercise).

Proposition 2.8.7 *Let K be a finite extension of F, and let w be a nontrivial valuation on K. Then an element of K_w that is algebraic over K is necessarily separable.*

Now, somewhat generalizing our setup, let L/K be a finite extension. Let B' be the integral closure of B in L. Note that B' is also the integral closure of A in L (the fact that the integral closure of A in L is contained in B' is obvious, and the reverse inclusion follows from the transitivity of integrality). Let $\mathfrak{p} \lhd B$ be a prime. Let

$$\mathfrak{p}B' = \mathfrak{P}_1^{e_1} \cdots \mathfrak{P}_r^{e_r} \tag{2.8.2}$$

be the prime decomposition of $\mathfrak{p}$ in B'. By an argument similar to the argument following (2.8.1), one deduces that a place w of L extends the place $v = \mathrm{ord}_{\mathfrak{p}}$ of K if and only if w is equivalent to $\mathrm{ord}_{\mathfrak{P}_i}$ for some $1 \leq i \leq r$. Let $L_{\mathfrak{P}_i}$ be the completion of L with respect to $\mathrm{ord}_{\mathfrak{P}_i}$, and $K_{\mathfrak{p}}$ be the completion of K with respect to $\mathrm{ord}_{\mathfrak{p}}$. For any $\alpha \in K$, we have $\mathrm{ord}_{\mathfrak{P}_i}(\alpha) = e_i \cdot \mathrm{ord}_{\mathfrak{p}}(\alpha)$. This implies that e_i is the ramification index of the extension $L_{\mathfrak{P}_i}/K_{\mathfrak{p}}$. We proved that $B'/\mathfrak{P}_i$ is the residue field of $L_{\mathfrak{P}_i}$; cf. (2.2.2). Thus, $f_i = [B'/\mathfrak{P}_i : B/\mathfrak{p}]$ is the residue degree of $L_{\mathfrak{P}_i}/K_{\mathfrak{p}}$.

Proposition 2.8.8 *Let v be a place of K, and let w be a place of L dividing v. Denote by $e(L_w/K_v)$ and $f(L_w/K_v)$ the ramification index and the residue degree of the extension L_w/K_v, respectively. Then*

$$[L : K] = \sum_{w|v} e(L_w/K_v) \cdot f(L_w/K_v).$$

In particular, for a prime $\mathfrak{p} \lhd B$ *with prime decomposition* $\mathfrak{p}B' = \mathfrak{P}_1^{e_1} \cdots \mathfrak{P}_r^{e_r}$ *in* B', *we have*

$$[L : K] = \sum_{i=1}^{r} e_i \cdot f_i, \qquad where \quad f_i = [B'/\mathfrak{P}_i : B/\mathfrak{p}].$$

Proof The formula is true in general, but we will prove it only under the assumption that either $K = F$ or L/K is separable. The reason is that we will use Theorem 2.8.5, which we proved for the extensions of F, but the same proof also works for L/K assuming it is separable. Now

$$[L : K] = \dim_{K_v}(L \otimes_K K_v) \tag{2.8.3}$$

$$= \sum_{w|v} [L_w : K_v] \qquad \text{(Theorem 2.8.5)}$$

$$= \sum_{w|v} e(L_w/K_v) \cdot f(L_w/K_v) \qquad \text{(Theorem 2.6.3)}.$$

$\square$

Definition 2.8.9 Let $\mathfrak{p}$ be a prime of B, and let $\mathfrak{p}B' = \mathfrak{P}_1^{e_1} \cdots \mathfrak{P}_r^{e_r}$ be its prime decomposition in B'. The prime $\mathfrak{p}$ is said to be *inert* (or to *remain prime*) in L if $\mathfrak{p}B'$ is prime in B', equiv., $r = 1$ and $e_1 = 1$. The prime $\mathfrak{p}$ is said to be *unramified* (respectively, *ramified*) in L if $e_i = 1$ for all $1 \le i \le r$ (respectively, $e_i > 1$ for some $1 \le i \le r$). The prime $\mathfrak{p}$ is said to be *totally ramified* in L if $r = 1$, $f_1 = 1$, and $e_1 = [L : K]$. The prime $\mathfrak{p}$ is said to *split completely* in L if $r = [L : K]$, equiv., $e_i = f_i = 1$ for all $1 \le i \le r$.

For the rest of this section, we assume that L/K is a Galois extension and denote $G = \mathrm{Gal}(L/K)$.

Theorem 2.8.10 *For any* $1 \le i, j \le r$, *there is* $\sigma \in G$ *such that* $\sigma \mathfrak{P}_i = \mathfrak{P}_j$.

Proof Denote $\mathfrak{P} = \mathfrak{P}_i$ and $\mathfrak{P}' = \mathfrak{P}_j$. It is clear that $\sigma \in G$ induces an isomorphism $B'/\mathfrak{P} \to B'/\sigma(\mathfrak{P})$, so $\sigma(\mathfrak{P}) \lhd B'$ is a maximal ideal. Assume $\mathfrak{P}' \ne \sigma(\mathfrak{P})$ for any $\sigma \in \mathrm{Gal}(L/K)$. By the Chinese Remainder Theorem 1.2.8, there is $\alpha \in B'$ such that $\alpha \in \mathfrak{P}'$, but $\alpha \equiv 1 \pmod{\sigma(\mathfrak{P})}$ for all σ. If $\alpha \equiv 1 \pmod{\sigma(\mathfrak{P})}$, then $\sigma^{-1}(\alpha) \notin \mathfrak{P}$. Therefore,

$$\mathrm{Nr}_{L/K}(\alpha) = \prod_{\sigma \in G} \sigma^{-1}(\alpha) \notin \mathfrak{P}.$$

On the other hand, because one of the factors in the above expression for $\mathrm{Nr}_{L/K}(\alpha)$ is α itself, we have

$$\mathrm{Nr}_{L/K}(\alpha) \in \mathfrak{P}' \cap B = \mathfrak{p} \subset \mathfrak{P}.$$

This leads to a contradiction.

$\square$

Because the Galois group G transitively permutes the primes in the decomposition (2.8.2) and because this decomposition is unique, we must have $e_i = e_j$ for all $1 \leq i, j \leq r$. In the previous proof, we also observed that $B'/\mathfrak{P}$ is isomorphic to $B'/\sigma(\mathfrak{P})$. Hence, $f_i = f_j$ for all $1 \leq i, j \leq r$. Overall, denoting the common values of the ramification indices and the residue degrees by e and f, respectively, the equality in Proposition 2.8.8 becomes

$$[L : K] = e \cdot f \cdot r. \tag{2.8.4}$$

For each prime $\mathfrak{P} \lhd B'$ lying over $\mathfrak{p}$, we define two subgroups of G: the *decomposition group*

$$D(\mathfrak{P}/\mathfrak{p}) = \{\sigma \in G \mid \sigma\mathfrak{P} = \mathfrak{P}\}$$

and the *inertia group*

$$I(\mathfrak{P}/\mathfrak{p}) = \{\sigma \in G \mid \sigma(\alpha) \equiv \alpha \pmod{\mathfrak{P}} \text{ for all } \alpha \in B'\}.$$

It is clear that $D(\mathfrak{P}/\mathfrak{p})$ and $I(\mathfrak{P}/\mathfrak{p})$ are subgroups of G. Moreover, $I(\mathfrak{P}/\mathfrak{p})$ is a subgroup of $D(\mathfrak{P}/\mathfrak{p})$ (for $\alpha \in \mathfrak{P}$ and $\sigma \in I(\mathfrak{P}/\mathfrak{p})$, the congruence $\sigma(\alpha) \equiv \alpha \equiv 0 \pmod{\mathfrak{P}}$ implies that $\sigma(\alpha) \in \mathfrak{P}$). We make a simple but important observation that for $\alpha \in B'$ and $\sigma \in D(\mathfrak{P}/\mathfrak{p})$, there is an equality $\operatorname{ord}_{\mathfrak{P}}(\alpha) = \operatorname{ord}_{\mathfrak{P}}(\sigma(\alpha))$, which can be seen by applying σ to the prime decomposition of the ideal (α). Since L is the field of fractions of B', this implies that $\operatorname{ord}_{\mathfrak{P}}(\alpha) = \operatorname{ord}_{\mathfrak{P}}(\sigma(\alpha))$ for all $\alpha \in L$. Now if $\{\alpha_i\}_{i \geq 0}$ is a Cauchy sequence in L with respect to the absolute value associated to $\operatorname{ord}_{\mathfrak{P}}$, then $\{\sigma(\alpha_i)\}_{i \geq 0}$ is also Cauchy for any $\sigma \in D(\mathfrak{P}/\mathfrak{p})$. Therefore, the action of $D(\mathfrak{P}/\mathfrak{p})$ on L naturally extends to an action on $L_{\mathfrak{P}}$. Thus, we get an injection $D(\mathfrak{P}/\mathfrak{p}) \hookrightarrow \operatorname{Gal}(L_{\mathfrak{P}}/K_{\mathfrak{p}})$. On the other hand, the orbit-stabilizer theorem from group theory and (2.8.4) imply that the order of $D(\mathfrak{P}/\mathfrak{p})$ is $e \cdot f = [L_{\mathfrak{P}} : K_{\mathfrak{p}}]$. Thus, the previous injection is an isomorphism:

Theorem 2.8.11 *There is a natural isomorphism*

$$D(\mathfrak{P}/\mathfrak{p}) \xrightarrow{\sim} \operatorname{Gal}(L_{\mathfrak{P}}/K_{\mathfrak{p}})$$

under which $I(\mathfrak{P}/\mathfrak{p})$ *maps isomorphically onto the inertia group of* $\operatorname{Gal}(L_{\mathfrak{P}}/K_{\mathfrak{p}})$.

Proof Recall that we defined the inertia subgroup of $\operatorname{Gal}(L_{\mathfrak{P}}/K_{\mathfrak{p}})$ to be the subgroup consisting of those elements that act trivially on $R_{\mathfrak{P}}/M_{\mathfrak{P}}$. On the other hand, $I(\mathfrak{P}/\mathfrak{p})$ can be equivalently defined as the subgroup of $D(\mathfrak{P}/\mathfrak{p})$ consisting of those elements that act trivially on $B'/\mathfrak{P}$. Thus, the second claim of the theorem follows from the canonical isomorphism $R_{\mathfrak{P}}/M_{\mathfrak{P}} \cong B'/\mathfrak{P}$; cf. (2.2.2). $\qquad\square$

Assume now that $\mathfrak{p}$ is unramified in L, so that $\mathfrak{p}B' = \mathfrak{P}_1 \cdots \mathfrak{P}_r$. Fix $\mathfrak{P} = \mathfrak{P}_i$ for some $1 \leq i \leq r$. Then $L_{\mathfrak{P}}/K_{\mathfrak{p}}$ is unramified, and we have an isomorphism $\operatorname{Gal}(L_{\mathfrak{P}}/K_{\mathfrak{p}}) \cong \operatorname{Gal}(l/k)$, where $l := B'/\mathfrak{P}$ and $k := B/\mathfrak{p}$. Therefore, $D(\mathfrak{P}/\mathfrak{p}) \cong$

$\mathrm{Gal}(l/k)$. The Galois group $\mathrm{Gal}(l/k)$ has a special generator, the Frobenius automorphism of l given by $x \mapsto x^{\#k}$. The corresponding element of $D(\mathfrak{P}/\mathfrak{p})$ will be denoted by $\mathrm{Frob}_{\mathfrak{P}}$ and is uniquely characterized by the property[12] that

$$\mathrm{Frob}_{\mathfrak{P}}(\alpha) \equiv \alpha^{\#k} \pmod{\mathfrak{P}} \qquad \text{for all } \alpha \in B'.$$

Note that for $\sigma \in G$ we have $D(\sigma(\mathfrak{P})/\mathfrak{p}) = \sigma D(\mathfrak{P}/\mathfrak{p})\sigma^{-1}$, which implies

$$\mathrm{Frob}_{\sigma(\mathfrak{P})} = \sigma \, \mathrm{Frob}_{\mathfrak{P}} \, \sigma^{-1}.$$

Since G transitively permutes the primes of B' over $\mathfrak{p}$, the conjugacy class of $\mathrm{Frob}_{\mathfrak{P}}$ is uniquely determined by $\mathfrak{p}$.

Theorem 2.8.12 *Assume $\mathfrak{p}$ is unramified in L. Then for any prime $\mathfrak{P} \lhd B'$ lying over $\mathfrak{p}$, the order of $\mathrm{Frob}_{\mathfrak{P}}$ in G is equal to $[L : K]/r$, where r is the number of distinct primes of B' over $\mathfrak{p}$. In particular, $\mathfrak{p}$ splits completely in L if and only if $\mathrm{Frob}_{\mathfrak{P}} = 1$.*

Proof The order of $\mathrm{Frob}_{\mathfrak{P}}$ is the residue degree $[l : k]$. On the other hand, by (2.8.4), we have $[l : k] = [L : K]/r$. □

If G is abelian, then $\mathrm{Frob}_{\sigma(\mathfrak{P})} = \sigma \, \mathrm{Frob}_{\mathfrak{P}} \, \sigma^{-1} = \mathrm{Frob}_{\mathfrak{P}}$, so $\mathrm{Frob}_{\mathfrak{P}}$ itself is uniquely determined by $\mathfrak{p}$. In that case, we denote $\mathrm{Frob}_{\mathfrak{P}}$ by $\mathrm{Frob}_{\mathfrak{p}}$. We arrive at the following:

Theorem 2.8.13 *Let L/K be an abelian extension. For each prime $\mathfrak{p} \lhd B$ that is unramified in L, there is an element $\mathrm{Frob}_{\mathfrak{p}} \in \mathrm{Gal}(L/K)$ depending only on $\mathfrak{p}$, which is uniquely characterized by the property that, for any prime $\mathfrak{P}$ lying over $\mathfrak{p}$, the following congruence holds:*

$$\mathrm{Frob}_{\mathfrak{p}}(\alpha) \equiv \alpha^{\#k} \pmod{\mathfrak{P}} \qquad \text{for all } \alpha \in B'.$$

Exercises

2.8.1 Assume L/K be an abelian extension, and $\mathfrak{p} \lhd B$ is unramified in L. Prove that $\mathrm{Frob}_{\mathfrak{p}}$ is uniquely characterized by the property that

$$\mathrm{Frob}_{\mathfrak{p}}(\alpha) \equiv \alpha^{\#k} \pmod{\mathfrak{p}B'} \qquad \text{for all } \alpha \in B'.$$

---

[12] Note that this property characterizes $\mathrm{Frob}_{\mathfrak{P}}$ uniquely also as an element of $\mathrm{Gal}(L/K)$. Indeed, if $\sigma \in \mathrm{Gal}(L/K)$ has the property that $\sigma(\alpha) \equiv \alpha^{\#k} \pmod{\mathfrak{P}}$ for all $\alpha \in B'$, then for $\alpha \in \mathfrak{P}$ we have $\sigma(\alpha) \equiv 0 \pmod{\mathfrak{P}}$, so $\sigma(\alpha) \in \mathfrak{P}$. It follows that $\sigma(\mathfrak{P}) = \mathfrak{P}$, and therefore, $\sigma \in D(\mathfrak{P}/\mathfrak{p})$.

2.8.2 Let K be a finite extension of $F = \mathbb{F}_q(T)$. Prove that if $\alpha \in K$, then

$$\mathrm{Nr}_{K/F}(\alpha) = \prod_{v|w} \mathrm{Nr}_{K_v/F_w}(\alpha),$$

where the product is over the set of all places v of K lying over w.

2.8.3 Let L/K be a finite extension of global fields.

(a) Assume L/K is Galois. Prove that if K has any inert primes, then L/K is cyclic.
    Prove the converse using Theorem 7.3.5.
(b) Prove that a prime of K splits completely in L if and only if it splits completely
    in the normal closure of L/K.

2.8.4 Let L/K be a Galois extension of global fields. Let $\mathfrak{p} \lhd B$ be a prime and
$\mathfrak{p}B' = \mathfrak{P}_1^e \cdots \mathfrak{P}_r^e$ be its prime decomposition in B'. Let $\mathfrak{P} \lhd B'$ be one of the primes
$\mathfrak{P}_i$. Let L^D and L^I be the fixed fields of $D(\mathfrak{P}/\mathfrak{p})$ and $I(\mathfrak{P}/\mathfrak{p})$, respectively. Prove
the following:

(a) The prime $\mathfrak{p}$ decomposes into a product $\mathfrak{p}O = \mathfrak{m}_1 \cdots \mathfrak{m}_r$ of r distinct primes in
    the integral closure O of A in L^D. Moreover, $f(\mathfrak{m}_i/\mathfrak{p}) = e(\mathfrak{m}_i/\mathfrak{p}) = 1$ for all
    $1 \le i \le r$, and, after reindexing, $\mathfrak{P}_i$ is the unique primes of B' lying over $\mathfrak{m}_i$.
(b) Each $\mathfrak{m}_i$ remains prime $\mathfrak{m}_i O' = \mathfrak{M}_i$ in the integral closure O' of A in L^I.
    Moreover, $f(\mathfrak{M}_i/\mathfrak{m}_i) = f(\mathfrak{P}/\mathfrak{p})$.
(c) Each $\mathfrak{M}_i$ totally ramifies $\mathfrak{M}_i B' = \mathfrak{P}_i^e$ in B'. In particular, $f(\mathfrak{P}_i/\mathfrak{M}_i) = 1$ and
    $e(\mathfrak{P}_i/\mathfrak{M}_i) = e(\mathfrak{P}/\mathfrak{p})$.

2.8.5 Let $F = \mathbb{F}_q(T)$, and let $\mathfrak{p}$ be a prime of F. Let $\widehat{K}$ be a finite separable
extension of $F_\mathfrak{p}$. Show that there is a finite separable extension K of F and prime $\mathfrak{P}$
of K such that $K_\mathfrak{P} = \widehat{K}$. Moreover, one can choose K so that $[K : F] = [\widehat{K} : F_\mathfrak{p}]$.

2.8.6 Let K be a global field. Prove that for all nonzero $\alpha \in K$ we have

$$\prod_v |\alpha|_v = 1,$$

where the product is over all places of K and $|\cdot|_v$ denotes the normalized absolute
value on K_v.

Chapter 3
Basic Properties of Drinfeld Modules

A Drinfeld module over a field K, our principal object of study in this book, is a field K equipped with an action of the ring of polynomials $\mathbb{F}_q[T]$, where $\mathbb{F}_q[T]$ acts via certain linearized polynomials in $K[x]$. In this chapter, we study the basic properties of Drinfeld modules which are valid over arbitrary fields. Later in the book we will be interested in the properties of Drinfeld modules defined over arithmetically interesting fields, such as finite fields, local fields, and global fields.

In Sect. 3.1, we discuss the $\mathbb{F}_q$-linear polynomials over a field K containing $\mathbb{F}_q$ as a subfield and the non-commutative ring $K\langle x \rangle$ that these polynomials form. A Drinfeld module over K is most conveniently given as a homomorphism $\phi \colon \mathbb{F}_q[T] \to K\langle x \rangle$ uniquely determined by the image $\phi_T(x)$ of T. The polynomial $\phi_T(x)$ in many respects plays the role of the Weierstrass equation defining an elliptic curve. The definition of a Drinfeld module is given in Sect. 3.2. The rest of the chapter is devoted to the study of homomorphisms between Drinfeld modules and their torsion submodules. In particular, in Sects. 3.3–3.6, we prove the important theorem that the homomorphisms between two given Drinfeld modules form a free finitely generated $\mathbb{F}_q[T]$-module, describe the torsion submodules of a Drinfeld module, and study the close relationship between the module of homomorphisms and the induced homomorphisms between the torsion submodules. In Sect. 3.7, we define a version of the determinant map on Drinfeld modules, which can be considered as an analogue of the Weil pairing for elliptic curves. In the final section, Sect. 3.8, we consider the problem of classifying Drinfeld modules up to isomorphism; this is the beginning of the theory of Drinfeld modular varieties.

Unless explicitly stated otherwise, the following notation will be fixed throughout the rest of the book:

- $\mathbb{F}_q$ is a finite field with q elements, where q is a power of the prime p.
- $A = \mathbb{F}_q[T]$ is the ring of polynomials in indeterminate T with coefficients in $\mathbb{F}_q$.
- $F = \mathbb{F}_q(T)$ is the fraction field of A.
- The ideals of A will be denoted by fraktur letters. Each nonzero ideal $\mathfrak{n} \lhd A$ has a unique monic generator, which, by abuse of notation, we will also denote by $\mathfrak{n}$.

© The Author(s), under exclusive license to Springer Nature Switzerland AG 2023
M. Papikian, *Drinfeld Modules*, Graduate Texts in Mathematics 296,
https://doi.org/10.1007/978-3-031-19707-9_3

It will always be clear from the context whether $\mathfrak{n}$ denotes an ideal or its monic generator.

- A_+ is the set of monic polynomials in A.
- We usually call a nonzero prime ideal of A simply a *prime of A*. Note that nonzero prime ideals of A are maximal.
- $\mathbb{F}_{\mathfrak{p}}$ is the quotient ring $A/\mathfrak{p}$ of A by a prime $\mathfrak{p}$; note that $\mathbb{F}_{\mathfrak{p}} \cong \mathbb{F}_{q^{\deg(\mathfrak{p})}}$ is a field since $\mathfrak{p}$ is maximal.
- $\mathbb{F}_{\mathfrak{p}^n}$ is the degree-n extension of $\mathbb{F}_{\mathfrak{p}}$.
- $F_{\mathfrak{p}}$ is the completion of F with respect to $\mathrm{ord}_{\mathfrak{p}}$, where $\mathfrak{p}$ is a prime of A.
- $A_{\mathfrak{p}}$ is the ring of integers of $F_{\mathfrak{p}}$; alternatively, $A_{\mathfrak{p}} = \varprojlim_n A/\mathfrak{p}^n$.

- F_∞ is the completion of F with respect to ord_∞.
- A_∞ is the ring of integers of F_∞.
- $\mathcal{M}_\infty$ is the maximal ideal of A_∞.
- $\mathbb{F}_\infty = A_\infty/\mathcal{M}_\infty \cong \mathbb{F}_q$.
- $\mathbb{C}_\infty$ is the completion of an algebraic closure of F_∞.
- $|\cdot|$ is the unique extension to $\mathbb{C}_\infty$ of the absolute value on F_∞ normalized by $|T| = q$. Note that the restriction of $|\cdot|$ to A can be equivalently defined by

$$|a| = \#A/(a) = q^{\deg(a)} \text{ for } 0 \neq a \in A, \text{ and } |0| = 0.$$

3.1   Additive Polynomials

Definition 3.1.1 Let K be a field, and let x, y be indeterminates. A polynomial $f(x) \in K[x]$ is *additive* if the equality

$$f(x + y) = f(x) + f(y)$$

holds in $K[x, y]$.

Lemma 3.1.2 *Let $f(x) \in K[x]$ be a polynomial.*

(1) If the characteristic of K is 0, then $f(x)$ is additive if and only if $f(x) = ax$ for some $a \in K$.

(2) If the characteristic of K is $p > 0$, then $f(x)$ is additive if and only if it is of the form

$$f(x) = \sum_{i=0}^{n} a_i x^{p^i}$$

for some $n \geq 0$ and $a_0, \ldots, a_n \in K$.

Proof According to (1), the additive polynomials over fields of characteristic 0 are not very interesting. We leave the proof of (1) as an exercise (Exercise 3.1.1) and assume for the rest of the proof that the characteristic of K is $p > 0$.

The claim that a polynomial of the form $\sum_{i=0}^{n} a_i x^{p^i}$ is additive follows from Example N.2, so we need to show that an arbitrary additive polynomial $f(x) \in K[x]$ has this form, i.e., only p-powers of x have nonzero coefficients in $f(x)$. Since $f(x)$ is also additive when considered as a polynomial in $\overline{K}[x]$, we can assume that K is algebraically closed.

We use induction on the degree of $f(x)$. If $f(x) = a + bx$ has degree 1, then $a + b(x + y) = a + bx + a + by$ implies $a = 0$, so f has the required form. Now assume that $f(x)$ is an additive polynomial of degree n and we have proved the claim for all additive polynomials of degree $\leq n - 1$. Let $f'(x) \in K[x]$ denote the derivative of $f(x)$. Consider the polynomial equality

$$f(x + y) = f(x) + f(y)$$

in $K[x, y] = K[y][x]$. Differentiating both sides with respect to x, we get $f'(x + y) = f'(x)$. Substituting $x = 0$ into this equality, we get $f'(y) = f'(0)$, which implies that $f'(x)$ is a constant. Therefore,

$$f(x) = ax + h(x)$$

for some $h(x) = \sum_{i=1}^{n} h_i x^i \in K[x]$ such that $h_i = 0$ if $p \nmid i$. Using our assumption that K is algebraically closed, we can write $h(x) = (\sum_{i=1}^{n} h_i^{1/p} x^{i/p})^p = g(x)^p$. Thus, $f(x) = ax + g(x)^p$. Since $f(x) - ax$ is additive, we have $g(x + y)^p = g(x)^p + g(y)^p$. But this implies

$$(g(x + y) - g(x) - g(y))^p = 0.$$

Since the Frobenius endomorphism on $K[x, y]$ is injective, we conclude that $g(x)$ is additive and has degree $< n$. Thus, by induction, $g(x)$, and therefore $g(x)^p$ and $f(x)$, have the required form. (A different proof of this lemma is indicated in Exercise 3.1.2.)   □

From now on, we assume that the finite field $\mathbb{F}_q$ is a subfield of K.

Definition 3.1.3 We say that a polynomial $f(x) \in K[x]$ is $\mathbb{F}_q$-*linear* if $f(x)$ is additive and satisfies $f(\alpha x) = \alpha f(x)$ for all $\alpha \in \mathbb{F}_q$.

Lemma 3.1.4 *A polynomial $f(x) \in K[x]$ is $\mathbb{F}_q$-linear if and only if it is of the form*

$$f(x) = \sum_{i=0}^{n} a_i x^{q^i}$$

for some $n \geq 0$ and $a_0, \ldots, a_n \in K$.

Proof That $f(x) = \sum_{i=0}^{n} a_i x^{q^i}$ is $\mathbb{F}_q$-linear is clear since it is additive, and $\alpha^{q^i} = \alpha$ for all $\alpha \in \mathbb{F}_q$ and $i \geq 0$. Conversely, if $f(x)$ is $\mathbb{F}_q$-linear, then it is additive. By Lemma 3.1.2, $f(x) = \sum_{j=0}^{m} b_j x^{p^j}$. The condition $f(\alpha x) = \alpha f(x)$ for all $\alpha \in \mathbb{F}_q$

implies that $b_j \alpha^{p^j} = b_j \alpha$ for all $0 \leq j \leq n$ and $\alpha \in \mathbb{F}_q$. Thus, if $b_j \neq 0$, then $\alpha^{p^j} = \alpha$ for all $\alpha \in \mathbb{F}_q$. On the other hand, this last equality implies that $\mathbb{F}_q$ is a subfield of $\mathbb{F}_{p^j}$ (see Sect. 1.6). Considering $\mathbb{F}_{p^j}$ as a vector space over $\mathbb{F}_q$, we get $p^j = q^i$ for some $i \geq 0$.   □

Let $f(x)$ be an $\mathbb{F}_q$-linear polynomial with coefficients in K. It is obvious that $\omega = 0$ is always a root of $f(x)$. Furthermore, if ω_1 and ω_2 are roots, then it is seen without difficulty that $\alpha_1 \omega_1 + \alpha_2 \omega_2$ are also roots for all $\alpha_1, \alpha_2 \in \mathbb{F}_q$. Hence the set of distinct roots of an $\mathbb{F}_q$-linear polynomial $f(x)$ forms a finite dimensional $\mathbb{F}_q$-vector subspace of $\overline{K}$. Conversely, we have the following:

Lemma 3.1.5 *Let W be a finite dimensional $\mathbb{F}_q$-vector subspace of K. Then*

$$f(x) = \prod_{w \in W} (x - w)$$

is $\mathbb{F}_q$-linear and separable.

Proof Denote $n = \dim_{\mathbb{F}_q} W$. Then $\deg(f) = q^n$ and f is monic, so

$$f(x) = x^{q^n} + \text{ lower degree terms.}$$

Moreover, since $f(x)$ has no multiple roots, it is separable.

Consider $f(x + y) - f(x) - f(y) =: H(x)$ as a polynomial in $K(y)[x]$. Since $(x + y)^{q^n} - x^{q^n} - y^{q^n} = 0$, we have $\deg H(x) < \deg f(x) = \#W$. On the other hand, for all $w \in W$, we have

$$H(w) = f(w + y) - f(w) - f(y)$$
$$= f(y) - 0 - f(y)$$
$$= 0,$$

where the second equality follows from the observation that

$$f(w + y) = \prod_{w' \in W} (y + w - w') = \prod_{w'' \in W} (y - w'') = f(y).$$

Since H cannot have more zeros than its degree, unless it is identically 0, we get

$$0 = H(x) = f(x + y) - f(x) - f(y).$$

Thus, $f(x)$ is additive. Next, if $\alpha \in \mathbb{F}_q^\times$, then

$$f(\alpha x) = \prod_{w \in W} (\alpha x - w) = \alpha^{q^n} \prod_{w \in W} (x - w/\alpha) = \alpha \prod_{w \in W} (x - w) = \alpha f(x),$$

where for the third equality we observe that $\alpha^q = \alpha$ and $w \mapsto w/\alpha$ is an automorphism of W. Therefore, $f(x)$ is $\mathbb{F}_q$-linear. □

Denote by $K\langle x\rangle$ the set of all $\mathbb{F}_q$-linear polynomials in $K[x]$. It is clear that $K\langle x\rangle$ is closed under the usual addition on $K[x]$, but $K\langle x\rangle$ is not closed under the usual multiplication of polynomials. Instead, we define a binary operation $\circ$ on $K\langle x\rangle$ via the composition of polynomials:

$$f \circ g = f(g(x)).$$

Thanks to (N.1), given $f \in K\langle x\rangle$, the polynomial f^{q^m} is also in $K\langle x\rangle$ for any $m \geq 0$, so $K\langle x\rangle$ is closed under composition. The composition of arbitrary functions is associative, so for $f, g, h \in K\langle x\rangle$, we have

$$f \circ (g \circ h) = (f \circ g) \circ h.$$

We also always have (not necessarily for $\mathbb{F}_q$-linear polynomials)

$$(f + g) \circ h = (f \circ h) + g \circ h.$$

Finally, since f is additive, we have

$$f \circ (g + h) = f \circ g + f \circ h.$$

Overall, $K\langle x\rangle$ is a ring with respect to the usual addition $+$ of polynomials and multiplication given by composition. The zero of this ring is the zero polynomial, and the multiplicative identity is the polynomial x. It is important to note that $K\langle x\rangle$ is generally not a commutative ring, i.e., $f \circ g = g \circ f$ does not always hold.

Example 3.1.6 Let $f(x) = ax + bx^q$ and $g(x) = x^q$. Then

$$f \cdot g = ax^{q+1} + bx^{2q} \notin K\langle x\rangle,$$

$$f \circ g = ax^q + bx^{q^2} \in K\langle x\rangle,$$

$$g \circ f = (ax + bx^q)^q = a^q x^q + b^q x^{q^2}.$$

Moreover, if a or b is not in $\mathbb{F}_q$, then $g \circ f \neq f \circ g$.

Remark 3.1.7 The ring $K\langle x\rangle$ was introduced and studied by Oystein Ore in [Ore33]; in fact, the main results of this section, e.g., Theorems 3.1.13 and 3.1.18, appear in the same paper. For that reason, the ring $K\langle x\rangle$ is sometimes called the *ring of Ore polynomials*. There is a more sophisticated interpretation of $K\langle x\rangle$ as the ring of $\mathbb{F}_q$-linear endomorphisms of the additive group-scheme $\mathbb{G}_{a,K}$ over K; see [Wat79, p. 65]. Although from a certain perspective this makes the definition of Drinfeld modules, as well as some of the properties of their torsion points, more natural, we will not use this in the book.

To study some of the ring-theoretic properties of $K\langle x\rangle$, it is convenient to introduce another version of the same ring.

Definition 3.1.8 Let R be a commutative $\mathbb{F}_q$-algebra. Let τ be an indeterminate, and let $R\{\tau\}$ be the set of polynomials $\sum_{i\geq 0}^{n} a_i\tau^i$, $n \geq 0$, $a_i \in R$. (By convention, $\tau^0 = 1$.) We define the addition on $R\{\tau\}$ to be the usual addition of polynomials

$$\sum_{i=0}^{n} a_i\tau^i + \sum_{i=0}^{n} b_i\tau^i = \sum_{i=0}^{n}(a_i + b_i)\tau^i.$$

To define the multiplication, we first put

$$(a\tau^i)(b\tau^j) \stackrel{\text{def}}{=} ab^{q^i}\tau^{i+j}$$

and then extend this to all polynomials by the distributive laws.

Example 3.1.9

$$(a + b\tau)(c + d\tau) = ac + (ad + bc^q)\tau + bd^q\tau^2.$$

Now define a map

$$\iota\colon K\{\tau\} \longrightarrow K\langle x\rangle$$

$$\sum_{i=0}^{n} a_i\tau^i \longmapsto \sum_{i=0}^{n} a_i x^{q^i}.$$

It is clear that ι is bijective and satisfies $\iota(0) = 0$, $\iota(1) = x$, $\iota(f + g) = \iota(f) + \iota(g)$. It is also straightforward, and left to the reader, to check that $\iota(fg) = \iota(f) \circ \iota(g)$. Thus, ι is an isomorphism. The ring $K\{\tau\}$ is called the *ring of twisted polynomials*. By slight abuse of notation, we will usually write $f(x)$ for $\iota(f)$.

Definition 3.1.10 Let $f = a_h\tau^h + a_{h+1}\tau^{h+1}\cdots + a_n\tau^n$, $0 \leq h \leq n$, $a_h \neq 0$, $a_n \neq 0$. The *height* of f is

$$\text{ht}(f) = h,$$

and the *degree* of f is

$$\deg(f) = n.$$

To distinguish the degree of f from the degree of $f(x)$ as a polynomial in x, we sometimes write $\deg_\tau(f)$ for $\deg(f)$. We formally put $\deg(0) = -\infty$ and $\text{ht}(0) = +\infty$. We say that f is *separable* if $\text{ht}(f) = 0$ and *inseparable* otherwise. This is equivalent to $f(x)$ being separable in the usual polynomial sense.

If $f \neq 0$ and $\mathrm{ht}(f) = h$, then we can always write

$$f = f_{\mathrm{sep}} \tau^h$$

for a uniquely determined separable $f_{\mathrm{sep}} \in K\{\tau\}$ of degree $n - h$. But there does not necessarily exist g such that $f = \tau^h g$ unless K is perfect. In terms of the corresponding $\mathbb{F}_q$-linear polynomials, the first decomposition $f = f_{\mathrm{sep}} \tau^h$ corresponds to writing $f(x) = a_h x^{q^h} + \cdots + a_n x^{q^n}$ as $f(x) = f_{\mathrm{sep}}(x^{q^h})$, where $f_{\mathrm{sep}}(x) = a_h x + \cdots + a_n x^{q^{n-h}}$, whereas $f = \tau^h g$ corresponds to $f(x) = g(x)^{q^h}$, where $g(x) = a_h^{1/q^h} x + \cdots + a_n^{1/q^h} x^{q^{n-h}}$.

The following is trivial to prove but is useful to have on record:

Lemma 3.1.11 *For any nonzero $f, g \in K\{\tau\}$, we have*

$$\mathrm{ht}(fg) = \mathrm{ht}(f) + \mathrm{ht}(g).$$

$$\mathrm{ht}(f + g) \geq \min(\mathrm{ht}(f), \mathrm{ht}(g)).$$

$$\deg(fg) = \deg(f) + \deg(g).$$

$$\deg(f + g) \leq \max(\deg(f), \deg(g)).$$

Definition 3.1.12 Define a homomorphism

$$\partial : K\{\tau\} \to K$$

$$\sum_{i=0}^{n} a_i \tau^i \mapsto a_0,$$

called the *derivative*. Note that $\partial(f) = f'(x)$ for any $f \in K\{\tau\}$.

Although $K\{\tau\}$ is not commutative, it does possess one of the crucial ring-theoretic properties of the usual polynomial ring, namely the (right) division algorithm (cf. Theorem 1.1.2):

Theorem 3.1.13 (Right Division Algorithm) *Given two polynomials $f, g \in K\{\tau\}$ with $g \neq 0$, there exist unique $h, r \in K\{\tau\}$ such that*

$$f = hg + r, \text{ and either } \deg(r) < \deg(g) \text{ or } r = 0.$$

Proof The proof is very similar to the proof of Theorem 1.1.2.

Suppose $g = b_m \tau^m + \cdots + b_0$, $m \geq 0$, $b_m \neq 0$. If $f = 0$ or $\deg(f) < \deg(g)$, then put $h = 0$ and $r = f$. Now assume

$$f = a_n \tau^n + \text{ lower degree terms}, \quad a_n \neq 0, \quad n \geq m,$$

and we have proved the claim for all polynomials f of degree $< n$. Put $h_1 = \frac{a_n}{b_m^{q^{n-m}}}\tau^{n-m}$. Then $\deg(f - h_1 g) < n$, so by induction we can find h_0 and r such that $\deg(r) < \deg(g)$ or $r = 0$, and

$$f - h_1 g = h_0 g + r.$$

Then $f = (h_0+h_1)g+r$, so we can take $h = h_0+h_1$. To see the uniqueness, suppose there is another pair $\tilde{h}, \tilde{r}$ satisfying the same conditions. Then $(h - \tilde{h})g = \tilde{r} - r$. Since $\deg(g) > \deg(\tilde{r} - r)$, we must have $\tilde{r} - r = 0$ and $h - \tilde{h} = 0$. □

Remark 3.1.14 It is important that h in Theorem 3.1.13 appears to the left of g. Assume K is not perfect and let $a \in K$ be such that $a \notin K^q$. Let $f = a\tau^2$ and $g = \tau$. Then, for $h = a\tau$, we have $f = hg$. On the other hand, there is no h such that $\deg(f - gh) < \deg(g)$, so Theorem 3.1.13 is false if we try to put h to the right of g. To see this, note that h has to have degree 1, so $h = b\tau + c$, and we need $\deg(a\tau^2 - b^q\tau^2) < 2$, which forces $a = b^q \in K^q$, contrary to our assumption.

Corollary 3.1.15 *Every left ideal of $K\{\tau\}$ is principal, i.e., if $I \subset K\{\tau\}$ is an additive subgroup such that $sf \in I$ for all $s \in K\{\tau\}$ and $f \in I$, then there is $g \in I$ such that $I = K\{\tau\}g$.*

Proof The proof is the same as the proof of the corresponding fact for Euclidean domains; cf. [DF04, p. 273]. In fact, if $m = \min\{\deg(f) \mid 0 \neq f \in I\}$, then any $g \in I$ with $\deg(g) = m$ is a generator of I thanks to the division algorithm. □

Corollary 3.1.16 *Let $f, g \in K\{\tau\}$ be such that $g \neq 0$ and $\mathrm{ht}(f) \geq \mathrm{ht}(g)$. Then the roots of $g(x)$ are roots of $f(x)$ if and only if there is $h \in K\{\tau\}$ such that $hg = f$. In that case, if f and g commute in $K\{\tau\}$, then we also have $gh = f$.*

Proof If $(h \circ g)(x) = f(x)$ and $\alpha \in \overline{K}$ is such that $g(\alpha) = 0$, then $f(\alpha) = h(g(\alpha)) = h(0) = 0$. Hence, the roots of $g(x)$ are roots of $f(x)$.

Conversely, suppose the roots of $g(x)$ are roots of $f(x)$. By Theorem 3.1.13, we can write $f(x) = h(g(x)) + r(x)$ for some $r \in K\{\tau\}$ such that either $\deg_\tau(r) < \deg_\tau(g)$ or $r = 0$. Let α be a root of $g(x)$. By our assumption,

$$0 = f(\alpha) = h(g(\alpha)) + r(\alpha) = h(0) + r(\alpha) = r(\alpha).$$

Hence, every root of $g(x)$ is a root of $r(x)$ without counting the multiplicities. This implies that $\deg_\tau(g_{\mathrm{sep}}) \leq \deg_\tau(r_{\mathrm{sep}})$. We also have

$$\mathrm{ht}(r) = \mathrm{ht}(f - hg) \geq \min\{\mathrm{ht}(f), \mathrm{ht}(hg)\} \geq \min\{\mathrm{ht}(f), \mathrm{ht}(g)\} = \mathrm{ht}(g),$$

where the last equality follows from the assumption that $\mathrm{ht}(f) \geq \mathrm{ht}(g)$. If $r \neq 0$, then we would have

$$\deg_\tau(r) = \deg_\tau(r_{\mathrm{sep}}) + \mathrm{ht}(r) \geq \deg_\tau(g_{\mathrm{sep}}) + \mathrm{ht}(g) = \deg_\tau(g),$$

contrary to $\deg_\tau(r) < \deg_\tau(g)$. Thus, $r = 0$, so we get $hg = f$.

Now assume $hg = f$ and $gf = fg$. Then $fg = gf = (gh)g$, which can be rewritten as $(f - gh)g = 0$. Since $K\{\tau\}$ has no zero-divisors, we must have $f = gh$.   □

We conclude this section with a discussion of the Moore determinant, which plays a role in the theory of $\mathbb{F}_q$-linear polynomials similar to the role of the Vandermonde determinant for the usual polynomials.[1] We will extensively use the Moore determinant in Sects. 3.7 and 5.4.

Definition 3.1.17 Let $K[x_1, \ldots, x_n]$ be the polynomial ring in indeterminates $x_1, \ldots, x_n$ with coefficients in K. The *Moore determinant* of $x_1, \ldots, x_n$ is the polynomial

$$M(x_1, \ldots, x_n) = \det \begin{pmatrix} x_1 & x_1^q & \cdots & x_1^{q^{n-1}} \\ x_2 & x_2^q & \cdots & x_2^{q^{n-1}} \\ \vdots & \vdots & \cdots & \vdots \\ x_n & x_n^q & \cdots & x_n^{q^{n-1}} \end{pmatrix}.$$

Theorem 3.1.18

$$M(x_1, \ldots, x_n) = \prod_{i=1}^{n} \prod_{c_{i-1} \in \mathbb{F}_q} \cdots \prod_{c_1 \in \mathbb{F}_q} (x_i + c_{i-1}x_{i-1} + \cdots + c_1 x_1).$$

Proof Denote

$$f(x_1, \ldots, x_n) = \prod_{i=1}^{n} \prod_{c_{i-1} \in \mathbb{F}_q} \cdots \prod_{c_1 \in \mathbb{F}_q} (x_i + c_{i-1}x_{i-1} + \cdots + c_1 x_1).$$

Fix some constants $c_1, \ldots, c_{i-1} \in \mathbb{F}_q$ and substitute $-(c_{i-1}x_{i-1} + \cdots + c_1 x_1)$ instead of x_i in the matrix of M. After doing this, the rows of the Moore matrix become linearly dependent (since $(c_{i-1}x_{i-1} + \cdots + c_1 x_1)^{q^s} = c_{i-1}x_{i-1}^{q^s} + \cdots + c_1 x_1^{q^s}$ for any $s \geq 0$); thus, the determinant becomes 0. Since $K[x_1, \ldots, x_n]$ is a unique factorization domain, this implies that $x_i + c_{i-1}x_{i-1} + \cdots + c_1 x_1$ divides $M(x_1, \ldots, x_n)$. Therefore, f divides M.

Next, since each factor in the product

$$f_i(x_1, \ldots, x_n) := \prod_{c_{i-1} \in \mathbb{F}_q} \cdots \prod_{c_1 \in \mathbb{F}_q} (x_i + c_{i-1}x_{i-1} + \cdots + c_1 x_1)$$

---

[1] One can consider $\tau \colon \overline{K} \to \overline{K}$ as an $\mathbb{F}_q$-linear operator. There is a subtle and deep analogy between τ and the differential operator d/dx from complex analysis. From this perspective, the Moore determinant also plays the role of the Wronskian; cf. [Gos96].

is linear, the degree of f_i as a multivariable polynomial is q^{i-1} (each x_j has degree 1). Thus,

$$\deg(f) = \sum_{i=1}^{n} \deg(f_i) = 1 + q + q^2 + \cdots + q^{n-1}.$$

On the other hand, if we compute M by the usual method of choosing and multiplying n arbitrary entries in the matrix lying in different rows and columns and then summing the resulting products with specific signs, each term of the resulting sum will have degree $1 + q + q^2 + \cdots + q^{n-1}$, as the entries in the i-th column of the matrix have degree q^{i-1}. Hence, $\deg(M) = \deg(f)$. This implies that $M = cf$ for some $c \in \mathbb{F}_q^{\times}$.

It remains to show that $c = 1$. The term $x_1 x_2^q \cdots x_n^{q^{n-1}}$ has coefficient 1 in M, as it arises as the product of the diagonal entries of the Moore matrix. Note that $x_i^{q^{i-1}}$ is one of the terms of f_i. Hence, $x_1 x_2^q \cdots x_n^{q^{n-1}}$ also has coefficient 1 in f. This implies that $c = 1$. $\square$

Corollary 3.1.19 *Let W be an $\mathbb{F}_q$-vector subspace of K of dimension n. Let $\{w_1, \ldots, w_n\}$ be a basis of W. Then*

$$\prod_{w \in W} (x - w) = M(w_1, w_2, \ldots, w_n, x)/M(w_1, w_2, \ldots, w_n).$$

Proof By Theorem 3.1.18,

$$M(w_1, w_2, \ldots, w_n, x) = M(w_1, w_2, \ldots, w_n) \prod_{c_n, \ldots, c_1 \in \mathbb{F}_q} (x + c_n w_n + \cdots + c_1 w_1)$$

$$= M(w_1, w_2, \ldots, w_n) \prod_{w \in W} (x - w).$$

     $\square$

Remark 3.1.20 Recall that the *Vandermonde determinant* is defined as

$$V(x_1, \ldots, x_n) = \det \begin{pmatrix} 1 & x_1 & x_1^2 & \cdots & x_1^{n-1} \\ 1 & x_2 & x_2^2 & \cdots & x_2^{n-1} \\ \vdots & \vdots & \vdots & \cdots & \vdots \\ 1 & x_n & x_n^2 & \cdots & x_n^{n-1} \end{pmatrix}.$$

It is a well-known fact that

$$V(x_1, \ldots, x_n) = \prod_{i > j} (x_i - x_j).$$

Hence, $V(\alpha_1, \ldots, \alpha_n) = 0$ for $\alpha_1, \ldots, \alpha_n \in K$ if and only if $\alpha_i = \alpha_j$ for some $i \neq j$. In comparison, $M(\alpha_1, \ldots, \alpha_n) = 0$ if and only if $\alpha_1, \ldots, \alpha_n$ are $\mathbb{F}_q$-linearly dependent.

Exercises

In all exercises, unless indicated otherwise, K is a field extension of $\mathbb{F}_q$.

3.1.1 Assume K is a field of characteristic 0. Prove that $f(x) \in K[x]$ is additive if and only if $f(x) = ax$ for some $a \in K$.

3.1.2 Let $n \geq 1$ be an integer.

(a) Write $n = p^s m$ with $(p, m) = 1$. Prove that in $K[x, y]$, we have

$$(x + y)^n = \sum_{i=0}^{m} \binom{m}{i} x^{p^s i} y^{p^s (m-i)}.$$

Deduce from this that if $m > 1$, then $(x + y)^n \neq x^n + y^n$.

(b) Give a different proof of Lemma 3.1.2 by expanding $f(x + y)$ and using (a).

3.1.3 Let W be an n-dimensional $\mathbb{F}_q$-vector subspace of K. Show that

$$\sum_{w \in W} w^k = 0$$

for all $0 \leq k < n$.

3.1.4 Let $f(x) = a_0 x + a_1 x^p + \cdots + a_n x^{p^n} \in K\langle x \rangle$ be an additive polynomial with $a_n \neq 0$. Prove the following:

(a) With the discriminant $\mathrm{disc}(f)$ defined by (1.1.5), we have

$$\mathrm{disc}(f) = (-1)^{p^n(p^n-1)/2} a_0^{p^n} a_n^{p^n-2}.$$

(b) If $a_0 \neq 0$, then

$$\frac{1}{f(x)} = \frac{1}{a_0} \sum_{i=1}^{p^n} \frac{1}{x - \lambda_i},$$

where $\lambda_1, \ldots, \lambda_{p^n}$ are the roots of $f(x)$ in $\overline{K}$.

3.1.5 Prove that the n-th power of $(x^q + x)$ in $K\langle x \rangle$ is equal to

$$\overbrace{(x^q + x) \circ (x^q + x) \circ \cdots \circ (x^q + x)}^{n}$$

$$= x^{q^n} + \binom{n}{n-1} x^{q^{n-1}} + \binom{n}{n-2} x^{q^{n-2}} + \cdots + \binom{n}{1} x^q + x.$$

3.1.6 Given an $\mathbb{F}_q$-vector subspace W of K, denote

$$f_W(x) = \prod_{w \in W} (x - w).$$

Let $\{w_1, \ldots, w_n\}$ be a basis of W over $\mathbb{F}_q$, so $\dim_{\mathbb{F}_q} W = n$. Given $1 \leq m \leq n$, let $W_m \subseteq W$ be the $\mathbb{F}_q$-span of $\{w_1, \ldots, w_m\}$; let $V_n = 0$ and, if $m < n$, let V_m be the $\mathbb{F}_q$-span of $\{f_{W_m}(w_{m+1}), \ldots, f_{W_m}(w_n)\}$. Prove the following equalities in $K\{\tau\}$:

(a)

$$f_W = \left(\tau - f_{W_{n-1}}(w_n)^{q-1}\right) \cdot f_{W_{n-1}}.$$

(b)

$$f_W = f_{V_m} \cdot f_{W_m}.$$

3.1.7 Let $f \in K\{\tau\}$ be a monic polynomial of degree n. Show that f can be decomposed in $\overline{K}\{\tau\}$ into linear factors

$$f = (\tau - \omega_1)(\tau - \omega_2) \cdots (\tau - \omega_n).$$

3.1.8 (Left Division Algorithm) Assume K is perfect. Prove that for any two polynomials $f, g \in K\{\tau\}$ with $g \neq 0$, there exist unique $h, r \in K\{\tau\}$ such that

$$f = gh + r \quad \text{and either } \deg(r) < \deg(g) \text{ or } r = 0.$$

(See Remark 3.1.14.)

3.1.9 For $f, g \in K\{\tau\}$, define their *greatest common divisor*, denoted $\gcd_\tau(f, g)$, to be the monic generator of the right ideal $(f, g) = K\{\tau\} f + K\{\tau\} g$. Let $\gcd(f(x), g(x))$ be the usual greatest common divisor of the polynomials $f(x)$ and $g(x)$ in the commutative polynomial ring $K[x]$. Prove that $\gcd_\tau(f, g)(x) = \gcd(f(x), g(x))$.

3.1.10 Prove that every finitely generated left $K\{\tau\}$-module M is isomorphic to

$$K\{\tau\}^{\oplus m} \oplus K\{\tau\}/K\{\tau\} f_1 \oplus \cdots \oplus K\{\tau\}/K\{\tau\} f_n$$

for some integers $m, n \geq 0$ and $f_i \in K\{\tau\}$ with $\deg(f_i) > 0$. Consequently, M is free over $K\{\tau\}$ if and only if it is torsion-free.

3.1.11 Show that

(a)

$$\prod_{c_1,\dots,c_n \in \mathbb{F}_q}' (c_1 x_1 + c_2 x_2 + \cdots + c_n x_n) = (-1)^n M(x_1, \dots, x_n)^{q-1},$$

where the product is over all n-tuples $(c_1, c_2, \dots, c_n) \neq (0, 0, \dots, 0)$.

(b) Let C be an $n \times n$ matrix over $\mathbb{F}_q$. Prove that

$$M\left(C\begin{pmatrix} x_1 \\ \vdots \\ x_n \end{pmatrix}\right) = \det(C) \cdot M(x_1, \dots, x_n).$$

3.1.12 Consider K as an $\mathbb{F}_q$-vector space. Prove the following:

(a) If K is finite, then every linear transformation of K is given by $\alpha \mapsto f(\alpha)$ for some $f(x) \in \mathbb{F}_q\langle x \rangle$.

(b) If K is infinite, then there are $\mathbb{F}_q$-linear transformations which are not given by any polynomial in $K\langle x \rangle$.

3.2 Definition of Drinfeld Modules

An *A-field* is a field K equipped with an $\mathbb{F}_q$-algebra homomorphism $\gamma : A \to K$. Since K is a field, there are two possibilities for the kernel of γ:

- $\ker(\gamma) = 0$, i.e., γ is injective. In this case, K is a field extension of F (not necessarily algebraic), and we say that K has *A-characteristic* 0, denoted $\mathrm{char}_A(K) = 0$.
- $\ker(\gamma) = \mathfrak{p}$ is a prime of A. In this case, γ factors through the quotient map $A \to A/\mathfrak{p}$, so K is a field extension of $\mathbb{F}_\mathfrak{p}$, and we say that K has *A-characteristic* $\mathfrak{p}$, denoted $\mathrm{char}_A(K) = \mathfrak{p}$.

Remark 3.2.1 It is important to observe that the same field can have different structures of an A-field. For example, $K = \mathbb{F}_q$ is an A-field via $\gamma_1 : A \to A/(T) \cong \mathbb{F}_q$ and also via $\gamma_2 : A \to A/(T-1) \cong \mathbb{F}_q$. An even more dramatic example is $K = \mathbb{F}_q(T)$ with $\gamma_1 : A \to K$ being the natural inclusion of A into its field of fractions and $\gamma_2 : A \to A/(T) = \mathbb{F}_q \hookrightarrow \mathbb{F}_q(T)$, so $\mathrm{char}_A(K) = 0$ in the first case and $\mathrm{char}_A(K) = (T)$ in the second case.

For the rest of this chapter, unless specified otherwise, K will be an A-field with a structure homomorphism $\gamma \colon A \to K$. We identify $\mathbb{F}_q \subset A$ with its isomorphic image in K.

Definition 3.2.2 A *Drinfeld module* of rank $r \geq 1$ over K is an $\mathbb{F}_q$-algebra homomorphism

$$\phi \colon A \longrightarrow K\{\tau\}$$

$$a \longmapsto \phi_a = \gamma(a) + g_1(a)\tau + \cdots + g_n(a)\tau^n,$$

such that for $a \neq 0$ we have

$$n = \deg_T(a) \cdot r \quad \text{and} \quad g_n(a) \neq 0.$$

Remark 3.2.3 Note that if $\deg_T(a) > 0$, then $\deg_\tau(\phi_a) > 0$, so the homomorphism ϕ is always injective. Thus, ϕ gives an embedding of the commutative ring A into the non-commutative ring $K\{\tau\}$.

Remark 3.2.4 The coefficients $g_1(a), \ldots, g_n(a) \in K$ of ϕ_a depend on a. On the other hand, if $a = \sum_{i=0}^{m} a_i T^i$, with $a_0, \ldots, a_m \in \mathbb{F}_q$, then

$$\phi_a = \sum_{i=0}^{m} a_i \phi_{T^i} = \sum_{i=0}^{m} a_i \phi_T^i.$$

Therefore, ϕ_T uniquely determines ϕ. Moreover, if $\deg_\tau(\phi_T) = r$ and the constant term of ϕ_T is $\gamma(T)$, then

$$\deg_\tau(\phi_a) = \deg_\tau(\phi_T^m) = m \deg_\tau(\phi_T) = \deg_T(a) \cdot r,$$

and the constant term of ϕ_a is

$$\sum_{i=0}^{m} a_i \gamma(T)^i = \sum_{i=0}^{m} a_i \gamma(T^i) = \gamma\left(\sum_{i=0}^{m} a_i T^i\right) = \gamma(a).$$

Hence, the conditions on ϕ_a in the definition of a Drinfeld module of rank r are automatically satisfied if they are satisfied for ϕ_T. Overall, we conclude that to define a Drinfeld module, one simply needs to choose $g_1, \ldots, g_r \in K$ such that $g_r \neq 0$ and put

$$\phi_T = \gamma(T) + g_1\tau + \cdots + g_r\tau^r.$$

From now on, for simplicity of notation, we denote

$$t = \gamma(T).$$

Definition 3.2.5 The *Carlitz module* is the Drinfeld module defined by $\phi_T = t + \tau$. This is the simplest possible Drinfeld module. We will distinguish the Carlitz module among all other Drinfeld modules by denoting it by C, i.e., $C_T = t + \tau$. The rank of C is 1.

Example 3.2.6 Suppose we want to compute C_a for a given a. For example, C_{T^2-T+1} can be computed by first computing

$$C_{T^2} = C_T C_T = (t + \tau)(t + \tau) = t^2 + (t + t^q)\tau + \tau^2$$

and then noting that

$$
\begin{aligned}
C_{T^2-T+1} &= C_{T^2} - C_T + 1 \\
&= (t^2 + (t + t^q)\tau + \tau^2) - (t + \tau) + 1 \\
&= (t^2 - t + 1) + (t + t^q - 1)\tau + \tau^2.
\end{aligned}
$$

But there is also a general recursive formula for computing the coefficients of

$$C_a = C_0(a) + C_1(a)\tau + \cdots + C_d(a)\tau^r, \qquad d = \deg(a).$$

To obtain this formula, first assume that C is defined over F, with $\gamma: A \to F$ being the natural embedding of A into its field of fractions. Comparing the coefficients of τ^m, $m \geq 1$, on both sides of the equation $C_a C_T = C_T C_a$, we get

$$C_m(a)T^{q^m} + C_{m-1}(a) = TC_m(a) + C_{m-1}(a)^q.$$

Therefore

$$C_m(a) = \frac{C_{m-1}(a)^q - C_{m-1}(a)}{T^{q^m} - T} \tag{3.2.1}$$

can be computed recursively starting with $C_0(a) = a$. For example, $C_1(a) = (a^q - a)/(T^q - T)$. Next, observe that $C_m(a)$ actually lies in A, even though it is written as a fraction of polynomials (this is clear from the fact that each $C_{T^n} = (C_T)^n$ has coefficients in A). Finally, if C is defined over a general A-field K, then we obtain the m-th coefficient of C_a by applying γ to the polynomial on the right-hand side of (3.2.1) (this works even if $t^{q^m} = t$). We will give another formula for the coefficients $C_m(a)$ in Corollary 5.4.4.

Remark 3.2.7 The Carlitz module was originally introduced by Carlitz in [Car35],[2] where he defined an analogue of the exponential function for A and proved a beautiful analogue of Euler's formula for $\sum_{n \geq 1} n^{-2m}$; see Sect. 5.4 for more details. Carlitz also showed in [Car38] that C gives rise to the correct analogue of cyclotomic polynomials over F; see Sect. 7.1. Unfortunately, [Car35] and [Car38] did not receive the attention they deserved from the larger mathematical community. This partly might be the result of the uninformative titles of many of the more than 750 papers by Carlitz[3] and the relative lack of good exposition/motivation in them.[4] It seems [Car35] and [Car38] were mostly forgotten until the 1970s, when Hayes [Hay74] (who was a student of Carlitz) used C to give an explicit description of the maximal abelian extension of F, similar to the Kronecker–Weber theorem for $\mathbb{Q}$. Note also that C is similar to a Lubin–Tate formal group law [LT65], which can be used to explicitly construct the totally ramified abelian extensions of local fields. Neither Lubin and Tate nor Drinfeld were aware of Carlitz's earlier work at the time of writing of [LT65] and [Dri74].

Definition 3.2.8 Note that K, being an A-algebra, has a natural A-module structure defined by $a * \beta = \gamma(a)\beta$, where $\gamma(a)\beta$ is just the multiplication in the field K. Via $\phi \colon A \to K\{\tau\}$, K acquires a new (twisted) A-module structure $a * \beta = \phi_a(\beta)$, where $\phi_a(\beta)$ is the $\mathbb{F}_q$-linear polynomial $\phi_a(x)$ evaluated at β; we will denote this A-module by $^\phi K$. The fact that ϕ gives a new A-module structure on K is the reason for calling ϕ a "module," although it is just a homomorphism.

Remark 3.2.9 If E is an elliptic curve over a field K, then the ring of integers $\mathbb{Z}$ acts on E via the group law on E: $n > 0$ maps a point $P \in E$ to the n-fold sum $P \oplus P \oplus \cdots \oplus P$. On the tangent space of E around the origin, which is isomorphic to K, the induced action is the usual multiplication by n. The significance of the condition $\partial(\phi_a) = \gamma(a)$ is that the induced action of A on the tangent space of $\mathbb{G}_{a,K}$ around the origin is the "old" action via γ.

Example 3.2.10 Let $K = A/(T) \cong \mathbb{F}_q$ with $\gamma \colon A \to A/(T)$ being the quotient map. Note that $^C K$ is a one-dimensional vector space over $\mathbb{F}_q$. Hence, it is isomorphic to $A/\mathfrak{p}$ for some prime $\mathfrak{p} \lhd A$ of degree 1. This prime $\mathfrak{p}$ is easy to find. Since $t = 0$, we have $C_T = \tau$, and T acts on $^C K$ by $T \circ \beta = \beta^q = \beta$. Thus, T acts as 1, so $^C K \cong A/(T - 1)$.

Since $\partial(\phi_a) = \gamma(a)$ for all $a \in A$, we see that if $\mathrm{char}_A(K) = 0$, then $\mathrm{ht}(\phi_a) = 0$ for all $0 \neq a \in A$, i.e., each ϕ_a is separable. On the other hand, if $\mathrm{char}_A(K) = \mathfrak{p} \neq 0$, then $\phi_\mathfrak{p}$ is not separable, so $\mathrm{ht}(\phi_\mathfrak{p}) \geq 1$. The next lemma extracts a more refined measure of inseparability of $\phi_\mathfrak{p}(x)$ from $\mathrm{ht}(\phi_\mathfrak{p})$.

---

[2] The module in [Car35] is actually the Drinfeld module defined by $\phi_T = t - \tau$.

[3] The titles of [Car35] and [Car38] are examples of this.

[4] For example, it is not mentioned anywhere in [Car35] that the formula for $\sum_{a \in A_+} a^{-(q-1)m}$ proved in that paper is similar to Euler's celebrated formula.

Lemma 3.2.11 *If* $\text{char}_A(K) = \mathfrak{p} \neq 0$ *and* ϕ *is a Drinfeld module of rank* r *over* K, *then there is an integer* $1 \leq H(\phi) \leq r$, *called the* **height** *of* ϕ, *such that for all* $0 \neq a \in A$ *we have*

$$\text{ht}(\phi_a) = H(\phi) \cdot \text{ord}_\mathfrak{p}(a) \cdot \deg(\mathfrak{p}).$$

Proof Let $h = \text{ht}(\phi_\mathfrak{p}) \geq 1$. We observe that $h \leq \deg(\phi_\mathfrak{p}) = r \cdot \deg(\mathfrak{p})$. Let $0 \neq a \in A$. Decompose $a = \mathfrak{p}^s \cdot b$ with $\mathfrak{p} \nmid b$. Since $\gamma(b) \neq 0$, we have $\text{ht}(\phi_b) = 0$, so by Lemma 3.1.11

$$\text{ht}(\phi_a) = \text{ord}_\mathfrak{p}(a) \cdot h.$$

Now it is enough to prove that h is divisible by $\deg(\mathfrak{p})$. We have

$$\phi_\mathfrak{p} = c_h \tau^h + \cdots + c_n \tau^n,$$

with $c_h \neq 0$. The commutation

$$\phi_a \phi_\mathfrak{p} = \phi_\mathfrak{p} \phi_a$$

implies that $\gamma(a) c_h = c_h \gamma(a)^{q^h}$. Thus, $\gamma(a)^{q^h} = \gamma(a)$. Since this is true for all $a \in A$, we conclude that $\gamma(A) = \mathbb{F}_\mathfrak{p} \cong \mathbb{F}_{q^{\deg(\mathfrak{p})}} \subseteq \mathbb{F}_{q^h}$. This implies that $\deg(\mathfrak{p})$ divides h. $\qquad\square$

Example 3.2.12 Let $q = 2$ and $K = A/\mathfrak{p}$, where $\mathfrak{p} = T^2 + T + 1$. Let ψ be the Drinfeld module of rank 2 defined by $\phi_T = t + \tau + \tau^2$. One computes that

$$\phi_\mathfrak{p} = (t^2 + t + 1) + (t^2 + t + 1)\tau + (t^2 + t + 1)(t^2 + t)\tau^2 + \tau^4$$
$$= \tau^4.$$

Hence, $H(\phi) = 2$.

Example 3.2.13 We use Lemma 3.2.11 to generalize Example 3.2.10 from $\mathfrak{p} = T$ to an arbitrary prime. So let $\mathfrak{p} \lhd A$ be a prime and let $K = A/\mathfrak{p}$, with γ being the quotient map. By Theorem 1.2.4, $^C K \cong A/(a_1) \oplus \cdots \oplus A/(a_s)$, for some $a_1 \mid a_2 \mid \cdots \mid a_s$. We have

$$q^{\deg(a_1) + \cdots + \deg(a_s)} = \#K = q^{\deg(\mathfrak{p})}.$$

On the other hand, every element of K is a root of $C_{a_s}(x)$ since a_s annihilates $^C K$. This implies that

$$\deg_x(C_{a_s}(x)) = q^{\deg(a_s)} \geq q^{\deg(\mathfrak{p})}.$$

Thus $s = 1$ and $\deg(a_s) = \deg(\mathfrak{p})$. Since the Carlitz module has rank 1, Lemma 3.2.11 implies that $H(C) = 1$; thus $C_{\mathfrak{p}} = \tau^{\deg(\mathfrak{p})}$. Since $\tau^{\deg(\mathfrak{p})}$ fixes all the elements of $\mathbb{F}_{\mathfrak{p}}$, we conclude that $\mathfrak{p} - 1$ annihilates $^C(A/\mathfrak{p})$. Therefore, a_1 divides $\mathfrak{p} - 1$. Since we proved that $\deg(a_1) = \deg(\mathfrak{p}) = \deg(\mathfrak{p} - 1)$, we get $a_1 = \mathfrak{p} - 1$. Overall, as A-modules,

$$^C(A/\mathfrak{p}) \cong A/(\mathfrak{p} - 1).$$

The definition of Drinfeld modules over fields can be easily generalized to arbitrary commutative rings. Let R be a commutative A-algebra, i.e., R is a ring equipped with a ring homomorphism $\gamma \colon A \to R$. We identity $\mathbb{F}_q \subset A$ with its isomorphic image in R. A *Drinfeld module* of rank $r \geq 1$ over R is a ring homomorphism $\phi \colon A \to R\{\tau\}$ satisfying the same conditions that we had imposed over fields, namely $\partial\phi_a = \gamma(a)$ and $\deg_\tau(\phi_a) = r \cdot \deg_T(a)$. Again, ϕ is uniquely determined by

$$\phi_T = t + g_1\tau + \cdots + g_r\tau^r,$$

where the coefficients $g_1, \ldots, g_r \in R$ can be chosen arbitrarily, except now g_r is not just nonzero, but it must be non-nilpotent (see Exercise 3.2.1).

Via ϕ, we get a new A-module structure $a * \beta = \phi_a(\beta)$ on R, denoted $^\phi R$. Passing from the A-algebra R to the A-module $^\phi R$ behaves well with respect to A-algebra homomorphisms: if $f \colon R \to S$ is an A-algebra homomorphism, then $f(\phi_a(\beta)) = \phi_a(f(\beta))$ for every $\beta \in R$, so $f \colon {}^\phi R \to {}^\phi S$ is an A-module homomorphism. (Thus, $R \mapsto {}^\phi R$ is a functor from the category of A-algebras to the category of A-modules.)

Later on, we will need a generalization of Lemma 3.2.11 valid for all A-algebras.

Lemma 3.2.14 *Let R be a commutative A-algebra with structure map $\gamma \colon A \to R$, and let $\phi \colon A \to R\{\tau\}$ be a Drinfeld module of rank r defined by*

$$\phi_a = \sum_{i=0}^{r\,\deg(a)} g_i(a)\tau^i, \qquad a \in A.$$

Let $\mathfrak{p} \lhd A$ be a prime of degree d. There is an integer $1 \leq H \leq r$ such that $g_i(\mathfrak{p}^n) \in \gamma(\mathfrak{p}^n)R$ for all $0 \leq i \leq Hd - 1$ and arbitrary $n \geq 1$.

Proof Consider the A-algebra $B = A[g_1, \ldots, g_r]$, where $g_1, \ldots, g_r$ are indeterminates. Let $\widetilde{\phi} \colon A \to B\{\tau\}$ be the Drinfeld module defined by $\widetilde{\phi}_T = T + g_1\tau + \cdots + g_r\tau^r$. First, we claim that there is an integer $1 \leq H \leq r$ such that, for all $n \geq 1$, we have

$$\widetilde{\phi}_{\mathfrak{p}^n} = \mathfrak{p}^n f_n + g_n$$

for some $f_n, g_n \in B\{\tau\}$ with $\deg_\tau(f_n) \leq Hd - 1$ and $\mathrm{ht}_\tau(g_n) \geq Hd$. (Here we include τ in the notation to avoid any confusion as what the degree and height stand for, since B itself is a polynomial ring.) We prove this claim by induction on n.

Consider the natural reduction map $\iota\colon B \to B' := \mathbb{F}_{\mathfrak{p}}[g_1, \ldots, g_r]$ obtained by reducing the coefficients of polynomials in B modulo $\mathfrak{p}$. This map can be extended to $B\{\tau\} \to B'\{\tau\}$. The composition $\bar\phi = \iota \circ \tilde\phi\colon A \to B'\{\tau\}$ is a Drinfeld module over B', which we can consider as a Drinfeld module over the field of fractions K of B'. Since $\mathrm{char}_A(K) = \mathfrak{p}$, Lemma 3.2.11 implies that $\mathrm{ht}(\bar\phi_{\mathfrak{p}}) = H \cdot d$ for some integer $1 \le H \le r$. Hence

$$\tilde\phi_{\mathfrak{p}} = \mathfrak{p} f_1 + g_1,$$

for some $f_1, g_1 \in B\{\tau\}$ with $\deg(f_1) \le Hd - 1$ and $\mathrm{ht}(g_1) \ge Hd$. This proves the claim for $n = 1$.

Now assume that we have proved that $\tilde\phi_{\mathfrak{p}^{n-1}} = \mathfrak{p}^{n-1} f_{n-1} + g_{n-1}$ for some $f_{n-1}, g_{n-1} \in B\{\tau\}$ with $\deg(f_{n-1}) \le Hd - 1$ and $\mathrm{ht}(g_{n-1}) \ge Hd$. Then

$$\tilde\phi_{\mathfrak{p}^n} = \tilde\phi_{\mathfrak{p}}\tilde\phi_{\mathfrak{p}^{n-1}}$$

$$= (\mathfrak{p} f_1 + g_1)(\mathfrak{p}^{n-1} f_{n-1} + g_{n-1})$$

$$= \mathfrak{p} f_1 \mathfrak{p}^{n-1} f_{n-1} + (\mathfrak{p} f_1 g_{n-1} + g_1 \mathfrak{p}^{n-1} f_{n-1} + g_1 g_{n-1}).$$

Let $g := \mathfrak{p} f_1 g_{n-1} + g_1 \mathfrak{p}^{n-1} f_{n-1} + g_1 g_{n-1}$. Since $\mathrm{ht}(g_1) \ge Hd$ and $\mathrm{ht}(g_{n-1}) \ge Hd$, we have $\mathrm{ht}(g) \ge Hd$ by the obvious extension of Lemma 3.1.11 to $B\{\tau\}$. Now observe that $\mathfrak{p} f_1 \mathfrak{p}^{n-1} f_{n-1} = \mathfrak{p}^n f$ for some $f \in B\{\tau\}$ (usually $f \ne f_1 f_{n-1}$). We can write $f = f_n + h_n$ with $\deg(f_n) \le Hd - 1$ and $\mathrm{ht}(h_n) \ge Hd$. Put $g_n = \mathfrak{p}^n h_n + g$. Then $\tilde\phi_{\mathfrak{p}^n} = \mathfrak{p}^n f_n + g_n$ with $\deg(f_n) \le Hd - 1$ and $\mathrm{ht}(g_n) \ge Hd$. This establishes the induction step.

To prove the claim of the lemma, observe that ϕ is obtained from $\tilde\phi$ through the homomorphism $B \to R$ mapping $a \in A$ to $\gamma(a)$ and g_i to $g_i(T)$. Therefore, $g_i(\mathfrak{p}^n) \subseteq \gamma(\mathfrak{p}^n) R$ for all $0 \le i \le Hd - 1$. $\qquad\square$

Exercises

3.2.1 Let R be a commutative A-algebra. Let $\phi\colon A \to R\{\tau\}$ be a homomorphism of $\mathbb{F}_q$-algebras defined by $\phi_T = t + g_1\tau + \cdots + g_r\tau^r$. Let $a = \sum_{i=0}^m a_i T^i$. Give a formula for the leading coefficient of ϕ_a. Conclude that $\deg_\tau(\phi_a) = r \cdot \deg_T(a)$ for all $0 \ne a \in A$ if and only if g_r is not nilpotent.

3.2.2 Let ϕ be a Drinfeld module over a field. Let $a, b \in A$ be nonzero. Prove the following:

(a) If $a \mid b$, then $\phi_a(x)$ divides $\phi_b(x)$ in $K[x]$.
(b) $\phi_a(x)$ and $\phi_b(x)$ have the same roots, counting multiplicities, if and only if $a = \alpha b$ for some $\alpha \in \mathbb{F}_q^\times$.

3.2.3 Let ϕ be a Drinfeld module of rank r over K. Assume $\mathrm{char}_A(K) = 0$. Let $a \in A$ be an element of positive degree. Prove that ϕ_a uniquely determines ϕ_T, and therefore ϕ. Moreover, if $\phi_a \in L\{\tau\}$ for a subfield $L \subseteq K$, then ϕ is defined over L. Give an example which shows that knowing ϕ_a might not be sufficient for determining ϕ if $\mathrm{char}_A(K) \neq 0$.

3.2.4 Let $a \in A$ have degree d. Let $C_m(a) \in A$ be the polynomial defined by (3.2.1). Prove that

$$\deg_T(C_m(a)) = q^m(d - m) \quad \text{for all} \quad 0 \le m \le d.$$

Moreover, if β is the leading coefficient of a, then $C_d(a) = \beta$.

3.3 Morphisms

Let $\gamma : A \to K$ be an A-field, and let ϕ be a Drinfeld module over K of rank r. As we have explained, ϕ induces an A-module structure on K. We denoted this module by ${}^\phi K$. We now turn to the study of homomorphisms between such modules.

First, as a motivation, recall the definition of homomorphisms between two modules M and N over a commutative ring R. An *R-module homomorphism* $f : M \to N$ is a map such that

(i) $f(x + y) = f(x) + f(y)$, for all $x, y \in M$;
(ii) $f(rx) = rf(x)$, for all $r \in R, x \in M$.

Now let ϕ and ψ be Drinfeld modules over a field K. If we apply the previous definition to ${}^\phi K$ and ${}^\psi K$, then we are lead to the following:

Definition 3.3.1 A *morphism* $u : \phi \to \psi$ of Drinfeld modules over K is a polynomial $u \in K\{\tau\}$ such that $u\phi_a = \psi_a u$ for all $a \in A$. In particular, u is a homomorphism ${}^\phi K \to {}^\psi K$ of A-modules, i.e., the diagram

$$
\begin{array}{ccc}
K & \xrightarrow{u} & K \\
\phi_a \downarrow & & \downarrow \psi_a \\
K & \xrightarrow{u} & K
\end{array}
$$

commutes. A nonzero morphism $u : \phi \to \psi$ is called an *isogeny*.[5] The group of all morphisms $\phi \to \psi$ over K is denoted by $\mathrm{Hom}_K(\phi, \psi)$. We also denote

$$\mathrm{End}_K(\phi) = \mathrm{Hom}_K(\phi, \phi) = \mathrm{Cent}_{K\{\tau\}}(\phi(A)).$$

---

[5] In the context of abelian varieties, an isogeny is a homomorphism with finite kernel.

The composition of morphisms makes $\text{End}_K(\phi)$ into a subring of $K\{\tau\}$, called the *endomorphism ring of* ϕ. The invertible elements of $\text{End}_K(\phi)$ form the *automorphism group of* ϕ, which is denoted by $\text{Aut}_K(\phi)$.

Since T generates the $\mathbb{F}_q$-algebra A, the commutation $u\phi_T = \psi_T u$ is sufficient to ensure the commutation $u\phi_a = \psi_a u$ for all $a \in A$. Therefore

$$\text{Hom}_K(\phi, \psi) = \{u \in K\{\tau\} \mid u\phi_T = \psi_T u\}.$$

For a field extension L/K, we define

$$\text{Hom}_L(\phi, \psi) = \{u \in L\{\tau\} \mid u\phi_T = \psi_T u\}.$$

Obviously, $\text{Hom}_K(\phi, \psi) \subset \text{Hom}_L(\phi, \psi)$, and, as we will see in the examples below, the inclusion can be strict.

Given $a \in A$ and $u \in \text{Hom}_K(\phi, \psi)$, we define

$$a \circ u = u\phi_a = \psi_a u.$$

Note that $a \circ u \in \text{Hom}_K(\phi, \psi)$, since

$$(a \circ u)\phi_T = u\phi_a \phi_T = u\phi_T \phi_a = \psi_T(a \circ u).$$

Therefore, $\text{Hom}_K(\phi, \psi)$ is an A-module. To simplify the notation, from now on, we write au instead of $a \circ u$. Since $K\{\tau\}$ has no zero-divisors, $\text{Hom}_K(\phi, \psi)$ has no torsion elements as an A-module. Furthermore, observe that with the given action of A, the ring $\text{End}_K(\phi)$ is an A-algebra (this is true because $\phi(A)$ lies in the center of $\text{End}_K(\phi)$), and $\text{Hom}_K(\phi, \psi)$ is an A-submodule of $\text{Hom}_L(\phi, \psi)$ for any field extension L/K.

Example 3.3.2 If $\text{char}_A(K) = 0$, then usually $\phi(A) = \text{End}_K(\phi)$. On the other hand, there are Drinfeld modules whose endomorphism rings are strictly larger than $\phi(A)$. One example of such a Drinfeld module is $\phi: A \to F\{\tau\}$ defined by $\phi_T = T + \tau^r$. Let $K = \mathbb{F}_{q^r} F$. It is clear that $\mathbb{F}_{q^r} \subset \text{End}_K(\phi)$, since any element of $\mathbb{F}_{q^r}$ is fixed by τ^r. Thus,

$$\mathbb{F}_{q^r}\phi(A) \cong \mathbb{F}_{q^r}[T] \subseteq \text{End}_K(\phi).$$

On the other hand, $\mathbb{F}_{q^r}[T]$ is the integral closure of A in K/F; in particular, it has rank r over A. It follows from Corollary 3.4.15 that

$$\text{End}_K(\phi) \cong \mathbb{F}_{q^r}[T].$$

Finally, we note that $\text{End}_F(\phi) = \phi(A) \subsetneq \text{End}_K(\phi)$; cf. Exercise 3.3.4.

Example 3.3.3 Let $K = \mathbb{F}_{q^2}$ and $\gamma: A \to A/(T) \cong \mathbb{F}_q \hookrightarrow K$. Let $\phi: A \to K\{\tau\}$ be the Drinfeld module of rank 2 defined by $\phi_T = t + \tau^2 = \tau^2$. In this case, as in the previous example, we have $\mathbb{F}_{q^2} \subset \operatorname{End}_K(\phi)$, but we also have $\tau \in \operatorname{End}_k(\phi)$. Thus, $K\{\tau\} \subseteq \operatorname{End}_K(\phi)$. Since, by definition, $\operatorname{End}_K(\phi) \subseteq K\{\tau\}$, we get $\operatorname{End}_K(\phi) = K\{\tau\}$. The ring $K\{\tau\}$ is not commutative: $\tau\alpha = \alpha^q\tau$ and $\alpha^q = \alpha$ if and only if $\alpha \in \mathbb{F}_q$. In fact, $K\{\tau\}$ is a subring of a quaternion division algebra over F. To see this, assume for simplicity that q is odd. Fix a non-square α in $\mathbb{F}_q^\times$, and let $j \in \mathbb{F}_{q^2}$ be such that $j^2 = \alpha$. The conjugate of j over $\mathbb{F}_q$ is $-j = j^q$. Thus, $\mathbb{F}_{q^2} \cong \mathbb{F}_q(j)$, and $\tau j = j^q\tau = -j\tau$. If we denote $i = \tau$, then $i^2 = \phi_T$, and we see that $K\{\tau\} \cong A[i,j]$, where $i^2 = T$, $j^2 = \alpha$, and $ij = -ji$. If we extend the coefficients from A to F, then we get the quaternion division algebra (cf. Example 1.7.3)

$$\Delta(T, \alpha) := F + Fi + Fj + Fij, \quad \text{with relations}$$

$$i^2 = T, \quad j^2 = \alpha, \quad ij = -ji.$$

The Drinfeld module ϕ is an example of what is called a "supersingular Drinfeld module." We will study this type of Drinfeld modules more extensively in Sect. 4.4.

Proposition 3.3.4 *Suppose* $u: \phi \to \psi$ *is an isogeny. We have:*

(1) *The rank of* ϕ *is equal to the rank of* ψ.
(2) $H(\phi) = H(\psi)$.
(3) *If* $\operatorname{char}_A(K) = 0$, *then* u *is separable.*
(4) *If* $\operatorname{char}_A(K) = \mathfrak{p} \neq 0$, *then* $\deg(\mathfrak{p})$ *divides* $\operatorname{ht}(u)$.
(5) *If* $\operatorname{char}_A(K) = 0$, *then*

$$\operatorname{Hom}_K(\phi, \psi) \longrightarrow K$$

$$u \longmapsto \partial u$$

is an injective homomorphism. In particular, in this case, $\operatorname{End}_K(\phi)$ *is a commutative ring.*

Proof Computing the degrees in τ of both sides of $u\phi_T = \psi_T u$, we see that $\deg(\phi_T) = \deg(\psi_T)$. Hence the rank of ϕ is equal to the rank of ψ. Similarly, computing the heights of both sides of $u\phi_\mathfrak{p} = \psi_\mathfrak{p} u$, we see that $\operatorname{ht}(\phi_\mathfrak{p}) = \operatorname{ht}(\psi_\mathfrak{p})$, which implies $H(\phi) = H(\psi)$; cf. Lemma 3.2.11. Next, comparing the lowest degree terms of $u\phi_T = \psi_T u$, we see that

$$t^{q^{\operatorname{ht}(u)}} = t.$$

As in the proof of Lemma 3.2.11, this implies (3) and (4). Finally, (5) is an immediate consequence of (3). $\square$

Definition 3.3.5 Let $u \in K\{\tau\}$. The *kernel of* u, denoted $\ker(u)$, is the kernel of the $\mathbb{F}_q$-linear transformation $\overline{K} \xrightarrow{u} \overline{K}$, i.e., the set of roots of $u(x)$ in $\overline{K}$, *without*[6] multiplicities. If ϕ is a Drinfeld module over K and $0 \neq a \in A$, then the set $\ker(\phi_a)$ is denoted by $\phi[a]$ and called the *a-torsion points* of ϕ. The $\mathbb{F}_q$-linear polynomial $\phi_a(x)$ is called the *a-division polynomial of* ϕ. Note that the polynomial $\phi_a(x)$ has coefficients in K. We denote by $K(\phi[a])$ the splitting field of $\phi_a(x)$, called the *a-division field of* ϕ. If $\mathrm{char}_A(K) \nmid a$, then $\phi_a(x)$ is separable, so the extension $K(\phi[a])/K$ is Galois.

Note that the degree of $\phi_a(x)$ is $|a|^r$, and if $\mathrm{char}_A(K) \nmid a$, then $\phi_a(x)$ is separable. Hence, if $\mathrm{char}_A(K) \nmid a$, then $\#\phi[a] = |a|^r$. Also, if $u: \phi \to \psi$ is a morphism defined over K, then u induces a homomorphism of A-modules $^\phi\overline{K} \to {}^\psi\overline{K}$, so u maps $\phi[a]$ into $\psi[a]$; explicitly, if $\beta \in \phi[a]$ for some $a \in A$, then

$$\psi_a u(\beta) = u\phi_a(\beta) = u(0) = 0,$$

so $u(\beta) \in \psi[a]$. As we will see, the study of torsion points of Drinfeld modules is intimately intertwined with the study of their isogenies.

Example 3.3.6 Suppose $\mathrm{char}_A(K) = \mathfrak{p} \neq 0$, and let $d = \deg(\mathfrak{p})$. Let $\pi_\mathfrak{p} := \tau^d \in K\{\tau\}$. Let ϕ be a Drinfeld module over K defined by $\phi_T = t + g_1\tau + \cdots + g_r\tau^r$. Define a Drinfeld module $\phi^{(\mathfrak{p}^n)}$ of the same rank r by

$$\phi_T^{(\mathfrak{p})} = t + g_1^{q^d}\tau + \cdots + g_r^{q^d}\tau^r.$$

Then $\pi_\mathfrak{p}: \phi \to \phi^{(\mathfrak{p})}$ is an isogeny, called the *Frobenius isogeny*, since

$$\pi_\mathfrak{p}\phi_T = \left(t^{q^d} + g_1^{q^d}\tau + \cdots + g_r^{q^d}\tau^r\right)\pi_\mathfrak{p}$$

$$= \left(t + g_1^{q^d}\tau + \cdots + g_r^{q^d}\tau^r\right)\pi_\mathfrak{p}$$

$$= \phi_T^{(\mathfrak{p})}\pi_\mathfrak{p}.$$

Iterating this isogeny n-times gives $\pi_\mathfrak{p}^n: \phi \to \phi^{(\mathfrak{p}^n)}$. Now let ϕ and ψ be Drinfeld modules and $u: \phi \to \psi$ be an isogeny over K. We can decompose $u = u_{\mathrm{sep}}\tau^{q^h}$, where u_{sep} is separable and $h = \mathrm{ht}(u)$. By Proposition 3.3.4, d divides h, so $\tau^{q^h} = \pi_\mathfrak{p}^n$, where $n = h/d$. Therefore, u is the composition of isogenies

$$\phi \xrightarrow{\pi_\mathfrak{p}^n} \phi^{(\mathfrak{p}^n)} \xrightarrow{u_{\mathrm{sep}}} \psi.$$

---

[6] One can certainly take into account the multiplicities of the roots of $u(x)$, i.e., consider $\ker(u)$ as a possibly non-reduced group-scheme. In fact, this is quite natural if one considers u as an endomorphism of the additive group-scheme $\mathbb{G}_{a,K}$. The reader will notice that we use the multiplicities of the roots of $u(x)$ when we use $\mathrm{ht}(u)$, as in Example 3.3.6.

Example 3.3.7 Let $\phi_T = T - \tau$ over $F = \mathbb{F}_q(T)$. Then $\phi[T]$ is the set of roots of $xT - x^q = x(T - x^{q-1})$. Since $\mathbb{F}_q^\times \subset F$, we have $F(\phi[T]) = F(\sqrt[q-1]{T})$ for a fixed root $\sqrt[q-1]{T}$ of $x^{q-1} = T$. Also, $\mathrm{Gal}(F(\phi[T])/F) \cong \mathbb{F}_q^\times$.

Example 3.3.8 Let $K = \mathbb{F}_2(y)$, where y is an indeterminate. Assume that $\gamma \colon A \to K$ factors through $A/(T) \cong \mathbb{F}_2$. Let $\phi_T(x) = yx^2 + x^4$. The extension $K(\phi[T])/K$ has degree 2 and is purely inseparable.

The next theorem implies that $\mathrm{Hom}_{K^{\mathrm{sep}}}(\phi, \psi)$ includes all morphisms $\phi \to \psi$ defined over an arbitrary (not necessarily algebraic) field extension of K. The proof of the theorem uses the torsion points of Drinfeld modules.

Theorem 3.3.9 *Let ϕ and ψ be Drinfeld modules defined over K. For any field L containing the separable closure of K as a subfield, we have*

$$\mathrm{Hom}_L(\phi, \psi) = \mathrm{Hom}_{K^{\mathrm{sep}}}(\phi, \psi).$$

Proof First, recall the Lagrange interpolation. Given n pairs $(x_1, y_1), \ldots, (x_n, y_n)$ of elements in a field K', where $x_1, \ldots, x_n$ are distinct, there is a polynomial $f(x) \in K'[x]$ of degree $\leq n - 1$ such that $f(x_i) = y_i$ for all $1 \leq i \leq n$. Explicitly, $f(x)$ is the polynomial

$$f(x) = \sum_{i=1}^n y_i f_i(x), \quad \text{where} \quad f_i(x) = \prod_{\substack{1 \leq j \leq n \\ j \neq i}} \frac{x - x_j}{x_i - x_j}.$$

Now, if K'' is any field extension of K' and $g(x) \in K''[x]$ is a polynomial of degree $\leq n - 1$ such that $g(x_i) = y_i$ for all $1 \leq i \leq n$, then $g(x) = f(x)$ since $x_1, \ldots, x_n$ are n distinct roots of the polynomial $f - g$ of degree $\leq n - 1$.

Let $u \in L\{\tau\}$ be an isogeny $\phi \to \psi$. Choose $a \in A$ not divisible by $\mathrm{char}_A(K)$ and such that $\#\phi[a] > \deg u(x)$. Since $\phi_a(x)$ and $\psi_a(x)$ are separable polynomials, the extension $K' = K(\phi[a], \psi[a])$ of K is separable. Since u maps $\phi[a]$ into $\psi[a]$, it is uniquely determined by its values on $\phi[a]$ and $u(x)$ is a polynomial in $K'[x]$, i.e., $u \in \mathrm{Hom}_{K'}(\phi, \psi)$. □

Definition 3.3.10 To simplify the notation, we will write

$$\mathrm{Hom}(\phi, \psi) := \mathrm{Hom}_{K^{\mathrm{sep}}}(\phi, \psi) = \mathrm{Hom}_{\overline{K}}(\phi, \psi),$$

$$\mathrm{End}(\phi) := \mathrm{End}_{K^{\mathrm{sep}}}(\phi) = \mathrm{End}_{\overline{K}}(\phi),$$

$$\mathrm{Aut}(\phi) := \mathrm{Aut}_{K^{\mathrm{sep}}}(\phi) = \mathrm{Aut}_{\overline{K}}(\phi).$$

Let ϕ and ψ be Drinfeld modules defined over K. Let $u \colon \phi \to \psi$ be an isogeny defined over some field extension of K. By Theorem 3.3.9, u is defined over K^{sep}. We observe the following:

(a) For any $\beta \in \ker(u)$ and $a \in A$, we have

$$u(\phi_a(\beta)) = \psi_a(u(\beta)) = 0.$$

Hence, $\phi(A)$ maps $\ker(u)$ to itself.
(b) By Proposition 3.3.4, $\mathrm{ht}(u) = 0$ if $\mathrm{char}_A(K) = 0$, and $\mathrm{ht}(u)$ is divisible by $\deg(\mathfrak{p})$ if $\mathrm{char}_A(K) = \mathfrak{p} \neq 0$.

Conversely, suppose we are given:

(a) A Drinfeld module ϕ defined over K and a finite dimensional $\mathbb{F}_q$-vector space $G \subset L$ in some field extension L of K, which is invariant under the action of ϕ_T, i.e., $\phi_T(\alpha) \in G$ for all $\alpha \in G$. (Note that because G is an $\mathbb{F}_q$-vector space, this condition is equivalent to G being invariant under the action of ϕ_a for all $a \in A$.)
(b) An integer $h \geq 0$, which is 0 if $\mathrm{char}_A(K) = 0$, and is divisible by $\deg(\mathfrak{p})$ if $\mathrm{char}_A(K) = \mathfrak{p} \neq 0$.

We would like to construct an isogeny u from ϕ to some other Drinfeld module such that $\ker(u) = G$ and $h = \mathrm{ht}(u)$.

Define the $\mathbb{F}_q$-linear polynomials (cf. Lemma 3.1.5)

$$w(x) = \prod_{\alpha \in G} (x - \alpha), \qquad u(x) = w(x)^{q^h}.$$

Obviously, $\ker(u) = \ker(w) = G$.

Proposition 3.3.11 *Let $K' = K(G) \subset L$ be the smallest field extension of K in L which contains all the elements of G. There are Drinfeld modules η and ψ defined over K' such that $w\phi_T = \eta_T w$ and $u\phi_T = \psi_T u$, i.e., w and u are isogenies from ϕ to these Drinfeld modules.*

Proof Note that $w \in K'\{\tau\}$. In terms of polynomials in $K'\{\tau\}$, we have $u = \tau^h w$. Suppose we proved the existence of η given by $\eta_T = t + g_1\tau + \cdots + g_r\tau^r \in K'\{\tau\}$. Define $\psi_T = t + g_1^{q^h}\tau + \cdots + g_r^{q^h}\tau^r \in K'\{\tau\}$. Since by assumption $t^{q^h} = t$, we have $\tau^h \eta_T = \psi_T \tau^h$. Thus, τ^h is an isogeny from η to ψ, and the composition $\phi \xrightarrow{w} \eta \xrightarrow{\tau^h} \psi$ is the isogeny u.

To prove the existence of η, consider $f := w\phi_T$. Since by assumption $\phi_T(\alpha) \in G$ for all $\alpha \in G$, we see that $f(\alpha) = w(\phi_T(\alpha)) = 0$, so G is a subset of the set of roots of f. By Corollary 3.1.16, we have

$$w\phi_T = gw \quad \text{for some} \quad g \in K'\{\tau\}.$$

The constant term of w is nonzero by construction, so the constant term of g is t. To get the desired Drinfeld module η, we put $\eta_T = g$. $\qquad \square$

Proposition 3.3.12 *Let $u: \phi \to \psi$ be an isogeny defined over K. There is an isogeny $\widehat{u}: \psi \to \phi$, also defined over K, such that $\widehat{u}u = \phi_a$ for some $a \in A$; such an isogeny $\widehat{u}$ is called a **dual** of u.*

Proof Let $\mathfrak{p} = \mathrm{char}_A(K)$ and $h = \mathrm{ht}(u)$. If $h \neq 0$, then $\mathfrak{p} \neq 0$ and $\deg(\mathfrak{p})$ divides h. In that case, $\phi_\mathfrak{p}$ is inseparable, and we can choose $n \geq 1$ such that $\mathrm{ht}(\phi_{\mathfrak{p}^n}) = n \cdot H(\phi) \cdot \deg(\mathfrak{p}) \geq h$. Let $G = \ker(u)$. By the earlier discussion, G is invariant under the action of $\phi(A)$. Since G is finite, there is some $0 \neq b \in A$ which annihilates G, i.e., ϕ_b vanishes on G, so by Corollary 3.1.16, there exists $g \in K\{\tau\}$ such that $\phi_{b\mathfrak{p}^n} = gu$.

We claim that g defines an isogeny $\psi \to \eta$ from ψ to another Drinfeld module η. To prove this, first note that

$$\mathrm{ht}(g) = \mathrm{ht}(\phi_{b\mathfrak{p}^n}) - \mathrm{ht}(u)$$

is divisible by $\deg(\mathfrak{p})$. Hence, by Proposition 3.3.11, it is enough to show that $\ker(g)$ is invariant under the action of $\psi(A)$. Let $\alpha \in \ker(g)$. There is some $\beta \in \overline{K}$ such that $u(\beta) = \alpha$. Now $\phi_{b\mathfrak{p}^n}(\beta) = g(u(\beta)) = g(\alpha) = 0$, so $\beta \in \phi[b\mathfrak{p}^n]$. Reversing this argument, we see that if $\beta \in \phi[b\mathfrak{p}^n]$, then $u(\beta) \in \ker(g)$. Therefore $\ker(g) = u(\phi[b\mathfrak{p}^n])$, i.e., $\ker(g)$ is the image of $\phi[b\mathfrak{p}^n]$ under u. Now let $\alpha \in \ker(g)$ and choose some $\beta \in \phi[b\mathfrak{p}^n]$ such that $\alpha = u(\beta)$. We have

$$g(\psi_T(\alpha)) = g(\psi_T(u(\beta))) = g(u(\phi_T(\beta)))$$
$$= \phi_{b\mathfrak{p}^n}\phi_T(\beta) = \phi_T\phi_{b\mathfrak{p}^n}(\beta) = \phi_T(0) = 0.$$

This implies that $\psi_T(\alpha) \in \ker(g)$, so $\ker(g)$ is indeed invariant under the action of $\psi(A)$.

If we show that $\eta = \phi$, then we can take $a = b\mathfrak{p}^n$ and $\widehat{u} = g$ to prove the proposition. On the one hand, we have

$$g\psi_T u = \eta_T gu = \eta_T \phi_{b\mathfrak{p}^n}.$$

On the other hand,

$$g\psi_T u = gu\phi_T = \phi_{b\mathfrak{p}^n}\phi_T = \phi_T\phi_{b\mathfrak{p}^n}.$$

Combining these equalities, we get $\eta_T\phi_{b\mathfrak{p}^n} = \phi_T\phi_{b\mathfrak{p}^n}$. Thus, $\eta_T = \phi_T$, as was required to be shown. □

A dual isogeny of u, as defined in Proposition 3.3.12, is not unique; for example, we can multiply $\widehat{u}$ by any ϕ_b, $0 \neq b \in A$, and obtain another dual isogeny:

$$(\phi_b\widehat{u})\psi_c = \phi_b\phi_c\widehat{u} = \phi_c(\phi_b\widehat{u}) \quad \text{for all } c \in A, \quad \text{and}$$
$$(\phi_b\widehat{u})u = \phi_b\widehat{u}u = \phi_{ba}.$$

On the other hand, among all dual isogenies, one could choose $\widehat{u}$ with minimal τ-degree and such that a in $\widehat{u}u = \phi_a$ is monic. It is easy to check that this $\widehat{u}$ is uniquely determined.[7] The next lemma gives a bound on the degree of this dual isogeny.

Lemma 3.3.13 *With notation and assumptions of Proposition 3.3.12, there is a dual isogeny $\widehat{u} \colon \psi \to \phi$ such that*

$$\deg(\widehat{u}) \le (r-1)\deg(u).$$

Proof We will use the notation from the proof of Proposition 3.3.12. Choose $b \in A_+$ of smallest degree such that ϕ_b annihilates $G = \ker(u)$, and let $n = \mathrm{ht}(u)/\deg_T(\mathfrak{p})$ (note that n is an integer). Then, as we showed in the proof of Proposition 3.3.12, there is $\widehat{u}$ such that

$$\deg_\tau(u) + \deg_\tau(\widehat{u}) = \deg_\tau \phi_{b\mathfrak{p}^n} \le r(\deg_T(b) + \mathrm{ht}(u)).$$

By Theorem 1.2.4, we have

$$G \cong A/b_1 A \times \cdots A/b_s A, \quad \text{for some monic } b_1 \mid b_2 \mid \cdots \mid b_s,$$

so $b = b_s$. Since $G \subseteq \phi[b]$ and, as we will show in Sect. 3.5, $\phi[b] \subseteq (A/bA)^r$, we have $s \le r$. We may assume that $s = r$ by assuming that some of the b_i's are equal to 1. Now,

$$\deg_\tau(u) = \deg_\tau(u_{\mathrm{sep}}) + \mathrm{ht}(u),$$
$$\deg_\tau(u_{\mathrm{sep}}) = \log_q \#G = \deg_T(b_1) + \cdots + \deg_T(b_r).$$

Thus,

$$\deg_\tau(\widehat{u}) \le \sum_{i=1}^{r-1}(\deg_T(b_r) - \deg_T(b_i)) + (r-1)\,\mathrm{ht}(u)$$
$$\le (r-1)\deg_T(b_r) + (r-1)\,\mathrm{ht}(u)$$
$$\le (r-1)(\deg_\tau(u_{\mathrm{sep}}) + \mathrm{ht}(u)) = (r-1)\deg_\tau(u).$$

$\square$

Definition 3.3.14 The *endomorphism algebra of ϕ* is

$$\mathrm{End}_K^\circ(\phi) := F \otimes_A \mathrm{End}_K(\phi).$$

---

[7] Still, this is not as "canonical" as the dual of an isogeny between two elliptic curves, which is defined using Picard functoriality; cf. [Sil09, p. 81]. For Drinfeld modules of rank 2, a dual isogeny $\widehat{u}$ can be defined through a similar functionality; see [PR03].

Since $\operatorname{End}_K(\phi)$ is an A-algebra, $\operatorname{End}_K^\circ(\phi)$ is an F-algebra.

Corollary 3.3.15 $\operatorname{End}_K^\circ(\phi)$ *is a division algebra. In particular, if* $\operatorname{char}_A(K) = 0$, *then* $\operatorname{End}_K^\circ(\phi)$ *is a field extension of* F.

Proof Let $0 \neq u \in \operatorname{End}_K^\circ(\phi)$. Every element of $F \otimes_A \operatorname{End}_K(\phi)$ can be scaled by a nonzero element of A to lie in $\operatorname{End}_K(\phi)$, so for some $b \in A$ we have $0 \neq bu \in \operatorname{End}_K(\phi)$. On the other hand, by Proposition 3.3.12, there exists $u' \in \operatorname{End}_K(\phi)$ such that $u'(bu) = a \in A - \{0\}$. Therefore, $(a^{-1}u'b)u = 1$. We also have (cf. Exercise 3.3.2)

$$u(a^{-1}u'b) = (ub)u'a^{-1} = aa^{-1} = 1,$$

so $a^{-1}u'b$ is the multiplicative inverse of u in $\operatorname{End}_K^\circ(\phi)$. Finally, if $\operatorname{char}_A(K) = 0$, then $\operatorname{End}_K^\circ(\phi)$ is commutative by Proposition 3.3.4 and so must be a field. □

Corollary 3.3.16 *Let* ϕ *and* ψ *be Drinfeld modules over* K. *There is an injective homomorphism of* A-*modules*

$$i \colon \operatorname{Hom}_K(\phi, \psi) \hookrightarrow \operatorname{End}_K(\phi),$$

such that, when $\operatorname{Hom}_K(\phi, \psi) \neq 0$, *the quotient* $\operatorname{End}_K(\phi)/i(\operatorname{Hom}_K(\phi, \psi))$ *is a torsion module.*

Proof We may assume that there exists $0 \neq u \in \operatorname{Hom}_K(\phi, \psi)$. Then the map

$$i \colon \operatorname{Hom}_K(\phi, \psi) \longrightarrow \operatorname{End}_K(\phi)$$

$$w \longmapsto \widehat{u}w$$

is the required injection. Note that $i' \colon \operatorname{End}_K(\phi) \to \operatorname{Hom}_K(\phi, \psi)$, $v \mapsto uv$, is also an injection. Since the composition $i'i$ is the multiplication by some $0 \neq a = u\widehat{u} \in A$ on $\operatorname{Hom}_K(\phi, \psi)$, the quotient $\operatorname{End}_K(\phi)/i(\operatorname{Hom}_K(\phi, \psi))$ is annihilated by a. □

Exercises

3.3.1 Let L/K be a field extension. Prove that $\operatorname{Hom}_L(\phi, \psi)/\operatorname{Hom}_K(\phi, \psi)$ is torsion-free as an A-module.

3.3.2 Let $u \colon \phi \to \psi$ and $\widehat{u} \colon \psi \to \phi$ be as in Proposition 3.3.12, so $\widehat{u}u = \phi_a$. Prove that we also have $u\widehat{u} = \psi_a$.

3.3.3 Give an example which shows that even if $u \colon \phi \to \psi$ is separable, there might not exist a separable dual isogeny $\widehat{u} \colon \psi \to \phi$.

3.3.4 Let ϕ be a Drinfeld module over F, where $\gamma \colon A \to F$ is the natural inclusion of A into its field of fractions. Prove that $\operatorname{End}_F(\phi) = A$.

3.4 Module of Morphisms

Let K be an A-field, and let ϕ and ψ be Drinfeld modules of rank r over K. To prove deeper results about the morphisms $\phi \to \psi$, we need a better understanding of the structure of the module $\mathrm{Hom}_K(\phi, \psi)$ of all morphisms. The main result of this section is the following fundamental fact.

Theorem 3.4.1 $\mathrm{Hom}_K(\phi, \psi)$ *is a free A-module of rank* $\leq r^2$.

We give two very different proofs of Theorem 3.4.1. The first proof is due to Anderson [And86] and the second is due to Drinfeld [Dri74]. Both proofs are important in their own right since they introduce important techniques into the theory.

3.4.1 Anderson Motive of a Drinfeld Module

Anderson's proof of Theorem 3.4.1 proceeds by first introducing a new category, equivalent to the category of Drinfeld modules, where the proof of Theorem 3.4.1 reduces to linear algebra.

Let $K[T, \tau]$ be the non-commutative polynomial ring in indeterminates T and τ subject to the commutation relations

$$T\tau = \tau T, \quad \alpha T = T\alpha, \quad \tau\alpha = \alpha^q \tau \quad \text{for all } \alpha \in K,$$

Note that

- $K[T, \tau] \cong K\{\tau\} \otimes_{\mathbb{F}_q} A$.
- $K[T, \tau] = K\{\tau\}[T]$ can be considered as the polynomial ring in T with coefficients in $K\{\tau\}$.
- $K[T]$ and $K\{\tau\}$ are subrings of $K[T, \tau]$, and $A = \mathbb{F}_q[T]$ is in the center of $K[T, \tau]$.

Definition 3.4.2 An *Anderson motive (of dimension 1)* over K is a left $K[T, \tau]$-module M with the following properties:

(i) As a $K\{\tau\}$-module, M is free of rank 1.
(ii) As a $K[T]$-module, M is free of finite rank.
(iii) $(T - t)m \in \tau\overline{M}$ for all $m \in \overline{M}$, where $\overline{M} = \overline{K} \otimes_K M$ is considered as a left $\overline{K}[T, \tau]$-module.

The *rank* of M is its rank as a $K[T]$-module. A *morphism* $f : M \to M'$ of Anderson motives is a homomorphism of $K[T, \tau]$-modules.

Remark 3.4.3 Definition 3.4.2 is a special case of Definition 1.2 in [And86]. In Anderson's original definition of what he calls "t-motive," M is only assumed to be finitely generated over $K\{\tau\}$ and (iii) is replaced by the requirement that

$(T - t)^n (\overline{M}/\tau\overline{M}) = 0$ for some integer $n > 0$. This leads to a higher dimensional generalization of Drinfeld modules. In this book, we only deal with the one dimensional case.

Given a Drinfeld module $\phi: A \to K\{\tau\}$ of rank $r \geq 1$, we associate with ϕ a left $K[T, \tau]$-module M_ϕ by taking $M_\phi = K\{\tau\}$ and defining

$$u \circ m = um \quad \text{for all } u \in K\{\tau\} \text{ and } m \in M_\phi,$$

$$a \circ m = m\phi_a \quad \text{for all } a \in A \text{ and } m \in M_\phi,$$

where um and $m\phi_a$ are multiplications in $K\{\tau\}$.

Lemma 3.4.4 M_ϕ *is an Anderson motive over K of rank r.*

Proof We claim that M_ϕ over $K[T]$ is freely generated by $1, \tau, \ldots, \tau^{r-1}$. Let $M' = K[T] + K[T]\tau + \cdots + K[T]\tau^{r-1}$ be the $K[T]$-submodule generated by these elements. Obviously, every element $m \in M$ with $\deg_\tau(m) \leq r - 1$ lies in M'. Assume by induction that the same is true for all elements of M_ϕ with τ-degree $\leq n$, where $n \geq r - 1$. Let $m \in M_\phi$ be such that $\deg_\tau(m) = n + 1$. By Theorem 3.1.13, we can write

$$m = h\phi_T + u = T \circ h + u,$$

where $h, u \in K\{\tau\}$, and $\deg_\tau(u) < r$ or $u = 0$. Since $\deg_\tau(h) \leq n$, we conclude that $h \in M'$ and hence also $m \in M'$. This completes the induction step, so M_ϕ is generated by $1, \tau, \ldots, \tau^{r-1}$.

Note that M_ϕ is torsion-free as a $K[T]$-module. By Theorem 1.2.4, M_ϕ is free of rank $n \leq r$. Now, consider $M_\phi/TM_\phi \cong K^n$, and note that the nonzero elements of $M'' = K + K\tau + \cdots + K\tau^{r-1} \subset M_\phi$ are not in $TM_\phi = M_\phi\phi_T$ since the nonzero elements of $M_\phi\phi_T$ have τ-degree $\geq r$. Thus, M'' maps injectively into M_ϕ/TM_ϕ, and therefore $n = \dim_K M_\phi/TM_\phi \geq r$. This implies that $n = r$, so M_ϕ is freely generated by $1, \tau, \ldots, \tau^{r-1}$.

Finally, let $m = m_0 + m_1 \circ \tau + \cdots + m_{r-1} \circ \tau^{r-1} \in M_\phi$, where $m_0, \ldots, m_{r-1} \in K[T]$. Then

$$T \circ m = m\phi_T = tm_0 + h\tau \quad \text{for some } h \in K\{\tau\}$$

$$= tm_0 + \tau g \quad \text{for some } g \in \overline{K}\{\tau\} \tag{3.4.1}$$

$$\equiv t \circ m \pmod{\tau\overline{M}_\phi}.$$

Hence, M_ϕ has the properties (i)–(iii) of Definition 3.4.2. (Note that if K is not perfect, then in (3.4.1) we cannot generally rewrite $h\tau$ as τg for some $g \in K\{\tau\}$; this is the reason for passing to the algebraic closure of K in (iii).) $\square$

Now let $M = K\{\tau\}$ be an Anderson motive of rank $r \geq 1$. Then $T \circ 1 = f$ for some $f \in K\{\tau\}$. By (iii), $\partial f = t$. Define $\phi: A \to K\{\tau\}$ by $\phi_T = f$. This is a Drinfeld module. To see that the rank of ϕ is r, note $M \cong M_\phi$, and $\text{rank}(\phi) = \text{rank}_{K[T]} M_\phi = \text{rank}_{K[T]} M = r$.

The module of morphism of Anderson motives $\text{Hom}_{K[T,\tau]}(M, M')$ is an A-module with the action of $a \in A$ defined by

$$(a \circ f)(x) := f(a \circ x) = a \circ f(x).$$

(We leave the easy verifications that this gives an A-module structure to the reader.)

Proposition 3.4.5 *Let ϕ and ψ be two Drinfeld modules over K. Then there is an isomorphism of A-modules*

$$\text{Hom}_K(\phi, \psi) \cong \text{Hom}_{K[T,\tau]}(M_\psi, M_\phi).$$

Note that on the right-hand side M_ψ and M_ϕ appear in the reverse order from ϕ and ψ. Hence the categories of Drinfeld modules and Anderson motives are anti-equivalent under the functor $\phi \mapsto M_\phi$.

Proof Given $u \in \text{Hom}_K(\phi, \psi)$, define a map

$$f_u : M_\psi \longrightarrow M_\phi$$

$$m \longmapsto mu,$$

where we identity both M_ϕ and M_ψ with $K\{\tau\}$. It is clear that f_u is $K\{\tau\}$-linear since $f_u(wm) = wmu = wf_u(m)$ for all $w, m \in K\{\tau\}$. Next, we have

$$f_u(T \circ m) = f_u(m\psi_T) = m\psi_T u = mu\phi_T = T \circ f_u(m).$$

Hence, f_u is also $\mathbb{F}_q[T]$-linear, so $f_u \in \text{Hom}_{K[T,\tau]}(M_\psi, M_\phi)$.

Conversely, given $f \in \text{Hom}_{K[T,\tau]}(M_\psi, M_\phi)$, set $u_f := f(1) \in K\{\tau\} = M_\phi$. By $K\{\tau\}$-linearity of f, we have $f(m) = mf(1) = mu_f$ for all $m \in M_\psi$. Now the T-linearity of f translates into

$$u_f\phi_T = T \circ f(1) = f(T \circ 1) = f(\psi_T) = \psi_T u_f.$$

Thus, $\psi_T u_f = u_f\phi_T$, so $u_f \in \text{Hom}_K(\phi, \psi)$.

It is easy to check that $u \mapsto f_u$ and $f \mapsto u_f$ are inverses of each other, so we have an isomorphism $\text{Hom}_K(\phi, \psi) \xrightarrow{\sim} \text{Hom}_{K[T,\tau]}(M_\psi, M_\phi)$ of abelian groups. Finally, for $a \in A$, we have

$$f_{a \circ u} = f_{u\phi_a} = a \circ f_u,$$

so the isomorphism is an isomorphism of A-modules. $\square$

Proposition 3.4.6 *Assume K is algebraically closed. Let M and M' be two Anderson motives over K. The evident map*

$$\mathrm{Hom}_{K[T,\tau]}(M, M') \otimes_{\mathbb{F}_q} K \longrightarrow \mathrm{Hom}_{K[T]}(M, M')$$

is injective.

Proof Suppose the claim is false. Let n be the smallest positive integer such that there exist $\mathbb{F}_q$-linearly independent $K[T, \tau]$-module homomorphisms $f_1, \ldots, f_n \colon M \to M'$ and $\alpha_1, \ldots, \alpha_n \in K$ such that

$$\sum_{i=1}^{n} \alpha_i f_i = 0, \qquad \alpha_1 = 1.$$

For an arbitrary $m \in M$, we have

$$(T - t) \left(\sum_{i=1}^{n} \alpha_i^q f_i(m) \right) = \sum_{i=1}^{n} \alpha_i^q f_i((T - t)m)$$

$$= \sum_{i=1}^{n} \alpha_i^q f_i(\tau m') \quad \text{for some } m' \in M$$

$$\text{(since } (T - t)m \in \tau M)$$

$$= \tau \left(\sum_{i=1}^{n} \alpha_i f_i(m') \right)$$

$$= 0.$$

Since M' is a torsion-free $K[T]$-module and $T \neq t$ (which really means that $\psi_T \neq t$), we get

$$\sum_{i=1}^{n} \alpha_i^q f_i = 0.$$

Consequently,

$$\sum_{i=2}^{n} (\alpha_i^q - \alpha_i) f_i = 0.$$

Since n is minimal, all the coefficients of this relation must vanish, i.e., $\alpha_i^q = \alpha_i$, hence $\alpha_1, \ldots, \alpha_n \in \mathbb{F}_q$. This gives an $\mathbb{F}_q$-linear relation among $f_1, \ldots, f_n$, leading to a contradiction. $\square$

Lemma 3.4.7 *Let k be a field, and let K/k be a field extension. Let M be a torsion-free $k[T]$-module. Then $M \otimes_k K$ is a free $K[T]$-module of rank d if and only if M is a free $k[T]$-module of rank d.*

Proof One direction is straightforward: if $M \cong k[T]^{\oplus d}$, then (cf. Proposition 1.5.1)

$$M \otimes_k K \cong (k[T] \otimes_k K)^{\oplus d} \cong K[T]^{\oplus d}.$$

Now assume $M \otimes_k K$ is a free $K[T]$-module of rank d. To show that M is free of rank d, it is enough to show that M is finitely generated over $k[T]$. Indeed, if M is finitely generated, then by Theorem 1.2.4 we have $M \cong k[T]^{\oplus d'}$ for some $d' \geq 0$. Since now $M \otimes_k K \cong K[T]^{\oplus d'} \cong K[T]^{\oplus d}$, from Theorem 1.2.4, we also get $d = d'$.

To prove that M is finitely generated, we argue by contradiction. Assume M is not finitely generated. Let M_1 be the $k[T]$-span of some nonzero $v_1 \in M$. Since M is not finitely generated, there is $v_2 \notin M_1$. Let $M_2 = k[T]v_1 + k[T]v_2$. Again there is some $v_3 \notin M_2$. Let M_3 be the $k[T]$-span of v_1, v_2, and v_3. Continuing in this manner, we find a strictly increasing chain of finitely generated $k[T]$-submodules

$$M_1 \subsetneq M_2 \subsetneq M_3 \subsetneq \cdots$$

We claim that $M_i \otimes_k K \subsetneq M_{i+1} \otimes_k K$ for all $i \geq 1$. Indeed, M_i is a k-vector subspace of M_{i+1}, so we can decompose $M_{i+1} = M_i \oplus M_i'$, where $M_i' \neq 0$. Now $M_{i+1} \otimes_k K = (M_i \otimes_k K) \oplus (M_i' \otimes_k K)$ and $M_i' \otimes_k K \neq 0$ since a nonzero vector space remains nonzero under extension of scalars. It is clear that each $M_i \otimes_k K$ is a finitely generated $K[T]$-submodule of $M \otimes_k K$ since it is invariant under the action of T. Thus, we get a strictly increasing infinite chain of submodules of $M \otimes_k K \cong K[T]^{\oplus d}$. On the other hand, any submodule of $K[T]^{\oplus d}$ is finitely generated (see Proposition 1.2.2), so $M \otimes_k K$ is Noetherian (see [DF04, p. 458]). But then there cannot exist an infinite increasing chain of submodules in $M \otimes_k K$, which leads to a contradiction. $\square$

Now, we are ready to prove the main theorem:

First Proof of Theorem 3.4.1 It is enough to prove the theorem assuming K is algebraically closed. Thanks to Proposition 3.4.5, it is enough to show that $\operatorname{Hom}_{K[T,\tau]}(M_\psi, M_\phi)$ has rank $\leq r^2$ over A. Furthermore, thanks to Lemma 3.4.7, it is enough to show that

$$\operatorname{Hom}_{K[T,\tau]}(M_\psi, M_\phi) \otimes_{\mathbb{F}_q} K,$$

as a module over $A \otimes_{\mathbb{F}_q} K \cong K[T]$, has rank $\leq r^2$.

Now, by Proposition 3.4.6, $\operatorname{Hom}_{K[T,\tau]}(M_\psi, M_\phi) \otimes_{\mathbb{F}_q} K$ is isomorphic to a $K[T]$-submodule of $\operatorname{Hom}_{K[T]}(M_\psi, M_\phi)$. But M_ψ and M_ϕ are free $K[T]$-modules of rank r, so $\operatorname{Hom}_{K[T]}(M_\psi, M_\phi) \cong \operatorname{Mat}_r(K[T])$ is a free $K[T]$-module of rank r^2. Finally, by Proposition 1.2.2, a submodule of the free module $\operatorname{Hom}_{K[T]}(M_\psi, M_\phi)$ is itself free of rank $\leq r^2$. $\square$

3.4.2 Embeddings into the Twisted Laurent Series Ring

In the second proof of Theorem 3.4.1, we will use the following interesting variant of the field of formal Laurent series; cf. Exercise 1.1.1. Assume K is perfect, and let $K((\tau^{-1}))$ be the ring of *twisted Lauren series* in τ^{-1} consisting of formal series of the form $\sum_{n \in \mathbb{Z}} \kappa_n \tau^n$ with $\kappa_n = 0$ for all sufficiently large *positive n*. The multiplication is defined so that

$$\tau^n \kappa = \kappa^{q^n} \tau^n \qquad \text{for all } \kappa \in K,$$

where $\kappa^{q^{-1}}$ is the unique element of K mapping to κ under the automorphism $\tau : K \to K$ (here we implicitly use the assumption that K is perfect).

Lemma 3.4.8 $K((\tau^{-1}))$ *is a division ring.*

Proof First, we note that because $K((\tau^{-1}))$ has no zero-divisors, it is enough to show that every nonzero element has a left inverse. Indeed, if $gf = 1$, then $g(fg) = (gf)g = g$, which implies $g(fg - 1) = 0$, so $fg = 1$.

Next, we show that $f = \sum_{n \geq 0} f_n \tau^{-n}$ with $f_0 \neq 0$ has a left inverse $g = \sum_{n \geq 0} g_n \tau^{-n}$. Formally expanding $gf = \sum_{n \geq 0} h_n \tau^{-n}$, we see that

$$h_n = g_n f_0^{q^{-n}} + g_{n-1} f_1^{q^{-(n-1)}} + \cdots + g_0 f_n, \qquad n \geq 0.$$

Considering $g_0, g_1, \ldots$ as indeterminates, we can recursively solve the equations $h_0 = g_0 f_0 = 1$ and $h_n = 0$ for all $n \geq 1$ because $f_0 \neq 0$. Thus $gf = 1$.

Finally, for general nonzero $f \in K((\tau^{-1}))$, there is a well-defined $m \in \mathbb{Z}$ such that $f = \tilde{f} \tau^m$ with $\tilde{f} = \sum_{n \geq 0} f_n \tau^{-n}$, $f_0 \neq 0$. Then $f^{-1} = \tau^{-m} \tilde{f}^{-1}$. □

Note that $K\{\tau\}$ is naturally a subring of $K((\tau^{-1}))$, and the map

$$- \deg_\tau : K\{\tau\} \longrightarrow \mathbb{Z} \cup \{+\infty\}$$

extends to a discrete valuation (cf. Definition 2.1.1)

$$\mathrm{ord}_{\tau^{-1}} : K((\tau^{-1})) \longrightarrow \mathbb{Z} \cup \{+\infty\}$$

defined as follows: given $f = \sum_{n \geq m} f_n \tau^{-n}$ with $f_m \neq 0$, put $\mathrm{ord}_{\tau^{-1}}(f) = m$, and put $\mathrm{ord}_{\tau^{-1}}(0) = +\infty$.

In the proof of Theorem 3.4.1, we will also need two auxiliary lemmas, which are important in their own right.

Lemma 3.4.9 *Let V be a finite dimensional normed vector space over F_∞, and let $\Lambda \subset V$ be a weakly discrete A-submodule. Then:*

(1) Λ is a free A-module of rank $\leq \dim(V)$.
(2) $\mathrm{rank}_A(\Lambda) = \dim(V)$ if and only if V/Λ is compact.

Proof

(1) After replacing V with its subspace $F_\infty \Lambda$ spanned over F_∞ by the elements of Λ, we may assume $V = F_\infty \Lambda$. We will prove that Λ is free of rank $n = \dim(V)$. Let $\{v_1, \ldots, v_n\} \subset \Lambda$ be a basis of V. We take $\{v_1, \ldots, v_n\}$ as the basis for the sup-norm, i.e., we define $\|\alpha_1 v_1 + \cdots + \alpha_n v_n\|_{\sup} = \max_{1 \le i \le n} |\alpha_i|$; see Example 2.3.3. By Proposition 2.3.4, the norm $\|\cdot\|$ on V is equivalent to $\|\cdot\|_{\sup}$. Hence Λ is weakly discrete with respect to $\|\cdot\|$ if and only if it is weakly discrete with respect to $\|\cdot\|_{\sup}$, so we may assume that $\|\cdot\| = \|\cdot\|_{\sup}$. (Note that, because F_∞ is local and V is finite dimensional, Λ is actually discrete in V according to Exercise 2.7.4, although we will not use this fact.)

The 1-cube

$$U = \{v \in V \: : \: \|v\| < 1\} \cong M_\infty^{\oplus n}$$

is compact; see Lemma 2.2.14. Let $\Lambda' = A v_1 + \cdots + A v_n$. Note that

$$F_\infty = A \oplus M_\infty$$

because, using $1/T$-expansions, we can uniquely decompose any $\alpha \in F_\infty$ as

$$\alpha = \sum_{i \ge d} \alpha_i (1/T)^i = \sum_{i=d}^{0} \alpha_i (1/T)^i + \sum_{i \ge 1} \alpha_i (1/T)^i \qquad \text{(all } \alpha_i \in \mathbb{F}_q\text{)}$$

$$= a + \beta \qquad \text{with } a \in A \text{ and } |\beta| < 1.$$

Therefore, every coset of Λ' in Λ has a representative in U. Thus, if Λ/Λ' is infinite, then $\Lambda \cap U$ is infinite. On the other hand, since U is open in V, the set $\Lambda \cap U$ is discrete with respect to the topology on U. But a discrete subset of a compact set must be finite, so Λ/Λ' is finite. This implies that Λ is finitely generated, thus free. As Λ/Λ' is finite, the rank of Λ must be equal to the rank of Λ', so $\mathrm{rank}_A \Lambda = n$. This proves (1).

(2) If $\mathrm{rank}_A(\Lambda) = \dim(V)$, then it follows from (1) that V/Λ is topologically isomorphic to U, which is compact. On the other hand, if $\mathrm{rank}_A(\Lambda) < \dim(V)$, then the positive dimensional F_∞-vector space $V/F_\infty \Lambda$ is a quotient of V/Λ. Since $V/F_\infty \Lambda$ is not compact (F_∞ is not compact), V/Λ cannot be compact either.

□

As we have observed earlier, $u \in \mathrm{Hom}_K(\phi, \psi)$ maps $\phi[a]$ into $\psi[a]$ for any $0 \ne a \in A$. Hence we get a map

$$\mathrm{Hom}_K(\phi, \psi) \to \mathrm{Hom}(\phi[a], \psi[a]),$$

which obviously factors through $\mathrm{Hom}_K(\phi, \psi) \otimes_A A/(a)$.

Lemma 3.4.10 *Assume* $a \in A$ *is not divisible by* $\mathrm{char}_A(K)$. *Then the natural homomorphism*

$$\mathrm{Hom}_K(\phi, \psi) \otimes_A A/(a) \longrightarrow \mathrm{Hom}_{A/(a)}(\phi[a], \psi[a])$$

is injective.

Proof Explicitly, the claim is that if $u \in \mathrm{Hom}_K(\phi, \psi)$ induces the zero map $\phi[a] \to \psi[a]$, then $u \in a \cdot \mathrm{Hom}_K(\phi, \psi)$. Suppose $\phi[a]$ is a subset of $\ker(u)$. By Corollary 3.1.16, we have $u = w\phi_a$ for some $w \in K\{\tau\}$. We need to show that $w \in \mathrm{Hom}_K(\phi, \psi)$. For any $b \in A$, we have $u\phi_b = w\phi_a\phi_b = w\phi_b\phi_a$ and $u\phi_b = \psi_b u = \psi_b w\phi_a$. Thus, $(w\phi_b - \psi_b w)\phi_a = 0$. Since $K\{\tau\}$ has no zero-divisors, we must have $w\phi_b = \psi_b w$, so $w \in \mathrm{Hom}_K(\phi, \psi)$. $\square$

Now, we are ready to prove the main theorem:

Second Proof of Theorem 3.4.1 First, we make two simplifying observations:

- Thanks to Corollary 3.3.16 and Proposition 1.2.2, it is enough to prove the theorem for $\mathrm{End}_K(\phi)$.
- Thanks to the natural injective A-module homomorphism $\mathrm{End}_K(\phi) \hookrightarrow \mathrm{End}_{\overline{K}}(\phi)$, we may assume that K is algebraically closed.

Now, using the Drinfeld module ϕ, we get a homomorphism

$$A \overset{\phi}{\hookrightarrow} K\{\tau\} \hookrightarrow K\left(\!\left(\tau^{-1}\right)\!\right), \tag{3.4.2}$$

which uniquely extends to a homomorphism $\phi \colon F \hookrightarrow K\left(\!\left(\tau^{-1}\right)\!\right)$. The valuation $\mathrm{ord}_{\tau^{-1}}$ induces a nontrivial valuation v on F. For any $a \in A$, we have

$$v(a) = \mathrm{ord}_{\tau^{-1}}(\phi_a) = -r \cdot \deg_T(a).$$

Therefore, $v = r \cdot \mathrm{ord}_\infty$. Since $K\left(\!\left(\tau^{-1}\right)\!\right)$ is complete with respect to $\mathrm{ord}_{\tau^{-1}}$ (cf. Example 2.2.6), the map ϕ uniquely extends to a homomorphism

$$\phi \colon F_\infty \hookrightarrow K\left(\!\left(\tau^{-1}\right)\!\right).$$

Thus, we can consider $K\left(\!\left(\tau^{-1}\right)\!\right)$ as a vector space over F_∞, with F_∞ acting by left multiplication. The absolute value on $K\left(\!\left(\tau^{-1}\right)\!\right)$ defined by

$$\|\cdot\| = q^{-\mathrm{ord}_{\tau^{-1}}(\cdot)/r}$$

is a norm on $K\left(\!\left(\tau^{-1}\right)\!\right)$ considered as an F_∞-vector space.

Note that $K\{\tau\}$, and therefore also

$$E := \mathrm{End}(\phi) = \mathrm{Cent}_{K\{\tau\}}(\phi(A)),$$

is a weakly discrete left A-submodule of $K((\tau^{-1}))$. Let v_1 be a nonzero element of E and put $W_1 = F_\infty v_1$. This is a one-dimensional F_∞-vector subspace of $K((\tau^{-1}))$. Now let $n \geq 1$, and assume that $v_1, \ldots, v_n \in E$ are such that

$$W_n = F_\infty v_1 + \cdots + F_\infty v_n$$

has dimension n, where we take the span in $K((\tau^{-1}))$. By Lemma 3.4.9 and its proof,

$$E_n := E \cap W_n$$

is a free A-module of rank n. If $E_n \neq E$, then there is $v_{n+1} \in E$ such that $W_{n+1} = W_n + F_\infty v_{n+1}$ has dimension $n + 1$ and E_{n+1} has rank $n + 1$.

Let $a \in A$ be nonzero. Note that $E_n \otimes A/(a) \cong (A/(a))^{\oplus n}$ injects into $E \otimes A/(a)$: if $av \in E_n$ for some $v \in E$, then $v \in W_n$ and so $v \in E_n$. Fix a prime $\mathfrak{p} \in A$ different from $\mathrm{char}_A(K)$. We will show in Sect. 3.5 that $\phi[\mathfrak{p}] \cong (A/\mathfrak{p})^{\oplus r}$; see Theorem 3.5.2. Hence,

$$\dim_{\mathbb{F}_\mathfrak{p}} \mathrm{End}_{A/\mathfrak{p}}(\phi[\mathfrak{p}]) = r^2.$$

By Lemma 3.4.10, $E \otimes_A A/\mathfrak{p}$ injects into $\mathrm{End}_{A/\mathfrak{p}}(\phi[\mathfrak{p}])$. Therefore, $E_n \otimes_A A/\mathfrak{p} \cong (\mathbb{F}_\mathfrak{p})^{\oplus n}$ injects into $\mathrm{End}_{A/\mathfrak{p}}(\phi[\mathfrak{p}])$. This implies that $n \leq r^2$. Thus, $E = E_n$ for some $n \leq r^2$. □

Corollary 3.4.11 *The embedding* $\mathrm{End}(\phi) = \mathrm{Cent}_{K\{\tau\}}(\phi(A)) \hookrightarrow K((\tau^{-1}))$ *naturally extends to an injective ring homomorphism*

$$F_\infty \otimes_A \mathrm{End}(\phi) \longrightarrow K((\tau^{-1})).$$

This implies the following:

(1) $F_\infty \otimes_A \mathrm{End}(\phi)$ *is a division algebra over* F_∞ *of dimension* $\leq r^2$.
(2) *The absolute value* $\|\cdot\|$ *on* $K((\tau^{-1}))$ *induces an absolute value on* $F_\infty \otimes_A \mathrm{End}(\phi)$, *which is an extension of the normalized absolute value* $|\cdot|$ *on* F_∞.

Proof Let $F_\infty \mathrm{End}(\phi)$ be the F_∞-span of $\mathrm{End}(\phi)$ in $K((\tau^{-1}))$. Since $\phi(F_\infty)$ commutes with $\mathrm{End}(\phi)$ in $K((\tau^{-1}))$, $F_\infty \mathrm{End}(\phi)$ is an algebra over F_∞. There is a natural A-bilinear map $F_\infty \times \mathrm{End}(\phi) \to F_\infty \mathrm{End}(\phi)$ given by $(\alpha, u) \mapsto \phi_\alpha u$, which induces a surjective homomorphism of F_∞-algebras $F_\infty \otimes_A \mathrm{End}(\phi) \to F_\infty \mathrm{End}(\phi)$. On the other hand,

$\dim_{F_\infty} F_\infty \otimes_A \mathrm{End}(\phi)$

$\quad = \mathrm{rank}_A \mathrm{End}(\phi)$ \qquad (because $\mathrm{End}(\phi)$ is a free A-module of finite rank)

$\quad = \dim_{F_\infty} F_\infty \mathrm{End}(\phi)$ \qquad\qquad (by Lemma 3.4.9).

Hence, the map $F_\infty \otimes_A \text{End}(\phi) \to F_\infty \text{End}(\phi)$ is an isomorphism.   □

Let

$$\Delta_\phi := \text{Cent}_{K((\tau^{-1}))} \left(\phi(F_\infty) \right).$$

Proposition 3.4.12 *The algebra Δ_ϕ is a central division algebra over F_∞ of dimension r^2.*

Proof Suppose ϕ is defined by $\phi_T = \sum_{i=0}^{r} g_i \tau^i$, where $g_0 = t$ and $g_r \neq 0$. We can find a series $u = \sum_{j \leq 0} u_j \tau^j \in K[\![\tau^{-1}]\!]^\times$ such that $u^{-1} \phi_T u = \tau^r$. To see this, consider the coefficients of u as indeterminates. Equating the coefficients of τ^{-m} on both sides of $\phi_T u = u \tau^r$ for all $m \geq -r$, one obtains an infinite sequence of polynomial equations

$$g_r u_0^{q^r} = u_0$$

$$g_r u_{-1}^{q^r} + g_{r-1} u_0^{q^{r-1}} = u_{-1}$$

$$g_r u_{-2}^{q^r} + g_{r-1} u_{-1}^{q^{r-1}} + g_{r-2} u_0^{q^{r-2}} = u_{-2}$$

$$\vdots$$

$$g_r u_{-m}^{q^r} + g_{r-1} u_{-m+1}^{q^{r-1}} + \cdots + g_0 u_{-m+r} = u_{-m}$$

$$\vdots$$

This system can be solved recursively starting with $u_0 \neq 0$ (the solution of the system is not unique, but it always exists).

Now, choose $1/T$ to be the uniformizer of F_∞. Then, for u such that $u^{-1} \phi_T u = \tau^r$, we have

$$u^{-1} \phi(F_\infty) u = \mathbb{F}_q \left((\tau^{-r}) \right)$$

and

$$u^{-1} \Delta_\phi u = \text{Cent}_{K((\tau^{-1}))} \left(\mathbb{F}_q \left((\tau^{-r}) \right) \right)$$

$$= \text{Cent}_{K((\tau^{-1}))} \left(\tau^{-r} \right)$$

$$= \mathbb{F}_{q^r} \left((\tau^{-1}) \right).$$

Since $\mathbb{F}_{q^r} \left((\tau^{-1}) \right)$ is a central division algebra over $\mathbb{F}_q \left((\tau^{-r}) \right)$ of dimension r^2, the proposition follows.   □

Definition 3.4.13 A finite extension L/F is called *totally imaginary* if there is a unique place of L over ∞. A finite extension L/F is called *totally real* if ∞ splits completely in L.

Remarks 3.4.14

(1) Note that L/F is totally imaginary if and only if $L \otimes_F F_\infty$ is a field. Similarly, L/F is totally real if and only if $L \otimes_F F_\infty \cong F_\infty^{[L:F]}$. Both of these claims follow from Theorem 2.8.5.

(2) A quadratic extension L/F is either totally imaginary or totally real. When discussing quadratic extensions, we will usually say "imaginary" instead if "totally imaginary."

(3) The terminology is motivated by the terminology from algebraic number theory. A number field L is called *totally imaginary* if it cannot be embedded into $\mathbb{R}$. A number field L is called *totally real* if for each embedding of L into $\mathbb{C}$ the image lies in $\mathbb{R}$. If one represents $L \cong \mathbb{Q}[x]/(f(x))$ for some irreducible $f(x) \in \mathbb{Q}[x]$, then L is totally imaginary (respectively, totally real) if $f(x)$ has no real roots (respectively, all the roots of $f(x)$ are real). In terms of tensor products, L is totally imaginary if and only if we have an $\mathbb{R}$-algebra isomorphism $L \otimes_{\mathbb{Q}} \mathbb{R} \cong \mathbb{C}^{[L:\mathbb{Q}]/2}$ (in particular, $[L:\mathbb{Q}]$ must be even), and L is totally real if and only if $L \otimes_{\mathbb{Q}} \mathbb{R} \cong \mathbb{R}^{[L:\mathbb{Q}]}$.

(4) Note that for a function field $L \cong F[x]/(f(x))$ to be totally imaginary, we require $f(x)$ to be irreducible over F_∞; this is a stronger condition that the requirement that $f(x)$ has no roots in F_∞.

Corollary 3.4.15 *Let ϕ be a Drinfeld module over K of rank r. Let L be a subfield of $\mathrm{End}^\circ(\phi)$ containing F. Then L/F is a totally imaginary extension of degree dividing r. In particular, if $\mathrm{char}_A(K) = 0$, then $\mathrm{End}^\circ(\phi)$ is a totally imaginary field extension of F of degree dividing r.*

Proof Note that

$$F_\infty \otimes_F L \hookrightarrow F_\infty \otimes_F \mathrm{End}^\circ(\phi) \hookrightarrow \Delta_\phi.$$

Hence $F_\infty \otimes_F L$ is a subfield of Δ_ϕ containing F_∞.

Because $F_\infty \otimes_F L$ is a field, L/F is totally imaginary. Next, since $[L : F] = [F_\infty \otimes_F L : F_\infty]$, Proposition 3.4.12 and Theorem 1.7.10 imply that $[L : F]$ divides r. (One can also prove that $[L : F]$ divides r using Anderson motives; see Corollary 3.6.11.) When $\mathrm{char}_A(K) = 0$, Corollary 3.3.15 says that $\mathrm{End}^\circ(\phi)$ itself is a field. $\square$

Remarks 3.4.16

(1) One cannot deduce Theorem 3.4.1 directly from Lemma 3.4.10. The problem is that, in principle, $\mathrm{End}(\phi)$ might contain a nonzero divisible submodule M without torsion elements, such as F. In that case, $M \otimes_A A/(a) = 0$, which obviously injects into $\mathrm{End}_{A/(a)}(\phi[a])$, but M is not a finitely generated A-

module. The fact that $\mathrm{End}(\phi)$ is a discrete submodule of $K((\tau^{-1}))$ implicitly rules out the possibility of the existence of such divisible submodules.

(2) The discreteness of $\mathrm{End}(\phi)$ in $K((\tau^{-1}))$ is also very important for Corollary 3.4.11. To see this, it is instructive to consider the following analogous situation. Assume K is perfect and $\mathrm{char}_A(K) = \mathfrak{p} \neq 0$. Let $\phi\colon A \to K\{\tau\}$ be a Drinfeld module over K. Similar to (3.4.2), we have an embedding

$$A \overset{\phi}{\hookrightarrow} K\{\tau\} \hookrightarrow K((\tau)),$$

where $K((\tau))$ is the ring of twisted Laurent series $\sum_{n\in\mathbb{Z}} \kappa_n \tau^n$ with $\kappa_n = 0$ for all sufficiently *negative* n and multiplication is defined by the same commutation rule $\tau^n \kappa = \kappa^{q^n} \tau^n$ as in $K((\tau^{-1}))$. The ring $K((\tau))$ is a division ring. Moreover, $K((\tau))$ is complete with respect to the valuation ord_τ defined analogously to $\mathrm{ord}_{\tau^{-1}}$ on $K((\tau^{-1}))$. The embedding $\phi\colon A \to K((\tau))$ extends to $\phi\colon F \to K((\tau))$, and the valuation ord_τ extends the valuation ht on $K\{\tau\}$; cf. Lemma 3.1.11. Therefore, because $\mathfrak{p} \neq 0$, the valuation ord_τ induces the valuation $H(\phi) \cdot \deg(\mathfrak{p}) \cdot \mathrm{ord}_\mathfrak{p}$ on F; see Lemma 3.2.11. This implies that ϕ extends to an embedding $\phi\colon F_\mathfrak{p} \to K((\tau))$. Thus, there is a natural ring homomorphism $F_\mathfrak{p} \otimes_A \mathrm{End}(\phi) \to K((\tau))$. But this homomorphism is generally not injective and the absolute value on $K((\tau))$ does not extend to $F_\mathfrak{p} \otimes_A \mathrm{End}(\phi)$. In fact, in Chap. 4, we will prove in case of finite A-fields that $F_\mathfrak{p} \otimes_A \mathrm{End}(\phi)$ is a division ring if and only if ϕ is supersingular.

(3) Let $d \geq 1$ be a positive integer, and let $L_d = \mathbb{F}_{q^d}((x))$ be the field of Laurent series in x with coefficients in $\mathbb{F}_{q^d}$. Let $L_d[\tau]$ be the non-commutative ring of polynomials in τ with coefficients in L_d such that the multiplication is defined under the commutation rules $\tau x = x\tau$ and $\tau\alpha = \alpha^q \tau$ for all $\alpha \in \mathbb{F}_{q^d}$. Let

$$D_{d,e} = L_d[\tau]/(\tau^d - x^e).$$

It is not hard to show that $D_{d,e}$ is a central division algebra over $\mathbb{F}_q((x))$ of dimension d^2. It turns out that every central division algebra over $\mathbb{F}_q((x))$ is isomorphic to some $D_{d,e}$, and the image of e/d in $\mathbb{Q}/\mathbb{Z}$, denoted $\mathrm{inv}(D_{d,e})$, determines the isomorphism class of $D_{d,e}$; see [Rei03, (31.8)]. The number $\mathrm{inv}(D_{d,e}) = e/d + \mathbb{Z}$ is called the *invariant* of $D_{d,e}$. Note that we proved in Proposition 3.4.12 that

$$\Delta_\phi \cong \mathbb{F}_{q^r}((\tau^{-1})) \cong D_{r,-1}.$$

Hence

$$\mathrm{inv}(\Delta_\phi) = -\frac{1}{r}. \tag{3.4.3}$$

By Corollary 3.4.15, when $\text{char}_A(K) = 0$, $\text{End}^\circ(\phi)$ is a totally imaginary field extension of F of degree dividing r. We give examples which show that generally all possibilities for the degree of $\text{End}^\circ(\phi)$ over F allowed by this result do actually occur.

Example 3.4.17 Let U denote a root of $x^r - T \in F[x]$, and let $K = F(U)$. Note that $[K : F] = r$. Let $\gamma: A \hookrightarrow F \hookrightarrow K$ be the natural embedding. Define a Drinfeld module over K by

$$\phi_T = (U + \tau)^r.$$

Note that $u = U + \tau \in K\{\tau\}$ lies in $\text{End}_K(\phi)$ since $u\phi_T = u^{r+1} = \phi_T u$. Moreover, $F(u) \subset \text{End}^\circ(\phi)$ is a field extension isomorphic to K since $u^r = \phi_T$. This implies that $\text{End}^\circ(\phi) = F(u)$ has degree r over F, i.e., has the maximal possible degree (see also Example 3.3.2). Moreover, it is not hard to show that $A[U]$ is the integral closure of A in K. Hence $\text{End}(\phi) = A[u]$.

Example 3.4.18 Consider the Drinfeld module over F defined by

$$\phi_T = T + \tau + \tau^2.$$

We claim that $\text{End}(\phi) = A$. To prove this, one can argue as follows. If $\text{End}(\phi)$ is strictly larger than A, then $\text{rank}_A \text{End}(\phi) = 2$ since ϕ has rank 2. On the other hand, the complete finite list of the j-invariants[8] of rank-2 Drinfeld modules over F for which $\text{End}^\circ(\phi)$ is imaginary quadratic is given in [Sch97, Thm. 6]; part of that list is reproduced in Remark 7.5.27. The j-invariant of ϕ, which is equal to 1, is not on that list, so $\text{End}(\phi) = A$.

Example 3.4.19 Consider the Drinfeld module of rank 4 over $K = F(U)$ defined by

$$\phi_T = (U + \tau + \tau^2)^2, \qquad \text{where } U = T^2.$$

This module contains $u = U + \tau + \tau^2$ in its endomorphism ring, so $\phi(A)[u] \subseteq \text{End}(\phi)$. Since $u^2 = \phi_T$ in $K\{\tau\}$, we get $\phi(A)[u] \cong A[U]$, so $\text{End}(\phi)$ is an $A[U]$-module. On the other hand, $A[U]$ is isomorphic to the polynomial ring $\mathbb{F}_q[U]$, so $\text{End}(\phi)$ is a free $A[U]$-module. The rank of $\text{End}(\phi)$ over $A[U]$ is 1 or 2 because

$$2\,\text{rank}_{A[U]}(\text{End}(\phi)) = \text{rank}_A(\text{End}(\phi)) \leq 4.$$

If $\text{rank}_A \text{End}(\phi) = 4$, then $\text{End}(\phi)$ contains an element $w \notin \phi(A)[u]$. In that case, w commutes with u, so w is in the endomorphism ring of the Drinfeld $\mathbb{F}_q[U]$-module ψ defined by $\psi_U = U + \tau + \tau^2$. But by the previous example, w must be in $\mathbb{F}_q[u]$, which leads to a contradiction. Hence, $\text{End}(\phi) \cong A[U]$ has rank 2 over A.

---

[8] The j-invariants will be defined in Sect. 3.8.

Definition 3.4.20 Let ϕ be a Drinfeld module of rank r over a field K of A-characteristic 0. If $\text{End}(\phi)$ has rank r over A (i.e., the maximal possible rank), then we say ϕ has *complex multiplication*, or *CM* for short, by the ring $\text{End}(\phi)$.

Remark 3.4.21 The terminology "complex multiplication" is used to emphasize the analogy with elliptic curves having complex multiplication, although complex numbers are not directly present in the setting of Drinfeld modules. Drinfeld modules with complex multiplication have many special properties, and they have important applications to class field theory of F. We will discuss them in detail in Sect. 7.5.

Definition 3.4.22 Let D be a finite dimensional algebra over F. An A-*order* (or just *order*) in D is an A-subalgebra O of D such that O is finitely generated as an A-module and contains an F-basis of D. Note that any A-order O in D is a free A-module of rank equal to $\dim_F D$. Moreover, if D is a field extension of F and B is the integral closure of A in D, then O is an A-order in D if and only if O is an A-subalgebra of B with the same unity element and B/O has finite cardinality (see Exercise 3.4.8).

Remark 3.4.23 Note that $\text{End}(\phi)$ is an A-order in $\text{End}^\circ(\phi)$, but generally it does not have to be maximal, i.e., $\text{End}(\phi)$ might be strictly contained in another A-order in $\text{End}^\circ(\phi)$; cf. Sect. 7.4.

In practice, it is useful to have explicit characterizations of totally imaginary extensions of F in terms of the polynomials defining those extensions. We conclude this section with two examples of this.

Example 3.4.24 The simplest example is when L/F is quadratic and the characteristic of F is odd. Hence, assume $f(x) = ax^2 + bx + c \in A[x]$ is irreducible and L is the splitting field of $f(x)$. If the characteristic of F is odd, then, by the quadratic formula, L is the splitting field of $x^2 - (b^2 - 4ac)$. Let αT^n be the leading term of the polynomial $b^2 - 4ac$. Since $\text{ord}_\infty(\sqrt{b^2 - 4ac}) = -n/2$, the extension L/F is ramified at ∞ if and only if n is odd. On the other hand, if n is even, then dividing both sides of $x^2 = b^2 - 4ac$ by T^n, we see that L is the splitting field of $x^2 - (\alpha + d)$, where $\text{ord}_\infty(d) \geq 0$. Applying Hensel's lemma, one deduces that ∞ is inert in L/F if and only if α is not a square in $\mathbb{F}_q$, and ∞ splits in L/F if and only if α is a square in $\mathbb{F}_q$. Thus, L/F is imaginary if and only if either the degree of $b^2 - 4ac$ is odd or the degree of $b^2 - 4ac$ is even, but its leading coefficient is not a square in $\mathbb{F}_q$. (For a generalization of this example, see Exercise 3.4.7.)

Example 3.4.25 In the previous example, we assumed that the characteristic of F is odd. If the characteristic of F is 2, then a quadratic extension L/F is either purely inseparable or an Artin–Schreier extension. In this example, we consider a more general version of this situation. Let L/F be a cyclic extension of degree p, where p is the characteristic of F.[9] By Theorem 1.5.21, L/F is the splitting field of an

---

[9] If L/F is purely inseparable, then L/F is totally imaginary since the place ∞ totally ramifies in L.

Artin–Schreier polynomial $f(x) = x^p - x - \frac{a}{b}$, where $a, b \in A$ are relatively prime. Using the division algorithm, we write $a = cb + d$ with $c, d \in A$, $\deg(d) < \deg(b)$. Then $\frac{a}{b} = c + \frac{d}{b}$. Suppose $\deg(c) > 0$ and $p \mid \deg(c)$. Then $c = \alpha T^{ps} +$ lower degree terms. Let $\beta \in \mathbb{F}_q$ be such that $\beta^p = \alpha$ in $\mathbb{F}_q$. After a change of variables $x \mapsto x - \beta T^s$, we obtain the Artin–Schreier polynomial $x^p - x - (\frac{d}{b} + c_1)$ with $\deg(c_1) < \deg(c)$ whose splitting field is L. Repeating this process finitely many times, we may assume that L/F is the splitting field of the polynomial

$$f(x) = x^p - x - \left(\frac{d}{b} + c \right),$$

where $\deg(d) < \deg(b)$, and either $c \in \mathbb{F}_q$ or $p \nmid \deg(c)$. Now, consider $f(x)$ over F_∞. Note that $d/b \in A_\infty$. If $c \in \mathbb{F}_q$, then reducing $f(x)$ modulo $(1/T)$, we obtain $x^p - x - c$ over $\mathbb{F}_q$. If $c \neq \alpha^p - \alpha$ for all $\alpha \in \mathbb{F}_q$, then $x^p - x - c$ is irreducible over $\mathbb{F}_q$, so ∞ is inert in L/F by Hensel's lemma. On the other hand, if $c = \alpha^p - \alpha$ for some $\alpha \in \mathbb{F}_q$, then $f(x)$ splits completely over $\mathbb{F}_q$, so ∞ splits completely in L/F. Finally, suppose $c \notin \mathbb{F}_q$ and $p \nmid \deg(c)$. The Newton polygon of $f(x)$ has one line segment joining $(0, -\deg(c))$ to $(p, 0)$. The slope is $\deg(c)/p$, so the valuation of any root of $f(x)$ is $-\deg(c)/p$. This implies that ∞ totally ramifies in L/F. Overall, we conclude that L/F is totally imaginary, unless $c = \alpha^p - \alpha$ for some $\alpha \in \mathbb{F}_q$.

Exercises

3.4.1 Prove the following:

(a) Any number field that is Galois over $\mathbb{Q}$ must be either totally real or totally imaginary.
(b) An abelian extension of $\mathbb{Q}$ is either totally real or is a CM field (i.e., a totally imaginary quadratic extension of a totally real field).

3.4.2 Let M be a left $K[T, \tau]$-module finitely generated both over $K[T]$ and $K\{\tau\}$. Prove that

$$\{m \in M \mid \dim_K(K[T]m) < \infty\} = \{m \in M \mid \dim_K(K\{\tau\}m) < \infty\}.$$

Conclude that M is free over $K[T]$ of finite rank if and only if it is free over $K\{\tau\}$ of finite rank.

3.4.3 Let ϕ and ψ be isogenous Drinfeld modules over K. Show that there is an isomorphism of division F-algebras $\mathrm{End}_K^\circ(\phi) \cong \mathrm{End}_K^\circ(\psi)$.

3.4.4 Show that $\mathrm{End}_K^\circ(\phi)$ is a central division algebra of dimension m^2 over a finite extension $\widetilde{F}$ of F and $m \cdot [\widetilde{F} : F]$ divides r.

3.4.5 Assume $\mathrm{char}_A(K) = 0$. Let ϕ be a Drinfeld module of rank r over K. Let $u \in \mathrm{End}(\phi)$ and $u_0 = \partial u \in \overline{K}$.

(a) Let $m_{u,F}(x)$ be the minimal polynomial of u over F. Show that $m_{u,F}(u_0) = 0$. Conclude that u_0 is integral over A and its minimal polynomial over A is $m_{u,F}(x)$.
(b) Prove that u_0 uniquely determines $\deg_\tau(u)$.
(c) Show that $u \in L\{\tau\}$, where $L = K(u_0)$.

3.4.6 Assume $\mathrm{char}_A(K) = 0$. Let ϕ be a Drinfeld module of rank r over K. Prove the following statements:

(a) $\mathrm{End}_K^\circ(\phi)$ is a finite separable extension of F if and only if $\mathrm{End}^\circ(\phi)$ is a finite separable extension of F.
(b) If the algebraic closure of F in K is separable over F, then $\mathrm{End}_K^\circ(\phi)$ is a finite separable extension of F.
(c) If $\mathrm{End}_K^\circ(\phi)$ is not separable over F, then ϕ cannot be defined over F.

3.4.7 Let $L = F(\sqrt[\ell]{d})$ be a Kummer extension of F of degree ℓ, where ℓ is a prime divisor of $q - 1$ and $d \in A$. Let $\delta = \deg(d)$, and let α be its leading coefficient. Prove the following:

(i)   The place ∞ is totally ramified in L/F if and only if $\ell \nmid \delta$.
(ii) The place ∞ is inert in L/F if and only if $\ell \mid \delta$ and $\alpha \notin (\mathbb{F}_q^\times)^\ell$.
(iii) The place ∞ splits completely in L/F if and only if $\ell \mid \delta$ and $\alpha \in (\mathbb{F}_q^\times)^\ell$.

3.4.8 Let L be a field extension of F of degree n and B be the integral closure of A in L.

(a) Prove that O is an A-order in L if and only if O is an A-subalgebra of B with the same unity element and B/O has finite cardinality.
(b) Assume $q = p^m$ and $m \geq 2$. Show that for any $a \in A$ of positive degree and $1 \leq s < m$, $\mathbb{F}_{p^s} + aB$ is a subring of B of finite index which is not an A-order.

3.4.9 Let L be a field extension of F of degree n, let B be the integral closure of A in L, and let O be an A-order in L.

(a) Deduce from Theorem 1.2.4 that there are uniquely determined monic polynomials

$$a_1, \ldots, a_{n-1} \in A \quad \text{such that}$$

$$B/O \cong A/a_1 A \oplus \cdots \oplus A/a_{n-1}A, \quad \text{and} \quad a_1 \mid a_2 \mid \cdots \mid a_{n-1}.$$

(b) The *conductor* of O in B is

$$\mathfrak{C} = \{c \in L \mid cB \subseteq O\}.$$

Prove that $\mathfrak{C}$ is the largest ideal in B which is also an ideal of O. Moreover, $\mathfrak{C} \cap A = (a_{n-1})$.

(c) Let $I = O\alpha_1 + \cdots + O\alpha_m$ for some $\alpha_1, \ldots, \alpha_m \in L$, i.e., I is a fractional ideal of O. Prove that

$$O' = \{c \in L \mid cI \subseteq I\}$$

is also an A-order in L. The ideal I is said to be *proper* if $O' = O$. Conclude that the conductor $\mathfrak{C}$ is not a proper ideal of O unless $O = B$.

(d) Let $0 \neq a \in A$. Show that $O = A + aB$ is an A-order, $\mathfrak{C} = aB$, but $\mathfrak{C}$ is not principal in O.

(e) Prove that if $n = 2$, then $O = A + aB$ for a uniquely determined monic $a \in A$.

(f) Let $L = \mathbb{F}_{q^2}(T)$, $B = \mathbb{F}_{q^2}[T]$, and $O = A + TB$. Let $\alpha \in \mathbb{F}_{q^2}$ be such that $\mathbb{F}_{q^2} = \mathbb{F}_q(\alpha)$. Show that $\mathfrak{C} = TA + T\alpha A$ is a maximal O-ideal and that $I = T^2 A + T\alpha A$ is a O-ideal contained in $\mathfrak{C}$ but I is not divisible by $\mathfrak{C}$, i.e., there is no O-ideal J such that $I = \mathfrak{C}J$.

3.5 Torsion Points

Let $\gamma: A \to K$ be an A-field, and let ϕ be a Drinfeld module over K of rank r. In this section, we determine the module structure of torsion points of ϕ. We then combine the $\mathfrak{p}$-power torsion points into an $A_{\mathfrak{p}}$-module $T_{\mathfrak{p}}(\phi)$, called the $\mathfrak{p}$-adic Tate module of ϕ, and consider the action of G_K and the endomorphism ring of ϕ on $T_{\mathfrak{p}}(\phi)$.

For $0 \neq a \in A$, we have

$$\phi_a = \gamma(a) + g_1(a)\tau + \cdots + g_n(a)\tau^n, \quad n = r \cdot \deg(a).$$

Recall that the a-division polynomial of ϕ is

$$\phi_a(x) = \gamma(a)x + g_1(a)x^q + \cdots + g_n(a)x^{q^n},$$

and $\phi[a]$ denotes the $\mathbb{F}_q$-vector space of the (distinct) roots of $\phi_a(x)$. Note that $\phi[a]$ is naturally an A-module via

$$b \circ \alpha := \phi_b(\alpha), \quad \text{where} \quad b \in A, \alpha \in \phi[a].$$

Indeed, to see that $b \circ \alpha \in \phi[a]$, we compute

$$\phi_a(b \circ \alpha) = \phi_a(\phi_b(\alpha)) = \phi_b(\phi_a(\alpha)) = \phi_b(0) = 0.$$

Thus, $\phi[a]$ is a finite A-module, and the basic problem that we want to solve first is to determine its elementary divisors.

Lemma 3.5.1 *Assume $a, b \in A$ are relatively prime nonzero elements. Then, as A-modules, we have*

$$\phi[ab] = \phi[a] \times \phi[b].$$

Proof Since ϕ_{ab} annihilates both $\phi[a]$ and $\phi[b]$, these are submodules of $\phi[ab]$. If $\alpha \in \phi[a] \cap \phi[b]$, then $\alpha \in \phi[c_1 a + c_2 b]$ for all $c_1, c_2 \in A$. Since a and b are assumed to be relatively prime, we can choose c_1 and c_2 such that $c_1 a + c_2 b = 1$. Hence $\alpha \in \phi[1] = 0$. This implies that $\phi[a] \times \phi[b]$ is a submodule of $\phi[ab]$.

Let $\mathfrak{p} = \mathrm{char}_A(K)$. The claim now follows by comparing the order of $\phi[ab]$ with the order of its submodule $\phi[a] \times \phi[b]$:

$$
\begin{aligned}
\log_q(\#\phi[ab]) &= r \deg(ab) - \mathrm{ht}(\phi_{ab}) \\
&= r \deg(ab) - H(\phi) \, \mathrm{ord}_{\mathfrak{p}}(ab) \deg(\mathfrak{p}) \\
&= (r \deg(a) - H(\phi) \, \mathrm{ord}_{\mathfrak{p}}(a) \deg(\mathfrak{p})) \\
&\quad + (r \deg(b) - H(\phi) \, \mathrm{ord}_{\mathfrak{p}}(b) \deg(\mathfrak{p})) \\
&= \log_q(\#\phi[a]) + \log_q(\#\phi[b]),
\end{aligned}
$$

where we used Lemma 3.2.11 to relate $\mathrm{ht}(\phi_{ab})$ to $\mathrm{ord}_{\mathfrak{p}}(ab)$.   □

It is easy to see that for $\beta \in \mathbb{F}_q^\times$ and $a \in A$, we have $\phi[\beta a] = \phi[a]$. Hence, given a nonzero ideal $\mathfrak{n} \lhd A$, one can define $\phi[\mathfrak{n}]$ to be the torsion points of any generator of $\mathfrak{n}$; by our usual convention, we choose the generator to be monic.

Theorem 3.5.2 *Let $\mathfrak{p}$ be a prime of A and $n \geq 1$ be an integer.*

(1) If $\mathfrak{p} \neq \mathrm{char}_A(K)$, then

$$\phi[\mathfrak{p}^n] \cong \prod_{i=1}^{r} A/\mathfrak{p}^n.$$

(2) If $\mathfrak{p} = \mathrm{char}_A(K)$, then

$$\phi[\mathfrak{p}^n] \cong \prod_{i=1}^{r-H(\phi)} A/\mathfrak{p}^n.$$

Proof By Theorem 1.2.4,

$$\phi[\mathfrak{p}^n] \cong A/\mathfrak{p}_1^{n_1} \times \cdots \times A/\mathfrak{p}_m^{n_m}$$

for some integer $m \geq 1$ and positive powers of primes $\mathfrak{p}_1^{n_1}, \ldots, \mathfrak{p}_m^{n_m}$ (not necessarily distinct). As in the proof of Lemma 3.5.1, we see that $\phi[\mathfrak{q}] \cap \phi[\mathfrak{p}^n] = 0$ if $\mathfrak{q}$ is a prime different from $\mathfrak{p}$. Hence no element of $\phi[\mathfrak{p}^n]$ is killed by a prime $\mathfrak{q} \neq \mathfrak{p}$. This implies that the only elementary divisors in the decomposition of $\phi[\mathfrak{p}^n]$ are powers of $\mathfrak{p}$:

$$\phi[\mathfrak{p}^n] \cong \prod_{i=1}^{m} A/\mathfrak{p}^{n_i}. \tag{3.5.1}$$

Since $\mathfrak{p}^n$ annihilates the left-hand side, we must have $n_1, \ldots, n_m \leq n$. Extracting the submodule killed by $\mathfrak{p}$ on both sides of (3.5.1), we get

$$\phi[\mathfrak{p}] = \prod_{i=1}^{m} A/\mathfrak{p}.$$

Now, comparing the orders of $\phi[\mathfrak{p}]$ and $\prod_{i=1}^{m} A/\mathfrak{p}$, we get $m = r - H$, where $H = 0$ if $\mathfrak{p} \neq \text{char}_A(K)$ and $H = H(\phi)$ if $\mathfrak{p} = \text{char}_A(K)$. Finally, we compute the orders of both sides of (3.5.1):

$$\log_q(\#\phi[\mathfrak{p}^n]) = n \cdot \deg(\mathfrak{p}) \cdot (r - H),$$

$$\log_q(\# \prod_{i=1}^{r-H} A/\mathfrak{p}^{n_i}) = \deg(\mathfrak{p}) \sum_{i=1}^{r-H} n_i.$$

Hence $\sum_{i=1}^{r-H} n_i = n(r - H)$. Since $1 \leq n_1, \ldots, n_m \leq n$, we must have $n_i = n$ for all i. $\qquad\square$

Corollary 3.5.3 *Assume $a \in A$ is not divisible by $\text{char}_A(K)$. Then*

$$\phi[a] \cong \prod_{i=1}^{r} A/aA.$$

The modules $\phi[\mathfrak{p}^n]$ can be fit together for varying n. Observe that the action by $\phi_{\mathfrak{p}}$ gives natural surjective homomorphisms

$$\phi[\mathfrak{p}^{n+1}] \xrightarrow{\phi_{\mathfrak{p}}} \phi[\mathfrak{p}^n], \quad \alpha \longmapsto \phi_{\mathfrak{p}}(\alpha).$$

Taking the inverse limit with respect to these maps, we obtain the $\mathfrak{p}$-*adic Tate module* of ϕ:

$$T_{\mathfrak{p}}(\phi) := \varprojlim_{n} \phi[\mathfrak{p}^n] \tag{3.5.2}$$

$$\cong \begin{cases} \varprojlim_{n}(A/\mathfrak{p}^n)^{\oplus r} \cong A_{\mathfrak{p}}^{\oplus r}, & \text{if } \mathfrak{p} \neq \text{char}_A(K); \\ \varprojlim_{n}(A/\mathfrak{p}^n)^{\oplus(r-H(\phi))} \cong A_{\mathfrak{p}}^{\oplus(r-H(\phi))}, & \text{if } \mathfrak{p} = \text{char}_A(K). \end{cases}$$

The inverse limit topology on $T_{\mathfrak{p}}(\phi)$ is equivalent to the $\mathfrak{p}$-adic topology it gains by being an $A_{\mathfrak{p}}$-module.

Let $u: \phi \to \psi$ be an isogeny of Drinfeld modules defined over K. Then u induces homomorphisms $\phi[\mathfrak{p}^n] \to \psi[\mathfrak{p}^n]$ for all $n \geq 1$, and hence it induces an $A_{\mathfrak{p}}$-linear map

$$u_{\mathfrak{p}}: T_{\mathfrak{p}}(\phi) \longrightarrow T_{\mathfrak{p}}(\psi).$$

We thus obtain a natural homomorphism

$$\mathrm{Hom}_K(\phi, \psi) \longrightarrow \mathrm{Hom}_{A_{\mathfrak{p}}}(T_{\mathfrak{p}}(\phi), T_{\mathfrak{p}}(\psi)). \qquad (3.5.3)$$

If $\phi = \psi$, then the map

$$\mathrm{End}_K(\phi) \longrightarrow \mathrm{End}_{A_{\mathfrak{p}}}(T_{\mathfrak{p}}(\phi))$$

is a homomorphism of rings. If $\mathfrak{p} \neq \mathrm{char}_A(K)$, then the homomorphism (3.5.3) is injective, because if $u_{\mathfrak{p}}: T_{\mathfrak{p}}(\phi) \to T_{\mathfrak{p}}(\psi)$ is the zero map, then $\phi[\mathfrak{p}^n]$ is a subset of $\ker(u)$ for all n, which is obviously false once $\#\phi[\mathfrak{p}^n]$ is larger than $\deg u(x)$. The next theorem says that the homomorphism (3.5.3) remains injective even after we extend it linearly to $A_{\mathfrak{p}}$.

Theorem 3.5.4 *Let ϕ and ψ be Drinfeld modules over K, and let $\mathfrak{p} \neq \mathrm{char}_A(K)$ be a prime of A. The natural map*

$$\mathrm{Hom}_K(\phi, \psi) \otimes_A A_{\mathfrak{p}} \longrightarrow \mathrm{Hom}_{A_{\mathfrak{p}}}(T_{\mathfrak{p}}(\phi), T_{\mathfrak{p}}(\psi)), \qquad u \longmapsto u_{\mathfrak{p}},$$

is injective. Moreover, the cokernel of this homomorphism is torsion-free.

Proof Let $u \in \mathrm{Hom}_K(\phi, \psi) \otimes_A A_{\mathfrak{p}}$. Suppose $u_{\mathfrak{p}} = 0$. Then u maps $\phi[\mathfrak{p}^n]$ to 0 for all $n \geq 1$. By Theorem 3.4.1, $\mathrm{Hom}_K(\phi, \psi)$ is free of finite rank, so we can choose an A-basis $\{v_1, \ldots, v_s\}$ of $\mathrm{Hom}_K(\phi, \psi)$, which is also an $A_{\mathfrak{p}}$-basis for $\mathrm{Hom}_K(\phi, \psi) \otimes_A A_{\mathfrak{p}}$. We expand $u = \alpha_1 v_1 + \cdots + \alpha_s v_s$ with $\alpha_1, \ldots, \alpha_s \in A_{\mathfrak{p}}$. Fix some $n \geq 1$ and choose $a_i \in A$ such that $a_i \equiv \alpha_i \pmod{\mathfrak{p}^n}$ for all $1 \leq i \leq s$. Then $u = w + \mathfrak{p}^n v$, where

$$w = a_1 v_1 + \cdots + a_s v_s \in \mathrm{Hom}_K(\phi, \psi),$$

and $v \in \mathrm{Hom}_K(\phi, \psi) \otimes_A A_{\mathfrak{p}}$. Obviously, $\mathfrak{p}^n v$ maps $\phi[\mathfrak{p}^n]$ to 0, so w maps $\phi[\mathfrak{p}^n]$ to 0. This implies that w maps to 0 under the reduction map $\mathrm{Hom}_K(\phi, \psi) \to \mathrm{Hom}_{A/\mathfrak{p}^n}(\phi[\mathfrak{p}^n], \psi[\mathfrak{p}^n])$. On the other hand, the natural homomorphism

$$\mathrm{Hom}_K(\phi, \psi) \otimes_A A/\mathfrak{p}^n \longrightarrow \mathrm{Hom}_{A/\mathfrak{p}^n}(\phi[\mathfrak{p}^n], \psi[\mathfrak{p}^n])$$

is injective by Lemma 3.4.10. Therefore, $a_i \equiv 0 \pmod{\mathfrak{p}^n}$ for all i, and hence $\alpha_i \in \mathfrak{p}^n A_{\mathfrak{p}}$. Since n was arbitrary, we conclude that $\alpha_i = 0$, and hence $u = 0$.

The proof of the second statement is similar. Let $u \in \mathrm{Hom}_K(\phi, \psi) \otimes_A A_{\mathfrak{p}}$. Suppose $u_{\mathfrak{p}} = \mathfrak{p}w_{\mathfrak{p}}$ for some $w_{\mathfrak{p}} \in \mathrm{Hom}_{A_{\mathfrak{p}}}(T_{\mathfrak{p}}(\phi), T_{\mathfrak{p}}(\psi))$. Then $u_{\mathfrak{p}}$ maps $\phi[\mathfrak{p}]$ to 0. As before, this implies that $\alpha_i \equiv 0 \pmod{\mathfrak{p}}$ for all i, so $u = \mathfrak{p}w$ for some $w \in \mathrm{Hom}_K(\phi, \psi) \otimes_A A_{\mathfrak{p}}$. $\qquad\qquad\qquad\qquad\qquad\qquad\qquad\qquad\qquad\qquad\qquad\qquad\qquad\quad\square$

Remark 3.5.5 In general, an injective homomorphism $M_1 \to M_2$ from a free finite rank A-module M_1 into a free finite rank $A_{\mathfrak{p}}$-module M_2 need not remain injective after extending M_1 to $M_1 \otimes_A A_{\mathfrak{p}}$. For example, let $M_1 = A \times A$ and $M_2 = A_{\mathfrak{p}}$. Fix some $\pi \in A_{\mathfrak{p}}$ such that $\pi \notin A_{\mathfrak{p}} \cap F$. Define a homomorphism $A \times A \to A_{\mathfrak{p}}$ by $(1, 0) \mapsto 1$, $(0, 1) \mapsto \pi$. It is easy to see that this is an injection, which is no longer an injection after tensoring the left-hand side with $A_{\mathfrak{p}}$. Thus, the statement of Theorem 3.5.4 is not a consequence of the injectivity of (3.5.3) and some general abstract algebra facts.

The $\mathfrak{p}$-adic Tate module can be extended to a vector space over $F_{\mathfrak{p}}$,

$$V_{\mathfrak{p}}(\phi) := T_{\mathfrak{p}}(\phi) \otimes_{A_{\mathfrak{p}}} F_{\mathfrak{p}}.$$

From Theorem 3.5.4, for $\mathfrak{p} \neq \mathrm{char}_A(\phi)$, we get an injection

$$\mathrm{End}_K^\circ(\phi) \otimes_F F_{\mathfrak{p}} \hookrightarrow \mathrm{End}_{F_{\mathfrak{p}}}(V_{\mathfrak{p}}(\phi)) \cong \mathrm{Mat}_r(F_{\mathfrak{p}}). \qquad (3.5.4)$$

Assume that $\mathrm{char}_A(K) \nmid a$. Recall that in this case the polynomial $\phi_a(x) \in K[x]$ is separable. The A-module $\phi[a]$ is naturally equipped with an action of the absolute Galois group $G_K := \mathrm{Gal}(K^{\mathrm{sep}}/K)$ of K since each element of G_K permutes the roots of $\phi_a(x)$. This action of G_K commutes with the action of A on $\phi[a]$, since for $\sigma \in G_K$, $b \in A$, $\alpha \in \phi[a]$, we have $\phi_b(\sigma\alpha) = \sigma(\phi_b(\alpha))$. Thus, we obtain a representation

$$\rho_{\phi,a} : G_K \longrightarrow \mathrm{Aut}_A(\phi[a]) \cong \mathrm{GL}_r(A/aA). \qquad (3.5.5)$$

The Galois group of $K(\phi[a])/K$ is isomorphic to the image of $\rho_{\phi,a}$.

Example 3.5.6 Let $q = 2$, and let ϕ be the Drinfeld module over F defined by $\phi_T = T + \tau + \tau^2$. Consider $\phi[T]$, i.e., the roots of $\phi_T(x) = x^4 + x^2 + Tx$. Let $f(x) = x^3 + x + T$, so that $\phi_T(x) = xf(x)$. The cubic $f(x)$ is irreducible by Gauss' lemma since $f(T) \neq 0$ and $f(1) \neq 0$. The *quadratic resolvent* of $f(x)$ is $x^2 + Tx + 1$, which is irreducible over F as $f(1) \neq 0$. Therefore, the Galois group of $f(x)$ is the symmetric group S_3 over a set of three elements. Finally, it is easy to show that $S_3 \cong \mathrm{GL}_2(\mathbb{F}_2)$. Thus,

$$\mathrm{Gal}(F(\phi[T])/F) \cong \mathrm{GL}_2(\mathbb{F}_2) \cong \mathrm{GL}_2(A/TA),$$

and the representation $\rho_{\phi,T}$ is surjective.

Now let $\mathfrak{p}$ be a prime different from $\mathrm{char}_A(K)$. The action of G_K on each $\phi[\mathfrak{p}^n]$ commutes with the action of $\phi_\mathfrak{p}$ used to form the inverse limit defining the $\mathfrak{p}$-adic Tate module of ϕ, so G_K also acts on $T_\mathfrak{p}(\phi)$ and $V_\mathfrak{p}(\phi)$. This gives a representation

$$\hat{\rho}_{\phi,\mathfrak{p}} : G_K \longrightarrow \mathrm{Aut}_{A_\mathfrak{p}}(T_\mathfrak{p}(\phi)) \cong \mathrm{GL}_r(A_\mathfrak{p}). \tag{3.5.6}$$

Next, given an isogeny $u \in \mathrm{Hom}_K(\phi, \psi)$, it is clear that the map $u_\mathfrak{p} : T_\mathfrak{p}(\phi) \to T_\mathfrak{p}(\psi)$ induced by u commutes with the action of G_K, so the image of (3.5.3) actually lies in the $A_\mathfrak{p}$-submodule $\mathrm{Hom}_{A_\mathfrak{p}}(T_\mathfrak{p}(\phi), T_\mathfrak{p}(\psi))$ of those homomorphisms $T_\mathfrak{p}(\phi) \to T_\mathfrak{p}(\psi)$ which commute with the action of G_K. Theorem 3.5.4 can be restated more precisely as the fact that there is a natural injective homomorphism

$$\mathrm{Hom}_K(\phi, \psi) \otimes_A A_\mathfrak{p} \longrightarrow \mathrm{Hom}_{A_\mathfrak{p}[G_K]}(T_\mathfrak{p}(\phi), T_\mathfrak{p}(\psi)) \tag{3.5.7}$$

with a torsion-free cokernel. In some important special cases, such as when K is a finite field or a finite extension of F, the homomorphism (3.5.7) is in fact a bijection (cf. [Yu95, Tag95b]); we will prove this for finite fields in Chap. 4. But the map (3.5.7) is certainly not always a bijection. For a trivial example, take $\phi = \psi$ over an algebraically closed field with $\mathrm{End}(\phi) = A$. Then the left-hand side of (3.5.7) is $A_\mathfrak{P}$, whereas the right-hand side has dimension r^2 over $A_\mathfrak{P}$.

We end this section by proving a group-theoretic fact which is useful when passing from the action of G_K or $\mathrm{End}_K(\phi)$ on $T_\mathfrak{p}(\phi)$ to their action on $\phi[\mathfrak{p}^n]$; cf. Exercise 3.5.2.

Given a matrix M in $\mathrm{GL}_r(A)$ or $\mathrm{GL}_r(A_\mathfrak{p})$, we can reduce its entries modulo $\mathfrak{p}^n$ to obtain a matrix $\overline{M}$ in $\mathrm{Mat}_r(A/\mathfrak{p}^n)$. From the explicit formulas describing the entries of the product of two matrices M_1 and M_2, it is easy to see that $\overline{M_1 \cdot M_2} = \overline{M_1} \cdot \overline{M_2}$. Therefore, the reduction modulo $\mathfrak{p}^n$ gives a homomorphism

$$\mathrm{GL}_r(A) \longrightarrow \mathrm{GL}_r(A/\mathfrak{p}^n), \tag{3.5.8}$$

and similarly for $\mathrm{GL}_r(A_\mathfrak{p}) \longrightarrow \mathrm{GL}_r(A/\mathfrak{p}^n)$.

Definition 3.5.7 Denote the kernel of the homomorphism (3.5.8) by

$$\Gamma(\mathfrak{p}^n) = \left\{ M \in \mathrm{GL}_r(A) \mid M \equiv 1 \ (\mathrm{mod} \ \mathfrak{p}^n) \right\}.$$

This group is the *principal congruence subgroup of level* $\mathfrak{p}^n$. By abuse of notation, we denote the kernel of $\mathrm{GL}_r(A_\mathfrak{p}) \longrightarrow \mathrm{GL}_r(A/\mathfrak{p}^n)$ also by $\Gamma(\mathfrak{p}^n)$.

Proposition 3.5.8 *Let G be a finite subgroup of $\Gamma(\mathfrak{p})$. Let p be the characteristic of A. Then*

(1) G is a p-group, i.e., the order of G is a power of p.
(2) For any $g \in G$, all the eigenvalues of g as an $r \times r$-matrix are equal to 1.
(3) When we consider G as acting on $F^{\oplus r}$ or $F_\mathfrak{p}^{\oplus r}$, the elements of G have a common eigenvector.

Proof The proof is the same whether $\Gamma(\mathfrak{p})$ is the principal congruence subgroup of $GL_r(A)$ or $GL_r(A_\mathfrak{p})$, so we only treat the first case.

(1) Let $g \in G$. Because G is a finite group, its elements have finite order, so $g^n = 1$ for some nonzero integer n (depending on g). We claim that the smallest such n is a power of p. If this is not the case, then writing $n = p^s n'$ with $n' > 1$ coprime to p and replacing g by g^{p^s}, we may assume that n is coprime to p. Since $g \in \Gamma(\mathfrak{p})$, we can write $g = 1 + \mathfrak{p}M$ for some $M \in \text{Mat}_r(A)$. Now

$$1 = (1 + \mathfrak{p}M)^n = 1 + \sum_{i=1}^{n} \binom{n}{i} \mathfrak{p}^i M^i$$

$$= 1 + n\mathfrak{p}M + \mathfrak{p}^2 M' \qquad \text{for some } M' \in \text{Mat}_r(A).$$

Considering this equality modulo $\mathfrak{p}^2$, we see that $M = \mathfrak{p}M_1$ for some $M_1 \in \text{Mat}_r(A)$ (observe that n is a unit in A). Hence,

$$1 = (1 + \mathfrak{p}M)^n = (1 + \mathfrak{p}^2 M_1)^n = 1 + n\mathfrak{p}^2 M_1 + \mathfrak{p}^4 M''$$

for some $M'' \subset \text{Mat}_r(A)$. Now, considering this equality modulo $\mathfrak{p}^4$, we conclude that $M_1 = \mathfrak{p}^2 M_2$ for some $M_2 \in \text{Mat}_r(A)$. Repeating this argument over and over, we get that $M \in \mathfrak{p}^c \text{Mat}_r(A)$ for all $c \geq 0$. This is possible only if $M = 0$, so $g = 1$. This proves that all elements of G have p-power orders, so G itself must be a p-group by Cauchy's theorem.

(2) Now suppose $g^{p^s} = 1$. Again writing $g = 1 + \mathfrak{p}M$, we have $1 = 1 + \mathfrak{p}^{p^s} M^{p^s}$. Thus, M is a nilpotent matrix. It is a well-known fact from linear algebra that a nilpotent matrix is conjugate to an upper-triangular matrix with zeros on the main diagonal (here one can consider g and M as elements of $\text{Mat}_r(F)$); cf. [DF04, p. 502]. Therefore, g is conjugate to an upper-triangular matrix with 1's on the main diagonal. From this, it is clear that all the eigenvalues of g are equal to 1.

(3) Let G be a finite p-group acting on a finite dimensional F-vector space V. Assume the eigenvalues of every $g \in G$ are all equal to 1. We use induction on the order of G and the dimension of V. If V has dimension 1 or $G = 1$, then the claim of (3) is clear. Suppose we proved (3) for any p-group G' acting on V with $\#G' < \#G$ or for G acting on any vector subspace $W \subset V$ with $\dim(W) < \dim(V)$. Since G is a p-group, its center $Z(G)$ is nontrivial (see [DF04, p. 188]). Let $1 \neq \gamma \in Z(G)$. Let W be the set of all eigenvectors of γ along with the zero vector. Note that γ has at least one eigenvector (the linear operator $\gamma - 1$ on V is not invertible since it is not invertible on $V \otimes_F \overline{F}$). Hence $W \neq 0$. Moreover, for all $w_1, w_2 \in W$ and $a, b \in F$, we have $\gamma(aw_1 + bw_2) = a\gamma w_1 + b\gamma w_2 = aw_1 + bw_2$ since all eigenvalues of γ are equal to 1. This implies that W is a subspace of V. If $W = V$, then $\gamma v = v$ for all $v \in V$, which means that γ acts as 1 on V, so the action of G on V factors through a smaller subgroup $G/\langle\gamma\rangle$, and we can apply the inductive hypothesis. Now suppose that

W is a proper nonzero subspace of V. We claim that W is G-invariant. Indeed, if $w \in W$ and $g \in G$, then $\gamma(gw) = g(\gamma w) = gw$, so $gw \in W$. By the inductive hypothesis, there is $w \in W$ such that $gw = w$ for all $g \in G$.

$\square$

Remark 3.5.9 $\Gamma(\mathfrak{p})$ does have elements of finite order, for example, $g = \begin{pmatrix} 1 & \mathfrak{p} \\ 0 & 1 \end{pmatrix} \in$ $\mathrm{GL}_2(A)$ has order p and lies in $\Gamma(\mathfrak{p})$. For stronger statements, see Lemma 3.5.11 and Exercise 3.5.5. It is worth pointing out that the principal congruence subgroups of $\mathrm{GL}_2(\mathbb{Z}_p)$ have no elements of finite order; see Exercise 3.5.6.

Definition 3.5.10 Let R be a ring. A matrix $M \in \mathrm{Mat}_r(R)$ is called *unipotent* if $M - 1$ is nilpotent, i.e., $(M - 1)^n = 0$ for some $n \geq 1$. Note that unipotent matrices are invertible since we can rewrite $(M - 1)^n = 0$ as

$$1 = M \sum_{i=0}^{n-1} \binom{n}{i} (-1)^{n-i} M^{n-i-1}.$$

Lemma 3.5.11 *If the characteristic of the ring R is a prime $p > 0$, then a unipotent matrix $M \in \mathrm{GL}_r(R)$ has order p^m for some $m \geq 0$.*

Proof Suppose $(M - 1)^n = 0$. Choose a power of p such that $p^m \geq n$. Then, by Example N.2, we have

$$0 = (M - 1)^{p^m} = M^{p^m} + (-1)^{p^m},$$

so $M^{p^m} = 1$.

$\square$

Exercises

3.5.1 Show that

$$T_{\mathfrak{p}}(\phi) \cong \mathrm{Hom}_{A_{\mathfrak{p}}}(F_{\mathfrak{p}}/A_{\mathfrak{p}}, {}^{\phi}K^{\mathrm{sep}}).$$

3.5.2 Let ϕ and ψ be two Drinfeld modules defined over K. Assume $\phi[\mathfrak{p}] \subset K$ and $\psi[\mathfrak{p}] \subset K$ for some prime $\mathfrak{p} \neq \mathrm{char}_A(K)$, and there is an isogeny $\phi \to \psi$ defined over some finite Galois extension K'/K. Prove that ϕ and ψ are isogenous over K.

3.5.3 Let ϕ be a Drinfeld module of rank r over K. Let L be a subfield of $\mathrm{End}_K^\circ(\phi)$ containing F. By Theorem 2.8.5,

$$L \otimes_F F_{\mathfrak{p}} \cong \prod_{\mathfrak{P}|\mathfrak{p}} L_{\mathfrak{P}},$$

where the product is over the primes of L lying over $\mathfrak{p}$.

(a) Show that $V_{\mathfrak{p}}(\phi)$ decomposes into a direct sum

$$V_{\mathfrak{p}}(\phi) = \bigoplus_{\mathfrak{P}|\mathfrak{p}} V_{\mathfrak{P}},$$

where each $V_{\mathfrak{P}}$ is a nonzero vector space over $L_{\mathfrak{P}}$.

(b) Let $d_{\mathfrak{P}} = \dim_{L_{\mathfrak{P}}} V_{\mathfrak{P}}$. Prove that

$$\sum_{\mathfrak{P}|\mathfrak{p}} d_{\mathfrak{P}} \cdot [L_{\mathfrak{P}} : F_{\mathfrak{p}}] = r.$$

(c) Using (b), conclude that $[L : F] \le r$. (This is a weaker version of Corollary 3.4.15.)

3.5.4 Prove that the reduction map $GL_r(A) \to GL_r(A/\mathfrak{p}^n)$ in (3.5.8) is not surjective if $n > 1$ or $\deg(\mathfrak{p}) > 1$.

3.5.5 Let $U \subset GL_r(A)$ be the subgroup of upper-triangular matrices with 1's on the diagonal. Let s be an integer such that $p^s \ge r$. Show that $g^{p^s} = 1$ for all $g \in U$. Conclude that the exponent of U is equal to $p^{\lceil \log_p r \rceil}$ by constructing an element of this order.

3.5.6 Let $p \ne 2$ be a prime. Denote by $\mathbb{Z}_p$ the ring of p-adic integers. Prove the following:

(a) If $1 \ne g \in GL_n(\mathbb{Z}_p)$ and $g \equiv 1 \pmod{p}$, then g has infinite order.
(b) If G is a finite subgroup of $GL_n(\mathbb{Z}_p)$, then the order of G divides

$$(p^n - p^{n-1})(p^n - p^{n-2}) \cdots (p^n - 1).$$

3.6 Torsion Points in Terms of Anderson Motives

In this section we study the torsion points of ϕ using its Anderson motive M_ϕ. This leads to a different proof of Theorem 3.5.2 and Corollary 3.4.15, as well as an interesting relationship between the characteristic polynomial of an isogeny and the A-module structure of its kernel.

The connection between $\phi[a]$ and M_ϕ is based on the simple observation that the elements of $\overline{M_\phi} \cong \overline{K}\langle x \rangle$ can be evaluated at $\alpha \in \phi[a]$, leading to a map $\overline{M_\phi} \to \overline{K}$. We make $\overline{K}$ into a $\overline{K}\{\tau\}$-module, denoted $\tilde{K}$, by defining

$$m \circ \alpha = m(\alpha), \qquad m \in \overline{K}\{\tau\}, \alpha \in \overline{K},$$

where $m(\alpha)$ is the evaluation of $m(x) \in \overline{K}\langle x \rangle$ at α. Then, for a fixed $\alpha \in \overline{K}$, the map $\overline{K}\{\tau\} \to \widetilde{K}$, $m \mapsto m(\alpha)$, is a homomorphism of $\overline{K}\{\tau\}$-modules, where $\overline{K}\{\tau\}$ acts on itself by left multiplication. Conversely, any $\overline{K}\{\tau\}$-module homomorphism $\overline{K}\{\tau\} \to \widetilde{K}$ is uniquely determined by the image of τ and thus corresponds to an evaluation map at some $\alpha \in \overline{K}$ (if $\tau \mapsto \beta$, then $\overline{K}\{\tau\} \to \widetilde{K}$ is the evaluation at $\beta^{1/q}$). Thus, we have a natural isomorphism of $\mathbb{F}_q$-vector spaces

$$\overline{K} \cong \mathrm{Hom}_{\overline{K}\{\tau\}}(\overline{K}\{\tau\}, \widetilde{K}) \tag{3.6.1}$$

$$\alpha \mapsto (m \mapsto m(\alpha)).$$

Note that on the left-hand side, we have $\overline{K}$, not $\widetilde{K}$, i.e., this is not an isomorphism of $\overline{K}\{\tau\}$-modules. On the other hand, if we make $\overline{K}$ into an A-module via ϕ, i.e., we define $a \circ \beta = \phi_a(\beta)$, and make $\mathrm{Hom}_{\overline{K}\{\tau\}}(\overline{K}\{\tau\}, \widetilde{K})$ into an A-module by defining $(a \circ f)(x) = f(x\phi_a)$, then the isomorphism (3.6.1) becomes an isomorphism of A-modules

$$\phi\overline{K} \cong \mathrm{Hom}_{\overline{K}\{\tau\}}(\overline{M}_\phi, \widetilde{K}). \tag{3.6.2}$$

Restricting the isomorphism (3.6.2) to the a-torsion submodules, we get

$$\phi[a] \cong \mathrm{Hom}_{\overline{K}\{\tau\}}(\overline{M}_\phi/a\overline{M}_\phi, \widetilde{K}). \tag{3.6.3}$$

Set

$$(\overline{M}_\phi/a\overline{M}_\phi)^\tau := \{m \in \overline{M}_\phi/a\overline{M}_\phi \mid \tau m = m\}.$$

Note that $(\overline{M}_\phi/a\overline{M}_\phi)^\tau$ is an A-submodule of $\overline{M}_\phi/a\overline{M}_\phi$, but it is not a $\overline{K}\{\tau\}$-submodule. We can extend $(\overline{M}_\phi/a\overline{M}_\phi)^\tau$ to a $\overline{K}[T, \tau]$-module by tensoring over $\mathbb{F}_q$ with $\overline{K}$, where $\overline{K}\{\tau\}$ acts on $(\overline{M}_\phi/a\overline{M}_\phi)^\tau \otimes_{\mathbb{F}_q} \widetilde{K}$ through its action on $\widetilde{K}$ and A acts through its action on $(\overline{M}_\phi/a\overline{M}_\phi)^\tau$.

Lemma 3.6.1 *Assume a is coprime to* $\mathrm{char}_A(K)$. *Then the evident map*

$$(\overline{M}_\phi/a\overline{M}_\phi)^\tau \otimes_{\mathbb{F}_q} \widetilde{K} \longrightarrow \overline{M}_\phi/a\overline{M}_\phi$$

is an isomorphism of $\overline{K}[T, \tau]$-modules.

Proof Let $V := \overline{M}_\phi/a\overline{M}_\phi$. Since $\overline{M}_\phi$ is a free $\overline{K}[T]$-module of finite rank, V is a finite dimensional $\overline{K}$-vector space (its dimension is equal to $r \cdot \deg_T(a)$). The map $\tau: V \to V$ is q-semilinear in the sense that $\tau(\alpha_1 v_1 + \alpha_2 v_2) = \alpha_1^q \tau(v_1) + \alpha_2^q \tau(v_2)$ for all $\alpha_1, \alpha_2 \in \overline{K}$ and $v_1, v_2 \in V$. Next, note that τ is injective. Indeed, if $\tau m = s\phi_a$ for some $m, s \in \overline{K}\{\tau\}$, then, applying ∂ to both sides of this equality, we get $\partial(s)\gamma(a) = 0$; by assumption, $\gamma(a) \neq 0$, so $s = \tau s'$ for some $s' \in \overline{K}\{\tau\}$, and thus $m \in a\overline{M}_\phi$. Using the fact that V is a vector space over an algebraically closed field, it is easy to check that the image and the kernel of a q-semilinear map are subspaces of V. Hence τ is also surjective.

The lemma will follow if we show that V has a basis fixed by τ. Let $\{e_1, \ldots, e_n\}$ be a basis of V. Let $\tau(e_i) = \sum_{j=1}^{n} \alpha_{i,j} e_j$, $1 \le i \le n$. Set $S = (\alpha_{ij}) \in \mathrm{GL}_n(\overline{K})$; note that here we use the fact that τ is a bijection. Let $\{e'_1, \ldots, e'_n\}$ be another basis, and let $B = (b_{ij}) \in \mathrm{GL}_n(A)$ be the matrix such that $Be_i = e'_i$ for all $1 \le i \le n$. Then

$$\tau e'_i = \tau B e_i = B^{(q)} \tau e_i = B^{(q)} S e_i = B^{(q)} S B^{-1} e'_i,$$

where $B^{(q)} := (b_{ij}^q)$. Thus, changing the basis of V corresponds to changing the matrix of τ from S to $B^{(q)} S B^{-1}$. The existence of a basis fixed by τ is equivalent to the existence of B such that $B^{(q)} S B^{-1} = I_n$, or equivalently $S = (B^{(q)})^{-1} B$. It is known[10] that the map $\mathrm{GL}_n(\overline{K}) \to \mathrm{GL}_n(\overline{K})$, $B \mapsto (B^{(q)})^{-1} B$, is surjective, so we can always find such B. $\qquad\square$

Applying Lemma 3.6.1 to (3.6.3), we get

$$\phi[a] \cong \mathrm{Hom}_{\overline{K}\{\tau\}}((\overline{M}_\phi/a\overline{M}_\phi)^\tau \otimes_{\mathbb{F}_q} \widetilde{K}, \widetilde{K}).$$

Now a $\overline{K}$-linear homomorphism

$$f: (\overline{M}_\phi/a\overline{M}_\phi)^\tau \otimes_{\mathbb{F}_q} \widetilde{K} \longrightarrow \widetilde{K}$$

is uniquely determined by its values on $m \otimes 1$, where $m \in (\overline{M}_\phi/a\overline{M}_\phi)^\tau$. If, in addition, f is equivariant with respect to the action of τ, then

$$f(m \otimes 1) = f(\tau(m \otimes 1)) = f(m \otimes 1)^q,$$

so $f(m \otimes 1) \in \mathbb{F}_q$. Conversely, given an $\mathbb{F}_q$-linear homomorphism $f: (\overline{M}_\phi/a\overline{M}_\phi)^\tau \to \mathbb{F}_q$, we can always extend it linearly to a homomorphism $(\overline{M}_\phi/a\overline{M}_\phi)^\tau \otimes_{\mathbb{F}_q} \widetilde{K} \to \widetilde{K}$ of $\overline{K}\{\tau\}$-modules. Therefore,

$$\phi[a] \cong \mathrm{Hom}_{\mathbb{F}_q}((\overline{M}_\phi/a\overline{M}_\phi)^\tau, \mathbb{F}_q). \qquad (3.6.4)$$

The isomorphism (3.6.4) can be used to give a different proof of Corollary 3.5.3.

---

[10] This is a special case of a well-known theorem of Lang [Lan56]. The proof is not that difficult using some basic algebraic geometry, but I do not know an elementary proof of this. Essentially, one considers the map $B \mapsto B^{(q)} B^{-1}$ as being induced by a morphism $f: \mathrm{GL}_n \to \mathrm{GL}_n$ of group varieties over $\overline{K}$. The Frobenius induces the zero map on the differentials, so f is an isomorphism on the space of differentials. This implies that f is smooth, and therefore its image is Zariski open in GL_n. Repeating the same argument for the map $g: B \mapsto B^{(q)} S B^{-1}$ for a fixed $S \in \mathrm{GL}_n(\overline{K})$, one concludes that the images of f and g intersect, as both are Zariski open and GL_n is connected. This implies that S is in the image of f.

Second proof of Corollary 3.5.3 Assume that a is coprime to $\mathrm{char}_A(K)$. As we have observed, $(\overline{M}_\phi/a\overline{M}_\phi)^\tau$ is an A-module. From the isomorphism (3.6.4), we get that $(\overline{M}_\phi/a\overline{M}_\phi)^\tau$ is a finite A-module since $\phi[a]$ is a finite A-module. Now it is enough to show that

$$(\overline{M}_\phi/a\overline{M}_\phi)^\tau \cong (A/aA)^r, \qquad (3.6.5)$$

or equivalently

$$(\overline{M}_\phi/a\overline{M}_\phi)^\tau \otimes_{\mathbb{F}_q} \overline{K} \cong (\overline{K}[T]/a\overline{K}[T])^r.$$

(It is easy to show that A/aA and its $\mathbb{F}_q$-dual $\mathrm{Hom}_{\mathbb{F}_q}(A/aA, \mathbb{F}_q)$ are isomorphic A-modules.)

(It is not hard to show that A/aA and its $\mathbb{F}_q$-dual $\mathrm{Hom}_{\mathbb{F}_q}(A/aA, \mathbb{F}_q)$ are isomorphic A-modules; see Lemma 3.6.2.) By Lemma 3.6.1, $(\overline{M}_\phi/a\overline{M}_\phi)^\tau \otimes_{\mathbb{F}_q} \overline{K} \cong \overline{M}_\phi/a\overline{M}_\phi$ as $\overline{K}[T]$-modules. Since $\overline{M}_\phi$ is a free $\overline{K}[T]$-module of rank r, we have $\overline{M}_\phi/a\overline{M}_\phi \cong (\overline{K}[T]/a\overline{K}[T])^r$, which completes the proof. $\qquad\square$

The isomorphism (3.6.4) relates $\phi[a]$ and $(\overline{M}_\phi/a\overline{M}_\phi)^\tau$ through an $\mathbb{F}_q$-duality. To relate these A-modules through an A-duality, one has to modify (3.6.2). The problem is that although $\overline{K}$ is naturally a module over $\overline{K}\{\tau\}$ and A via evaluations of linearized polynomials, these two actions do not commute, so $\overline{K}$ is not a $\overline{K}[T, \tau]$-module. Anderson's idea for rectifying the situation is to separate the two actions by replacing $\overline{K}$ by $\mathrm{Hom}_{\mathbb{F}_q}(A, \overline{K})$ given the left $\overline{K}[T, \tau]$-module structure:

$$(\beta \circ f)(x) = \beta f(x),$$

$$(a \circ f)(x) = f(ax),$$

$$(\tau \circ f)(x) = f(x)^q,$$

for all $\beta \in \overline{K}$, $f \in \mathrm{Hom}_{\mathbb{F}_q}(A, \overline{K})$, and $a \in A$. With this definition, there is a map

$$\overline{K} \longrightarrow \mathrm{Hom}_{\overline{K}[T,\tau]}(\overline{M}_\phi, \mathrm{Hom}_{\mathbb{F}_q}(A, \overline{K})) \qquad (3.6.6)$$

$$\beta \longmapsto (m \longmapsto (a \longmapsto m(\phi_a(\beta)))),$$

which induces a map

$$\phi[a] \longrightarrow \mathrm{Hom}_{\overline{K}[T,\tau]}(\overline{M}_\phi/a\overline{M}_\phi, \mathrm{Hom}_{\mathbb{F}_q}(A/aA, \overline{K})). \qquad (3.6.7)$$

By an argument similar to the argument by which we deduced (3.6.4), we get

$$\mathrm{Hom}_{\overline{K}[T,\tau]}(\overline{M}_\phi/a\overline{M}_\phi, \mathrm{Hom}_{\mathbb{F}_q}(A/aA, \overline{K}))$$

$$\cong \mathrm{Hom}_{\overline{K}[T,\tau]}((\overline{M}_\phi/a\overline{M}_\phi)^\tau \otimes_{\mathbb{F}_q} \widetilde{K}, \mathrm{Hom}_{\mathbb{F}_q}(A/aA, \overline{K}))$$

$$\cong \mathrm{Hom}_A((\overline{M}_\phi/a\overline{M}_\phi)^\tau, \mathrm{Hom}_{\mathbb{F}_q}(A/aA, \mathbb{F}_q)).$$

It is easy to check that (3.6.7) is injective. On the other hand, by (3.6.5), $(\overline{M}_\phi/a\overline{M}_\phi)^\tau \cong (A/aA)^r$. Denote

$$(A/aA)^\vee = \mathrm{Hom}_{\mathbb{F}_q}(A/aA, \mathbb{F}_q).$$

Treating the elements of $(A/aA)^\vee$ as functions $A/aA \to \mathbb{F}_q$, the set $(A/aA)^\vee$ acquires a natural A-module structure via $(b \circ f)(x) = f(bx)$, $b \in A$.

Lemma 3.6.2 *There is an isomorphism of A-modules $(A/aA)^\vee \cong A/aA$.*

Proof As a finite dimensional $\mathbb{F}_q$-vector space, $(A/aA)^\vee$ is just the dual of A/aA. Thus, $\#(A/aA)^\vee = \#A/aA$. Next, by Theorem 1.2.4,

$$(A/aA)^\vee \cong A/a_1A \oplus \cdots \oplus A/a_mA, \tag{3.6.8}$$

where $a_1 \mid a_2 \mid \cdots \mid a_m$. Obviously a annihilates $(A/aA)^\vee$, so $a_m \mid a$. On the other hand, $f(a_m) = 0$ for all $f \in (A/aA)^\vee$. If $\deg(a_m)$ were strictly smaller than $n = \deg(a)$, then we could choose a_m as part of an $\mathbb{F}_q$-basis $\{e_1 = a_m, e_2, \ldots, e_n\}$ of A/aA and define $f \in (A/aA)^\vee$ by $f(\alpha_1 e_1 + \cdots \alpha_n e_n) = \alpha_1$. Since $f(a_m) = 1$, we get a contradiction. Therefore, $a_m A = aA$, and, by comparing the cardinalities of the A-modules on both sides of (3.6.8), we obtain an isomorphism $(A/aA)^\vee \cong A/aA$. $\qquad\square$

Since $(A/aA)^\vee$ is (non-canonically) isomorphic to A/aA, we get

$$\mathrm{Hom}_A((\overline{M}_\psi/a\overline{M}_\psi)^\tau, (A/aA)^\vee) \cong (A/aA)^r.$$

Therefore, (3.6.7) is an isomorphism and induces the A-module isomorphism

$$\phi[a] \xrightarrow{\sim} \mathrm{Hom}_{A/aA}((\overline{M}_\phi/a\overline{M}_\phi)^\tau, (A/aA)^\vee) \tag{3.6.9}$$

that we were looking for.

Now let $\mathfrak{p} \lhd A$ be a prime distinct from $\mathrm{char}_A(K)$. For any $n \geq 0$, the multiplication by $\mathfrak{p}$ induces a surjective homomorphism $(\overline{M}_\phi/\mathfrak{p}^{n+1}\overline{M}_\phi)^\tau \to (\overline{M}_\phi/\mathfrak{p}^n\overline{M}_\phi)^\tau$. We form the inverse limit with respect to these maps

$$H^1_\mathfrak{p}(\phi) := \varprojlim_n (\overline{M}_\phi/\mathfrak{p}^n\overline{M}_\phi)^\tau. \tag{3.6.10}$$

We also have the surjective homomorphisms

$$\mathrm{Hom}_{\mathbb{F}_q}(A/\mathfrak{p}^{n+1}, \mathbb{F}_q) \longrightarrow \mathrm{Hom}_{\mathbb{F}_q}(A/\mathfrak{p}^n, \mathbb{F}_q),$$

$$f(x) \longmapsto f(\mathfrak{p}x)$$

and denote

$$A_{\mathfrak{p}}^{\vee} = \varprojlim_{n} \operatorname{Hom}_{\mathbb{F}_q}(A/\mathfrak{p}^{n+1}, \mathbb{F}_q).$$

Note that $A_{\mathfrak{p}}^{\vee} \cong \varprojlim_{n} A/\mathfrak{p}^{n} \cong A_{\mathfrak{p}}$, so taking the inverse limits in (3.6.9), one arrives at the following:

Theorem 3.6.3 *There is an isomorphism of $A_{\mathfrak{p}}$-modules*

$$T_{\mathfrak{p}}(\phi) \cong \operatorname{Hom}_{A_{\mathfrak{p}}}(H_{\mathfrak{p}}^1(\phi), A_{\mathfrak{p}}^{\vee}).$$

Remarks 3.6.4

(1) Suppose E is an elliptic curve over a field K. Then its n-torsion points $E[n]$ form a group isomorphic to $\mathbb{Z}/n\mathbb{Z} \times \mathbb{Z}/n\mathbb{Z}$, assuming n is relatively prime to the characteristic of K. If p is a prime distinct from the characteristic of K, then taking the inverse limit over the maps $E[p^{n+1}] \xrightarrow{p} E[p^n]$ one obtains the p-adic Tate module $T_p(E) \cong \mathbb{Z}_p \times \mathbb{Z}_p$ of E. Now $T_p(E)$ can be interpreted as the first homology group of E, either as an algebraic variety or as a complex manifold if $K = \mathbb{C}$. Essentially, the multiplication by p^n gives an unramified finite cover $E \to E$, so it corresponds to a quotient of the étale fundamental group $\pi_1(E)$ of E and Grothendieck's theory of these groups gives an isomorphism $\pi_1(E) \otimes \mathbb{Z}_p \cong T_p(E)$. When E is defined over $\mathbb{C}$, its corresponding complex manifold is isomorphic to $\mathbb{C}/\Lambda$ for some lattice $\Lambda \subset \mathbb{C}$; in that case, Λ is isomorphic to the first singular homology group $H_1(E, \mathbb{Z})$ of E and $T_p(E) \cong \Lambda \otimes \mathbb{Z}_p$. The dual $\operatorname{Hom}(T_p(E), \mathbb{Z}_p)$ is then the first (étale or singular) cohomology group $H^1(E, \mathbb{Z}_p)$ of E with $\mathbb{Z}_p$ coefficients. Within the analogies of Drinfeld modules and elliptic curves, $T_{\mathfrak{p}}(\phi)$ is clearly the analogue of $T_p(E)$. Therefore, from the duality of Theorem 3.6.3, one could think of $H_{\mathfrak{p}}^1(\phi)$ as the first cohomology group of ϕ with $A_{\mathfrak{p}}$ coefficients, which is the reason for the notation (3.6.10).

(2) Given an algebraic variety X, Grothendieck's vision of motives, very vaguely, posits the existence of an object which recovers the natural cohomology groups of X upon tensoring with $\mathbb{Q}_p$ or $\mathbb{C}$; cf. [Mil12b]. Now note that $H_{\mathfrak{p}}^1(\phi)$ is related to $M_\phi \otimes A_{\mathfrak{p}}$. This is the reason for calling M_ϕ a "motive."

Next, we use Theorem 3.6.3 to study the action of endomorphisms of ϕ on $T_{\mathfrak{p}}(\phi)$.

Definition 3.6.5 Let $\mathfrak{p}$ be a prime distinct from $\operatorname{char}_A(K)$. Through the map of Theorem 3.5.4, we can associate with every endomorphism $u \in \operatorname{End}_K(\phi)$ a matrix $u_{\mathfrak{p}} \in \operatorname{Mat}_r(A_{\mathfrak{p}})$. The *characteristic polynomial of u* is

$$P_u(x) = \det(x - u_{\mathfrak{p}}).$$

Theorem 3.6.6 *With notation and assumptions of Definition 3.6.5, the polynomial $P_u(x)$ has coefficients in A which do not depend on the choice of $\mathfrak{p}$.*

Proof By Theorem 3.6.3, the endomorphism u acts on $H^1_{\mathfrak{p}}(\phi)$ by the transpose of $u_{\mathfrak{p}}$, which has the same characteristic polynomial as $u_{\mathfrak{p}}$. Now the action of u on $H^1_{\mathfrak{p}}(\phi)$ is induced by its action on the A-modules $(\overline{M}_\phi/\mathfrak{p}^n\overline{M}_\phi)^\tau \cong (A/\mathfrak{p}^n)^r$, $n \geq 0$. Thus,

$$\det\left(x - u\big|(\overline{M}_\phi/\mathfrak{p}^n\overline{M}_\phi)^\tau\right) = (P_u(x) \mod \mathfrak{p}^n).$$

(By definition, $P_u(x) \in A_{\mathfrak{p}}[x]$.) On the other hand,

$$\det{}_{A/\mathfrak{p}^n}\left(x - u\big|(\overline{M}_\phi/\mathfrak{p}^n\overline{M}_\phi)^\tau\right) = \det{}_{\overline{K}[T]/\mathfrak{p}^n\overline{K}[T]}\left(x - u\big|(\overline{M}_\phi/\mathfrak{p}^n\overline{M}_\phi)^\tau \otimes_{\mathbb{F}_q} \overline{K}\right)$$

$$= \det{}_{\overline{K}[T]/\mathfrak{p}^n\overline{K}[T]}\left(x - u\big|(\overline{M}_\phi/\mathfrak{p}^n\overline{M}_\phi)\right)$$

$$= \det{}_{\overline{K}[T]}\left(x - u\big|\overline{M}_\phi\right) \mod \mathfrak{p}^n\overline{K}[x].$$

(In above congruences, we consider $\mathfrak{p}$ as a polynomial in $\overline{K}[T]$.) Since $\overline{M}_\phi$ is a free $\overline{K}[T]$-module of rank r,

$$\hat{P}_u(x) := \det{}_{\overline{K}[T]}\left(x - u\big|\overline{M}_\phi\right)$$

is a monic polynomial of degree r with coefficients in $\overline{K}[T]$ which does not depend on $\mathfrak{p}$. On the other hand, these coefficients modulo $\mathfrak{p}^n$ lie in $A/\mathfrak{p}^n \subset \overline{K}[T]/\mathfrak{p}^n$ for all $n \geq 0$. This is possible if and only if $\hat{P}_u(x)$ lies in $A[x]$. Since

$$\hat{P}_u(x) \equiv P_u(x) \pmod{\mathfrak{p}^n} \quad \text{for all } n \geq 0,$$

we conclude that $P_u(x) = \hat{P}_u(x)$, so $P_u(x)$ has coefficient in A which do not depend on $\mathfrak{p}$. $\qquad\square$

We record separately the following fact established in the proof of the previous theorem:

Proposition 3.6.7 *For any* $\mathfrak{p} \neq \operatorname{char}_A(K)$ *and* $u \in \operatorname{End}_K(\phi)$, *we have*

$$P_u(x) = \det{}_{A_{\mathfrak{p}}}(x - u\big|T_{\mathfrak{p}}(\phi)) = \det{}_{\overline{K}[T]}(x - u\big|\overline{M}_\phi).$$

Definition 3.6.8 The *determinant* of $u \in \operatorname{End}_K(\phi)$ is

$$\det(u) = \det{}_{\overline{K}[T]}\left(u\big|\overline{M}_\phi\right).$$

Since $u_{\mathfrak{p}}$ is invertible if $u \neq 0$, Proposition 3.6.7 implies that $\det(u) = (-1)^r P_u(0) \in A$ and $\det(u) = 0$ if and only if $u = 0$. We extend det linearly to $\operatorname{End}^\circ_K(\phi) = \operatorname{End}_K(\phi) \otimes_A F$:

$$\det(u) = \det{}_{\overline{K}[T]}\left(u\big|\overline{M}_\phi \otimes_A F\right).$$

Proposition 3.6.9 *The function* det: $\text{End}_K^{\circ}(\phi) \to F$ *is a norm form.*

Proof This follows from the standard linear algebra properties of the determinant.
□

Now, we give a different proof of part of the claim of Corollary 3.4.15.

Theorem 3.6.10 *Let Z be the center of $D := \text{End}_K^{\circ}(\phi)$. Then $\sqrt{[D:Z]} \cdot [Z:F]$ divides r.*

Proof We know that D is a division algebra over F, so it is a central division algebra over Z. By Proposition 3.6.9, det: $D \to F$ is a norm form. By the existence of the canonical norm form (see Theorem 1.7.20), we must have

$$\det = \left(\text{Nr}_{Z/F} \circ \text{nr}_{D/Z}\right)^m$$

for some $m \geq 1$. Evaluating both sides on $a \in A$, we get

$$a^r = a^{\sqrt{[D:Z]} \cdot [Z:F] \cdot m}.$$

Thus, $\sqrt{[D:Z]} \cdot [Z:F] \cdot m = r$, so $\sqrt{[D:Z]} \cdot [Z:F]$ divides r.
□

Corollary 3.6.11 *Let L be a subfield of $\text{End}_K^{\circ}(\phi)$ containing F. Then $[L:F]$ divides r.*

Proof Any subfield of $D := \text{End}_K^{\circ}(\phi)$ containing F is itself contained in some maximal subfield of D. Hence it is enough to prove the claim assuming L is maximal. On the other hand, a maximal subfield contains the center Z of D and has degree $\sqrt{[D:Z]}$ over Z; see Theorem 1.7.10. Thus,

$$[L:F] = [L:Z][Z:F] = \sqrt{[D:Z]} \cdot [Z:F],$$

which divides r by Theorem 3.6.10.
□

Given a nonzero endomorphism $u \in \text{End}_K(\phi)$, let

$$\ker(u) := \{\alpha \in \overline{K} \mid u(\alpha) = 0\},$$

which is the set of distinct roots of the polynomial $u(x) \in K\langle x \rangle$. This is a finite A-module, with the action of A defined by $a \circ \alpha = \phi_a(\alpha)$.

Definition 3.6.12 The *norm of u* is

$$\mathfrak{N}(u) = \begin{cases} \chi(\ker u), & \text{if } \text{char}_A(K) = 0, \\ \mathfrak{p}^{\text{ht}(u)/\deg(\mathfrak{p})} \cdot \chi(\ker u), & \text{if } \text{char}_A(K) = \mathfrak{p} \neq 0. \end{cases}$$

We remind the reader that for a finite A-module M we defined $\chi(M)$ as the product of the invariant factors of M, which are assumed to be monic; cf.

Definition 1.2.5. Also, by our convention at the beginning on this chapter, we denote the monic generator of a maximal ideal $\mathfrak{p} \lhd A$ by the same symbol. And finally, by Proposition 3.3.4, if $\mathrm{char}_A(K) = \mathfrak{p} \neq 0$, then $\deg(\mathfrak{p})$ divides $\mathrm{ht}(u)$. Thus, $\mathfrak{N}(u) \in A_+$.

It is immediate from the definition that

$$\deg_\tau(u) = \deg_T \mathfrak{N}(u). \tag{3.6.11}$$

Theorem 3.6.13 *For any isogeny $u \in \mathrm{End}_K(\phi)$, we have the equality of ideals in* A:

$$(\det u) = (P_u(0)) = (\mathfrak{N}(u)).$$

Proof The following argument is motivated by the proof of Theorem 5.1 in [Gek91].

The first equality $(\det u) = (P_u(0))$ immediately follows from Proposition 3.6.7. Let $\mathfrak{p} = \mathrm{char}_A(K)$, which we allow to be 0. Let $\mathfrak{l}$ be a prime different from $\mathfrak{p}$. Denote by $\ker(u)_\mathfrak{l}$ the $\mathfrak{l}$-primary part of $\ker(u)$ and by $|\cdot|_\mathfrak{l}$ the normalized absolute value on $F_\mathfrak{l}$.

The isogeny u defines is an injective $A_\mathfrak{l}$-linear transformation $u_\mathfrak{l} \colon T_\mathfrak{l}(\phi) \to T_\mathfrak{l}(\phi)$. Note that $T_\mathfrak{l}(\phi)$ is a free $A_\mathfrak{l}$-module of rank r and the valuation on $A_\mathfrak{l}$ makes $A_\mathfrak{l}$ into a Euclidean domain. We have

$$
\begin{aligned}
|P_u(0)|_\mathfrak{l} &= |\det u_\mathfrak{l}|_\mathfrak{l} && \text{(by definition of } P_u\text{)} \\
&= \#\left(T_\mathfrak{l}(\phi)/u_\mathfrak{l}T_\mathfrak{l}(\phi)\right) && \text{(by Lemma 1.2.6)} \\
&= \#(\ker u)_\mathfrak{l} && \text{(by Exercise 3.6.2)} \\
&= |\mathfrak{N}(u)|_\mathfrak{l} && \text{(by the definition of } \mathfrak{N}(u)\text{)}.
\end{aligned}
$$

Next, note that $-\deg_T \det$ and $-\deg_\tau$ are valuations on $D = \mathrm{End}_K^\circ(\phi)$ and both are equivalent with the ∞-adic valuation; cf. Sect. 3.4. The proportionality factor comes out by evaluating on $a \in A$:

$$\deg_T(\det(a)) = \deg_T(a^r) = r \cdot \deg_T(a) = \deg_\tau(\phi_a).$$

Thus,

$$\deg_T \det(u) = \deg_\tau(u) \quad \text{for all } u \in \mathrm{End}_k(\phi). \tag{3.6.12}$$

Now

$$
\begin{aligned}
\deg_T P_u(0) &= \deg_T \det(u) \\
&= \deg_\tau(u) && \text{by (3.6.12)} \\
&= \deg_T(\mathfrak{N}(u)) && \text{by (3.6.11)}.
\end{aligned}
$$

Therefore, $P_u(0)$ and $\mathfrak{N}(u)$ are two elements of A of the same degree which have the same $\mathfrak{l}$-adic valuation for any $\mathfrak{l} \neq \mathfrak{p}$. This implies that $P_u(0)$ and $\mathfrak{N}(u)$ are equal, up to $\mathbb{F}_q^\times$-multiples, so $(P_u(0)) = (\mathfrak{N}(u))$. $\square$

Exercises

3.6.1 Prove that (3.6.6) is an isomorphism.

3.6.2 Let ϕ and ψ be Drinfeld modules over K. Let $u: \phi \to \psi$ be an isogeny, and let

$$\ker(u) := \{\alpha \in \overline{K} \mid u(\alpha) = 0\}.$$

We make $\ker(u)$ into an A-module by $a \circ \alpha = \phi_a(\alpha)$. For a prime $\mathfrak{p}$, different from $\mathrm{char}_A(K)$, denote by $\ker(u)_\mathfrak{p}$ the $\mathfrak{p}$-primary part of $\ker(u)$. Prove that

$$\#T_\mathfrak{p}(\psi)/u(T_\mathfrak{p}(\phi)) = \#\ker(u)_\mathfrak{p} = |\chi(\ker(u))|_\mathfrak{p},$$

where $|\cdot|_\mathfrak{p}$ is the normalized absolute value on $F_\mathfrak{p}$ and χ was defined in Sect. 1.2.

3.6.3 Let $u, w \in \mathrm{End}_K(\phi)$ be isogenies. Prove the following:

(a) $\mathfrak{N}(u \cdot w) = \mathfrak{N}(u) \cdot \mathfrak{N}(w)$.
(b) There is a polynomial $f(x) \in A[x]$ such that $(\mathfrak{N}(au + w)) = (f(a))$ for all $a \in A$. For example, if $u, w \in A$, then $f(x) = (xu + w)^r$.

3.7   Weil Pairing

The Weil pairing is an important tool in the study of elliptic curves. It is a pairing on the torsion points of an elliptic curve with values in roots of unity, which is useful, for example, in the study of the action of Galois groups on torsion points; cf. [Sil09, III.8]. In this section we construct two versions of this pairing in the context of Drinfeld modules.

Let ϕ be a Drinfeld module of rank r over K defined by

$$\phi_T = t + g_1\tau + \cdots + g_r\tau^r,$$

and let ψ be the Drinfeld module of rank 1 over K defined by

$$\psi_T = t + (-1)^{r-1}g_r\tau. \tag{3.7.1}$$

The main result of this section is the existence for each $a \in A$, $\text{char}_A(K) \nmid a$, of a
map

$$W_a = W_{\phi,a} : \prod_{i=1}^{r} \phi[a] \longrightarrow \psi[a],$$

with the following properties:

(1) $W_a(x_1, \ldots, x_r)$ is A-multilinear, i.e., for each fixed i and fixed elements $\alpha_j \in$
$\phi[a]$, $j \neq i$, the map

$$\phi[a] \to \psi[a] \quad \text{defined by} \quad \beta \mapsto W_a(\alpha_1, \ldots, \alpha_{i-1}, \beta, \alpha_{i+1}, \ldots, \alpha_r)$$

is an A-module homomorphism.
(2) $W_a(x_1, \ldots, x_r)$ is alternating, i.e.,

$$W_a(\alpha_1, \ldots, \alpha_r) = 0 \text{ if } \alpha_i = \alpha_j \text{ for some } i \neq j.$$

(3) $W_a(x_1, \ldots, x_r)$ is surjective and nondegenerate, i.e.,

$$\text{if } W_a(\beta_1, \ldots, \beta_{i-1}, \alpha_i, \beta_{i+1}, \ldots, \beta_r) = 0$$
$$\text{for all } \beta_1, \ldots, \beta_{i-1}, \beta_{i+1}, \ldots, \beta_r \in \phi[a], \text{ then } \alpha_i = 0.$$

(4) $W_a(x_1, \ldots, x_r)$ is G_K-equivariant, i.e.,

$$\sigma W_a(\alpha_1, \ldots, \alpha_r) = W_a(\sigma \alpha_1, \ldots, \sigma \alpha_r),$$
$$\text{for all } \alpha_1, \ldots, \alpha_r \in \phi[a] \text{ and } \sigma \in G_K.$$

(5) $W_a(x_1, \ldots, x_r)$ satisfies the compatibility relation

$$\psi_b(W_{ab}(\alpha_1, \ldots, \alpha_r)) = W_a(\phi_b(\alpha_1), \ldots, \phi_b(\alpha_r)),$$

for all $a, b \in A$ not divisible by $\text{char}_A(K)$ and for all $\alpha_1, \ldots, \alpha_r \in \phi[ab]$.

Let $\mathfrak{p}$ be a prime different from $\text{char}_A(K)$. The last property of the W_a's implies
that the diagram

$$
\begin{array}{ccc}
\prod_{i=1}^{r} \phi[\mathfrak{p}^{n+1}] & \xrightarrow{\ W_{\mathfrak{p}^{n+1}}\ } & \psi[\mathfrak{p}^{n+1}] \\
\Big\downarrow{\scriptstyle \prod_{i=1}^{r} \phi_\mathfrak{p}} & & \Big\downarrow{\scriptstyle \psi_\mathfrak{p}} \\
\prod_{i=1}^{r} \phi[\mathfrak{p}^{n}] & \xrightarrow[\ W_{\mathfrak{p}^{n}}\ ]{} & \psi[\mathfrak{p}^{n}]
\end{array}
$$

is commutative. Thus, the maps W_{p^n} are compatible with taking the inverse limits, and we can combine them to create a map on the Tate modules,

$$W_{p^\infty} : \prod_{i=1}^{r} T_p(\phi) \longrightarrow T_p(\psi),$$

which inherits the properties (1)-(4) of the W_a's.

The map W_a plays a role in the theory of Drinfeld modules similar to the Weil pairing on the torsion points of elliptic curves. Therefore, we call W_a the *Weil pairing for $\phi[a]$*, although, strictly speaking, it is a pairing only when $r = 2$.

The existence of the Weil pairing implies the following theorem about Galois representations, which we will use in Chap. 4.

Theorem 3.7.1 *Let ϕ and ψ be Drinfeld modules over K of rank r and 1, respectively, defined by*

$$\phi_T = t + g_1\tau + \cdots + g_r\tau^r, \qquad \psi_T = t + (-1)^{r-1}g_r\tau.$$

(1) Assume $a \in A$ is not divisible by $\text{char}_A(K)$. Let

$$\rho_{\phi,a} : G_K \to \text{Aut}_A(\phi[a]) \cong \text{GL}_r(A/aA)$$

be the representation (3.5.5), and let $\rho_{\psi,a} : G_K \to \text{Aut}_A(\psi[a]) \cong (A/aA)^\times$ be the corresponding representation for ψ. Then

$$\det(\rho_{\phi,a}(\sigma)) = \rho_{\psi,a}(\sigma) \quad \text{for all} \quad \sigma \in G_K.$$

(2) $K(\psi[a]) \subseteq K(\phi[a])$.
(3) Assume p is a prime not equal to $\text{char}_A(K)$. Let

$$\hat{\rho}_{\phi,p} : G_K \to \text{Aut}_{A_p}(T_p(\phi)) \cong \text{GL}_r(A_p)$$

be the representation (3.5.6), and let $\hat{\rho}_{\psi,p} : G_K \to \text{Aut}_{A_p}(T_p(\psi)) \cong A_p^\times$ be the corresponding representation for ψ. Then

$$\det \hat{\rho}_{\phi,p}(\sigma) = \hat{\rho}_{\psi,p}(\sigma) \quad \text{for all} \quad \sigma \in G_K.$$

Proof The G_K-equivariance of W_a means that

$$W_a\left(\rho_{\phi,a}(\sigma)(\alpha_1), \ldots, \rho_{\phi,a}(\sigma)(\alpha_r)\right) = \rho_{\psi,a}(\sigma) \cdot W_a(\alpha_1, \ldots, \alpha_r). \qquad (3.7.2)$$

On the other hand, the fact that W_a is A-multilinear and alternating implies that

$$W_a\left(\rho_{\phi,a}(\sigma)(\alpha_1), \ldots, \rho_{\phi,a}(\sigma)(\alpha_r)\right) = \det(\rho_{\phi,a}(\sigma)) \cdot W_a(\alpha_1, \ldots, \alpha_r). \qquad (3.7.3)$$

(This last equation is essentially the "coordinate-free" definition of the determinant; see [DF04, pp. 436-437].) Since W_a is surjective, comparing (3.7.2) and (3.7.3), we get $\det(\rho_{\phi,a}(\sigma)) = \rho_{\psi,a}(\sigma)$, which is (1).

The proof of (3) is similar to the proof of (1), except that it uses $W_{\mathfrak{p}^\infty}$ instead of W_a, and (2) is an immediate consequence of (1).   □

3.7.1 Exterior Product of Anderson Motives

Now we discuss the existence of W_a with the desired properties. Note that W_a is supposed to induce an isomorphism of A-modules

$$\bigwedge_{A/aA}^{r} \phi[a] \xrightarrow{\sim} \psi[a]. \tag{3.7.4}$$

This observation suggests a natural plan for the construction of W_a, which was carried out by Van der Heiden [vdH04]; there are also earlier closely related results by Taguchi [Tag95a] and Hamahata [Ham93].

Let M_ϕ be the Anderson motive associated with ϕ. Since M_ϕ is a free $K[T]$-module of rank r, we can consider its exterior power $\bigwedge_{K[T]}^{r} M_\phi$, which is a free $K[T]$-module of rank 1. If we knew that $\bigwedge_{K[T]}^{r} M_\phi$ is also a free $K\{\tau\}$-module of rank 1, then, by the equivalence of categories in Proposition 3.4.5, it would correspond to the Anderson motive of a rank-1 Drinfeld module ψ. Under this construction, (3.7.4) should be a natural consequence. The problem with this approach is that $\bigwedge_{K[T]}^{r} M_\phi$ is, by definition, the quotient of the tensor product $M_\phi \otimes_{K[T]} M_\phi \otimes \cdots \otimes_{K[T]} M_\phi$ (r factors), so first one needs to consider this module. But it is clear that $\bigotimes_{K[T]}^{r} M_\phi$ does not have rank 1 over $K\{\tau\}$ if $r > 1$, so even though the starting and ending objects in $M_\phi \rightsquigarrow M_\psi = \bigwedge^{r} M_\phi$ are in the category of Drinfeld modules, the actual construction has to take place in a larger category[11] of left $K[T, \tau]$-modules, where one can form tensor products and quotients.

We will not deal with the technical issues in the construction of $M := \bigwedge_{K[T]}^{r} M_\phi$ and will simply assume that M exists and has the usual properties of an exterior product. In particular, M is free of rank 1 over $K\{\tau\}$. Then $M = M_\psi$ for a Drinfeld module of rank 1 defined by $\psi_T = t + c\tau$ for some $c \in K^\times$. Let $\phi_T = t + g_1\tau + \cdots + g_r\tau^r$. Since M_ϕ is freely generated by $1, \tau, \ldots, \tau^{r-1}$ over $K[T]$ (see the proof of Lemma 3.4.4), M_ψ is freely generated by $e = 1 \wedge \tau \wedge \cdots \wedge \tau^{r-1}$ over $K[T]$. On the one hand, since $\tau = c^{-1}(T - t)$ on M_ψ, we have

$$\tau e = c^{-1}(T - t)e. \tag{3.7.5}$$

---

[11] This category is the category of $K[T, \tau]$-modules which are free of finite rank over both $K[T]$ and $K\{\tau\}$, satisfy $(T - t)^n(\overline{M/\tau M}) = 0$ for $n \gg 0$, and an extra technical condition called "purity"; cf. [And86].

On the other hand,

$$
\tau e = \tau \left(1 \wedge \tau \wedge \cdots \wedge \tau^{r-1} \right)
$$

$$
= \tau \wedge \tau^2 \wedge \cdots \wedge \tau^{r-1} \wedge \left(g_r^{-1}(T - t - g_1\tau - \cdots - g_{r-1}\tau^{r-1}) \right)
$$

$$
= \tau \wedge \tau^2 \wedge \cdots \wedge \tau^{r-1} \wedge \left(g_r^{-1}(T - t) \right)
$$

$$
= (-1)^{r-1} \left(g_r^{-1}(T - t) \right) \left(1 \wedge \tau \wedge \cdots \wedge \tau^{r-1} \right)
$$

$$
= (-1)^{r-1} \left(g_r^{-1}(T - t) \right) e. \tag{3.7.6}
$$

Comparing (3.7.5) and (3.7.6), we get

$$
c = (-1)^{r-1} g_r.
$$

Thus, ψ is defined by (3.7.1).

For the rest of this subsection, we follow [vdH04, §§5-7]. Assume $a \in A$ is relatively prime to $\mathrm{char}_A(K)$.

Lemma 3.7.2 *With above notation, there is a canonical isomorphism*

$$
(\overline{M}_\psi / a\overline{M}_\psi)^\tau \cong \bigwedge_{K[T]}^r (\overline{M}_\phi / a\overline{M}_\phi)^\tau.
$$

Proof We leave the proof of this lemma as an exercise (see Exercise 3.7.2). □

Let

$$
\mathrm{Det}:\ \prod_{i=1}^r \mathrm{Hom}_{A/aA}((\overline{M}_\phi / a\overline{M}_\phi)^\tau, (A/aA)^\vee)
$$

$$
\longrightarrow \mathrm{Hom}_{A/aA} \left(\bigwedge_{K[T]}^r (\overline{M}_\phi / a\overline{M}_\phi)^\tau, ((A/aA)^\vee)^{\otimes r} \right)
$$

be defined by

$$
(f_1, \ldots, f_r)
$$

$$
\longmapsto \left(m_1 \wedge \cdots \wedge m_r \longmapsto \sum_{\sigma \in S_r} \mathrm{sgn}(\sigma) \left(f_1(m_{\sigma(1)}) \otimes \cdots \otimes f_r(m_{\sigma(r)}) \right) \right),
$$

where S_r is the group of permutations of $\{1, \ldots, r\}$. Since (cf. Lemma 3.6.2)

$$
(A/aA)^\vee = \mathrm{Hom}_{\mathbb{F}_q}(A/aA, \mathbb{F}_q) \cong A/aA,
$$

we have

$$\mathrm{Hom}_{A/aA}\left(\bigwedge_{K[T]}^{r}(\overline{M}_{\phi}/a\overline{M}_{\phi})^{\tau},\,((A/aA)^{\vee})^{\otimes r}\right)$$

$$\cong \mathrm{Hom}_{A/aA}\left(\bigwedge_{K[T]}^{r}(\overline{M}_{\phi}/a\overline{M}_{\phi})^{\tau},\,(A/aA)^{\vee}\right)\otimes_{A/aA}((A/aA)^{\vee})^{\otimes(r-1)}$$

$$\cong \mathrm{Hom}_{A/aA}\left((\overline{M}_{\psi}/a\overline{M}_{\psi})^{\tau},\,(A/aA)^{\vee}\right)\otimes_{A/aA}((A/aA)^{\vee})^{\otimes(r-1)},$$

where the last isomorphism is due to Lemma 3.7.2. Now, using (3.6.9), we get a map

$$\mathrm{Det}:\ \prod_{i=1}^{r}\phi[a]\longrightarrow \psi[a]\otimes_{A/aA}((A/aA)^{\vee})^{\otimes(r-1)}.$$

Finally, fixing an isomorphism $(A/aA)^{\vee}\cong A/aA$, we obtain a map

$$W_{a}:\ \prod_{i=1}^{r}\phi[a]\longrightarrow \psi[a].$$

Theorem 3.7.3 *The map W_a has the properties (1)–(5) listed at the beginning of this section.*

Proof This follows from the well-known properties of the determinant since Det is defined similarly to the usual determinant. We leave the details of the verification to the reader.
□

Remark 3.7.4 Let B be an abelian variety over a field K. Let m be a positive integer not divisible by the characteristic of K. The dual of the multiplication-by-m isogeny $m_B: B \to B$ is the multiplication-by-m on the dual abelian variety $m_{B^\vee}: B^\vee \to B^\vee$, and the kernel $B[m] := \ker(m_B)$ is the Cartier dual of $B^\vee[m] := \ker(m_{B^\vee})$. This leads to a pairing

$$e_m:\ B[m] \times B^\vee[m] \to \mathbb{G}_{m,K}[m],$$

which is the classical Weil pairing; cf. [Mil86, §16].

Now the map W_a can be similarly interpreted as a pairing between the a-torsion points of ϕ and its "dual" $\phi^\vee = \wedge^{r-1}\phi$, where $\phi^\vee$ is defined as the Anderson module corresponding to the motive $\bigwedge_{K[T]}^{r-1} M_\phi$. It is important to note that $\phi^\vee$ is an Anderson module of dimension[12] $r - 1$, so it is a Drinfeld module if and

---

[12] This essentially means that $\phi^\vee$ defines an embedding $A \to \mathrm{Mat}_{r-1}(K\{\tau\})$.

only if $r = 2$. This is the perspective on W_a in [Tag95a], where Taguchi shows that W_a is an incarnation of an analogue of Cartier duality for Anderson modules. This analogy was pushed even further by Papanikolas and Ramachandran in [PR03] who constructed a group extension $\text{Ext}^1(\phi, C)$ and showed that it is essentially isomorphic to $\phi^\vee$; the analogy with abelian varieties becomes apparent once one recalls that $B^\vee \cong \text{Ext}^1(B, \mathbb{G}_m)$ (see [Mil86, §11]) and that the Carlitz module C in the Drinfeld context plays the role of the multiplicative group.

Example 3.7.5 The following instructive example is due to Van der Heiden [vdH04, §7.1] (see also [Ham93, §4]).

Assume $r = 2$. We take 1 and τ as a basis of $\overline{M}_\phi$ over $\overline{K}[T]$ and $1 \wedge \tau$ as the generator of $\wedge^2 \overline{M}_\phi$. The isomorphism $\wedge^2 \overline{M}_\phi \cong \overline{M}_\psi$ is given by $1 \wedge \tau \mapsto 1$. Let $\alpha, \beta \in \phi[a]$, and let f_α, f_β be the corresponding elements of $\text{Hom}_{A/aA}\left((\overline{M}_\phi/a\overline{M}_\phi)^\tau, (A/aA)^\vee\right)$. Then

$$\text{Det}(f_\alpha, f_\beta)(1 \wedge \tau) = f_\alpha(1) \otimes f_\beta(\tau) - f_\alpha(\tau) \otimes f_\beta(1).$$

Next, from the definition of the map (3.6.6), we have

$$f_\alpha(m) = (b \mapsto m(\phi_b(\alpha))).$$

Let $n = \deg_T(a)$. Let $e_0, e_1, \ldots, e_{n-1}$ be a basis of the $\mathbb{F}_q$-vector space $(A/aA)^\vee = \text{Hom}_{\mathbb{F}_q}(A/aA, \mathbb{F}_q)$ such that $e_i(T^j) = \delta_{i,j} =$ the Kronecker delta. Then

$$f_\alpha(1) = \sum_{i=0}^{n-1} \phi_{T^i}(\alpha) \otimes e_i,$$

$$f_\alpha(\tau) = \sum_{i=0}^{n-1} \phi_{T^i}(\alpha)^q \otimes e_i.$$

We have the same for f_β. Therefore,

$$\text{Det}(f_\alpha, f_\beta)(1 \wedge \tau) = \sum_{i,j=0}^{n-1} \left(\phi_{T^i}(\alpha)\phi_{T^j}(\beta)^q - \phi_{T^i}(\alpha)^q \phi_{T^j}(\beta)\right) \otimes e_i \otimes e_j$$

$$= \sum_{j=0}^{n-1} \left(\sum_{i=0}^{n-1} \left(\phi_{T^i}(\alpha)\phi_{T^j}(\beta)^q - \phi_{T^i}(\alpha)^q \phi_{T^j}(\beta)\right) \otimes e_i\right) \otimes e_j$$

$$= \sum_{j=0}^{n-1} g_j \otimes e_j,$$

where

$$g_j : \overline{M}_\psi \longrightarrow \overline{K} \otimes (A/aA)^\vee$$

is given by

$$g_j(1) = \sum_{i=0}^{n-1} \left(\phi_{T^i}(\alpha)\phi_{T^j}(\beta)^q - \phi_{T^i}(\alpha)^q \phi_{T^j}(\beta) \right) \otimes e_i.$$

Using (3.6.7), we conclude that g_j corresponds to the element $w_j \in \psi[a]$ with

$$\psi_{T^i}(w_j) = \phi_{T^i}(\alpha)\phi_{T^j}(\beta)^q - \phi_{T^i}(\alpha)^q \phi_{T^j}(\beta).$$

In particular,

$$w_j = \psi_1(w_j) = \alpha\phi_{T^j}(\beta)^q - \alpha^q \phi_{T^j}(\beta).$$

Thus, we get a map

$$\phi[a] \times \phi[a] \longrightarrow \psi[a] \otimes_{A/aA} (A/aA)^\vee \qquad (3.7.7)$$

$$(\alpha, \beta) \longmapsto \sum_{j=0}^{n-1} w_j \otimes e_j.$$

Let $a = a_0 + a_1 T + \cdots + a_{n-1}T^{n-1} + a_n T^n$ with $a_n = 1$. Observe that

$$T \circ e_0 = -a_0 e_{n-1}, \qquad T \circ e_i = e_{i-1} - a_i e_{n-1}, \quad 1 \le i \le n - 1.$$

This implies that

$$e_i = \left(\sum_{j=0}^{n-i-1} a_{i+j+1}T^j \right) \circ e_{n-1}.$$

In particular, e_{n-1} generates $(A/aA)^\vee$ over A/aA. We define an isomorphism

$$(A/aA)^\vee \xrightarrow{\sim} A/aA$$

$$e_{n-1} \longmapsto 1.$$

Under this isomorphism, (3.7.7) becomes

$$W_a : \phi[a] \times \phi[a] \longrightarrow \psi[a] \qquad (3.7.8)$$

$$(\alpha, \beta) \longmapsto \sum_{i=0}^{n-1} \left(\sum_{j=0}^{n-i-1} a_{i+j+1}T^j \right) w_l.$$

In particular, for $a = T^n$, we have

$$W_{T^n}(\alpha, \beta) = \sum_{i=0}^{n-1} \psi_{T^{n-1-i}} \left(\alpha \phi_{T^i}(\beta)^q - \alpha^q \phi_{T^i}(\beta) \right).$$

3.7.2 Weil Pairing via Explicit Formulas

In this subsection, we discuss an alternative approach to the Weil pairing on Drinfeld modules based on explicit formulas. This approach is much more elementary than the approach via Anderson motives but has the disadvantage of being less conceptual. One essentially writes down an explicit polynomial $W_a(x_1, \ldots, x_r)$ in r variables which when evaluated on $\prod_{i=1}^{r} \phi[a]$ gives a map $\prod_{i=1}^{r} \phi[a] \to \psi[a]$ with the desired properties (1)-(5). Here we will do this only for $r = 2$. A generalization of this to higher ranks is indicated in Exercise 3.7.3 for $a = T^n$; this was further extended to arbitrary a and r by Katen [Kat21].

For the rest of this section, ϕ will be a Drinfeld module of rank 2 over K defined by $\phi_T = t + g_1 \tau + g_2 \tau^2$, and ψ will be the Drinfeld module of rank 1 over K defined by $\psi_T = t - g_2 \tau$. For each nonzero $a \in A$, we define a polynomial $W_a(x, y) \in K[x, y]$, using the Moore determinant and the division polynomials of ϕ. Let

$$M(x, y) = xy^q - x^q y$$

be the 2×2 Moore determinant, considered now as a polynomial in $K[x, y]$. For $n \geq 0$, define a polynomial $W_{T^n}(x, y) \in K[x, y]$ by

$$W_1(x, y) = 0,$$

and

$$W_{T^n}(x, y) = M(x, \phi_{T^{n-1}}(y)) + M(\phi_T(x), \phi_{T^{n-2}}(y))$$
$$+ \cdots + M(\phi_{T^{n-1}}(x), y), \quad n \geq 1.$$

Now, for $a = a_0 + a_1 T + \cdots + a_n T^n \in A$, define

$$W_a(x, y) = \sum_{i=0}^{n} a_i W_{T^i}(x, y). \tag{3.7.9}$$

Lemma 3.7.6 *The polynomials W_a have the following properties:*

(1) They are additive in a's:

$$W_{a+b}(x, y) = W_a(x) + W_b(y) \quad \text{for all } a, b \in A.$$

(2) They are $\mathbb{F}_q$-bilinear:

$$W_a(\alpha x, y) = W_a(x, \alpha y) = \alpha W_a(x, y) \quad \text{for all } \alpha \in \mathbb{F}_q,$$

$$W_a(x_1 + x_2, y) = W_a(x_1, y) + W_a(x_2, y),$$

$$W_a(x, y_1 + y_2) = W_a(x, y_1) + W_a(x, y_2).$$

(3) They are alternating:

$$W_a(x, x) = 0.$$

In particular, $W_a(x, y) = -W_a(y, x)$.
(4) For $n \geq 1$, the degree of $W_a(x, y)$ as a polynomial in x is $q^{2\deg(a)-1}$, and its leading coefficient is αy for some $\alpha \in K^\times$.

Proof

(1) This property is immediate from the definition.
(2) The $\mathbb{F}_q$-bilinearity easily follows from the fact that $\phi_{T^i}(x)$ is $\mathbb{F}_q$-linear and $M(x, y)$ is $\mathbb{F}_q$-bilinear.
(3) Note that $M(x, y) = -M(y, x)$ and $M(x, x) = 0$. Thus, for any $n \geq 1$ and $i \neq (n-1)/2$

$$M(\phi_{T^i}(x), \phi_{T^{n-i-1}}(x)) + M(\phi_{T^{n-i-1}}(x), \phi_{T^i}(x)) = 0,$$

and for odd n,

$$M(\phi_{T^{(n-1)/2}}(x), \phi_{T^{(n-1)/2}}(x)) = 0.$$

Now, from the definition of $W_{T^n}(x, y)$, it follows that $W_{T^n}(x, x) = 0$ and so $W_a(x, x) = 0$.
(4) Note that

$$\deg_x M(\phi_{T^i}(x), \phi_{T^{n-i-1}}(y)) = \deg_x((\phi_{T^i}(x))^q)$$

$$= q^{1+2i}$$

$$> \deg_x(M(\phi_{T^{i-1}}(x), \phi_{T^{n-i}}(y))).$$

Hence, from the definition of W_{T^n} as a sum of Moore determinants, we get

$$\deg_x(W_{T^n}(x, y)) = \deg_x M(\phi_{T^{n-1}}(x), y) = q^{2n-1}.$$

Moreover, the leading coefficient of $W_{T^n}(x, y)$ is the same as the leading coefficient of $M(\phi_{T^{n-1}}(x), y)$ as a polynomial in x, which is y times the leading coefficient of $\phi_{T^{n-1}}(x)$ (the latter is an element of $K^\times$). Now, from the definition of $W_a(x, y)$, we see that the leading term of $W_a(x, y)$ as a polynomial in x comes from the leading term of $W_{T^n}(x, y)$, and this implies the claim.

$\square$

Lemma 3.7.7 *We have*

$$\psi_T(W_a(x, y)) = W_a(\phi_T(x), y) + (\phi_a(x)^q y - x^q \phi_a(y)).$$

Proof First, we prove this for $a = T^n$, $n \geq 0$. For $n = 0$, the claim is trivial since

$$\phi_{T^0}(x)^q y - x^q \phi_{T^0}(y) = x^q y - x^q y = 0.$$

For $n \geq 1$, we will need the auxiliary formula

$$\psi_T(M(x, y)) = \phi_T(x)y^q - x^q \phi_T(y),$$

which can be easily proved by expanding the left-hand side. Now, using this formula and the $\mathbb{F}_q$-linearity of ψ_T, we get

$$\psi_T(W_{T^n}(x, y)) = \sum_{i=0}^{n-1} \psi_T(M(\phi_{T^i}(x), \phi_{T^{n-1-i}}(y)))$$

$$= \sum_{i=0}^{n-1} \left(\phi_{T^{i+1}}(x)\phi_{T^{n-1-i}}(y)^q - \phi_{T^i}(x)^q \phi_{T^{n-i}}(y)\right).$$

Denote $f_i = \phi_{T^{i+1}}(x)\phi_{T^{n-1-i}}(y)^q - \phi_{T^i}(x)^q \phi_{T^{n-i}}(y)$. For each $0 \leq i \leq n - 2$, by grouping together the first summand in f_i with the second summand in f_{i+1}, we can rewrite the previous expression as

$$\sum_{i=1}^{n-1} M(\phi_{T^i}(x), \phi_{T^{n-i}}(y)) + (\phi_{T^n}(x)y^q - x^q \phi_{T^n}(y))$$

$$= W_{T^n}(\phi_T(x), y) - M(\phi_{T^n}(x), y) + (\phi_{T^n}(x)y^q - x^q \phi_{T^n}(y))$$

$$= W_{T^n}(\phi_T(x), y) + (\phi_{T^n}(x)^q y - x^q \phi_{T^n}(y)).$$

For general $a = a_0 + a_1 T + \cdots + a_n T^n \in A$, using the already proved formula for T^n, we get

$$\psi_T(W_a(x, y)) = \sum_{i=0}^{n} a_i \psi_T W_{T^i}(x, y)$$

$$= \sum_{i=0}^{n} a_i W_{T^i}(\phi_T(x), y) + \sum_{i=0}^{n} a_i(\phi_{T^i}(x)^q y - x^q \phi_{T^i}(y))$$

$$= W_a(\phi_T(x), y) + \left(\sum_{i=0}^{n} a_i \phi_{T^i}(x)\right)^q y - x^q \sum_{i=0}^{n} a_i \phi_{T^i}(y)$$

$$= W_a(\phi_T(x), y) + (\phi_a(x)^q y - x^q \phi_a(y)).$$

$\square$

Proposition 3.7.8 *Assume* $x, y \in \phi[a]$. *Then, for any* $b \in A$, *we have*

$$\psi_b(W_a(x, y)) = W_a(\phi_b(x), y),$$
$$\psi_b(W_a(x, y)) = W_a(x, \phi_b(y)).$$

Hence, $W_a : \phi[a] \times \phi[a] \to \psi[a]$ *is* A-*bilinear and alternating.*

Proof If $x, y \in \phi[a]$, then $\phi_a(x) = \phi_a(y) = 0$, so by Lemma 3.7.7 we have

$$\psi_T(W_a(x, y)) = W_a(\phi_T(x), y).$$

Write $b = b_0 + b_1 T + \cdots + b_m T^m$. Since $\psi_b(x) = \sum_{i=0}^{m} b_i \psi_{T^i}$, iterating the previous identity, we obtain

$$\psi_b(W_a(x, y)) = \sum_{i=0}^{m} b_i W(\phi_{T^i}(x), y).$$

On the other hand, $W_a(x, y)$ is $\mathbb{F}_q$-bilinear, so

$$\sum_{i=0}^{m} b_i W(\phi_{T^i}(x), y) = W_a\left(\sum_{i=0}^{m} b_i \phi_{T^i}(x), y\right) = W_a(\phi_b(x), y).$$

The A-linearity in the second variable follows from

$$\psi_b(W_a(x, y)) = \psi_b(-W_a(y, x)) = -\psi_b(W_a(y, x))$$
$$= -W_a(\phi_b(y), x) = W_a(x, \phi_b(y)).$$

$\square$

Proposition 3.7.9 *Let $a, b \in A$ and $x \in \phi[a]$. Then*

$$W_{ab}(x, y) = W_a(x, \phi_b(y)).$$

Proof First, note that

$$
\begin{aligned}
W_{T^n}(x, y) &= M(x, \phi_{T^{n-1}}(y)) + M(\phi_T(x), \phi_{T^{n-2}}(y)) + \cdots + M(\phi_{T^{n-1}}(x), y) \\
&= M(x, \phi_{T^{n-2}}(\phi_T(y))) + M(\phi_T(x), \phi_{T^{n-3}}(\phi_T(y))) \\
&\quad + \cdots + M(\phi_{T^{n-2}}(x), \phi_T(y)) + M(\phi_{T^{n-1}}(x), y) \\
&= W_{T^{n-1}}(x, \phi_T(y)) + M(\phi_{T^{n-1}}(x), y).
\end{aligned}
$$

Using this observation, we obtain

$$
\begin{aligned}
W_{aT}(x, y) &= \sum_{i=0}^{n} a_i W_{T^{i+1}}(x, y) \\
&= \sum_{i=0}^{n} a_i (W_{T^i}(x, \phi_T(y)) + M(\phi_{T^i}(x), y)) \\
&= W_a(x, \phi_T(y)) + M\left(\sum_{i=0}^{n} a_i \phi_{T^i}(x), y\right) \\
&= W_a(x, \phi_T(y)) + M(\phi_a(x), y) \\
&= W_a(x, \phi_T(y)),
\end{aligned}
$$

where in the last equality we have used the assumption that $\phi_a(x) = 0$.

Now write $b = b_0 + b_1 T + \cdots + b_m T^m$. Using (1) and (2) of Lemma 3.7.6, we get

$$
\begin{aligned}
W_{ab}(x, y) &= \sum_{j=0}^{m} b_j W_{aT^j}(x, y) \\
&= \sum_{j=0}^{m} b_j W_a\left(x, \phi_{T^j}(y)\right) \\
&= W_a(x, \sum_{j=0}^{m} b_j \phi_{T^j}(y)) \\
&= W_a(x, \phi_b(y)),
\end{aligned}
$$

as was claimed.   □

Let $a \in A$. By Proposition 3.7.8, if $x, y \in \phi[a]$, then $\psi_a(W_a(x, y)) = W_a(\phi_a(x), y) = W_a(0, y) = 0$. This allows us to define a pairing

$$W_a : \phi[a] \times \phi[a] \longrightarrow \psi[a],$$

$$(x, y) \longmapsto W_a(x, y).$$

Now, we are ready to prove the main theorem of this section:

Theorem 3.7.10 *The pairing* $W_a : \phi[a] \times \phi[a] \rightarrow \psi[a]$ *has the following properties:*

(1) It is A-bilinear and alternating.
(2) It is nondegenerate and surjective, assuming $\text{char}_A(K)$ *does not divide a.*
(3) It is Galois equivariant:

$$\sigma W_a(x, y) = W_a(\sigma x, \sigma y) \quad \text{for all } \sigma \in G_K.$$

(4) If $a, b \in A$ *and* $x, y \in \phi[ab]$, *then*

$$\psi_b(W_{ab}(x, y)) = W_a(\phi_b(x), \phi_b(y)).$$

Proof

(1) This is Proposition 3.7.8.
(2) Recall from Lemma 3.7.6 that $W_a(x, y)$, as a polynomial in x, has degree $q^{2 \deg(a)-1}$ and its leading coefficient is αy for some $\alpha \in K^{\times}$. Also, recall that the number of elements of $\phi[a]$ is $q^{2 \deg(a)}$, assuming $\text{char}_A(K)$ does not divide a. If for some fixed y we have $W_a(x, y) = 0$ for all $x \in \phi[a]$, then we have a polynomial of degree $q^{2 \deg(a)-1}$, which has $q^{2 \deg(a)}$ distinct roots. This is possible if and only if the polynomial is zero. In particular, its leading coefficient must be 0, so $y = 0$. This proves that the Weil pairing is nondegenerate.
   The image of W_a is an A-submodule of $\psi[a] \cong A/aA$. If it is a proper submodule, then there is b such that $b \mid a$, $\deg(b) < \deg(a)$, and

$$0 = \psi_b(W_a(x, y)) = W_a(\phi_b(x), y) \quad \text{for all } x, y \in \phi[a].$$

   The nondegeneracy of the W_a-pairing implies that $\phi_b(x) = 0$. This leads to a contradiction since x is arbitrary and $\deg(b) < \deg(a)$. Thus, the Weil pairing is surjective.
(3) The Galois equivariance is obvious since $W_a(x, y)$, as a polynomial, has coefficients in K.
(4) If $x \in \phi[ab]$, then $\phi_b(x) \in \phi[a]$. Hence,

$$\psi_b(W_{ab}(x, y)) = W_{ab}(\phi_b(x), y) = W_a(\phi_b(x), \phi_b(y)),$$

where the first equality follows from Proposition 3.7.8 and the second from Proposition 3.7.9.

$\square$

3.7.3 Adjoint of a Drinfeld Module

In this subsection, we discuss another, completely different, "Weil pairing" for Drinfeld modules constructed by Poonen [Poo96]. This construction is based on the concept of adjoint of a Drinfeld module, which essentially goes back to Ore [Ore33].

Assume K is perfect, and let $K\{\tau^{-1}\}$ be the ring of *twisted polynomials in* τ^{-1} with coefficients in K; the multiplication in this ring is given by the rule

$$\tau^{-1}a = a^{1/q}\tau^{-1} \quad \text{for all } a \in K.$$

Note that $K\{\tau^{-1}\}$ is naturally a subring of the ring $K((\tau^{-1}))$ that we defined in §3.4.2. Given a polynomial $f = \sum_{i=0}^{n} a_i \tau^i \in K\{\tau\}$ of degree $n \geq 0$, the *adjoint of* f is

$$f^* := \sum_{i=0}^{n} a_i^{1/q^i} \tau^{-i} \in K\{\tau^{-1}\}.$$

One easily verifies that for all $f, g \in K\{\tau\}$, we have $(f + g)^* = f^* + g^*$ and

$$(fg)^* = g^* f^*. \tag{3.7.10}$$

Hence, the map $K\{\tau\} \to K\{\tau^{-1}\}$, $f \mapsto f^*$, is an anti-isomorphism.

Note that f^* defines an $\mathbb{F}_q$-linear operator on $\overline{K}$, and we denote by $\ker(f^*)$ the kernel of this operator:

$$\ker(f^*) = \left\{ \beta \in \overline{K} \;\middle|\; \sum_{i=0}^{n} a_i^{1/q^i} \beta^{1/q^i} = 0 \right\}.$$

Assume $f \in K\{\tau\}$ is nonzero. Then $\ker(f)$ is a finite dimensional vector space over $\mathbb{F}_q$, and we claim that $\ker(f^*)$ has the same dimension over $\mathbb{F}_q$. This follows from the following two observations:

- $\beta \in \ker(f^*)$ if and only if $\beta \in \ker(\tau^n \cdot f^*)$.
- $\tau^n \cdot f^* \in K\{\tau\}$ and $\deg(\tau^n \cdot f^*) - \mathrm{ht}(\tau^n \cdot f^*) = \deg(f) - \mathrm{ht}(f)$.

Remark 3.7.11 There are explicit formulas due to Ore [Ore33] for a basis of $\ker(f^*)$ in terms of a basis of $\ker(f)$; see Exercise 3.7.7.

Let $\alpha \in \ker f$ and $\beta \in \ker f^*$. Then $f \cdot \alpha$ vanishes on the elements of $\mathbb{F}_q$ since $(f \cdot \alpha)(s) = s \cdot f(\alpha) = 0$ for all $s \in \mathbb{F}_q$. (Here $f \cdot \alpha$ denotes the product of f and α as elements of $K\{\tau\}$.) Since $\ker(1 - \tau) = \mathbb{F}_q$, applying Corollary 3.1.16, we may write

$$f \cdot \alpha = g_\alpha \cdot (1 - \tau)$$

for a uniquely determined $g_\alpha \in K\{\tau\}$. The *Elkies–Poonen pairing* for f is the map

$$\langle \cdot, \cdot \rangle_f : \ker f \times \ker f^* \longrightarrow \overline{K}$$

given by

$$\langle \alpha, \beta \rangle_f = g_\alpha^*(\beta).$$

This pairing was independently defined by Poonen [Poo96] and Elkies [Elk99].

Proposition 3.7.12 *The pairing $\langle \cdot, \cdot \rangle_f$ takes values in $\mathbb{F}_q$. Moreover, $\langle \cdot, \cdot \rangle_f$ is $\mathbb{F}_q$-bilinear, non-degenerate, and G_K-equivariant, i.e., $\langle \sigma(\alpha), \sigma(\beta) \rangle_f = \langle \alpha, \beta \rangle_f$ for all $\sigma \in G_K$, $\alpha \in \ker f$, $\beta \in \ker f^*$.*

Proof Let $\alpha \in \ker f$ and $\beta \in \ker f^*$. Taking the adjoints of both sides of $f \cdot \alpha = g_\alpha \cdot (1 - \tau)$ and using (3.7.10), we get

$$\alpha \cdot f^* = (1 - \tau^{-1}) \cdot g_\alpha^*.$$

Evaluating both sides of the above equation at β gives

$$\alpha \cdot f^*(\beta) = 0 = \left((1 - \tau^{-1}) \cdot g_\alpha^* \right)(\beta) = g_\alpha^*(\beta) - g_\alpha^*(\beta)^{1/q}.$$

Thus, $g_\alpha^*(\beta)^q = g_\alpha^*(\beta)$, which implies that $g_\alpha^*(\beta) \in \mathbb{F}_q$.

Next, note that for $\alpha, \alpha' \in \ker(f)$ and $s \in \mathbb{F}_q$, we have $g_{\alpha + \alpha'} = g_\alpha + g_{\alpha'}$ and $g_{s\alpha} = sg_\alpha$ because $f \cdot (s\alpha) = sf \cdot \alpha$. Thus, $\langle \cdot, \cdot \rangle_f$ is $\mathbb{F}_q$-linear in the first argument. The $\mathbb{F}_q$-linearity in the second argument follows from the $\mathbb{F}_q$-linearity of g_α^*.

If $\langle \alpha, \beta \rangle_f = 0$ for all $\beta \in \ker f^*$, with α being fixed, then $\dim \ker g_\alpha^* \geq \dim \ker f^*$. On the other hand, if $\alpha \neq 0$, then

$$\dim \ker g_\alpha^* = \dim \ker g_\alpha = \dim(\ker f) - 1 = \dim(\ker f^*) - 1,$$

which is a contradiction. Thus, the kernel of the map

$$\ker f \longrightarrow \operatorname{Hom}(\ker f^*, \mathbb{F}_q), \quad \alpha \longmapsto \langle \alpha, \cdot \rangle_f,$$

is trivial. Since $\dim \ker f = \dim \ker f^*$, a standard fact from linear algebra (see Exercise 3.7.5) implies that the kernel of the map

$$\ker f^* \longrightarrow \mathrm{Hom}(\ker f, \mathbb{F}_q), \quad \beta \longmapsto \langle \cdot, \beta \rangle_f,$$

is also trivial. This means that the pairing $\langle \cdot, \cdot \rangle_f$ is non-degenerate.

Finally, because f has coefficients in K, for $\sigma \in G_K$, we have $f \cdot \sigma(\alpha) = \sigma(f \cdot \alpha) = \sigma(g_\alpha \cdot (1 - \tau)) = \sigma(g_\alpha) \cdot (1 - \tau)$. Also, $(\sigma(g_\alpha))^* = \sigma(g_\alpha^*)$. Thus, $\sigma(g_\alpha^*) = g_{\sigma(\alpha)}^*$. Using this observation, we compute

$$\langle \sigma(\alpha), \sigma(\beta) \rangle_f = g_{\sigma(\alpha)}^*(\sigma(\beta)) = (\sigma(g_\alpha^*))(\sigma(\beta)) = \sigma(g_\alpha^*(\beta)) = g_\alpha^*(\beta) = \langle \alpha, \beta \rangle_f,$$

where the equality $\sigma(g_\alpha^*(\beta)) = g_\alpha^*(\beta)$ follows from the fact that $g_\alpha^*(\beta) \in \mathbb{F}_q$. $\square$

Definition 3.7.13 Given a Drinfeld module $\phi: A \to K\{\tau\}$, the *adjoint of* ϕ is the map

$$\phi^*: A \longrightarrow K\left\{\tau^{-1}\right\}$$

$$a \longmapsto \phi_a^* := (\phi_a)^*.$$

The adjoint ϕ^* is an $\mathbb{F}_q$-linear homomorphism because

$$\phi_{ab}^* = (\phi_{ab})^* = (\phi_b \phi_a)^* = (\phi_a)^*(\phi_b)^* = \phi_a^* \phi_b^*.$$

Denote $\phi^*[a] = \ker \phi_a^*$. It is trivial to check that the set $\phi^*[a]$ has a natural structure of an A-module, with the action of A defined by $b \circ \beta = \phi_b^*(\beta)$ for all $b \in A$ and $\beta \in \phi^*[a]$.

Lemma 3.7.14 *Let a and b be nonzero elements of A. Then for all $\alpha \in \phi[a]$ and $\beta \in \phi^*[a]$ we have*

$$\langle \phi_b(\alpha), \beta \rangle_{\phi_a} = \langle \alpha, \phi_b^*(\beta) \rangle_{\phi_a}.$$

Proof Since $\phi_b(\alpha) \in \phi[a]$, we can write

$$\phi_a \cdot \alpha = g_\alpha \cdot (1 - \tau),$$

$$\phi_a \cdot \phi_b(\alpha) = g_{\phi_b(\alpha)} \cdot (1 - \tau).$$

Multiply the first of the above equations by ϕ_b on the left and then subtract the second:

$$\phi_a \cdot (\phi_b \cdot \alpha - \phi_b(\alpha)) = \left(\phi_b \cdot g_\alpha - g_{\phi_b(\alpha)}\right) \cdot (1 - \tau).$$

Since $\phi_b \cdot \alpha - \phi_b(\alpha)$ kills 1,

$$\phi_b \cdot \alpha - \phi_b(\alpha) = u \cdot (1 - \tau)$$

for some $u \in K\{\tau\}$. Substitute and cancel $(1 - \tau)$:

$$\phi_a \cdot u = \phi_b \cdot g_\alpha - g_{\phi_b(\alpha)}.$$

Take adjoints and evaluate both sides at $\beta \in \phi^*[a]$ to get

$$0 = g_\alpha^*(\phi_b^*(\beta)) - g_{\phi_b(\alpha)}^*(\beta)$$
$$= \langle \alpha, \phi_b^*(\beta) \rangle_{\phi_a} - \langle \phi_b(\alpha), \beta \rangle_{\phi_a}$$

as desired. □

Definition 3.7.15 Given $0 \neq a \in A$, the *Poonen–Weil pairing* $[\cdot, \cdot]_a$ is the pairing

$$\phi[a] \times \phi^*[a] \longrightarrow (A/aA)^\vee$$
$$\alpha, \beta \longmapsto [\alpha, \beta]_a = \left(b \longmapsto \langle \phi_b(\alpha), \beta \rangle_{\phi_a}\right).$$

Theorem 3.7.16 *The Poonen–Weil pairing $\phi[a] \times \phi^*[a] \to (A/aA)^\vee$ is a perfect G_K-equivariant pairing of finite A-modules, with G_K acting trivially on $(A/aA)^\vee$.*

Proof First, we check that $[\cdot, \cdot]_a$ is a pairing of A-modules. Treating $[\alpha, \beta]_a$ as a function $A/uA \to \mathbb{F}_q$, the previous sentence explicitly means checking that for all $b \in A$, we have

$$(c \circ [\alpha, \beta]_a)(x) := [\alpha, \beta]_a(cx) = [\phi_c(\alpha), \beta]_a(x) = [\alpha, \phi_c^*(\beta)]_a(x).$$

The last equality follows from Lemma 3.7.14. To show the middle equality, we compute that for any $b \in A$ we have

$$[\phi_c(\alpha), \beta]_a(b) = \langle \phi_b(\phi_c(\alpha)), \beta \rangle_{\phi_a} = \langle \phi_{bc}(\alpha), \beta \rangle_{\phi_a} = [\alpha, \beta]_a(cb).$$

The pairing $[\cdot, \cdot]_a$ induces a homomorphism of A-modules

$$\phi^*[a] \longrightarrow \mathrm{Hom}_A(\phi[a], (A/aA)^\vee), \quad \beta \longmapsto [\cdot, \beta]_a. \tag{3.7.11}$$

To show that the pairing is perfect, one needs to check that the above homomorphism is an isomorphism. But we know that $(A/aA)^\vee \cong A/aA$ (see Lemma 3.6.2) and $\phi[a]$ is a free A/aA-module. Thus, $\#\mathrm{Hom}_A(\phi[a], (A/aA)^\vee) = \#\phi[a]$. Since $\#\phi^*[a] = \#\phi[a]$, to prove the perfectness of $[\cdot, \cdot]_a$, it is enough to show that (3.7.11) is injective. This injectivity is equivalent to the claim that for any nonzero $\beta \in \phi^*[a]$, there is some $\alpha \in \phi[a]$ such that $[\alpha, \beta]_a$ is not the zero function. But this easily follows from the non-degeneracy of the pairing $\langle \cdot, \cdot \rangle_{\phi_a}$ established in Proposition 3.7.12. The G_K-equivariance also follows from the corresponding property of $\langle \cdot , \cdot \rangle_{\phi_a}$.   □

Corollary 3.7.17 *We have an isomorphism* $\phi[a] \cong \phi^*[a]$ *of A-modules and an equality* $K(\phi[a]) = K(\phi^*[a])$ *of field extensions.*

The Poonen–Weil pairing has the following compatibility property, whose proof we leave as an exercise:

Proposition 3.7.18 *Assume* $a, b \in A$ *are not nonzero. Then, for all* $\alpha \in \phi[ab]$ *and* $\beta \in \phi^*[ab]$, *we have*

$$b \circ [\alpha, \beta]_{ab} = [\phi_b(\alpha), \phi_b^*(\beta)]_a.$$

Let $\mathfrak{p}$ be a prime different from the A-characteristic of K. The natural maps

$$\phi^*[\mathfrak{p}^{i+1}] \xrightarrow{\phi_\mathfrak{p}^*} \phi^*[\mathfrak{p}^i]$$

fit into a projective limit $T_\mathfrak{p}(\phi^*)$, the Tate module of ϕ^*. Similar to the case of W_a pairing, Theorem 3.7.16 and Proposition 3.7.18 imply the existence of a perfect G_K-equivariant pairing of $A_\mathfrak{p}$-modules:

$$[\cdot , \cdot]_{\mathfrak{p}^\infty} : T_\mathfrak{p}(\phi) \times T_\mathfrak{p}(\phi^*) \to A_\mathfrak{p},$$

which is analogous to the Weil pairing between the Tate module of an abelian variety and its dual abelian variety; cf. Remark 3.7.4.

Exercises

3.7.1 Let $a \in A$ and $\alpha \in \mathrm{Aut}(\phi)$. For the pairing (3.7.9), show that $W_a(\alpha x, \alpha y) = \alpha^{1+q} W_a(x, y)$; cf. Lemma 3.8.2.

3.7.2 Prove Lemma 3.7.2.

3.7.3 Let ϕ and ψ be Drinfeld modules over K defined by $\phi_T = t + g_1 \tau + \cdots + g_r \tau^r$ and $\psi_T = t + (-1)^{r-1} g_r \tau$. Consider the Moore determinant $M(x_1, \ldots, x_r)$ as a polynomial in $K[x_1, \ldots, x_r]$. Prove the following statements:

(a)

$$M(x_1, \ldots, x_r) = \sum_{i=1}^{r} (-1)^{i-1} x_i M(x_1, x_2, \ldots, \widehat{x_i}, \ldots, x_r)^q$$

$$= (-1)^{r-1} \sum_{i=1}^{r} (-1)^{i-1} x_i^{q^{r-1}} M(x_1, x_2, \ldots, \widehat{x_i}, \ldots, x_r),$$

where $\widehat{x_i}$ indicates that x_i does not appear in the corresponding Moore determinant.

(b)

$$\psi_T(M(x_1,\ldots,x_r)) = \det \begin{pmatrix} \phi_T(x_1) \; x_1^q \cdots x_1^{q^{r-1}} \\ \phi_T(x_2) \; x_2^q \cdots x_2^{q^{r-1}} \\ \vdots \quad\; \vdots \; \cdots \; \vdots \\ \phi_T(x_r) \; x_r^q \cdots x_r^{q^{r-1}} \end{pmatrix}.$$

(c) $W_T(x_1,\ldots,x_r) = M(x_1,\ldots,x_r)$ gives a map $W_T\colon \prod_{i=1}^r \phi[T] \longrightarrow \psi[T]$, which is $\mathbb{F}_q$-linear in each variable and Galois equivariant. Moreover, if $T \neq \mathrm{char}_A(K)$, then W_T is nondegenerate and surjective.

(d) For $n \geq 1$, define

$$W_{T^n}(x_1,\ldots,x_r) = \sum_{\substack{0 \leq i_1,\ldots,i_r \leq n-1 \\ i_1+\cdots+i_r=(n-1)(r-1)}} M(\phi_{T^{i_1}}(x_1),\ldots,\phi_{T^{i_r}}(x_r)).$$

If $x_1,\ldots,x_r \in \phi[T^n]$, then

$$\psi_T(W_{T^n}(x_1,\ldots,x_r)) = W_{T^{n-1}}(\phi_T(x_1),\ldots,\phi_T(x_r)).$$

(e) Deduce that W_{T^n} gives a map $W_{T^n}\colon \prod_{i=1}^r \phi[T^n] \longrightarrow \psi[T^n]$, which is $\mathbb{F}_q$-linear in each variable and Galois equivariant. Moreover, if $T \neq \mathrm{char}_A(K)$, then W_{T^n} is nondegenerate and surjective.

3.7.4 Show that the pairings $\phi[a] \times \phi[a] \to \psi[a]$ defined by (3.7.8) and (3.7.9) are the same.

3.7.5 Let V and W be finite dimensional vector spaces over a field K. Let $\langle\cdot,\cdot\rangle\colon V \times W \to K$ be a bilinear form. Consider the linear maps:

$$f\colon V \longrightarrow \mathrm{Hom}_K(W,K), \quad f(v)(w) = \langle v, w\rangle$$

$$g\colon W \longrightarrow \mathrm{Hom}_K(V,K), \quad g(w)(v) = \langle v, w\rangle.$$

Show that any two of the following statements imply the third:

(i) f is injective.
(ii) g is injective.
(iii) $\dim(V) = \dim(W)$.

The bilinear form $\langle\cdot,\cdot\rangle$ is called *non-degenerate* if it satisfies these three conditions.

3.7.6 Assume K is perfect, and let $f \in K\{\tau\}$ be a nonzero element. Show that $\beta \in \overline{K}^\times$ lies in $\ker f^*$ if and only if

$$f^* = g^* \cdot \left(\beta^{(1-q)/q} - \tau^{-1} \right)$$

for some $g \in K\{\tau\}$.

3.7.7 Assume $f \in K\{\tau\}$ is separable of degree n. Let $\{w_1, \ldots, w_n\}$ be a basis of $\ker f$ over $\mathbb{F}_q$. For $1 \le i \le n$, let

$$v_i := \frac{M(w_1, \ldots, \widehat{w_i}, \ldots, w_n)}{M(w_1, \ldots, w_n)},$$

where $\widehat{w_i}$ means that w_i is omitted from the list $w_1, \ldots, w_n$. Prove that $\{v_1^q, \ldots, v_n^q\}$ is an $\mathbb{F}_q$-basis of $\ker f^*$.

3.7.8 Let f and h be nonzero elements of $K\{\tau\}$. Show that for all $\alpha \in \ker(f \cdot h)$ and $\beta \in \ker f^* \subseteq \ker(f \cdot h)^*$, we have

$$\langle \alpha, \beta \rangle_{f \cdot h} = \langle h(\alpha), \beta \rangle_f.$$

Similarly, show that for all $\alpha \in \ker f \subseteq \ker(h \cdot f)$ and $\beta \in \ker(h \cdot f)^*$, we have

$$\langle \alpha, \beta \rangle_{h \cdot f} = \langle \alpha, h^*(\beta) \rangle_f.$$

3.7.9 Prove Proposition 3.7.18.

3.8 Isomorphisms

In this section we consider the question of classifying Drinfeld modules up to isomorphism. This classification problem is of central importance in the arithmetic of function fields because its solution is given by an algebraic variety, called the *Drinfeld modular variety*, and this variety and its generalizations are the meeting ground of Galois representations and automorphic forms. This eventually leads to the proof of the Langlands conjectures over function fields, which is a much more advanced topic than those discussed in this book. For a comprehensive discussion of Drinfeld modular varieties and their applications to the Langlands program, the reader can consult [Lau96] and [Lau97].

In this section, we consider the simplest version of the classification problem of Drinfeld modules up to isomorphism—the version which does not involve "level structures." We will reconsider the same problem in Chap. 5 from the analytic perspective of uniformization of Drinfeld modules.

Let ϕ and ψ be Drinfeld modules over K. An isogeny $u: \phi \to \psi$ over K is an *isomorphism* if it has an inverse in $K\{\tau\}$, i.e., there is $v \in K\{\tau\}$ such that $uv = vu = 1$. Computing the degrees $\deg_\tau(uv) = \deg_\tau(u) + \deg_\tau(v) = 0$, we see that an isomorphism $\phi \to \psi$ over K is given by a nonzero constant $c \in K$ such that

$$c\phi_T c^{-1} = \psi_T. \tag{3.8.1}$$

If

$$\phi_T = t + g_1\tau + \cdots + g_r\tau^r,$$
$$\psi_T = t + h_1\tau + \cdots + h_r\tau^r,$$

then (3.8.1) is equivalent to

$$g_i = h_i c^{q^i - 1} \quad \text{for all} \quad 1 \le i \le r. \tag{3.8.2}$$

We say that ϕ and ψ are isomorphic over a field extension L of K, if there is $c \in L^\times$ satisfying (3.8.2).

Example 3.8.1 Suppose ϕ and ψ have rank 1. Let c be a root of $x^{q-1} = g_1/h_1$. Then $c\phi_T c^{-1} = \psi_T$, so ϕ and ψ are isomorphic over $K(\sqrt[q-1]{g_1/h_1})$. This implies that, up to isomorphism, the Carlitz module $C_T = t + \tau$ is the only Drinfeld module of rank 1 over K^{sep}. But note that $x^{q-1} = g_1/h_1$ might not have roots in K, so ϕ and ψ might not be isomorphic over K.

Lemma 3.8.2 *Let* $\phi_T = t + g_1\tau + \cdots + g_r\tau^r$ *be a Drinfeld module of rank r over K. Let*

$$m = \gcd\{i \mid 1 \le i \le r, \quad g_i \ne 0\}.$$

Then

$$\text{Aut}(\phi) \cong \mathbb{F}_{q^m}^\times.$$

(Note that m divides r, since $g_r \ne 0$ by assumption.)

Proof Let $0 \ne c \in \text{Aut}(\phi) \subset K^{\text{sep}}$. From (3.8.2), we get $c^{q^i-1} = 1$ if $g_i \ne 0$. Hence, if $g_i \ne 0$, then $c \in \mathbb{F}_{q^i}^\times$. Since $\mathbb{F}_{q^i} \cap \mathbb{F}_{q^j} = \mathbb{F}_{q^{\gcd(i,j)}}$, we must have $c \in \mathbb{F}_{q^m}^\times$. Conversely, any $c \in \mathbb{F}_{q^m}^\times$ satisfies (3.8.2). Therefore, $\text{Aut}(\phi) \cong \mathbb{F}_{q^m}^\times$. $\square$

Definition 3.8.3 (Gekeler [Gek83]) Assume that the rank of ϕ is 2 and that

$$\phi_T = t + g_1\tau + g_2\tau^2.$$

The *j-invariant of* ϕ is

$$j(\phi) = g_1^{q+1}/g_2.$$

(The terminology is motivated by the terminology from the theory of elliptic curves, where the j-invariant of an elliptic curve determines the isomorphism class of the curve.)

Lemma 3.8.4 *Two Drinfeld modules* ϕ *and* ψ *of rank* 2 *are isomorphic over* K^{sep} *if and only if* $j(\phi) = j(\psi)$.

Proof If c satisfies (3.8.2), then

$$j(\phi) = \frac{g_1^{q+1}}{g_2} = \frac{h_1^{q+1} c^{(q-1)(q+1)}}{h_2 c^{q^2-1}} = \frac{h_1^{q+1}}{h_2} = j(\psi),$$

where $\phi_T = t + g_1\tau + g_2\tau^2$ and $\psi_T = t + h_1\tau + h_2\tau^2$.

Conversely, assume $j(\phi) = j(\psi)$. If $g_1 = 0$, then $h_1 = 0$. In that case, if we let c be a root of $x^{q^2-1} = g_2/h_2$, then $c^{-1}\psi c = \phi$. If $g_1 \neq 0$, then let c be a root of $x^{q-1} = g_1/h_1$, so that $h_1 c^{q-1} = g_1$. The equality $j(\phi) = j(\psi)$ implies $h_1 c^{q^2-1}/g_2 = h_1/h_2$. Thus we also have $h_2 c^{q^2-1} = g_2$; thus (3.8.2) is satisfied, and thus ϕ and ψ are isomorphic.   □

Remark 3.8.5 Similar to Example 3.8.1, two Drinfeld modules ϕ and ψ of rank 2 with $j(\phi) = j(\psi)$ might not be isomorphic over K, e.g., $\phi_T = t + t\tau^2$ and $\psi_T = t + \tau^2$ both have j-invariant 0, but they are isomorphic over K if and only if t is a $(q^2 - 1)$-th power in K.

Corollary 3.8.6 *Let* ϕ *be a Drinfeld module of rank* 2 *over* K. *Then*

$$\mathrm{Aut}(\phi) = \begin{cases} \mathbb{F}_{q^2}^{\times} & \text{if } j(\phi) = 0; \\ \mathbb{F}_q^{\times} & \text{if } j(\phi) \neq 0. \end{cases}$$

For each $j \in K$, there is a Drinfeld module ϕ of rank 2 defined over $\mathbb{F}_q(t, j)$ with $j(\phi) = j$. Indeed,

$$\phi_T = t + \tau^2 \quad \text{has} \quad j(\phi) = 0, \tag{3.8.3}$$

and for $j \in K^{\times}$,

$$\phi_T = t + \tau + j^{-1}\tau^2 \quad \text{has} \quad j(\phi) = j. \tag{3.8.4}$$

Example 3.8.1 shows that over a separably closed field K (i.e., $K^{\mathrm{sep}} = K$), there is a unique Drinfeld module of rank 1. Similarly, Lemma 3.8.4 and Eqs. (3.8.3) and (3.8.4) imply that the isomorphism classes of Drinfeld modules of rank 2 over a

separably closed field K are in bijection with the elements of K. One would like
to extend this classification to Drinfeld modules of rank ≥ 3. It might appear that
over a separably closed field K, the map that sends the Drinfeld module $\phi_T =
t + g_1\tau + \cdots + g_r\tau^r$ to

$$
j(\phi) = \left(\frac{g_1^{(q^r-1)/(q-1)}}{g_r}, \frac{g_2^{(q^r-1)/(q^2-1)}}{g_r}, \ldots, \frac{g_{r-1}^{(q^r-1)/(q^{r-1}-1)}}{g_r} \right)
$$

gives a bijection between the isomorphism classes of Drinfeld modules of rank r
over K and K^{r-1}. We leave it to the reader to figure out the (easy) reason why this
does not work for $r \geq 3$.

The correct generalization of the j-invariant is more complicated and is due to
Potemine [Pot98]:

Definition 3.8.7 Let ϕ be a Drinfeld module of rank r defined over K by $\phi_T =
t + g_1\tau + \cdots + g_r\tau^r$. We consider a multi-index $(k_1, \ldots, k_l)$ with $1 \leq k_1 < \cdots <
k_l \leq r - 1$ which admits integers $s_1, \ldots, s_l, s_r$ such that

(a) $\quad 0 \leq s_i \leq \dfrac{q^r - 1}{q^{\gcd(i,r)} - 1}, \quad 1 \leq i \leq l;$

(b) $\quad \gcd(s_1, \ldots, s_l, s_r) = 1;$

(c) $\quad s_1(q^{k_1} - 1) + \cdots + s_l(q^{k_l} - 1) = s_r(q^r - 1).$

For such a multi-index $(k_1, \ldots, k_l)$ and a collection of integers $s_1, \ldots, s_l, s_r$
satisfying the above conditions, the $j_{k_1,\ldots,k_l}^{s_1,\ldots,s_l}$-invariant of ϕ is

$$
j_{k_1,\ldots,k_l}^{s_1,\ldots,s_l}(\phi) = \frac{g_{k_1}^{s_1} \cdots g_{k_l}^{s_l}}{g_r^{s_r}}.
$$

A *j-invariant* of ϕ is one of the $j_{k_1,\ldots,k_l}^{s_1,\ldots,s_l}$-invariants.

Example 3.8.8 If $r = 2$, then $k_1 = 1$ is the only possible index. The integers s_1 and
s_2 must be coprime and satisfy $s_1(q - 1) = s_2(q^2 - 1)$, $0 \leq s_1 \leq q + 1$. The only
such integers are $s_1 = q + 1$ and $s_2 = 1$. Therefore, the only j-invariant of ϕ is
$j_1^{q+1}(\phi)$, which is the j-invariant from Definition 3.8.3.

Example 3.8.9 Let $r = 3$. Then, the possible multi-indices are (1), (2), and (1, 2).
For $(k_1) = (1)$, the only possible s_i's are $s_1 = q^2 + q + 1$ and $s_3 = 1$. This gives

$$
j_1(\phi) := \frac{g_1^{q^2+q+1}}{g_3}
$$

as a j-invariant of ϕ. Similarly, for $(k_1) = (2)$, the only possible s_i's are $s_1 = q^2 + q + 1$ and $s_3 = q + 1$, which gives

$$j_2(\phi) := \frac{g_2^{q^2+q+1}}{g_3^{q+1}}$$

as a j-invariant. For $(k_1, k_2) = (1, 2)$, we have

$$j_{1,2}(\phi) := \frac{g_1^{q^2} g_2}{g_3}$$

as a j-invariant, but there are also others, such as $g_1 g_2^q / g_3$ (see also Exercise 3.8.2).

Example 3.8.10 For general r, the following $r - 1$ elements of K,

$$j_k(\phi) := \frac{g_k^{(q^r-1)/(q^{\gcd(k,r)}-1)}}{g_r^{(q^k-1)/(q^{\gcd(k,r)}-1)}}, \quad 1 \le k \le r - 1,$$

are j-invariants of ϕ. These are called *basic j-invariants* of ϕ.

Theorem 3.8.11 *Two Drinfeld modules ϕ and ψ of rank r over a separably closed field K are isomorphic if and only if they have the same j-invariants, i.e., $j_{k_1,\dots,k_l}^{s_1,\dots,s_l}(\phi) = j_{k_1,\dots,k_l}^{s_1,\dots,s_l}(\psi)$ for all the multi-indices $(k_1,\dots,k_l)$ and integers $s_1,\dots,s_l, s_r$ satisfying the conditions of Definition 3.8.7*

Proof If ϕ and ψ are isomorphic, then the equalities $j_{k_1,\dots,k_l}^{s_1,\dots,s_l}(\phi) = j_{k_1,\dots,k_l}^{s_1,\dots,s_l}(\psi)$ follow from (3.8.2). We will prove the converse only for $r = 3$. This special case already includes the key ideas of the general proof.

For $r = 3$, among all the j-invariants, it is enough to consider the three j-invariants j_1, j_2, and $j_{1,2}$ from Example 3.8.9. Assume these three j-invariants of ϕ and ψ are equal. Let ϕ and ψ be defined by $\phi_T = t + g_1\tau + g_2\tau^2 + g_3\tau^3$ and $\psi_T = t + h_1\tau + h_2\tau^2 + h_3\tau^3$. We proceed as in the proof of Lemma 3.8.4.

If $g_1 \ne 0$, then from $j_1(\phi) = j_1(\psi)$ we see that $h_1 \ne 0$. Let c be a root of $x^{q-1} = g_1/h_1$. The equality $j_1(\phi) = j_1(\psi)$ implies $g_3 = c^{q^3-1}h_3$, and $j_{1,2}(\phi) = j_{1,2}(\psi)$ implies $g_2 = c^{q^2-1}h_2$. Hence (3.8.2) holds.

If $g_1 = g_2 = 0$, then $h_1 = h_2 = 0$. Let c be a root of $x^{q^3-1} = g_3/h_3$. Then $c^{-1}\psi c = \phi$.

If $g_1 = 0$ but $g_2 \ne 0$, then $h_1 = 0$ and $h_2 \ne 0$. Let c be a root of $x^{q^2-1} = g_2/h_2$. From $j_2(\phi) = j_2(\psi)$, we get $(c^{q^3-1}h_3)^{q+1} = g_3^{q+1}$. Hence, $g_3 = \zeta^{-1}c^{q^3-1}h_3$ for some ζ satisfying $\zeta^{q+1} = 1$. Replacing ψ by the isomorphic module $c^{-1}\psi c$, we may assume that $\psi_T = t + g_2\tau^2 + \zeta g_3\tau^3$. For $\alpha \in \mathbb{F}_{q^2}^\times$, we have

$$\alpha^{-1}\psi_T\alpha = t + g_2\tau^2 + \alpha^{q^3-1}\zeta g_3\tau^3.$$

Since $\mathbb{F}_{q^2}^{\times}$ is cyclic, the kernel of the homomorphism $\mathbb{F}_{q^2}^{\times} \to \mathbb{F}_{q^2}^{\times}, \alpha \mapsto \alpha^{q^3-1}$, is the unique subgroup of $\mathbb{F}_{q^2}^{\times}$ of order $\gcd(q^2-1, q^3-1) = q-1$, i.e., the kernel is $\mathbb{F}_q^{\times}$, and the image is the subgroup of order $q+1$. Hence, we can find $\alpha \in \mathbb{F}_{q^2}^{\times}$ such that $\alpha^{q^3-1} = \zeta^{-1}$. Replacing ψ by $\alpha^{-1}\psi\alpha$, we get ϕ. $\square$

Remarks 3.8.12

(1) The j-invariants are useful for the study of geometric properties of Drinfeld modular varieties. For example, in the rank-2 case, the existence of the j-invariant implies that the Drinfeld modular variety is the affine line $\mathrm{Spec}(K[j])$. Similarly, in higher ranks, the existence of the j-invariants implies the existence of a geometric variety whose points are in natural bijection with the isomorphism classes of Drinfeld modules; the j-invariants are the coordinate functions on the modular variety.[13] When $r \geq 3$, the Drinfeld modular variety looks like the affine space of dimension $r - 1$, but with some singularities. We refer to Potemine's paper [Pot98] for more details, where he uses the j-invariants to describe the modular variety as a toric variety. For example, he shows that when $r = 3$, the modular variety is defined by the system of q equations:

$$\begin{cases} X_q^{q+2} = X_{q+1}X_{q-1}, \\ X_i^2 = X_{i+1}X_{i-1}, \quad 1 \leq i \leq q-1, \end{cases}$$

where the X_i's are related to the j-invariants.

The modular varieties provide a powerful geometric tool for the study of Drinfeld modules since one is able to deduce facts about all Drinfeld modules simultaneously from the geometric properties of a single algebraic variety. For example, this approach is extremely useful in the study of rational torsion submodules of Drinfeld modules; cf. Sect. 7.3.

(2) There is a different approach to the j-invariants Drinfeld modules in ranks ≥ 3 due to Breuer and Rück [BR16]. The set of their j-invariants is less explicit than Potemine's, but the key property of these invariants, namely that they distinguish the isomorphism classes of Drinfeld modules, is easier to prove. Let $B = A[x_1, \ldots, x_{r-1}]$, $A' = \mathbb{F}_{q^r}[T]$, and $B' = A'[x_1, \ldots, x_{r-1}]$. Define an action of the group $\mathbb{F}_{q^r}^{\times}$ on B' via an action on the indeterminates

$$\mu \circ x_i = \mu^{q^i-1}x_i, \quad 1 \leq i \leq r-1, \quad \mu \in \mathbb{F}_{q^r}^{\times}.$$

Let $(B')^{\mathbb{F}_{q^r}^{\times}}$ be the subring of B' consisting of invariant polynomials under this action, and let

$$C = B \cap (B')^{\mathbb{F}_{q^r}^{\times}}.$$

---

[13] Potemine proved that $\mathrm{Spec}\, A\left[\left\{j_{k_1,\ldots,k_l}^{s_1,\ldots,s_l}\right\}\right]$ is the coarse moduli scheme of Drinfeld modules of rank r.

It is not hard to show that if $\mathbb{F}_{q^r}^\times$ fixes a polynomial in B, then it fixes each monomial of that polynomial. Hence C consists of those polynomials in B whose monomials are of the form

$$f(T)x_1^{e_1}x_2^{e_2}\cdots x_{r-1}^{e_{r-1}}, \quad \text{where} \quad f(T) \in A,$$

$$\text{and} \quad \sum_{i=1}^{r-1} e_i(q^i - 1) \equiv 0 \pmod{q^r - 1}.$$

Breuer and Rück prove that, for every separably closed A-field K, there is a canonical bijection between the set of isomorphism classes of rank-r Drinfeld modules over K and ring homomorphisms $\Gamma \colon C \to K$ satisfying $\Gamma|_A = \gamma$. This implies that $\operatorname{Spec} A\left[\left\{j_{k_1,\ldots,k_l}^{s_1,\ldots,s_l}\right\}\right] = \operatorname{Spec}(C)$, so the j-invariants of a Drinfeld module ϕ can be defined as the elements of C evaluated at the coefficients of ϕ_T.

Definition 3.8.13 Let ϕ be a Drinfeld module over K and let L be an A-subfield of K. We say that L is a *field of definition for* ϕ if there is a Drinfeld module ψ defined over L which becomes isomorphic to ϕ over $\overline{K}$.

By Theorem 3.8.11, a field of definition of ϕ contains the field

$$\mathbb{F}_q\left(t, \left\{j_{k_1,\ldots,k_l}^{s_1,\ldots,s_l}(\phi)\right\}\right)$$

obtained by adjoining all the j-invariants of ϕ to $\mathbb{F}_q(t) \subseteq K$. Moreover, from Eqs. (3.8.3) and (3.8.4), it is clear that a Drinfeld module ϕ of rank 2 over K can actually be defined over $\mathbb{F}_q(t, j(\phi))$. One may then ask whether the same is true in higher ranks, i.e., whether ϕ can be defined over $\mathbb{F}_q\left(t, \left\{j_{k_1,\ldots,k_l}^{s_1,\ldots,s_l}(\phi)\right\}\right)$. We will see next that the answer is positive,[14] but the proof is more subtle. The proof[15] is interesting in its own right because it introduces important techniques from Galois cohomology into the study of isomorphism classes of Drinfeld modules; see also Exercise 3.8.4. (A version of this theorem is also proved in [Hay79, Thm. 6.5] by a different argument.)

---

[14] In a fancier terminology, this means that the *field of moduli* of a Drinfeld module is a field of definition. In some other classification problems, such as the classification of abelian surfaces with quaternionic multiplication, the field of moduli can be strictly smaller than the field of definition. Geometrically, this means that the modular variety classifying the isomorphism classes of certain algebraic objects might have an L-rational point but none of the algebraic objects corresponding to that point can be defined over L.

[15] The proof is motivated by an argument of Shimura for abelian varieties; see [Shi75, Thm. 9.5].

Theorem 3.8.14 *Let ϕ be a Drinfeld module over K. Then $\mathbb{F}_q\left(t, \left\{j_{k_1,\ldots,k_l}^{s_1,\ldots,s_l}(\phi)\right\}\right)$ is a field of definition for ϕ. In particular, $\mathbb{F}_q\left(t, \left\{j_{k_1,\ldots,k_l}^{s_1,\ldots,s_l}(\phi)\right\}\right)$ is the smallest field of definition of ϕ.*

Proof Assume ϕ is given by $\phi_T = t + g_1\tau + \cdots + g_r\tau^r$. After adjoining to K a root α of the separable polynomial $x^{q^r-1} - g_r$ and replacing ϕ by the isomorphic module $\alpha\phi\alpha^{-1}$, we may assume that $g_r = 1$. Now the j-invariants $j_k(\phi)$ of Example 3.8.10 give the equations $j_k(\phi) = g_k^{(q^r-1)/(q^{\gcd(k,r)}-1)}$, $1 \le k \le r-1$, from which it follows that $g_1, \ldots, g_{r-1}$ lie in a finite separable extension of $L := \mathbb{F}_q\left(t, \left\{j_{k_1,\ldots,k_l}^{s_1,\ldots,s_l}(\phi)\right\}\right)$. Passing to the Galois closure of that separable extension, we may assume that K is a finite Galois extension of L. Moreover, using either Theorem 3.3.9 or Lemma 3.8.2, we may also assume that all the automorphisms of ϕ are defined over K.

Denote $G = \mathrm{Gal}(K/L)$. For $\sigma \in G$, let ϕ^σ be the Drinfeld module defined by

$$\phi_T^\sigma = t + \sigma(g_1)\tau + \cdots + \sigma(g_r)\tau^r = \sigma(\phi_T).$$

Clearly,

$$j_{k_1,\ldots,k_l}^{s_1,\ldots,s_l}(\phi^\sigma) = \sigma\left(j_{k_1,\ldots,k_l}^{s_1,\ldots,s_l}(\phi)\right) = j_{k_1,\ldots,k_l}^{s_1,\ldots,s_l}(\phi).$$

Hence, by Theorem 3.8.11, ϕ^σ is isomorphic to ϕ. This means that for each $\sigma \in G$, there is some $\lambda_\sigma \in K^\times$ such that $\phi_T^\sigma = \lambda_\sigma\phi_T\lambda_\sigma^{-1}$. Now, for $\sigma, \delta \in G$, we have

$$\lambda_{\sigma\delta}\phi_T\lambda_{\sigma\delta}^{-1} = \phi_T^{\sigma\delta} = (\phi_T^\delta)^\sigma = (\lambda_\delta\phi_T\lambda_\delta^{-1})^\sigma = \sigma(\lambda_\delta)\phi_T^\sigma\sigma(\lambda_\delta)^{-1} \qquad (3.8.5)$$

$$= \sigma(\lambda_\delta)\lambda_\sigma\phi_T\lambda_\sigma^{-1}\sigma(\lambda_\delta)^{-1}$$

Therefore

$$\lambda_{\sigma\delta} = \sigma(\lambda_\delta)\lambda_\sigma\alpha_{\sigma,\delta} \qquad \text{for some} \qquad \alpha_{\sigma,\delta} \in \mathrm{Aut}(\phi) \subset K^\times.$$

Let $\mathrm{Aut}(\phi) \cong \mathbb{F}_{q^m}^\times$. Denote $h = q^m - 1$ and $\mu_\sigma = \lambda_\sigma^h$. Then

$$\mu_{\sigma\delta} = \sigma(\mu_\delta)\mu_\sigma \qquad \text{for all} \quad \sigma, \delta \in G.$$

This means that the map $G \to K^\times$ defined by $\sigma \mapsto \mu_\sigma$ is a 1-*cocycle*. At this point, we apply the famous Hilbert's Theorem 90 (see [DF04, p. 814]), which implies that there is $b \in K^\times$ such that $\mu_\sigma = b/\sigma(b)$ for all $\sigma \in G$. Let a be an element of K^{sep} such that $a^h = b$. Let K' be the Galois closure of $K(a)$ over L. Put $G^* = \mathrm{Gal}(K'/L)$, and let $\pi: G^* \to G$ be the natural quotient homomorphism from Galois theory. For every $\sigma \in G^*$, we see that $\lambda_{\pi(\sigma)}\sigma(a)/a$ is an h-th root of unity. Hence, there is a unique $\alpha_\sigma \in \mathrm{Aut}(\phi)$ such that $\lambda_{\pi(\sigma)}\alpha_\sigma = a/\sigma(a)$.

Put $c_\sigma = \lambda_{\pi(\sigma)}\alpha_\sigma = a/\sigma(a)$ and $\psi = a\phi a^{-1}$. Then, for any $\sigma \in G^*$, we have

$$c_\sigma \phi_T c_\sigma^{-1} = \lambda_{\pi(\sigma)}\alpha_\sigma \phi_T \alpha_\sigma^{-1}\lambda_{\pi(\sigma)}^{-1} = \lambda_{\pi(\sigma)}\phi_T \lambda_{\pi(\sigma)}^{-1} = \phi_T^{\pi(\sigma)} = \phi_T^\sigma,$$

where the last equality follows from the fact that ϕ_T has coefficients in K. Next, we have

$$\psi_T^\sigma = \sigma(a\phi_T a^{-1}) = \sigma(a)\phi_T^\sigma \sigma(a)^{-1} = \sigma(a)c_\sigma \phi_T c_\sigma^{-1}\sigma(a)^{-1} = a\phi_T a^{-1} = \psi_T.$$

Now, Galois theory implies that ψ, which is isomorphic to ϕ, is defined over L since the coefficients of ψ_T are fixed by G^*.   □

Exercises

3.8.1 Let $\phi_T = t + g_1\tau + g_2\tau^2$ and $\psi_T = t + h_1\tau + h_2\tau^2$ be two Drinfeld modules over K. Prove that ϕ and ψ are isomorphic over K if and only if

(a) $j(\phi) = j(\psi)$
(b) h_1/g_1 is a $(q-1)$-th power in K (if $j = j(\phi) = j(\psi) \neq 0$) and h_2/g_2 is a (q^2-1)-th power in K (if $j = 0$).

3.8.2 Assume ϕ is a Drinfeld module of rank 3.

(a) List all the j-invariants of ϕ, assuming $q = 2$.
(a) List all the j-invariants of ϕ, assuming $q = 3$.

3.8.3 Let $K = \mathbb{F}_{q^n}$ equipped with some $\gamma: A \to K$. Prove that

(a) The number of rank-2 Drinfeld modules over K is $q^n(q^n - 1)$.
(b) The number of K-isomorphism classes of such modules is

$$(q^n - 1)(q - 1) + \#(K^\times/K^{\times q^2-1}).$$

(c)

$$\sum \frac{1}{\# \mathrm{Aut}_K(\phi)} = q^n,$$

where the sum is over the K-isomorphism classes of Drinfeld modules of rank 2.

3.8.4 Let ϕ be a Drinfeld module of rank r defined over K. A *K-form* of ϕ is a Drinfeld module ψ over K which becomes isomorphic to ϕ over K^{sep}.

(a) Let $\lambda \in K^{\text{sep}}$ be such that $\lambda \phi \lambda^{-1} = \psi$. Show that for any $\sigma \in G_K$,

$$\sigma(\lambda)/\lambda \in \text{Aut}(\phi).$$

(b) Denote $c_{\psi,\lambda} \colon G_K \to \text{Aut}(\phi)$, $\sigma \mapsto \sigma(\lambda)/\lambda$, the map constructed in (a). Prove that

$$c_{\psi,\lambda}(\sigma_1 \sigma_2) = c_{\psi,\sigma_2(\lambda)}(\sigma_1) \cdot c_{\psi,\lambda}(\sigma_2).$$

(c) Let $\mu \in K^{\text{sep}}$ be another isomorphism $\mu \phi \mu^{-1} = \psi$. Show that there is $u \in \text{Aut}(\phi)$ such that

$$c_{\psi,\lambda}(\sigma) = c_{\psi,\mu}(\sigma) \frac{\sigma(u)}{u}.$$

Conclude that $c_\psi := c_{\psi,\lambda}$ is a well-defined element of $H^1(G_K, \text{Aut}(\phi))$, which does not depend on the choice of an isomorphism λ. Here, $H^1(G_K, \text{Aut}(\phi))$ denotes the first continuous cohomology group of the G_K-module $\text{Aut}(\phi)$. For the definition of this group, we refer to [Sil09, Appendix B].

(d) Prove that $c_\psi = 1$ if and only if ϕ and ψ are isomorphic over K. Hence, if we denote by $\text{Twist}(\phi/K)$ the set of K-isomorphism classes of K-forms of ϕ, then $\psi \mapsto c_\psi$ defines an injection

$$\text{Twist}(\phi/K) \longrightarrow H^1(G_K, \text{Aut}(\phi)).$$

(e) Prove that the map $\text{Twist}(\phi/K) \to H^1(G_K, \text{Aut}(\phi))$ is surjective.

(f) Prove that if $\text{Aut}(\phi) \cong \mathbb{F}_{q^m}^\times$, then $\text{Twist}(\phi/K)$ is in bijection with $K^\times/(K^\times)^{q^m-1}$. (This requires a nontrivial fact from Galois cohomology, cf. Proposition 2.5 in [Sil09, Appendix B].)

(g) Assume K is a finite field. Show that

$$\sum_{\psi \in \text{Twist}(\phi/K)} \frac{1}{\#\text{Aut}_K(\psi)} = 1.$$

3.8.5 The reader with some background in algebraic geometry can try to show that the isomorphism classes of Drinfeld modules of rank r over an algebraically closed field correspond bijectively to points in the weighted projective space $\mathbb{P}(q-1, q^2 - 1, \ldots, q^r - 1)$ with nonzero last coordinate. When $r = 2$, this is isomorphic to the usual projective line minus a point. A nice introduction to the theory of weighted projective spaces can be found in [BR86].

Chapter 4
Drinfeld Modules over Finite Fields

In this chapter we study Drinfeld modules defined over a finite field $k = \mathbb{F}_{q^n}$. What distinguishes the theory of these Drinfeld modules from the general theory is that the Frobenius $\pi := \tau^n$ commutes with every other element of $k\{\tau\}$, hence $A[\pi]$ is a subring of $\mathrm{End}_k(\phi)$ for any Drinfeld module ϕ; this simple observation is the starting point of the main results of this chapter.

In Sect. 4.1, we construct a non-commutative division algebra $k(\tau)$, which can be considered as the field of fractions of $k\{\tau\}$. The center of $k(\tau)$ is $\mathbb{F}_q(\pi)$. Then we describe $\mathrm{End}_k^\circ(\phi)$ as a central division algebra over $F(\pi)$. To do this, we use the fact that $F = \mathbb{F}_q(T)$ and $\mathbb{F}_q(\pi)$ can be directly related to each other as subalgebras of $k(\tau)$. In particular, it turns out that the dimension of $\mathrm{End}_k^\circ(\phi)$ over F is determined by the degree of the field extension $F(\pi)/F$.

In Sect. 4.2, we consider π as an $A_{\mathfrak{l}}$-linear operator on the Tate module $T_{\mathfrak{l}}(\phi) \cong A_{\mathfrak{l}}^{\oplus r}$ for a prime $\mathfrak{l} \neq \mathrm{char}_A(k)$. It follows from the results in Chap. 3 that the characteristic polynomial $P_\phi(x)$ of π has coefficients in A, which do not depend on $\mathfrak{l}$. We proved this using the motive of ϕ. Here we give two other proofs of this result which use specific aspects of the theory of Drinfeld modules over finite fields, and thus can be instructive on their own. In fact, this alternative approach produces stronger results in this setting. Then we prove a version of the Riemann Hypothesis for ϕ as a statement about the absolute values of the roots of $P_\phi(x)$ in $\mathbb{C}_\infty$, and show that there is an equality of ideals $(\chi(^\phi k)) = (P_\phi(1))$. These results fit nicely into the context of the similarities between Drinfeld modules and elliptic curves since they are the analogues of the results of Hasse for elliptic curves defined over finite fields.

In Sect. 4.3, we show that two Drinfeld modules ϕ and ψ are isogenous over k if and only if $P_\phi(x) = P_\psi(x)$. This leads to a classification of the isogeny classes of Drinfeld modules in terms of algebraic elements over F with certain properties.

In the last section, Sect. 4.4, we study Drinfeld modules over k with the largest possible endomorphism algebras, i.e., such that $\dim_F \mathrm{End}^\circ(\phi) = r^2$. These Drinfeld modules, called supersingular Drinfeld modules, have several special properties. For

© The Author(s), under exclusive license to Springer Nature Switzerland AG 2023
M. Papikian, *Drinfeld Modules*, Graduate Texts in Mathematics 296,
https://doi.org/10.1007/978-3-031-19707-9_4

example, there are only finitely many of them (up to $\bar{k}$-isomorphisms), they are all isogenous over $\bar{k}$, and they have no nonzero $\mathfrak{p}$-torsion points for $\mathfrak{p} = \operatorname{char}_A(k)$.

The following notation will be used throughout this chapter:

- $0 \neq \mathfrak{p} \lhd A$ is a prime of degree d.
- $k = \mathbb{F}_{q^n}$ is a finite extension of $A/\mathfrak{p} = \mathbb{F}_\mathfrak{p} \cong \mathbb{F}_{q^d}$. In particular, d divides n.
- $\gamma: A \to A/\mathfrak{p} \hookrightarrow k$ is the A-field structure on k, so $\operatorname{char}_A(k) = \mathfrak{p}$.
- $\pi = \tau^n$.

4.1 Endomorphism Algebras

Let $\phi: A \to k\{\tau\}$ be a Drinfeld module of rank r defined over k. We know from the general theory of morphisms of Drinfeld modules developed in Sect. 3.3 that $\operatorname{End}_k(\phi)$ is an algebra over $\phi(A) \cong A$, which is free of rank $\leq r^2$ as an A-module. Note that π commutes with every element of $k\{\tau\}$ because $\pi\alpha = \alpha^{q^n}\pi = \alpha\pi$ if $\alpha \in k \cong \mathbb{F}_{q^n}$, and obviously $\pi\tau = \tau\pi$. In particular, π commutes with the elements of $\phi(A)$, and so $\pi \in \operatorname{End}_k(\phi)$. The main goal of this section is to relate π as an integral element over $\phi(A)$ to the size of $\operatorname{End}_k(\phi)$. As long as ϕ is fixed, we write "A" for the subring $\phi(A)$ of $k\{\tau\}$.

First, we construct a version of the fraction field of an integral domain for the non-commutative polynomial ring $k\{\tau\}$ where our arguments will take place. Consider $\mathbb{F}_q[\pi]$ as the ring of polynomials over $\mathbb{F}_q$ in the indeterminate π. Since $\mathbb{F}_q[\pi]$ is in the center of $k\{\tau\}$, we can consider $k\{\tau\}$ as an $\mathbb{F}_q[\pi]$-algebra. It is easy to check that $k\{\tau\}$ is a free $\mathbb{F}_q[\pi]$-module of rank n^2. Indeed,

$$k\{\tau\} = k[\pi] \oplus k[\pi]\tau \oplus \cdots \oplus k[\pi]\tau^{n-1}$$

and

$$k[\pi] = \mathbb{F}_q[\pi] \oplus \mathbb{F}_q[\pi]\zeta \oplus \cdots \oplus \mathbb{F}_q[\pi]\zeta^{n-1},$$

where ζ is a primitive element of k over $\mathbb{F}_q$, i.e., $k = \mathbb{F}_q[\zeta]$. Hence

$$\{\zeta^i\tau^j \mid 0 \leq i, j \leq n - 1\}$$

is a basis of $k\{\tau\}$ over $\mathbb{F}_q[\pi]$. Let

$$K = \mathbb{F}_q(\pi)$$

be the fraction field of $\mathbb{F}_q[\pi]$ and define

$$k(\tau) = k\{\tau\} \otimes_{\mathbb{F}_q[\pi]} K.$$

Consider $k(\pi) = \mathbb{F}_{q^n}(\pi)$ as the degree-n extension of $K = \mathbb{F}_q(\pi)$ obtained by the extension of constants $\mathbb{F}_{q^n}/\mathbb{F}_q$. Let σ be the generator of $\mathrm{Gal}(k(\pi)/K) \cong \mathbb{Z}/n\mathbb{Z}$ which acts on

$$\alpha = a_0 + a_1\pi + \cdots + a_s\pi^s \in k[\pi] \quad \text{by} \quad \sigma(\alpha) = a_0^q + a_1^q\pi + \cdots + a_s^q\pi^s.$$

Then we can describe $k(\tau)$ in terms of generators and relations as follows:

$$k(\tau) = k(\pi) \oplus k(\pi)\tau \oplus \cdots \oplus k(\pi)\tau^{n-1}, \qquad (4.1.1)$$

$$\tau^n = \pi, \quad \tau\alpha = \sigma(\alpha)\tau \quad \text{for all } \alpha \in k(\pi).$$

Hence, $k(\tau) \cong (\mathbb{F}_{q^n}(\pi)/\mathbb{F}_q(\pi), \sigma, \pi)$ is a cyclic algebra; in particular, $k(\tau)$ is a central simple algebra over K of dimension n^2; cf. Example 1.7.5. In fact, it is easy to check directly that the center of $k(\tau)$ is K, and in the next proposition we show that $k(\tau)$ is a division algebra.

Proposition 4.1.1 *We have:*

(1) $k(\tau)$ *is a central division algebra over* K *of dimension* n^2.
(2) $k(\tau) \otimes_K \mathbb{F}_q((\pi)) \cong k((\tau))$.
(3) $k(\tau) \otimes_K \mathbb{F}_q((\pi^{-1})) \cong k((\tau^{-1}))$.
(4) $k(\tau) \otimes_K K_v \cong \mathrm{Mat}_n(K_v)$ *for any place* v *of* K *distinct from* (π) *and* $(1/\pi)$, *where* K_v *be the completion of* K *at* v.

Proof Using (4.1.1) and the observation that $k(\pi) \otimes_{\mathbb{F}_q(\pi)} \mathbb{F}_q((\pi)) \cong k((\pi))$, we get

$$k(\tau) \otimes_K \mathbb{F}_q((\pi)) \cong k((\tau)).$$

This proves both (2) and (1) because we already know that $k((\tau))$ is a division algebra; cf. Sect. 3.4. The proof of (3) is similar to the proof of (2).

(4) This is not essential for our purposes, so we only indicate the main steps of the proof. The proof requires some facts from the theory of cyclic algebras. Let $L := \mathbb{F}_{q^n}(\pi)$, let w be a place of L over v, and let L_w be the completion of L at w. Let $s = [L_w : K_v]$. Considering $\sigma^{n/s}$ as a generator of $\mathrm{Gal}(L_w/K_v)$, one shows that (see [Rei03, Thm. 30.8])

$$(L/K, \sigma, \pi) \otimes K_v \cong \mathrm{Mat}_s((L_w/K_v, \sigma^{n/s}, \pi)).$$

Next, because π is a unit in the ring of integers of K_v, and the field L_w contains the n-th roots of 1, it is not hard to show that $\pi = \mathrm{Nr}_{L_w/K_v}(c)$ for some $c \in L_w$. Finally, one concludes that $(L/K, \sigma, \pi) \otimes K_v \cong \mathrm{Mat}_n(K_v)$ using Exercise 1.7.2. $\qquad\square$

Remarks 4.1.2

(1) In the terminology of central simple algebras over global fields, Proposition 4.1.1 says that $k(\tau)$ *splits* at all places of K, except at (π) and $(1/\pi)$, where

it *ramifies*. Moreover, the so-called *local invariants* of $k(\tau)$ are (cf. (3.4.3))

$$\operatorname{inv}_v(k(\tau)) = \begin{cases} 1/n & \text{if } v = (\pi); \\ -1/n & \text{if } v = (1/\pi); \\ 0 & \text{if } v \neq (\pi), (1/\pi). \end{cases}$$

It is known that the local invariants of a central division algebra over $\mathbb{F}_q(\pi)$ determine the isomorphism class of that algebra; see [Rei03, §32].

(2) Assume $n = 2$ and q is odd. In this case, as follows from (4.1.1), $k(\tau)$ is isomorphic to the quaternion algebra $\Delta(\pi, \beta)$ over K, where $\beta \in \mathbb{F}_q^\times$ is a non-square; cf. Example 1.7.3. One can check that $\Delta(\pi, \beta)$ splits at a place v of K, $v \neq (\pi), (1/\pi)$, using the observation we made in Example 1.7.3: namely, $\Delta(\pi, \beta)$ splits at v if and only if the quadratic form

$$\operatorname{nr}_{\Delta/K} = x_1^2 - \pi x_2^2 - \beta x_3^2 + \beta \pi x_4^2$$

has nontrivial zeros in K_v. The reduction of $\operatorname{nr}_{\Delta/K}$ modulo the maximal ideal of the ring of integers of K_v is a quadratic form in *four* variables over the residue field of K_v. On the other hand, the Chevalley-Warning theorem (cf. [Lan02, p. 213]) implies that a quadratic form in three or more variables over a finite field always has nonzero solutions. Now, using a multivariable generalization of Hensel's Lemma (see Exercise 2.4.3), one concludes that $\operatorname{nr}_{\Delta/K} = 0$ has nonzero solutions in K_v.

(3) Observe that a division ring with a subring isomorphic to $k\{\tau\}$ must contain a subring isomorphic to K, so it must also contain a subring isomorphic to $k(\tau)$. Hence, $k(\tau)$ is the smallest division ring with a subring isomorphic to $k\{\tau\}$ in the sense that any homomorphism $s_1 \colon k\{\tau\} \to D$ into a division ring extends to a homomorphism $s_2 \colon k(\tau) \to D$ such that $s_1 = s_2 \circ \iota$, where $\iota \colon k\{\tau\} \hookrightarrow k\{\tau\} \otimes_{\mathbb{F}_q[\pi]} K$ is the natural embedding. From this perspective, $k(\tau)$ can be considered as the ring of fractions of the twisted polynomial ring $k\{\tau\}$. Given a general A-field L one might construct the ring of fractions $L(\tau)$ of $L\{\tau\}$ as the smallest division ring containing $L\{\tau\}$ in $L(\!(\tau)\!)$; see also the exercises for an alternative construction.

Now returning to the Drinfeld module ϕ, we get an embedding $A \xrightarrow{\phi} k\{\tau\} \hookrightarrow k(\tau)$. Since every nonzero element of $k(\tau)$ has a multiplicative inverse, the embedding $A \hookrightarrow k(\tau)$ extends to an embedding $F \hookrightarrow k(\tau)$. Identifying F with its image in $k(\tau)$ (which depends on ϕ), we can consider $\widetilde{F} := F(\pi)$ as a field extension of F inside $k(\tau)$. By definition,

$$\operatorname{End}_k^\circ(\phi) = \text{Centralizer of } \phi(F) \text{ in } k(\tau)$$

$$= \text{Centralizer of } \widetilde{F} \text{ in } k(\tau),$$

where the second equality is a consequence of the fact that π is in the center of $k(\tau)$. We know that

$$D := \text{End}_k^\circ(\phi)$$

is a division algebra over F. Since $\widetilde{F}$ is in the center of D, we can consider D as a division algebra over $\widetilde{F}$.

Theorem 4.1.3 *With previous notation and assumptions, the following hold:*

(1) $[\widetilde{F} : F]$ divides r.
(2) $\widetilde{F}$ is a totally imaginary extension of F.
(3) D is a central division algebra over $\widetilde{F}$ of dimension $\left(\frac{r}{[\widetilde{F}:F]}\right)^2$. In particular,

$$\dim_F(D) = \frac{r^2}{[\widetilde{F} : F]} \geq r.$$

(4) There are exactly two places in $\widetilde{F}$ where π has nonzero valuation. One of these places is the unique place $\widetilde{\infty}$ in $\widetilde{F}$ over ∞, and the other place, denoted $\widetilde{\mathfrak{p}}$, lies over $\mathfrak{p}$.
(5) $D \otimes_{\widetilde{F}} \widetilde{F}_{\widetilde{\mathfrak{p}}}$ and $D \otimes_{\widetilde{F}} \widetilde{F}_{\widetilde{\infty}}$ are division algebras.

Proof The key idea of the proof is to compare the extensions $\widetilde{F}/F$ and $\widetilde{F}/K$ inside $k(\tau)$:

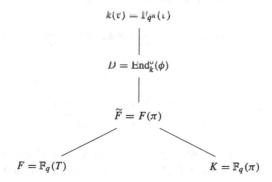

$$k(\tau) = \mathbb{F}_{q^n}'(\iota)$$

$$D = \text{End}_k^v(\phi)$$

$$\widetilde{F} = F(\pi)$$

$$F = \mathbb{F}_q(T) \qquad\qquad K = \mathbb{F}_q(\pi)$$

Since $k(\tau) \otimes_K K_\pi$ and $k(\tau) \otimes_K K_{1/\pi}$ are division algebras by Proposition 4.1.1, $\widetilde{F} \otimes_K K_\pi$ and $\widetilde{F} \otimes_K K_{1/\pi}$ are fields.[1] By Corollary 2.8.6, there is a unique place $\widetilde{\infty}$ in $\widetilde{F}$ over $(1/\pi)$ and a unique place $\widetilde{\mathfrak{p}}$ over (π). Obviously the only places of K where π has nonzero valuations are (π) and $(1/\pi)$, so $\widetilde{\infty}$ and $\widetilde{\mathfrak{p}}$ are the only places of $\widetilde{F}$ where π has nonzero valuation.

---

[1] If Δ is an algebra over a field L, $\Delta' \subset \Delta$ is a subalgebra, and L'/L is a field extension, then $\Delta' \otimes_L L'$ is a subalgebra of the L'-algebra $\Delta \otimes_L L'$ since L' is a *flat* L-module; see [DF04, p. 400].

The normalized valuation $\mathrm{ord}_{1/\pi}$ on K is equivalent to the restriction of the valuation $-\deg_\tau$ on $k(\tau)$ to K, i.e., $\mathrm{ord}_{1/\pi}(\cdot) = c \cdot \deg_\tau(\cdot)$ for some $c \in \mathbb{Q}^\times$. Hence, the normalized valuation $\mathrm{ord}_{\widetilde{\infty}}$ is also equivalent to the restriction of $-\deg_\tau$ to $\widetilde{F} \subset k(\tau)$. On the other hand, the restriction of the valuation $-\deg_\tau$ on $k(\tau)$ to F is equivalent to $-\deg_T$ on F since for any $a \in A$ we have $\deg_\tau(\phi_a) = r \cdot \deg_T(a)$. We proved in Sect. 3.3 that the place ∞ of F does not split in the extension $\widetilde{F}/F$, and from what we just discussed it follows that the unique place of $\widetilde{F}$ over ∞ is in fact $\widetilde{\infty}$. This proves (2).

Let $m_1 := [\widetilde{F} : F]$ and $m_2 := [\widetilde{F} : K]$. Since ∞ and $1/\pi$ do not split in $\widetilde{F}$, we have

$$m_1 = [\widetilde{F}_{\widetilde{\infty}} : F_\infty] = f_1 \cdot e_1$$

$$m_2 = [\widetilde{F}_{\widetilde{\infty}} : K_{1/\pi}] = f_2 \cdot e_2,$$

where f_i (respectively, e_i), $i = 1, 2$, is the residue degree (respectively, the ramification index) of the corresponding extension of the completions (see Proposition 2.8.8). The residue field at both ∞ and $1/\pi$ is $\mathbb{F}_q$, which implies $f_1 = f_2$. To determine the proportionality factor of $\deg_\tau$ and $\mathrm{ord}_{\widetilde{\infty}}$, we compute that for any nonzero $a \in A$ we have

$$\deg_\tau(\phi_a) = r \cdot \deg_T(a) = -r \cdot \mathrm{ord}_\infty(a) = -\frac{r}{e_1} \cdot \mathrm{ord}_{\widetilde{\infty}}(a).$$

Hence, $\deg_\tau = -\frac{r}{e_1} \cdot \mathrm{ord}_{\widetilde{\infty}}$, and plugging in π we get

$$n = \deg_\tau(\pi) = -\frac{r}{e_1} \cdot \mathrm{ord}_{\widetilde{\infty}}(\pi) = \frac{r}{e_1} e_2. \tag{4.1.2}$$

Thus,

$$\frac{[\widetilde{F} : F]}{[\widetilde{F} : K]} = \frac{m_1}{m_2} = \frac{e_1 f_1}{e_2 f_2} = \frac{e_1}{e_2} = \frac{r}{n}. \tag{4.1.3}$$

Now consider $k(\tau)$ as a central division algebra over K and $\widetilde{F}/K$ as a finite extension, and apply Theorem 1.7.22. Then D is a central division algebra over $\widetilde{F}$ and

$$n^2 = [k(\tau) : K] = [D : K] \cdot [\widetilde{F} : K] = [D : \widetilde{F}] \cdot [\widetilde{F} : K]^2.$$

Combining this equation with (4.1.3), we get

$$[D : \widetilde{F}] = \left(\frac{n}{[\widetilde{F} : K]}\right)^2 = \left(\frac{r}{[\widetilde{F} : F]}\right)^2.$$

This proves (1) and (3).

To prove (4), it remains to prove that $\widetilde{\mathfrak{p}}$ lies over $\mathfrak{p}$. For that, observe that for $0 \neq a \in A$

$$\mathrm{ord}_{\widetilde{\mathfrak{p}}}(a) \neq 0 \quad \Longleftrightarrow \quad \mathrm{ord}_{\widetilde{\mathfrak{p}}}(a^s) \neq 0 \text{ for all } s \geq 1$$

$$\Longleftrightarrow \quad \pi \text{ divides } \phi_{a^s} \text{ in } k\{\tau\} \text{ for some } s \geq 1.$$

On the other hand, $\pi = \tau^n$ divides ϕ_{a^s} for some $s \geq 1$ if and only if ϕ_a is inseparable, which itself is equivalent to $\mathfrak{p}$ dividing a. Thus, $\mathrm{ord}_{\mathfrak{p}}(a) > 0$ if and only if $\mathrm{ord}_{\widetilde{\mathfrak{p}}}(a) > 0$, so $\widetilde{\mathfrak{p}}$ divides $\mathfrak{p}$.

Finally, to prove (5), observe that

$$D \otimes_{\widetilde{F}} \widetilde{F}_{\widetilde{\mathfrak{p}}} = D \otimes_{\widetilde{F}} (\widetilde{F} \otimes_K K_\pi) = D \otimes_K K_\pi$$

is a subalgebra of $k(\tau) \otimes_K K_\pi$, which is a division algebra by Proposition 4.1.1. Similarly,

$$D \otimes_{\widetilde{F}} \widetilde{F}_{\widetilde{\infty}} = D \otimes_K K_{1/\pi} \subset k(\tau) \otimes_K K_{1/\pi}$$

is a division algebra.   □

Corollary 4.1.4 *Suppose r, the rank of ϕ, is prime. Then either*

- $[\widetilde{F} : F] = r$ and $D = \widetilde{F}$ or
- $\widetilde{F} = F$ and $\dim_F D = r^2$.

Remarks 4.1.5

(1) It is important to note that Theorem 4.1.3 *does not* claim that $\widetilde{\mathfrak{p}}$ is the only place of $\widetilde{F}$ over $\mathfrak{p}$, it only claims that $\widetilde{\mathfrak{p}}$ lies over $\mathfrak{p}$. Generally, $\mathfrak{p}$ splits in $\widetilde{F}$. The case when $\mathfrak{p}$ does not split is rather special (see Corollary 4.1.11).

(2) Using Remark 4.1.2, one can show that D, as a central division algebra over $\widetilde{F}$, splits at all places of $\widetilde{F}$, except at $\widetilde{\mathfrak{p}}$ and $\widetilde{\infty}$, where its invariants are

$$[\widetilde{F} : F]/r \quad \text{and} \quad -[\widetilde{F} : F]/r,$$

respectively. Up to isomorphism, this uniquely characterizes D among the central division algebras over the field $\widetilde{F}$.

(3) It follows from the results in Sect. 4.3 that all the possibilities for D allowed by Theorem 4.1.3 do actually occur. More precisely, given r, an extension $\widetilde{F}/F$ satisfying certain restrictions, and a $(r/d)^2$-dimensional central division algebra D over $\widetilde{F}$ with appropriate local invariants, there is a finite extension k of $\mathbb{F}_\mathfrak{p}$ and a Drinfeld module ϕ of rank r over k such that $\mathrm{End}_k^\circ(\phi) \cong D$. In particular, if $r \geq 2$, then D is always strictly larger than F, D can be non-commutative, and D can have the largest possible dimension r^2 allowed by Theorem 3.4.1.

(4) Continuing (3), suppose K is an A-field with $\mathrm{char}_A(K) = 0$ and ϕ is a Drinfeld module of rank r over K. Then we proved that $\mathrm{End}_K^\circ(\phi)$ is a totally imaginary

field extension of F of degree dividing r. Conversely, we will show in Sect. 5.3 that if L/F is a totally imaginary extension, then there is a Drinfeld module ϕ over $\mathbb{C}_\infty$ such that $\mathrm{End}^\circ(\phi) \cong L$.

(5) The case when $\mathrm{char}_A(K) = \mathfrak{p} \neq 0$ but ϕ over K cannot be defined over a finite extension of $\mathbb{F}_\mathfrak{p}$ lies somewhere in between the cases (3) and (4): D can be equal to F, but it also can be non-commutative as Example 4.1.6 below demonstrates (although $\dim_F D = r^2$ is not possible because supersingular Drinfeld modules are defined over $\overline{\mathbb{F}}_\mathfrak{p}$). As of 2023, a complete classification of algebras that occur as endomorphism algebras of Drinfeld modules in this case seems to be lacking.

Example 4.1.6 This example is based on Example 5.2 in [Pin06]. Let $\mathfrak{p} \lhd A$ be a prime, let K be a transcendental extension of $A/\mathfrak{p}$, and let γ be the composition $A \to A/\mathfrak{p} \hookrightarrow K$. Let $\phi \colon A \to K\{\tau\}$ be a Drinfeld module of rank r. Let Z be the center of $D := \mathrm{End}^\circ(\phi)$. By Theorem 3.6.10, $\sqrt{[D:Z]} \cdot [Z:F]$ divides r. In addition, we have the following fact (see [Pin06, Thm. 6.1]):

Theorem 4.1.7 *If ϕ cannot be defined over a finite extension of $\mathbb{F}_\mathfrak{p}$, then*

$$\frac{r}{\sqrt{[D:Z]} \cdot [Z:F]} > 1.$$

In particular, if r is prime, then $\mathrm{End}(\phi) = A$.

Now let $K = \mathbb{F}_{q^2}(x)$ with x transcendental over $\mathbb{F}_q$. Let γ be the composition $A \to A/(T) \cong \mathbb{F}_q \hookrightarrow K$. Let ϕ be the Drinfeld module of rank 3 over K defined by

$$\phi_T = x\tau + \tau^3.$$

This Drinfeld module cannot be defined over a finite extension of $\mathbb{F}_T$ because its j-invariant $j_1(\phi) = x^{q^2+q+1}$ is transcendental over $\mathbb{F}_q$; see Example 3.8.10. By Theorem 4.1.7, we have $\mathrm{End}(\phi) = A$. Next, let φ be the Drinfeld module of rank 6 over K defined by

$$\varphi_T = (x\tau + \tau^3)(x\tau + \tau^3).$$

Again, this Drinfeld module cannot be defined over a finite extension of $\mathbb{F}_T$ because its j-invariant $j_2(\phi) = x^{(q+1)(q^4+q^2+1)}$ is transcendental over $\mathbb{F}_q$. Let $D' := \mathrm{End}^\circ(\varphi)$ and let Z' be the center of D'. Obviously $u = (x\tau + \tau^3) \in \mathrm{End}_K(\varphi)$. We must have $Z' = F$, since $z \in Z'$ commutes with u, and so lies in $D = \mathrm{End}^\circ(\phi) = F$. Thus, by Theorem 4.1.7, D' is a central division algebra over F such that $\sqrt{[D':F]} = 2$ or 3. Note that for any $\alpha \in \mathbb{F}_{q^2}$ we have $u\alpha = \alpha^q u$, so α commutes with $\varphi_T = u^2$. Thus, the non-commutative polynomial ring $\mathbb{F}_{q^2}\{u\}$ with $u\alpha = \alpha^q u$ is a subring of $\mathrm{End}_K(\phi)$. This implies that $\sqrt{[D':F]}$ is divisible by 2, so D' is a quaternion division algebra over F. Moreover, because $\mathbb{F}_{q^2}\{u\}$ is a maximal

A-order in the quaternion algebra over F ramified at T and ∞, we conclude that
$\mathbb{F}_{q^2}\{u\} = \mathrm{End}_K(\phi) = \mathrm{End}(\phi)$.

Now we return to the initial setting of this section.

Proposition 4.1.8 *Let $\widetilde{\mathfrak{p}}$ be the unique prime of $\widetilde{F}$ such that $\mathrm{ord}_{\widetilde{\mathfrak{p}}}(\pi) > 0$. Let $H(\phi)$ be the height of ϕ. Then we have*

$$H(\phi) = \frac{r}{[\widetilde{F} : F]} \cdot [\widetilde{F}_{\widetilde{\mathfrak{p}}} : F_{\mathfrak{p}}].$$

Proof We use an argument similar to the proof of Theorem 4.1.3.

Since $\widetilde{\mathfrak{p}}$ is the only prime of $\widetilde{F}$ over π, we have

$$[\widetilde{F} : K] = [\widetilde{F}_{\widetilde{\mathfrak{p}}} : K_\pi] = e_1 \cdot f_1, \tag{4.1.4}$$

where now e_1 and f_1 denote the ramification index and the residue degree of the extension $\widetilde{F}_{\widetilde{\mathfrak{p}}}/K_\pi$. Let e_2 and f_2 denote the ramification index and the residue degree of $\widetilde{F}_{\widetilde{\mathfrak{p}}}/F_{\mathfrak{p}}$, so that

$$[\widetilde{F}_{\widetilde{\mathfrak{p}}} : F_{\mathfrak{p}}] = e_2 \cdot f_2. \tag{4.1.5}$$

Since the residue field of K_π is $\mathbb{F}_q$, we have

$$f_1 = [\mathbb{F}_{\widetilde{\mathfrak{p}}} : \mathbb{F}_q] = [\mathbb{F}_{\widetilde{\mathfrak{p}}} : \mathbb{F}_{\mathfrak{p}}] \cdot [\mathbb{F}_{\mathfrak{p}} : \mathbb{F}_q] = f_2 \cdot d, \tag{4.1.6}$$

where we denote by $\mathbb{F}_{\widetilde{\mathfrak{p}}}$ the residue field of $\widetilde{F}_{\widetilde{\mathfrak{p}}}$. The restriction of ht on $k\{\tau\}$ to $\mathbb{F}_q[\pi]$ satisfies

$$\mathrm{ht} = n \cdot \mathrm{ord}_\pi = \frac{n}{e_1}\, \mathrm{ord}_{\widetilde{\mathfrak{p}}}, \tag{4.1.7}$$

where ord_π (respectively, $\mathrm{ord}_{\widetilde{\mathfrak{p}}}$) is the normalized valuation on $\mathbb{F}_q[\pi]$ (respectively, $\widetilde{F}$) associated with the prime π (respectively, $\widetilde{\mathfrak{p}}$). On the other hand, by Lemma 3.2.11, the restriction of ht to A satisfies

$$\mathrm{ht} = H(\phi) \cdot d \cdot \mathrm{ord}_{\mathfrak{p}} = \frac{H(\phi) \cdot d}{e_2}\, \mathrm{ord}_{\widetilde{\mathfrak{p}}}. \tag{4.1.8}$$

Combining (4.1.4)–(4.1.8), we get

$$\frac{[\widetilde{F} : K]}{[\widetilde{F}_{\widetilde{\mathfrak{p}}} : F_{\mathfrak{p}}]} = \frac{e_1 f_1}{e_2 f_2} = \frac{n}{H(\phi)}. \tag{4.1.9}$$

On the other hand, by (4.1.3),

$$[\widetilde{F} : K] = \frac{n}{r}[\widetilde{F} : F].$$ (4.1.10)

Plugging (4.1.10) into (4.1.9) gives the desired formula for $H(\phi)$. □

Definition 4.1.9 A Drinfeld module ϕ of rank r over k is called *ordinary* if $H(\phi) = 1$ and *supersingular* if $H(\phi) = r$. Since the height of ϕ does not change under field extensions, ϕ remains ordinary (respectively, supersingular) over any extension k'/k if it is ordinary (respectively, supersingular) over k. Note that, because k has nonzero A-characteristic, $1 \le H(\phi) \le r$. Hence, a Drinfeld module of rank 2 is either ordinary or supersingular. When $r \ge 3$, there are Drinfeld modules which are neither ordinary nor supersingular.

Remark 4.1.10 The terminology in the previous definition could be justified as follows. The modular variety X classifying Drinfeld modules of rank r over $\bar{k}$, up to isomorphism, is an algebraic variety over $\bar{k}$ of dimension $r-1$. One can show that the set of points on X corresponding to ordinary Drinfeld modules is Zariski dense (so Drinfeld modules are usually "ordinary"), whereas the set of points corresponding to the supersingular Drinfeld modules is finite (so supersingular Drinfeld modules are "extra special"); cf. [Boy99].

Corollary 4.1.11 *We have:*

(1) ϕ is supersingular if and only if $\mathfrak{p}$ does not split in $\widetilde{F}/F$, i.e., $\widetilde{\mathfrak{p}}$ is the only prime of $\widetilde{F}$ over $\mathfrak{p}$.
(2) ϕ is ordinary if and only if $[\widetilde{F} : F] = r$ and $\widetilde{F}_{\widetilde{\mathfrak{p}}} = F_{\mathfrak{p}}$.

Proof By Proposition 2.8.8, $\widetilde{\mathfrak{p}}$ is the only prime over $\mathfrak{p}$ if and only if $[\widetilde{F} : F] = [\widetilde{F}_{\widetilde{\mathfrak{p}}} : F_{\mathfrak{p}}]$. On the other hand, by Proposition 4.1.8, we have $[\widetilde{F} : F] = [\widetilde{F}_{\widetilde{\mathfrak{p}}} : F_{\mathfrak{p}}]$ if and only if $H(\phi) = r$. This proves (1). The proof of (2) similarly follows from Proposition 4.1.8 since $[\widetilde{F} : F]$ divides r. □

Corollary 4.1.12 *We have:*

(1) If $H(\phi) = r/[\widetilde{F} : F]$, then $\widetilde{F}/F$ is separable.
(2) If ϕ is ordinary, then $\widetilde{F}/F$ is a separable extension of degree r and $D = \widetilde{F}$.

Proof

(1) If $H(\phi) = r/[\widetilde{F} : F]$, then by Proposition 4.1.8 we have $\widetilde{F}_{\widetilde{\mathfrak{p}}} = F_{\mathfrak{p}}$. On the other hand, if $\widetilde{F}/F$ is inseparable, then every prime of F ramifies in $\widetilde{F}$, so $\widetilde{F}_{\widetilde{\mathfrak{p}}} \ne F_{\mathfrak{p}}$; see Exercise 1.5.8.
(2) If ϕ is ordinary, then $[\widetilde{F} : F] = r$ by Corollary 4.1.11. Moreover, $\widetilde{F}/F$ is separable since $H(\phi) = 1 = r/[\widetilde{F} : F]$. Finally, the equality $D = \widetilde{F}$ follows from Theorem 4.1.3, since according to that theorem D has dimension $(r/[\widetilde{F} : F])^2$ over $\widetilde{F}$.

□

Exercises

In all exercises ϕ is a Drinfeld module defined over k of rank r and $D := \mathrm{End}_k^\circ(\phi)$.

4.1.1 Use Theorem 1.7.10 to prove that D contains a field extension F'/F of degree r. Conclude that r divides $[D : F]$. In particular, $\mathrm{End}_k(\phi) \neq A$ when $r \geq 2$.

4.1.2 Prove that $D = k(\tau)$ if and only if $\phi_T \in \mathbb{F}_q[\pi]$. In particular, if $D = k(\tau)$, then $d = 1$.

4.1.3 Suppose ϕ is a Drinfeld module over k of rank r such that $\mathrm{End}_{\bar{k}}(\phi) = \mathrm{End}_k(\phi)$, i.e., all endomorphisms of ϕ are defined over k. Prove that $\widetilde{F}/F$ is separable.

4.1.4 Suppose $\widetilde{F}/F$ is purely inseparable. Prove that ϕ is supersingular.

4.1.5 Assume ϕ is a Drinfeld module over k. Prove that if $\widetilde{F_{\bar{\mathfrak{p}}}}/F_{\mathfrak{p}}$ is unramified, then $\phi(A)[\pi]$ contains an element α with $\mathrm{ord}_{\bar{\mathfrak{p}}}(\alpha) = 1$. Conversely, show that if $[k : \mathbb{F}_{\mathfrak{p}}] \geq H(\phi)$ and $\phi(A)[\pi]$ contains an element α with $\mathrm{ord}_{\bar{\mathfrak{p}}}(\alpha) = 1$, then $\widetilde{F_{\bar{\mathfrak{p}}}}/F_{\mathfrak{p}}$ is unramified.

4.1.6 Let ϕ be a Drinfeld module over k. Let $L := \mathbb{F}_q(\mathfrak{p}, \pi)$ be the subfield of $k(\tau)$ generated over $\mathbb{F}_q$ by $\phi_{\mathfrak{p}}$ and π. Let v be the unique place of L that extends the place π of $\mathbb{F}_q(\pi)$.

(a) Show that the minimal polynomial $f(x) \in L[x]$ of $T \in \widetilde{F}$ over L divides
    $\mathfrak{p}(x) - \mathfrak{p}$, where $\mathfrak{p} = p_0 + p_1 T + \cdots + T^d$ and $\mathfrak{p}(x) = p_0 + p_1 x + \cdots + x^d$.
(b) Prove that $f(x)$ modulo v is separable.
(c) Conclude that v does not ramify in the extension $\widetilde{F}/L$.

Exercises 4.1.7–4.1.9 give the construction of the ring of fractions of a domain (i.e., a ring without zero-divisors) satisfying certain assumptions;[2] cf. [Coh95]. This construction is due to Ore.

4.1.7 Let R be a domain. A *right division ring of fractions of R* is a ring Q together with a homomorphism $\theta : R \to Q$ such that

(i) For any $0 \neq r \in R$, $\theta(r)$ is a unit in Q.
(ii) For any $q \in Q$, $q = \theta(r_1)\theta(r_2)^{-1}$ for some $r_1, r_2 \in R$.

A left division ring of fractions of R is defined similarly, with the condition in (ii) being $q = \theta(r_1)^{-1}\theta(r_2)$.

(a) Prove that if a right (respectively, left) division ring of fractions of R exists, then it is unique up to isomorphism. Because of this uniqueness property, we can talk about *the* right (respectively, left) division ring of fractions of R, and will denote this ring by $D_r(R)$ (respectively, $D_l(R)$).
(b) Prove that if both $D_r(R)$ and $D_l(R)$ exist, then $D_r(R) \cong D_l(R)$.

---

[2] Some assumptions are necessary since not every domain can be embedded into a division ring; see [Coh95, p. 9].

4.1.8 We say that R satisfies the *right (respectively, left) Ore condition* if for all $a, b \in R$, both nonzero, there exist $a', b' \in R$, both nonzero, such that $aa' = bb'$ (respectively, $a'a = b'b$).

Assume R satisfies the right Ore condition. Define a relation on

$$\widetilde{R} = \{(a, b) : a, b \in R, \ b \neq 0\}$$

by $(a, b) \sim (c, d)$ if $ab' = cd'$ when $bb' = dd'$. Prove that this is an equivalence relation.

Let $D = \widetilde{R}/\sim$ be the set of equivalence classes. Denote the equivalence class of (a, b) by $\frac{a}{b}$, and define addition and multiplication on D by

$$\frac{a}{b} + \frac{c}{d} = \frac{ab' + cd'}{bb'}, \quad \text{where } bb' = dd',$$

$$\frac{a}{b} \cdot \frac{c}{d} = \frac{ab'}{dc'}, \quad \text{where } bb' = cc'.$$

(a) Check that addition and multiplication on D are well-defined, and D is a division ring. Prove that $D \cong D_r(R)$. Prove that the same claims hold if R satisfies the left Ore condition, and this construction produces $D_l(R)$.

(b) Prove that if $D_r(R)$ (respectively, $D_l(R)$) exists, then R satisfies the right (respectively, left) Ore condition.

4.1.9 Prove the following:

(a) Given an A-field L, the twisted polynomial ring $L\{\tau\}$ has a left division ring of fractions. Moreover, if L is perfect, then $L\{\tau\}$ also has a right division ring of fractions.

(b) $D_l(k\{\tau\}) \cong D_r(k\{\tau\})$ are isomorphic to the division algebra $k(\tau)$ constructed at the beginning of this section.

(c) Assume L is perfect and denote by $L(\tau)$ the division ring of fractions $D_l(L\{\tau\}) \cong D_r(L\{\tau\})$ of $L\{\tau\}$. Then, a Drinfeld module $\phi: A \to L\{\tau\}$ extends uniquely to $\phi: F \to L(\tau)$.

4.2   Characteristic Polynomial of the Frobenius

Let $\phi: A \to k\{\tau\}$ be a Drinfeld module of rank r. Let $\mathfrak{l}$ be a prime not equal to $\mathfrak{p} = \operatorname{char}_A(k)$, let $T_{\mathfrak{l}}(\phi) \cong A_{\mathfrak{l}}^{\oplus r}$ be the $\mathfrak{l}$-adic Tate module of ϕ, and let

$$V_{\mathfrak{l}}(\phi) = T_{\mathfrak{l}}(\phi) \otimes_{A_{\mathfrak{l}}} F_{\mathfrak{l}} \cong F_{\mathfrak{l}}^{\oplus r}.$$

Recall that the Galois group $G_k = \operatorname{Gal}(\bar{k}/k)$ acts on $T_{\mathfrak{l}}(\phi)$, cf. (3.5.6),

$$\hat{\rho}_{\phi,\mathfrak{l}}: G_k \longrightarrow \operatorname{Aut}_{A_{\mathfrak{l}}}(T_{\mathfrak{l}}(\phi)) \cong \operatorname{GL}_r(A_{\mathfrak{l}}).$$

Let $\mathrm{Fr}_k \in G_k$ be the Frobenius automorphism of $\overline{k}$:

$$\mathrm{Fr}_k(\alpha) = \alpha^{\#k} = \alpha^{q^n}, \qquad \alpha \in \overline{k}.$$

Let

$$P_{\phi,\mathfrak{l}}(x) = \det\left(x - \hat{\rho}_{\phi,\mathfrak{l}}(\mathrm{Fr}_k)\right) \in A_{\mathfrak{l}}[x]$$

be the characteristic polynomial of $\hat{\rho}_{\phi,\mathfrak{l}}(\mathrm{Fr}_k) \in \mathrm{GL}_r(A_{\mathfrak{l}})$, which we call the *characteristic polynomial of the Frobenius*. In this section we study the properties of the polynomial $P(x) := P_{\phi,\mathfrak{l}}(x)$ and their relationship to the arithmetic of ϕ.

Remark 4.2.1 To motivate and give some context for the results of this section, we recall the central facts of the theory of elliptic curves over finite fields. Let E be an elliptic curve defined over $\mathbb{F}_q$. Let $P_E(x) \in \mathbb{Z}_\ell[x]$ be the characteristic polynomial of the Frobenius automorphism in $\mathrm{Gal}(\overline{\mathbb{F}}_q/\mathbb{F}_q)$ acting on the ℓ-adic Tate module $T_\ell(E) \cong \mathbb{Z}_\ell \times \mathbb{Z}_\ell$, where ℓ is a prime different from the characteristic of $\mathbb{F}_q$. Then the following is true:

(i) $P_E(x) = x^2 - ax + b$ has coefficients in $\mathbb{Z}$ which do not depend on ℓ.
(ii) $b = q$ and the roots of $P_E(x)$ have complex absolute values equal to $q^{1/2}$.
(iii) The number $\#E(\mathbb{F}_q)$ of $\mathbb{F}_q$-rational points on E is equal to $P_E(1) = q + 1 - a$.

These statements were conjectured by Emil Artin and proved by Helmut Hasse in the 1930s; the reader can find the proofs in Chapter V of [Sil09]. They can be reformulated as a version of the Riemann Hypothesis for the zeta function of E. In the 1940s, André Weil extended Hasse's result to abelian varieties defined over $\mathbb{F}_q$, and made a series of remarkable conjectures concerning the number of points on varieties defined over finite fields. Weil's conjectures had a tremendous impact on mathematics. For example, they motivated the development of several cohomology theories in arithmetic geometry. The deepest part of Weil's conjectures, proved by Deligne in 1973, is a version of the Riemann Hypothesis for the zeta functions of varieties defined over finite fields.

In Theorem 4.2.2, we will prove that $P(x)$ has coefficients in A that do not depend on $\mathfrak{l}$, i.e., the analogue of (i). Theorem 4.2.7 gives $P(0)$ and the absolute values of the roots of $P(x)$ in $\mathbb{C}_\infty$, i.e., the analogue of (ii). Theorem 4.2.6 says that there is an equality of ideals $(\chi(^\phi k)) = (P(1))$, where $\chi(^\phi k)$ is the product of the invariant factors of the A-module $^\phi k$. To see that this last statement is the analogue of (iii), note that for a finite abelian group the product of its invariant factors as a $\mathbb{Z}$-module is the order of the group, so $\chi(E(\mathbb{F}_q)) = \#E(\mathbb{F}_q) = P_E(1)$.

We start proving the theorems mentioned in the previous remark by first relating $P(x)$ to the minimal polynomial of π as an endomorphism of ϕ. Let $D := \mathrm{End}_k^\circ(\phi)$. We proved that there is a natural injective homomorphism (see (3.5.7))

$$\mathrm{End}_k(\phi) \otimes_A A_{\mathfrak{l}} \hookrightarrow \mathrm{End}_{A_{\mathfrak{l}}[G_k]}(T_{\mathfrak{l}}(\phi)). \qquad (4.2.1)$$

After tensoring with F_l, one obtains an injective[3] homomorphism

$$D \otimes_F F_l \hookrightarrow \operatorname{End}_{F_l[G_k]}(V_l(\phi)). \tag{4.2.2}$$

For all $s \in \mathbb{Z}_{>0}$, the Frobenius endomorphism $\pi : \phi \to \phi$ induces the map $\phi[l^s] \to \phi[l^s], \alpha \mapsto \alpha^{\#k}$. Clearly the action of π on $T_l(\phi)$ and $V_l(\phi)$ coincides with the action of Fr_k. Hence, $P(x)$ is also the characteristic polynomial of π acting on $V_l(\phi)$. The advantage of working with π is that it is an integral element over $\phi(A) \cong A$, so there is another polynomial naturally associated with π, namely the minimal polynomial $m(x) := m_{\pi,\phi(F)}(x) \in A[x]$ of π over F. The next theorem is a stronger version of Theorem 3.6.6 for π.

Theorem 4.2.2 *Let* $\widetilde{F} := F(\pi)$. *Then*

$$P(x) = m(x)^{r/[\widetilde{F}:F]}.$$

In particular, $P(x)$ has coefficients in A which do not depend on the choice of l.

Proof We have $\widetilde{F} \cong F[x]/(m(x))$. By Theorem 2.8.5,

$$\widetilde{F}_l := \widetilde{F} \otimes_F F_l \cong F_l[x]/(m(x)) \tag{4.2.3}$$

$$\cong \prod_{\mathfrak{L}|l} \widetilde{F}_{\mathfrak{L}},$$

where the product is over the primes of $\widetilde{F}$ dividing l. Similarly, we have

$$D_l := D \otimes_F F_l \cong D \otimes_{\widetilde{F}} (\widetilde{F} \otimes_F F_l)$$

$$\cong D \otimes_{\widetilde{F}} \prod_{\mathfrak{L}|l} \widetilde{F}_{\mathfrak{L}}$$

$$\cong \prod_{\mathfrak{L}|l} D_{\mathfrak{L}}, \qquad \text{where } D_{\mathfrak{L}} := D \otimes_{\widetilde{F}} \widetilde{F}_{\mathfrak{L}},$$

and

$$V_l(\phi) \cong V_l(\phi) \otimes_{\widetilde{F}_l} (\widetilde{F}_l \otimes_{F_l} F_l)$$

$$\cong V_l(\phi) \otimes_{\widetilde{F}_l} \prod_{\mathfrak{L}|l} \widetilde{F}_{\mathfrak{L}}$$

$$\cong \prod_{\mathfrak{L}|l} V_{\mathfrak{L}}, \qquad \text{where } V_{\mathfrak{L}} := V_l(\phi) \otimes_{\widetilde{F}_l} \widetilde{F}_{\mathfrak{L}}.$$

---

[3] F_l is flat over A_l; cf. [DF04, p. 400].

The injective F_l-linear map (4.2.2) is the product over $\mathcal{L} \mid \mathfrak{l}$ of the injective $\widetilde{F}_{\mathcal{Q}}$-linear maps

$$D_{\mathcal{Q}} \longhookrightarrow \operatorname{End}_{\widetilde{F}_{\mathcal{Q}}}(V_{\mathcal{Q}}). \qquad (4.2.4)$$

Let

$$d_{\mathcal{Q}} := \dim_{\widetilde{F}_{\mathcal{Q}}}(V_{\mathcal{Q}}).$$

From the previous injection we get

$$\dim_{\widetilde{F}_{\mathcal{Q}}}(D_{\mathcal{Q}}) \leq d_{\mathcal{Q}}^2 \quad \text{for all} \quad \mathcal{L} \mid \mathfrak{l}.$$

On the other hand,

$$\dim_{\widetilde{F}_{\mathcal{Q}}}(D_{\mathcal{Q}}) = \dim_{\widetilde{F}_{\mathcal{Q}}}(D \otimes_{\widetilde{F}} \widetilde{F}_{\mathcal{Q}})$$
$$= \dim_{\widetilde{F}}(D)$$
$$= \left(r/[\widetilde{F} : F]\right)^2,$$

where the last equality follows from Theorem 4.1.3. Therefore,

$$d_{\mathcal{Q}} \geq \frac{r}{[\widetilde{F} : F]}. \qquad (4.2.5)$$

We also have

$$r = \dim_{F_{\mathfrak{l}}}(V_{\mathfrak{l}}(\phi)) = \sum_{\mathcal{L} \mid \mathfrak{l}} \dim_{F_{\mathfrak{l}}}(V_{\mathcal{Q}})$$
$$= \sum_{\mathcal{L} \mid \mathfrak{l}} \dim_{\widetilde{F}_{\mathcal{Q}}}(V_{\mathcal{Q}}) \cdot [\widetilde{F}_{\mathcal{Q}} : F_{\mathfrak{l}}]$$
$$= \sum_{\mathcal{L} \mid \mathfrak{l}} d_{\mathcal{Q}} \cdot [\widetilde{F}_{\mathcal{Q}} : F_{\mathfrak{l}}], \qquad (4.2.6)$$

and (see Proposition 2.8.8)

$$[\widetilde{F} : F] = \sum_{\mathcal{L} \mid \mathfrak{l}} [\widetilde{F}_{\mathcal{Q}} : F_{\mathfrak{l}}]. \qquad (4.2.7)$$

Comparing (4.2.5), (4.2.6), (4.2.7), we deduce that

$$d_{\mathcal{Q}} = \frac{r}{[\widetilde{F} : F]} \quad \text{for all} \quad \mathcal{L} \mid \mathfrak{l}. \qquad (4.2.8)$$

In particular, all $d_\mathfrak{L}$ are equal, so $V_\mathfrak{l}(\phi)$ is a free $\widetilde{F}_\mathfrak{l}$-module. Thus, we have an isomorphism of $F_\mathfrak{l}[x]$-modules

$$V_\mathfrak{l}(\phi) \cong \widetilde{F}_\mathfrak{l}^{\oplus \frac{r}{[\widetilde{F}:F]}} \cong (F_\mathfrak{l}[x]/m(x))^{\oplus \frac{r}{[\widetilde{F}:F]}} , \tag{4.2.9}$$

where x acts on $V_\mathfrak{l}(\phi)$ via π. Now Exercise 1.2.7 implies that the characteristic polynomial of π acting on the vector space $V_\mathfrak{l}(\phi)$ is equal to $m(x)^{r/[\widetilde{F}:F]}$, that is, $P(x) = m(x)^{r/[\widetilde{F}:F]}$ as was claimed.   □

Remarks 4.2.3

(1) In the previous proof we have established the important fact that $V_\mathfrak{l}(\phi)$ is a free module of rank $r/[\widetilde{F}:F]$ over $\widetilde{F}_\mathfrak{l}$. But it is generally not true that $T_\mathfrak{l}(\phi)$ is free over $\left(\widetilde{F} \cap \mathrm{End}_k(\phi)\right) \otimes_A A_\mathfrak{l}$.

(2) It follows from (4.2.9) and Exercise 1.2.7 that $m(x)$ is also the minimal polynomial of π acting on the vector space $V_\mathfrak{l}(\phi)$; keep in mind that $m(x)$ was originally defined as the minimal polynomial of π over A.

The equality (4.2.8) has another important consequence: it implies that the injection (4.2.4) is an isomorphism $D_\mathfrak{L} \xrightarrow{\sim} \mathrm{End}_{\widetilde{F}_\mathfrak{L}}(V_\mathfrak{L})$. Hence, (4.2.2) is also an isomorphism (because the subgroup generated by Fr_k is dense in G_k):

$$D_\mathfrak{l} \xrightarrow{\sim} \mathrm{End}_{\widetilde{F}_\mathfrak{l}}(V_\mathfrak{l}(\phi)) = \mathrm{End}_{F_\mathfrak{l}[\mathrm{Fr}_k]}(V_\mathfrak{l}(\phi)) = \mathrm{End}_{F_\mathfrak{l}[G_k]}(V_\mathfrak{l}(\phi)).$$

Corollary 4.2.4 *The natural map*

$$\mathrm{End}_k(\phi) \otimes_A A_\mathfrak{l} \xrightarrow{\sim} \mathrm{End}_{A_\mathfrak{l}[G_k]}(T_\mathfrak{l}(\phi))$$

is an isomorphism.

Proof We know that the map in question is injective. Let N be its cokernel. Tensoring with $F_\mathfrak{l}$, one obtains the exact sequence

$$D_\mathfrak{l} \longrightarrow \mathrm{End}_{F_\mathfrak{l}[G_k]}(V_\mathfrak{l}(\phi)) \longrightarrow N \otimes_{A_\mathfrak{l}} F_\mathfrak{l} \longrightarrow 0.$$

But $D_\mathfrak{l} \to \mathrm{End}_{F_\mathfrak{l}[G_k]}(V_\mathfrak{l}(\phi))$ is an isomorphism, so $N \otimes_{A_\mathfrak{l}} F_\mathfrak{l} = 0$. Since N is a free $A_\mathfrak{l}$-module of finite rank by Theorem 3.5.4, the vanishing $N \otimes_{A_\mathfrak{l}} F_\mathfrak{l} = 0$ is possible if and only if $N = 0$.   □

Next, we generalize Theorem 4.2.2 to arbitrary endomorphisms of ϕ; this gives an alternative proof of Theorem 4.2.2. Recall that we defined the characteristic polynomial of $u \in \mathrm{End}_k(\phi)$ as

$$P_u(x) = \det(x - u_\mathfrak{l}),$$

where $u_\mathfrak{l} \in \mathrm{Mat}_r(A_\mathfrak{l})$ is a matrix by which u acts on $T_\mathfrak{l}(\phi)$. Let $m_u(x) := m_{u,\phi(F)}(x) \in A[x]$ be the minimal polynomial of u over F. Since D is a division algebra, the F-subalgebra $F(u)$ generated by u is a field; in fact, $F(u) \cong F[x]/(m_u(x))$. Note that $F(u)$ is contained in some maximal subfield of D, and any such maximal field has degree r over F. Thus, $[F(u) : F]$ divides r.

Theorem 4.2.5 *For any $u \in \mathrm{End}_k(\phi)$, we have*

$$P_u(x) = m_u(x)^{r/[F(u):F]}.$$

In particular, $P_u(x)$ has coefficients in A which do not depend on the choice of $\mathfrak{l}$.

Proof We follow the argument in [Gek91, §3].

Since D is a central division algebra over $\widetilde{F}$, we can define a map $N \colon D \to F$ as the composition

$$N = \mathrm{Nr}_{\widetilde{F}/F} \circ \mathrm{nr}_{D/\widetilde{F}},$$

where $\mathrm{nr}_{D/\widetilde{F}}$ is the reduced norm and $\mathrm{Nr}_{\widetilde{F}/F}$ is the field norm. Let L be a maximal subfield of D. Then $[L : F] = r$. Under the injective homomorphism

$$i_\mathfrak{l} \colon D \otimes_F F_\mathfrak{l} \longrightarrow \mathrm{End}_{F_\mathfrak{l}}(V_\mathfrak{l}(\phi)) \cong \mathrm{Mat}_r(F_\mathfrak{l}), \tag{4.2.10}$$

the $F_\mathfrak{l}$-algebra $L_\mathfrak{l} := L \otimes_F F_\mathfrak{l}$ maps to a maximal commutative semisimple $F_\mathfrak{l}$-subalgebra of $\mathrm{Mat}_r(F_\mathfrak{l})$. (Note that $L_\mathfrak{l}$ is not necessarily a field, but it is isomorphic to a direct product $\prod_{\mathfrak{l}\mathfrak{l}} L_\mathfrak{l}$ of fields, similar to (4.2.3).) Such a subalgebra of $\mathrm{Mat}_r(F_\mathfrak{l})$, after possibly extending the coefficient to $\overline{F}_\mathfrak{l}$, is conjugate to the subalgebra consisting of the diagonal matrices. Hence the norm map $\mathrm{Nr}_{L_\mathfrak{l}/F_\mathfrak{l}}$ on $L_\mathfrak{l}$ agrees with $\det \circ i_\mathfrak{l}$. From the definition of the norm on finite dimensional algebras it clear that under $L \to L \otimes_F F_\mathfrak{l}, \alpha \mapsto \alpha \otimes 1$, we have $\mathrm{Nr}_{L/F}(\alpha) = \mathrm{Nr}_{L_\mathfrak{l}/F_\mathfrak{l}}(\alpha \otimes 1)$. (Choose a basis $\{e_1, \ldots, e_r\}$ of L. Since $\{e_1 \otimes 1, \ldots, e_r \otimes 1\}$ is an $F_\mathfrak{l}$-basis of $L_\mathfrak{l}$, the matrix by which α acts on L with respect to the basis $\{e_1, \ldots, e_r\}$ is the same as the matrix by which $\alpha \otimes 1$ acts on $L_\mathfrak{l}$ with respect to the basis $\{e_1 \otimes 1, \ldots, e_r \otimes 1\}$.) On the other hand, by Corollary 1.7.17, the restriction of N to L is the field norm $\mathrm{Nr}_{L/F}$. Thus, $N|_L = \det \circ i_\mathfrak{l}$. But L was an arbitrary maximal subfield of D and every element of D is contained in a such field, so

$$N = \det \circ i_\mathfrak{l}. \tag{4.2.11}$$

Now we are ready to prove the theorem. Since any nonzero polynomial has finitely many roots, it suffices to show that $P_u(a) = m_u(a)^{r/[F(u):F]}$ for all $a \in A$.

Let L be a maximal subfield of D containing u, so that there is a tower of extensions $F \subset F(u) \subset L$. We have

$$P_u(a) = (\det \circ i_1)(a - u)$$

$$= N(a - u) \qquad \qquad \text{(by (4.2.11))}$$

$$= \mathrm{Nr}_{L/F}(a - u) \qquad \qquad \text{(by Corollary 1.7.17)}$$

$$= \mathrm{Nr}_{F(u)/F}\left(\mathrm{Nr}_{L/F(u)}(a - u)\right) \qquad \qquad \text{(by Lemma 1.4.5)}$$

$$= \mathrm{Nr}_{F(u)/F}\left((a - u)^{r/[F(u):F]}\right) \qquad \qquad \text{(since } a - u \in F(u))$$

$$= \left(\mathrm{Nr}_{F(u)/F}(a - u)\right)^{r/[F(u):F]} \qquad \qquad \text{(by (1.4.1))}$$

$$= (-1)^r \cdot m_{a-u}(0)^{r/[F(u):F]} \qquad \text{(by Corollary 1.4.4 with } D = F(u))$$

$$= m_u(a)^{r/[F(u):F]},$$

where the last equality follows from the observation that $m_{a-u}(x) = (-1)^{[F(u):F]}m_u(a - x)$. $\qquad \square$

Recall that $^\phi k$ denotes the A-module whose underlying group is $(k, +)$ and on which $a \in A$ acts by $a \circ \beta = \phi_a(\beta)$. Since k is a finite A-module, by Theorem 1.2.4 we have an isomorphism

$$^\phi k \cong A/(a_1) \times \cdots \times A/(a_s), \qquad a_1 \mid a_2 \mid \cdots \mid a_s, \qquad (4.2.12)$$

for uniquely determined monic polynomials $a_1, \ldots, a_s$ of positive degrees. (Of course, $a_1, \ldots, a_s$ depend on ϕ although it is not explicitly indicated in the notation.) Note that $^\phi k$ is annihilated by a_s, so, using Theorem 3.5.2, one obtains the inclusions

$$^\phi k \subseteq \phi[a_s] \subseteq (A/(a_s))^{\oplus r}.$$

This implies that $s \leq r$. By definition

$$\chi(^\phi k) \doteq \prod_{i=1}^{s} a_i.$$

Theorem 4.2.6 *We have an equality of ideals*

$$(\chi(^\phi k)) = (P(1)).$$

Proof The kernel of $\pi - 1$ acting on $\bar{k}$ consists of the elements fixed by π. Thus, $\ker(\pi - 1) = {}^\phi k$ and

$$\chi(^\phi k) = \mathfrak{N}(\pi - 1),$$

where $\Re(\pi - 1)$ was defined in Definition 3.6.12. On the other hand, by Theorem 3.6.13,

$$(\Re(\pi - 1)) = (P_{\pi-1}(0)) = (P_\pi(1)).$$

$\square$

Now we proceed to the proof of the "Riemann Hypothesis" for Drinfeld modules. For the statement of the next theorem, recall that $n = [k : \mathbb{F}_q]$ and $d = \deg(\mathfrak{p})$.

Theorem 4.2.7 *Let*

$$P(x) = x^r + a_{r-1}x^{r-1} + \cdots + a_1 x + a_0 \in A[x]$$

be the characteristic polynomial of the Frobenius of a rank-r Drinfeld module ϕ. Then:

(1) For any root α of $P(x)$ in $\mathbb{C}_\infty$ we have

$$|\alpha| = q^{n/r} = (\#k)^{1/r},$$

where $|\cdot|$ denotes the unique extension of the normalized absolute value on F_∞ to $\mathbb{C}_\infty$.

(2) For any $0 \le i \le r - 1$, we have

$$\deg(a_i) \le \frac{(r-i)n}{r}.$$

(3) If $\phi_T = t + g_1\tau + \cdots + g_r\tau^r \in k\{\tau\}$, then

$$a_0 = (-1)^{rn-r-n} \cdot \mathrm{Nr}_{k/\mathbb{F}_q}(g_r)^{-1} \cdot \mathfrak{p}^{n/d}.$$

In particular, if $g_r = 1$, then

$$a_0 = \begin{cases} \mathfrak{p}^{n/d}, & \text{if } r \text{ and } n \text{ are even;} \\ -\mathfrak{p}^{n/d}, & \text{otherwise.} \end{cases}$$

Proof From Theorem 4.2.2 we know that $P(x) = m(x)^{r/[\widetilde{F}:F]}$, where $m(x)$ is the minimal polynomial of π over A. Hence, the roots of $P(x)$ are the roots of $m(x)$ with multiplicity $r/[\widetilde{F} : F]$. If the roots of $m(x)$ had different valuations in $\mathbb{C}_\infty$, then $m(x)$ would decompose over F_∞ into at least two irreducible factors, where the roots of each irreducible factor have the same valuations; see Proposition 2.5.4. Theorem 2.8.5 then would imply that $\widetilde{F}$ has at least two places over ∞, contradicting Theorem 4.1.3. Thus, all the roots of $P(x)$ have the same ∞-adic absolute value.

Let $\widetilde{\infty}$ be the unique place of $\widetilde{F}$ over ∞; cf. Theorem 4.1.3. By Definition 2.6.1, the unique extension of $|\cdot| = q^{-\mathrm{ord}_\infty(\cdot)}$ on F_∞ to $\widetilde{F}_{\widetilde{\infty}}$ is $|\cdot| = q^{-\frac{1}{e}\mathrm{ord}_{\widetilde{\infty}}(\cdot)}$, where e

is the ramification index of $\widetilde{F}_{\widetilde{\infty}}/F_\infty$ and $\mathrm{ord}_{\widetilde{\infty}}(\cdot)$ is the normalized valuation at $\widetilde{\infty}$. By (4.1.2), we have $\mathrm{ord}_{\widetilde{\infty}}(\pi) = -ne/r$. Since π is a root of $P(x)$, we get

$$|\alpha| = |\pi| = q^{-\frac{1}{e}\,\mathrm{ord}_{\widetilde{\infty}}(\pi)} = q^{n/r},$$

which is the first claim.

Next, since a_i, up to ± 1, is the $(r-i)$-th elementary symmetric polynomial in the roots of $P(x)$, from the triangle inequality we get

$$\deg(a_i) = -\,\mathrm{ord}_\infty(a_i)$$

$$\leq -(r-i)\frac{1}{e}\,\mathrm{ord}_{\widetilde{\infty}}(\pi)$$

$$= \frac{(r-i)n}{r},$$

for any $1 \leq i \leq r-1$, and also

$$\deg(a_0) = n.$$

To determine a_0, note that, by the definition of $P(x)$,

$$a_0 = (-1)^r \det(\hat{\rho}_{\phi,\mathfrak{l}}(\mathrm{Fr}_k)) \qquad (4.2.13)$$

for a prime $\mathfrak{l} \neq \mathfrak{p}$. On the other hand, according to Theorem 3.7.1,

$$\det(\hat{\rho}_{\phi,\mathfrak{l}}(\mathrm{Fr}_k)) = \hat{\rho}_{\psi,\mathfrak{l}}(\mathrm{Fr}_k), \qquad (4.2.14)$$

where $\psi_T = t + (-1)^{r-1}g_r\tau$. Since $\mathfrak{p}^{n/d} = T^n +$ lower degree terms, the leading term of $\psi_{\mathfrak{p}^{n/d}}$ is equal to the leading term of ψ_{T^n}, which is

$$((-1)^{r-1}g_r\tau)^n = (-1)^{(r-1)n}g_r \cdot g_r^q \cdots g_r^{q^{n-1}}\tau^n$$

$$= (-1)^{(r-1)n} \cdot \mathrm{Nr}_{k/\mathbb{F}_q}(g_r) \cdot \tau^n$$

$$= (-1)^{(r-1)n} \cdot \mathrm{Nr}_{k/\mathbb{F}_q}(g_r) \cdot \pi.$$

Since $H(\psi) \geq 1$ and ψ has rank 1, Lemma 3.2.11 implies that $\mathrm{ht}(\psi_{\mathfrak{p}^{n/d}}) = n$, so

$$\psi_{\mathfrak{p}^{n/d}} = (-1)^{(r-1)n}\,\mathrm{Nr}_{k/\mathbb{F}_q}(g_r) \cdot \pi.$$

This means that π (considered as an endomorphism of ψ) acts on $T_\mathfrak{l}(\psi)$ as the scalar multiplication by $(-1)^{(r-1)n} \cdot \mathrm{Nr}_{k/\mathbb{F}_q}(g_r)^{-1} \cdot \mathfrak{p}^{n/d}$. Since Fr_k acts on $T_\mathfrak{l}(\psi)$ as π, we conclude that

$$\hat{\rho}_{\psi,\mathfrak{l}}(\mathrm{Fr}_k) = (-1)^{(r-1)n} \cdot \mathrm{Nr}_{k/\mathbb{F}_q}(g_r)^{-1} \cdot \mathfrak{p}^{n/d}.$$

Substituting this expression into (4.2.14) and then into (4.2.13), we obtain the desired formula for a_0. (It is possible to deduce the formula for a_0 without using the full force of Theorem 3.7.1. This alternative argument is sketched in the exercises.)   □

Remark 4.2.8 The constant term of the characteristic polynomial of the Frobenius for an elliptic curve over k is always #k. The situation is more complicated in the case of Drinfeld modules since $P(0)$ is equal to an $\mathbb{F}_q^{\times}$-multiple of the monic polynomial $\mathfrak{p}^{n/d}$, so $P(0)$ varies. As is clear from the proof of Theorem 4.2.7, this is closely related to the fact that the value group of the Weil pairing on Drinfeld modules varies, whereas the value group of the Weil pairing on elliptic curves is always the group of roots of unity.

Second Proof of Theorem 4.2.7 (1) We give a different proof using motives. This argument is due to Anderson [And86]. The starting point is the following simple lemma:

Lemma 4.2.9 *Let K be a non-Archimedean field and let R be its ring of integers. Let V be a vector space over K of dimension n. An R-**lattice** $\Lambda \subset V$ is a free R-submodule of rank n which contains a basis of V. Let S be a linear transformation of V such that $S\Lambda = \Lambda$. Then all eigenvalues of S have absolute value 1.*

Proof Fix an R-basis of Λ, which is a basis of V over K. Since S maps Λ to itself, the matrix of S with respect to this basis lies in $\mathrm{Mat}_n(R)$. Hence the characteristic polynomial $\chi_S(x) = \prod_{i=1}^{n}(x - \alpha_i)$ of S is a monic polynomial in $R[x]$ of degree n, where $\alpha_1, \ldots, \alpha_n \in \bar{K}$ are the eigenvalues of S. For any $1 \leq i \leq n$, the minimal polynomial of α_i over K divides $\chi_S(x)$, so by Gauss' Lemma (see Corollary 1.1.7) α_i is integral over R. Thus, $|\alpha_i| \leq 1$. On the other hand, by Lemma 1.2.6,

$$\prod_{i=1}^{n} |\alpha_i| = |\det(S)| = |\chi(\Lambda/S\Lambda)| = 1.$$

This implies that $|\alpha_1| = \cdots = |\alpha_n| = 1$.   □

Let M_ϕ be the Anderson motive of ϕ over k. Recall that $M_\phi = k\{\tau\}$ is a left $k[T, \tau]$-module, freely generated by $1, \tau, \ldots, \tau^{r-1}$ over $k[T]$. Denote $K = k(\!(1/T)\!)$, $R = k[\![1/T]\!]$, and

$$V = M_\phi \otimes_{k[T]} K.$$

We define an action of τ on V by

$$\tau\left(m \otimes \left(\sum \alpha_i (1/T)^i\right)\right) = \tau m \otimes \left(\sum \alpha_i^q (1/T)^i\right).$$

This equips V with a structure of a left $k[T, \tau]$-module for which M_ϕ is a submodule of V. The Frobenius endomorphism π acts $k[T]$-linearly on M_ϕ, and extends to a linear transformation of V as a vector space over K. Note that K is isomorphic to the unramified extension of F_∞ of degree n.

By Proposition 3.6.7,

$$P_\pi(x) = \det_{k[T]} \left(x - \pi \,|\, M_\phi \right)$$
$$= \det_K \left(x - \pi \,|\, V \right).$$

Let $e_i := \tau^i \otimes 1$ for $i = 0, \ldots, r-1$, and let

$$\Lambda = Re_0 \oplus \cdots \oplus Re_{r-1}.$$

Then Λ is a R-lattice since $e_0, \ldots, e_{r-1}$ generate V over K. We will show that

$$\tau^r \Lambda = T\Lambda. \tag{4.2.15}$$

Assuming this for the moment, we get

$$\pi^r \Lambda = T^n \Lambda.$$

Therefore, $(\pi^r / T^n)\Lambda = \Lambda$. By Lemma 4.2.9, all eigenvalues of π^r / T^n have absolute value 1, so the eigenvalues of π have absolute value $q^{n/r}$.

It remains to prove (4.2.15). Let

$$\phi_T = t + g_1\tau + \cdots + g_r\tau^r, \qquad g_r \neq 0.$$

The generators of the lattice $\tau\Lambda$ are $\tau e_0 = e_1, \cdots, \tau e_{r-2} = e_{r-1}$, and

$$\tau e_{r-1} = \tau^r \otimes 1 = g_r^{-1}(\phi_T - t - g_1\tau - \cdots - g_{r-1}\tau^{r-1}) \otimes 1$$
$$\in (\phi_T - t)Re_0 + Re_1 + \cdots + Re_{r-1}.$$

Therefore, $\tau\Lambda$ contains $e_1, \ldots, e_{r-1}$. Moreover, since $(T-t)^{-1} \in R$, we see that e_0 also can be written as an R-linear combination of the elements of $\tau\Lambda$. Thus, $\Lambda \subset \tau\Lambda$, and, by iterating this inclusion,

$$\Lambda \subset \tau^r\Lambda. \tag{4.2.16}$$

The generators of $\tau^r\Lambda$ are $e_r := \tau^r \otimes 1, \ldots, e_{2r-1} := \tau^{2r-1} \otimes 1$, so

$$Te_0 = \phi_T \otimes 1 = te_0 + g_1e_1 + \cdots + g_{r-1}e_{r-1} + g_re_r \tag{4.2.17}$$

$$Te_1 = \tau\phi_T \otimes 1 = t^qe_1 + g_1^qe_2 + \cdots + g_{r-1}^qe_r + g_r^qe_{r+1} \tag{4.2.18}$$

$$\vdots$$

$$Te_{r-1} = \tau^{r-1}\phi_T \otimes 1 = t^{q^{r-1}}e_{r-1} + g_1^{q^{r-1}}e_r + \cdots + g_{r-1}^{q^{r-1}}e_{2r-2} + g_r^{q^{r-1}}e_{2r-1}.$$

From these equations it follows that each $Te_i, 0 \leq i \leq r-1$, lies in $\Lambda + \tau^r \Lambda = \tau^r \Lambda$, where the last equality follows from (4.2.16). Hence

$$T\Lambda \subseteq \tau^r \Lambda. \tag{4.2.19}$$

On the other hand, using (4.2.17), we can express e_r as a k-linear combination of Te_0 and $e_0, \ldots, e_{r-1}$. Next, using (4.2.18), we can express e_{r+1} as a k-linear combination of Te_1 and $e_1, \ldots, e_r$, and therefore as a k-linear combination of Te_0, Te_1 and $e_1, \ldots, e_{r-1}$. Applying a similar recursive argument to $e_{r+2}, \ldots, e_{2r-1}$, we conclude that $e_{r+i} \in \Lambda + T\Lambda, 0 \leq i \leq r - 1$. Because $1/T \in R$, we have $\Lambda \subset T\Lambda$, so $\Lambda + T\Lambda = T\Lambda$. Hence

$$\tau^r \Lambda \subseteq T\Lambda. \tag{4.2.20}$$

Comparing (4.2.19) and (4.2.20), we get (4.2.15). □

Example 4.2.10 Let $q = 3$, $\mathfrak{p} = T^7 - T^2 + 1$, and ϕ be the Drinfeld module of rank 3 over $\mathbb{F}_\mathfrak{p}$ defined by

$$\phi_T = t + (t^2 + 1)\tau + t\tau^2 + \tau^3.$$

By a method that will be discussed at the end of this section, one computes

$$P(x) = x^3 + (-T + 1)x^2 + (T^3 + T - 1)x - \mathfrak{p}.$$

We see that

$$a_0 = -\mathfrak{p}, \quad \deg(a_1) = 3 < \frac{14}{3}, \quad \deg(a_2) = 1 < \frac{7}{3},$$

which is in agreement with Theorem 4.2.7.

Example 4.2.11 Let $A = \mathbb{F}_5[T]$, $\mathfrak{p} = T^4 + 4T^2 + 4T + 2$, and $k = \mathbb{F}_\mathfrak{p}$. Let ϕ be the Drinfeld module of rank 2 over k defined by

$$\phi_T = t + (t + 1)\tau + t^3\tau^2.$$

Then

$$P(x) = x^2 + (4T + 2)x + 2\mathfrak{p}$$

and

$$\phi_k \cong A/(T^4 + 4T^2 + T + 1).$$

One easily computes that $P(1) = 2(T^4 + 4T^2 + T + 1) = 2\chi(^\phi k)$, which agrees with Theorem 4.2.6.

Now let ϕ be the Drinfeld module of rank 2 over k defined by

$$\phi_T = t + (t^2 + 1)\tau + t\tau^2.$$

In this case

$$P(x) = x^2 + (2T^2 + T + 3)x + 3\mathfrak{p}$$

and

$$^\phi k \cong A/(T + 3) \times A/(T(T + 3)(T + 4)).$$

Hence, $P(1) = 3(T(T + 3)^2(T + 4)) = 3 \cdot \chi(^\phi k)$.

Example 4.2.12 Let $C_T = t + \tau$ be the Carlitz module over k. By Theorem 4.2.7,

$$P(x) = x - \mathfrak{p}^{n/d}.$$

Thus, $P(1) = 1 - \mathfrak{p}^{n/d}$. Now using Theorem 4.2.6, we get $(\chi(^C k)) = (\mathfrak{p}^{n/d} - 1)$. On the other hand, since $r = 1$, $^C k \cong A/(a)$ is a cyclic A-module. Hence,

$$^C k \cong A/(\mathfrak{p}^{n/d} - 1).$$

We have already seen a special case of this statement in Example 3.2.13. Note that the Carlitz module is supersingular over finite fields since $H(C) = 1 = r$. From this perspective, when $k = \mathbb{F}_\mathfrak{p}$, this example itself is a special case of a more general fact stated in Exercise 4.2.2.

Let

$$\overline{P}(x) = x^r + \overline{a}_{r-1}x^{r-1} + \cdots + \overline{a}_0 \in \mathbb{F}_\mathfrak{p}[x]$$

be the reduction of $P(x)$ modulo $\mathfrak{p}$, where $\overline{a}_i$, $0 \leq i \leq r - 1$, denotes the image of a_i under the canonical homomorphism $A \to A/\mathfrak{p}$. By Theorem 4.2.7, $\overline{a}_0 = 0$. Hence, x divides $\overline{P}(x)$. It turns out that the exact power of x that divides $\overline{P}(x)$ is the height of ϕ, so $P(x)$ also encodes the height of ϕ.

Theorem 4.2.13 *Let $H(\phi)$ be the height of ϕ. Then*

$$\overline{P}(x) = x^{H(\phi)} \cdot g(x), \quad \text{where} \quad g(x) \in \mathbb{F}_\mathfrak{p}[x] \text{ and } g(0) \neq 0.$$

Proof When we consider $m(x)$ as a polynomial in $\mathbb{F}_\mathfrak{p}[x]$, it decomposes as a product

$$m(x) = m_1(x) \cdot m_2(x) \cdots m_s(x) \tag{4.2.21}$$

of distinct monic irreducibles $m_i(x) \in A_{\mathfrak{p}}[x]$; cf. Theorem 2.8.5 and its proof. (The polynomials $m_i(x)$ have coefficients in $A_{\mathfrak{p}}$ by Gauss' Lemma.) By Proposition 2.5.4, the roots of $m_i(x)$ have the same (non-negative) valuation with respect to the extension of $\mathrm{ord}_{\mathfrak{p}}$ to $\overline{F}_{\mathfrak{p}}$. Since $m_i(0)$ is ± 1 times the product of the roots of $m_i(x)$, the roots of $m_i(x)$ will have positive valuations if and only if $\mathrm{ord}_{\mathfrak{p}}(m_i(0)) > 0$.

On the other hand,

$$\widetilde{F} \otimes_F F_{\mathfrak{p}} \cong F_{\mathfrak{p}}[x]/(m(x)) \cong \prod_{i=1}^{s} F_{\mathfrak{p}}[x]/(m_i(x)) \cong \prod_{i=1}^{s} \widetilde{F}_{\mathfrak{P}_i},$$

where the product is over the primes of $\widetilde{F}$ lying over $\mathfrak{p}$. The primes $\mathfrak{P}_1, \ldots, \mathfrak{P}_s$ correspond to embeddings $\widetilde{F} = F(\pi) \hookrightarrow \widetilde{F}_{\mathfrak{P}_i} \hookrightarrow \overline{F}_{\mathfrak{p}}$ given by mapping π to a root of $m_i(x)$. Therefore, the roots of $m_i(x)$ have positive valuations in $\overline{F}_{\mathfrak{p}}$ if and only if $\mathrm{ord}_{\mathfrak{P}_i}(\pi) > 0$.

According to Theorem 4.1.3, there is a unique place $\widetilde{\mathfrak{p}}$ in $\widetilde{F}$ where π has positive valuation and that prime lies over $\mathfrak{p}$. We may assume $\widetilde{\mathfrak{p}} = \mathfrak{P}_1$. Denote by $\overline{m}_i(x)$ the reduction of $m_i(x)$ modulo $\mathfrak{p}$. By the previous paragraph, $\overline{m}_1(0) = 0$ but $\overline{m}_i(0) \neq 0$ for $2 \leq i \leq s$. On the other hand, by Exercise 2.4.1, $\overline{m}_i(x) = g_i(x)^{d_i}$ for some irreducible polynomial $g_i(x) \in \mathbb{F}_{\mathfrak{p}}[x]$ and $d_i \in \mathbb{Z}_{>0}$ (this is a consequence of Hensel's Lemma). Thus, $\overline{m}_1(x) = x^{[\widetilde{F}_{\widetilde{\mathfrak{p}}}:F_{\mathfrak{p}}]}$, but x does not divide $\overline{m}_i(x)$ for $2 \leq i \leq s$. Finally,

$$\overline{P}(x) = \overline{m}(x)^{r/[\widetilde{F}:F]}$$

$$= (m_1(x) \cdots \overline{m}_s(x))^{r/[\widetilde{F}:F]}$$

$$= x^{r[\widetilde{F}_{\widetilde{\mathfrak{p}}}:F_{\mathfrak{p}}]/[\widetilde{F}:F]} g(x), \quad \text{with} \quad g(0) \neq 0,$$

$$= x^{H(\phi)} g(x),$$

where in the last equality we have used Proposition 4.1.8. □

Corollary 4.2.14 *The Drinfeld module ϕ is supersingular if and only if $\mathfrak{p}$ divides each non-leading coefficient of $P(x)$.*

Example 4.2.15 Let $q = 5$, $\mathfrak{p} = T^4 + T^2 + 2$, and $k = \mathbb{F}_{\mathfrak{p}}$. Let ϕ_T be the Drinfeld module of rank 3 over k defined by $\phi_T = t + \tau + t\tau^2 + \tau^3$. Then $H(\phi) = 2$, $P(x) = x^3 + x^2 - \mathfrak{p}$, and $P(x)$ modulo $\mathfrak{p}$ is equal to $x^2(1 + x)$.

Remark 4.1 A more conceptual proof of Theorem 4.2.13 can be obtained by using the de Rham or crystalline cohomology of Drinfeld modules; cf. [AnglesMM97]. Moreover, one can show that $g(x) \in \mathbb{F}_{\mathfrak{p}}[x]$ in the statement of that theorem is the characteristic polynomial of π acting on $\phi[\mathfrak{p}] \cong \mathbb{F}_{\mathfrak{p}}^{r-H(\phi)}$.

We end this section with a discussion of some methods for computing $P_\phi(x)$ and ϕ_k; see [GP20] for more details. We will assume $k = \mathbb{F}_p$, which simplifies the problem.

Computing $P(x)$

Finding the characteristic polynomial of the Frobenius of an elliptic curve E over a finite field k amounts to finding the number of solutions of a defining equation of E over k: if $P_E(x) = x^2 + ax + b$, then $b = \#k$ and $a = \#E(k) - b - 1$. Computing $\#E(k)$ is relatively easy for small k since one can just plug all the pairs $(\alpha, \beta) \in k \times k$ into the cubic equation defining E and check if they give a solution.

The calculation of the characteristic polynomial of the Frobenius for Drinfeld modules has to be approached differently since it is not related to point-counting on an algebraic curve. Let

$$P(x) = x^r + a_{r-1}x^{r-1} + \cdots + a_1x + a_0 \in A[x].$$

By Theorem 4.2.7, $\deg(a_i) < d$ for all $1 \le i \le r - 1$, so $a_1, \ldots, a_{r-1}$ are uniquely determined by their residues modulo $\mathfrak{p}$. (Here we use the assumption that $k = \mathbb{F}_p$.) We also know from Theorem 4.2.7 that $a_0 = c \cdot \mathfrak{p}$, where $c \in \mathbb{F}_q^\times$ is given by a simple explicit formula, so we need to find $a_1, \ldots, a_{r-1}$. Now, since $P(\pi) = 0$, we have

$$\tau^{dr} + \phi_{a_{r-1}}\tau^{d(r-1)} + \cdots + \phi_{a_1}\tau^d + \phi_{a_0} = 0.$$

Denote

$$f_i = \phi_{a_i}\tau^{di} + \phi_{a_{i-1}}\tau^{d(i-1)} + \cdots + \phi_{a_0}$$

and

$$f_i^\dagger = \tau^{dr} + \phi_{a_{r-1}}\tau^{d(r-1)} + \cdots + \phi_{a_{i+1}}\tau^{d(i+1)}.$$

Note that $\deg_\tau \phi_{a_j}\tau^{dj} \ge dj$, so the coefficient of $\tau^{d(i+1)}$ in $f_i^\dagger$ is the constant term of $\phi_{a_{i+1}}$, i.e., $\gamma(a_{i+1})$. Therefore,

$$\gamma(a_{i+1}) = -\text{coefficient of } \tau^{d(i+1)} \text{ in } f_i.$$

Since we know f_0 explicitly, we can compute all a_j recursively, where $a_0, a_1, \ldots, a_{i-1}$ are used to calculate a_i. This algorithm is fairly easy to implement on a computer, and the amount of work involved in the calculation of $P(x)$ using this algorithm is polynomial in r and d.

Remark 4.2.16 When $k = \mathbb{F}_p$ but $r = 2$, there is a different recursive procedure due to Gekeler for computing $P(x)$ based on the properties of Eisenstein series;

see [Gek08, Prop. 3.7]. Let ϕ be a Drinfeld module of rank 2 over $\mathbb{F}_p$ defined by $\phi_T = t + g_1\tau + g_2\tau^2$. Let $P(x) = x^2 - ax + b$. By Theorem 4.2.7,

$$b = (-1)^d \operatorname{Nr}_{\mathbb{F}_p/\mathbb{F}_q}(g_2)^{-1}\mathfrak{p}.$$

For $n \in \mathbb{Z}_{>0}$, put $[n] := T^{q^n} - T \pmod{\mathfrak{p}}$, and define recursively $f_0 = 1$, $f_1 = g_1$,

$$f_n = -[n - 1] \cdot f_{n-2} \cdot g_2^{q^{n-2}} + f_{n-1} \cdot g_1^{q^{n-1}} \qquad (n \geq 2).$$

Then the Frobenius trace $a \in A$ is determined through

$$a \equiv (-1)^d \operatorname{Nr}_{\mathbb{F}_p/\mathbb{F}_q}(g_2)^{-1} f_d \pmod{\mathfrak{p}} \quad \text{and} \quad \deg(a) \leq d/2.$$

In practice, Gekeler's algorithm seems to have the same efficiency as what was presented above.

Computing $\phi\mathbb{F}_p$
Similar to (4.2.12), let

$$\phi\mathbb{F}_p \cong A/(b_1) \times \cdots \times A/(b_r),$$

where $b_1 \mid b_2 \mid \cdots \mid b_r$ are monic, but we allow some of the b_i's to be equal to 1. We call b_r the *exponent* of $\phi\mathbb{F}_p$ since $b_r \in A$ is the monic polynomial of smallest degree such that ϕ_{b_r} annihilates $\mathbb{F}_p$. We treat $\mathbb{F}_p$ as an $\mathbb{F}_q$-vector space with basis $1, t, t^2, \ldots, t^{d-1}$. Put

$$\mathfrak{b} = x_0 + x_1 T + \cdots + x_{d-1}T^{d-1} + T^d$$

for a hypothetical annihilator of $\phi\mathbb{F}_p$, i.e., $\phi_{\mathfrak{b}}$ acts as zero on $\mathbb{F}_p$. For a fixed $0 \leq s \leq d - 1$, compute

$$\phi_{T^i}(t^s) = \alpha_{i,1} + \alpha_{i,2}t + \cdots + \alpha_{i,d}t^{d-1} \qquad (4.2.22)$$

for $i = 1, 2, \ldots, d$, which can be easily done by a repeated application of ϕ_T. Let M_s be the $d \times d$ matrix whose first row consists of zeros except at position $s + 1$ where it is 1, and the $(i + 1)$-th row is $[\alpha_{i,1}, \alpha_{i,2}, \ldots, \alpha_{i,d}]$ for $1 \leq i \leq d - 1$. Let N_s be the column vector $-[\alpha_{d,1}, \ldots, \alpha_{d,d}]$. Then $\phi_{\mathfrak{b}}$ acts as 0 on $\mathbb{F}_p$ if and only if $\phi_{\mathfrak{b}}(t^s) = 0$ for all $s = 0, \ldots, d - 1$, which itself is equivalent to

$$[x_0, \ldots, x_{d-1}]M_s = N_s \quad \text{for all} \quad s = 0, \ldots, d - 1.$$

This system of linear equations always has a solution (since $\phi\mathbb{F}_p$ has an exponent). Find a particular solution $\mathbf{x}$ and find a basis $\mathbf{b}_1, \ldots, \mathbf{b}_h$ for the intersection of the null-spaces of all M_s, so that every other solution is of the form $\mathbf{x} + \operatorname{span}(\mathbf{b}_1, \ldots, \mathbf{b}_h)$. It is easy to see that the exponent of $\phi\mathbb{F}_p$ is the greatest common divisor of $f_{\mathbf{x}}, f_{\mathbf{x}+\mathbf{b}_1}, \ldots, f_{\mathbf{x}+\mathbf{b}_h}$, where $f_{\mathbf{y}} := y_0 + y_1 T \cdots + y_{d-1} T^{d-1} + T^d$ for $\mathbf{y} = (y_0, \ldots, y_{d-1})$. Finding the greatest common divisor of a set of polynomials is a routine calculation thanks to the division algorithm (Theorem 1.1.2). Computing all $\phi_{T^i}(t^s)$ can be done in polynomial time in d; solving the system of linear equations also can be done in polynomial time in d. Hence, we can find the exponent b_r of $\phi \mathbb{F}_{\mathfrak{p}}$ in polynomial time in d. Since $b_1 \mid b_2 \mid \cdots \mid b_r$ and $\deg(b_1 \cdot b_2 \cdots b_r) = d$, this gives us finitely many possibilities for these invariants. Let $(c_1, \ldots, c_r)$ be a possible r-tuple for the invariants. Each ϕ_{c_i} is an $\mathbb{F}_q$-linear transformation of $\mathbb{F}_{\mathfrak{p}} = \mathbb{F}_q + \mathbb{F}_q t + \cdots + \mathbb{F}_q t^{d-1}$. An explicit matrix for its action with respect to the given basis can be computed by writing $c_i = \beta_0 + \beta_1 T^1 + \cdots + T^m$, $m \le d$, and using (4.2.22). This allows us to compute the dimension of the null-space of ϕ_{c_i} acting on $\mathbb{F}_{\mathfrak{p}}$. Then $c_1, \ldots, c_r$ are the actual invariant factors of $\phi \mathbb{F}_{\mathfrak{p}}$ if and only if the dimension of the null-space of ϕ_{c_i} is equal to $\deg(c_1) + \deg(c_2) + \cdots + \deg(c_{i-1}) + (r - i + 1) \cdot \deg(c_i)$ for all $1 \le i \le r$.

The explicit examples in this section were calculated using the above algorithms implemented in Magma.

Exercises

In all exercises ϕ is a Drinfeld module defined over k and r denotes the rank of ϕ.

4.2.1 Consider k as an A-module via γ. Prove that $\chi(k) = \mathfrak{p}^{n/d}$.

4.2.2 Assume $k = \mathbb{F}_{\mathfrak{p}}$ and ϕ is supersingular. Let $\phi_T = t + g_1 \tau + \cdots + g_r \tau^r$. Prove that

$$P(x) = x^r + (-1)^{rd-d-r} g_r^{-(q^d-1)/(q-1)} \cdot \mathfrak{p}$$

4.2.3 Assume $k = \mathbb{F}_{\mathfrak{p}}$. Prove that $[\widetilde{F} : F] = r$ and $D = \widetilde{F}$.

4.2.4 Prove that $\frac{\mathfrak{p}^{n/d}}{\pi} \in \mathrm{End}_k(\phi)$.

4.2.5 Let $\mathfrak{n} \lhd A$ be an ideal not divisible by $\mathfrak{p}$, so that $\phi[\mathfrak{n}] \cong (A/\mathfrak{n})^r$. The Frobenius Frob_k acts on $\phi[\mathfrak{n}]$ by an A-linear automorphism, so can be represented by a matrix in $\mathrm{GL}_r(A/\mathfrak{n})$, well-defined up to conjugation. Let $P_{\mathfrak{n}}(x) \in (A/\mathfrak{n})[x]$ be the characteristic polynomial of the matrix of Frob_k; this is well-defined. Prove that $P_{\mathfrak{n}}(x)$ is equal to the polynomial that one obtains by reducing the coefficients of $P(x)$ modulo $\mathfrak{n}$.

4.2.6 Assume $k = \mathbb{F}_{\mathfrak{p}}$ and r is not divisible by the characteristic of k. Suppose there is a prime $\mathfrak{l} \ne \mathfrak{p}$ for which $\phi[\mathfrak{l}]$ is rational over $\mathbb{F}_{\mathfrak{p}}$, or, in other words, $\phi_{\mathfrak{l}}(x)$ splits completely over $\mathbb{F}_{\mathfrak{p}}$. Prove that ϕ must be ordinary.

4.2.7 Assume $r = 2$, ϕ is defined by $\phi_T = t + g_1 \tau + g_2 \tau^2$, and ψ is defined by $\psi_T = t + c g_1 \tau + c^{q+1} g_2 \tau^2$. Let $P_\phi = x^2 + a_1 x + a_0$ be the characteristic polynomial of the Frobenius of ϕ. Prove that the characteristic polynomial of the Frobenius of ψ is

$$P_\psi = x^2 + b^{-1}a_1 x + b^{-2}a_0,$$

where $b = \mathrm{Nr}_{k/\mathbb{F}_q}(c)$.

4.2.8 Let $m(x) = m_1(x) \cdots m_s(x)$ be the decomposition into irreducibles of the minimal polynomial of π over $F_\mathfrak{p}$; cf. (4.2.21). Each factor $m_i(x)$ corresponds to prime $\mathfrak{P}_i$ of $\widetilde{F}$ lying over $\mathfrak{p}$. The restriction of $\mathrm{ord}_{\mathfrak{P}_i}$ to $\mathbb{F}_q(\pi)$ is a valuation, hence corresponds to a unique irreducible monic polynomial $h_i(\pi) \in \mathbb{F}_q[\pi]$. Assume $\mathfrak{P}_1 = \widetilde{\mathfrak{p}}$, so that $h_1(\pi) = \pi$.

(a) Some of the polynomials $h_i(\pi)$ might coincide for different indices $2 \le i \le s$, but show that they are all coprime to π.

(b) Denote by $\overline{m}_i(x)$ the reduction of $m_i(x)$ modulo $\mathfrak{p}$. Then, as in the proof of Theorem 4.2.13, $\overline{m}_i(x) = g_i(x)^{d_i}$ for some irreducible polynomial $g_i(x) \in \mathbb{F}_\mathfrak{p}[x]$ and $d_i \in \mathbb{Z}_{>0}$. Prove that $g_i(x)$ divides $h_i(x)$ when both are considered as polynomials in $\mathbb{F}_\mathfrak{p}[x]$. [Hint: Note that $g_i(x)$ is the minimal polynomial of $\pi \pmod{\mathfrak{P}_i}$ over $\mathbb{F}_\mathfrak{p}$, and that $h_i(\pi) \in \mathfrak{P}_i$.]

(c) Conclude that x does not divide $\overline{m}_2(x) \cdots \overline{m}_s(x)$, which is a key step in the proof of Theorem 4.2.13.

Exercises 4.2.9–4.2.12 give an alternative proof of the formula for $a_0 = P(0)$ from Theorem 4.2.7. It is based on an argument of Hsia and Yu [HY00], and Gekeler [Gek08].

4.2.9 By definition, $\hat{\rho}_{\phi,1}(\mathrm{Fr}_k) \in \mathrm{GL}_r(A_1)$. Deduce from this that the ideal (a_0) must be a power of $\mathfrak{p}$. Use the fact that $n = \deg(a_0)$ to conclude that $a_0 = c \cdot \mathfrak{p}^{n/d}$ for some $c \in \mathbb{F}_q^\times$.

4.2.10 Assume $\mathfrak{p} \ne T$. (If $\mathfrak{p} = T$ one can apply the arguments of Exercises 4.2.10–4.2.12 with $T + 1$ playing the role of T.)

(a) Consider $W := \phi[T]$ as an $\mathbb{F}_q$-vector space. Prove that

$$(-1)^r \det(\pi \mid W) = a_0 \pmod{T} = c \cdot p_0^{n/d},$$

where p_0 is the constant term of the monic polynomial $\mathfrak{p}$.

(b) Choose a basis $\{w_1, \ldots, w_r\}$ of W and denote $\delta(W) = (-1)^r M(w_1, \ldots, w_r)^{q-1}$. Prove that

$$\delta(W) = \frac{t}{g_r}.$$

(c) Prove that

$$\det(\pi \mid W) = \left((-1)^r \delta(W)\right)^{\frac{q^n-1}{q-1}}.$$

(d) Deduce that

$$c = (-1)^{r(n+1)} t^{\frac{q^n-1}{q-1}} (p_0^{-1})^{n/d} \left(g_r^{-1}\right)^{\frac{q^n-1}{q-1}}.$$

4.2.11 The $(q-1)$-th power residue symbol for $a \in A$ not divisible by the prime $\mathfrak{p}$ is defined as

$$\left(\frac{a}{\mathfrak{p}}\right)_{q-1} \equiv a^{(q^{\deg(\mathfrak{p})}-1)/(q-1)} \pmod{\mathfrak{p}}.$$

Prove that

$$t^{\frac{q^n-1}{q-1}} = \left(\frac{T}{\mathfrak{p}}\right)_{q-1}^{n/d} \quad \text{and} \quad p_0 = \left(\frac{\mathfrak{p}}{T}\right)_{q-1}.$$

4.2.12 For two monic primes $\mathfrak{p} \neq \mathfrak{q}$, there is a reciprocity law:

$$\left(\frac{\mathfrak{q}}{\mathfrak{p}}\right)_{q-1} \left(\frac{\mathfrak{p}}{\mathfrak{q}}\right)_{q-1}^{-1} = (-1)^{\deg(\mathfrak{p})\cdot\deg(\mathfrak{q})}.$$

(See [Ros02, Ch. 3] for a proof.) Use this reciprocity law to prove that

$$c = (-1)^{nr-r-n} \cdot \mathrm{Nr}_{k/\mathbb{F}_q}(g_r)^{-1}.$$

4.2.13 This exercise describes an algorithm for computing $P(x)$ for $r = 2$ and arbitrary k. Let ϕ be a Drinfeld module of rank 2 over k defined by $\phi_T = t + g_1\tau + g_2\tau^2$. Let $P(x) = x^2 - ax + b$ be its Frobenius characteristic polynomial. From Theorem 4.2.7 we know that

$$b = (-1)^n \mathrm{Nr}_{\mathbb{F}_{q^n}/\mathbb{F}_q}(g_2)^{-1} \mathfrak{p}^{n/d},$$

and $\deg(a) \leq [n/2] =: m$. Write $a = \sum_{i=0}^m \alpha_i T^i$, where $\alpha_0, \ldots, \alpha_m$ are unknowns which we need to determine.

(a) Show that the coefficients of $\phi_a \in k[\alpha_0, \ldots, \alpha_m]\{\tau\}$, as a polynomial in τ, are linear expressions in $\alpha_0, \ldots, \alpha_m$.

(b) Show that $P(\pi) = 0$ leads to a system of linear equations in $\alpha_0, \ldots, \alpha_m$ over k, which has a unique solution.

4.3 Isogeny Classes

Isogenies of Drinfeld modules over a given A-field K define an equivalence relation. Indeed, it is clear that a composition of isogenies $\phi_1 \to \phi_2 \to \phi_3$ is an isogeny, so the relation is transitive; moreover, thanks to Proposition 3.3.12, each isogeny $u : \phi \to \psi$ has a dual isogeny $\hat{u} : \psi \to \phi$ also defined over K, so the relation is symmetric. We will show in this section that isogeny classes of Drinfeld modules over finite fields are uniquely determined by their characteristic polynomials of the Frobenius.

An important problem in the theory of Drinfeld modular varieties is the problem of counting Drinfeld modules over finite fields, up to isomorphism. This is directly related to the problem of calculating the Weil zeta functions of Drinfeld modular varieties; see [Dri77, Lau96]. It turns out that for this counting, it is more convenient to first classify Drinfeld modules up to isogeny, and then count the number of isomorphism classes in each isogeny class. In turn, the classification of the isogeny classes reduces to the classification of characteristic polynomials of the Frobenius, which, as we will see, can be accomplished fairly explicitly.

For a given Drinfeld module ϕ over k, denote by $\widetilde{F}_\phi$ the field generated by $\pi \in k(\tau)$ over $\phi(F) \subset k(\tau)$, and by $m_\phi(x)$ the minimal polynomial of π over $\phi(F) \cong F$; recall that $m_\phi(x)$ is monic with coefficients in A. Let $P_\phi(x)$ be the characteristic polynomial of the Frobenius, and let $D_\phi := \mathrm{End}_k^\circ(\phi)$.

In Theorem 4.3.2, as in the proof of Theorem 4.1.3, we consider $\widetilde{F}_\phi$ as a finite extension of $K = \mathbb{F}_q(\pi)$ in $k(\tau)$. Thus, $\widetilde{F}_\phi$ is the quotient of $K[x]$ by the ideal generated by some irreducible polynomial. It turns out that this irreducible polynomial is essentially $m_\phi(\pi)$ viewed as a polynomial in T. More precisely, consider T and π as two independent indeterminates of the polynomial ring $\mathbb{F}_q[T, \pi]$. Then consider $m_\phi(\pi) \in \mathbb{F}_q[T, \pi] = \mathbb{F}_q[\pi][T]$ as a polynomial $\widetilde{m}_\phi(T)$ in indeterminate T with coefficients in $\mathbb{F}_q[\pi]$.

Lemma 4.3.1 *The polynomial $\widetilde{m}_\phi(T)$ has degree $[\widetilde{F}_\phi : K]$ in T and is irreducible in $K[T]$. The leading coefficient of $\widetilde{m}_\phi(T)$ is in $\mathbb{F}_q^\times$.*

Proof In the proof we omit the subscript ϕ from notation, so, for example, $\widetilde{m}(T)$ stands for $\widetilde{m}_\phi(T)$. By Theorem 4.2.7, the degree in T of $m(0)$ is strictly larger than the degrees of the other coefficients of $m(x)$. Hence, the leading term of $\widetilde{m}(T)$ is the leading term of $m(0) \in A$, so its leading coefficient is in $\mathbb{F}_q^\times$. Moreover, by Theorems 4.2.2 and 4.2.7, up to an $\mathbb{F}_q^\times$-multiple, $m(0)$ is equal to $\mathfrak{p}^{\frac{[\widetilde{F}:F] \cdot n}{r \cdot d}}$. Using (4.1.3), we get

$$(m(0)) = \mathfrak{p}^{\frac{[\widetilde{F}:K]}{d}}.$$

Thus, $\deg_T \widetilde{m}(T) = [\widetilde{F} : K]$. Since $\widetilde{F}$ is obtained by adjoining to K a root of $\widetilde{m}(T)$, the equality $\deg_T \widetilde{m}(T) = [\widetilde{F} : K]$ implies that $\widetilde{m}(T)$ is irreducible over K. □

Theorem 4.3.2 *For two Drinfeld modules ϕ and ψ of rank r over k, the following statements are equivalent:*

(1) ϕ *and* ψ *are isogenous over* k.
(2) $m_\phi(x) = m_\psi(x)$.
(3) $P_\phi(x) = P_\psi(x)$.

Proof
$(1) \Rightarrow (2)$: Let $u \in k\{\tau\}$ be an isogeny $\phi \to \psi$. Let

$$m_\phi(x) = x^s + a_{s-1}x^{s-1} + \cdots + a_1 x + a_0 \in A[x].$$

By definition, $m_\phi(\pi) = 0$ means that in $k\{\tau\}$ we have the equality

$$\pi^s + \phi_{a_{s-1}}\pi^{s-1} + \cdots + \phi_{a_0} = 0.$$

Multiplying both sides by u and using the commutation rule $u\phi_a = \psi_a u$, we obtain

$$0 = u\pi^s + u\phi_{a_{s-1}}\pi^{s-1} + \cdots + u\phi_{a_0} = \pi^s u + \psi_{a_{s-1}}\pi^{s-1}u + \cdots + \psi_{a_0}u$$
$$= (\pi^s + \psi_{a_{s-1}}\pi^{s-1} + \cdots + \psi_{a_0})u.$$

Hence, $\pi^s + \psi_{a_{s-1}}\pi^{s-1} + \cdots + \psi_{a_0} = 0$. This implies that $m_\phi(x)$ annihilates $\pi \in \mathrm{End}_k(\psi)$, which then implies that $m_\psi(x)$ divides $m_\phi(x)$. Reversing the roles of ϕ and ψ we conclude that $m_\phi(x)$ divides $m_\psi(x)$. Therefore, $m_\phi(x) = m_\psi(x)$, as both polynomials are monic.
$(2) \Leftrightarrow (3)$: Since $m_\phi(x)$ and $m_\psi(x)$ are irreducible over F, and $P_\phi(x)$ and $P_\psi(x)$ have the same degree r, the equivalence follows from Theorem 4.2.2.
$(2) \Rightarrow (1)$: If $m_\phi(x) = m_\psi(x)$, then $\widetilde{m}_\phi(T) = \widetilde{m}_\psi(T)$. It follows from Lemma 4.3.1 that, as K-subalgebras of $k(\tau)$, we have $\widetilde{F}_\phi \cong K[T]/(\widetilde{m}_\phi(T))$ and $\widetilde{F}_\psi \cong K[T]/(\widetilde{m}_\psi(T))$. Thus, $\widetilde{F}_\phi \cong \widetilde{F}_\psi$ as K-subalgebras of $k(\tau)$. Using Theorem 1.7.23, we conclude that there exists $0 \neq u \in k(\tau)$ such that $\widetilde{F}_\psi = u\widetilde{F}_\phi u^{-1}$. Since conjugate elements in $k(\tau)$ have the same minimal polynomials over K, the element $u\phi_T u^{-1}$ is a root of $\widetilde{m}_\psi$. But ψ_T is also a root of this polynomial and $\widetilde{F}_\psi \cong K[T]/(\widetilde{m}_\psi(T))$, so there is a K-automorphism of $\widetilde{F}_\psi$ which maps $u\phi_T u^{-1}$ to ψ_T. Applying Theorem 1.7.23 again, there is $w \in k(\tau)$ such that $\psi_T = wu\phi_T u^{-1}w^{-1}$. Thus, after replacing u by wu, we may assume that $u\phi_T u^{-1} = \psi_T$. Finally, since $k(\tau) = k\{\tau\} \otimes_{\mathbb{F}_q[\pi]} K$, after multiplying u by an element from the center K, we may assume that $u \in k\{\tau\}$. This u gives an isogeny $\phi \to \psi$. $\square$

Lemma 4.3.3 *Given two Drinfeld modules ϕ and ψ of rank r over k, the following two statements are equivalent:*

(1) $\widetilde{F}_\phi \cong \widetilde{F}_\psi$ *are isomorphic K-subalgebras of $k(\tau)$.*
(2) $D_\phi \cong D_\psi$ *are isomorphic K-subalgebras of $k(\tau)$.*

These statements hold if ϕ and ψ are isogenous over k.

Proof

(1)⇒(2): If $\widetilde{F}_\phi \cong \widetilde{F}_\psi$ as K-subalgebras of $k(\tau)$, then by Theorem 1.7.23 there exists $0 \neq u \in k(\tau)$ such that $\widetilde{F}_\psi = u\widetilde{F}_\phi u^{-1}$. Now

$$D_\psi = \mathrm{Cent}_{k(\tau)}(\widetilde{F}_\psi) = \mathrm{Cent}_{k(\tau)}(u\widetilde{F}_\phi u^{-1}) = u\,\mathrm{Cent}_{k(\tau)}(\widetilde{F}_\phi)u^{-1} = uD_\phi u^{-1}.$$

Therefore, $D_\phi \cong D_\psi$ as K-subalgebras of $k(\tau)$.

(2)⇒(1): Suppose $D_\phi \cong D_\psi$ as K-algebras. Since both are division algebras containing K, Theorem 1.7.23 ensures the existence of $0 \neq u \in k(\tau)$ such that $D_\psi = uD_\phi u^{-1}$. The conjugation by u maps the center of D_ϕ, which is $\widetilde{F}_\phi$, onto the center of D_ψ, which is $\widetilde{F}_\psi$. Hence, $\widetilde{F}_\psi = u\widetilde{F}_\phi u^{-1}$.

We already showed in the proof of Theorem 4.3.2 that if ϕ and ψ are isogenous Drinfeld modules, then $\widetilde{F}_\phi$ and $\widetilde{F}_\psi$ are isomorphic as K-subalgebras of $k(\tau)$. □

Remark 4.3.4 It is important to note that the fields $\widetilde{F}_\phi$ and $\widetilde{F}_\psi$ can be isomorphic over K without ϕ and ψ being isogenous over k. For example, take $\mathfrak{p} = T, n = 2$, q odd, and define $\phi_T = \tau^2$ and $\psi_T = -\tau^2$. Then $m_\phi = x - T$ and $m_\psi = x + T$. By Theorem 4.3.2, ϕ and ψ are not isogenous over k (this is also easy to see directly), although $\widetilde{F}_\phi = \widetilde{F}_\psi = F = K$ and $D_\phi = D_\psi = k(\tau)$. (But note that ϕ and ψ become isomorphic over a quadratic extension of k.)

According to Theorem 4.3.2, classifying the isogeny classes of Drinfeld modules over k is equivalent to classifying the possible minimal polynomials $m(x) := m_\phi(x)$ of π over $\phi(F)$ for Drinfeld modules ϕ of rank r defined over k. Let us summarize the properties of these polynomials that we proved so far. Let $r_1 := \deg(m(x))$.

(1) $m(x) \in A[x]$ is monic.
(2) $\frac{r}{r_1}$ and $\frac{nr_1}{dr}$ are integers.
(3) $m(0) = c\mathfrak{p}^{\frac{nr_1}{dr}}$ for some $c \in \mathbb{F}_q^\times$.
(4) $m(x)$ remains irreducible over F_∞.
(5) $m(x) = m_1(x) \cdot m_2(x)$ over $F_\mathfrak{p}$, with $m_1(x), m_2(x) \in A_\mathfrak{p}[x]$ monic, such that

    (i) $m_1(x)$ is irreducible over $F_\mathfrak{p}$.
    (ii) $m_1(x) \equiv x^h \pmod{\mathfrak{p}}$ for some $1 \leq h \leq r_1$.
    (iii) $m_2(0) \in A_\mathfrak{p}^\times$.

Definition 4.3.5 We call a polynomial a (k, r)-*Weil polynomial* if it satisfies (1)-(5).

Example 4.3.6 This example is based on [Yu95, Prop. 4]. Let $s = n/d = [k : \mathbb{F}_\mathfrak{p}]$. Assume the characteristic of F is odd. Let us show that a $(k, 2)$-Weil polynomial $m(x)$ is one of the following:

(i) $x^2 + ax + c\mathfrak{p}^s$, $\mathfrak{p} \nmid a, c \in \mathbb{F}_q^\times$, and either $\deg(a^2 - 4c\mathfrak{p}^s)$ is odd or $\deg(a^2 - 4c\mathfrak{p}^s)$ is even but the leading coefficient of $a^2 - 4c\mathfrak{p}^s$ is not a square in $\mathbb{F}_q^\times$.

(ii) $x^2 + c_1\mathfrak{p}^{s/2}x + c\mathfrak{p}^s$, s is even, $c_1, c \in \mathbb{F}_q, c \neq 0, c_1^2 - 4c$ is not a square in $\mathbb{F}_q^\times$, and d is odd.

(iii) $x^2 + c\mathfrak{p}^s$, s is odd, $c \in \mathbb{F}_q^\times$, and either d is odd or d is even and $-c$ is not a
    square in $\mathbb{F}_q^\times$.

(iv) $x + c\mathfrak{p}^{s/2}$, s is even, $c \in \mathbb{F}_q^\times$.

It is clear that (iv) is a $(k, 2)$-Weil polynomial, and every $(k, 2)$-Weil polynomial
of degree one must be a polynomial of that form. Hence, from now on we assume
that $m(x)$ is quadratic. Note that $m(x)$ satisfies (1)–(3) if and only if $m(x) = x^2 +$
$ax + c\mathfrak{p}^s$, with $a \in A$ and $c \in \mathbb{F}_q^\times$, so we assume $m(x)$ has this form. We need to
analyze conditions (4) and (5).

If $\mathfrak{p} \nmid a$, then $m(x)$ satisfies (5) by Hensel's Lemma (see Theorem 2.4.2) because
modulo $\mathfrak{p}$ it factors as $x(x + \alpha)$ for some $0 \neq \alpha \in \mathbb{F}_\mathfrak{p}$. The polynomial $m(x)$
is irreducible over F_∞ if and only if $\widetilde{F} = F[x]/(m(x))$ is a totally imaginary
extension of F, so the necessary and sufficient conditions for this are the same as in
Example 3.4.24. This gives (i).

Now suppose $\mathfrak{p} \mid a$. Then (5) forces $m(x)$ to be irreducible over $F_\mathfrak{p}$ (if $m(x) =$
$(x + b_1)(x + b_2)$, with $\mathfrak{p} \mid b_1$, $\mathfrak{p} \nmid b_2$, then $a = b_1 + b_2$ is not divisible by
$\mathfrak{p}$). Since $m(x)$ is irreducible over F_∞, we must have $2 \deg(a) \leq s \cdot \deg(\mathfrak{p})$; cf.
Example 3.4.24. This implies that $v := \mathrm{ord}_\mathfrak{p}(a) \leq s/2$. If $v < s/2$ and $a \neq 0$,
then making a change of variables $x \mapsto \mathfrak{p}^v x$ and dividing $\mathfrak{p}^{2v} x^2 + a\mathfrak{p}^v x + c\mathfrak{p}^s$ by
$\mathfrak{p}^{2v}$, we obtain $x^2 + b_1 x + b_2$ with $b_1, b_2 \in A_\mathfrak{p}$, $\mathrm{ord}_\mathfrak{p}(b_1) = 0$, $\mathrm{ord}_\mathfrak{p}(b_2) > 0$. By
Hensel's Lemma, this last polynomial is reducible over $F_\mathfrak{p}$, which is a contradiction.
Therefore, either $a = 0$ or $a = c_1 \mathfrak{p}^{s/2}$ for some $c_1 \in \mathbb{F}_q^\times$.

Suppose s is even. Write $a = c_1 \mathfrak{p}^{s/2}$ for some $c_1 \in \mathbb{F}_q$ ($c_1 = 0$ is allowed). By
the above argument, $x^2 + c_1 x + c$ must be irreducible over $\mathbb{F}_q$ and $\mathbb{F}_\mathfrak{p}$. A necessary
and sufficient condition for this is that $c_1^2 - 4c$ is not a square in $\mathbb{F}_q$ and d is odd.
This gives (ii).

Now suppose s is odd. Then necessarily $a = 0$. The polynomial $x^2 + c\mathfrak{p}^s$ is
irreducible over $F_\mathfrak{p}$. It is irreducible over F_∞ if and only if either d is odd or d is
even but $-c$ is not a square in F_∞. This gives (iii).

Note that $\mathfrak{p}$ splits in $\widetilde{F}$ in case (i), remains inert in case (ii), ramifies in case
(iii), and $F = \widetilde{F}$ in case (iv). Therefore, if $m(x) = m_\phi(x)$, then ϕ is ordinary in
case (i), and supersingular in cases (ii), (iii), (iv); cf. Corollary 4.1.11. Moreover,
by Theorem 4.1.3, $D_\phi = \widetilde{F}$ in cases (i), (ii), (iii), and D_ϕ is a quaternion division
algebra over F in case (iv).

Theorem 4.3.2 implies that the map $\phi \to m_\phi(x)$ gives an injection

$$\{\text{isogeny classes of Drinfeld modules of rank } r \text{ over } k\} \qquad (4.3.1)$$

$$\hookrightarrow \{(k, r)\text{-Weil polynomials}\}.$$

It turns out that this map is also surjective, so any polynomial satisfying the
necessary conditions (1)–(5) is in fact the minimal polynomial of the Frobenius
endomorphism of some Drinfeld module of rank r defined over k. This is an
analogue for Drinfeld modules of the Honda-Tate theorem for abelian varieties. The
proof of this theorem for Drinfeld modules is considerably simpler than the proof

of the corresponding fact for abelian varieties. This is due to the existence of the ambient algebra $k(\tau)$ where the constructions take place. Still, the proof is not easy and relies on some of the deepest results from the theory of central division algebras.

Theorem 4.3.7 *The map* (4.3.1) *is bijective.*

Proof Let $m(x)$ be a (k, r)-Weil polynomial. Denote $K = \mathbb{F}_q(\pi)$ and $\widetilde{F} = F[\pi]/(m(\pi))$, where now we consider π as an indeterminate. Since $\widetilde{F}/F$ is a finite extension, the transcendence degree of $\widetilde{F}$ over $\mathbb{F}_q$ is 1. On the other hand, from the properties of $m(x)$, it is not hard to see that π is not algebraic over $\mathbb{F}_q$. Hence, the natural homomorphism $\mathbb{F}_q[\pi] \to F[\pi]/(m(\pi))$ is injective, so we can consider K as a subfield of $\widetilde{F}$. Since K has transcendence degree 1 over $\mathbb{F}_q$, the extension $\widetilde{F}/K$ is algebraic; in particular, T is algebraic over K. But $K(T) = \widetilde{F}$, so $\widetilde{F}/K$ is a finite extension.

Let v be a valuation on $\widetilde{F}$. If $\mathrm{ord}_v(\pi) > 0$, then $m(0) = c\mathfrak{p}^{\frac{nr_1}{dr}}$ implies that $\mathrm{ord}_v(\mathfrak{p}) > 0$, so v lies over $\mathfrak{p}$. On the other hand, condition (5) implies that there is a unique place $\widetilde{\mathfrak{p}}$ of $\widetilde{F}$ over $\mathfrak{p}$ where $\mathrm{ord}_{\widetilde{\mathfrak{p}}}(\pi) > 0$; see the proof of Theorem 4.2.13. Thus, $\mathrm{ord}_v(\pi) \neq 0$ only at $v = \widetilde{\mathfrak{p}}$ or at v lying over ∞ in F. Condition (4) implies that there is a unique place $\widetilde{\infty}$ in $\widetilde{F}$ over ∞, so π has nonzero valuation at exactly two places $\widetilde{\mathfrak{p}}$ and $\widetilde{\infty}$ of $\widetilde{F}$. Hence the places (π) and $(1/\pi)$ of K do not split in $\widetilde{F}/K$, and the corresponding places of $\widetilde{F}$ over (π) and $(1/\pi)$ are $\widetilde{\mathfrak{p}}$ and $\widetilde{\infty}$, respectively.

Now, as in the proof of Theorem 4.1.3, we have

$$[\widetilde{F} : F] = [\widetilde{F}_\infty : F_\infty] = f_1 \cdot e_1,$$

$$[\widetilde{F} : K] = [\widetilde{F}_{\widetilde{\infty}} : K_{1/\pi}] = f_2 \cdot e_2,$$

where f_i (respectively, e_i), $i = 1, 2$, is the residue degree (respectively, the ramification index) of the corresponding extension of the completions. The residue field at both ∞ and $1/\pi$ is $\mathbb{F}_q$. Hence $f_1 = f_2$. Since $m(x)$ is irreducible over F_∞, all its roots have the same $\widetilde{\infty}$-adic valuation. Thus,

$$\mathrm{ord}_{\widetilde{\infty}}(\pi) = \frac{1}{r_1} \mathrm{ord}_{\widetilde{\infty}}(m(0))$$

$$= \frac{1}{r_1 e_1} \mathrm{ord}_\infty(m(0))$$

$$= -\frac{1}{r_1 e_1} \frac{nr_1}{dr} d$$

$$= -\frac{1}{e_1} \frac{n}{r}.$$

On the other hand,

$$\mathrm{ord}_{\widetilde{\infty}}(\pi) = \frac{1}{e_2} \mathrm{ord}_{1/\pi}(\pi) = -\frac{1}{e_2}.$$

Combining these calculations, we get

$$\frac{n}{[\widetilde{F} : K]} = \frac{r}{[\widetilde{F} : F]} = \frac{r}{r_1}.$$

Since by assumption, r/r_1 is an integer, we conclude that $[\widetilde{F} : K]$ divides n.

At this point we need a fact from the theory of central division algebras which says that a field extension L/K embeds into $k(\tau)$, i.e., there is a homomorphism of K-algebras $L \hookrightarrow k(\tau)$, if and only if $[L : K]$ divides n and the places (π) and $(1/\pi)$ do not split in L. (Note that $k(\tau)$ splits at all places of K, except at (π) and $(1/\pi)$; cf. Remark 4.1.2.) The necessity of these conditions is easy to prove, but the sufficiency (which is actually what we need) lies much deeper; we refer the interested reader to [Rei03, §32]. Assuming this fact, we get an embedding $\widetilde{F} \hookrightarrow k(\tau)$. Thus, A embeds into $k(\tau)$ via $A \hookrightarrow \widetilde{F} \hookrightarrow k(\tau)$. Using an observation made in the proof of the implication $(2) \Rightarrow (4)$ of Theorem 4.3.2, namely that the leading coefficient of $\widetilde{m}(T)$ is in $\mathbb{F}_q^\times$, we conclude that T is integral over $\mathbb{F}_q[\pi]$. Now another fact from the theory of central division algebras[4] implies that, after an appropriate conjugation in $k(\tau)$, we get an embedding $\phi \colon A \to k\{\tau\}$. Since π is in the kernel of $\partial \colon k\{\tau\} \to k$ and π divides some power of $\mathfrak{p}$ by construction, the composition $\partial\phi \colon A \to k$ factors through the quotient map $\gamma \colon A \to A/\mathfrak{p}$. Hence ϕ is a Drinfeld module over k. Also, by construction, the minimal polynomial over A of the Frobenius endomorphism of ϕ is $m(x)$. The fact that ϕ has rank r can be deduced from $m(0)$. $\square$

Lemma 4.3.8 *Let r be an integer coprime to n. Let $c \in \mathbb{F}_q^\times$. Let $a_1, a_2, \ldots, a_{r-1} \in A$ be polynomials such that $\deg(a_i) \le (r-i)n/r$ for all $1 \le i \le r-1$. In addition, assume that $\mathfrak{p} \nmid a_1$. Then*

$$P(x) = x^r + a_{r-1}x^{r-1} + \cdots + a_1 x + c\mathfrak{p}^{n/d}$$

is the characteristic polynomial of the Frobenius of some ordinary Drinfeld module of rank r over k.

Proof Because $\deg(a_i) \le (r-i)n/r$, the point $(i, \operatorname{ord}_\infty(a_i)) = (i, -\deg(a_i))$ for $a_i \ne 0$ lies on or above the line segment joining $(0, -n)$ to $(r, 0)$ in $\mathbb{R}^2$. Therefore, the Newton polygon of $P(x)$ over F_∞ consists of a single line segment joining $(0, -n)$ to $(r, 0)$. This implies that all the roots of $P(x)$ have ∞-adic valuation equal to $-n/r$. Since r and n are assumed to be coprime, $P(x)$ is irreducible over F_∞ by Remark 2.5.7. Next, using the assumption that $\mathfrak{p} \nmid a_1$, we get that $P(x)$ modulo $\mathfrak{p}$ splits as $xg(x)$ with $g(0) \ne 0$. Therefore, by Hensel's Lemma, $P(x)$ splits over $F_\mathfrak{p}$ as a product of two monic polynomials $P(x) = P_1(x)P_2(x)$ with $\deg(P_1) = 1$, $P_1(x) \equiv x \pmod{\mathfrak{p}}$, and $P_2(0) \in A_\mathfrak{p}^\times$. Now Theorem 4.3.7 implies that $P(x)$ is the characteristic polynomial of the Frobenius of some Drinfeld modules ϕ of rank

---

[4] The key point here is that, up to conjugation, $k\{\tau\}$ is the unique maximal $\mathbb{F}_q[\pi]$-order in $k(\tau)$; see [Rei03, §21].

r over k (in our case, $P(x) = m(x)$ since $P(x)$ is irreducible over F). Moreover, $H(\phi) = 1$ by Theorem 4.2.13. □

Corollary 4.3.9 *Let $b \in A$ be a monic polynomial of degree n. There exists an ordinary Drinfeld module ϕ over k such that $\chi(^\phi k) = b$.*

Proof It is enough to show that there is a polynomial $P(x) \in A[x]$ satisfying the assumptions of Lemma 4.3.8, and such that $P(1) = b$; cf. Theorem 4.2.6. In fact, we will prove the slightly stronger statement that such a polynomial exists for any $b \in A$ of degree n, not necessarily monic.

Let c be the leading coefficient of b. Note that $b' := b - c\mathfrak{p}^{n/d}$ has degree $\leq n-1$. If $b' = 0$, then choose $r \geq 2$ relatively prime to n and let

$$P(x) = x^r - x + c\mathfrak{p}^{n/d}.$$

In this case, we obviously have $P(1) = b$ and the assumptions of Lemma 4.3.8 are satisfied.

If $b' \neq 0$, then let β be the leading coefficient of b'. Choose $r \geq \max(n, 3)$ relatively prime to n. If $\mathfrak{p} \neq T$, then put $a_1 = \beta T^{\deg(b')}$. If $\mathfrak{p} = T$, then put $a_1 = \beta(T - 1)^{\deg(b')}$. Now $b'' := b' - a_1$ has degree $\leq n - 2$. (If $n = 1$, then $b'' = 0$.) Put $a_2 = b'' - 1$, and

$$P(x) = x^r + a_2 x^2 + a_1 x + c\mathfrak{p}^{n/d}.$$

Because $r \geq n$, we have $n - i \leq (r - i)n/r$, so the assumptions of Lemma 4.3.8 are satisfied. Also, by construction, $P(1) = b$. □

Remark 4.3.10 Corollary 4.3.9 is the Drinfeld module analogue of a result of Howe and Kedlaya [HK21], which says that, given a natural number $n \geq 1$, there exists an ordinary abelian variety A over $\mathbb{F}_2$ such that $n = P_A(1) = \#A(\mathbb{F}_2)$; here $P_A(x) \in \mathbb{Z}[x]$ is the characteristic polynomial of the Frobenius automorphism of $\overline{\mathbb{F}}_2$ acting on the Tate module $T_\ell(A)$ of A for some prime $\ell \neq 2$. The dimension of A has to increase with n due to the Weil bounds on the number of rational points on an abelian variety a finite field. The proof in [HK21] proceeds by constructing an appropriate polynomial $P_A(x)$ and showing that it is the characteristic polynomial of the Frobenius for some ordinary abelian variety using the Honda-Tate theory. Building on the techniques of Howe and Kedlaya, Van Bommel et al. [vBCL$^+$21] have shown that, for any fixed q, every sufficiently large positive integer is the order of an ordinary abelian variety over $\mathbb{F}_q$.

Exercises

4.3.1 Show that, given two Drinfeld modules ϕ and ψ of rank r over k, the homomorphism of Theorem 3.5.4 is an isomorphism:

$$\mathrm{Hom}_k(\phi, \psi) \otimes_A A_{\mathfrak{l}} \overset{\sim}{\longrightarrow} \mathrm{Hom}_{A_{\mathfrak{l}}[G_k]}(T_{\mathfrak{l}}(\phi), T_{\mathfrak{l}}(\psi)).$$

This is a generalization of Corollary 4.2.4.

4.3.2 Let ϕ be a Drinfeld module of rank r over k. Let $\widetilde{\mathfrak{p}}$ be the unique prime in $\widetilde{F} = F(\pi)$ over π of $K = F_q(\pi)$. Let

$$A[\pi]_{\widetilde{\mathfrak{p}}} := A[\pi] \otimes_{\mathbb{F}_q[\pi]} \mathbb{F}_q[\![\pi]\!] .$$

Prove the following:

(a) $A[\pi]_{\widetilde{\mathfrak{p}}}$ is a local ring with maximal ideal $\mathcal{M}$ generated by π and $\mathfrak{p}$.
(b) $A[\pi]_{\widetilde{\mathfrak{p}}}/\mathcal{M} \cong \mathbb{F}_{\mathfrak{p}}$.
(c) The completion $A_{\mathfrak{p}}$ is a subring of $A[\pi]_{\widetilde{\mathfrak{p}}}$.

4.3.3 Let $\widetilde{F}_{\widetilde{\mathfrak{p}}}$ be the completion of $\widetilde{F}$ at $\widetilde{\mathfrak{p}}$ and let $B_{\widetilde{\mathfrak{p}}}$ be the ring of integers of $\widetilde{F}_{\widetilde{\mathfrak{p}}}$. Let

$$[\widetilde{F}_{\widetilde{\mathfrak{p}}} : K_\pi] = e_K \cdot f_K ,$$

$$[\widetilde{F}_{\widetilde{\mathfrak{p}}} : F_{\mathfrak{p}}] = e_F \cdot f_F ,$$

where e and f denote the ramification index and the residue degree of the corresponding extension. We say that $A[\pi]$ is *locally maximal* at π if $A[\pi]_{\widetilde{\mathfrak{p}}} = B_{\widetilde{\mathfrak{p}}}$. Prove that $A[\pi]$ is locally maximal at π if and only if one of the following holds:

- $f_F = 1$ and $e_F = 1$.
- $f_F = 1$ and $e_K = 1$.

4.3.4 Let H be the height of ϕ. Show that

$$\left\lceil \frac{n}{H \cdot d} \right\rceil \le \frac{[\widetilde{F} : K]}{d} ,$$

with an equality if and only if $A[\pi]$ is locally maximal at π; here $\lceil \alpha \rceil$ denotes the smallest integer $\ge \alpha$ (for example, $\lceil 5/3 \rceil = 2$). Conclude that if $[\widetilde{F} : F] = r$, then $A[\pi]$ is locally maximal at π if and only if either ϕ is ordinary or $k = \mathbb{F}_{\mathfrak{p}}$.

4.3.5 Let $\mathfrak{p} = T$, $n = 3$, and let ϕ be defined by $\phi_T = \tau^2$. Show that $A[\pi]_{\widetilde{\mathfrak{p}}} = A_{\mathfrak{p}}[\mathfrak{p}\sqrt{\mathfrak{p}}]$. Conclude that $A[\pi]$ is not locally maximal at π.

4.3.6 Let ϕ and ψ be Drinfeld modules over k such that there is a separable isogeny $u : \phi \to \psi$ defined over k. Prove that there is a separable isogeny $w : \psi \to \phi$ defined over k. Thus, separable isogenies give an equivalence relation between Drinfeld modules over finite fields. (Compare this with Exercise 3.3.3.)

4.3.7 Call $\pi \in \overline{F}$ a (k, r)-*Weil number* if

(1) π is integral over A.
(2) $[F(\pi) : F]$ divides r.

(3) $F(\pi)/F$ is totally imaginary and $|\pi| = q^{n/r}$, where $|\cdot|$ is the unique extension of the normalized absolute value of F corresponding to ∞.
(4) There is only one place of $F(\pi)$ which is a zero of π, and this place lies above $\mathfrak{p}$.

Prove that the map which associates to a (k, r)-Weil number its minimal polynomial gives a bijection between the conjugacy classes of (k, r)-Weil numbers and (k, r)-Weil polynomials.

4.3.8 A $(\mathfrak{p}, r)$-*Weil pair* is a pair $(\widetilde{F}, \widetilde{\mathfrak{p}})$ consisting of a finite extension $\widetilde{F}$ of F and a place $\widetilde{\mathfrak{p}}$ of $\widetilde{F}$ over $\mathfrak{p}$ such that

(1) $[\widetilde{F} : F]$ divides r.
(2) $\widetilde{F}/F$ is totally imaginary.
(3) $\widetilde{F} = F(\varpi)$ for any ϖ such that $\mathrm{ord}_{\widetilde{\mathfrak{p}}}(\varpi) > 0$, $\mathrm{ord}_{\widetilde{\infty}}(\varpi) < 0$, and $\mathrm{ord}_v(\varpi) = 0$ for all places $v \neq \widetilde{\mathfrak{p}}, \widetilde{\infty}$, where $\widetilde{\infty}$ is the unique place of $\widetilde{F}$ over ∞.

Two $(\mathfrak{p}, r)$-Weil pairs $(\widetilde{F}, \widetilde{\mathfrak{p}})$ and $(\widetilde{F}', \widetilde{\mathfrak{p}}')$ are said to be *isomorphic* if there is an isomorphism $\iota\colon \widetilde{F} \to \widetilde{F}'$ over F such that $\iota(\widetilde{\mathfrak{p}}) = \widetilde{\mathfrak{p}}'$.

(a) Let π be a (k, r)-Weil number for some finite $k/\mathbb{F}_\mathfrak{p}$. Let $\widetilde{F} = \bigcap_{m \geq 1} F(\pi^m)$, and $\widetilde{\mathfrak{p}}$ be the prime of $\widetilde{F}$ dividing π. Prove that $(\widetilde{F}, \widetilde{\mathfrak{p}})$ is a $(\mathfrak{p}, r)$-Weil pair; moreover, conjugate Weil numbers produce isomorphic Weil pairs.
(b) Prove that the construction in (a) gives a natural bijection between the isogeny classes of Drinfeld modules of rank r over $\overline{\mathbb{F}}_\mathfrak{p}$ and the isomorphism classes of $(\mathfrak{p}, r)$-Weil pairs.
(c) Prove that a $(\mathfrak{p}, r)$-Weil pair $(\widetilde{F}, \widetilde{\mathfrak{p}})$ corresponds to a supersingular Drinfeld module if and only if $(\widetilde{F}, \widetilde{\mathfrak{p}}) = (F, \mathfrak{p})$. Conclude that any two supersingular Drinfeld modules of rank r over $\overline{\mathbb{F}}_\mathfrak{p}$ are isogenous.

4.3.9 Show that the statement of Corollary 4.3.9 is false if we restrict to ordinary Drinfeld modules of rank 2 and assume $n \geq 2$.

4.4 Supersingular Drinfeld Modules

In this section we study in more detail the supersingular Drinfeld modules. We start by giving alternative characterizations of these modules:

Theorem 4.4.1 *Let ϕ be a Drinfeld module of rank r over k. The following statements are equivalent:*

(1) ϕ is supersingular in the sense of Definition 4.1.9, i.e., $H(\phi) = r$.
(2) $\phi[\mathfrak{p}^s] = 0$ for some $s \geq 1$.
(3) $\phi[\mathfrak{p}^s] = 0$ for all $s \geq 1$.
(4) $\pi^m \in A$ for some $m \geq 1$.
(5) $\dim_F(\mathrm{End}^\circ(\phi)) = r^2$.

Proof

(1) $\Leftrightarrow$ (2) $\Leftrightarrow$ (3): By Lemma 3.2.11, for a given $s \geq 1$, we have $\mathrm{ht}(\phi_{\mathfrak{p}^s}) = H(\phi) \cdot \deg(\mathfrak{p}^s)$. Therefore,

$$\phi[\mathfrak{p}^s] = 0 \quad \Longleftrightarrow \quad \mathrm{ht}(\phi_{\mathfrak{p}^s}) = \deg_\tau(\phi_{\mathfrak{p}^s})$$
$$\Longleftrightarrow \quad \mathrm{ht}(\phi_{\mathfrak{p}^s}) = r \cdot \deg(\mathfrak{p}^s)$$
$$\Longleftrightarrow \quad H(\phi) = r.$$

(3) $\Rightarrow$ (4): Suppose $\phi[\mathfrak{p}^{n/d}] = 0$. Then $\phi_{\mathfrak{p}^{n/d}} = c\pi^r$ for some $c \in k^\times$. Denote $f = q^{nr}$. Because $\pi^r c = c^f \pi^r$, for any $e \geq 1$ we have

$$\phi_{\mathfrak{p}^{ne/d}} = (c\pi^r)^e = c^{1+f+f^2+\cdots+f^{e-1}} \pi^{re}.$$

Note that $c^{f-1} = 1$ since $f - 1$ is divisible by $\#k^\times = q^n - 1$. On the other hand,

$$1 + f + f^2 + \cdots + f^{e-1} \equiv e \pmod{f - 1}.$$

Hence, if we choose e divisible by $f - 1$, then $\phi_{\mathfrak{p}^{ne/d}} = \pi^{re}$. This implies that $\pi^{r(f-1)} \in A$.

(4) $\Rightarrow$ (2): If $\pi^m \in A$ for some $m \geq 1$, then there is $a \in A$ such that $\pi^m = \phi_a$. Decomposing $a = \mathfrak{p}^s \cdot b$ with $\mathfrak{p} \nmid b$, we get

$$\mathrm{ht}(\phi_{\mathfrak{p}^s}) + \mathrm{ht}(\phi_b) = \mathrm{ht}(\phi_a) = \deg_\tau(\phi_a) = \deg_\tau(\phi_{\mathfrak{p}^s}) + \deg_\tau(\phi_b).$$

Since $\mathrm{ht} \leq \deg_\tau$ and $\mathrm{ht}(\phi_b) = 0$ (as $\gamma(b) \neq 0$), we see that $\deg_\tau(\phi_b) = 0$. Thus, $b \in \mathbb{F}_q^\times$ and $s \geq 1$. Now it is clear that $\phi[\mathfrak{p}^s] = \ker(\pi^m) = 0$.

(4) $\Leftrightarrow$ (5): Since $\mathrm{End}(\phi)$ is a finitely generated A-module, there is a finite extension k'/k such that $\mathrm{End}(\phi) = \mathrm{End}_{k'}(\phi)$. Let $\pi' = \pi^{[k':k]}$ be the Frobenius of k'. By Theorem 4.1.3, we have

$$\dim_F(\mathrm{End}_{k'}^\circ(\phi)) = r^2 \quad \Longleftrightarrow \quad F(\pi') = F \quad \Longleftrightarrow \quad \pi' \in F.$$

On the other hand, π is integral over A, so $\pi' \in F \Longleftrightarrow \pi' \in A$. $\square$

Remarks 4.4.2

(1) An alternative proof of (5) $\Rightarrow$ (3) is the following: By passing to a finite extension of k, without loss of generality we may assume that $\mathrm{End}_k(\phi) = \mathrm{End}(\phi)$. Now $D_\mathfrak{p} := \mathrm{End}_k^\circ(\phi) \otimes_F F_\mathfrak{p}$ acts on $V_\mathfrak{p}(\phi)$. On the other hand, by Theorem 4.1.3, $D_\mathfrak{p}$ is a division algebra, so $V_\mathfrak{p}(\phi)$ is a vector space over the division algebra $D_\mathfrak{p}$. This implies that $\dim_{F_\mathfrak{p}} V_\mathfrak{p}(\phi) \leq r - 1$ is a multiple of $\dim_{F_\mathfrak{p}} D_\mathfrak{p} = r^2$. Thus, $V_\mathfrak{p}(\phi) = 0$, so $\phi[\mathfrak{p}^s] = 0$ for all $s \geq 1$.

(2) Let ϕ be a supersingular Drinfeld module over k. Let k' be a finite extension of k such that $\mathrm{End}(\phi) = \mathrm{End}_{k'}(\phi)$. In this case, $F = F(\pi')$, so by Remark 4.1.5,

$\mathrm{End}^\circ(\phi)$ is the central division algebra over F with invariants

$$\mathrm{inv}_v(\mathrm{End}^\circ(\phi)) = \begin{cases} \frac{1}{r}, & \text{if } v = \mathfrak{p}; \\ -\frac{1}{r}, & \text{if } v = \infty; \\ 0, & \text{if } v \neq \mathfrak{p}, \infty. \end{cases}$$

With more tools from the theory of central simple algebras one can show that not only $\mathrm{End}(\phi) \otimes_A F$ has the largest possible dimension over F, but the endomorphism ring $\mathrm{End}(\phi)$ itself is as large as possible, in the sense that it is a maximal A-order in $\mathrm{End}^\circ(\phi)$; see [Gek91]. Moreover, the number of supersingular Drinfeld modules over $\overline{k}$, up to isomorphism, is equal to the class number of the division algebra $\mathrm{End}^\circ(\phi)$.

Lemma 4.4.3 *Any two supersingular Drinfeld modules of rank r over $\overline{\mathbb{F}}_\mathfrak{p}$ are isogenous.*

Proof Note that this statement is a special case of Exercise 4.3.8. We give a more direct argument for supersingular Drinfeld modules.

Let ϕ and ψ be supersingular Drinfeld modules of rank r over a finite field k. After passing to a finite extension of k, we may assume that all endomorphisms of both ϕ and ψ are defined over k. Thus, by Theorem 4.4.1, $\dim_F \mathrm{End}_k^\circ(\phi) = \dim_F \mathrm{End}_k^\circ(\psi) = r^2$. Theorem 4.1.3 then implies that $m_\phi(x) = x - c_1 \mathfrak{p}^{n/dr}$ and $m_\psi(x) = x - c_2 \mathfrak{p}^{n/dr}$ for some $c_1, c_2 \in \mathbb{F}_q^\times$. Over the extension k' of k of degree $q - 1$ the Frobenius is π^{q-1}, so its minimal polynomial in $\mathrm{End}_{k'}(\phi)$ is $x - (c_1 \mathfrak{p}^{n/dr})^{q-1} = x - \mathfrak{p}^{\frac{n(q-1)}{dr}}$. The same applies to ψ. Thus, the minimal polynomials are the same, so ϕ and ψ are isogenous over k' by Theorem 4.3.2. $\square$

Remark 4.4.4 If $s = [k : \mathbb{F}_\mathfrak{p}]$ is divisible by r, then $m(x) = x + c \cdot \mathfrak{p}^{s/r}$ is a (k, r)-Weil polynomial for any $c \in \mathbb{F}_q^\times$. By Theorem 4.3.7, this polynomial is the minimal polynomial of the Frobenius of a supersingular Drinfeld module of rank r defined over k. The k-isogeny classes corresponding to $x + c \cdot \mathfrak{p}^{s/r}$ for different c are distinct. When we make an extension of k of degree $q-1$ all these classes fall together, so any two supersingular Drinfeld modules of rank r are isogenous over the degree $q - 1$ extension of a field where all their endomorphisms are defined. But the extension which identifies these $q - 1$ classes also creates new isogeny classes. There are at least $q - 1$ classes at each stage, even though any two fixed supersingular Drinfeld modules eventually become isogenous.

Example 4.4.5 Consider the Drinfeld module of rank r defined by

$$\phi_T = t + \tau^r.$$

Assume r is a prime number. We claim that ϕ is supersingular if and only if $d = \deg(\mathfrak{p})$ is not divisible by r. Let $D = \mathrm{End}^\circ(\phi)$. Obviously the elements of $\mathbb{F}_{q^r} \subset \overline{k}\{\tau\}$ are in the centralizer of $\phi(A)$, so the field $L := \mathbb{F}_{q^r}(T)$ is an F-subalgebra of D.

Suppose ϕ is supersingular. By Theorems 4.1.3 and 4.4.1, D is a central division algebra over F of dimension r^2 such that $D \otimes_F F_\mathfrak{p}$ is still a division algebra. This implies that $L \otimes_F F_\mathfrak{p}$ is a field, being a commutative subalgebra of $D \otimes_F F_\mathfrak{p}$. Let $f(x) \in F_q[x]$ be an irreducible polynomial of degree r, so that F_{q^r} and L are the splitting fields of $f(x)$ over F_q and F, respectively. By Theorem 2.8.5 and its proof, $f(x)$ remains irreducible in $F_\mathfrak{p}[x]$. Then $f(x)$ is also irreducible as an element of the subring $F_\mathfrak{p}[x]$ of $F_\mathfrak{p}[x]$. But according to Exercise 1.6.4 an irreducible polynomial $f(x) \in F_q[x]$ remains irreducible over $F_\mathfrak{p} \cong F_{q^d}$ if and only if the degree of $f(x)$ is coprime to d. Hence $r \nmid d$. (Note that so far we have not used the assumption that r is prime.)

Now suppose $r \nmid d$. The Drinfeld module ϕ is defined over $F_\mathfrak{p}$. By Exercise 4.2.3, $L' := \mathrm{End}^\circ_{F_\mathfrak{p}}(\phi)$ is a field extension of F of degree r, which we can consider as an F-subalgebra of D. If $r \nmid d$, then F_{q^r} is not contained in $F_\mathfrak{p}\{\tau\}$, and therefore $L \neq L'$. This implies that $\dim_F D > r$. Because r is prime, Corollary 4.1.4 forces $\dim_F D = r^2$. Thus, ϕ is supersingular by Theorem 4.4.1.

To give a different perspective on the previous example, consider the Drinfeld module Φ over F defined by $\Phi_T = T + \tau^r$. We can reduce the coefficients of Φ_T modulo a prime $\mathfrak{p}$ to obtain a Drinfeld module $\Phi \otimes F_\mathfrak{p}$ over $F_\mathfrak{p}$. The previous example implies that the reduction $\Phi \otimes F_\mathfrak{p}$ is supersingular for infinitely many $\mathfrak{p}$ (in fact, the set of such primes has positive density). The next example, which is due to Poonen [Poo98], gives the completely opposite situation, where a Drinfeld module over F has no supersingular reductions at all.

Example 4.4.6 Assume $\mathrm{char}_A(k) \neq T$. Let ϕ be the Drinfeld module of rank r defined by

$$\phi_T = t(1 - \tau)^r.$$

We claim that ϕ is not supersingular, unless r is a p-th power. Moreover, if $p \nmid r$, then ϕ is ordinary. Before giving the proof, we explain the significance of this fact.

Consider the Drinfeld module ϕ over F defined by $\Phi_T = T(1 - \tau)^r$ and assume that $p \nmid r$. For any prime $\mathfrak{p} \neq T$, reduce the coefficients of Φ_T modulo $\mathfrak{p}$ to obtain a Drinfeld module $\Phi \otimes F_\mathfrak{p}$ over $F_\mathfrak{p}$. According to our claim, $\Phi \otimes F_\mathfrak{p}$ is ordinary, so there are no primes where ϕ has supersingular reduction. (At $\mathfrak{p} = T$, Φ has "bad" reduction; we will define this notion rigorously in Chap. 6.) Poonen's discovery was somewhat unexpected since a well-known theorem of Elkies says that any elliptic curve over $\mathbb{Q}$ has supersingular reduction at infinitely many primes of $\mathbb{Z}$. Thus, the situation is more nuanced in the case of Drinfeld modules. It was proved by Brown [Bro92] that any Drinfeld module ϕ of rank 2 over F for which $j(\phi)T$ is not a square in F_∞ has infinitely many supersingular reductions,[5] and Poonen's result shows that this assumption is necessary (see [Poo98] for a small correction of Brown's result).

---

[5] It seems quite challenging to generalize Elkies' method to higher rank Drinfeld modules. As far as I know, currently there are no known examples of Drinfeld modules ϕ of rank ≥ 3 over F with $\mathrm{End}(\phi) = A$ having infinitely many supersingular reductions.

To prove that ϕ is not supersingular, we will bound its height. Let $\alpha_1, \alpha_2, \ldots, \alpha_r \in \bar{k}$ be a set of elements which satisfy

$$\alpha_1 = 1, \qquad \alpha_i - \alpha_i^q = \alpha_{i-1}, \quad 2 \le i \le r.$$

Let $W_i \subset \bar{k}$ be the $\mathbb{F}_q$-linear span of $\{\alpha_1, \ldots, \alpha_i\}$. The linear operator $1 - \tau$ gives an exact sequence of $\mathbb{F}_q$-vector spaces:

$$0 \longrightarrow W_1 \longrightarrow W_i \overset{1-\tau}{\longrightarrow} W_{i-1} \longrightarrow 0.$$

Hence, inductively, $\dim_{\mathbb{F}_q} W_i = i$ and $W_i = \ker(1-\tau)^i$. In particular,

$$\phi[T] = \bigoplus_{i=1}^{r} \mathbb{F}_q \alpha_i.$$

The action of τ on $\phi[T]$ with respect to this basis is given by the unipotent matrix

$$S = \begin{pmatrix} 1 & -1 & 0 & \cdots & 0 \\ 0 & 1 & -1 & \cdots & 0 \\ \vdots & & \ddots & \ddots & -1 \\ 0 & 0 & 0 & \cdots & 1 \end{pmatrix}.$$

We can consider ϕ as being defined over $\mathbb{F}_\mathfrak{p}$. The Frobenius $\pi = \tau^d$ of $\mathbb{F}_\mathfrak{p}$ acts on $\phi[T]$ by the matrix S^d, which is also unipotent. Thus, the characteristic polynomial $P(x) = a_0 + a_1 x + \cdots + a_{r-1} x^{r-1} + x^r$ of π is congruent to $(x-1)^r$ modulo T. If we write $r = r_0 p^s$ with $p \nmid r_0$, then $(x-1)^r = (x^{p^s} - 1)^{r_0}$. The coefficient of x^{p^s} in this latter polynomial is $\pm r_0 \neq 0$. Hence, the coefficient of x^{p^s} in $P(x)$ cannot be zero. Let $h := H(\phi)$. By Theorem 4.2.13, the coefficients $a_0, \ldots, a_{h-1}$ are divisible by $\mathfrak{p}$. On the other hand, by Theorem 4.2.7, $\deg(a_i) < d$ for $1 \le i \le r-1$. Therefore, $a_1 = a_2 = \cdots = a_{h-1} = 0$. Contrasted with our earlier conclusion that $a_{p^s} \neq 0$, we get the bound $H(\phi) \le p^s$. This implies our initial claim because ϕ is supersingular (respectively, ordinary) if $H(\phi) = r$ (respectively, $H(\phi) = 1$).

Lemma 4.4.7 *A supersingular Drinfeld module of rank r over $\overline{\mathbb{F}}_\mathfrak{p}$ is isomorphic to a supersingular Drinfeld module defined over the degree r extension $\mathbb{F}_{\mathfrak{p}^r}$ of $\mathbb{F}_\mathfrak{p}$.*

Proof Suppose ϕ is given by

$$\phi_T = t + g_1 \tau + \cdots + g_r \tau^r.$$

After possibly replacing ϕ by an isomorphic Drinfeld module, we can assume $g_r = 1$. Then $\phi_\mathfrak{p} = \tau^{dr}$. The commutation $\phi_T \phi_\mathfrak{p} = \phi_\mathfrak{p} \phi_T$ implies that $g_i^{q^{rd}} = g_i$ for all $1 \le i \le r-1$; thus, $g_i \in \mathbb{F}_{\mathfrak{p}^r}$ and ϕ is defined over $\mathbb{F}_{\mathfrak{p}^r}$. $\qquad\square$

The previous lemma implies that, up to isomorphism, there are only finitely many supersingular Drinfeld modules over $\overline{\mathbb{F}}_{\mathfrak{p}}$ of a given rank (indeed, the number of polynomials in $\mathbb{F}_{\mathfrak{p}'}\{\tau\}$ of degree r is finite, thus the number of possible $\phi_T \in \mathbb{F}_{\mathfrak{p}'}\{\tau\}$ is finite). A natural question then is whether it is possible to explicitly describe all these Drinfeld modules, or at least count their number. We consider this question for rank-2 Drinfeld modules.

Let ϕ be a Drinfeld module over $\overline{\mathbb{F}}_{\mathfrak{p}}$ of rank 2. After possibly replacing ϕ by an isomorphic Drinfeld module, we assume that ϕ is defined by

$$\phi_T = t + \lambda\tau + \tau^2, \qquad j(\phi) = \lambda^{q+1}.$$

Then the coefficients of

$$\phi_{\mathfrak{p}} = \sum_{i \geq 0} g_i(\lambda)\tau^i$$

are in $\overline{\mathbb{F}}_{\mathfrak{p}}[\lambda]$. We have $g_i(\lambda) = 0$ for $i > 2d$ and $g_{2d}(\lambda) = 1$. Moreover, by Lemma 3.2.11, $g_i(\lambda) = 0$ for $0 \leq i < d$, and

$$\phi \text{ is supersingular if and only if } H_{\mathfrak{p}}(\lambda) := g_d(\lambda) = 0.$$

Thus, our questions about the isomorphism classes of supersingular Drinfeld modules of rank 2 reduce to questions about the roots of the polynomial $H_{\mathfrak{p}}(\lambda)$.

Remark 4.4.8 In the context of analogies between Drinfeld modules and elliptic curves, the polynomials $H_{\mathfrak{p}}(\lambda)$ have a well-known counterpart. For a prime p, the *Deuring polynomial* $H_p(x) \in \mathbb{F}_p[x]$ is defined by

$$H_p(x) = \sum_{i=0}^{(p-1)/2} \binom{m}{i}^2 x^i.$$

Then $\lambda \in \overline{\mathbb{F}}_p$, $\lambda \neq 0, 1$, is a root of $H_p(x)$ if and only if the elliptic curve in the Legendre form

$$y^2 = x(x-1)(x-\lambda)$$

is supersingular; see [Sil09, §V.4].

Example 4.4.9 If $\mathfrak{p} = T + \alpha$ is a prime of degree 1, then

$$H_{\mathfrak{p}}(\lambda) = \lambda.$$

If $\mathfrak{p} = T^2 + \alpha T + \beta$ is a prime of degree 2, then

$$\phi_{\mathfrak{p}} = \phi_T \phi_T + \alpha \phi_T + \beta$$
$$= (\tau^2 + \lambda\tau + t)(\tau^2 + \lambda\tau + t) + \alpha(\tau^2 + \lambda\tau + t) + \beta$$
$$= \tau^4 + (\lambda + \lambda^{q^2})\tau^3 + (t + \lambda^{q+1} + t^{q^2} + \alpha)\tau^2 + \lambda(t + t^q + \alpha)\tau + t^2 + \alpha t + \beta$$
$$= \tau^4 + (\lambda + \lambda^{q^2})\tau^3 + (2t + \alpha + \lambda^{q+1})\tau^2,$$

where for the last equality we observe that $t^{q^2} = t$ (since $\mathbb{F}_{\mathfrak{p}} \cong \mathbb{F}_{q^2}$), $t^q + t = -\alpha$ (since t and t^q are the roots of $\mathfrak{p}$), and $t^2 + \alpha t + \beta = 0$ (since t is a root of $\mathfrak{p}$). Therefore,

$$H_{\mathfrak{p}}(\lambda) = \lambda^{q+1} + (2t + \alpha).$$

These calculations imply that for $d = 1$ the only supersingular j-invariant is $j = 0$, and for $d = 2$ the only supersingular j-invariant is $j = -(2t + \alpha)$.

Example 4.4.10 Let $q = 2$.

If $\mathfrak{P} = T^3 + T + 1$, then $H_{\mathfrak{P}}(\lambda) = \lambda^7 + (t^2 + t)\lambda^4 + t^2\lambda$.

If $\mathfrak{P} = T^4 + T + 1$, then $H_{\mathfrak{P}}(\lambda) = \lambda^{15} + (t^2 + t)\lambda^{12} + \lambda^9 + (t^2 + t + 1)\lambda^3 + 1$.

Theorem 4.4.11 (Gekeler [Gek83]) *The polynomial $H_{\mathfrak{p}}(\lambda)$ has the following properties:*

(1) $H_{\mathfrak{p}}(\lambda)$ has no multiple roots.
(2) $H_{\mathfrak{p}}(\lambda)$ is a monic polynomial of degree $(q^d - 1)/(q - 1)$.
(3) For even (respectively, odd) d, the polynomial $H_{\mathfrak{p}}(\lambda)$ is a polynomial in λ^{q+1} (respectively, λ times a polynomial in λ^{q+1}).

Proof

(1) By comparing the coefficients of both sides of the equation $\phi_{\mathfrak{p}}\phi_T = \phi_T\phi_{\mathfrak{p}}$, we get

$$g_i(\lambda)(t^{q^i} - t) + g_{i-1}(\lambda)\lambda^{q^{i-1}} - g_{i-1}(\lambda)^q\lambda + g_{i-2}(\lambda) - g_{i-2}(\lambda)^{q^2} = 0.$$

Because $t^{q^i} \neq t$ for $d < i < 2d$, this equation recursively implies that $g_d = H_{\mathfrak{p}}$ divides all g_i for $d \leq i < 2d$. For $i = 2d + 1$, the equation becomes

$$\lambda^{q^{2d}} - \lambda + g_{2d-1}(\lambda) - g_{2d-1}(\lambda)^{q^2} = 0.$$

Taking the derivative with respect to λ, we get

$$g'_{2d-1}(\lambda) = 1.$$

This implies that g_{2d-1}, and hence also $H_{\mathfrak{p}}$, has no multiple roots.

(2) Given $n \geq 1$, consider

$$\phi_{T^n} = b_{n,0}(\lambda) + b_{n,1}(\lambda)\tau + \cdots + b_{n,2n-1}(\lambda)\tau^{2n-1} + b_{n,2n}(\lambda)\tau^{2n}.$$

Put $b_{n,m} = 0$ for $m < 0$ and $2n < m$. From $\phi_{T^{n+1}} = \phi_{T^n}\phi_T$, we obtain the recursive formula

$$b_{n+1,m}(\lambda) = b_{n,m}(\lambda)t^{q^m} + b_{n,m-1}(\lambda)\lambda^{q^{m-1}} + b_{n,m-2}(\lambda).$$

Using this formula, one easily proves by induction on n that

(i) $\deg_\lambda(b_{n,m}) \leq \frac{q^m-1}{q-1}$.

(ii) $b_{n,n}$ is monic of degree $\frac{q^n-1}{q-1}$.

(iii) $\deg_\lambda(b_{n,m}) < \frac{q^m-1}{q-1}$ if $n < m$.

Now let $\mathfrak{p} = T^d + p_{d-1}T^{d-1} + \cdots + p_0$. Then

$$H_{\mathfrak{p}}(\lambda) = b_{d,d}(\lambda) + p_{d-1}b_{d-1,d}(\lambda) + \cdots + p_0 b_{0,d}(\lambda),$$

which shows that the leading term of $H_{\mathfrak{p}}$ is the leading term of $b_{d,d}$. Therefore, $H_{\mathfrak{p}}(\lambda)$ is monic of degree $(q^d - 1)/(q - 1)$.

(3) Let $0 \neq \alpha \in \overline{\mathbb{F}}_q$ be a root of $H_{\mathfrak{p}}$. Let $\zeta_1, \ldots, \zeta_{q+1} \in \overline{\mathbb{F}}_q$ be the roots of $x^{q+1} - 1 = 0$. The Drinfeld module $\phi_T = t + \alpha\tau + \tau^2$ is supersingular and isomorphic to $\psi_T = t + \zeta_i\alpha\tau + \tau^2$ for any $1 \leq i \leq q+1$, as both have the same j-invariant, namely α^{q+1}. But then ψ_T is also supersingular, so $\zeta_i\alpha$ is a root of $H_{\mathfrak{p}}(\lambda)$. By grouping the roots $\zeta_i\alpha$ of $H_{\mathfrak{p}}(\lambda)$ together, we conclude that $H_{\mathfrak{p}}(\lambda)$ is divisible by $\prod_{i=1}^{q+1}(\lambda - \zeta_i\alpha) = \lambda^{q+1} - \alpha^{q+1}$. To deduce (3), it remains to recall from Example 4.4.5 that $\phi_T = t + \tau^2$ corresponding to $\lambda = 0$ is supersingular if and only if d is odd, and so 0 is a root of $H_{\mathfrak{p}}(\lambda)$ if and only if d is odd.

$\square$

Corollary 4.4.12 *The number of supersingular Drinfeld modules of rank 2 over $\overline{\mathbb{F}}_{\mathfrak{p}}$, up to isomorphism, is*

$$S(2, \mathfrak{p}) = \begin{cases} \frac{q^d-1}{q^2-1} & \text{if } d \text{ is even;} \\ \frac{q^d-q}{q^2-1} + 1 & \text{if } d \text{ is odd.} \end{cases}$$

Example 4.4.13 This example is essentially due to Schweizer [Sch95]. Let ϕ be the rank 2 Drinfeld module over $k = \overline{\mathbb{F}}_{\mathfrak{p}}$ defined by $\phi_T = t + \tau^2$. Put $\psi_T = t + \lambda\tau + \tau^2$, where $\lambda \in k$ is to be determined. We construct isogenies $\phi \to \psi$ by considering the equation

$$(\hat{\alpha} + \tau)(t + \tau^2) = (t + \lambda\tau + \tau^2)(\alpha + \tau).$$

Expanding both sides of the equation, we obtain

$$\lambda\alpha^q = (t^q - t) \quad \text{and} \quad \lambda = \alpha - \alpha^{q^2}.$$

If $d = 1$, then $\lambda = 0$. Now assume $d \geq 2$ and $\alpha \neq 0$. If we denote $\beta = \alpha^{1+q}$, then the above equations give $t^q - t = \beta - \beta^q$. Thus, $(t + \beta)^q = (t + \beta)$. This implies that $t + \beta \in \mathbb{F}_q$, so $\beta = c - t$ for some $c \in \mathbb{F}_q$. Now

$$j(\psi) = \lambda^{q+1} = \frac{(t^q - t)^{q+1}}{\beta^q} = \frac{(t^q - t)^{q+1}}{(c - t)^q}.$$

The values of $(t^q - t)^{q+1}/(c - t)^q$ are distinct and nonzero as c varies over $\mathbb{F}_q$, so we obtain q non-isomorphic Drinfeld modules ψ, all isogenous to ϕ.

If d is odd, then ϕ is supersingular by Example 4.4.5, so ψ is also supersingular. Thus, we obtain $q + 1$ supersingular Drinfeld modules of rank 2 (including $j(\phi) = 0$), all of which can be defined over $\mathbb{F}_p$ (since all $j(\psi) \in \mathbb{F}_p$). By Corollary 4.4.12, when $d = 3$, this gives a representative of each isomorphism class of supersingular Drinfeld modules over $\overline{\mathbb{F}}_p$.

Remarks 4.4.14

(1) The polynomial $H_p(\lambda)$ has a natural generalization to arbitrary $r \geq 2$. Assume for simplicity of notation that $r = 3$ (the idea of the general construction will be clear from this special case). Assume ϕ is a Drinfeld module of rank 3 over $\overline{\mathbb{F}}_p[x, y]$ defined by

$$\phi_T = t + x\tau + y\tau^2 + \tau^3,$$

where x and y as indeterminates. Let

$$\phi_p = \sum_{i=0}^{3d} g_i(x, y)\tau^i.$$

Since the A-characteristic of $\overline{\mathbb{F}}_p[x, y]$ is p, the coefficients $g_i(x, y)$, $0 \leq i < d$, are zero; see Lemma 3.2.14. Denote

$$H_1(x, y) = g_d(x, y), \qquad H_2(x, y) = g_{2d}(x, y).$$

If $d = 1$, then $H_1(x, y) = x$ and $H_2(x, y) = y$. For $d = 2$ these polynomials are also easy to compute:

$$H_1(x, y) = (2t + a)y + x^{q+1}, \qquad H_2(x, y) = y^{q^2+1} + (x^{q^3} + x),$$

where a is the coefficient of T in $\mathfrak{p}$. But generally, $H_1(x, y)$ and $H_2(x, y)$ are quite complicated.

If for some $\alpha, \beta \in \overline{\mathbb{F}}_\mathfrak{p}$ we have $H_1(\alpha, \beta) = 0$, then all $g_i(\alpha, \beta) = 0$ for $d \leq i \leq 2d - 1$. If moreover $H_2(\alpha, \beta) = 0$, then all $g_i(\alpha, \beta) = 0$ for $0 \leq i \leq 3d - 1$. In that case, $\phi_T = t + \alpha \tau + \beta \tau^2 + \tau^3$ defines a supersingular Drinfeld module of rank 3 over $\overline{\mathbb{F}}_\mathfrak{p}$. Thus, the number of solutions of the system of polynomial equations $H_1(x, y) = H_2(x, y) = 0$ is closely related to the number of isomorphism classes of supersingular Drinfeld modules of rank 3 over $\overline{\mathbb{F}}_\mathfrak{p}$. But there are several subtle technical problems that one has to overcome to actually deduce from this approach an explicit formula for the number $S(r, \mathfrak{p})$ of isomorphism classes of supersingular Drinfeld modules of rank r over $\overline{\mathbb{F}}_\mathfrak{p}$, similar to the formula for $S(2, \mathfrak{p})$ in Corollary 4.4.12. This was done by Gekeler in [Gek92]. For example, if r is prime, then

$$S(r, \mathfrak{p}) = \begin{cases} \prod_{i=1}^{r-1} \frac{q^{id}-1}{q^{i+1}-1} & \text{if } r \mid d; \\ \prod_{i=1}^{r-1} \frac{q^{id}-1}{q^{i+1}-1} + \frac{q^r-q}{q^r-1} & \text{if } r \nmid d. \end{cases}$$

(2) There are other versions of "Deuring polynomials" for Drinfeld modules of rank 2 other than $H_\mathfrak{p}(\lambda)$. These different versions arise from different normalizations of the defining equation ϕ_T. For example, in [BB18], Bassa and Beelen consider Drinfeld modules defined by

$$\phi_T^\Delta = t - (\Delta + t)\tau + \Delta\tau^2 = (\Delta\tau - t)(\tau - 1),$$

where $0 \neq \Delta \in \overline{\mathbb{F}}_\mathfrak{p}$. One easily checks that every Drinfeld module of rank 2 over $\overline{\mathbb{F}}_\mathfrak{p}$ is isomorphic to ϕ_T^Δ for some Δ. Let $h_\mathfrak{p}(\Delta) := (-1)^d g_d(\Delta)$, where $\phi_\mathfrak{p}^\Delta = \sum_{i \geq 0} g_i(\Delta)\tau^i$. By arguments similar to those that we used for $H_\mathfrak{p}(\lambda)$, one shows that $h_\mathfrak{p}(\Delta)$ is a separable monic polynomial of degree $(q^d - 1)/(q - 1)$. Moreover, for a given $\Delta \in \overline{\mathbb{F}}_\mathfrak{p}$, we have $h_\mathfrak{p}(\Delta) = 0$ if and only if ϕ_T^Δ is supersingular. The advantage of this version of the Deuring polynomial is that $h_\mathfrak{p}(\Delta)$ can be computed using recursive formulas as follows. Define the polynomials $f_n(x) \in A[x]$ by

$$f_{-1}(x) = 0, \quad f_0(x) = 1,$$

and for $n \geq 0$,

$$f_{n+1}(x) = (x + T^q)^{q^n} f_n(x) - x^{q^n}(T^{q^n} - T)f_{n-1}(x).$$

Then $h_{\mathfrak{p}}(x)$ is the reduction of $f_d(x)$ modulo $\mathfrak{p}$, assuming $\mathfrak{p} \neq T$. In [BB18], Bassa and Beelen deduce this fact from their calculations of the equations of certain Drinfeld modular curves.

(3) A completely different method for producing the set $S(\mathfrak{p})$ of all j-invariants of supersingular Drinfeld modules of rank 2 over $\overline{\mathbb{F}}_{\mathfrak{p}}$ was developed by Schweizer [Sch97]. Let $\mathscr{S}(n)$ be the set of all j-invariants of Drinfeld modules of rank 2 over $\mathbb{C}_\infty$ having complex multiplication by an order of conductor $\mathfrak{c}$ in $\mathbb{F}_{q^2}[T]$ with $\deg(\mathfrak{c}) \leq n$ (these j-invariants are integral over A; cf. Sect. 7.5). Schweizer proved that, for any prime $\mathfrak{p}$ of degree $2n + 1$, the reduction modulo $\mathfrak{p}$ gives a bijection $\mathscr{S}(n) \to S(\mathfrak{p})$. Thus, $\mathscr{S}(n)$ is a set of "universal" supersingular j-invariants. Schweizer also gave a similar construction for primes of even degrees.

Exercises

4.4.1 Assume d and r are relatively prime, and $q > 2$. Show that there are at least two distinct isogeny classes of supersingular Drinfeld modules of rank r over $\mathbb{F}_{\mathfrak{p}}$.

4.4.2 Assume d is even and $q \equiv 1 \pmod 4$. Prove that the Drinfeld module defined by $\phi_T = t + g\tau + \tau^2$ is not supersingular for any $g \in \mathbb{F}_{\mathfrak{p}}$.

4.4.3 Assume q is odd. Let ϕ be the Drinfeld module over $\mathbb{F}_{\mathfrak{p}}$ defined by

$$\phi_T = t + (\sqrt{t} + \sqrt{t^q})\tau + \tau^2.$$

Let $p_0 \in \mathbb{F}_q$ be the constant term of $\mathfrak{p}$. Prove that ϕ is supersingular if and only if

$$(-1)^{\frac{q-1}{2}d} p_0^{\frac{q-1}{2}} \neq 1.$$

4.4.4 Prove the "mass-formula"

$$\sum \frac{1}{\#\operatorname{Aut}(\phi)} = \frac{\#\mathbb{F}_{\mathfrak{p}} - 1}{(q^2 - 1)(q - 1)},$$

where the sum is over the isomorphism classes of supersingular Drinfeld modules of rank 2 over $\overline{\mathbb{F}}_{\mathfrak{p}}$.

4.4.5 Let ϕ be a Drinfeld module over k of rank 2. Let g_d be the coefficient of τ^d in $\phi_{\mathfrak{p}}$. Note that $g_d = 0$ if and only if ϕ is supersingular. Let $P_\phi(x) = x^2 + a_1 x + a_2$ be the characteristic polynomial of the Frobenius of ϕ. Let c be the leading coefficient of a_2. Prove that

$$c \cdot \operatorname{Nr}_{k/\mathbb{F}_{\mathfrak{p}}}(g_d) \equiv -a_1 \pmod{\mathfrak{p}}.$$

4.4.6 Let $\mathfrak{l} \neq \mathfrak{p}$ be a prime.

(a) Let ϕ be a supersingular Drinfeld module of rank r over k such that $\pi \in \phi(A)$. Prove that $\rho_{\phi,\mathfrak{l}}(\mathrm{Fr}_k) \in \mathrm{GL}_r(\mathbb{F}_\mathfrak{l})$ is a scalar matrix. Conclude that if ϕ has a nontrivial $\mathfrak{l}$-torsion point over k, then all $\mathfrak{l}$-torsion points of ϕ are defined over k.

(b) Conversely, suppose ϕ is a Drinfeld module of rank r over k such that $\rho_{\phi,\mathfrak{l}}(\mathrm{Fr}_k) \in \mathrm{GL}_r(\mathbb{F}_\mathfrak{l})$ is a scalar matrix. Prove that there is a natural number $N(k, r)$ depending only on k and r such that

$$\deg(\mathfrak{l}) > N(k, r) \quad \Longrightarrow \quad \phi \text{ is supersingular.}$$

4.4.7 Let ϕ be a Drinfeld module of rank $r \geq 2$ over F, let $L = F(\phi[\mathfrak{l}])$, and let $\mathfrak{p}$ be a prime that splits completely in L. Assume that the coefficients of ϕ_T have non-negative valuations at $\mathfrak{p}$, so that we can reduce these coefficients modulo $\mathfrak{p}$. Furthermore, assume that $\phi \otimes \mathbb{F}_\mathfrak{p}$ is a Drinfeld module of rank r. Prove that $\phi \otimes \mathbb{F}_\mathfrak{p}$ is ordinary. Using Theorem 7.3.5, deduce from this that a Drinfeld module over F of rank ≥ 2 has infinitely many ordinary reductions.

Chapter 5
Analytic Theory of Drinfeld Modules

The origins of the theory of elliptic curves lie in analysis. It was the insight of Abel and Weiertrass that meromorphic functions that are periodic with respect to a lattice $\Lambda \subset \mathbb{C}$ are intimately related to elliptic curves. In particular, such functions provide an explicit description of an elliptic curve E over $\mathbb{C}$ as a quotient $\mathbb{C}/\Lambda$.

Drinfeld modules also made their first appearance in an analytic context. In the 1930s, Carlitz introduced an analogue of the exponential function over $\mathbb{C}_\infty$, which is an entire function $e_C(z)$ on $\mathbb{C}_\infty$ satisfying $e_C(z+\lambda) = e_C(z)$ for all $\lambda \in \pi_C A$, for a certain $\pi_C \in \mathbb{C}_\infty$. Carlitz's motivation was to prove a version of Euler's celebrated formula for the special values of the Riemann zeta function at even integers. Carlitz showed that $e_C(z)$ satisfies the functional equation $e_C(T \cdot z) = C_T(e_C(z))$, where on the right-hand side we have the defining equation $C_T(z) = Tz + z^q$ of the Carlitz module. Unlike the complex numbers, $\mathbb{C}_\infty$ contains lattices of arbitrary finite rank and the associated periodic functions lead to general Drinfeld modules over $\mathbb{C}_\infty$. This was independently discovered by Drinfeld in the 1970s. The ability to study Drinfeld modules over $\mathbb{C}_\infty$ using their associated lattices and exponential functions is a powerful tool. This chapter is devoted to the study of A-lattices in $\mathbb{C}_\infty$, their associated Carlitz–Drinfeld exponential functions, and the applications of these to the theory of Drinfeld modules.

In Sect. 5.1, we combine additive polynomials with power series. More precisely, we define the non-commutative ring of additive power series $\mathbb{C}_\infty\langle\langle x \rangle\rangle$. For our purposes, the most important example of such series is the Carlitz–Drinfeld exponential function e_Λ associated to a discrete $\mathbb{F}_q$-submodule Λ of $\mathbb{C}_\infty$, which we define and study in this section. We also introduce the multiplicative inverse $\log_\Lambda$ of e_Λ in $\mathbb{C}_\infty\langle\langle x \rangle\rangle$. The interplay between e_Λ and $\log_\Lambda$ will be useful in some of the arguments that follow in later sections.

In Sect. 5.2, we further assume that the lattice Λ is an A-module. In that case, e_Λ gives rise to a Drinfeld module ϕ over $\mathbb{C}_\infty$. The coefficients of ϕ_a can be expressed in terms of Eisenstein series of Λ. We then show that every Drinfeld module over $\mathbb{C}_\infty$ arises from this construction.

M. Papikian, *Drinfeld Modules*, Graduate Texts in Mathematics 296,
https://doi.org/10.1007/978-3-031-19707-9_5

In Sect. 5.3, we give applications of analytic uniformization of Drinfeld modules. In particular, we give a relatively simple analytic construction of the Weil pairing, show how to construct Drinfeld modules over $\mathbb{C}_\infty$ with given endomorphism rings, and give an analytic solution to the problem of classifying Drinfeld modules over $\mathbb{C}_\infty$, up to isomorphism.

In Sect. 5.4, we review the results in Carlitz's paper [Car35], especially his beautiful analogue of Euler's formula.

In the final section, Sect. 5.5, we consider the field extensions of F_∞ generated by the lattices of Drinfeld modules over F_∞. This can be viewed as either a more detailed study of the lattice of a Drinfeld module or the study of its division fields.

5.1 Additive Power Series

Let R be a commutative $\mathbb{F}_q$-algebra, and let $R\langle\!\langle x \rangle\!\rangle$ be the set of power series of the form

$$f(x) = \sum_{n \geq 0} a_n x^{q^n}, \quad a_n \in R.$$

For such power series, just as in the case of additive polynomials, we have

$$f \circ (g + h) = f \circ g + f \circ h.$$

Therefore, $R\langle\!\langle x \rangle\!\rangle$ is a non-commutative ring with the usual addition but multiplication $\circ$, cf. Sect. 2.7.5. In particular, $f(x) \in R\langle\!\langle x \rangle\!\rangle$ is $\mathbb{F}_q$-linear, i.e., for two indeterminates x, y, we have

$$f(x + y) = f(x) + f(y)$$

and

$$f(\alpha x) = \alpha \cdot f(x) \text{ for all } \alpha \in \mathbb{F}_q.$$

Sometimes it will be more convenient to work with the ring of twisted power series,

$$R\{\!\{\tau\}\!\} = \left\{ \sum_{n \geq 0} a_n \tau^n \,\middle|\, a_n \in R \right\},$$

which is isomorphic to $R\langle\!\langle x \rangle\!\rangle$. The addition and multiplication on $R\{\!\{\tau\}\!\}$ are defined by

$$\sum_{n\geq 0} a_n \tau^n + \sum_{n\geq 0} b_n \tau^n = \sum_{n\geq 0} (a_n + b_n)\tau^n,$$

$$\left(\sum_{n\geq 0} a_n \tau^n\right)\left(\sum_{n\geq 0} b_n \tau^n\right) = \sum_{n\geq 0} \left(\sum_{i=0}^{n} a_i b_{n-i}^{q^i}\right)\tau^n,$$

and the isomorphism $R\{\!\{\tau\}\!\} \to R\langle\!\langle x \rangle\!\rangle$ is given by

$$f = \sum_{n\geq 0} a_n \tau^n \longmapsto f(x) = \sum_{n\geq 0} a_n x^{q^n}.$$

Similar to Lemma 2.7.21, we have:

Lemma 5.1.1 $f = \sum_{n\geq 0} a_n \tau^n \in R\{\!\{\tau\}\!\}^{\times}$ *if and only if $a_0 \in R^{\times}$.*

For the rest of this section, we will be concerned with additive power series over complete non-Archimedean fields of positive characteristic. Let K be a field complete with respect to a non-trivial non-Archimedean absolute value $|\cdot|$. Assume $\mathbb{F}_q$ is a subfield of K. Let

$R = \{z \in K \ : \ |z| \leq 1\}$ be the ring of integers in K,
$M = \{z \in K \ : \ |z| < 1\}$ be the maximal ideal of R,
$k = R/M$ be the residue field,
$\mathbb{C}_K$ be the completion of an algebraic closure of K.

Definition 5.1.2 Let Λ be a discrete $\mathbb{F}_q$-vector subspace of $\mathbb{C}_K$. The *Carlitz–Drinfeld exponential* of Λ is

$$e_\Lambda(x) = x \prod_{\lambda \in \Lambda}{}' \left(1 - \frac{x}{\lambda}\right). \tag{5.1.1}$$

Proposition 5.1.3 *Let Λ be a discrete $\mathbb{F}_q$-vector subspace of $\mathbb{C}_K$. Then:*

(1) $e_\Lambda(x)$ is an entire function.
(2) $e_\Lambda(x) \in \mathbb{C}_K\langle\!\langle x \rangle\!\rangle$.
(3) If $f(x) = \sum_{n\geq 0} a_n x^{q^n} \in \mathbb{C}_K\langle\!\langle x \rangle\!\rangle$ is entire and $a_0 \neq 0$, then all zeros of $f(x)$ are simple, i.e., have multiplicity 1 and the set of zeros $Z(f) \subset \mathbb{C}_K$ is a discrete $\mathbb{F}_q$-vector subspace. Thus,

$$f(x) = a_0 \cdot e_{Z(f)}(x).$$

Moreover, if $f(x)$ is defined over K, then $Z(f) \subset K^{\mathrm{sep}}$ and $Z(f)$ is G_K-invariant.

Proof

(1) The fact that $e_\Lambda(x)$ is entire is a special case of Lemma 2.7.15.
(2) Let $n \geq 1$ be an integer and $\Lambda_n = \{\lambda \in \Lambda \; : \; |\lambda| \leq n\}$. The strong triangle inequality implies that Λ_n is a finite dimensional $\mathbb{F}_q$-vector subspace of $\mathbb{C}_K$. Let

$$e_{\Lambda_n}(x) = x \prod_{\lambda \in \Lambda_n}' \left(1 - \frac{x}{\lambda}\right)$$

$$= \sum_{i=1}^{m} e_i(\Lambda_n) x^i, \quad m = \#\Lambda_n.$$

Expand $e_\Lambda(x) = \sum_{i \geq 0} e_i(\Lambda) x^i$. Then, for each $i \geq 0$,

$$e_i(\Lambda_n) \to e_i(\Lambda) \quad \text{as} \quad n \to \infty.$$

On the other hand, by Proposition 3.1.2 and Lemma 3.1.5, $e_{\Lambda_n}(x)$ is an $\mathbb{F}_q$-linear polynomial, so $e_i(\Lambda_n) = 0$ unless i is a q-th power. Hence, the same is true for the coefficients of $e_\Lambda(x)$. This implies that $e_\Lambda(x) \in \mathbb{C}_K \langle\!\langle x \rangle\!\rangle$.

(3) We already proved (cf. Theorem 2.7.16) that the multiset of zeros $Z := Z(f)$ of f is discrete and

$$f(x) = a_0 \cdot x \prod_{\lambda \in Z}' \left(1 - \frac{x}{\lambda}\right).$$

Define $Z_n = \{\lambda \in Z \; : \; |\lambda| \leq n\}$ and $f_n(x) = a_0 \cdot x \prod_{\lambda \in Z_n}' (1 - \frac{x}{\lambda})$. As in (2), the coefficients of $f_n(x) = \sum_{m \geq 1} a_{n,m} x^m$ converge to the coefficients of $f(x)$ as $n \to \infty$. Moreover, for any m, there is an integer $N(m)$ such that $|a_m| = |a_{n,m}|$ if $n > N(m)$ (this is because Z is discrete). If some $\lambda \in Z$ has multiplicity > 1, then the polynomials $f_n(x)$ for all large enough n cannot be $\mathbb{F}_q$-linear since an $\mathbb{F}_q$-linear polynomial with nonzero first coefficient is separable. This implies that $a_m \neq 0$ for some m which is not a power of q in contradiction to the assumption that $f(x) \in \mathbb{C}_K \langle\!\langle x \rangle\!\rangle$. Thus, each zero of $f(x)$ has multiplicity 1 and all the $f_n(x)$ are $\mathbb{F}_q$-linear. Since the roots of a separable $\mathbb{F}_q$-linear polynomial form an $\mathbb{F}_q$-vector subspace in $\mathbb{C}_K$, we see that Z is a discrete $\mathbb{F}_q$-vector subspace. Finally, assume $f(x)$ is defined over K. Using Proposition 2.7.12 and its proof, we observe that (i) each $\lambda \in Z$ is a root of a polynomial $g(x) \in K[x]$, (ii) all the roots of $g(x)$ are in Z, and (iii) $g(x)$ is separable because every element of Z has multiplicity 1. Hence, $Z \subset K^{\text{sep}}$ and Z is G_K-invariant.

$\square$

Lemma 5.1.4 *Let $\Lambda \subset \Lambda'$ be discrete $\mathbb{F}_q$-vector subspaces of $\mathbb{C}_K$. Assume that the dimension of Λ'/Λ is finite over $\mathbb{F}_q$. Let $z_1, \ldots, z_m$ be a set of coset representatives of Λ in Λ', where $m = \#\Lambda'/\Lambda$.*

(1) The set $\{e_\Lambda(z_1), \ldots, e_\Lambda(z_m)\}$ is a vector space over $\mathbb{F}_q$ that does not depend on the choice of coset representatives.
(2) Without loss of generality, assume $z_1 \in \Lambda$ so that $e_\Lambda(z_1) = 0$. Put

$$P_{\Lambda'/\Lambda}(x) = x \prod_{i=2}^{m} \left(1 - \frac{x}{e_\Lambda(z_i)}\right).$$

The polynomial $P_{\Lambda'/\Lambda}(x)$ is $\mathbb{F}_q$-linear.
(3) We have

$$e_{\Lambda'}(x) = P_{\Lambda'/\Lambda}(e_\Lambda(x)).$$

Proof

(1) Suppose $y_1, \ldots, y_m$ is another set of coset representatives. After reindexing, we may assume that $y_l = z_i + \lambda_i$ for some $\lambda_i \in \Lambda$, $1 \le i \le m$. Using the fact that e_Λ is $\mathbb{F}_q$-linear, we compute

$$e_\Lambda(y_i) = e_\Lambda(z_i + \lambda_i) = e_\Lambda(z_i) + e_\Lambda(\lambda_i) = e_\Lambda(z_i).$$

Thus $\{e_\Lambda(z_1), \ldots, e_\Lambda(z_m)\}$ is independent of the choice of $z_1, \ldots, z_m$. Next, for any $1 \le i, j \le m$, we can write $z_i + z_j = z_k + \lambda$ for some $\lambda \in \Lambda$. Therefore,

$$e_\Lambda(z_i) + e_\Lambda(z_j) = e_\Lambda(z_i + z_j)$$
$$= e_\Lambda(z_k + \lambda)$$
$$= e_\Lambda(z_k) + e_\Lambda(\lambda)$$
$$= e_\Lambda(z_k).$$

Similarly, for any $\alpha \in \mathbb{F}_q$ and any z_i, we can write $\alpha z_i = z_j + \lambda$ for some $1 \le j \le m$ and $\lambda \in \Lambda$. As above, one checks that $\alpha \cdot e_\Lambda(z_i) = e_\Lambda(\alpha z_i) = e_\Lambda(z_j)$. We conclude that $W = \{e_\Lambda(z_1), \ldots, e_\Lambda(z_m)\}$ is the set of elements of an $\mathbb{F}_q$-vector space; note that

$$e_\Lambda(z_i) = e_\Lambda(z_j) \implies e_\Lambda(z_i - z_j) = 0$$
$$\implies z_i - z_j \in \Lambda \implies z_i = z_j,$$

so the listed elements are distinct.

(2) Since the roots of $P_{\Lambda'/\Lambda}(x)$ are precisely the elements of W, the polynomial $P_{\Lambda'/\Lambda}(x)$ is $\mathbb{F}_q$-linear by Lemma 3.1.5.

(3) Below we use $\overset{\circ}{=}$ to denote that both sides are equal up to a constant multiple. We have

$$P_{\Lambda'/\Lambda}(e_\Lambda(x)) \overset{\circ}{=} e_\Lambda(x) \prod_{i=2}^{m} (e_\Lambda(z_i) - e_\Lambda(x))$$

$$= e_\Lambda(x) \prod_{i=2}^{m} e_\Lambda(z_i - x)$$

$$\overset{\circ}{=} x \prod_{i=1}^{m} \prod_{\lambda \in \Lambda + z_i}' \left(1 - \frac{x}{\lambda}\right)$$

$$= e_{\Lambda'}(x),$$

where in the last equality we use the fact that $\coprod_{i=1}^{m}(z_i + \Lambda) = \Lambda'$. Thus, $P_{\Lambda'/\Lambda}(e_\Lambda(x)) \overset{\circ}{=} e_{\Lambda'}(x)$. Since the coefficient of x in the power series expansions of both $P_{\Lambda'/\Lambda}(e_\Lambda(x))$ and $e_{\Lambda'}(x)$ is 1, we conclude that these power series are actually equal, as was claimed.

$\square$

Thanks to Proposition 5.1.3, we can consider $e_\Lambda(x)$ as an element $e_\Lambda = \sum_{n\geq 0} e_n \tau^n$ of $\mathbb{C}_K\{\!\{\tau\}\!\}$. Since $e_0 = 1$ is nonzero, e_Λ has a multiplicative inverse $\log_\Lambda$ in $\mathbb{C}_K\{\!\{\tau\}\!\}$, called the *Carlitz–Drinfeld logarithm of* Λ. This logarithm can be explicitly described as follows. Put

$$\log_\Lambda = \sum_{n\geq 0} b_n \tau^n,$$

and consider the system of equations in variables b_i, $i \geq 0$, resulting from $\log_\Lambda \cdot e_\Lambda = 1$:

$$b_0 e_0 = 1,$$

$$\sum_{i+j=n} b_i \cdot e_j^{q^i} = 0, \quad n \geq 1.$$

There is a unique solution given by the recursive formulas

$$b_0 = 1,$$

$$b_n = -\sum_{i=0}^{n-1} b_i e_{n-i}^{q^i}, \quad n \geq 1. \tag{5.1.2}$$

To consider $\log_\Lambda$ as a function $\log_\Lambda(x) = \sum_{n\geq 0} b_n x^{q^n}$ on $\mathbb{C}_K$, it is important to take into account the issue of convergence. Denote

$$m(\Lambda) = \min_{0\neq\lambda\in\Lambda} |\lambda|.$$

Lemma 5.1.5 *We have*

$$\rho(\log_\Lambda) = m(\Lambda),$$

so $\log_\Lambda$ *converges on the open disk* $\{z \in \mathbb{C}_K : |z| < m(\Lambda)\}$.

Proof Denote $e_\Lambda(x) = \sum_{n\geq 0} e_n(\Lambda)x^{q^n}$. For any $c \in \mathbb{C}_K^\times$, by expanding the product defining $e_{c\Lambda}(x)$, we see that

$$e_n(c\Lambda) = c^{1-q^n} e_n(\Lambda) \qquad \text{for all} \quad n \geq 0. \tag{5.1.3}$$

Next, denote $\log_\Lambda(x) = \sum_{n\geq 0} b_n(\Lambda)x^{q^n}$. We claim that $b_n(c\Lambda) = c^{1-q^n}b_n(\Lambda)$. This is obviously true for $n = 0$. Assume by induction that we proved this equality for $b_0, b_1, \ldots, b_{n-1}$. Then, using (5.1.2) and (5.1.3), we get

$$b_n(c\Lambda) = -\sum_{i=0}^{n-1} b_i(c\Lambda)e_{n-i}^{q^i}(c\Lambda)$$

$$= -\sum_{i=0}^{n-1} c^{1-q^i}b_i(\Lambda)c^{q^i-q^n}e_{n-i}^{q^i}(\Lambda)$$

$$= c^{1-q^n}b_n(\Lambda).$$

In particular,

$$\rho(\log_{c\Lambda})^{-1} = \varlimsup_{n\to\infty} |b_n(c\Lambda)|^{1/q^n}$$

$$= \varlimsup_{n\to\infty} |c|^{(1-q^n)/q^n} |b_n(\Lambda)|^{1/q^n}$$

$$= |c|^{-1} \rho(\log_\Lambda)^{-1}.$$

Since $m(c\Lambda) = |c|\, m(\Lambda)$, we see that it is enough to prove $\rho(\log_\Lambda) = m(\Lambda)$ assuming $m(\Lambda) = 1$.

If $m(\Lambda) = 1$, then $|\lambda_1 \cdots \lambda_n| \geq 1$ for any nonzero $\lambda_1, \ldots, \lambda_n \in \Lambda$. Since $e_n(\Lambda)$ is the (infinite) sum of inverses of such products, we get $|e_n(\Lambda)| \leq 1$ for all n. Then, using (5.1.2) and induction, we also get $|b_n(\Lambda)| \leq 1$ for all n. Now,

$$\rho(\log_\Lambda)^{-1} = \varlimsup_{n\to\infty} |b_n|^{1/q^n} \leq 1.$$

Thus, $1 \leq \rho(\log_\Lambda)$. To prove that this is an equality, one could use Theorem 2.7.24. Suppose on the contrary that $1 < \rho(\log_\Lambda)$. Since $m(\Lambda) = 1$, there is $\lambda \in \Lambda$ such that $|\lambda| = 1$. Then

$$\max_n \{ \left| e_n(\Lambda) \lambda^{q^n} \right| \} = \max_n \{ |e_n(\Lambda)| \} \leq 1 < \rho(\log_\Lambda).$$

Hence the assumptions of Theorem 2.7.24 are satisfied for $f = \log_\Lambda$, $g = e_\Lambda$, and $\alpha = \lambda$, so we must have

$$\lambda = \log_\Lambda(e_\Lambda(\lambda)) = \log_\Lambda(0) = 0.$$

This is a contradiction, so $1 = \rho(\log_\Lambda)$. $\qquad\qquad\qquad\qquad\qquad\qquad\qquad\square$

Remark 5.1.6 There is an alternative argument for proving $1 = \rho(\log_\Lambda)$, after reducing to the case $m(\Lambda) = 1$. Goss [Gos96, p. 124] attributes this proof to Poonen. As we showed above, all the coefficients of $e_\Lambda(x)$ and $\log_\Lambda(x)$ have valuations ≤ 1. Moreover, since $e_\Lambda(x)$ is entire, all but finitely many coefficients of $e_\Lambda(x)$ are in the maximal ideal $\mathcal{M}_{\mathbb{C}_K}$ of $\mathbb{C}_K$. Therefore, reducing the coefficients of $e_\Lambda(x)$ modulo $\mathcal{M}_{\mathbb{C}_K}$, one obtains a polynomial $\bar{e}_\Lambda(x)$ in $\bar{k}\langle x \rangle$. From the product expansion of $e_\Lambda(x)$, one sees that $\deg_x(\bar{e}_\Lambda(x)) \geq q$ since $\bar{e}_\Lambda(x)$ has a nonzero root, namely λ modulo $\mathcal{M}_{\mathbb{C}_K}$, where $\lambda \in \Lambda$ is such that $|\lambda| = 1$. On the other hand, the reduction $\overline{\log}_\Lambda(x)$ of $\log_\Lambda(x)$ modulo $\mathcal{M}_{\mathbb{C}_K}$ is still the formal inverse of $\bar{e}_\Lambda(x)$ in $k\langle\!\langle x \rangle\!\rangle$. The inverse of a polynomial of degree > 1 cannot be a polynomial, so infinitely many coefficients of $\overline{\log}_\Lambda(x)$ are nonzero, implying that infinitely many $b_n(\Lambda)$ satisfy $|b_n(\Lambda)| = 1$. Then, $\rho(\log_\Lambda)^{-1} = \overline{\lim}_{n\to\infty} |b_n|^{1/q^n} = 1$.

The series $e_\Lambda(x)$ does not have an inverse in the ring $\mathbb{C}_K[\![x]\!]$ since its constant term is 0. On the other hand, the constant term of $e_\Lambda(x)/x$ is 1, so this power series does have a multiplicative inverse $x/e_\Lambda(x)$ in $\mathbb{C}_K[\![x]\!]$. Let

$$\frac{x}{e_\Lambda(x)} = \sum_{n \geq 0} c_n(\Lambda) x^n. \qquad\qquad (5.1.4)$$

Next, we show how to explicitly describe the coefficients $c_n = c_n(\Lambda)$ of this power series in terms of the lattice Λ.

Definition 5.1.7 Let $n \geq 1$ be an integer. The *Eisenstein series of Λ of weight n*, denoted $E_n(\Lambda)$, is the series

$$E_n(\Lambda) = {\sum_{\lambda \in \Lambda}}' \frac{1}{\lambda^n}.$$

Note that this series is convergent since Λ is discrete.

Lemma 5.1.8 *If $n \not\equiv 0 \pmod{q-1}$, then $E_n(\Lambda) = 0$.*

Proof Suppose $E_n(\Lambda) \neq 0$. Since Λ is an $\mathbb{F}_q$-vector space, we have $\alpha\Lambda = \Lambda$ for all $\alpha \in \mathbb{F}_q^\times$. Thus,

$$E_n(\Lambda) = E_n(\alpha\Lambda) = \alpha^{-n} E_n(\Lambda).$$

This implies that $\alpha^n = 1$ for all $\alpha \in \mathbb{F}_q^\times$. Now, since $\mathbb{F}_q^\times$ is cyclic, Lagrange's theorem implies that n is divisible by $q - 1$. □

Theorem 5.1.9 *For all $n \geq 1$, we have $c_n(\Lambda) = -E_n(\Lambda)$, i.e.,*

$$\frac{x}{e_\Lambda(x)} = 1 - \sum_{n\geq 1} E_n(\Lambda)x^n.$$

Proof Observe that coefficient-wise $x/e_\Lambda(x)$ is the limit of $x/e_{\Lambda_n}(x)$. On the other hand, $e_{\Lambda_n}(x)$ is an additive polynomial, so its derivative $e'_{\Lambda_n}(x)$ is $e_0(\Lambda_n) = 1$. Thus, by Exercise 3.1.4,

$$\frac{1}{e_{\Lambda_n}(x)} = \frac{e'_{\Lambda_n}(x)}{e_{\Lambda_n}(x)} = \sum_{\lambda\in\Lambda_n} \frac{1}{x-\lambda}.$$

We expand

$$\frac{x}{e_{\Lambda_n}(x)} = \sum_{\lambda\in\Lambda_n} \frac{1}{1 - \frac{\lambda}{x}}$$

$$= 1 - {\sum_{\lambda\in\Lambda_n}}' \frac{\frac{x}{\lambda}}{1 - \frac{x}{\lambda}}$$

$$= 1 - {\sum_{\lambda\in\Lambda_n}}' \sum_{i\geq 1} \left(\frac{x}{\lambda}\right)^i.$$

Letting $n \to \infty$, we obtain

$$\frac{x}{e_\Lambda(x)} = 1 - {\sum_{\lambda\in\Lambda}}' \sum_{i\geq 1} \left(\frac{x}{\lambda}\right)^i \tag{5.1.5}$$

$$= 1 - \sum_{i\geq 1} \left({\sum_{\lambda\in\Lambda}}' \frac{1}{\lambda^i}\right) x^i.$$

Therefore, $c_i = -E_i(\Lambda)$ for all $i \geq 1$, as was claimed. □

Corollary 5.1.10 *We have a recursive formula for the coefficients of the exponential function $e_\Lambda(x) = \sum_{n \geq 0} e_n(\Lambda)x^{q^n}$ in terms of the Eisenstein series:*

$$e_n(\Lambda) = E_{q^n-1}(\Lambda) + \sum_{i=1}^{n-1} e_i(\Lambda) \cdot E_{q^{n-i}-1}(\Lambda)^{q^i}.$$

Proof For simplicity, denote $e_n = e_n(\Lambda)$, $c_n = c_n(\Lambda)$, and $E_n = E_n(\Lambda)$. From $e_\Lambda(x)(x/e_\Lambda(x)) = x$, we get

$$\left(\sum_{i \geq 0} e_i x^{q^i-1}\right)\left(\sum_{j \geq 0} c_j x^j\right) = 1.$$

The coefficient of x^{q^m-1}, $m \geq 1$, on the left-hand side of the above equation is $\sum_{i=0}^{m} e_i c_{q^m-q^i}$, which is therefore equal to zero:

$$\sum_{i=0}^{m} e_i c_{q^m-q^i} = 0, \qquad \text{for all } m \geq 1. \tag{5.1.6}$$

From the definition of the Eisenstein series, it is easy to see that

$$E_{q^i(q^{m-i}-1)} = E_{q^{m-i}-1}^{q^i} \qquad \text{for all } m > i \geq 0. \tag{5.1.7}$$

Therefore,

$$e_n = -\sum_{i=0}^{n-1} e_i c_{q^n-q^i} \qquad\qquad \text{(by (5.1.6))}$$

$$= \sum_{i=0}^{n-1} e_i E_{q^n-q^i} \qquad\qquad \text{(by Theorem 5.1.9)}$$

$$= \sum_{i=0}^{n-1} e_i E_{q^{n-i}-1}^{q^i} \qquad\qquad \text{(by (5.1.7)).}$$

$\square$

Somewhat surprisingly, the coefficients of $x/e_\Lambda(x)$ are related to the coefficients of $\log_\Lambda$.

Proposition 5.1.11 *Given integers s and m such that $0 \leq s \leq m$, we have*

$$c_{q^m-q^s}(\Lambda) = b_{m-s}(\Lambda)^{q^s}.$$

Proof Denote $n = q^m - q^s$. If $m = 0$, then $n = 0$, and the claim reduces to $c_0 = b_0 = 1$. Now let $m \geq 1$ and assume that the claim is proved for all integers strictly less than m.

Because

$$c_n = -E_n(\Lambda) = -E_{q^s(q^{m-s}-1)}$$
$$= -E_{q^{m-s}-1}^{q^s} \qquad\qquad \text{(by (5.1.7))}$$
$$= c_{q^{m-s}-1}^{q^s} \qquad\qquad \text{(by Theorem 5.1.9)},$$

it is enough to prove the claim assuming $s = 0$. By the induction hypothesis, for $0 < i \leq m$, we have $c_{q^m-q^i} = c_{q^{m-i}-1}^{q^i} = b_{m-i}^{q^i}$. Therefore, by (5.1.6),

$$c_{q^m-1} + \sum_{i=1}^{m} e_i b_{m-i}^{q^i} = 0. \qquad\qquad (5.1.8)$$

On the other hand, $e_\Lambda \cdot \log_\Lambda = 1$ implies that, besides (5.1.2), we also have the relation

$$\sum_{i=0}^{m} e_i b_{m-i}^{q^i} = 0 \qquad\qquad (5.1.9)$$

between the coefficients of v_Λ and $\log_\Lambda$. Comparing (5.1.8) and (5.1.9), we get $c_{q^m-1} = b_m$. $\qquad\qquad\qquad\qquad\qquad\qquad\qquad\qquad\qquad\qquad\qquad\square$

Finally, considering $x/e_\Lambda(x)$ as a function on $\mathbb{C}_K$ again raises the question of convergence. As with $\log_\Lambda$, we have the equality:

Lemma 5.1.12 $\rho(x/e_\Lambda(x)) = m(\Lambda)$.

Proof On the one hand, $\rho(x/e_\Lambda(x)) \leq m(\Lambda)$ since $x/e_\Lambda(x)$ does converge at any $0 \neq \lambda \in \Lambda$ (if $x/e_\Lambda(x)$ converges at $\lambda \in \Lambda$, then by Corollary 2.7.4 we have $\lambda = e_\Lambda(\lambda) \cdot (\lambda/e_\Lambda(\lambda)) = 0 \cdot (\lambda/e_\Lambda(\lambda)) = 0$.) On the other hand, by the strong triangle inequality, we have $|E_n(\Lambda)| \leq m(\Lambda)^{-n}$. Hence, using Theorem 5.1.9, we get

$$\rho(x/e_\Lambda(x))^{-1} = \varlimsup_{n\to\infty} \sqrt[n]{|c_n(\Lambda)|} = \varlimsup_{n\to\infty} \sqrt[n]{|E_n(\Lambda)|} \leq m(\Lambda)^{-1}.$$

Therefore, $\rho(x/e_\Lambda(x)) = m(\Lambda)$. $\qquad\qquad\qquad\qquad\qquad\qquad\qquad\qquad\square$

Exercises

5.1.1 Let K be a field complete with respect to a nontrivial non-Archimedean absolute value, let R be its ring of integers, and let M be the maximal ideal of R. There is a variant of the Weierstrass Preparation Theorem for $R\{\{\tau\}\}$. A polynomial $\sum_{n=0}^{d} b_n \tau^n \in R\{\tau\}$ is called *distinguished* if $b_0, \ldots, b_{d-1} \in M$ and $b_d = 1$.

Let $f = \sum_{n \geq 0} a_n \tau^n \in R\{\{\tau\}\}$ be such that $a_0, \ldots, a_{d-1} \in M$ and $a_d \in R^\times$. Prove that there is a unit $u \in R\{\{\tau\}\}^\times$ and a distinguished polynomial $g \in R\{\tau\}$ of degree d such that $f = ug$. Moreover, u and g with these properties are uniquely determined by f.

5.1.2 Let $\lambda \in \mathbb{C}_\infty^\times$ and $\Lambda = \mathbb{F}_q \cdot \lambda$. Show that

$$\log_\Lambda(z) = \sum_{i \geq 0} \frac{z^{q^i}}{\lambda^{q^i-1}}.$$

5.1.3 Prove that on the disk $\{z \in \mathbb{C}_K \ : \ |z| < m(\Lambda)\}$, we have

$$\left|\exp_\Lambda(z)\right| = |z| = \left|\log_\Lambda(z)\right|.$$

5.1.4 A power series $f(x) \in \mathbb{C}_K[\![x]\!]$ is $\mathbb{F}_q$-*linear* if for indeterminates x and y we have an equality $f(x + y) = f(x) + f(y)$ in the ring $\mathbb{C}_K[\![x, y]\!]$ and $f(\alpha x) = \alpha \cdot f(x)$ for all $\alpha \in \mathbb{F}_q$. Prove that $f(x) \in \mathbb{C}_K[\![x]\!]$ is $\mathbb{F}_q$-linear if and only if $f(x) = \sum_{n \geq 0} a_n x^{q^n}$.

5.1.5 Let Λ be a discrete $\mathbb{F}_q$-vector subspace of $\mathbb{C}_K$. Note that in the proof of Theorem 5.1.9, we showed that

$$u_\Lambda(x) := \frac{1}{e_\Lambda(x)} = \sum_{\lambda \in \Lambda} \frac{1}{x - \lambda}.$$

Prove that for any integer $m \geq 1$, there is a unique polynomial $G_{m,\Lambda}(x) \in \mathbb{C}_K[x]$ such that

$$\sum_{\lambda \in \Lambda} \frac{1}{(x - \lambda)^m} = G_{m,\Lambda}(u_\Lambda(x)).$$

Moreover, $G_{m,\Lambda}(x)$ is monic of degree m. The polynomials $G_{m,\Lambda}(x)$ are called *Goss polynomials*; they play an important role in the theory of Drinfeld modular forms; cf. [Gos80, Gek88].

5.2 Lattices and Drinfeld Modules

We specialize the setup of the previous section to the case $K = F_\infty$. In particular, $\mathbb{C}_K = \mathbb{C}_\infty$. Let $|\cdot|$ be the unique extension to $\mathbb{C}_\infty$ of the absolute value on F_∞ normalized by $|T| = q$.

For the rest of this chapter, we consider $\mathbb{C}_\infty$ as an A-field via the natural inclusions

$$\gamma : A \hookrightarrow F \hookrightarrow F_\infty \hookrightarrow \mathbb{C}_\infty.$$

To simplify the notation, we denote $\gamma(a) \in \mathbb{C}_\infty$ by a.

Definition 5.2.1 An A-*lattice* of rank r in $\mathbb{C}_\infty$ is a discrete subgroup $\Lambda \subset \mathbb{C}_\infty$ which is a finitely generated (thus free) A-submodule of $\mathbb{C}_\infty$ of rank r.

We will usually call an A-lattice simply a "lattice." The main result of this section is a natural bijection between the set of Drinfeld modules over $\mathbb{C}_\infty$ and the set of lattices.

Let Λ be a lattice of rank r. Fix $0 \neq a \in A$ and consider

$$a^{-1}\Lambda = \{\lambda/a \mid \lambda \in \Lambda\}.$$

It is clear that $a^{-1}\Lambda$ is a lattice of rank r and $\Lambda \subset a^{-1}\Lambda$ is a sublattice. Moreover,

$$a^{-1}\Lambda/\Lambda \cong \prod_{i=1}^{r} A/aA.$$

From the definition of the exponential function $e_\Lambda(x)$, cf. (5.1.1), it readily follows that for any $c \in \mathbb{C}_\infty^\times$ we have

$$e_{c\Lambda}(cx) = ce_\Lambda(x).$$

In particular,

$$e_\Lambda(ax) = ae_{a^{-1}\Lambda}(x).$$

Denote the polynomial $P_{a^{-1}\Lambda/\Lambda}$ in Lemma 5.1.4 by P_a^Λ. Combining Lemma 5.1.4 with the above expression for $e_\Lambda(ax)$, we get

$$e_\Lambda(ax) = a \cdot P_a^\Lambda(e_\Lambda(x)).$$

Proposition 5.2.2 *Let Λ be a lattice of rank r. The map*

$$\phi^\Lambda : A \longrightarrow \mathbb{C}_\infty\{\tau\}$$

$$a \longmapsto a P_a^\Lambda$$

is a Drinfeld module of rank r.

Proof Since $a P_a^\Lambda(x) = ax$ + higher degree terms, we need to check that ϕ^Λ is a ring homomorphism. Since $e_\Lambda(ax) = \phi_a^\Lambda(e_\Lambda(x))$, we have

$$\begin{aligned}
\phi_{a+b}^\Lambda(e_\Lambda(x)) &= e_\Lambda((a+b)x) \\
&= e_\Lambda(ax) + e_\Lambda(bx) \\
&= \phi_a^\Lambda(e_\Lambda(x)) + \phi_b^\Lambda(e_\Lambda(x)) \\
&= (\phi_a^\Lambda + \phi_b^\Lambda)(e_\Lambda(x)).
\end{aligned}$$

Since $\phi_{a+b}^\Lambda(x)$ and $\phi_a^\Lambda(x) + \phi_b^\Lambda(x)$ are polynomials, this implies that $\phi_{a+b}^\Lambda(x) = \phi_a^\Lambda(x) + \phi_b^\Lambda(x)$. (By Proposition 2.7.12, $e_\Lambda(x)$ is a surjective function from $\mathbb{C}_\infty$ to $\mathbb{C}_\infty$. Using Lemma 2.7.23, we conclude that the polynomials $\phi_{a+b}^\Lambda(x)$ and $\phi_a^\Lambda(x) + \phi_b^\Lambda(x)$ are equal on all of $\mathbb{C}_\infty$, and thus are equal.) Similarly,

$$\begin{aligned}
\phi_{ab}^\Lambda(e_\Lambda(x)) &= e_\Lambda(abx) \\
&= \phi_a^\Lambda(e_\Lambda(bx)) \\
&= \phi_a^\Lambda(\phi_b^\Lambda(e_\Lambda(x))),
\end{aligned}$$

which implies that $\phi_{ab}^\Lambda(x) = \phi_a^\Lambda(\phi_b^\Lambda(x))$. Finally, the degree of $P_a^\Lambda(x)$ is $\#(a^{-1}\Lambda/\Lambda) = q^{\deg(a)r}$. Hence, ϕ^Λ is a Drinfeld module of rank r.   □

Proposition 5.2.3 *Let Λ be a lattice of rank r. For $a \in A$, let*

$$\phi_a^\Lambda = a + g_1^\Lambda(a)\tau + \cdots + g_n^\Lambda(a)\tau^n, \quad n = r \cdot \deg(a).$$

The coefficients $g_i^\Lambda(a)$ can be computed recursively as follows:

$$g_m^\Lambda(a) = (a^{q^m} - a) \cdot E_{q^m-1}(\Lambda) + \sum_{i=1}^{m-1} E_{q^i-1}(\Lambda) \cdot g_{m-i}^\Lambda(a)^{q^i}, \quad 1 \le m \le n.$$

Proof To simplify the notation, we write g_i for $g_i^\Lambda(a)$ and put $g_0 = a$ and $g_i = 0$ for $i > n$. Let

$$\log_\Lambda = \sum_{i \geq 0} b_i \tau^i$$

be the inverse of e_Λ in $\mathbb{C}_\infty\{\{\tau\}\}$. In $\mathbb{C}_\infty\{\{\tau\}\}$, we have $e_\Lambda \cdot a = \phi_a^\Lambda \cdot e_\Lambda$. This identity, combined with $e_\Lambda \cdot \log_\Lambda = \log_\Lambda \cdot e_\Lambda = 1$, gives

$$a \cdot \log_\Lambda = \log_\Lambda \cdot e_\Lambda \cdot a \cdot \log_\Lambda \qquad (5.2.1)$$
$$= \log_\Lambda \cdot \phi_a^\Lambda \cdot e_\Lambda \cdot \log_\Lambda$$
$$= \log_\Lambda \cdot \phi_a^\Lambda.$$

Hence

$$a b_m = \sum_{i=0}^{m} b_i g_{m-i}^{q^i}. \qquad (5.2.2)$$

On the other hand, by Theorem 5.1.9 and Proposition 5.1.11,

$$b_i = -E_{q^i-1}(\Lambda), \quad i \geq 1.$$

This implies the desired formula for g_m. ⌐

Remark 5.2.4 Proposition 5.2.3 gives a formula for $g_i^\Lambda(a)$ in terms of the coefficients $b_m(\Lambda)$ of the series expansion of $\log_\Lambda(x)$. Conversely, note that from (5.2.2) one may recursively compute $b_m(\Lambda)$ using $g_i^\Lambda(a)$'s.

Example 5.2.5 We will show next that every Drinfeld module of rank r over $\mathbb{C}_\infty$ arises as ϕ^Λ for some lattice Λ of rank r. Let $\Lambda_C = \pi_C A$ be the rank-1 lattice corresponding to the Carlitz module $C_T = T + \tau$, where $\pi_C \in \mathbb{C}_\infty$ is well-defined up to an $\mathbb{F}_q^\times$-multiple. The element π_C^{q-1} is well-defined, and Proposition 5.2.3 gives an explicit formula for π_C^{q-1}. Indeed, according to that proposition,

$$1 = g_1^{\Lambda_C}(T) = (T^q - T)E_{q-1}(\Lambda_C) = (T^q - T)\pi_C^{1-q}E_{q-1}(A).$$

Hence,

$$\pi_C^{q-1} = (T^q - T){\sum_{a \in A}}' \frac{1}{a^{q-1}} = -(T^q - T)\sum_{a \in A_+} \frac{1}{a^{q-1}}. \qquad (5.2.3)$$

This is a special case of a more general formula due to Carlitz; see Theorem 5.4.10. (In Sect. 5.4, we will deduce two more explicit expressions for π_C^{q-1} by different methods; see Corollary 5.4.9.) Note that (5.2.3) implies that $\pi_C^{q-1} \in F_\infty$ and $\left| \pi_C^{q-1} \right| = |T^q - T| = q^q$. On the other hand, one can show that π_C^{q-1} is transcendental over F; see Remark 5.4.17.

Example 5.2.6 Since ϕ^Λ is uniquely determined by ϕ_T^Λ, to determine the Drinfeld module corresponding to a given lattice Λ, it is enough to know $g_1^\Lambda(T), \ldots, g_r^\Lambda(T)$. In this example, we give explicit formulas for $r = 3$. Let $\phi_T^\Lambda = T + g_1\tau + g_2\tau^2 + g_3\tau^3$ and $E_m := E_m(\Lambda)$. Then,

$$g_1 = (T^q - T) \cdot E_{q-1},$$

$$g_2 = (T^{q^2} - T) \cdot E_{q^2-1} + g_1^q \cdot E_{q-1}$$

$$= (T^{q^2} - T) \cdot E_{q^2-1} + (T^{q^2} - T^q) \cdot E_{q-1}^{q+1},$$

$$g_3 = (T^{q^3} - T) \cdot E_{q^3-1} + g_2^q \cdot E_{q-1} + g_1^{q^2} \cdot E_{q^2-1}$$

$$= (T^{q^3} - T) \cdot E_{q^3-1} + (T^{q^3} - T) \cdot E_{q^2-1}^q \cdot E_{q-1}$$

$$+ (T^{q^3} - T^{q^2}) \cdot E_{q-1}^{q^2+q+1} + (T^{q^3} - T^{q^2}) \cdot E_{q-1}^{q^2} \cdot E_{q^2-1}.$$

Note that the formulas for g_1, g_2, and g_3 are universal in the sense that the first three coefficients of ϕ_T^Λ are given by these formulas for any $r \geq 3$.

Starting with a lattice Λ, we have produced a Drinfeld module ϕ^Λ. In this construction, the Carlitz–Drinfeld exponential function e_Λ plays an important role of a steppingstone:

$$\Lambda \rightsquigarrow e_\Lambda \rightsquigarrow \phi^\Lambda.$$

Next, we will produce a lattice Λ_ϕ associated with a Drinfeld module ϕ over $\mathbb{C}_\infty$ by reversing the arrows in the above diagram: given a Drinfeld module ϕ, we first construct an entire $\mathbb{F}_q$-linear function $e_\phi(x)$ satisfying $e_\phi(ax) = \phi_a(e_\phi(x))$, and then the lattice corresponding to ϕ turns out to be the set of zeros of e_ϕ.

Let ϕ be a Drinfeld module of rank r over $\mathbb{C}_\infty$ given by

$$\phi_T = T + g_1\tau + \cdots + g_r\tau^r \in \mathbb{C}_\infty\{\tau\}, \quad g_r \neq 0.$$

We want to find $e_i \in \mathbb{C}_\infty$, $i \geq 0$, such that

$$e_\phi = e_0 + e_1\tau + e_2\tau^2 + \cdots$$

satisfies

$$e_\phi T = \phi_T e_\phi$$

in $\mathbb{C}_\infty\{\{\tau\}\}$. Expanding both sides leads to a system of equations

$$(T^{q^n} - T) \cdot e_n = e_{n-1}^q \cdot g_1 + e_{n-2}^{q^2} \cdot g_2 + \cdots + e_{n-r}^{q^r} \cdot g_r, \qquad n \geq 0, \qquad (5.2.4)$$

where we put $e_i = 0$ for $i < 0$. If we put $e_0 = 1$, then every other e_n can be uniquely determined from the above recursive formulas. Thus, there exists a unique $e_\phi \in \mathbb{C}_\infty\{\{\tau\}\}$ with constant term 1 such that $e_\phi T = \phi_T e_\phi$. Since $\phi: A \to \mathbb{C}_\infty\{\tau\}$ is a homomorphism, it is easy to show that in fact the equality

$$e_\phi a = \phi_a e_\phi \qquad (5.2.5)$$

holds in $\mathbb{C}_\infty\{\{\tau\}\}$ for all $a \in A$.

Proposition 5.2.7 *Let ϕ be a Drinfeld module of rank r over $\mathbb{C}_\infty$.*

(1) The power series $e_\phi(x) \in \mathbb{C}_\infty\langle\langle x\rangle\rangle$ defines an entire function on $\mathbb{C}_\infty$.
(2) The multiset of zeros Λ_ϕ of $e_\phi(x)$ is a lattice of rank r.

Proof Denote $c_n = |e_n|^{1/q^n}$. By Lemma 2.7.6, showing that $e_\phi(x)$ is entire is equivalent to showing that $c_n \to 0$. For any $\varepsilon > 0$ there is $N = N(\varepsilon)$ such that $\max_{1 \leq i \leq r} |g_i|^{1/q^n} < 1 + \varepsilon$ for all $n > N$. Then, for $n > N + r$, from (5.2.4), we get

$$c_n \leq \frac{1 + \varepsilon}{q} \max(c_{n-1}, c_{n-2}, \ldots, c_{n-r}).$$

This easily implies that there is a constant B such that $0 \leq c_{N+sr} < B/q^s$ for any $s \geq 1$. Thus, $c_n \to 0$, which proves (1).

By Proposition 5.1.3, the zeros of e_ϕ are simple and $\Lambda_\phi := Z(e_\phi)$ is a discrete $\mathbb{F}_q$-vector subspace of $\mathbb{C}_\infty$. Moreover, for $\lambda \in \Lambda_\phi$ and $a \in A$, we have (cf. Corollary 2.7.25)

$$e_\phi(a\lambda) = \phi_a(e_\phi(\lambda)) = \phi_a(0) = 0.$$

Hence, $a\lambda \in \Lambda_\phi$, so Λ_ϕ is an A-submodule of $\mathbb{C}_\infty$. Note that Λ_ϕ is a torsion-free A-module, since $\mathbb{C}_\infty$ is a torsion-free A-module. We will show now that Λ_ϕ is finitely generated, thus free.

Suppose $\lambda_1, \lambda_2, \ldots, \lambda_s \in \Lambda_\phi$ are linearly independent over F_∞. Let $V = F_\infty\lambda_1 + \cdots + F_\infty\lambda_s$ Be the F_∞-subspace of $\mathbb{C}_\infty$ generated by these elements, and let $\Lambda' := \Lambda_\phi \cap V$. By Lemma 3.4.9, Λ' is a free A-module of rank s. With an appropriate choice of $\lambda_1, \ldots, \lambda_s$, we can assume that these elements form an A-basis of Λ'.

Observe that for all $1 \leq i \leq s$,

$$\phi_T(e_\phi(\lambda_i/T)) = e_\phi(T(\lambda_i/T)) = e_\phi(\lambda_i) = 0.$$

Thus, $e_\phi(\lambda_i/T) \in \phi[T]$, which is a vector space of dimension r over $\mathbb{F}_q$. We claim that the elements

$$\{e_\phi(\lambda_1/T), \ldots, e_\phi(\lambda_s/T)\}$$

are linearly independent over $\mathbb{F}_q$; this will imply that $s \le r$. Suppose by contradiction that there are $\alpha_1, \ldots, \alpha_s \in \mathbb{F}_q$, not all zero, such that

$$0 = \sum_{i=0}^{s} \alpha_i \cdot e_\phi(\lambda_i/T) = e_\phi\left(\frac{\alpha_1\lambda_1 + \cdots + \alpha_s\lambda_s}{T}\right).$$

This implies that $\lambda' := \alpha_1\lambda_1 + \cdots + \alpha_s\lambda_s \in T\Lambda_\phi \cap V = T\Lambda'$. Without loss of generality, assume that $\alpha_1 \neq 0$. Then $\{\lambda', \lambda_2, \ldots, \lambda_s\}$ is an A-basis of Λ', but $\lambda' \in T\Lambda'$, which leads to a contradiction.

We conclude that the F_∞-vector space in $\mathbb{C}_\infty$ spanned by the elements of Λ_ϕ is finite dimensional. Let $\lambda_1, \lambda_2, \ldots, \lambda_s \in \Lambda_\phi$ be a basis of this vector space. Using Lemma 3.4.9, one concludes that $\Lambda_\phi = \Lambda'$ is a free A-module of rank s. Finally, to show that $s = r$, one can simply use the earlier construction that associates a Drinfeld module with a lattice. From that construction, it is clear that $\phi^{\Lambda_\phi} = \phi$; hence, the rank of ϕ is equal to the rank of Λ_ϕ. $\square$

Propositions 5.2.2 and 5.2.7 together give the following:

Theorem 5.2.8 *The maps $\Lambda \mapsto \phi^\Lambda$ and $\phi \mapsto \Lambda_\phi$ are inverses of each other and give a bijection between the set of lattices of rank r in $\mathbb{C}_\infty$ and the set of Drinfeld modules of rank r over $\mathbb{C}_\infty$.*

Since an entire function on $\mathbb{C}_\infty$ is surjective, we obtain from (5.2.5) and Proposition 5.2.7 a short exact sequence of A-modules

$$0 \longrightarrow \Lambda_\phi \longrightarrow \mathbb{C}_\infty \xrightarrow{e_\phi} {}^\phi\mathbb{C}_\infty \longrightarrow 0. \qquad (5.2.6)$$

One says that Drinfeld modules over $\mathbb{C}_\infty$ have "analytic uniformization" since the above exact sequence can be thought of as saying that $\mathbb{C}_\infty/\Lambda_\phi$ is isomorphic to ${}^\phi\mathbb{C}_\infty$.

Remarks 5.2.9

(1) The sequence (5.2.6) should be compared with the sequence

$$0 \longrightarrow 2\pi i\mathbb{Z} \longrightarrow \mathbb{C} \xrightarrow{e^x} \mathbb{C}^\times \longrightarrow 0,$$

where e^x is the usual exponential. Note that e^x induces a homomorphism of $\mathbb{Z}$-modules

$$e^x: (\mathbb{C}, +) \longrightarrow (\mathbb{C}^\times, \times),$$

and its kernel is a lattice in $\mathbb{C}$.

(2) For the reader familiar with the theory of elliptic curves, it might also be instructive to compare the analytic uniformization of elliptic curves with the analytic uniformization of Drinfeld modules. We refer to Chapter VI of [Sil09] for a detailed account of the analytic uniformization of elliptic curves.

Let $\Lambda = \mathbb{Z}\omega_1 + \mathbb{Z}\omega_2 \subset \mathbb{C}$ be a lattice, where ω_1 and ω_2 are complex numbers linearly independent over $\mathbb{R}$. The Weierstrass $\wp$-function (relative to Λ) is defined by the series

$$\wp_\Lambda(z) = \frac{1}{z^2} + \sideset{}{'}\sum_{\lambda \in \Lambda} \left(\frac{1}{(z-\lambda)^2} - \frac{1}{\lambda^2} \right).$$

The Weierstrass function $\wp_\Lambda$ is a meromorphic function on $\mathbb{C}$ satisfying $\wp_\Lambda(z + \lambda) = \wp_\Lambda(z)$ for all $z \in \mathbb{C}$ and $\lambda \in \Lambda$. In fact, any function $f(z)$ that is meromorphic on $\mathbb{C}$ and satisfies $f(z + \lambda) = f(z)$ for all $z \in \mathbb{C}$ and $\lambda \in \Lambda$ is a rational combination of $\wp_\Lambda(z)$ and its derivative $\wp'_\Lambda(z)$. Some elementary calculus leads to the Laurent expansion

$$\wp_\Lambda(z) = \frac{1}{z^2} + \sum_{k=1}^{\infty} (2k+1) E_{2k+2}(\Lambda) z^{2k}$$

for $\wp_\Lambda(z)$ around the origin, where

$$E_k(\Lambda) = \sideset{}{'}\sum_{\lambda \in \Lambda} \frac{1}{\lambda^k}, \qquad k > 2.$$

The Weierstrass function $\wp_\Lambda(z)$ and its derivative play a role in the setting of elliptic curves similar to the role of the Carlitz–Drinfeld exponential $e_\Lambda(x)$ for Drinfeld modules.[1] Recall that $e_\Lambda(x)$ can be expanded into power series whose coefficients are also related to Eisenstein series (see Corollary 5.1.10), although by more complicated formulas than the coefficients of $\wp_\Lambda(z)$. Let E_Λ be the elliptic curve over $\mathbb{C}$ defined by the equation

$$Y^2 Z = 4X^3 - 60E_4(\Lambda)XZ^2 - 140E_6(\Lambda)Z^3.$$

---

[1] One notable difference between $e_\Lambda(x)$ and $\wp_\Lambda(z)$ is that $e_\Lambda(x)$ is entire. An entire doubly periodic functions on $\mathbb{C}$ is bounded and hence is a constant.

The map $\varepsilon(z) \colon \mathbb{C} \longrightarrow E_\Lambda(\mathbb{C})$ defined by

$$
z \longmapsto \begin{cases} (\wp_\Lambda(z), \wp'_\Lambda(z), 1) & \text{for } z \notin \Lambda; \\ (0, 1, 0) & \text{for } z \in \Lambda \end{cases}
$$

induces an exact sequence of $\mathbb{Z}$-modules

$$
0 \longrightarrow \Lambda \longrightarrow \mathbb{C} \xrightarrow{\ \varepsilon(z)\ } E_\Lambda(\mathbb{C}) \longrightarrow 0,
$$

i.e., $\varepsilon(nz) = [n]\varepsilon(z)$ for all $n \in \mathbb{Z}$, where on the right-hand side we have the group operation on E_Λ. This is another analogue of (5.2.6). The fact that the coefficients of ϕ_T^Λ can be expressed in terms of Eisenstein series is similar to the above cubic equation for E_Λ.

Next, every elliptic curve E over $\mathbb{C}$ is isomorphic to E_Λ for some lattice Λ. In [Sil09, Thm. VI.5.1], this is proved as a trivial consequence of the fact from complex analysis that for any $a, b \in \mathbb{C}$ satisfying $4a^3 - 27b^2 \neq 0$ there exists a unique lattice $\Lambda \subset \mathbb{C}_\infty$ such that $60E_4(\Lambda) = a$ and $140E_6(\Lambda) = b$ (the proof of this fact can be found in [Sil94, Ch. I]). Alternatively (cf. [Mum70, p. 1]), if one considers $E(\mathbb{C})$ as a compact connected complex Lie group, then the exponential map

$$
\exp \colon \operatorname{Tan}_{0,E} \longrightarrow E(\mathbb{C})
$$

from the tangent space $\operatorname{Tan}_{0,E} \cong \mathbb{C}$ at the identity point of $E(\mathbb{C})$ to $E(\mathbb{C})$ is surjective. The kernel[2] of exp is the sought after lattice Λ. This is more in line with our construction for Drinfeld modules.

Finally, one shows that

$$
\operatorname{Hom}(E_\Lambda, E_{\Lambda'}) \cong \operatorname{Hom}(\Lambda, \Lambda'),
$$

where $\operatorname{Hom}(\Lambda, \Lambda')$ is defined similarly to Definition 5.2.10. We prove this for Drinfeld modules next.

(3) We point out that $\mathbb{C}$ only has enough space for lattices of rank 1 and rank 2 since $[\mathbb{C} : \mathbb{R}] = 2$. Thus, only the multiplicative group and elliptic curves can be analytically uniformized as quotients of $\mathbb{C}$. In contrast, $\mathbb{C}_\infty$ is infinite dimensional over F_∞, so $\mathbb{C}_\infty$ contains lattices of arbitrary ranks; these, through the analytic uniformization, produce Drinfeld modules of arbitrary ranks. From this perspective, Drinfeld modules of rank ≥ 3 have no proper classical analogues.

---

[2] Since $\mathbb{C}$ is the universal covering space of $E(\mathbb{C})$, the kernel of exp is canonically isomorphic to the homology group $H_1(E(\mathbb{C}), \mathbb{Z})$. Because of this, in some problems in the theory of Drinfeld modules, the lattice Λ_ϕ associated with a Drinfeld module ϕ is thought of as its first homology group, although ϕ does not define an actual topological object.

The bijection in Theorem 5.2.8 in fact provides an equivalence of categories of Drinfeld modules and lattices, that is, the morphisms of Drinfeld modules correspond to morphisms of lattices, with the latter notion appropriately defined:

Definition 5.2.10 Let Λ and Λ' be lattices of the same rank. A *morphism* $\Lambda \to \Lambda'$ is an element $c \in \mathbb{C}_\infty$ such that $c\Lambda \subseteq \Lambda'$. The set of all morphism $\Lambda \to \Lambda'$ is denoted

$$\mathrm{Hom}(\Lambda, \Lambda') = \{c \in \mathbb{C}_\infty \mid c\Lambda \subseteq \Lambda'\}.$$

Note that $\mathrm{Hom}(\Lambda, \Lambda')$ is an A-module, where $a \in A$ acts on $c \in \mathrm{Hom}(\Lambda, \Lambda')$ by $a \circ c = ac$, i.e., by the usual multiplication in $\mathbb{C}_\infty$.

Theorem 5.2.11 *Let ϕ and ψ be Drinfeld modules over $\mathbb{C}_\infty$, and let Λ_ϕ and Λ_ψ be their associated lattices. Then, there is an isomorphism of A-modules*

$$\mathrm{Hom}(\phi, \psi) \cong \mathrm{Hom}(\Lambda_\phi, \Lambda_\psi)$$

$$u \mapsto \partial u.$$

Moreover, for $0 \neq u \in \mathrm{Hom}(\phi, \psi)$, there is an isomorphism of A-modules

$$\ker(u) \cong \Lambda_\psi / \partial(u)\Lambda_\phi.$$

Proof Let $u \colon \phi \to \psi$ be an isogeny. Recall that this means that u is a nonzero element of $\mathbb{C}_\infty\{\tau\}$ such that $u\phi_T = \psi_T u$. Let $u_0 = \partial u$ be the constant term of the twisted polynomial u. Since $\mathrm{char}_A(\mathbb{C}_\infty) = 0$, we have $u_0 \neq 0$; see Lemma 3.3.4. Consider $u\phi_T = \psi_T u$ as an equality in $\mathbb{C}_\infty\{\tau\}$ and multiply both sides by $e_\phi u_0^{-1}$ from the right. Using the equation $\phi_T e_\phi = e_\phi T$, we obtain

$$\psi_T u e_\phi u_0^{-1} = u\phi_T e_\phi u_0^{-1} = u e_\phi T u_0^{-1} = u e_\phi u_0^{-1} T.$$

On the other hand, from (5.2.4), we know that $e_\psi \in \mathbb{C}_\infty\{\tau\}$ is the unique solution of the equation $\psi_T e = eT$ in $\mathbb{C}_\infty\{\tau\}$, assuming e has constant term 1. By our previous calculation, $u e_\phi u_0^{-1}$ satisfies this equation and has constant term 1. Hence,

$$e_\psi u_0 = u e_\phi. \tag{5.2.7}$$

Considering the sets of zeros of both power series, we get $u_0 \Lambda_\phi \subseteq \Lambda_\psi$. Thus, $u_0 \in$ $\mathrm{Hom}(\Lambda_\phi, \Lambda_\psi)$. Moreover, the following diagram of A-modules is commutative:

$$
\begin{array}{ccccccccc}
0 & \longrightarrow & \Lambda_\phi & \longrightarrow & \mathbb{C}_\infty & \xrightarrow{\ e_\phi\ } & {}^\phi\mathbb{C}_\infty & \longrightarrow & 0 \\
& & \downarrow{\scriptstyle u_0} & & \downarrow{\scriptstyle u_0} & & \downarrow{\scriptstyle u} & & \\
0 & \longrightarrow & \Lambda_\psi & \longrightarrow & \mathbb{C}_\infty & \xrightarrow{\ e_\psi\ } & {}^\psi\mathbb{C}_\infty & \longrightarrow & 0.
\end{array}
$$

Since multiplication by $u_0 \neq 0$ on $\mathbb{C}_\infty$ is an isomorphism, the Snake Lemma 5.2.13 gives an isomorphism of A-modules $\ker(u) \cong \Lambda_\psi / u_0 \Lambda_\phi$.

Conversely, suppose $c\Lambda_\phi \subseteq \Lambda_\psi$ for some $c \in \mathbb{C}_\infty^\times$. We have

$$ e_{c\Lambda_\phi} = c e_\phi c^{-1}, $$

since the constant term of the entire function on the right-hand side is 1 and its set of zeros is $c\Lambda_\phi$. By Lemma 5.1.4, there is $P_{\Lambda_\psi / c\Lambda_\phi} \in \mathbb{C}_\infty\{\tau\}$ with constant term 1 such that

$$ e_\psi = P_{\Lambda_\psi / c\Lambda_\phi} c e_\phi c^{-1}. $$

Let $u = P_{\Lambda_\psi / c\Lambda_\phi} c$. Note that $u_0 = c$ and $e_\psi u_0 = u e_\phi$. Now,

$$ e_\psi u_0 = u e_\phi \implies e_\psi u_0 T = u e_\phi T $$
$$ \implies \psi_T e_\psi u_0 = u \phi_T e_\phi. $$

From the last equality, using $e_\psi u_0 = u e_\phi$ once more, we obtain $\psi_T u e_\phi = u \phi_T e_\phi$. Thus,

$$ (\psi_T u - u \phi_T) e_\phi = 0. $$

Since $\mathbb{C}_\infty\{\!\{\tau\}\!\}$ has no zero divisors, we must have $\psi_T u = u \phi_T$, i.e., $u \in \mathrm{Hom}(\phi, \psi)$. $\square$

Corollary 5.2.12 *Two Drinfeld modules ϕ and ψ over $\mathbb{C}_\infty$ are isomorphic if and only if there is $c \in \mathbb{C}_\infty^\times$ such that $c\Lambda_\phi = \Lambda_\psi$.*

Lemma 5.2.13 (Snake Lemma) *Let R be a commutative ring, and let*

$$
\begin{array}{ccccccc}
M_1 & \xrightarrow{g} & M_2 & \longrightarrow & M_3 & \longrightarrow & 0 \\
\downarrow{\scriptstyle f_1} & & \downarrow{\scriptstyle f_2} & & \downarrow{\scriptstyle f_3} & & \\
0 & \longrightarrow & N_1 & \longrightarrow & N_2 & \xrightarrow{h} & N_3
\end{array}
$$

be a commutative diagram of R-modules with exact rows. Then, there is an exact sequence of modules

$$
0 \longrightarrow \ker(g) \longrightarrow \ker(f_1) \longrightarrow \ker(f_2) \longrightarrow \ker(f_3)
$$

$$
\mathrm{coker}(f_1) \longleftarrow \mathrm{coker}(f_2) \longrightarrow \mathrm{coker}(f_3) \longrightarrow \mathrm{coker}(h) \longrightarrow 0.
$$

Proof This is given as an exercise in [DF04]; the proof can be found in [Lan02, Sec. III.9]. The non-obvious map is the "connecting" homomorphism $\ker(f_3) \to \mathrm{coker}(f_1)$. $\qquad\square$

Exercises

5.2.1 Let Λ be an A-lattice of rank r. Assign weight m to the Eisenstein series $E_m(\Lambda)$. Prove that $g_i^{\Lambda}(a)$, $1 \le i \le r \cdot \deg(a)$, is a polynomial in E_m's of homogeneous weight $q^i - 1$ for any a.

5.2.2 Let Λ be a lattice of rank r, let $\{\omega_1, \ldots, \omega_r\}$ be an A basis of Λ, and let ψ^{Λ} be the associated Drinfeld module of rank r. Let $\Delta(\Lambda)$ be the leading coefficient of $\phi_T^{\Lambda}(x)$. Prove that

$$
\Delta(\Lambda) = T \prod_{\alpha_1,\ldots,\alpha_r \in \mathbb{F}_q}{}' e_\Lambda\left(\frac{\alpha_1\omega_1 + \cdots \alpha_r\omega_r}{T}\right)^{-1}.
$$

5.2.3 Let Λ be a lattice of rank $r \ge 1$ and $\Lambda' = A\lambda + \Lambda$ a lattice of rank $r + 1$. Prove that

$$
e_{\Lambda'}(z) = e_\Lambda(z) \prod_{a \in A}{}' \frac{e_\Lambda(z) + e_\Lambda(a\lambda)}{e_\Lambda(a\lambda)}.
$$

5.2.4 Let Λ be a lattice of rank r, and let ϕ^Λ be its associated Drinfeld module. Show that for any $0 \neq a \in A$, we have the following relationship between the coefficients of $e_\Lambda(x) = \sum_{n \geq 0} e_n(\Lambda) x^{q^n}$ and the coefficients of $\phi_a^\Lambda(x) = \sum_{n \geq 0} g_n^\Lambda(a) x^{q^n}$, where $g_0^\Lambda(a) = a$, $g_n^\Lambda(a) = 0$ for $n > r \deg(a)$, and $e_n(\Lambda) = 0$ for $n < 0$:

$$(a^{q^n} - a) \cdot e_n(\Lambda) = g_n^\Lambda(a) + \sum_{i=1}^{n-1} g_i^\Lambda(a) \cdot e_{n-i}(\Lambda)^{q^i}.$$

Conclude that if $d := \deg(a) \geq 1$, then each $e_n(\Lambda)$ is a polynomial in $g_0^\Lambda(a), g_1^\Lambda(a), \ldots, g_{rd}^\Lambda(a)$.

5.2.5 Let $\Lambda = A \subset \mathbb{C}_\infty$. Assume $z \in \mathbb{C}_\infty$ is such that $|z| = \mathfrak{I}(z) = q^{d-\varepsilon}$ for some $0 \leq \varepsilon < 1$ and $d \in \mathbb{Z}_{>0}$; here $\mathfrak{I}(z)$ was defined as in Exercise 2.3.1. Prove that

$$|e_A(z)| = |z|^{q^d} / \prod_{\deg(a) < d} |a|.$$

5.2.6 Let ϕ be a Drinfeld module over $\mathbb{C}_\infty$. Verify the following statements:

(a) $\phi_a(x) = \exp_\phi(a \log_\phi(x))$.
(b) $a \log_\phi(x/a) \to x$ as $\deg(a) \to \infty$.
(c) $\displaystyle\lim_{\deg(a) \to \infty} \phi_a(x/a) = \exp_\phi(x)$.

5.2.7 Let ϕ be a Drinfeld module over $\mathbb{C}_\infty$.

(a) Prove that there are polynomials $f_{\phi,n}(y) \in \mathbb{C}_\infty[y]$, $n \geq 0$, such that

$$\phi_a(x) = \sum_{n \geq 0} f_{\phi,n}(a) x^{q^n}.$$

(b) Compute $\deg f_{\phi,n}(y)$.

5.2.8 Let ϕ be a Drinfeld module of rank 2, and let $\Lambda_\phi = A\lambda_1 + A\lambda_2$ be its corresponding lattice, where $\{\lambda_1, \lambda_2\}$ is an A-basis of Λ. Prove that $\mathrm{End}(\phi) \neq A$ if and only if $F(\lambda_1/\lambda_2)$ is an imaginary quadratic extension of F, in which case $\mathrm{End}^\circ(\phi) = F(\lambda_1/\lambda_2)$.

5.2.9 Show that there are A-lattices in $\mathbb{C}_\infty$ of infinite countable rank. Explore the possibility of defining a generalized Drinfeld module $\phi^\Lambda : A \to \mathbb{C}_\infty\{\{\tau\}\}$ associated with such a lattice Λ using its exponential function e_Λ.

5.3 Applications of Analytic Uniformization

Analytic uniformization of Drinfeld modules can be used to give alternative analytic proofs of some of the results that we proved in Chap. 3 using algebraic methods.

Before discussing the details of this approach, we remark on the "Lefschetz principle" in this setting which allows the application of analytic methods to Drinfeld modules over arbitrary fields of A-characteristic 0. Suppose ϕ is a Drinfeld module defined by $\phi_T = t + g_1\tau + \cdots + g_r\tau^r$ over a field K of A-characteristic 0. Note that t is transcendental over $\mathbb{F}_q$, so F can be identified with $\mathbb{F}_q(t)$. Then ϕ is actually defined over $K' = F(g_1, \ldots, g_r)$, which has finite transcendence degree over F. Since $\mathbb{C}_\infty$ has infinite transcendence degree over F and is algebraically closed, by general field theory, it contains a subfield isomorphic to K'. Identifying K' with such a subfield of $\mathbb{C}_\infty$, in many problems concerning ϕ, one may assume that ϕ is defined over $\mathbb{C}_\infty$.

Proposition 5.3.1 *Let ϕ be a Drinfeld module of rank r over $\mathbb{C}_\infty$, and let Λ be the lattice corresponding to ϕ. Then, for any $0 \neq a \in A$, we have*

(1) $\phi[a] \cong \Lambda/a\Lambda \cong (A/aA)^r$.
(2) $T_\mathfrak{p}(\phi) \cong \Lambda \otimes_A \Lambda_\mathfrak{p}$ for any prime $\mathfrak{p}$.
(3) $\phi[a] = \{e_\phi(\lambda/a) \mid \lambda \in \Lambda\}$.

Proof

(1) By Theorem 5.2.11, we have $\phi[a] \cong \Lambda/a\Lambda$. Since Λ is a free A-module of rank r, we have $\Lambda/a\Lambda \cong (A/aA)^r$, which implies the claim.[3]
(2) By part (1), we have $\phi[\mathfrak{p}^n] = \Lambda/\mathfrak{p}^n\Lambda \cong \Lambda \otimes_A A/\mathfrak{p}^n$. Moreover, under this isomorphism, the transition map $\phi[\mathfrak{p}^{n+1}] \xrightarrow{\phi_\mathfrak{p}} \phi[\mathfrak{p}^n]$ corresponds to multiplication by $\mathfrak{p}$ on $\Lambda \otimes_A A/\mathfrak{p}^n$. Taking inverse limits, we obtain

$$T_\mathfrak{p}(\phi) \cong \varprojlim_n (\Lambda \otimes_A A/\mathfrak{p}^n) \cong \Lambda \otimes_A \left(\varprojlim_n A/\mathfrak{p}^n \right) \cong \Lambda \otimes_A A_\mathfrak{p}.$$

(Tensor products do not always commute with inverse limits, but they do in this case because Λ is a free A-module of finite rank.)
(3) By definition, $\phi[a] = \{z \in \mathbb{C}_\infty \mid \phi_a(z) = 0\}$. Since $e_\phi \colon \mathbb{C}_\infty \to \mathbb{C}_\infty$ is surjective, for each $z \in \mathbb{C}_\infty$, there is some $\beta \in \mathbb{C}_\infty$ such that $z = e_\phi(\beta)$. Since $\phi_a(e_\phi(\beta)) = e_\phi(a \cdot \beta)$ and $Z(e_\phi) = \Lambda$, we conclude that $z \in \phi[a]$ if and only if $\beta \in \{\lambda/a \mid \lambda \in \Lambda\}$. Thus, $\phi[a] = \{e_\phi(\lambda/a) \mid \lambda \in \Lambda\}$, as was claimed.
$\square$

---

[3] The argument here might appear circular since we used $\phi[T] \cong (A/TA)^r$ in the proof of Proposition 5.2.7, but that can be avoided since in that proof we only need to know that $\phi[T]$ is finite, which is obvious.

Let ϕ be a Drinfeld module of rank r over $\mathbb{C}_\infty$, let $u \in \text{End}(\phi)$, and let $\mathfrak{p}$ be any prime. Recall that we defined (Definition 3.6.5) the characteristic polynomial of u as

$$P_u(x) = \det \left(x - u \mid T_\mathfrak{p}(\phi) \right).$$

We then proved using Anderson motives that $P_u(x)$ has coefficients in A which do not depend on the choice of $\mathfrak{p}$; see Theorem 3.6.6. In fact, we proved this over an arbitrary base field. Next, we show that there is an easier proof over $\mathbb{C}_\infty$.

Theorem 5.3.2 *With previous notation and assumptions, the polynomial $P_u(x)$ has coefficients in A which do not depend on the choice of $\mathfrak{p}$. Moreover, we have an equality of ideals*

$$(P_u(0)) = (\chi(\Lambda/(\partial u)\Lambda)),$$

where Λ is the lattice of ϕ.

Proof The action of u on $\phi[a]$ corresponds to the action of $u_0 = \partial u$ on Λ under the isomorphism $\phi[a] \cong \Lambda \otimes A/aA$. Hence $u_\mathfrak{p}$ is the matrix by which u_0 acts on $\Lambda \cong A^r$. Now it is clear that $P_u(x) = \det(x - u_0 \mid \Lambda)$ has coefficients in A which do not depend on the choice of $\mathfrak{p}$. Next,

$$(P_u(0)) = (\det(u_0 \mid \Lambda)) = (\chi(\Lambda/u_0\Lambda)),$$

where the second equality follows Lemma 1.2.6.   □

Next, we give an analytic construction of the Weil pairing discussed in Sect. 3.7. Let ϕ be a Drinfeld module of rank r over $\mathbb{C}_\infty$, and let ψ be a Drinfeld module of rank 1. Fix an A-basis $\omega_1, \ldots, \omega_r \in \mathbb{C}_\infty$ of the lattice Λ_ϕ of ϕ and an A-basis $\pi_\psi \in \mathbb{C}_\infty$ of the lattice Λ_ψ of ψ, i.e.,

$$\Lambda_\phi = A\omega_1 + \cdots + A\omega_r \quad \text{and} \quad \Lambda_\psi = A\pi_\psi.$$

For $\lambda \in \Lambda_\phi$, denote by $[\lambda]_\omega$ the column vector in A^r consisting of the coefficients of the expansion $\lambda = a_1\omega_1 + \cdots + a_r\omega_r$ of λ with respect to the basis $\{\omega_1, \ldots, \omega_r\}$. Let $0 \neq a \in A$ and $x_1, \ldots, x_r \in \phi[a]$. By Proposition 5.3.1, we can write $x_i = e_\phi(\lambda_i/a)$ for some $\lambda_i \in \Lambda$, $1 \leq i \leq r$. Define

$$W_a(x_1, x_2, \ldots, x_r) = e_\psi \left(\frac{\pi_\psi \cdot \det([\lambda_1]_\omega, \ldots, [\lambda_r]_\omega)}{a} \right).$$

Proposition 5.3.3 *We have:*

(1) W_a *does not depend on the choice of* $\lambda_i \in \Lambda$ *such that* $x_i = e_\phi(\lambda_i/a)$.
(2) *A different choice of a basis of* Λ_ϕ *or* Λ_ψ *changes* W_a *by an* $\mathbb{F}_q^\times$-*multiple.*
(3) $W_a(x_1, x_2, \ldots, x_r) \in \psi[a]$.

(4) $W_a\colon \prod_{i=1}^{r} \phi[a] \longrightarrow \psi[a]$ *is A-multilinear and alternating.*
(5) W_a *is surjective and non-degenerate.*
(6) If $a, b \in A$ *are nonzero and* $x_1, \ldots, x_r \in \phi[ab]$, *then*

$$\psi_b(W_{ab}(x_1, \ldots, x_r)) = W_a(\phi_b(x_1), \ldots, \phi_b(x_r)).$$

Before giving the proof, we remark that the rank-1 module ψ in Sect. 3.7 was a specific module constructed from ϕ, whereas in the present setting ψ is arbitrary. This is explained by the fact that all Drinfeld modules of rank 1 are isomorphic over $\mathbb{C}_\infty$, whereas this is not the case over general A-fields. The choice of ψ in Sect. 3.7 is important for the Galois equivariance of W_a, but this property is not applicable over $\mathbb{C}_\infty$.

Proof

(1) If $x_i = e_\phi(\lambda_i'/a)$, then $\lambda_i - \lambda_i' = a\mu_i$ for some $\mu_i \in \Lambda_\phi$. By the A-multilinearity of the determinant, we have

$$\det\left([\lambda_1']_\omega, \ldots, [\lambda_r']_\omega\right) = \det\left([\lambda_1]_\omega, \ldots, [\lambda_r]_\omega\right) + ac$$

for some $c \in A$. Therefore,

$$e_\psi\left(\frac{\pi_\psi \cdot \det\left([\lambda_1']_\omega, \ldots, [\lambda_r']_\omega\right)}{a}\right) = e_\psi\left(\frac{\pi_\psi \cdot \det\left([\lambda_1]_\omega, \ldots, [\lambda_r]_\omega\right)}{a}\right)$$
$$+ e_\psi(\pi_\psi c).$$

Since $e_\psi(\pi_\psi c) = 0$, the claim follows.

(2) If π_ψ' is another generator of Λ_ψ, then $\pi_\psi' = \alpha \pi_\psi$ for some $\alpha \in \mathbb{F}_q^\times$, and by the $\mathbb{F}_q$-linearity of Carlitz–Drinfeld exponential functions we have $e_\psi(\alpha x) = \alpha e_\psi(x)$. Now, let $\{\omega_1', \ldots, \omega_r'\}$ be another basis of Λ_ϕ. By linear algebra, $([\lambda_1]_{\omega'}, \ldots, [\lambda_r]_{\omega'}) = S \cdot ([\lambda_1]_\omega, \ldots, [\lambda_r]_\omega)$ for some $S \in \mathrm{GL}_r(A)$; cf. Lemma 1.2.6. Since

$$\det([\lambda_1]_{\omega'}, \ldots, [\lambda_r]_{\omega'}) = \det(S) \cdot \det([\lambda_1]_\omega, \ldots, [\lambda_r]_\omega)$$

and $\det(S) \in \mathbb{F}_q^\times$, the claim again follows from the $\mathbb{F}_q$-linearity of e_ψ.
(3) We have

$$\psi_a(W_a(x_1, x_2, \ldots, x_r)) = \psi_a e_\psi\left(\frac{\pi_\psi \cdot \det([\lambda_1]_\omega, \ldots, [\lambda_r]_\omega)}{a}\right)$$
$$= e_\psi\left(a\frac{\pi_\psi \cdot \det([\lambda_1]_\omega, \ldots, [\lambda_r]_\omega)}{a}\right)$$
$$= e_\psi\left(\pi_\psi \cdot \det([\lambda_1]_\omega, \ldots, [\lambda_r]_\omega)\right) = 0.$$

(4) W_a is alternating because the determinant is alternating. We show that W_a is A-linear in x_1; the argument for the other coordinates is similar. By definition, for $b \in A$, we have $b \circ x_1 = \phi_b(x_1) = \phi_b(e_\phi(\lambda_i/a)) = e_\phi(b\lambda_i/a)$. On the other hand,

$$e_\psi \left(\frac{\pi_\psi \cdot \det([b\lambda_1]_\omega, \ldots, [\lambda_r]_\omega)}{a} \right) = e_\psi \left(\frac{\pi_\psi \cdot \det(b[\lambda_1]_\omega, \ldots, [\lambda_r]_\omega)}{a} \right)$$

$$= e_\psi \left(b \frac{\pi_\psi \cdot \det([\lambda_1]_\omega, \ldots, [\lambda_r]_\omega)}{a} \right)$$

$$= \psi_b \left(e_\psi \left(\frac{\pi_\psi \cdot \det([\lambda_1]_\omega, \ldots, [\lambda_r]_\omega)}{a} \right) \right).$$

Thus, $W_a(b \circ x_1, x_2, \ldots, x_r) = b \circ W_a(x_1, x_2, \ldots, x_r)$.

(5) This is an easy consequence of Proposition 5.3.1 applied to the a-torsion of ψ and the fact that $\det([\lambda_1]_\omega, \ldots, [\lambda_r]_\omega)$ assumes every value in A as we vary the λ_i's in Λ_ϕ.

(6) Let $x_1, \ldots, x_r \in \phi[ab]$. Write $x_i = e_\phi(\lambda_i/ab)$ for some $\lambda_i \in \Lambda_\phi$, $1 \le i \le r$. Then $\phi_b(x_i) = \phi_b(e_\phi(\lambda_i/ab)) = e_\phi(b\lambda_i/ab) = e_\phi(\lambda_i/a)$. Thus,

$$W_a(\phi_b(x_1), \ldots, \phi_b(x_r)) = W_a(e_\phi(\lambda_1/a), \ldots, e_\phi(\lambda_r/a))$$

$$= e_\psi \left(\frac{\pi_\psi \cdot \det([\lambda_1]_\omega, \ldots, [\lambda_r]_\omega)}{a} \right)$$

$$= e_\psi \left(b \frac{\pi_\psi \cdot \det([\lambda_1]_\omega, \ldots, [\lambda_r]_\omega)}{ab} \right)$$

$$= \psi_b e_\psi \left(\frac{\pi_\psi \cdot \det([\lambda_1]_\omega, \ldots, [\lambda_r]_\omega)}{ab} \right)$$

$$= \psi_b(W_{ab}(x_1, \ldots, x_r)).$$

$\square$

Now we return to the endomorphism rings of Drinfeld modules, armed with analytic tools for their study. First, we give an analytic proof Corollary 3.4.15.

Proposition 5.3.4 *Let ϕ be a Drinfeld module of rank r over $\mathbb{C}_\infty$. Then $\mathrm{End}^\circ(\phi)$ is a totally imaginary field extension of F of degree dividing r.*

Proof Let Λ be the lattice corresponding to ϕ. By Theorem 5.2.11, we have

$$\mathrm{End}(\phi) \cong \{c \in \mathbb{C}_\infty \mid c\Lambda \subseteq \Lambda\}.$$

Let $F\Lambda$ be the F-vector subspace of $\mathbb{C}_\infty$ spanned by Λ. Then $F\Lambda$ has dimension r over F because a linear relation $\alpha_1\lambda_1 + \cdots + \alpha_r\lambda_r = 0$ with $\lambda_1, \ldots, \lambda_r \in \Lambda$ and $\alpha_1, \ldots, \alpha_r \in F$ can be scaled to give a linear relation over A. It is also not hard to check that

$$\text{End}^\circ(\phi) \cong \{c \in \mathbb{C}_\infty \mid c(F\Lambda) \subseteq F\Lambda\}.$$

(We leave this as an exercise.) Suppose $c \neq 0$ and $c(F\Lambda) \subseteq F\Lambda$. Clearly, multiplication by c gives an injective F-linear map $F\Lambda \to F\Lambda$. Since $F\Lambda$ is finite dimensional, we conclude that $c(F\Lambda) = F\Lambda$. Hence, every nonzero element of $\text{End}^\circ(\phi)$ is invertible. This implies that $L = \text{End}^\circ(\phi)$ is a field extension of F which embeds into $\mathbb{C}_\infty$. Observe that $F\Lambda$ is a vector space over L. Hence,

$$r = \dim_F F\Lambda = [L : F] \cdot \dim_L F\Lambda,$$

so $[L : F]$ divides r.

It remains to show that L is totally imaginary. After replacing ϕ by an isomorphic module, we may assume that $1 \in \Lambda$ (see Corollary 5.2.12). Then $L \subseteq F\Lambda$. Choose a basis of L as an F-vector space, and extend it to a basis $\{\omega_1, \ldots, \omega_r\}$ of $F\Lambda$. Proposition 5.3.8 implies that the elements $\{\omega_1, \ldots, \omega_r\}$ are linearly independent over F_∞. Hence, $[F_\infty L : F_\infty] = [L : F]$. On the other hand, by Theorem 1.5.19, $L \cong F[x]/(f(x))$ for some irreducible polynomial $f(x) \in F[x]$. The equality $[F_\infty L : F_\infty] = [L : F]$ implies that $f(x)$ remains irreducible over F_∞; thus, there is a unique place in L over ∞.   □

Example 5.3.5 Let us specialize the previous proposition and its proof to $r = 2$. After replacing ϕ by an isomorphic Drinfeld module, we may assume that $\Lambda = A + Az$, where $z \notin F_\infty$. Assume $\text{End}(\phi)$ is strictly larger than A. Let $u \notin A$ be an endomorphism of ϕ. Then $u = a + bz$ and $uz = c + dz$ for some $a, b, c, d \in A$ such that $b \neq 0$. This implies that z is a root of the quadratic equation $bx^2 + (a-d)x - c = 0$, and moreover $F(z) = F(u)$. Hence, $F(u)$ is a quadratic extension of F, which is imaginary since $z \notin F_\infty$. Since u was an arbitrary element of $\text{End}(\phi)$ not in A, we see that $\text{End}^\circ(\phi) = F(u) = F + Fz = F\Lambda$ is an imaginary quadratic extension of F.

The analytic uniformization of Drinfeld modules also allows us to give a constructive converse of Proposition 5.3.4 in the following sense: given an order O in a totally imaginary field extension L/F of degree dividing r, one could ask whether there exists a Drinfeld module ϕ of rank r over $\mathbb{C}_\infty$ with $\text{End}(\phi) = O$. This was not addressed in Chap. 3. In fact, it is difficult to give a positive answer to this question using purely algebraic methods. The analytic approach on the other hand provides a natural setting for a constructive proof.

Example 5.3.6 Assume L/F is a totally imaginary extension with $[L : F] = r$. Let O be an A-order in L. Then

$$O = A + A\omega_2 + \cdots + A\omega_r,$$

for some $1, \omega_2, \ldots, \omega_r \in O$ which form a basis of L as an F-vector space. Fix an embedding $L \hookrightarrow \mathbb{C}_\infty$. Because L/F is totally imaginary, as in the proof of Proposition 5.3.4, the elements $1, \omega_2, \dots, \omega_r$ are linearly independent over F_∞. Hence, by Proposition 5.3.8, O is a lattice in $\mathbb{C}_\infty$. Since O is a ring, we obviously have $O \subseteq \operatorname{End}(O) := \operatorname{Hom}(O, O)$, where $\operatorname{Hom}(O, O)$ is defined as in Definition 5.2.10. On the other hand, if $c \in \mathbb{C}_\infty$ belongs to $\operatorname{End}(O)$, then $c = c \cdot 1 \in O$. Thus $\operatorname{End}(O) = O$. By Theorem 5.2.11, for the Drinfeld module ϕ of rank r corresponding to the lattice O, we have

$$\operatorname{End}(\phi) \cong \operatorname{End}(O) = O.$$

Example 5.3.7 At the other extreme, suppose we want to construct a lattice Λ of rank $r \geq 1$ with $\operatorname{End}(\Lambda) = A$. The field $\mathbb{C}_\infty$ has infinite transcendence degree over F_∞, so we can choose $\omega_2, \dots, \omega_r \in \mathbb{C}_\infty$ such that $F_\infty(\omega_2, \dots, \omega_r)/F_\infty$ is a purely transcendental extension of transcendence degree $r - 1$. By Proposition 5.3.8,

$$\Lambda = A + A\omega_2 + \cdots + A\omega_r$$

is a lattice of rank r. Let $c \in \operatorname{End}(\Lambda)$. On the one hand, $c \cdot 1 = c \in \Lambda \subset F(\omega_2, \dots, \omega_r)$. On the other hand, by the Cayley–Hamilton theorem, c is a root of a monic polynomial of degree r with coefficients in A, so it is integral over A. Since the elements of A are the only elements of Λ which are algebraic over F, we conclude that $\operatorname{End}(\Lambda) = A$.

The final problem that we readdress using analytic methods is the problem of classifying Drinfeld modules up to isomorphism that we discussed in Sect. 3.8. We call two lattices $\Lambda, \Lambda' \subset \mathbb{C}_\infty$ *homothetic* if $\Lambda' = c\Lambda$ for some $c \in \mathbb{C}_\infty^\times$. By Corollary 5.2.12, the isomorphism classes of Drinfeld modules of rank r over $\mathbb{C}_\infty$ are in bijection with the homothety classes of lattices of rank r in $\mathbb{C}_\infty$. Hence, using the analytic uniformization, we reinterpret our problem as the problem of classifying lattices of rank r in $\mathbb{C}_\infty$, up to homothety. To proceed with this approach, one needs to understand how lattices in $\mathbb{C}_\infty$ are formed.

The obvious way to construct an A-submodule of $\mathbb{C}_\infty$ of finite rank is to fix some elements $v_1, \dots, v_r \in \mathbb{C}_\infty$ and take the span

$$\Lambda = Av_1 + \cdots + Av_r.$$

Obviously, Λ will have rank r over A if and only if $v_1, \dots, v_r$ are linearly independent over F, i.e.,

$$a_1 v_1 + \cdots + a_r v_r = 0 \quad \text{with} \quad a_1, \dots, a_r \in F$$

if and only if $a_1 = \cdots = a_r = 0$. But besides this restriction, there should be some other condition related to the discreteness of Λ in $\mathbb{C}_\infty$. This problem arises also for the usual $\mathbb{Z}$-lattices in $\mathbb{C}$; for example, 1 and $\pi \approx 3.14$ are linearly independent over

Q, but $\mathbb{Z} + \mathbb{Z}\pi$ is not a lattice since it is not discrete.[4] It is well-known, and not hard to prove, that, given two complex numbers v_1 and v_2, their $\mathbb{Z}$-span is a lattice in $\mathbb{C}$ of rank 2 if and only if v_1 and v_2 are linearly independent over $\mathbb{R}$. Since $\mathbb{C}$ has dimension 2 over $\mathbb{R}$, this also implies that there are no lattices in $\mathbb{C}$ of rank greater than 2. It turns out that the same principle applies to lattices in $\mathbb{C}_\infty$.

Proposition 5.3.8 *Given $v_1, \ldots, v_r \in \mathbb{C}_\infty$, their A-span*

$$\Lambda = Av_1 + \cdots + Av_r$$

is a lattice of rank r if and only if $v_1, \ldots, v_r$ are linearly independent over F_∞.

Proof Suppose Λ is a lattice of rank r. Let $V = F_\infty v_1 + \cdots + F_\infty v_r$. Since Λ is discrete in $\mathbb{C}_\infty$, Λ is a discrete A-submodule of V. By Lemma 3.4.9,

$$r = \text{rank}_A \, \Lambda \leq \dim_{F_\infty} V \leq r.$$

Therefore, $\dim_{F_\infty} V = r$, so $v_1, \ldots, v_r$ are linearly independent over F_∞.

Conversely, assume $v_1, \ldots, v_r$ are linearly independent over F_∞. Then $v_1, \ldots, v_r$ are automatically linearly independent over F, so Λ is a free A-module of rank r. Suppose, contrary to the claim of the proposition, Λ is not a lattice, i.e., Λ is not discrete. This means that there is $n > 0$ such that

$$\Lambda_n = \{\lambda \in \Lambda \, : \, |\lambda| \leq n\}$$

is infinite. Thus, there is an infinite sequence

$$\lambda_s = a_1^{(s)} v_1 + \cdots + a_r^{(s)} v_r, \qquad a_i^{(s)} \in A,$$

of distinct elements in Λ all with $|\lambda_s| \leq n$. Since the set $\{\lambda_s\}$ is infinite, we can choose a subsequence which, after possibly reindexing $v_1, \ldots, v_r$, satisfies $\left| a_1^{(s)} \right| \geq \left| a_i^{(s)} \right|$ for all s and $2 \leq i \leq r$. Without loss of generality, we assume that the original sequence $\{\lambda_s\}$ has this property. Now, consider

$$\frac{\lambda_s}{a_1^{(s)}} = v_1 + b_2^{(s)} v_2 + \cdots + b_r^{(s)} v_r. \tag{5.3.1}$$

Since $|\lambda_s| \leq n$ is bounded whereas $\left| a_1^{(s)} \right| \to \infty$ (as there are only finitely many elements in A with $|a| < N$ for any fixed N), we get $\lambda_s / a_1^{(s)} \to 0$. On the other hand, $b_2^{(s)}, \ldots, b_r^{(s)}$ are in A_∞, which is compact. Hence, we can choose a

---

[4] In fact, $\mathbb{Z} + \mathbb{Z}\pi$ is dense in $\mathbb{R}$.

subsequence of $\left\{\lambda_s / a_1^{(s)}\right\}_{s \geq 0}$ such that the sequences $\left\{b_i^{(s)}\right\}_{s \geq 0}$ are convergent in F_∞ for all $2 \leq i \leq r$. Denote $b_i = \lim_{s \to \infty} b_i^{(s)}$. Taking $s \to \infty$ in (5.3.1), we get

$$0 = v_1 + b_2 v_2 + \cdots + b_r v_r,$$

where $b_2, \ldots, b_r \in F_\infty$. This implies that $v_1, \ldots, v_r$ are linearly dependent over F_∞, contrary to our assumption. $\square$

Corollary 5.3.9 *For any $r \geq 1$, there is a lattice in $\mathbb{C}_\infty$ of rank r.*

Proof This follows from the previous proposition and the fact that $\mathbb{C}_\infty$ has infinite degree over F_∞, so for any $r \geq 1$ it is possible to choose $v_1, \ldots, v_r \in \mathbb{C}_\infty$ which are linearly independent over F_∞. Of course, the same conclusion can be reached using Theorem 5.2.8 since there are obviously Drinfeld modules over $\mathbb{C}_\infty$ of arbitrary rank r. $\square$

Let

$$S = \{(c_1, \ldots, c_r) \mid c_1, \ldots, c_r \in \mathbb{C}_\infty \text{ not all } 0\}.$$

Define an equivalence relation $\sim$ on S by declaring two r-tuples $(c_1, \ldots, c_r)$, $(c_1', \ldots, c_r')$ as being equivalent if there exists $c \in \mathbb{C}_\infty^\times$ such that $c_i' = c c_i$ for all $1 \leq i \leq r$. By definition, the $(r-1)$-dimensional projective space $\mathbb{P}^{r-1}(\mathbb{C}_\infty) = S / \sim$ is the quotient of S by this equivalence relation. Denote the image of $(c_1, \ldots, c_r)$ in $\mathbb{P}^{r-1}(\mathbb{C}_\infty)$ by $[c_1, \ldots, c_r]$. An F_∞-*rational hyperplane* in $\mathbb{P}^{r-1}(\mathbb{C}_\infty)$ is the set of solutions of a linear equation

$$\alpha_1 x_1 + \cdots + \alpha_r x_r = 0, \quad \text{where } \alpha_1, \ldots, \alpha_r \in F_\infty \text{ are not all zero.}$$

The *Drinfeld symmetric space* is

$$\Omega^r = \mathbb{P}^{r-1}(\mathbb{C}_\infty) - \bigcup_{F_\infty\text{-rational}} H,$$

where the union is over the F_∞-rational hyperplanes; equivalently, Ω^r is the set of all $[c_1, \ldots, c_r]$ such that $c_1, \ldots, c_r$ are linearly independent over F_∞.

Note that $\mathrm{GL}_r(F_\infty)$ acts on S from the left as on column vectors in $\mathbb{C}_\infty^r$ (or from the right as on row vectors). This action is $\mathbb{C}_\infty$-linear, so it preserves the equivalence relation on S that we defined above. Therefore, $\mathrm{GL}_r(F_\infty)$ also naturally acts on $\mathbb{P}^{r-1}(\mathbb{C}_\infty)$. Write $\vec{c} = (c_1, \ldots, c_r)$. For $M \in \mathrm{GL}_r(F_\infty)$ and two vectors $\vec{\alpha}$ and $\vec{c}$ in $\mathbb{C}_\infty^r$, we have

$$\vec{\alpha}^\top \cdot \vec{c} := \sum_{i=1}^r \alpha_i c_i = \left((M^\top)^{-1} \vec{\alpha}\right)^\top \cdot (M \vec{c}).$$

Obviously, $\vec{\alpha} \in F_\infty^r$ if and only if $(M^\top)^{-1}\vec{\alpha} \in F_\infty^r$. Therefore, the action of $\mathrm{GL}_r(F_\infty)$ on $\mathbb{P}^{r-1}(\mathbb{C}_\infty)$ preserves Ω^r.

Let $r \geq 2$ and $z = [z_1, \ldots, z_r] \in \Omega^r$. Because $z_1, \ldots, z_r$ are linearly independent over F_∞, we can normalize the projective coordinates of z by assuming $z_r = 1$. With this normalization, the group $\mathrm{GL}_r(F_\infty)$ acts on Ω^r by linear fractional transformations: for $\gamma = (\gamma_{i,j}) \in \mathrm{GL}_r(F_\infty)$ and $z = [z_1, \ldots, z_{r-1}, z_r = 1] \in \Omega^r$, we have

$$
\gamma z = \left[\frac{\gamma_{1,1}z_1 + \cdots + \gamma_{1,r-1}z_{r-1} + \gamma_{1,r}}{\gamma_{r,1}z_1 + \cdots + \gamma_{r,r-1}z_{r-1} + \gamma_{r,r}}, \right.
$$
$$
\frac{\gamma_{2,1}z_1 + \cdots + \gamma_{2,r-1}z_{r-1} + \gamma_{2,r}}{\gamma_{r,1}z_1 + \cdots + \gamma_{r,r-1}z_{r-1} + \gamma_{r,r}}, \ldots,
$$
$$
\left. \frac{\gamma_{r-1,1}z_1 + \cdots + \gamma_{r-1,r-1}z_{r-1} + \gamma_{r-1,r}}{\gamma_{r,1}z_1 + \cdots + \gamma_{r,r-1}z_{r-1} + \gamma_{r,r}}, 1 \right].
$$

(Note that scalar matrices act trivially on Ω^r.)

Let $\Lambda = A\lambda_1 + \cdots + A\lambda_r \subset \mathbb{C}_\infty$ be a lattice of rank r. Since Λ is discrete in $\mathbb{C}_\infty$, the elements $\lambda_1, \ldots, \lambda_r$ are linearly independent over F_∞. So if we put $z_i = \lambda_i/\lambda_r$, $1 \leq i \leq r$, then $z = [z_1, \ldots, z_r = 1] \in \Omega^r$.

Theorem 5.3.10 *Let ϕ^Λ and $\phi^{\Lambda'}$ be Drinfeld modules of rank r over $\mathbb{C}_\infty$, which correspond to $\Lambda = A\lambda_1 + \cdots + A\lambda_r$ and $\Lambda' = A\lambda_1' + \cdots + A\lambda_r'$, respectively. For these lattices, define $z, z' \in \Omega^r$ as above. Then, the following are equivalent:*

(1) ϕ^Λ is isogenous (respectively, isomorphic) to $\phi^{\Lambda'}$.
(2) There exists an element $\gamma \in \mathrm{GL}_r(F)$ (respectively, $\mathrm{GL}_r(A)$) such that $\gamma z = z'$.

Proof

(1) $\Rightarrow$ (2): Assume ϕ^Λ is isogenous to $\phi^{\Lambda'}$. By Theorem 5.2.11, there exists $c \in \mathbb{C}_\infty^\times$ such that $c\Lambda' \subseteq \Lambda$. Then, for each $1 \leq i \leq r$, we can write

$$
c\lambda_i' = \sum_{j=1}^r a_{i,j}\lambda_j, \quad \text{for some } a_{i,j} \in A.
$$

Let $\gamma = (a_{i,j})$. The previous equations imply that $z' = \gamma z$. Note that the F-span of $\{c\lambda_1', \ldots, c\lambda_r'\}$ and $\{\lambda_1, \ldots, \lambda_r\}$ in $\mathbb{C}_\infty$ is the same F-vector subspace of $\mathbb{C}_\infty$ since both sets are linearly independent over F. This implies that $\det(\gamma) \neq 0$. Hence $\gamma \in \mathrm{GL}_r(F)$.

If ϕ^Λ is isomorphic to $\phi^{\Lambda'}$, then there exists $c \in \mathbb{C}_\infty^\times$ such that $c\Lambda' = \Lambda$. Now one can use Lemma 1.2.6 to conclude that $\det(a_{i,j}) \in \mathbb{F}_q^\times$, i.e., $\gamma \in \mathrm{GL}_r(A)$.

(2) $\Rightarrow$ (1): Conversely, suppose $\gamma z = z'$ for some $\gamma \in \mathrm{GL}_r(F)$. After replacing Λ and Λ' by homothetic lattices, we may assume that $\Lambda = Az_1 + \cdots + Az_r$ and $\Lambda' = Az_1' + \cdots + Az_r'$, where $z = [z_1, \ldots, z_r = 1]$, $z' = [z_1', \ldots, z_r' = 1]$.

We can write $\gamma = d^{-1}(a_{i,j})$ for some $d, a_{i,j} \in A$. Then $\gamma z = z'$ if and only if $(a_{i,j})z = z'$. Hence we may also assume that $\gamma = (a_{i,j})$ has entries in A. Let

$$c := a_{r,1}z_1 + \cdots + a_{r,r-1}z_{r-1} + a_{r,r}.$$

Now the assumption $\gamma z = z'$ translates into

$$cz'_i = a_{i,1}z_1 + \cdots + a_{i,r-1}z_{r-1} + a_{i,r} \quad \text{for all } 1 \leq i \leq r. \tag{5.3.2}$$

Hence, $c\Lambda' \subseteq \Lambda$. By Theorem 5.2.11, this implies that ϕ^Λ is isogenous to $\phi^{\Lambda'}$. If $\gamma \in \mathrm{GL}_r(A)$, then (5.3.2) and Lemma 1.2.6 imply that $c\Lambda' = \Lambda$, so ϕ^Λ and $\phi^{\Lambda'}$ are isomorphic.

$\square$

Corollary 5.3.11 *The set of isomorphism classes of Drinfeld modules of rank r over $\mathbb{C}_\infty$ is in natural bijection with the set of orbits $\mathrm{GL}_r(A) \setminus \Omega^r$.*

Remarks 5.3.12 The quotient $\mathrm{GL}_r(A) \setminus \Omega^r$ is much more than just a set. In [Dri74], Drinfeld showed that $\mathrm{GL}_r(A) \setminus \Omega^r$ has a structure of an analytic manifold over F_∞. Moreover, he proved that $\mathrm{GL}_r(A) \setminus \Omega^r$ is algebraizable, in the sense that it is the analytic space corresponding to an affine algebraic variety over F_∞. For more about analytic geometry over non-Archimedean fields, the reader might consult [FvdP04].

For the rest of this section, we specialize to $r = 2$. In this case, $\mathbb{P}^{r-1}(\mathbb{C}_\infty) = \mathbb{P}^1(\mathbb{C}_\infty)$ consists of $[1,0]$ and $[z,1]$, $z \in \mathbb{C}_\infty$. It is easy to see that

$$\Omega^2 = \{[z,1] \mid z \notin F_\infty\}$$
$$= \mathbb{C}_\infty - F_\infty.$$

After identifying Ω^2 with $\mathbb{C}_\infty - F_\infty$, $\mathrm{GL}_2(F_\infty)$ acts on Ω^2 by linear fractional transformations

$$\begin{pmatrix} a & b \\ c & d \end{pmatrix} z = \frac{az+b}{cz+d}.$$

To a point $z \in \Omega^2$, we associate the lattice $\Lambda_z = A + Az$ of rank 2. As in Definition 5.1.7, define the Eisenstein series

$$E_k(z) = E_k(\Lambda_z) = {\sum_{n,m \in A}}' \frac{1}{(m+nz)^k},$$

which is an entire function on Ω^2. Furthermore, let $\phi^z := \phi^{\Lambda_z}$ be the Drinfeld module associated with Λ_z. From Example 5.2.6, we have

$$\phi_T^z = T + g_1(z)\tau + g_2(z)\tau^2,$$

where

$$g_1(z) = (T^q - T)E_{q-1}(z),$$ (5.3.3)

$$g_2(z) = (T^{q^2} - T)E_{q^2-1}(z) + (T^{q^2} - T^q)E_{q-1}(z)^{q+1}.$$

Lemma 5.3.13 *Let* $\gamma = \begin{pmatrix} a & b \\ c & d \end{pmatrix} \in GL_2(A).$ *Then*

$$E_k(\gamma z) = (cz + d)^k E_k(z).$$

Proof We have

$$E_k(\gamma z) = \sideset{}{'}\sum_{n,m \in A} \frac{1}{\left(m + n\frac{az+b}{cz+d}\right)^k}$$

$$= (cz+d)^k \sideset{}{'}\sum_{n,m \in A} \frac{1}{((md+nb) + (mc+na)z)^k}.$$

The assumption that $\gamma \in GL_2(A)$ is equivalent to the assumption that the rows of γ form an A-basis of $A \oplus A$. Hence, for any $(m', n') \in A \oplus A$, there are unique $m, n \in A$ such that

$$(m', n') = m(c, d) + n(a, b) = (mc + na, md + nb).$$

Hence, the above sum can be rewritten as

$$(cz+d)^k \sideset{}{'}\sum_{m',n' \in A} \frac{1}{(m' + n'z)^k} = (cz+d)^k E_k(z).$$

$\square$

It easily follows from the previous lemma and (5.3.3) that

$$g_1(\gamma z) = (cz + d)^{q-1} g_1(z), \qquad g_2(\gamma z) = (cz + d)^{q^2-1} g_2(z).$$

Hence,

$$j(z) = g_1(z)^{q+1}/g_2(z)$$

is $GL_2(A)$-invariant. In fact, a stronger statement holds:

Proposition 5.3.14 *For* $z, z' \in \Omega^2$, *we have* $j(z) = j(z')$ *if and only if* $z' = \gamma z$ *for some* $\gamma \in GL_2(A).$

Proof Note that $j(z)$ is the j-invariant of ϕ^{Λ_z}. Since $\mathbb{C}_\infty$ is algebraically closed, $j(z) = j(z')$ if and only if $\phi^{\Lambda_z} \cong \phi^{\Lambda_{z'}}$. The proposition now follows from Theorem 5.3.10.   □

Since for any $c \in \mathbb{C}_\infty$ there is a Drinfeld module of rank 2 over $\mathbb{C}_\infty$ with j-invariant c, we conclude that the map $j \colon \Omega^2 \to \mathbb{C}_\infty$ is surjective and induces a bijection

$$j \colon \mathrm{GL}_2(A) \setminus \Omega^2 \xrightarrow{\sim} \mathbb{C}_\infty. \tag{5.3.4}$$

Remarks 5.3.15 The bijection (5.3.4) is strikingly similar to the parametrization of complex elliptic curves up to isomorphism. Note that Ω^2 is the function field analogue of $\mathbb{C} - \mathbb{R} = \mathbb{H}^\pm$, which consists of the upper and lower half-planes:[5]

$$\mathbb{H}^+ = \{z \in \mathbb{C} \mid \Im(z) > 0\},$$

$$\mathbb{H}^- = \{z \in \mathbb{C} \mid \Im(z) < 0\}.$$

The group $\mathrm{GL}_2(\mathbb{R})$ acts on $\mathbb{H}^\pm$ by linear fractional transformations

$$\begin{pmatrix} a & b \\ c & d \end{pmatrix} z = \frac{az + b}{cz + d}.$$

Since $\begin{pmatrix} 0 & 1 \\ 1 & 0 \end{pmatrix}$ switches $\mathbb{H}^+$ and $\mathbb{H}^-$, we have $\mathrm{GL}_2(\mathbb{Z}) \setminus \mathbb{H}^\pm = \mathrm{SL}_2(\mathbb{Z}) \setminus \mathbb{H}$, where $\mathbb{H} := \mathbb{H}^+$. Now, the classical j-function

$$j(z) = e^{-2\pi i z} + 744 + 196884 \cdot e^{2\pi i z} + \cdots$$

gives a complex analytic isomorphism of Riemann surfaces

$$j \colon \mathrm{SL}_2(\mathbb{Z}) \setminus \mathbb{H} \xrightarrow{\sim} \mathbb{C}.$$

Explicitly, to $z \in \mathbb{H}$, we associate the lattice $\Lambda_z = \mathbb{Z} + \mathbb{Z}z$, to this lattice we associate the elliptic curve $E_z = \mathbb{C}/\Lambda_z$, and $j(z) = j(E_z)$ is the j-invariant of this elliptic curve.

Finally, one can show that the region

$$\mathcal{F} = \{z \in \mathbb{H} : |\Re(z)| \leq 1/2, |z| \geq 1\} \tag{5.3.5}$$

is a "fundamental domain" for the action of $\mathrm{SL}_2(\mathbb{Z})$ on $\mathbb{H}$, in the sense that $\mathcal{F}$ contains a point from each orbit of the action of $\mathrm{SL}_2(\mathbb{Z})$ on $\mathbb{H}$, and this point is

---

[5] Unlike $\mathbb{H}^\pm$, the Drinfeld upper half-plane Ω^2 is connected as a rigid-analytic manifold.

unique if it lies in the interior of $\mathcal{F}$; cf. [Ser73, p. 77]. Thus, the restriction of j to $\mathcal{F}$ is essentially a bijection. The analogue of such a domain for the action of $GL_r(A)$ on Ω^r is described in Exercise 5.5.7.

Exercises

5.3.1 Let ϕ be a Drinfeld module over $\mathbb{C}_\infty$ and let Λ be its lattice. Prove that

$$\text{End}^\circ(\phi) \cong \{c \in \mathbb{C}_\infty \mid c(F\Lambda) \subseteq F\Lambda\},$$

as was claimed in the proof of Proposition 5.3.4.

5.3.2 Assume $\text{End}(\phi) = O$ is the integral closure of A in an imaginary quadratic extension K of F. Let $I \lhd O$ be a nonzero ideal. Let $I^{-1} = \{\alpha \in K \mid \alpha I \subseteq O\}$ be the inverse of I in the group of fractional ideals of O. Define

$$\phi[I] = \{x \in \mathbb{C}_\infty \mid i(x) = 0 \text{ for all } i \in I\},$$

$$I^{-1}\Lambda = \{a_1\lambda_1 + \cdots + a_n\lambda_n \mid a_i \in I^{-1}, \lambda_i \in \Lambda, n \in \mathbb{Z}_{>0}\}.$$

Show that

(a) $\phi[I]$ is a submodule of $^\phi\mathbb{C}_\infty$ of finite order.
(b) $\phi[I] \cong I^{-1}\Lambda/\Lambda$, where Λ is the lattice of ϕ.
(c) $\phi/\phi[I] \cong \mathbb{C}_\infty/I^{-1}\Lambda$, i.e., the lattice corresponding to the Drinfeld module ψ obtained through an isogeny $\phi \to \psi$ with kernel $\phi[I]$ is homothetic to $I^{-1}\Lambda$; cf. Proposition 3.3.11.

5.3.3 Let O be an A-order in a totally imaginary extension L of F of degree dividing r. Combine the constructions in Examples 5.3.6 and 5.3.7 to show that there is a lattice $\Lambda \subset \mathbb{C}_\infty$ of rank r such that $\text{End}(\Lambda) = O$.

5.3.4 Let $r \geq 2$. For $z = [z_1, \ldots, z_r = 1] \in \Omega^r$, denote $\Lambda_z = Az_1 + \cdots + Az_r$ and

$$E_k(z) = \sum_{\lambda \in \Lambda_z}' \lambda^{-k},$$

$$\phi_T^{\Lambda_z}(x) = Tx + g_1(z)x^q + \cdots + g_r(z)x^{q^r},$$

$$e_{\Lambda_z}(x) = \sum_{n \geq 0} e_n(z)x^{q^n}.$$

Prove that for $\gamma \in \mathrm{GL}_r(A)$, we have

$$E_k(\gamma z) = j(\gamma, z)^k E_k(z),$$

$$g_n(\gamma z) = j(\gamma, z)^{q^n - 1} g_n(z), \quad 1 \leq n \leq r,$$

$$e_n(\gamma z) = j(\gamma, z)^{q^n - 1} e_n(z), \quad 1 \leq n,$$

where

$$j(\gamma, z) = \sum_{j=1}^{r} \gamma_{r,j} z_j.$$

This type of functions on Ω^r are called (weak) Drinfeld modular forms; cf. [BBP21].

5.3.5 Assume $r \geq 2$. Let $z = [z_1, \ldots, z_r = 1] \in \Omega^r$, $\Lambda = A z_1 + \cdots + A z_r$, and $\phi = \phi^\Lambda$. Prove that $\mathrm{End}(\phi)$ is strictly larger than A if and only if there exists a non-scalar $\gamma \in \mathrm{GL}_r(F)$ such that $\gamma z = z$.

5.3.6 We say that a linear form $\ell(x_1, \ldots, x_r) = \sum_{i=1}^{r} \alpha_i x_i$ on F_∞^r is *unimodular* if

$$\max_{1 \leq i \leq r} |a_i| = 1.$$

For each F_∞-rational hyperplane $H \subset \mathbb{P}^{r-1}(\mathbb{C}_\infty)$, choose a unimodular linear form ℓ_H that defines it. Let $\|\cdot\|$ be the sup-norm on $\mathbb{C}_\infty^r$ with respect to the standard basis, i.e.,

$$\|(c_1, \ldots, c_r)\| = \max_{1 \leq i \leq r} |c_i|.$$

For $z = [z_1, \ldots, z_r] \in \Omega^r$, define

$$\Im(z) = \frac{1}{\|z\|} \cdot \inf\{|\ell_H(z)| : H \text{ is } F_\infty\text{-rational hyperplane}\}.$$

This measures the distance from z to the boundary of Ω^r in $\mathbb{P}^{r-1}(\mathbb{C}_\infty)$; cf. Exercise 2.3.1.

(a) Show that $\Im(z)$ is well-defined, i.e., does not depend on the choice of ℓ_H and a representative of z in $\mathbb{C}_\infty^r$.

(b) Assume $r = 2$, $z = [z, 1]$, and $\gamma = \begin{pmatrix} a & b \\ c & d \end{pmatrix} \in \mathrm{GL}_2(F_\infty)$. Show that

$$\Im(\gamma z) = \frac{|\det(\gamma)|}{|j(\gamma, z)|^2} \cdot \Im(z).$$

(c) Prove that for any $\gamma \in \mathrm{GL}_r(F_\infty)$, there is a constant c_γ depending only on γ such that for all $z \in \Omega^r$ we have

$$\Im(\gamma z) \geq c_\gamma \cdot \Im(z).$$

5.4 Carlitz Module and Zeta-Values

The goal of this section is to give explicit formulas for the coefficients of the Carlitz module, the coefficients of its exponential and logarithmic functions, as well as explicit formulas for the generator of its lattice. All these results go back to Carlitz [Car35, Car38], whose motivation was to construct over A the analogues of the classical exponential function and the natural logarithm and to derive from this a formula for $\sum_{a \in A_+} a^{-n}$ similar to Euler's formula for the values of the Riemann zeta function at even integers.

For $n \in \mathbb{Z}_{>0}$, denote

$$[n] = T^{q^n} - T, \tag{5.4.1}$$

$$L_0 = 1, \qquad L_n = [n][n-1]\cdots[1],$$

$$D_0 = 1, \qquad D_n = [n][n-1]^q \cdots [1]^{q^{n-1}}.$$

Let C be the Carlitz module over C_∞; recall that this is the Drinfeld module of rank 1 defined by $C_T = T + \tau$. Let $e_C \in C_\infty\{\{\tau\}\}$ be the exponential function of C and let $\log_C \in C_\infty\{\{\tau\}\}$ be the multiplicative inverse of e_C.

Proposition 5.4.1 *We have*

$$e_C = \sum_{n \geq 0} \frac{1}{D_n} \tau^n,$$

$$\log_C = \sum_{n \geq 0} \frac{(-1)^n}{L_n} \tau^n.$$

Proof Recall that $e_C = \sum_{n \geq 0} a_n \tau^n$ is uniquely determined by $a_0 = 1$ and the functional equation $e_C T = C_T e_C$. This leads to the recursive formula $a_n = a_{n-1}^q / [n]$ for the coefficients, which then easily implies that $a_n = D_n^{-1}$ for all $n \geq 0$.

Next, if we put $\log_C = \sum_{n \geq 0} b_n \tau^n$, then the functional equation (5.2.1),

$$T \log_C = \log_C C_T,$$

implies that $b_n = -b_{n-1}/[n]$. Since $b_0 = 1$, we deduce that $b_n = (-1)^n/L_n$ for all $n \geq 0$. □

For $n \in \mathbb{Z}_{>0}$, let

$$A_{<n} = \{a \in A \mid \deg(a) \leq n - 1\},$$

and

$$f_{<n}(x) = \prod_{a \in A_{<n}} (x - a) \in A[x].$$

Lemma 5.4.2 *We have*

$$f_{<n}(x) = \sum_{i=0}^{n} (-1)^{n-i} \frac{D_n}{D_i L_{n-i}^{q^i}} x^{q^i}.$$

Proof We consider $A_{<n}$ as an $\mathbb{F}_q$-vector subspace of F of dimension n and choose

$$\{1, T, \ldots, T^{n-1}\}$$

as a basis of $A_{<n}$. By Corollary 3.1.19, we can express $f_{<n}(x)$ in terms of the Moore determinant

$$f_{<n}(x) = M(1, T, \ldots, T^{n-1}, x)/M(1, T, \ldots, T^{n-1}). \tag{5.4.2}$$

Now note that

$$M(1, T, \ldots, T^{n-1}) = \det \begin{pmatrix} 1 & 1 & \cdots & 1 \\ T & T^q & \cdots & T^{q^{n-1}} \\ T^2 & (T^q)^2 & \cdots & (T^{q^{n-1}})^2 \\ \vdots & \vdots & \cdots & \vdots \\ T^{n-1} & (T^q)^{n-1} & \cdots & (T^{q^{n-1}})^{n-1} \end{pmatrix}$$

is a Vandermonde determinant. From the formula in Remark 3.1.20, we get

$$M(1, T, \ldots, T^{n-1}) = \prod_{0 \leq j < i \leq n-1} (T^{q^i} - T^{q^j}).$$

Since

$$D_i = (T^{q^i} - T)(T^{q^i} - T^q) \cdots (T^{q^i} - T^{q^{i-1}}),$$

we see that

$$M(1, T, \ldots, T^{n-1}) = D_1 \cdots D_{n-1}. \tag{5.4.3}$$

Next, expanding the determinant $M(1, T, \ldots, T^{n-1}, x)$ along its last row, one obtains

$$M(1, T, \ldots, T^{n-1}, x) = \sum_{i=0}^{n} (-1)^{n-i} \theta_i \cdot x^{q^i}, \qquad (5.4.4)$$

where

$$\theta_i = \det \begin{pmatrix} 1 & 1 & \cdots & 1 & \cdots & 1 \\ T & T^q & \cdots & T^{q^i} & \cdots & T^{q^n} \\ T^2 & (T^q)^2 & \cdots & (T^{q^i})^2 & \cdots & (T^{q^n})^2 \\ \vdots & \vdots & \cdots & \vdots & \cdots & \vdots \\ T^{n-1} & (T^q)^{n-1} & \cdots & (T^{q^i})^{n-1} & \cdots & (T^{q^n})^{n-1} \end{pmatrix}$$

(the $(i + 1)$-th column is crossed out). This is again a Vandermonde determinant. Hence,

$$\theta_i = \prod_{\substack{0 \le j < k \le n \\ j, k \ne i}} (T^{q^k} - T^{q^j})$$

$$= \frac{\prod_{0 \le j < k \le n} (T^{q^k} - T^{q^j})}{\prod_{0 \le j \le i-1} (T^{q^i} - T^{q^j}) \prod_{i+1 \le k \le n} (T^{q^k} - T^{q^i})}$$

$$= \frac{D_1 \cdots D_n}{D_i L_{n-i}^{q^i}}.$$

The lemma follows from combining this formula with (5.4.2), (5.4.3), and (5.4.4).

□

Given $a \in A$ and $n \in \mathbb{Z}_{\ge 0}$, denote by $\left\{ \begin{smallmatrix} a \\ n \end{smallmatrix} \right\}$ the coefficient of x^{q^n} in $C_a(x)$:

$$C_a(x) = \sum_{n \ge 0} \left\{ \begin{matrix} a \\ n \end{matrix} \right\} x^{q^n}.$$

In particular, $\left\{ \begin{smallmatrix} a \\ 0 \end{smallmatrix} \right\} = a$, $\left\{ \begin{smallmatrix} a \\ \deg(a) \end{smallmatrix} \right\} \in \mathbb{F}_q^\times$ is the leading coefficient of a, and $\left\{ \begin{smallmatrix} a \\ n \end{smallmatrix} \right\} = 0$ for $n > \deg(a)$.

Proposition 5.4.3 *For $a \in A$ and $n \in \mathbb{Z}_{>0}$, we have*

$$\left\{ \begin{matrix} a \\ n \end{matrix} \right\} = \frac{f_{<n}(a)}{D_n}.$$

Proof Let y be an indeterminate. We formally expand

$$e_C(y \log_C(x)) = \sum_{n \geq 0} \frac{y^{q^n} \log_C(x)^{q^n}}{D_n}$$

$$= \sum_{n \geq 0} \frac{y^{q^n}}{D_n} \left(\sum_{m \geq 0} (-1)^m \frac{x^{q^m}}{L_m} \right)^{q^n} \qquad \text{(by Proposition 5.4.1)}$$

$$= \sum_{n \geq 0} \frac{y^{q^n}}{D_n} \sum_{m \geq 0} (-1)^m \frac{x^{q^{m+n}}}{L_m^{q^n}}$$

$$= \sum_{n \geq 0} x^{q^n} \sum_{i=0}^{n} (-1)^{n-i} \frac{y^{q^i}}{D_i L_{n-i}^{q^i}}$$

$$= \sum_{n \geq 0} x^{q^n} \frac{f_{<n}(y)}{D_n} \qquad \text{(by Lemma 5.4.2).}$$

On the other hand, we have $e_C(a \log_C(x)) = C_a(e_C(\log_C(x))) = C_a(x)$. Comparing this with the previous equation gives the desired formula for the coefficients of $C_a(x)$. (Note that $f_{<n}(a) = 0$ for $n > \deg(a)$ since then $a \in A_{<n}$.) $\square$

Corollary 5.4.4 *For all $n \geq 0$, we have*

$$D_n = \prod_{\substack{a \in A_+ \\ \deg(a)=n}} a.$$

Hence,

$$\left\{ \begin{matrix} a \\ n \end{matrix} \right\} = \left(\prod_{\substack{b \in A \\ \deg(b)<n}} (a-b) \right) \left(\prod_{\substack{b \in A_+ \\ \deg(b)=n}} b \right)^{-1}.$$

Proof The leading coefficient of $C_{T^n}(x)$ is 1. On the other hand, by Proposition 5.4.3, this coefficient is equal to $f_{<n}(T^n)/D_n$. Since

$$f_{<n}(T^n) = \prod_{\substack{a \in A \\ \deg(a) \leq n-1}} (T^n - a) = \prod_{\substack{a \in A_+ \\ \deg(a)=n}} a,$$

the first claim follows. Now, the second claim immediately follows from Proposition 5.4.3. $\square$

Remarks 5.4.5

(1) It is possible to deduce the formula for D_n in Corollary 5.4.4 directly by comparing the prime decompositions of both sides; see Exercise 5.4.5.

(2) As we will see in Sect. 7.1, the splitting fields of $C_a(x)$ over F are in many respects similar to the cyclotomic extensions of $\mathbb{Q}$. In this spirit, one observes that the coefficients of $C_a(x)$ are similar to the binomial coefficients. Indeed, the binomial coefficient $\binom{m}{n}$ is the value at m of the polynomial

$$\frac{x(x-1)\cdots(x-(n-1))}{1\cdot 2\cdots(n-1)\cdot n},$$

whereas the n-th coefficient of $C_a(x)$ is the value at a of the polynomial

$$\prod_{\substack{b\in\Lambda \\ \deg(b)\leq n-1}}(x-b)\Big/\prod_{\substack{b\in A_+ \\ \deg(b)=n}}b.$$

For a much deeper discussion of this analogy, the reader may consult [Tha04, Chapter 4].

(3) Continuing the line of thought of the previous remark, D_n looks like a factorial, or rather the factorial of q^n, since it is the denominator of the coefficient of x^{q^n} in the expansion of $C_a(x)$. Carlitz generalized this to a "factorial" of any $m\in\mathbb{Z}_{\geq 0}$ using the q-adic expansion of m

$$m = c_0 + c_1 q + \cdots + c_s q^s, \qquad 0\leq c_i < q.$$

In [Car37], he defined what is nowadays is called the *Carlitz factorial of* m as

$$\Pi(m) = D_0^{c_0} D_1^{c_1}\cdots D_s^{c_s}.$$

In particular, $\Pi(q^n) = D_n$. The Carlitz factorial satisfies many divisibility results analogous to those of usual factorials. The interested reader should consult [Gos96, Sec. 9.1], [Gos78], [Car37], [Car40]. For an elementary example of such divisibility, we refer to Exercise 5.4.10.

Next, we want to determine the lattice Λ_C of C. Since Λ_C has rank 1,

$$\Lambda_C = \pi_C \cdot A \quad\text{for some } \pi_C \in \mathbb{C}_\infty,$$

so we need to determine a generator π_C of Λ_C. Such a generator is called a *Carlitz period*; it is well-defined up to an $\mathbb{F}_q^\times$-multiple. We already obtained a series expansion for π_C^{q-1} in Example 5.2.5. Here we obtain a product expansion for π_C by comparing e_C with the exponential e_A of the lattice $A\subset\mathbb{C}_\infty$.

Let

$$e_{<n}(x) := x \prod_{a \in A_{<n}}{}' \left(1 - \frac{x}{a}\right). \qquad (5.4.5)$$

By considering the coefficient of x in Lemma 5.4.2, we get

$$(-1)^{q^n-1} \prod_{a \in A_{<n}}{}' a = (-1)^n D_n/L_n. \qquad (5.4.6)$$

On the other hand,

$$(-1)^{q^n-1} f_{<n}(x) = e_{<n}(x) \prod_{a \in A_{<n}}{}' a.$$

Hence, we deduce from Lemma 5.4.2 that

$$e_{<n}(x) = \sum_{i=0}^{n}(-1)^i \frac{x^{q^i}}{D_i} \cdot \frac{L_n}{L_{n-i}^{q^i}}. \qquad (5.4.7)$$

For $i \geq 0$, denote

$$\beta_i = [1]^{(q^i-1)/(q-1)},$$

$$\varpi_i = \beta_i/L_i. \qquad (5.4.8)$$

It is easy to show that

$$\frac{L_n}{L_{n-i}^{q^i}} = \frac{\beta_i \cdot \varpi_{n-i}^{q^i}}{\varpi_n}.$$

Hence,

$$e_{<n}(x) = \frac{1}{\varpi_n} \sum_{i=0}^{n}(-1)^i \frac{x^{q^i}}{D_i} \cdot \beta_i \varpi_{n-i}^{q^i}.$$

Note that $[i+1] - [i] = [1]^{q^i}$. Thus,

$$\varpi_{n+1} - \varpi_n = -\frac{[n]}{[n+1]}\varpi_n. \qquad (5.4.9)$$

Using induction, one deduces from this

$$\varpi_n = \prod_{i=1}^{n-1} \left(1 - \frac{[i]}{[i+1]} \right). \tag{5.4.10}$$

Lemma 5.4.6 *The sequence* $\{\varpi_n\}_{n\geq 1}$ *is Cauchy in* F_∞.

Proof From (5.4.8) and (5.4.9), we get

$$\mathrm{ord}_\infty(\varpi_{n+1} - \varpi_n) = q^{n+1} - q^n + \mathrm{ord}_\infty\left([1]^{(q^n-1)/(q-1)}\right) - \mathrm{ord}_\infty(L_n)$$

$$= q^{n+1} - q^n.$$

This implies that $|\varpi_{n+1} - \varpi_n| \to 0$, so the sequence is Cauchy by Exercise 2.2.8.
□

Denote the limit of $\{\varpi_n\}_{n\geq 1}$ by ϖ. By (5.4.10), this limit has a product expansion

$$\varpi = \prod_{i=1}^{\infty} \left(1 - \frac{[i]}{[i+1]} \right). \tag{5.4.11}$$

As in the proof of Proposition 5.1.3, the coefficients of $e_{<n}(x)$ converge to the coefficients of the power series expansion of

$$e_A(x) := x \prod_{0 \neq a \in A} \left(1 - \frac{x}{a} \right).$$

Therefore,

$$e_A(x) = \frac{1}{\varpi} \sum_{i=0}^{\infty} (-1)^i \frac{x^{q^i}}{D_i} \cdot \beta_i \varpi^{q^i}.$$

Fix a $(q-1)$-th root of $-[1]$ in $\mathbb{C}_\infty$, denoted $\mathbf{i}$,

$$\mathbf{i} := (-[1])^{1/(q-1)}.$$

Since $(-1)^i = (-1)^{(q^i-1)/(q-1)}$, we have

$$(-1)^i \beta_i = (-[1])^{(q^i-1)/(q-1)} = \mathbf{i}^{q^i-1}.$$

Therefore,

$$e_A(x) = \frac{1}{\varpi \mathbf{i}} \sum_{i=0}^{\infty} \frac{(\varpi \mathbf{i} \cdot x)^{q^i}}{D_i}.$$

Comparing this with Proposition 5.4.1, one obtains

$$\varpi \mathbf{i} \cdot e_A(x) = e_C(\varpi \mathbf{i} \cdot x).$$

Since for any $c \in \mathbb{C}_\infty$ we have $e_{cA}(cx) = ce_A(x)$, this implies that

$$\pi_C := \varpi \mathbf{i} \tag{5.4.12}$$

is a Carlitz period.

Theorem 5.4.7 *We have $\Lambda_C = (\varpi \mathbf{i}) \cdot A$.*

Next, we give another formula for π_C proved by Gekeler in [Gek86, (4.10)].

Proposition 5.4.8 *Fix a $(q-1)$-th root of $-T$, denoted $(-T)^{1/(q-1)}$. Then*

$$\pi_C = (-T)^{1/(q-1)} T \lim_{n \to \infty} \prod_{a \in A_{<n}}{}' \left(a/T^{\deg(a)} \right).$$

Proof The Drinfeld module corresponding to the lattice $A \subset \mathbb{C}_\infty$ is defined by $\phi_T(x) = Tx + ux^q$ for some $u \in \mathbb{C}_\infty^\times$. Let w be a $(q-1)$-th root of u, i.e., $w^{q-1} = u$. Then $w\phi_T(w^{-1}x) = C_T(x)$. Under this isomorphism, $\Lambda_C = w\Lambda_\phi$. Thus, $\phi_T(x) = Tx + \pi_C^{q-1}x^q$. On the other hand,

$$\phi_T(x) = Tx \prod_{\alpha \in \phi[T]}{}' \left(1 - \frac{x}{\alpha} \right).$$

Hence,

$$\pi_C^{q-1} = T \prod_{\alpha \in \phi[T]}{}' \frac{1}{\alpha}. \tag{5.4.13}$$

By Proposition 5.3.1, we have

$$\prod_{\alpha \in \phi[T]}{}' \alpha = \prod_{c \in \mathbb{F}_q^\times} e_A(c/T) \tag{5.4.14}$$

$$= \prod_{c \in \mathbb{F}_q^\times} \left(\frac{c}{T} \prod_{a \in A}{}' \left(1 - \frac{c}{aT} \right) \right)$$

$$= T^{1-q} \lim_{n \to \infty} \prod_{c \in \mathbb{F}_q^\times} \left(c \prod_{a \in A_{<n}}{}' \left(\frac{aT - c}{aT} \right) \right).$$

Note that

$$\prod_{c\in\mathbb{F}_q^\times}\left(c\prod_{a\in A_{<n}}'(aT-c)\right) = -\left(\prod_{b\in A_{<n+1}}'b\right)\left(\prod_{a\in A_{<n}}'aT\right)^{-1}.$$

Hence,

$$\prod_{c\in\mathbb{F}_q^\times}\left(c\prod_{a\in A_{<n}}'\left(\frac{aT-c}{aT}\right)\right) = -\left(\prod_{b\in A_{<n+1}}'b\right)\left(\prod_{a\in A_{<n}}'aT\right)^{-q}. \qquad (5.4.15)$$

Denote $\langle a\rangle = a/T^{\deg(a)}$. Since the numerator and the denominator of each factor on the left-hand side of (5.4.15) have the same degree (as polynomials in T), the degrees of the numerator and the denominator on the right-hand side are also the same, so we can rewrite that expression as

$$\left(\prod_{b\in A_{<n+1}}'\langle b\rangle\right)\left(\prod_{a\in A_{<n}}'\langle a\rangle\right)^{-q} - \left(\prod_{a\in A_{<n}}'\langle a\rangle\right)^{1-q}\prod_{\substack{b\in A\\\deg(b)=n+1}}\langle b\rangle. \qquad (5.4.16)$$

It is not hard to show that

$$\lim_{n\to\infty}\prod_{\substack{b\in A\\\deg(b)=n}}\langle b\rangle = 1 \qquad (5.4.17)$$

Taking the limit in (5.4.16) as $n\to\infty$, and combining the result with (5.4.13) and (5.4.14), proves the proposition. □

Corollary 5.4.9 *We have the following four formulas for* π_C^{q-1}:

$$\pi_C^{q-1} = -[1]\sum_{a\in A_+}\frac{1}{a^{q-1}} \qquad (Example\ 5.2.5);$$

$$\pi_C^{q-1} = -[1]\prod_{n\geq 1}\left(1-\frac{[n]}{[n+1]}\right)^{q-1} \qquad (Equation\ (5.4.12));$$

$$\pi_C^{q-1} = -T^q\lim_{n\to\infty}\prod_{a\in A_{<n}}'\left(a/T^{\deg(a)}\right)^{q-1} \qquad (Proposition\ 5.4.8);$$

$$\pi_C^{q-1} = -T^q\prod_{n\geq 1}\left(1-\frac{1}{T^{q^n-1}}\right)^{-(q-1)} \qquad (Exercise\ 5.4.12).$$

Some of our earlier calculations in this section have beautiful number-theoretic applications to the so-called Carlitz zeta-values. For $n \geq 1$, the n-th *Carlitz zeta-value* is

$$\zeta_C(n) := \sum_{a \in A_+} \frac{1}{a^n}.$$

(In this context, the set of monic polynomials, A_+, should be thought of as the function field analogue of the set of natural numbers $\mathbb{Z}_{>0}$.) The series $\zeta_C(n)$ converges in F_∞ since $|a| \to 0$. Let

$$\frac{x}{e_C(x)} = \sum_{n \geq 0} c_n x^n$$

be the power series (5.1.4) specialized to the Carlitz module.

Theorem 5.4.10 *Let $n \geq 1$ be an integer. If n is divisible by $q - 1$, then*

$$\zeta_C(n) = c_n \pi_C^n.$$

Moreover, if $n = q^m - 1$, then

$$\zeta_C(n) = (-1)^m \frac{\pi_C^n}{L_m}.$$

Proof By Theorem 5.1.9, we have

$$c_n = -E_n(\pi_C A)$$

$$= -\sum_{\lambda \in \pi_C A}' \frac{1}{\lambda^n}$$

$$= -\frac{1}{\pi_C^n} \sum_{a \in A}' \frac{1}{a^n}$$

$$= -\frac{1}{\pi_C^n} \left(\sum_{\alpha \in \mathbb{F}_q^\times} \frac{1}{\alpha^n} \right) \left(\sum_{a \in A_+}' \frac{1}{a^n} \right)$$

$$= \frac{1}{\pi_C^n} \zeta_C(n),$$

where the last equality follows from the observation that if $q - 1$ divides n, then

$$\sum_{\alpha \in \mathbb{F}_q^\times} \alpha^{-n} = \sum_{\alpha \in \mathbb{F}_q^\times} \alpha^n = \sum_{\alpha \in \mathbb{F}_q^\times} 1 = q - 1 = -1.$$

This proves the first formula. The formula for $\zeta_C(q^m - 1)$ now follows from Proposition 5.1.11, and the explicit expression for the coefficients of $\log_C$ in Proposition 5.4.1. $\qquad\square$

It is natural to consider polynomial power sums in F,

$$S_n(i) := \sum_{\substack{a \in A_+ \\ \deg(a)=n}} a^i, \quad i \in \mathbb{Z}, \quad n \in \mathbb{Z}_{\geq 0},$$

for both their intrinsic interest and their applications to Carlitz zeta-values.

Obviously, $S_0(i) = 1$ for all i. We examine $S_n(i)$, $n \geq 1$, using the logarithmic derivative of $f_{<n}(x)$ (cf. Exercise 3.1.4):

$$\frac{f'_{<n}(x)}{f_{<n}(x)} = \sum_{a \in A_{<n}} \frac{1}{x - a}. \tag{5.4.18}$$

Since $f_{<n}$ is $\mathbb{F}_q$-linear, $f'_{<n}(x)$ is equal to the constant term of $f_{<n}(x)/x$. Using Lemma 5.4.2, we get

$$f'_{<n}(x) = (-1)^n \frac{D_n}{L_n}. \tag{5.4.19}$$

Now note that

$$f_{<n}(x - T^n) = f_{<n}(x) - f_{<n}(T^n) \qquad \text{(because } f_{<n}(x) \text{ is additive)}$$

$$\tag{5.4.20}$$

$$= f_{<n}(x) - \prod_{\substack{a \in A \\ \deg(a) \leq n-1}} (T^n - a)$$

$$= f_{<n}(x) - \prod_{\substack{a \in A_+ \\ \deg(a)=n}} a$$

$$= f_{<n}(x) - D_n \qquad \text{(by Corollary 5.4.4)}.$$

Substituting $x - T^n$ for x in (5.4.18) and using (5.4.19) and (5.4.20), we obtain

$$(-1)^n \frac{D_n}{L_n} \frac{1}{f_{<n}(x) - D_n} = \sum_{\substack{a \in A_+ \\ \deg(a)=n}} \frac{1}{x - a}. \tag{5.4.21}$$

By Lemma 5.4.2,

$$f_{<n}(x)/D_n = \sum_{i=0}^{n} (-1)^{n-i} \frac{1}{D_i L_{n-i}^{q^i}} x^{q^i},$$

so (5.4.21) can be rewritten as

$$\frac{(-1)^n}{L_n} \frac{1}{-1 + \frac{(-1)^n}{L_n} x + x^q(\cdots)} = \sum_{\substack{a \in A_+ \\ \deg(a)=n}} \frac{1}{x - a}. \tag{5.4.22}$$

Now there are two ways to proceed from here. The first is to expand both sides of (5.4.22) in the formal Laurent series ring $\mathbb{C}_\infty((x))$. In this ring, we have

$$\frac{1}{x - a} = -\frac{1}{a} \cdot \frac{1}{1 - \frac{x}{a}} = -\sum_{i \geq 0} \frac{x^i}{a^{i+1}}.$$

Hence, the right-hand side of (5.4.22) is equal to

$$-\sum_{i \geq 0} x^i \sum_{\substack{a \in A_+ \\ \deg(a)=n}} \frac{1}{a^{i+1}} = -\sum_{i \geq 0} S_n(-(i+1)) x^i. \tag{5.4.23}$$

On the other hand, the first q terms of the formal expansion of the left-hand side of (5.4.22) are

$$-\frac{(-1)^n}{L_n} \left(1 + \frac{(-1)^n}{L_n} x + \left(\frac{(-1)^n}{L_n} x \right)^2 + \cdots + \left(\frac{(-1)^n}{L_n} x \right)^{q-1} \right). \tag{5.4.24}$$

Proposition 5.4.11 *For $1 \leq i \leq q$, we have*

$$S_n(-i) = \left(\frac{(-1)^n}{L_n} \right)^i.$$

Proof For $1 \leq i \leq q - 1$, the formula of the proposition follows by comparing (5.4.23) and (5.4.24). The claim for $i = q$ follows by observing that $S_n(-q) = S_n(-1)^q$.   □

As a second approach to analyzing (5.4.21), one could expand both sides of this formula in terms of $1/x$ in $\mathbb{C}_\infty(\!(1/x)\!)$. In this ring, we have

$$\frac{1}{x-a} = \frac{1}{x} \cdot \frac{1}{1-\frac{a}{x}} = \sum_{i \geq 0} \frac{a^i}{x^{i+1}}.$$

Hence, the right-hand side of (5.4.22) is equal to

$$\sum_{i \geq 0} \frac{1}{x^{i+1}} \sum_{\substack{a \in A_+ \\ \deg(a)=n}} a^i = \sum_{i \geq 0} \frac{S_n(i)}{x^{i+1}}. \qquad (5.4.25)$$

On the other hand,

$$(f_{<n}(x) - D_n)^{-1} = x^{-q^n} + \alpha x^{-q^n-1} + \cdots \qquad (5.4.26)$$

By comparing (5.4.25) to (5.4.26), we obtain

Proposition 5.4.12

$$S_n(i) = 0, \quad 0 \leq i \leq q^n - 2,$$

$$S_n(q^n - 1) = (-1)^n \frac{D_n}{L_n}.$$

Definition 5.4.13 The *m-th Carlitz polylogarithm* is the power series

$$\mathrm{Li}_m(x) = \sum_{n=0}^{\infty} \left(\frac{(-1)^n}{L_n} \right)^m x^{q^n}. \qquad (5.4.27)$$

The series $\mathrm{Li}_m(x)$ converges for $|x| < q^{mq/(q-1)}$ (see Exercise 5.4.4). Observe that, by Proposition 5.4.1, $\mathrm{Li}_1(x) = \log_C(x)$.

Theorem 5.4.14 *For* $1 \leq m \leq q$, *we have*

$$\zeta_C(m) = \mathrm{Li}_m(1).$$

In particular,

$$\zeta_C(1) = \log_C(1).$$

Proof

$$\zeta_C(m) = \sum_{\substack{n \geq 0}} \sum_{\substack{a \in A_+ \\ \deg(a)=n}} \frac{1}{a^m}$$

$$= \sum_{n \geq 0} \left(\frac{(-1)^n}{L_n} \right)^m \qquad \text{(by Proposition 5.4.11)}$$

$$= \mathrm{Li}_m(1).$$

$\square$

Note that Proposition 5.4.12 implies that, for any $m \leq 0$,

$$\zeta_C(m) = \sum_{n \geq 0} S_n(-m)$$

is a finite sum that lies in A. This allows one to extend the domain of $\zeta_C(m)$ from $\mathbb{Z}_{>0}$ to the whole $\mathbb{Z}$ just by grouping the summands of the same degree; this is a discrete analogue of the meromorphic continuation of the Riemann zeta function to $\mathbb{C}$. We also have the following analogue of the vanishing of the Riemann zeta function at negative even integers (these are the so-called trivial zeros of the zeta function).

Theorem 5.4.15 *If $m \in \mathbb{Z}_{<0}$ and $m \equiv 0 \pmod{q-1}$, then $\zeta_C(m) = 0$.*

Proof By Proposition 5.4.12, there is N depending on m such that $S_n(-m) = 0$ for all $n \geq N$. If $(q-1) \mid m$, then $\sum_{\alpha \in \mathbb{F}_q} \alpha^m = -1$. Thus,

$$\zeta_C(m) = \sum_{n \geq 0} S_n(-m) = \sum_{n=0}^{N-1} S_n(-m) = - \sum_{a \in A_{<N}} a^{-m}.$$

Since $A_{<N}$ is an N-dimensional $\mathbb{F}_q$-vector subspace of $\mathbb{C}_\infty$ and we may assume that $N > -m$, the vanishing of $\sum_{a \in A_{<N}} a^{-m}$ follows from Exercise 3.1.3. $\square$

Example 5.4.16 Note that

$$\zeta_C(0) = \sum_{n \geq 0} S_n(0) = \sum_{n \geq 0} q^n = 1.$$

Moreover, one can show that $\zeta_C(m) \neq 0$ if $m \in \mathbb{Z}_{<0}$ but $m \not\equiv 0 \pmod{q-1}$; see Exercise 5.4.11.

We conclude this section by comparing known results about the Carlitz zeta-values to known and conjectural results about the Riemann zeta function.

Remarks 5.4.17

(1) The Carlitz period π_C is a good analogue of $2\pi i \in \mathbb{C}$ since $\pi_C \cdot A$ is the kernel of the exponential function $e_C(x)$, just like $2\pi i\mathbb{Z}$ is the kernel of the usual exponential function e^x (here $i = \sqrt{-1}$ is the usual imaginary unit); cf. Remark 5.2.9.

(2) The formula for $\zeta_C(n)$ in Theorem 5.4.10 is similar to Euler's famous formula

$$\zeta(2m) = \sum_{n=1}^{\infty} \frac{1}{n^{2m}} = -\frac{B_{2m}}{(2m)!} \frac{(2\pi i)^{2m}}{2}, \qquad m \in \mathbb{Z}_{>0}, \qquad (5.4.28)$$

where $\frac{B_n}{n!}$ is the coefficient of x^n in the expansion

$$\frac{x}{e^x - 1} = \sum_{n=0}^{\infty} \frac{B_n}{n!} x^n. \qquad (5.4.29)$$

For example, $\zeta(2) = \pi^2/6$.

Note that $x/e_C(x)$ is the analogue of $\frac{x}{e^x-1}$ over A, since the poles of $\frac{x}{e^x-1}$ are the elements of $2\pi i\mathbb{Z}$. To further highlight the similarity between Euler's and Carlitz's formulas, one can rewrite

$$\zeta_C(n) = -c_n \frac{\pi_C^n}{q - 1},$$

and note that $q - 1 = \#A^{\times}$ is the analogue of $2 = \#\mathbb{Z}^{\times}$.

In fact, the proof of Theorem 5.4.10 itself is similar to the usual proof of (5.4.28). This latter proof is based on the product expansion of the hyperbolic sine function,

$$\frac{e^x - e^{-x}}{2} = x \prod_{n=1}^{\infty} \left(1 + \frac{x^2}{(\pi n)^2}\right), \qquad (5.4.30)$$

which can be considered as an analogue of the product expansion of $e_C(x)$ over $\pi_C A$. Having this product expansion, Euler's formula follows by comparing the logarithmic derivatives of both sides of (5.4.30). Note that this idea is also present in the proof of Theorem 5.1.9.

(3) The numbers B_n on the right-hand side of (5.4.29) are called the *Bernoulli numbers*. The first few of these numbers are

$$B_0 = 1, \quad B_1 = -\frac{1}{2}, \quad B_2 = \frac{1}{6}, \quad B_3 = 0, \quad B_4 = -\frac{1}{30},$$

$$B_5 = 0, \quad B_6 = \frac{1}{42}, \quad B_7 = 0, \quad B_8 = -\frac{1}{30}, \quad B_9 = 0,$$

$$B_{10} = \frac{5}{66}, \quad B_{11} = 0, \quad B_{12} = -\frac{691}{2730}.$$

(All Bernoulli numbers with odd indices, except for B_1, are equal to 0.) The Bernoulli numbers have been much studied in number theory and occur in some central results, such as the famous theorem of Kummer, which says that if a given odd prime p does not divide any of the numerators of $B_2, B_4, \ldots, B_{p-3}$, then $x^p + y^p = z^p$ has no solutions in positive integers, i.e., Fermat's Last Theorem is true for such p.

In Remark 5.4.5, we mentioned Carlitz's analogue of the usual factorial. Using this factorial, Carlitz defined in [Car37] what are now called the *Bernoulli–Carlitz numbers*:

$$BC_n = c_n \cdot \Pi(n).$$

The reader can find some explicit examples of these numbers in [Gos96, Sec. 9.2], such as, if $q = 3$, then

$$BC_{12} = \frac{T^6 + T^4 + T^2 + 1}{T^3 + 2T}.$$

Remarkably, the Bernoulli–Carlitz numbers have properties similar to the properties of the usual Bernoulli numbers; for example, there is a version of the classical von Staudt–Clausen theorem for these numbers (see [Car37, Car40, Gos78]), and they are related to the usual class groups and the Taelman class modules of Carlitz cyclotomic extensions similar to the classical cyclotomic theory (see [Oka91, Ang01, Tae12a, ANDTR20]).

(4) The formulas (3) and (4) in Corollary 5.4.9 are analogous to the classical Wallis' formula:

$$\pi = 2 \prod_{n \geq 1} \left(1 - \frac{1}{4n^2}\right)^{-1} = \frac{4}{3} \cdot \frac{16}{15} \cdot \frac{36}{35} \cdots$$

(Note that this formula easily follows by evaluating both sides of (5.4.30) at $\pi i/2$.)

(5) Wade [Wad41] proved that π_C is transcendental over F, just like π is transcendental over $\mathbb{Q}$. Hence, even though the Carlitz module C is defined over F, its

lattice is transcendental over F. Wade was actually interested in the construction of transcendental elements over F. He proved the following: if α is nonzero and algebraic over F, then $e_C(\alpha)$ is transcendental over F. In particular,

$$e_C(1) = \sum_{i \geq 1} 1/D_i$$

is transcendental over F. This number can be considered as an analogue of $e \approx 2.718$, the base of the natural logarithm, which is well-known to be transcendental. The transcendence of π_C follows from the observation that $e_C(\pi_C/T) \in C[T] \subset \overline{F}$.

(6) The transcendence of π_C over F is a special case of a general principle established by Jing Yu [Yu86] that if ϕ is a Drinfeld module defined over $\overline{F}$, then every $0 \neq \lambda \in \Lambda_\phi$ is transcendental over F. (This is a stronger version of a result of Siegel [Sie32] for elliptic curves: if E is an elliptic curve defined over $\overline{\mathbb{Q}}$ with lattice $\mathbb{Z}\omega_1 + \mathbb{Z}\omega_2$, then at least one of the two complex numbers ω_1 and ω_2 is transcendental.) Chang and Papanikolas [CP12] proved that for a Drinfeld module ϕ of rank r defined over $\overline{F}$, the transcendence degree of the extension $\overline{F}(\Lambda_\phi)/\overline{F}$ is r, assuming $\mathrm{End}(\phi) = A$. The analogue of this remarkable statement for elliptic curves over $\overline{\mathbb{Q}}$ is a long standing open problem.

(7) By Theorem 5.4.10, the transcendence of π_C implies that $\zeta_C(n)$ is transcendental over F if $n \equiv 0 \pmod{q-1}$. Anderson and Thakur [AT90] deduced a formula for $\zeta_C(m)$ for arbitrary $m \geq 1$ using certain objects which can be considered as the tensor powers of the Carlitz module. Using this result, Jing Yu [Yu91] proved that $\zeta_C(n)$ is transcendental over F for all positive integers n. The transcendence of the Riemann zeta-values $\zeta(n)$ for odd $n \geq 3$ is a major open problem in number theory.

(8) The Riemann zeta function has a meromorphic continuation to $\mathbb{C}$ with a pole at 1. In analogy with this, Goss [Gos79] showed that the domain of ζ_C can be extended from $\mathbb{Z}$ to $\mathbb{C}_\infty^\times \times \mathbb{Z}_p$ by defining (the *Carlitz–Goss zeta function*)

$$\zeta_{CG}(x, y) := \sum_{n \geq 0} x^{-n} \sum_{\substack{a \in A_+ \\ \deg(a) = n}} \left(\frac{a}{T^n}\right)^{-y}.$$

Note that for any $m \in \mathbb{Z}$ we have

$$\zeta_{CG}(T^m, m) = \sum_{n \geq 0} S_n(-m) = \zeta_C(m).$$

Moreover, the following analogue of the Riemann Hypothesis for the Carlitz–Goss zeta function is known by the work of Sheats [She98] and others: for a fixed $y \in \mathbb{Z}_p$, the zeros of $\zeta_{CG}(x, y)$, as a function of $x \in \mathbb{C}_\infty^\times$, lie in F_∞, i.e., all lie on the same "real line."

(9) As we will explain in Sect. 7.6, the formula $\zeta_C(1) = \log_C(1)$ of Theorem 5.4.14 is a special case of a more general formula due to Taelman, which itself is an analogue of the class number formula for the zeta functions of number fields; see Theorem 7.6.23.

Exercises

5.4.1 Show that for all $n \geq 1$, we have

$$\sum_{i=0}^{n}(-1)^{n-i}\frac{1}{D_i L_{n-i}^{q^i}} = \sum_{i=0}^{n}(-1)^{i}\frac{1}{L_i D_{n-i}^{q^i}} = 0.$$

5.4.2 Assume $q = 3$. Find the first three nonzero terms in the $1/T$-adic expansion of π_C^{q-1}. This will give an analogue of the approximation $\pi \approx 3.14$ of the usual π.

5.4.3 Prove that $F_\infty(\pi_C)$ is the splitting field of $x^{q-1} + T \in F_\infty[x]$.

5.4.4
(a) Prove that the radius of convergence of Li_m is $q^{qm/(q-1)}$.
(b) Show that $|\pi_C| = q^{q/(q-1)}$. Compare these calculations with the statement of Lemma 5.1.5.

5.4.5
(a) Show that $\deg(D_n) = nq^n$. Deduce from this and Proposition 5.4.1 that $e_C(x)$ is entire on $\mathbb{C}_\infty$.
(b) Show that

$$D_n = \prod_{\substack{a \in A_+ \\ \deg(a)=n}} a,$$

by showing that the exact power with which a monic irreducible polynomial in A of degree d divides either side of this equation is

$$\sum_{i=1}^{\lfloor \frac{n}{d} \rfloor} q^{n-id}.$$

(c) Prove that

$$e_{<n}(x) + D_n = \prod_{\substack{a \in A_+ \\ \deg(a)=n}} (x + a).$$

5.4.6 Show that L_n is the least common multiple of all monic polynomials in A of degree n.

5.4.7 Let $\mathfrak{p}$ be a prime of degree n. Prove the following congruences:

(a) $D_n/\mathfrak{p} \equiv -1 \pmod{\mathfrak{p}}$.
(b) $D_n/L_n \equiv (-1)^{n-1} \pmod{\mathfrak{p}}$.
(c) $L_n/\mathfrak{p} \equiv (-1)^n \pmod{\mathfrak{p}}$.

5.4.8 Verify (5.4.17), which was used in the proof of Proposition 5.4.8.

5.4.9 Let $\mathfrak{p} \lhd A$ be a prime. Let $\mathbb{C}_\mathfrak{p}$ be the completion of an algebraic closure of $F_\mathfrak{p}$. Because the coefficients of $e_C(x)$ and $\log_C(x)$ are in F, one can consider these as elements of $\mathbb{C}_\mathfrak{p}[\![x]\!]$. Prove the following:

(a) $e_C(x)$ converges in $\mathbb{C}_\mathfrak{p}$ for all $x \in \mathbb{C}_\mathfrak{p}$ with $\mathrm{ord}_\mathfrak{p}(x) > \frac{1}{q^{\deg(\mathfrak{p})}-1}$.
(b) $\log_C(x)$ converges in $\mathbb{C}_\mathfrak{p}$ for all $x \in \mathbb{C}_\mathfrak{p}$ with $\mathrm{ord}_\mathfrak{p}(x) > 0$.

5.4.10 It is a well-known fact that the prime factorization of the usual factorial $n!$ is given by

$$n! = \prod_p p^{\sum_{s>1} \lfloor n/p^s \rfloor},$$

where the product is over all prime numbers. The Carlitz factorial $\Pi(n)$ has a similar factorization

$$\Pi(n) = \prod_\mathfrak{p} \mathfrak{p}^{\sum_{s \geq 1} \lfloor n/|\mathfrak{p}|^s \rfloor},$$

where the product is over all monic irreducibles of A; this formula is attributed to Sinnott in [Gos96]. The reader should try to prove both formulas.

5.4.11 Let m be a negative integer not divisible by $q - 1$. Prove that $\zeta_C(m) \neq 0$.

5.4.12 This exercise outlines the proof of (4) of Corollary 5.4.9; the formula is due to Anderson and Thakur [AT90, Cor. 2.5.9].
    Consider the power series

$$H(x) := \sum_{n \geq 0} e_C\left(\pi_C/T^{n+1}\right) x^n \in \mathbb{C}_\infty[\![x]\!].$$

(a) Use the expansion of $e_C(x)$ to show that in a small neighborhood of T,

$$H(x) = -\frac{\pi_C}{x - T} + g(x),$$

where $g(x)$ is holomorphic.
(b) For $f(x) = \sum_{n \geq 0} a_n x^n \in \mathbb{C}_\infty[\![x]\!]$, let $f^{(1)}(x) := \sum_{n \geq 0} a_n^q x^n$. Prove that the functional equation $f^{(1)}(x) = (x - T)f(x)$ has at most one solution in $\mathbb{C}_\infty[\![x]\!]$,

by showing that the coefficients of $f(x)$ can be uniquely recovered from this equation.

(c) Show that $H(x)$ satisfies $H^{(1)}(x) = (x - T)H(x)$.

(d) Show that

$$h(x) = (-T)^{1/(q-1)} \prod_{n=0}^{\infty} \left(1 - \frac{x}{T^{q^n}}\right)^{-1}$$

also satisfies $h^{(1)}(x) = (x - T)h(x)$.

(e) Compute the residue of $h(x)$ at $x = T$.

5.5   Fields Generated by Lattices of Drinfeld Modules

In this section we are interested in the extension $F_\infty(\Lambda_\phi)$ generated by the lattice of a Drinfeld module ϕ defined over F_∞. This problem is closely related to the problem of analyzing the extensions of F_∞ generated by the torsion points of ϕ.

There are different approaches to studying $F_\infty(\Lambda_\phi)/F_\infty$. Since Λ_ϕ is the set of zeros of the entire function e_ϕ, analyzing the Newton polygon of e_ϕ is one of these approaches. Another approach is based on the observation that $F_\infty(\Lambda_\phi)/F_\infty$ is the same extension as the extension obtained by attaching T^n-torsion points of ϕ to F_∞ for all $n \geq 1$. Hence, analyzing the Newton polygons of $\phi_{T^n}(x)$ provides information about $F_\infty(\Lambda_\phi)/F_\infty$. The third approach, the one that we take in this section, is based on the analysis of a special type basis of Λ_ϕ, called successive minimum basis (SMB). The advantage of the SMB approach is that it gives more information about $F_\infty(\Lambda_\phi)$ than the Newton polygon method, and it also has other interesting applications: for example, an SMB can be used to define for ϕ an analogue of the area of a fundamental domain for the lattice of an elliptic curve.

Some of the results in this section were obtained independently by different mathematicians: Chen and Lee [CL13] (using the Newton polygon of e_ϕ), Maurischat [Mau19] (using the Newton polygons of the iterates of ϕ_T), Gekeler [Gek19a] (using SMB), Gardeyn [Gar02] (using SMB and the Newton polygon of e_ϕ), Taguchi [Tag92] (using the Newton polygon of ϕ_T). We follow Gekeler's approach [Gek19a], which is the only one that shows that the degree of $\overline{\mathbb{F}}_q \cap F_\infty(\Lambda_\phi)$ over $\mathbb{F}_q$ is bounded by a constant depending only on the rank of ϕ.

Let K be a finite extension of F_∞, and let ϕ be a Drinfeld module of rank r over K. By Theorem 5.2.8, there is a lattice $\Lambda := \Lambda_\phi$ of rank r associated with ϕ. Let $K(\Lambda)$ be the field generated over K by the elements of Λ. Let

$$\text{tor}(\phi) = \bigcup_{0 \neq a \in A} \phi[a] \cong (F/A)^r$$

be the torsion submodule of $^\phi C_\infty$, and let $K(\text{tor}(\phi))$ be the field extension of K generated by $\text{tor}(\phi)$.

Proposition 5.5.1 (Maurischat [Mau19]) *With previous notation and assumption, we have:*

(1) For any $a \in A$ of positive degree,

$$\bigcup_{n \geq 1} K(\phi[a^n]) = K(\mathrm{tor}(\phi)) = K(\Lambda).$$

(2) The extension $K(\Lambda)/K$ is finite and Galois.
(3) The lattice Λ is invariant under the action of $\mathrm{Gal}(K(\Lambda)/K)$.

Proof

(1) Since Λ is finitely generated over A, after choosing an A-basis $\{\lambda_1, \ldots, \lambda_r\}$ of Λ, we have $K(\Lambda) = K(\lambda_1, \ldots, \lambda_r)$. On the other hand, by Proposition 2.7.12, the elements of Λ are algebraic over K. Hence, $K(\Lambda)$ is a finite extension of K. This implies that $K(\Lambda)$ is complete with respect to the unique extension of the absolute value on K to $\mathbb{C}_\infty$. By the recursive equations (5.2.4), the coefficients of $e_\phi(x)$ are in K. By Proposition 5.3.1, for any $a \in A$ of positive degree, we have $\phi[a] = \{e_\phi(\lambda/a) \mid \lambda \in \Lambda\}$. Since $e_\phi(\lambda/a)$ converges in $K(\Lambda)$, we get

$$\bigcup_{n \geq 1} K(\phi[a^n]) \subseteq K(\mathrm{tor}(\phi)) \subseteq K(\Lambda). \qquad (5.5.1)$$

Now let $\lambda_0 \in \Lambda$ be a nonzero element. For any $a \in A$ of positive degree, we can find $n \geq 1$ such that $|\lambda_0/a^n| < \min_{0 \neq \lambda \in \Lambda} |\lambda|$. Assume n is chosen so that this inequality holds. Then, by Lemma 5.1.5, the logarithm $\log_\phi$ of Λ converges on $e_\phi(\lambda_0/a^n)$ because

$$\left| e_\phi(\lambda_0/a^n) \right| = \left| \frac{\lambda_0}{a^n} \right| \prod_{\lambda \in \Lambda}' \left| 1 - \frac{\lambda_0}{a^n \lambda} \right|$$

$$= \left| \lambda_0/a^n \right|$$

$$< \min_{0 \neq \lambda \in \Lambda} |\lambda|.$$

On the other hand, from the recursive equations (5.1.2), it is clear that the coefficients of $\log_\phi$ are in K. Therefore,

$$\lambda_0/a^n = \log_\phi(e_\phi(\lambda_0/a^n)) \in K(e_\phi(\lambda_0/a^n)).$$

Since $K(e_\phi(\lambda_0/a^n)) \subseteq K(\phi[a^n])$, this implies that $\lambda_0 \in K(\phi[a^n])$. Thus,

$$K(\Lambda) \subseteq \bigcup_{n \geq 1} K(\phi[a^n]).$$

Combining this with (5.5.1), we get the first claim.

(2) Since each extension $K(\phi[a])/K$ is Galois, the extension $K(\Lambda)/K$ is also Galois. Since we already observed that $K(\Lambda)/K$ is a finite extension, we get (2).

(3) Let $\sigma \in \mathrm{Gal}(K(\Lambda)/K)$, and let $0 \neq \lambda \in \Lambda$. We have $e_\phi(\sigma(\lambda)/T) = \sigma(e_\phi(\lambda/T))$. Since $\phi[T]$ is invariant under the action of $\mathrm{Gal}(K(\Lambda)/K)$, we see that $e_\phi(\sigma(\lambda)/T) \in \phi[T]$. Thus,

$$e_\phi(\sigma(\lambda)/T) = e_\phi(\lambda'/T) \quad \text{for some } \lambda' \in \Lambda \quad \Longrightarrow$$

$$e_\phi((\sigma(\lambda) - \lambda')/T) = 0 \quad \Longrightarrow$$

$$(\sigma(\lambda) - \lambda')/T \in \Lambda \quad \Longrightarrow$$

$$\sigma(\lambda) \in \Lambda.$$

$\square$

Lemma 5.5.2 *If the rank of ϕ is 1, then $\mathrm{Gal}(K(\Lambda)/K)$ is isomorphic to a subgroup of $\mathbb{F}_q^\times$. In particular, $[K(\Lambda) : K]$ divides $q - 1$.*

Note that by Exercise 5.4.3, for the Carlitz module C, we have $[F_\infty(\Lambda_C) : F_\infty] = q - 1$, so the bound in the lemma is sharp.

Proof Let λ be a generator of Λ. Then $K(\Lambda) = K(\lambda)$. By Proposition 5.5.1, $\sigma(\lambda)$ is also a generator of Λ for any $\sigma \in \mathrm{Gal}(K(\Lambda)/\Lambda)$. Thus, $\sigma(\lambda) = \alpha_\sigma \lambda$ for some $\alpha_\sigma \in \mathbb{F}_q^\times$. It is easy to check that $\mathrm{Gal}(K(\Lambda)/K) \to \mathbb{F}_q^\times$, $\sigma \mapsto \alpha_\sigma$, is an injective homomorphism. $\square$

It is natural to ask whether the degree of $K(\Lambda)/K$ is universally bounded as Λ varies over the lattices of Drinfeld modules over K of fixed rank $r \geq 2$. We will show that when $r \geq 2$, the ramification index of $K(\Lambda)/K$ can be arbitrarily large, whereas the residue degree remains universally bounded. (It is not hard to show that if r is allowed to vary, then the residue degree also can become arbitrarily large; see Exercise 5.5.10.) We follow Gekeler's arguments in [Gek19a]. The idea is to relate the valuations of the elements of a special basis of Λ to the valuations of the coefficients of ϕ_T; see Proposition 5.5.8. We will do this assuming $r = 2$; this is a reasonable special case in which the main ideas of [Gek19a] are present, minus some of the technicalities.

Definition 5.5.3 An ordered A-basis $\{\lambda_1, \ldots, \lambda_r\}$ of the A-lattice Λ in $\mathbb{C}_\infty$ is a *successive minimum basis* (SMB) if

(i) λ_1 has the minimal absolute value among the nonzero elements of Λ;
(ii) for each $2 \leq i \leq r$, the element λ_i has the minimal absolute value among all $\lambda \in \Lambda$ not in the A-span of $\{\lambda_1, \ldots, \lambda_{i-1}\}$.

Remarks 5.5.4

(1) It is clear that an SMB $\{\lambda_1, \ldots, \lambda_r\}$ exists for any lattice (since the lattice is discrete), and

$$|\lambda_1| \leq |\lambda_2| \leq \cdots \leq |\lambda_r|.$$

(2) An SMB of Λ is not necessarily unique, but the r-tuple $(|\lambda_1|, \ldots, |\lambda_r|)$ is an invariant of Λ. For $r = 2$, this can be seen as follows (the proof for the general case is not much different). Since λ_1 is an element of Λ with minimal nonzero valuation, $|\lambda_1|$ is an invariant of Λ. Now, let $\mathbb{D}(n) = \{x \in \mathbb{C}_\infty : |x| \leq n\}$ be a disk of smallest radius such that $\mathbb{D}(n) \cap \Lambda$ contains an element which is not an A-multiple of λ_1. Then $|\lambda_2| = n$ is also an invariant of Λ.

(3) SMB was first introduced by Taguchi in [Tag93] for the purpose of defining a height of a Drinfeld module over F_∞ (see Exercise 5.5.14) and was further developed by Gekeler for applications to the theory of Drinfeld modular forms (see [Gek17], [Gek19b]).

An SMB has the following important "orthogonality" property:

Proposition 5.5.5 *Let* $\{\lambda_1, \ldots, \lambda_r\}$ *be an SMB of* Λ. *Then for all* $a_1, \ldots, a_r \in F_\infty$,

$$|a_1\lambda_1 + \cdots + a_r\lambda_r| = \max_{1 \leq i \leq r} |a_i||\lambda_i|.$$

Proof Let $M := \max_{1 \leq i \leq r} |a_i||\lambda_i|$. Let $i_1 < \cdots < i_n$ be the indices for which $|a_i\lambda_i| = M$. By the strong triangle inequality,

$$|a_1\lambda_1 + \cdots + a_r\lambda_r| \leq M.$$

Suppose the inequality is strict. Then $n \geq 2$ and

$$\left| a_{i_1} a_{i_n}^{-1} \lambda_{i_1} + \cdots + a_{i_{n-1}} a_{i_n}^{-1} \lambda_{i_{n-1}} + \lambda_{i_n} \right| < \left| a_{i_n}^{-1} \right| M = \left| \lambda_{i_n} \right|.$$

For $1 \leq j \leq n$, we have $|a_{i_j}\lambda_{i_j}| = |a_{i_n}\lambda_{i_n}|$ but $|\lambda_{i_j}| \leq |\lambda_{i_n}|$. Hence $|a_{i_j}a_{i_n}^{-1}| \geq 1$ for all $1 \leq j \leq n-1$, so we can decompose $a_{i_j}a_{i_n}^{-1} = b_j + c_j$, where $b_j \in A$ is nonzero and $\left| c_j \right| < 1$. Now

$$\left| b_1\lambda_{i_1} + \cdots + b_{n-1}\lambda_{i_{n-1}} + \lambda_{i_n} \right|$$

$$= \left| a_{i_1} a_{i_n}^{-1} \lambda_{i_1} + \cdots + a_{i_{n-1}} a_{i_n}^{-1} \lambda_{i_{n-1}} + \lambda_{i_n} - c_1\lambda_{i_1} + \cdots - c_{n-1}\lambda_{i_{n-1}} \right|$$

$$\leq \max(\left| a_{i_1} a_{i_n}^{-1} \lambda_{i_1} + \cdots + a_{i_{n-1}} a_{i_n}^{-1} \lambda_{i_{n-1}} + \lambda_{i_n} \right|, \left| c_1\lambda_{i_1} \right|, \ldots, \left| c_{n-1}\lambda_{i_{n-1}} \right|)$$

$$< \left| \lambda_{i_n} \right|.$$

After replacing λ_{i_n} by $b_1\lambda_{i_1} + \cdots + b_{n-1}\lambda_{i_{n-1}} + \lambda_{i_n}$, we obtain a new A-basis of Λ, such that

$$\left| b_1\lambda_{i_1} + \cdots + b_{n-1}\lambda_{i_{n-1}} + \lambda_{i_n} \right| < \left| \lambda_{i_n} \right|.$$

But this contradicts the assumption that $\{\lambda_1, \ldots, \lambda_r\}$ is an SMB.   □

Assume from now on that $r = 2$. Let $\{\lambda_1, \lambda_2\}$ be an SMB of Λ. Denote

$$\mu_1 = e_\Lambda(\lambda_1/T) \qquad \text{and} \qquad \mu_2 = e_\Lambda(\lambda_2/T).$$

By Proposition 5.3.1, the elements $\{\mu_1, \mu_2\}$ form an $\mathbb{F}_q$-basis of $\phi[T]$.

For the statement of the next lemma, we need the following function:

$$s(n) := \deg \prod_{a \in A_{<n}}{}' a.$$

It is easy to check either directly or by using (5.4.6) that

$$s(n) = (q-1)\sum_{m=1}^{n-1} m q^m \tag{5.5.2}$$

$$= nq^n - (q^n + \cdots + q)$$

$$= (n-1)q^n - \frac{q^n - 1}{q - 1} + 1.$$

(By convention, an empty product is equal to 1, so $s(0) = 0$.)

Lemma 5.5.6 *In addition to above notation and assumptions, let*

$$n = \max\{i \in \mathbb{Z}_{\geq 0} \;:\; |\lambda_2/\lambda_1| \geq q^i\} \tag{5.5.3}$$

be the integral part of $\log_q |\lambda_2/\lambda_1|$. *Then*

(1)

$$|\mu_1| = \frac{|\lambda_1|}{q},$$

$$|\mu_2| = |\lambda_1| \left| \frac{\lambda_2}{\lambda_1} \right|^{q^n} q^{-(q^n + s(n))}.$$

(2) $|\mu_1| \leq |\mu_2|$ *with equality if and only if* $|\lambda_1| = |\lambda_2|$.

(3) For any $\alpha, \beta \in \mathbb{F}_q$, we have

$$|\alpha\mu_1 + \beta\mu_2| = \begin{cases} 0, & \text{if } \alpha = \beta = 0; \\ |\mu_1|, & \text{if } \alpha \neq 0, \beta = 0; \\ |\mu_2|, & \text{if } \beta \neq 0. \end{cases}$$

(4) Among the nonzero elements of $\phi[T]$, $q-1$ elements have absolute value $|\mu_1|$, and the other $q^2 - q$ elements have absolute value $|\mu_2|$.

Proof

(1) We have

$$|\mu_1| = |e_\Lambda(\lambda_1/T)| = \left|\frac{\lambda_1}{T}\right| \prod_{\lambda \in \Lambda}' \left|1 - \frac{\lambda_1}{T\lambda}\right|.$$

By definition, λ_1 has the minimal nonzero absolute value among the elements of Λ. Hence, $|T\lambda| > |\lambda_1|$ for all $0 \neq \lambda \in \Lambda$. This implies that $|1 - \lambda_1/T\lambda| = 1$ for all $0 \neq \lambda \in \Lambda$. Thus,

$$|\mu_1| = |\lambda_1/T| = |\lambda_1|/q. \tag{5.5.4}$$

Similarly,

$$|\mu_2| = \left|\frac{\lambda_2}{T}\right| \prod_{\lambda \in \Lambda}' \left|1 - \frac{\lambda_2}{T\lambda}\right|.$$

If $|T\lambda| > |\lambda_2|$, then $|1 - \lambda_2/T\lambda| = 1$. If $|T\lambda| < |\lambda_2|$, then $|1 - \lambda_2/T\lambda| = |\lambda_2/T\lambda|$. If $|T\lambda| = |\lambda_2|$, then $|1 - \lambda_2/T\lambda| \leq 1$, with strict inequality if and only if $|T\lambda - \lambda_2| < |T\lambda| = |\lambda_2|$. If we expand λ in terms of the basis $\{\lambda_1, \lambda_2\}$,

$$\lambda = a\lambda_1 + b\lambda_2, \quad \text{with } a, b \in A,$$

then the coefficient of λ_2 in the expansion of $T\lambda - \lambda_2$ is $Tb - 1$, which is obviously nonzero. Hence $T\lambda - \lambda_2$ does not lie in the sublattice of Λ spanned by λ_1. The inequality $|T\lambda - \lambda_2| < |\lambda_2|$ then would contradict the SMB property of $\{\lambda_1, \lambda_2\}$, so $|1 - \lambda_2/T\lambda| = 1$. Overall, we obtain the formula

$$|\mu_2| = \left|\frac{\lambda_2}{T}\right| \prod_{\substack{\lambda \in \Lambda \\ |T\lambda| \leq |\lambda_2|}}' \left|\frac{\lambda_2}{T\lambda}\right|. \tag{5.5.5}$$

To proceed, note that, by Proposition 5.5.5, we have

$$|T\lambda| = |aT\lambda_1 + bT\lambda_2| = \max(|aT\lambda_1|, |bT\lambda_2|).$$

The inequality $|T\lambda| \leq |\lambda_2|$ is satisfied if and only if $b = 0$ and $|aT| \leq |\lambda_2/\lambda_1|$, in which case $|T\lambda| = |aT\lambda_1|$. Since the inequality $|aT| \leq |\lambda_2/\lambda_1|$ is equivalent to $a \in A_{<n}$, we get

$$
\begin{aligned}
|\mu_2| &= \left| \frac{\lambda_2}{T} \right| \prod_{a \in A_{<n}}' \left| \frac{\lambda_2}{aT\lambda_1} \right| \\
&= \left| \frac{\lambda_2}{T} \right| \left| \frac{\lambda_2}{T\lambda_1} \right|^{q^n-1} q^{-s(n)} \\
&= |\lambda_1| \left| \frac{\lambda_2}{\lambda_1} \right|^{q^n} q^{-(q^n+s(n))}.
\end{aligned}
$$

(2) Comparing (5.5.4) with (5.5.5), it is clear that $|\mu_1| \leq |\mu_2|$, with equality if and only if $|\lambda_1| = |\lambda_2|$.

(3) Let $\alpha, \beta \in \mathbb{F}_q$, and consider $\mu = \alpha\mu_1 + \beta\mu_2$. If $\alpha = \beta = 0$, then obviously $|\mu| = 0$. Now assume one of these elements is nonzero. Note that $\mu = e_\Lambda\left(\lambda'/T\right)$, where $\lambda' = \alpha\lambda_1 + \beta\lambda_2$. The same argument that gives (5.5.5) also shows that

$$
|\mu| = \left| \frac{\lambda'}{T} \right| \prod_{\substack{\lambda \in \Lambda \\ |T\lambda| \leq |\lambda'|}}' \left| \frac{\lambda'}{T\lambda} \right|.
$$

Because $\{\lambda_1, \lambda_2\}$ is an SMB, Proposition 5.5.5 implies that $\left|\lambda'\right| = |\lambda_1|$ (respectively, $\left|\lambda'\right| = |\lambda_2|$), if $\beta = 0$ (respectively, $\beta \neq 0$). This proves (3).

(4) This is an immediate consequence of (3).

$\square$

Lemma 5.5.7 *Let*

$$
\phi_T(x) = Tx + g_1 x^q + g_2 x^{q^2}
$$

be the T-division polynomial of ϕ.

(1) If $|\lambda_1| = |\lambda_2|$, then

$$
|g_2| = q^{q^2} |\lambda_1|^{1-q^2}.
$$

(2) If $|\lambda_1| < |\lambda_2|$, then

$$
|g_1| = q^q |\lambda_1|^{1-q},
$$

$$
|g_2| = q^{q+q(q-1)(q^n+s(n))} |\lambda_1|^{1-q^2} \left| \frac{\lambda_2}{\lambda_1} \right|^{(1-q)q^{n+1}}.
$$

Proof After decomposing

$$\phi_T(x) = Tx \prod_{\mu \in \phi[T]}' \left(1 - \frac{x}{\mu}\right),$$

we see that

$$|g_2| = |T| \prod_{\mu \in \phi[T]}' |\mu|^{-1}$$

$$= q \, |\mu_1|^{1-q} \, |\mu_2|^{q-q^2}$$

$$= q^{q+q(q-1)(q^n+s(n))} \, |\lambda_1|^{1-q^2} \left|\frac{\lambda_2}{\lambda_1}\right|^{(1-q)q^{n+1}} \qquad \text{(by Lemma 5.5.6).}$$

This equation for $|g_2|$ is valid whether the inequality $|\lambda_1| \le |\lambda_2|$ is strict or not (in the latter case, $n = 0$ and $s(0) = 0$).

In addition, if $|\lambda_1| < |\lambda_2|$, then

$$|g_1| = |T| \, |\mu_1|^{1-q} = q^q \, |\lambda_1|^{1-q} \, .$$

This is true because, up to sign, g_1/T is equal to the sum of $(\mu_{i_1} \dots \mu_{i_{q-1}})^{-1}$, $\mu_{i_j} \in \phi[T]$, and the product of the $q - 1$ "short" elements of $\phi[T]$ is strictly smaller than any other product of $q - 1$ elements of $\phi[T]$. $\qquad \square$

Proposition 5.5.8 *If $|\lambda_1| < |\lambda_2|$, then*

$$\log_q |j(\phi)| = (q - 1)q^{n+1} \left(\frac{1}{q-1} + \log_q \left|\frac{\lambda_2}{\lambda_1}\right| - n\right).$$

Proof Since $|j(\phi)| = |g_1|^{q+1}/|g_2|$, from Lemma 5.5.7, we get

$$|j(\phi)| = \left|\frac{\lambda_2}{\lambda_1}\right|^{q^{n+1}(q-1)} q^{q+(q^n-1+s(n))(q-q^2)}.$$

The formula of the corollary follows by taking $\log_q$ of both sides and noting that (cf. (5.5.2))

$$q + (q^n - 1 + s(n))(q - q^2) = q^{n+1} - nq^{n+1}(q - 1).$$

$$\square$$

Lemma 5.5.9 $\log_q |j(\phi)| > q$ *if and only if $|\lambda_2| > |\lambda_1|$.*

Proof The Newton polygon of $\phi_T(x)/x = T + g_1 x^{q-1} + g_2 x^{q^2-1}$ is the lower convex hull of the set of points

$$P_0 = (0, -1), \qquad P_{q-1} = (q - 1, \operatorname{ord}_\infty(g_1)), \qquad P_{q^2-1} = (q^2 - 1, \operatorname{ord}_\infty(g_2)).$$

$\operatorname{NP}(\phi_T(x)/x)$ has two line segments if and only if the slope of the line segment joining P_0 with P_{q-1} is strictly smaller than the slope of the line segment joining P_0 with P_{q^2-1}, i.e., if and only if

$$\frac{\operatorname{ord}_\infty(g_1) + 1}{q - 1} < \frac{\operatorname{ord}_\infty(g_2) + 1}{q^2 - 1}.$$

It is easy to check that this inequality is equivalent to

$$q < \operatorname{ord}_\infty(g_2) - (q + 1)\operatorname{ord}_\infty(g_1)$$
$$= -\operatorname{ord}_\infty(j(\phi)) = \log_q |j(\phi)|.$$

On the other hand, by Theorem 2.5.2, $\operatorname{NP}(\phi_T(x)/x)$ has two line segments if and only if $|\mu_1| < |\mu_2|$. Since, by Lemma 5.5.6, the inequality $|\mu_1| < |\mu_2|$ is equivalent to $|\lambda_1| < |\lambda_2|$, we conclude that

$$\log_q |j(\phi)| > q \quad \Longleftrightarrow \quad |\lambda_2| > |\lambda_1|.$$

(Note that the implication $|\lambda_2| > |\lambda_1| \Longrightarrow \log_q |j(\phi)| > q$ also easily follows from Proposition 5.5.8.)   □

Finally, we are ready to prove the main result of this section:

Theorem 5.5.10 *Let ϕ be a Drinfeld module of rank 2 over K, and let Λ be its associated lattice.*

(1) *If $\log_q |j(\phi)| \leq q$, then $K(\Lambda) = K(\phi[T])$ and $[K(\Lambda) : K]$ divides $(q^2 - 1)(q^2 - q)$.*
(2) *Suppose $\log_q |j(\phi)| > q$. Let $n \geq 0$ be the largest integer such that $\log_q |j(\phi)| \geq q^{n+1}$. Then $[K(\Lambda) : K]$ divides $q^{n+1}(q - 1)^2$.*
(3) *Assume $\log_q |j(\phi)| > q$ and the numerator of $\log_q |j(\phi)|$ is not divisible by p. Let p^e, $e \geq 0$, be the largest power of p that divides the ramification index of K/F_∞. If $q^{n+1} \geq p^e$, then the ramification index of $K(\Lambda)/K$ is divisible by q^{n+1}/p^e.*
(4) *$K(\Lambda)$ may have arbitrarily large ramification index over K. In particular, the degree $[K(\Lambda) : K]$ is unbounded as ϕ varies.*
(5) *The residue degree of $K(\Lambda)$ over K is bounded by $(q^2 - 1)(q^2 - q)$.*

Proof Before proving each individual claim, we make some general observations. By Proposition 5.5.1, the Galois group $G := \operatorname{Gal}(K(\Lambda)/K)$ acts on Λ by A-linear automorphisms, which gives an embedding $G \hookrightarrow \operatorname{Aut}_A(\Lambda) \cong \operatorname{GL}_2(A)$. For an SMB $\{\lambda_1, \lambda_2\}$ of Λ and $\sigma \in G$, we have

$$\sigma(\lambda_1) = a_{1,1}\lambda_1 + a_{2,1}\lambda_2,$$

$$\sigma(\lambda_2) = a_{1,2}\lambda_1 + a_{2,2}\lambda_2$$

with $a_{i,j} \in A$. By Lemma 2.3.8, we also have $|\sigma(\lambda_i)| = |\lambda_i|$. On the other hand, by Proposition 5.5.5,

$$|a_{1,j}\lambda_1 + a_{2,j}\lambda_2| = \max\left(|a_{1,j}| |\lambda_1|, |a_{2,j}| |\lambda_2|\right).$$

Therefore,

$$|a_{1,1}| \le 1, \quad |a_{2,1}| \le 1, \quad |a_{2,2}| \le 1, \quad |a_{1,2}| \le \frac{|\lambda_2|}{|\lambda_1|}.$$

This implies that $a_{1,1}, a_{2,1}, a_{2,2} \in \mathbb{F}_q$, $\deg(a_{1,2}) \le \log_q |\lambda_2/\lambda_1|$, and $a_{2,1} = 0$ if $|\lambda_1| < |\lambda_2|$.

There are two separate cases:

(i) $|\lambda_1| = |\lambda_2|$. In this case, G is isomorphic to a subgroup of $\mathrm{GL}_2(\mathbb{F}_q)$, so its order divides $(q^2 - 1)(q^2 - q)$.

(ii) $|\lambda_1| < |\lambda_2|$. In this case, G is isomorphic to subgroup of

$$B = \left\{ \begin{pmatrix} a & b \\ 0 & d \end{pmatrix} \middle| a, d \in \mathbb{F}_q^\times, b \in A \text{ such that } \deg(b) \le \log_q |\lambda_2/\lambda_1| \right\},$$

so its order divides $(q - 1)^2 q^{n+1}$, where n is the integer such that

$$0 \le \log_q \left| \frac{\lambda_2}{\lambda_1} \right| - n < 1. \tag{5.5.6}$$

By Lemma 5.5.9, these two cases correspond to $\log_q |j(\phi)| \le q$ and $\log_q |j(\phi)| > q$, respectively. Moreover, in case (ii), Proposition 5.5.8 implies that

$$q^{n+1} \le \log_q |j(\phi)| < q^{n+2},$$

so n in (5.5.6) is the same n that appears in the statement of the theorem.

Now we prove each claim of the theorem:

(1) With the given assumption, we are in case (i). Then $|\lambda_1/T| = |\lambda_2/T| < |\lambda|$ for all $0 \ne \lambda \in \Lambda$. Hence, as in the proof of Proposition 5.5.1, $\log_\phi$ converges on $e_\phi(\lambda_1/T)$ and $e_\phi(\lambda_2/T)$, so $\lambda_1, \lambda_2 \in K(\phi[T])$. This then implies $K(\Lambda) \subseteq K(\phi[T])$. Thus, by Proposition 5.5.1, $K(\Lambda) = K(\phi[T])$. We already observed that in this case $[K(\Lambda) : K]$ divides $(q^2 - 1)(q^2 - q)$.

(2) With the given assumption, we are in case (ii). We already proved the claim in the first paragraph.

(3) Assume we are in case (ii) and the numerator of $\log_q |j(\phi)|$ is not divisible by p. Then Proposition 5.5.8 implies that q^{n+1} divides the denominator of

$\log_q |\lambda_2/\lambda_1|$. Therefore, $1/q^{n+1} \in \mathrm{ord}_\infty(K(\Lambda))$, so the ramification index of $K(\Lambda)/F_\infty$ is divisible by q^{n+1}. On the other hand,

$$e(K(\Lambda)/F_\infty) = e(K(\Lambda)/K) \cdot e(K/F_\infty),$$

so if $q^{n+1} \geq p^e$, then $e(K(\Lambda)/K)$ must be divisible by q^{n+1}/p^e.

(4) Choose ϕ such that $\log_q |j(\phi)| = q^{n+1} + 1$, where n is any positive integer. Then (3) implies that $e(K(\Lambda)/K)$ is divisible by q^{n+1}/p^e, so it can be made arbitrarily large.

(5) Let $f(K(\Lambda)/K)$ be the residue degree of $K(\Lambda)$ over K. In case (i), we have

$$f(K(\Lambda)/K) \leq [K(\Lambda) : K] \leq (q^2 - 1)(q^2 - q).$$

Now assume we are in case (ii), so that G is isomorphic to a subgroup of B. Let U be the subgroup of B consisting of unipotent matrices with $a = d = 1$. Note that U is normal in B and $B/U \cong \mathbb{F}_q^\times \times \mathbb{F}_q^\times$. Let $L := K(\Lambda)^{G \cap U}$ be the fixed field of the subgroup $G \cap U \subseteq G$. Then $[L : K]$ divides $(q-1)^2$. We will show that $f(K(\Lambda)/L)$ is either 1 or p. Since residue degrees are multiplicative in towers of field extensions, this implies

$$f(K(\Lambda)/K) \leq (q^2 - 1)p \leq (q^2 - 1)(q^2 - q).$$

Note that $U \cong (\mathbb{Z}/p\mathbb{Z})^{[\mathbb{F}_q : \mathbb{F}_p]}$ is a p-elementary abelian group since each element of U, except the identity, has order p. Let L'/L be the maximal unramified subextension of $K(\Lambda)/L$. Then L'/L is a cyclic extension and $f(K(\Lambda)/L) = [L' : L]$. Since a maximal cyclic subgroup of U is isomorphic to $\mathbb{Z}/p\mathbb{Z}$, we conclude that $[L' : L] = 1$ or p.

$\square$

Exercises

5.5.1 Let ϕ be a Drinfeld module of rank r over K. Let $\{\lambda_1, \ldots, \lambda_r\}$ be an SMB of its lattice Λ. Assume $|\lambda_1| = \cdots = |\lambda_r|$. Prove that $\mathrm{Gal}(K(\Lambda)/K)$ is isomorphic to a subgroup of $\mathrm{GL}_r(\mathbb{F}_q)$.

5.5.2 With notation and assumptions of Lemma 5.5.6, show that the pairs $(|\lambda_1|, |\lambda_2|)$ and $(|\mu_1|, |\mu_2|)$ determine each other.

5.5.3 Let $z \in \mathbb{C}_\infty - F_\infty$. Show that $\{1, z\}$ is an SMB of $A + Az$ if and only if $|z| = \Im(z) \geq 1$, where the imaginary part $\Im(z)$ is defined in Exercise 2.3.1.

5.5.4 Let K be an algebraically closed field equipped with a nontrivial valuation. Assume $\mathbb{F}_q$ is a subfield of K. Let $f(x) \in K\langle x \rangle$ be a separable $\mathbb{F}_q$-linear polynomial. Let $W = \ker(f)$.

(a) Consider W as an $\mathbb{F}_q$-vector subspace of K. Show that W has an SMB $\{w_1, \ldots, w_n\}$, where SMB is defined as in Definition 5.5.3, except in (ii) one takes the $\mathbb{F}_q$-span instead of the A-span.
(b) Prove that the valuation of each w_i is the negative of the slope of the segment of $\mathrm{NP}(f)$ above the interval $[q^{i-1}, q^i]$.

5.5.5 Prove the converse of Proposition 5.5.5: If $\{\lambda_1, \ldots, \lambda_r\}$ is an A-basis of $\Lambda \subset \mathbb{C}_\infty$ such that

$$|a_1\lambda_1 + \cdots + a_r\lambda_r| = \max_{1 \le i \le r} |a_i||\lambda_i|$$

for all $a_1, \ldots, a_r \in F_\infty$, then $\{\lambda_1, \ldots, \lambda_r\}$ is an SMB.

5.5.6 Let ϕ be a Drinfeld module over K defined by $\phi_T(x) = Tx + g_1 x^q + \cdots + g_r x^{q^r}$, and let Λ be the lattice associated with ϕ. Let $\{\lambda_1, \ldots, \lambda_r\}$ be an SMB of Λ. Define $\mu_i = e_\Lambda(\lambda_i/T)$, $1 \le i \le r$.

(a) Generalize (5.5.5) to prove that $|\mu_i| = |\mu_{i+1}|$ for some $1 \le i < r$ if and only if $|\lambda_i| = |\lambda_{i+1}|$.
(b) Suppose $g_i = 0$ for some $1 \le i < r$. Prove that $|\lambda_i| = |\lambda_{i+1}|$.

5.5.7 Let Ω^r be the Drinfeld symmetric space. Let

$$\mathcal{F} = \left\{ [c_1, c_2, \ldots, c_r] \in \Omega^r \,\middle|\, \{c_1, c_2 \ldots, c_r\} \text{ is an SMB of } \sum_{i=1}^{r} Ac_i \right\}.$$

Prove that each $[z_1, \ldots, z_r] \in \Omega^r$ is $\mathrm{GL}_r(A)$-equivalent with at least one and at most finitely many elements of $\mathcal{F}$. Essentially, $\mathcal{F}$ is a "fundamental domain" for the action of $\mathrm{GL}_r(A)$ on Ω^r similar to the domain (5.3.5) for $\mathrm{SL}_2(\mathbb{Z})$ acting on the upper half-plane $\mathbb{H}$.

5.5.8 Let $\phi_T = T + g_1\tau + g_2\tau^2$ be a Drinfeld module of rank 2 over F_∞ such that $\mathrm{ord}_\infty(g_1) = 0$ and $\mathrm{ord}_\infty(g_2) = q^n - q^{n-1} - 1$, $n \ge 1$. Let $\{\lambda_1, \lambda_2\}$ be an SMB of the lattice Λ of ϕ.

(a) Prove that $\mathrm{ord}_\infty(\lambda_1) = -1$ and

$$\mathrm{ord}_\infty(\lambda_2) = -\left(n + \frac{q^{n-1} - 1}{q^{n-1}(q-1)}\right).$$

(b) Conclude that the ramification index of $F_\infty(\Lambda)/F_\infty$ is $q^{n-1}(q-1)$.

5.5.9 Let $\phi_T = T + g_1\tau + g_2\tau^2 + \cdots + g_r\tau^r$ be a Drinfeld module of rank r over F_∞ such that

$$\text{ord}_\infty(g_r) = 0,$$

$$\text{ord}_\infty(g_{r-1}) = -n(q^r - q^{r-1}) + 1, \quad n \geq 1,$$

$$\text{ord}_\infty(g_{r-1}) \leq \text{ord}_\infty(g_i) \text{ for all } 1 \leq i \leq r - 1.$$

Prove that $\phi_T(x)$ has $q^r - q^{r-1}$ roots μ with

$$\text{ord}_\infty(\mu) = -n + \frac{1}{q^r - q^{r-1}},$$

and all other nonzero roots have non-negative valuations.

5.5.10

(a) Give an explicit example of a Drinfeld module of rank $r \geq 1$ over F_∞ such that $F_\infty(\phi[T]) = \mathbb{F}_{q^r} F_\infty$. Conclude that a general bound on the residue degree of $K(\Lambda)$ over K has to depend on the rank of the Drinfeld module ϕ.

(b) Show that there is an explicit constant $C(q, r)$, depending only on q and r, such that for any Drinfeld module ϕ of rank r over K the prime-to-p part of $[K(\Lambda_\phi) : K]$ is bounded by $C(q, r)$.

5.5.11 Let $\phi_T = T + g_1\tau + g_2\tau^2$ be a Drinfeld module of rank 2 over F_∞, and let Λ be the lattice associated with ϕ. Prove the following:

(a) If $\log_q |j(\phi)| \leq q$, then $F_\infty(\Lambda)/F_\infty$ is unramified if and only if

$$\log_q |g_2| \equiv 1 \pmod{q^2 - 1}.$$

(b) If $\log_q |j(\phi)| > q$, then $F_\infty(\Lambda)/F_\infty$ is unramified if and only if

$$\log_q |g_1| \equiv 1 \pmod{q - 1} \text{ and}$$

$$\log_q |j(\phi)| \equiv q^{n+1} \pmod{q^{n+1}(q - 1)}.$$

5.5.12 Let ϕ be a Drinfeld module of rank 2 over K and Λ be its associated lattice. Let $\{\lambda_1, \lambda_2\}$ be an SMB of Λ. Let $G = \text{Gal}(K(\Lambda)/K)$. Assume $|\lambda_1| < |\lambda_2|$.

(a) Let B_1 be the subgroup of the group B defined in the proof of Theorem 5.5.10 consisting of matrices with $a = 1$. Prove that $K(\Lambda)^{G \cap B_1} = K(\lambda_1)$.

(b) Let $e_{\lambda_1}(x)$ be the exponential function of the lattice $A\lambda_1 \subset \mathbb{C}_\infty$ generated by λ_1. Prove that $K(\Lambda)^{G \cap U} = K\left(e_{\lambda_1}(\lambda_2)\right)$.

5.5.13 Let ϕ be a Drinfeld module of rank 2 over K. Assume $\log_q |j(\phi)| \leq q$.

(a) Let $0 \neq z_1 \in \phi[T]$. Prove that $\mathrm{ord}_\infty(z_1) = -\frac{\mathrm{ord}_\infty(g_2)+1}{q^2-1}$, so all nonzero elements of $\phi[T]$ have the same valuation.

(b) Consider the Newton polygon of $\phi_T(x) - z_1$ to show that there is $z_2 \in \phi[T^2]$ such that $z_2 \in K(\phi[T])$, $\phi_T(z_2) = z_1$, and $\mathrm{ord}_\infty(z_2) = 1 + \mathrm{ord}_\infty(z_1)$.

(c) Using induction, show that for redany $n \geq 2$ there is $z_n \in \phi[T^n]$ such that $z_n \in K(\phi[T])$, $\phi_T(z_n) = z_{n-1}$, and $\mathrm{ord}_\infty(z_n) = 1 + \mathrm{ord}_\infty(z_{n-1})$.

(d) Deduce from (c) and Proposition 5.5.1 that $K(\Lambda) = K(\phi[T])$.

This gives an alternative proof of Part (1) of Theorem 5.5.10. Parts (2) and (3) of the same theorem can be proved by a more complicated version of this argument. We refer to Maurischat's paper [Mau19] for the details.

5.5.14 Let $\Lambda \subset \mathbb{C}_\infty$ be a lattice of rank r and $\{\lambda_1, \ldots, \lambda_r\}$ be an SMB of Λ. Following [Tag93], we attach an invariant to Λ similar to the area of the fundamental parallelogram of an elliptic curve:

$$\mathrm{Vol}(\Lambda) = \prod_{i=1}^{r} |\lambda_i|.$$

(a) Prove that

$$\mathrm{Vol}(\Lambda) \leq q^r \prod_{i=1}^{r} |\mu_i|,$$

where $\mu_i = e_\Lambda(\lambda_1/T)$ for all $1 \leq i \leq r$.

(b) As was discussed in Sect. 5.3, the group $\mathrm{GL}_r(F_\infty)$ acts on the set of lattices of rank r in $\mathbb{C}_\infty$. Prove that

$$\mathrm{Vol}(g(\Lambda)) = |\det(g)| \cdot \mathrm{Vol}(\Lambda)$$

for any $g \in \mathrm{GL}_r(F_\infty)$.

(c) Let $u : \phi \to \psi$ be an isogeny given on lattices by multiplication by $c \in \mathbb{C}_\infty^\times$ (cf. Theorem 5.2.11). Show that

$$\# \ker(u) = |c|^r \cdot \mathrm{Vol}(\Lambda_\phi)/\mathrm{Vol}(\Lambda_\psi).$$

5.5.15 For an elliptic curve defined over the complex numbers, there is a special isomorphism representative which corresponds to a point z in the standard fundamental domain of the upper half-plane; cf. (5.3.5). The analogous concept for Drinfeld modules is a Drinfeld module whose corresponding lattice is *reduced*, i.e., has an SMB corresponding to a point in the region $\mathcal{F}$ in Exercise 5.5.7. We call a Drinfeld module ϕ over $\mathbb{C}_\infty$ *reduced* if its associated lattice is reduced. Every Drinfeld module is isomorphic over $\mathbb{C}_\infty$ to a reduced Drinfeld module.

(a) Show that a lattice Λ is reduced if and only if $1 \in \Lambda$ and $|\lambda| \geq 1$ for all nonzero $\lambda \in \Lambda$.

(b) Let $u \colon \phi \to \phi'$ be an isogeny of reduced Drinfeld modules, corresponding to reduced lattices Λ and Λ', respectively. Prove that

$$\mathrm{Vol}(\Lambda)/\mathrm{Vol}(\Lambda') \leq \#\ker(u).$$

(c) Let $\hat{u} \colon \phi' \to \phi$ be a dual isogeny. Prove that

$$-\log_q \#\ker(\hat{u}) \leq \log_q \mathrm{Vol}(\Lambda) - \log_q \mathrm{Vol}(\Lambda') \leq \log_q \#\ker(u).$$

Chapter 6
Drinfeld Modules over Local Fields

In this chapter we study Drinfeld modules defined over a field K which is complete with respect to a discrete valuation. More precisely, we assume that K is a finite extension of $F_\mathfrak{p}$ for a prime $\mathfrak{p} \lhd A$, and consider K as an A-field via the natural embeddings

$$\gamma : A \lhook\joinrel\longrightarrow F \lhook\joinrel\longrightarrow F_\mathfrak{p} \lhook\joinrel\longrightarrow K.$$

To simplify the notation, we denote $\gamma(a) \in K$ by a. In addition, we use the following notation and assumptions:

- The valuation v on K is *normalized*, i.e., $v(K^\times) = \mathbb{Z}$.
- $R = \{x \in K \mid v(x) \geq 0\}$ is the ring of integers of K.
- $M = \{x \in K \mid v(x) > 0\}$ is the maximal ideal of R.
- π is a fixed uniformizer for R, i.e., $M = \pi R$.
- $k = R/M$ is the residue field of R.
- $|\alpha| = (\#k)^{-v(\alpha)}$ is the normalized absolute value on K associated with the valuation v.
- $\mathbb{C}_K$ is the completion of an algebraic closure of K with respect to the unique extension of the absolute value on K to $\overline{K}$, which again will be denoted by $|\cdot|$.
- $R_{\mathbb{C}_K} = \{z \in \mathbb{C}_K : |z| \leq 1\}$ is the ring of integers of $\mathbb{C}_K$.
- $M_{\mathbb{C}_K} = \{z \in \mathbb{C}_K : |z| < 1\}$ is the maximal ideal of $R_{\mathbb{C}_K}$.
- The reduction modulo M, and everything derived from it, are denoted by $\bar{*}$. For example, given $f \in R\{\tau\}$, we denote by $\bar{f} \in k\{\tau\}$ the polynomial obtained by reducing the coefficients of f modulo M.
- Notice that we have $\mathfrak{p} = A \cap M$ and $v(\mathfrak{p}) = e(K/F_\mathfrak{p})$ is the ramification index of the extension $K/F_\mathfrak{p}$.

In Sect. 6.1, we introduce a special operation that becomes available for Drinfeld modules over K, namely the reduction of a Drinfeld module modulo M. Through

© The Author(s), under exclusive license to Springer Nature Switzerland AG 2023   345
M. Papikian, *Drinfeld Modules*, Graduate Texts in Mathematics 296,
https://doi.org/10.1007/978-3-031-19707-9_6

this operation the theory of Drinfeld modules over k comes to bear on the theory of Drinfeld modules over K.

In Sect. 6.2, we prove a theorem which essentially says that a Drinfeld module with bad reduction can be uniquely decomposed into a Drinfeld module with good reduction and a lattice. This introduces analytic techniques into the study of Drinfeld modules over K, similar to those in Chap. 5.

In Sect. 6.3, we relate the action of G_K on the torsion points of a Drinfeld module ϕ over K to the reduction properties of ϕ. This leads to an analogue of the Néron-Ogg-Shafarevich criterion for elliptic curves. We also show that the behaviors of $\mathfrak{p}$-primary and $\mathfrak{l}$-primary torsion points of ϕ differ drastically under the action of G_K (here $\mathfrak{l}$ is a prime of A different from $\mathfrak{p}$).

In Sect. 6.4, using the Newton polygons of the division polynomials of Drinfeld modules, we show that only finitely many torsion points of ϕ over K are fixed by G_K, i.e., $(^{\phi}K)_{\text{tor}}$ is finite; this is in contrast to the situation over F_{∞}, where we showed that all torsion points of ϕ are defined over a finite extension of F_{∞}.

In Sect. 6.5, we define the formal completion at $\mathfrak{p}$ of a Drinfeld module over K, and use this theory to reprove the finiteness of $(^{\phi}K)_{\text{tor}}$.

6.1 Reductions of Drinfeld Modules

Let $\phi: A \to K\{\tau\}$ be a Drinfeld module of rank r defined by

$$\phi_T = T + g_1\tau + \cdots + g_r\tau^r.$$

We say that ϕ is *defined over R* (or *has integral coefficients*) if $\phi_a \in R\{\tau\}$ for all $a \in A$.

Lemma 6.1.1 *The following are equivalent:*

(1) ϕ is defined over R.
(2) $\phi_T \in R\{\tau\}$.
(3) $\phi_a \in R\{\tau\}$ for some $a \in A$ with $\deg(a) \geq 1$.

Proof The implications $(1) \Leftrightarrow (2) \Rightarrow (3)$ are obvious.

$(3) \Rightarrow (2)$. Assume ϕ_T does not have integral coefficients. Let $1 \leq m \leq r$ be the maximal index such that $v(g_m) < 0$. Using $\phi_{T^n} = \phi_T\phi_{T^{n-1}}$ and induction, it is easy to show that the maximal index of a coefficient of ϕ_{T^n} with negative valuation is $m + (n-1)r$ and the valuation of that coefficient is $q^{r(n-1)}v(g_m)$. Now let $a \in A$ be a polynomial of degree $n \geq 1$, which we write as $a = a_0 + a_1T + \cdots a_nT^n$ with $a_0, \ldots, a_n \in \mathbb{F}_q$. From $\phi_a = \sum_{i=0}^{n} a_i\phi_{T^i}$, it is clear that the $(m + (n-1)r)$-th coefficient of ϕ_a has valuation $q^{r(n-1)}v(g_m) < 0$.   □

We can always find a Drinfeld module ψ isomorphic to ϕ which is defined over R. Indeed, it is possible to choose $c \in K^{\times}$ such that $(q^i - 1)v(c) + v(g_i) \geq 0$ for all $1 \leq i \leq r$. Then

$$\psi_T := c^{-1}\phi_T c = T + \sum_{i=1}^{r} c^{q^i-1} g_i \tau^i \in R\{\tau\}$$

defines a Drinfeld module ψ over R which is isomorphic to ϕ. Reducing the coefficients of ψ modulo $\mathcal{M}$ we obtain a homomorphism $\overline{\psi} : A \to k\{\tau\}$. The problem is that $\overline{\psi}$ might be a Drinfeld module of strictly smaller rank than ψ or not be a Drinfeld module at all if $\overline{\psi}(A) \subset k$.

Example 6.1.2 Let $\phi_T = T + \frac{1}{\pi}\tau + \tau^2$. To make the coefficients of $\psi = c^{-1}\phi c$ integral we must choose c with valuation $v(c) \geq 1$. In that case, $\psi_T = T + \frac{c^{q-1}}{\pi}\tau + c^{q^2-1}\tau^2$. Since $v(c^{q^2-1}) > 0$, we see that $\overline{\psi}$ never has rank 2. Moreover, if $q > 2$, then $v(c^{q-1}/\pi) > 0$, so $\overline{\psi}$ is not a Drinfeld module.

Definition 6.1.3 We say that ϕ has *stable reduction* (respectively, *good reduction*) over K if there is $c \in K^{\times}$ such that $\psi = c^{-1}\phi c$ is defined over R and $\overline{\psi}$ is a Drinfeld module (respectively, Drinfeld module of rank r). We say that ϕ has *potentially stable reduction* (respectively, *potentially good reduction*) if ϕ has stable reduction (respectively, good reduction) when considered as a Drinfeld module over some finite extension L of K. Note that in the $r = 1$ case the notions of stable and good reduction coincide. If ϕ does not have good reduction, then we say that it has *bad reduction*.

Definition 6.1.4 Let

$$e(\phi) = \min_{1 \leq i \leq r} \frac{v(g_i)}{q^i - 1} \tag{6.1.1}$$

and

$$r'(\phi) = \max\{1 \leq i \leq r \mid e(\phi) = v(g_i)/(q^i - 1)\}.$$

Lemma 6.1.5 *Let ϕ be a Drinfeld module over K, and let L/K be a finite extension.*

(1) ϕ has stable reduction over L if and only if $e(\phi) \in v(L)$.
(2) Suppose ϕ has stable reduction over L. Let $c \in L^{\times}$ be such that $\overline{\psi} = \overline{c^{-1}\phi c}$ is a Drinfeld module. Then the rank of $\overline{\psi}$ is $r'(\phi)$.

Proof It is easy to see that ϕ has stable reduction over L if and only if there is $c \in L^{\times}$ such that $e(c^{-1}\phi c) = 0$. On the other hand,

$$e(c^{-1}\phi c) = e(\phi) + v(c).$$

Combining these two facts gives (1).

To prove (2), note that if ϕ is such that $e(\phi) = 0$, then the rank of the Drinfeld module $\bar{\phi}$ is the largest index $1 \le i \le r$ such that $v(g_i) = 0$, which itself is equal to $r'(\phi)$.   □

Proposition 6.1.6 *(1) and (2) follow from Lemma 6.1.5 and its proof. (3) essentially follows from the argument in the proof of Lemma 6.1.7.*

(1) *ϕ has stable reduction over a totally tamely ramified extension L/K of degree equal to the denominator of $e(\phi)$ reduced to lowest terms.*

(2) *Any stable reduction of ϕ has rank $r'(\phi)$, so this rank does not depend on the extension L/K over which ϕ acquires stable reduction.*

(3) *ϕ has potentially good reduction if and only if for **some** $a \in A$ of positive degree we have*

$$\frac{v(g_n(a))}{(q^n - 1)} \le \frac{v(g_i(a))}{(q^i - 1)} \quad \text{for all} \quad 1 \le i \le n = r \cdot \deg(a),$$

where $\phi_a = a + \sum_{i=1}^{n} g_i(a)\tau^i$.

Proof This follows from Lemma 6.1.5 and its proof.   □

Lemma 6.1.7 *Let ϕ be a Drinfeld module defined over K. Let $a \in A$ be an element of positive degree and assume that $\mathfrak{p} \nmid a$. If the slope of the first line segment of $\mathrm{NP}(\phi_a(x)/x)$ is an integer, then ϕ has stable reduction.*

Proof Let

$$\phi_a(x) = \sum_{i=0}^{n} g_i(a)x^{q^i}, \qquad g_0(a) = a, \quad n = r \cdot \deg(a).$$

Since the constant term a is a unit in R the minimal slope of $\mathrm{NP}(\phi_a(x)/x)$ is the slope of the line segment joining $(0, 0)$ to $(q^m - 1, v(g_m(a)))$ for some $0 < m \le n$. Since the slope of the line segment joining $(0, 0)$ to $(q^i - 1, v(g_i(a)))$ is $v(g_i(a))/(q^i - 1)$, we have

$$\frac{v(g_m(a))}{q^m - 1} \le \frac{v(g_i(a))}{q^i - 1} \quad \text{for all } 1 \le i \le n.$$

By assumption, $v(g_m(a))/(q^m - 1)$ is an integer. Choose some $c \in K^\times$ with $v(c) = v(g_m(a))/(q^m - 1)$, and put $\psi = c\phi c^{-1}$. Then

$$\psi_a(x) = a + c^{1-q}g_1(a)x^q + \cdots + c^{1-q^n} g_n(a)x^{q^n}.$$

For all $1 \leq i \leq n$ we have

$$v(c^{1-q^i} g_i(a)) = v(g_i(a)) - (q^i - 1)v(c)$$

$$= v(g_i(a)) - \frac{q^i - 1}{q^m - 1} v(g_m(a))$$

$$\geq 0.$$

By Lemma 6.1.1, this implies that $\psi_T \in R\{\tau\}$. Moreover, since $v(c^{1-q^m} g_m(a)) = 0$, at least one of the coefficients of ψ_T, besides T, must be a unit. Thus, ψ has stable reduction. $\qquad\square$

Corollary 6.1.8 *Let ϕ be a Drinfeld module over K. Let $a \in A$ be an element of positive degree and assume that $\mathfrak{p} \nmid a$. Then ϕ acquires stable reduction over $K(\phi[a])$.*

Proof By considering ϕ as being defined over $K(\phi[a])$, it is enough to show that ϕ has stable reduction assuming $\phi[a] \subset K$. On the other hand, if $\phi[a] \subset K$, then the valuations of all its elements are integers. This implies that all the slopes of $\text{NP}(\phi_a(x)/x)$ are integers, so ϕ has stable reduction by Lemma 6.1.7. $\qquad\square$

Next, we study the reductions of isogenies of Drinfeld modules. To start, we make a simple observation. Let ϕ and ψ be Drinfeld modules over K. Let $\alpha, \beta \in K^\times$. If $u: \phi \to \psi$ is an isogeny, then $\beta u \alpha^{-1}: \alpha \phi \alpha^{-1} \to \beta \psi \beta^{-1}$ is also an isogeny. Hence,

$$\text{Hom}_K(\phi, \psi) \cong \text{Hom}_K(\alpha \phi \alpha^{-1}, \beta \psi \beta^{-1}).$$

This allows us in questions concerning isogenies between two Drinfeld modules ϕ and ψ to assume that both ϕ and ψ are defined over R, and moreover $\bar{\phi}$ (and/or $\bar{\psi}$) is a Drinfeld module if ϕ (and/or ψ) has stable reduction.

Lemma 6.1.9 (Gekeler [Gek83]) *Let $f, g \in R\{\tau\}$, and let $0 \neq u \in K\{\tau\}$ be such that $uf = gu$. If $\deg(\bar{g}) > 0$, then $u \in R\{\tau\}$.*

Proof Let $u = \sum u_i \tau^i$, $f = \sum f_i \tau^i$, $g = \sum g_i \tau^i$. Then for $m \geq 0$ we have

$$\sum_{i+j=m} \left(u_i f_j^{q^i} - u_i^{q^j} g_j \right) = 0. \tag{6.1.2}$$

Let i_0 be the largest index i with maximal $|u_i|^{q^{-i}}$, and let j_0 be the largest index j with $|g_j| = 1$. By assumption, $j_0 \geq 1$. If $u \notin R\{\tau\}$, then $|u_{i_0}| > 1$. In that case, for $m = i_0 + j_0 = i + j$ the following is true:

$$\left| u_{i_0}^{q^{j_0}} g_{j_0} \right| = \left| u_{i_0}^{q^{-i_0}} \right|^{q^m} > \left| u_i^{q^{-i}} \right|^{q^m} |g_j| = \left| u_i^{q^j} g_j \right|, \quad i \neq i_0. \tag{6.1.3}$$

Furthermore, for $i \leq m$ with $|u_i| > 1$ and $i + j = m$, we have

$$\left| u_{i_0}^{q^{j_0}} g_{j_0} \right| = \left| u_{i_0}^{q^{-i_0}} \right|^{q^m} \overset{(*)}{\geq} \left| u_i^{q^{-i}} \right|^{q^m} \overset{(\dagger)}{\geq} |u_i| \geq \left| u_i f_j^{q^i} \right|. \tag{6.1.4}$$

An equality in $(\dagger)$ implies $i = m$; an equality in $(*)$ implies $\left| u_{i_0}^{q^{-i_0}} \right| = \left| u_i^{q^{-i}} \right|$. By assumption, $i_0 \geq i$, so equalities in both $(*)$ and $(\dagger)$ imply $m \geq i_0 \geq i = m$. Thus, $i_0 = m$, which contradicts $j_0 = m - i_0 \geq 1$. On the other hand, because of the strong triangle inequality, strict inequalities in (6.1.3) and (6.1.4) contradict (6.1.2). Thus, $|u_{i_0}| \leq 1$, so $u \in R\{\tau\}$. $\qquad\square$

Proposition 6.1.10 *Let ϕ and ψ be Drinfeld modules over R. Denote*

$$\mathrm{Hom}_R(\phi, \psi) = \{u \in R\{\tau\} \mid u\phi_a = \psi_a u \text{ for all } a \in A\}.$$

If $\bar{\psi}$ is a Drinfeld module, then

$$\mathrm{Hom}_K(\phi, \psi) = \mathrm{Hom}_R(\phi, \psi).$$

Proof Apply Lemma 6.1.9 to $f = \phi_T$ and $g = \psi_T$. $\qquad\square$

Proposition 6.1.11 *Let ϕ and ψ be Drinfeld modules over R. Denote*

$$\mathrm{Hom}_k(\bar{\phi}, \bar{\psi}) = \left\{u \in k\{\tau\} \mid u\bar{\phi}_a = \bar{\psi}_a u \text{ for all } a \in A\right\}.$$

If $\bar{\phi}$ is a Drinfeld module, then the natural homomorphism

$$\mathrm{Hom}_R(\phi, \psi) \xrightarrow{\bmod \mathcal{M}} \mathrm{Hom}_k(\bar{\phi}, \bar{\psi})$$

is injective.

Proof We argue by contradiction. Suppose $u \in \mathrm{Hom}_R(\phi, \psi)$ is such that $u \neq 0$, but $\bar{u} = 0$. We can write $u = \pi^m w$, where $m > 0$, $w \in R\{\tau\}$, and $\bar{w} \neq 0$. From $u\phi_{\mathfrak{p}} = \psi_{\mathfrak{p}} u$, we get $\pi^m w \phi_{\mathfrak{p}} = \psi_{\mathfrak{p}} \pi^m w$. If $\psi_{\mathfrak{p}} = \sum_{n=0}^{r \cdot \deg(\mathfrak{p})} g_n(\mathfrak{p}) \tau^n$, then

$$\psi_{\mathfrak{p}} \pi^m = \left(\sum_{n=0}^{r \cdot \deg(\mathfrak{p})} g_n(\mathfrak{p}) \tau^n \right) \pi^m$$

$$= \sum_{n=0}^{r \cdot \deg(\mathfrak{p})} g_n(\mathfrak{p}) \pi^{m q^n} \tau^n$$

$$= \pi^m f,$$

where $\bar{f} = 0$ because $g_0(\mathfrak{p}) = \mathfrak{p} \in M$. Substituting, we find $\pi^m w\phi_\mathfrak{p} = \pi^m f w$, and so $w\phi_\mathfrak{p} = fw$. Reducing modulo M, we get $\bar{w}\bar{\phi}_\mathfrak{p} = \bar{f}\bar{w} = 0$. Since $\bar{\phi}$ is a Drinfeld module we know that $\bar{\phi}_\mathfrak{p} \neq 0$. It follows that $\bar{w} = 0$, which is a contradiction. □

Corollary 6.1.12 *If ϕ and ψ have stable reduction over K, then there is a natural injective homomorphism*

$$\mathrm{Hom}_K(\phi, \psi) \longrightarrow \mathrm{Hom}_k(\bar{\phi}, \bar{\psi}).$$

Moreover, if ϕ has good reduction, then the degree in τ of a morphism $u : \phi \to \psi$ is preserved under the reduction.

Proof After possibly replacing ϕ and ψ by isomorphic modules over K, we may assume that both are defined over R and $\bar{\phi}, \bar{\psi}$ are Drinfeld modules. By Proposition 6.1.10, we have $\mathrm{Hom}_K(\phi, \psi) = \mathrm{Hom}_R(\phi, \psi)$. By Proposition 6.1.11, the reduction map

$$\mathrm{Hom}_K(\phi, \psi) = \mathrm{Hom}_R(\phi, \psi) \xrightarrow{\mathrm{mod}\, M} \mathrm{Hom}_k(\bar{\phi}, \bar{\psi})$$

is injective.

Suppose there is an isogeny $u : \phi \to \psi$ defined over K. If ϕ has good reduction, so that $\deg(\phi_T) = \deg(\bar{\phi}_T)$, then ψ also has good reduction. Indeed, since $\bar{u} \neq 0$, computing the degrees of both sides of the equality $\bar{u}\bar{\phi}_T = \bar{\psi}_T\bar{u}$, we see that $\deg_\tau(\bar{\phi}_T) = \deg_\tau(\bar{\psi}_T)$. Hence,

$$\deg_\tau(\psi_T) = \deg_\tau(\bar{\psi}_T) = \deg_\tau(\bar{\phi}_T) = \deg_\tau(\bar{\psi}_T).$$

Let g_r and g'_r be the leading coefficients of ϕ_T and ψ_T. Let $u_n\tau^n$ be the leading term of u. Then $u\phi_T = \psi_T u$ implies that $u_n g_r^{q^n} = u_n^{q^r} g'_r$, which itself implies that $u_n \in R^\times$ because $g_r, g'_r \in R^\times$ and $r > 0$. □

Exercises

6.1.1 Let ϕ be a Drinfeld module of rank 2 defined over K. Prove that ϕ has potentially good reduction if and only if $v(j(\phi)) \geq 0$.

6.1.2 Let ϕ be Drinfeld module of rank 2 over K defined by $\phi_T = T + g_1\tau + g_2\tau^2$. Assume $j(\phi) \neq 0$. Let $\delta(\phi)$ be the image of $g_1 \in K^\times$ in $K^\times/(K^\times)^{q-1}$. Prove the following:

(a) Two Drinfeld modules ϕ, ψ are isomorphic over K if and only if $j(\phi) = j(\psi)$ and $\delta(\phi) = \delta(\psi)$.
(b) ϕ has stable reduction such that $\bar{\phi} \cong \bar{C}$ over k if and only if $v(j(\phi)) < 0$ and $\delta(\phi) = 1$.

6.1.3 We say that ϕ is *minimal* if $v(g_r)$ is minimal among those Drinfeld modules which are defined over R and are isomorphic to ϕ. Prove that ϕ is minimal if and only if $0 \leq e(\phi) < 1$.

6.1.4 Let ϕ be a Drinfeld module over K with potentially good reduction. Prove that ϕ has good reduction over K if one of the following conditions holds:

(a) There is $a \in A$ of positive degree such that the leading coefficient of ϕ_a is in $R^\times$.
(b) ϕ has a nonzero K-rational a-torsion point for some $a \in A$ such that $\mathfrak{p} \nmid a$.

6.1.5 Let $f, g \in R\{\tau\}, 0 \neq u \in K\{\tau\}$ with $uf = gu$.

(a) Give an example where $\deg(\bar{f}) > 0$ while $u \notin R\{\tau\}$.
(b) Prove that if $\deg(\bar{f}) > 0$ and $\deg(\bar{g}) > 0$, then $\bar{u} \neq 0$.

6.1.6 Let ϕ be a Drinfeld module over R with good reduction. Assume $\mathfrak{n} \in A$ is not divisible by $\mathfrak{p}$.

(a) Prove that $\phi[\mathfrak{n}]$, under the reduction modulo $\mathfrak{p}$, maps isomorphically onto $\bar{\phi}[\mathfrak{n}]$.
(b) Without using Corollary 6.1.12, show that if $u \in \mathrm{End}_R(\phi)$ maps to $0 \in \mathrm{End}_k(\bar{\phi})$, then $u(\alpha) = 0$ for all $\alpha \in \phi[\mathfrak{n}]$.
(c) Conclude from (a) and (b) that the map $\mathrm{End}_R(\phi) \xrightarrow{\bmod M} \mathrm{End}_k(\bar{\phi})$ is injective.

6.1.7 Let ϕ be a Drinfeld module over R with good reduction. Let $u \in \mathrm{End}_K(\phi)$. Assume $\mathfrak{n} \in A$ is not divisible by $\mathfrak{p}$.

(a) Show that $v(\alpha) = 0$ for all nonzero $\alpha \in \phi[\mathfrak{n}]$.
(b) Suppose $\bar{u} = g\bar{\phi}_\mathfrak{n}$ for some $g \in k\{\tau\}$. Prove that $u = w\phi_\mathfrak{n}$ for some $w \in R\{\tau\}$ with $\bar{w} = g$.
(c) Show that w from (b) is in $\mathrm{End}_K(\phi)$.
(d) Conclude that the torsion of the cokernel of the injective homomorphism of A-modules

$$\mathrm{End}_K(\phi) \hookrightarrow \mathrm{End}_k(\bar{\phi})$$

is $\mathfrak{p}$-primary.

6.1.8 For a polynomial $f(x) = \sum_{i \geq 1} f_i x^i \in K[x]$ with zero constant term and $r \in \mathbb{R}$, define a function $V_f : \mathbb{R} \to \mathbb{R}$ by

$$V_f(r) = \min\{v(f_i) + ir \mid i \geq 1\}.$$

Prove the following:

(a) For two polynomials with zero constant terms f and g, we have $V_{f \circ g} = V_f \circ V_g$.
(b) If $f : \phi \to \phi'$ is an isogeny of Drinfeld modules over K, then $V_f(e(\phi)) = e(\phi')$.

6.2 Tate Uniformization

Since K is a complete non-Archimedean field, one might attempt to uniformize Drinfeld modules over K by A-lattices in $\mathbb{C}_K$, as we did over $\mathbb{C}_\infty$ in Chap. 5. Unfortunately, this approach immediately fails because $\mathbb{C}_K$ has no nontrivial A-lattices. Indeed, if $\Lambda \subset \mathbb{C}_K$ is any nonzero A-submodule and $0 \neq \lambda \in \Lambda$, then $\mathfrak{p}^n \lambda \in \Lambda$ for all $n \geq 0$ and $\mathfrak{p}^n \lambda \to 0$, so 0 is an accumulation point of Λ. Hence $\mathbb{C}_K$ contains no discrete A-submodules. Fortunately, one is able to remedy the situation by considering a different A-module structure on $\mathbb{C}_K$ given by a Drinfeld module.

Definition 6.2.1 Let ϕ be a Drinfeld module over K. A *ϕ-lattice of rank d* is a free A-submodule $\Lambda \subset {}^\phi K^{\mathrm{sep}}$ of rank d, which is invariant under the action of G_K and discrete in $\mathbb{C}_K$. As before, "discrete" in this context means that only finitely many elements of Λ are contained in any closed disc of finite radius in $\mathbb{C}_K$.

Example 6.2.2 Assume ϕ is a Drinfeld module defined over R with good reduction. Fix some $\omega \in K^\times$ and consider $\Lambda = \{\phi_a(\omega) \mid a \in A\}$. Obviously, Λ is an A-submodule of ${}^\phi K$, and Λ is G_K-invariant. If $\omega \notin {}^\phi K_{\mathrm{tor}}$, then Λ is free of rank 1. Using the assumptions on ϕ, one easily computes

$$v(\phi_a(\omega)) = \begin{cases} v(\omega)q^{r \deg(a)}, & \text{if } v(\omega) < 0, \\ \geq 0, & \text{if } v(\omega) \geq 0. \end{cases}$$

This implies that Λ is a ϕ-lattice of rank 1 if and only if $v(\omega) < 0$ and $\omega \notin {}^\phi K_{\mathrm{tor}}$. On the other hand, since the torsion points of ϕ are integral over R, the assumption $v(\omega) < 0$ already implies $\omega \notin {}^\phi K_{\mathrm{tor}}$. Thus, Λ is a ϕ-lattice of rank 1 if and only if $v(\omega) < 0$.

Remark 6.2.3 The reader familiar with the theory of Tate curves will recognize the similarity of the above construction with the construction of discrete subgroups $\omega^{\mathbb{Z}}$ of $\mathbb{Q}_p^\times$ generated by elements with $|\omega|_p > 1$; cf. [Sil94, §V.3]. Note that $(\mathbb{Q}_p, +)$ has no discrete subgroups other than 0, so the change $(\mathbb{Q}_p, +) \rightsquigarrow (\mathbb{Q}_p^\times, \times)$ plays a role similar to $K \rightsquigarrow {}^\phi K$.

Let Λ be a ϕ-lattice. As in Chap. 5, we define

$$e_\Lambda(x) = x \prod_{\lambda \in \Lambda}{}' \left(1 - \frac{x}{\lambda}\right).$$

Since Λ is a discrete $\mathbb{F}_q$-subspace of $\mathbb{C}_K$, by Proposition 5.1.3, $e_\Lambda(x)$ is entire and $\mathbb{F}_q$-linear. Moreover, due to the G_K-invariance of Λ, the coefficients of e_Λ are in K.

Given $0 \neq a \in A$, define

$$\Lambda' := \phi_a^{-1}\Lambda = \{z \in \mathbb{C}_K \mid \phi_a(z) \in \Lambda\}.$$

It is clear that Λ' is an A-submodule of $^\phi\mathbb{C}_K$ such that $\phi_a(\Lambda') = \Lambda$, and $\Lambda \subset \Lambda'$. But Λ' is not a ϕ-lattice since it contains the torsion elements $\phi[a] \subset {}^\phi(\mathbb{C}_K)_{\mathrm{tor}}$. To prove that Λ' is discrete, we use the following well-known fact:

Lemma 6.2.4 (Kernel-Cokernel Lemma) *Every pair of homomorphisms*

$$M_1 \xrightarrow{f} M_2 \xrightarrow{g} M_3$$

of modules over a commutative ring gives rise to an exact sequence

$$0 \longrightarrow \ker(f) \longrightarrow \ker(g \circ f) \longrightarrow \ker(g) \longrightarrow \mathrm{coker}(f) \longrightarrow$$
$$\mathrm{coker}(g \circ f) \longrightarrow \mathrm{coker}(g) \to 0.$$

Proof This is an easy consequence of the Snake Lemma (Lemma 5.2.13) applied to the commutative diagram

$$
\begin{array}{ccccccc}
 & M_1 & \xrightarrow{f} & M_2 & \longrightarrow & \mathrm{coker}\, f & \longrightarrow & 0 \\
 & \downarrow{\scriptstyle g\circ f} & & \downarrow{\scriptstyle g} & & \downarrow & & \\
0 & \longrightarrow & M_3 & \xrightarrow{\mathrm{id}} & M_3 & \longrightarrow & 0.
\end{array}
$$

$\square$

Applying the previous lemma to

$$\Lambda \xhookrightarrow{\iota} \Lambda' \xrightarrow{\phi_a} \Lambda,$$

one obtains the exact sequence of A-modules

$$0 \longrightarrow \ker(\iota) \longrightarrow \ker(\phi_a\iota) \longrightarrow \ker(\phi_a) \longrightarrow \mathrm{coker}(\iota)$$
$$\longrightarrow \mathrm{coker}(\phi_a\iota) \longrightarrow \mathrm{coker}(\phi_a) \to 0.$$

Now $\phi_a\iota: \Lambda \to \Lambda$ is injective (since Λ is a ϕ-lattice) and $\phi_a: \Lambda' \to \Lambda$ is surjective, so the previous exact sequence degenerates to the short-exact sequence

$$0 \longrightarrow \phi[a] \longrightarrow \Lambda'/\Lambda \longrightarrow \Lambda/\phi_a\Lambda \longrightarrow 0. \qquad (6.2.1)$$

In particular, Λ has finite index in Λ', which implies that Λ' is discrete. Thus, $e_{\Lambda'}(x)$ is an $\mathbb{F}_q$-linear entire function on $\mathbb{C}_K$.

Lemma 6.2.5 *The set of zeros of $e_\Lambda(\phi_a(x))$ is Λ'.*

Proof Denote $h(x) = e_\Lambda(\phi_a(x))$. By Corollary 2.7.25, $h(z) = e_\Lambda(\phi_a(z))$ for all $z \in \mathbb{C}_K$. Therefore, $h(z) = 0$ if and only if $\phi_a(z) \in \Lambda$, i.e., if and only if $z \in \Lambda'$.

Because $h(x)$ is $\mathbb{F}_q$-linear with nonzero first coefficient, its zeros are simple. Hence, $Z(h) = \Lambda'$. □

Using Lemma 6.2.5 and Theorem 2.7.16, one deduces that $a e_{\Lambda'}(x) = e_\Lambda(\phi_a(x))$. On the other hand, by Lemma 5.1.4, there is an $\mathbb{F}_q$-linear polynomial $P_a(x)$ such that $e_{\Lambda'}(x) = P_a(e_\Lambda(x))$. Denote $\psi_a(x) = a P_a(x)$. Then we get

$$e_\Lambda(\phi_a(x)) = \psi_a(e_\Lambda(x)). \tag{6.2.2}$$

Moreover, since e_Λ and ϕ_a have coefficients in K, the same is true for ψ_a, i.e., $\psi_a \in K\{\tau\}$.

Proposition 6.2.6 *With above notation and assumptions, the map*

$$\psi : A \longrightarrow K\{\tau\},$$
$$a \longmapsto \psi_a$$

is a Drinfeld module over K of rank $r + d$, where r is the rank of ϕ and d is the rank of Λ.

Proof The proof is the same as the proof of Proposition 5.2.2. Note that

$$\begin{aligned} \operatorname{rank}(\psi) &= \deg_\tau \psi_T \\ &= \log_q(\#(\phi_T^{-1}\Lambda/\Lambda)) \\ &= \log_q(\#\phi[T]) + \log_q(\#\Lambda/\psi_T\Lambda) \\ &= r + d. \end{aligned}$$

□

One can summarize the relationship between e_Λ, ϕ, and ψ into the following exact sequence of A-modules

$$0 \longrightarrow \Lambda \longrightarrow {}^\phi\mathbb{C}_K \xrightarrow{e_\Lambda} {}^\psi\mathbb{C}_K \longrightarrow 0, \tag{6.2.3}$$

which is the analogue of (5.2.6). If ϕ has good reduction over K, then (6.2.3) is called the *Tate uniformization of ψ*.

The sequence (6.2.3) more explicitly means that the diagram

$$\begin{CD} 0 @>>> \Lambda @>>> \mathbb{C}_K @>e_\Lambda>> \mathbb{C}_K @>>> 0 \\ @. @VV\phi_a V @VV\phi_a V @VV\psi_a V @. \\ 0 @>>> \Lambda @>>> \mathbb{C}_K @>e_\Lambda>> \mathbb{C}_K @>>> 0 \end{CD} \tag{6.2.4}$$

is commutative for all $a \in A$. Therefore, the Snake Lemma (Lemma 5.2.13) implies that there is a short-exact sequence of A-modules

$$0 \longrightarrow \phi[a] \longrightarrow \psi[a] \longrightarrow \Lambda/\phi_a\Lambda \longrightarrow 0. \tag{6.2.5}$$

Moreover, since e_Λ, ϕ_a, and ψ_a have coefficients in K, the diagram (6.2.4) is compatible with the action of G_K, so (6.2.5) is also an exact sequence of G_K-modules.

Note that $\psi[a]$ can be explicitly described as

$$\psi[a] = \left\{ e_\Lambda(z) \mid z \in \phi_a^{-1}\Lambda \right\}, \tag{6.2.6}$$

and that (6.2.5) and (6.2.6) constitute the analogue of Proposition 5.3.1 in this setting (the proof of (6.2.6) is similar to the proof of that proposition and is left to the reader). Thus, $e_\Lambda(x)$ gives an isomorphism of A-modules

$$e_\Lambda : \phi_a^{-1}\Lambda/\Lambda \xrightarrow{\sim} \psi[a].$$

Example 6.2.7 Let $C_T = T + \tau$ be the Carlitz module. Fix $\omega \in K$ with $v(\omega) < 0$ and let $\Lambda = \{ C_a(\omega) \mid a \in A \}$ be the corresponding C-lattice of rank 1; cf. Example 6.2.2. Let ψ be the Drinfeld module of rank 2 corresponding to the pair (C, Λ). Let $\log_\Lambda = \sum_{n \geq 0} b_n \tau^n$ be the inverse of e_Λ in $K\{\{\tau\}\}$. By Lemma 5.1.11,

$$b_n = -E_{q^n-1}(\Lambda) = -\sum_{\lambda \in \Lambda}' \frac{1}{\lambda^{q^n-1}} = \sum_{a \in A_+} \frac{1}{C_a(\omega)^{q^n-1}}, \qquad n \geq 1.$$

Since $v(C_a(\omega)^{q^n-1}) = (q^n - 1)q^{\deg(a)}v(\omega) < 0$, we have

$$v(b_n) = -v(C_1(\omega)^{q^n-1}) = -(q^n - 1)v(\omega) > 0. \tag{6.2.7}$$

Let

$$\psi_T = T + g_1\tau + g_2\tau^2.$$

From the logarithmic version of the functional equation relating C and ψ,

$$C_T \log_\Lambda = \log_\Lambda \psi_T,$$

we deduce

$$g_1 = 1 + (T - T^q)b_1$$
$$g_2 = (T - T^{q^2})b_2 - b_1 + b_1^q + (T^{q^2} - T^q)b_1^{q+1}.$$

These formulas, combined with (6.2.7), imply $\bar{g}_1 = 1$ and $\bar{g}_2 = 0$. Hence,

$$\bar{\psi}_T = \bar{T} + \bar{g}_1 \tau + \bar{g}_2 \tau^2 = \bar{T} + \tau = \bar{C}_T,$$

so ψ has bad stable reduction of rank 1, which in fact coincides with the reduction of C. Also note that $v(g_1) = 0$ and $v(g_2) = v(b_1) = -(q-1)v(\omega) > 0$, so

$$v(j(\psi)) = v(g_1^{q+1}/g_2) = (q-1)v(\omega) < 0.$$

So far, starting with a Drinfeld module ϕ and a ϕ-lattice Λ, we have constructed another Drinfeld module ψ. As in Chap. 5, the exponential function e_Λ has played a crucial role of a steppingstone between the pair (ϕ, Λ) and ψ:

$$(\phi, \Lambda) \rightsquigarrow e_\Lambda \rightsquigarrow \psi.$$

In Chap. 5, we were able to reverse the arrows in a similar diagram to produce a lattice associated with a given Drinfeld module. The lattice was the set of zeros of an exponential function e_ϕ, which we found by simply solving the equations resulting from the functional equation $e_\phi T = \phi_T e_\phi$. In the current situation, given a Drinfeld module ψ over K, if we want to find a Drinfeld module ϕ of smaller rank and a ϕ-lattice Λ such that ${}^\psi \mathbb{C}_K \cong {}^\phi \mathbb{C}_K/\Lambda$, then solving the functional equation $e_\Lambda \phi_T = \psi_T e_\Lambda$ will not be possible as we have two unknown quantities, ϕ_T and e_Λ. Nevertheless, Example 6.2.7 gives a hint on how one might proceed. If we start with a Drinfeld module ψ with stable bad reduction, then at least we might be able to recover the reduction of ϕ as $\bar{\psi}$. Once we know $\bar{\phi}_i$ we might be able to find e_Λ modulo $\mathcal{M}^2$ by solving the reduction of the function equation modulo $\mathcal{M}^2$, and then continue this process with higher powers of $\mathcal{M}$. This suggest a possibility of finding e_Λ and ϕ simultaneously by successive approximation.

First, we need a few preliminary lemmas. The first lemma is a variant of Lemma 5.1.1.

Lemma 6.2.8 *Let B be an $\mathbb{F}_q$-algebra. Suppose $u = \sum_{i=0}^n u_i \tau^i \in B\{\tau\}$ is such that u_0 is invertible and $u_1, \ldots, u_n$ are nilpotent. Then u is invertible in $B\{\tau\}$.*

Proof After replacing u by $u_0^{-1}u$, we may assume $u_0 = 1$. It is enough to show that $u^{p^s} - 1 = (u-1)^{p^s} = 0$ for some $s \geq 1$ since then $u^{p^s-1}u = uu^{p^s-1} = 1$, so $u^{p^s-1} = u^{-1}$. Note that we can write

$$(u-1)^m = w_m \tau^{m-1}(u_1 \tau + \cdots + u_n \tau^n)$$

for some $w_m \in B\{\tau\}$. For all large enough m, we have $u_i^{q^{m-1}} = 0$ for all $i = 1, \ldots, n$, so $(u-1)^m = 0$. $\qquad\square$

Lemma 6.2.9 *Let B be an $\mathbb{F}_q$-algebra and let $d > 0$ be an integer. Suppose $f = \sum_{i=0}^n f_i \tau^i \in B\{\tau\}$ is such that $f_d \in B^\times$ and $f_{d+1}, \ldots, f_n$ are nilpotent. Then there exists a unique $u = \sum_{j \geq 0} u_j \tau^j \in B\{\tau\}$ with the following properties:*

(i) $u_0 = 1$.

(ii) u_j are nilpotent for $j \geq 1$.

(iii) $g = u^{-1} f u = \sum_{i=1}^{d} g_i \tau^i$ has degree d and $g_d \in B^\times$.

Note that (iii) implicitly assumes that u is invertible, but, thanks to Lemma 6.2.8, the properties (i) and (ii) imply that u is indeed invertible.

Proof First, we prove the existence of u. Let $\mathcal{N}$ be the ideal of B generated by $f_{d+1}, \ldots, f_n$. It is easy to check that this ideal is nilpotent, i.e., $\mathcal{N}^s = 0$ for some $s \geq 1$. If $s = 1$, then $f_{d+1} = \cdots = f_n = 0$ and we can take $u = 1$.

We prove the case $s = 2$ separately. If $n = d$, then there is nothing to prove, as in this case we can take $u = 1$. Let $N \geq 1$ be an arbitrary integer. Assume by induction that we proved the lemma for $0 \leq n - d < N$. Now suppose $n - d = N$. Let $w = 1 + \frac{f_n}{f_d^{q^{n-d}}} \tau^{n-d}$. This is a unit since it has an inverse:

$$w^{-1} = 1 - \frac{f_n}{f_d^{q^{n-d}}} \tau^{n-d}.$$

Moreover,

$$\tilde{f} = w^{-1} f w = f - \frac{f_n}{f_d^{q^{n-d}}} \left(f_0^{q^{n-d}} + f_1^{q^{n-d}} \tau + \cdots + f_d^{q^{n-d}} \tau^d \right) \tau^{n-d}.$$

Note that $\tilde{f} = \sum_{i=0}^{n-1} \tilde{f}_i \tau^i$ has degree $\leq n - 1$ and all the coefficients $\tilde{f}_i$ with $d + 1 \leq i \leq n - 1$ are in $\mathcal{N}$. Since $f_d + \alpha$ is a unit for $\alpha \in \mathcal{N}$ (its inverse being $f_d^{-1} - \alpha/f_d^2$), we also see that $\tilde{f}_d$ is a unit. By the induction hypothesis, since $(n-1) - d = N - 1$, there exists $\tilde{u}$ satisfying (i), (ii), and such that $g = \tilde{u}^{-1} \tilde{f} \tilde{u}$ has degree d with $g_d \in B^\times$. Let $u = w\tilde{u}$. Then $u_0 = 1$ since the constant terms of both $\tilde{u}$ and w are 1. The coefficients u_j are nilpotent for $j \geq 1$ since the coefficients $\tilde{u}_j$ and w_j are nilpotent for $j \geq 1$. Finally, $g = u^{-1} f u$ by construction. This proves the claim for $n - d = N$.

Now suppose $s = M \geq 3$. Assume by induction that we proved the statement for $s = 1, \ldots, M - 1$. Let $B' := B/\mathcal{N}^{M-1}$ and f' be the image of f in $B'\{\tau\}$. The coefficients of f' generate the ideal $\mathcal{N}' = \mathcal{N}/\mathcal{N}^{M-1}$, and obviously $(\mathcal{N}')^{M-1} = 0$. By the induction hypothesis, we can find $u' = \sum_j u'_j \tau^j \in B'\{\tau\}$ with properties (i)–(iii). Let $z = \sum_{i \geq 0} z_i \tau^i$ be any element of $B\{\tau\}$ which maps to u'. It is an easy exercise to show that the coefficients $z_1, z_2, \ldots$ are nilpotent (since $u'_1, u'_2, \ldots$ are nilpotent and $\mathcal{N}^{M-1}$ is nilpotent). Let $g' = z^{-1} f z = \sum_{i=0}^{m} g'_i \tau^i$. It is another easy exercise to show that g'_d is a unit and the ideal I generated by $g'_{d+1}, \ldots, g'_m$ lies in $\mathcal{N}^{M-1}$. Hence, $I^2 = 0$. Applying the argument from the $s = 2$ case, we find $y = \sum_{j \geq 0} y_j \tau^j$ such that $y_1, y_2, \ldots$ are in I (hence also in $\mathcal{N}^{M-1}$) and $g = y^{-1} g' y \in B\{\tau\}$ satisfies (iii). By construction, $u = zy$ has properties (i)–(iii).

To prove the uniqueness of u, assume there is another υ with properties (i)–(iii). Let $h = \upsilon^{-1} f \upsilon$. We have $g = (\upsilon^{-1} u)^{-1} h (\upsilon^{-1} u)$. Since g and h have degree d and invertible leading coefficients, to prove that $u = \upsilon$ it is enough to prove that $u = 1$ if $f_{d+1} = \cdots = f_n = 0$. We argue by contradiction, so suppose $f_{d+1} = \cdots = f_n = 0$ but $\deg(u) = m > 0$. From $fu = ug$, we get $u_m g_d^{q^m} = u_m^{q^d} f_d$. Since f_d and g_d are units, this implies that $u_m = \alpha u_m^{q^d}$ for some $\alpha \in B^\times$. Let $N > 1$ be the smallest integer such that $u_m^N = 0$. If $q^d \geq N$, then $u_m^{q^d} = 0$, and so $u_m = 0$; this contradicts the assumption that $\deg(u) = m$. If $q^d < N$, then

$$0 = u_m^N = u_m^{N - q^d} u_m^{q^d} = u_m^{N - q^d} (\alpha^{-1} u_m) = \alpha^{-1} u_m^{N - q^d + 1}.$$

Thus, $u_m^{N - q^d + 1} = 0$. Since $N - q^d + 1 < N$, we get a contradiction to the assumption that N is the smallest positive integer for which $u_m^N = 0$. $\qquad\square$

Lemma 6.2.10 *Let $f \in R\{\tau\}$ with $d = \deg(\bar{f}) > 0$. There exists a unique $u = u_f \in R\{\{\tau\}\}$ with the following properties:*

(i) $u = 1 + \sum_{i \geq 1} \alpha_i \tau^i$, $|\alpha_i| < 1$, $\alpha_i \to 0$.
(ii) $g = u^{-1} f u$ *lies in* $R\{\tau\}$, *and* $\deg(g) = \deg(\bar{g}) = d$.
(iii) $u(x)$ *is an entire function.*

Proof For $m \geq 1$, let $R_m = R/\mathcal{M}^m$ and let $\bar{f}_m \in R_m\{\tau\}$ be the reduction of f modulo $\mathcal{M}^m$. By Lemma 6.2.9, there exists a unique $u_m \in R_m\{\tau\}^\times$ such that the constant term of u_m is 1, and $u_m^{-1} \bar{f}_m u_m$ is a polynomial of degree d whose leading coefficient is in $R_m^\times$. The uniqueness implies that $u_m \equiv u_{m-1} \pmod{\mathcal{M}^{m-1}}$. Hence, taking the inverse limit of the coefficients, we obtain an element $u = \varprojlim u_m$ of $R\{\{\tau\}\}$. By construction, u satisfies (i) and (ii), and is unique with these properties.

It remains to show that $u(x)$ is entire. Let $f = \sum_{i=0}^n f_i \tau^i$ and $g = \sum_{j=0}^d g_j \tau^j$. For $m > n$, computing the coefficient of τ^m on both sides of the equation $ug = fu$ gives

$$\alpha_m g_0^{q^m} + \alpha_{m-1} g_1^{q^{m-1}} + \cdots + \alpha_{m-d} g_d^{q^{m-d}} = \alpha_m f_0 + \alpha_{m-1}^q f_1 + \cdots + \alpha_{m-n}^{q^n} f_n,$$

which we rewrite as

$$\alpha_{m-d}(g_d^{q^{m-d}} - f_d \alpha_{m-d}^{q^d - 1}) = -\sum_{i=0}^{d-1} \alpha_{m-i} g_i^{q^{m-i}} + \sum_{\substack{0 \leq j \leq n \\ j \neq d}} f_j \alpha_{m-j}^{q^j}. \qquad (6.2.8)$$

Note that $g_d \in R^\times$ is a unit and $f_d \alpha_{m-d}^{q^d - 1} \in \mathcal{M}$, so we have

$$|\alpha_{m-d}| \cdot \left| (g_d^{q^{m-d}} - f_d \alpha_{m-d}^{q^d - 1}) \right| = |\alpha_{m-d}|.$$

Using the fact that g and f have integral coefficients, the triangle inequality applied to (6.2.8) implies

$$|\alpha_{m-d}| \leq \max \left(|\alpha_{m-d+1}|, \ldots, |\alpha_m|, \right.$$

$$\left. |\alpha_{m-n}|^{q^n}, |\alpha_{m-n+1}|^{q^{n-1}}, \ldots, \widehat{|\alpha_{m-d}|^{q^d}}, \ldots |\alpha_m| \right),$$

where $\widehat{|\alpha_{m-d}|^{q^d}}$ means that term has been removed. Since $|\alpha_{m-j}| < 1$ for $0 \leq j \leq d - 1$, we have $|\alpha_{m-j}|^{q^j} < |\alpha_{m-j}|$, so some of the terms of the above inequality can be omitted. After doing so, one concludes that for $i > s := n - d$

$$|\alpha_i| \leq \max \left(|\alpha_{i-s}|^{q^{d+s}}, \ldots, |\alpha_{i-1}|^{q^{d+1}}, |\alpha_{i+1}|, \ldots, |\alpha_{i+d}| \right). \tag{6.2.9}$$

Denote the set on the right-hand side of this inequality by S_i. Now execute the following iterative process. Initially, put $S := S_i$. If $|\alpha_i|^{q^\ell} \in S$ for some $\ell \geq 1$, then delete that element from S. Next, replace each $|\alpha_j|^{q^\ell} \in S$ with $j > i$ by $S_j^{q^\ell}$, where $S_j^{q^\ell}$ denotes the set of elements of S_j raised to power q^ℓ; call the resulting set S. Repeat the same process for this new S. It is easy to see that with each iteration, either the elements $|\alpha_j|$ appear in S to higher powers of q than before or $|\alpha_j|$ has larger index than the elements in the previous S. Thanks to (6.2.9), at each step of the process we have $|\alpha_i| \leq \max S$. On the other hand, since $0 \leq |\alpha_j| < 1$ for all $j > 0$ and $|\alpha_j| \to 0$ as $j \to \infty$, the maximum of the elements in S with indices greater than i will tend to 0. Therefore,

$$|\alpha_i| \leq \max \left(|\alpha_{i-s}|^{q^{d+s}}, \ldots, |\alpha_{i-1}|^{q^{d+1}} \right). \tag{6.2.10}$$

If we denote $\beta_j = |\alpha_j|^{1/q^j}$, $j \geq 1$, then (6.2.10) implies

$$\beta_i \leq \max (\beta_{i-s}, \ldots, \beta_{i-1})^{q^d}. \tag{6.2.11}$$

Hence

$$\beta_{i+1} \leq \max (\beta_{i+1-s}, \ldots, \beta_i)^{q^d}$$

$$\leq \max \left(\beta_{i+1-s}, \ldots, \beta_{i-1}, \beta_{i-s}^{q^d}, \ldots, \beta_{i-1}^{q^d} \right)^{q^d} \qquad \text{(by (6.2.11))}$$

$$\leq \max \left(\beta_{i-s}^{q^d}, \beta_{i+1-s}, \ldots, \beta_{i-1} \right)^{q^d}.$$

Iterating this argument, one shows that $\beta_{i+2} \leq \max\left(\beta_{i-s}^{q^d}, \beta_{i+1-s}^{q^d}, \ldots, \beta_{i-1}\right)^{q^d}$ and so on, eventually arriving at

$$\beta_{i+js} \leq \max(\beta_{i-s}, \ldots, \beta_{i-1})^{q^{(j+1)d}} \qquad \text{for all } j \geq 0.$$

Since $0 \leq \max(\beta_{i-s}, \ldots, \beta_{i-1}) < 1$, this implies that $\beta_j \to 0$ as $j \to \infty$, which is a condition equivalent to $u(x)$ being entire (see Lemma 2.7.6). $\qquad \square$

Theorem 6.2.11 *The Tate uniformization* $(\phi, \Lambda) \rightsquigarrow \psi$ *is a bijection from the set of pairs consisting of a Drinfeld module ϕ over R of rank r with good reduction and a ϕ-lattice Λ of rank d to the set of Drinfeld modules over R of rank $r + d$ with reduction of rank r. Moreover, $\bar{\phi} = \bar{\psi}$.*

Proof Let $\phi_T = T + g_1\tau + \cdots + g_r\tau^r \in R\{\tau\}$, where $g_r \in R^\times$. Let Λ be a ϕ-lattice of rank d. By Proposition 6.2.6, the Tate uniformization produces a Drinfeld module ψ of rank $r + d$. If there is $0 \neq \lambda \in \Lambda$ satisfying $|\lambda| \leq 1$, then $|\phi_a(\lambda)| \leq 1$ for all $a \in A$, contradicting the discreteness of Λ. Hence, for any $0 \neq \lambda \in \Lambda$ we have $|\lambda| > 1$. This implies that $e_\Lambda \in R\{\{\tau\}\}$ and $e_\Lambda \equiv 1 \pmod{\mathcal{M}}$. Now the functional equation $e_\Lambda\phi_T = \psi_T e_\Lambda$ implies that $\psi_T \in R\{\tau\}$. By reducing the functional equation modulo $\mathcal{M}$, we see that $\bar{\phi}_T = \bar{\psi}_T$. Thus, ψ has reduction of rank r.

Conversely, suppose ψ is a Drinfeld module of rank $r + d$ defined over R such that $\bar{\psi}$ has rank r. Lemma 6.2.10 applied to $f = \psi_T$ implies the existence of a unique power series

$$e = 1 + \sum_{i \geq 1} \alpha_i \tau^i \in 1 + \mathcal{M}\{\{\tau\}\}\tau$$

such that

(i) $\phi_T := e^{-1}\psi_T e \in R\{\tau\}$ has degree r;
(ii) $\bar{\phi}_T = \bar{\psi}_T$;
(iii) $e(x) \in R\langle\langle x\rangle\rangle$ is entire.

Note that ϕ_T defines a Drinfeld module ϕ over R of rank r, and (ii) implies that ϕ has good reduction. We will show that the set $\Lambda := Z(e(x))$ of zeros of $e(x)$ is a ϕ-lattice of rank d.

Proposition 5.1.3 implies that Λ is a discrete $\mathbb{F}_q$-linear subspace of $\mathbb{C}_K$; moreover, it lies in K^{sep} and is G_K-invariant.

The relation $e\phi_T = \psi_T e$ implies that $e\phi_a = \psi_a e$ for all $a \in A$. Substituting λ into this equation and using Corollary 2.7.25, we see that $\phi_a(\lambda) \in \Lambda$ for all $a \in A$. Hence, Λ is a discrete $\phi(A)$-module. Next, note that for $0 \neq \lambda \in \Lambda$ we have $|\lambda| > 1$; indeed, if $|\lambda| \leq 1$, then the fact that the coefficients of e are in $\mathcal{M}$ implies that $e(\lambda) \in 1 + \mathcal{M}$, hence cannot be zero. Since the coefficients of $\phi_a(x)$ are in R

and the leading coefficient is a unit in R, we get $|\phi_a(\lambda)| = |\lambda|^{q^{r \cdot \deg(a)}}$. This implies that Λ is a torsion-free $\phi(A)$-module.

Choose some $a \in A$ of positive degree. The functional equation $e\phi_a = \psi_a e$ gives a commutative diagram similar to (6.2.4). From that diagram, as in (6.2.5), we get the short exact sequence

$$0 \longrightarrow \phi[a] \longrightarrow \psi[a] \longrightarrow \Lambda/\phi_a\Lambda \longrightarrow 0,$$

which then implies that $\Lambda/\phi_a\Lambda \cong (A/aA)^d$. Fix a finite subset S of Λ which maps surjectively onto $\Lambda/\phi_a\Lambda$. We claim that S generates Λ. If we assume this for the moment, then it follows from Theorem 1.2.4 that Λ is a free $\phi(A)$-module of finite rank; moreover, the rank has to be d because $\Lambda/\phi_a\Lambda \cong (A/aA)^d$. Thus, to ψ we have associated a pair (ϕ, Λ), and, by the uniqueness of e, this pair is unique. This proves the bijection of the theorem.

It remains to show that S generates Λ. Let $M = \max_{\lambda \in S} |\lambda|$ and let $m = \min_{0 \neq \lambda \in \Lambda} |\lambda|$. Both of these numbers are well-defined and strictly greater than 1 (the second number is well-defined because Λ is discrete). Let $\lambda \in \Lambda$ be arbitrary. We can write $\lambda = \lambda' + \phi_a\lambda_1$, where $\lambda' \in S$ and $\lambda_1 \in \Lambda$. Note that $|\lambda'| \leq M$ while, if $\lambda_1 \neq 0$, then $|\phi_a\lambda_1| \geq m^{q^{r \deg(a)}}$. If $\lambda_1 \neq 0$, then we can write $\lambda_1 = \lambda'' + \phi_a\lambda_2$, where $\lambda'' \in S$ and $\lambda_2 \in \Lambda$, so that

$$\lambda = \lambda' + \phi_a\lambda'' + \phi_{a^2}\lambda_2.$$

Note that $|\lambda' + \phi_a\lambda''| \leq M^{q^{r \deg(a)}}$ while, if $\lambda_2 \neq 0$, $|\phi_{a^2}\lambda_2| \geq m^{q^{2r \deg(a)}}$. If $\lambda_2 \neq 0$, then we iterate this process, and keep doing so while $\lambda_n \neq 0$. If at some stage $\lambda_n = 0$, then λ has been written as an A-span of the elements of S. On the other hand, if $\lambda_n \neq 0$, then λ has been written as a sum $\mu + \phi_{a^n}\lambda_n$ such that $|\mu| \leq M^{q^{(n-1)r \deg(a)}}$ and $|\phi_{a^n}\lambda_n| \geq m^{q^{nr \deg(a)}}$. We are free to choose a, so we pick an element such that $\log M / \log m < q^{r \deg(a)}$. Since $M, m > 1$, we have $\log M / \log m > 0$, so with the previous choice of a we get $M^{q^{(n-1)r \deg(a)}} < m^{q^{nr \deg(a)}}$. This implies that $|\lambda| = |\phi_{a^n}\lambda_n| \geq m^{q^{nr \deg(a)}}$. Since λ is fixed and $m > 1$, this inequality cannot be true for arbitrary n, so eventually λ_n must be 0.   □

Theorem 6.2.12 *Let ψ_1 and ψ_2 be two Drinfeld modules over R with stable reduction. Let (ϕ_1, Λ_1) and (ϕ_2, Λ_2) be the Tate uniformizations of ψ_1 and ψ_2, respectively. Then*

$$\mathrm{Hom}_K(\psi_1, \psi_2) \overset{\sim}{\longrightarrow} \{u \in \mathrm{Hom}_K(\phi_1, \phi_2) \mid u\Lambda_1 \subseteq \Lambda_2\}.$$

If e_1 and e_2 are the exponential functions associated with Λ_1 and Λ_2, respectively, then the above isomorphism is explicitly given by mapping $w \in \mathrm{Hom}_K(\psi_1, \psi_2)$ to $e_2^{-1} w e_1$.

Proof If ψ_1 and ψ_2 have different ranks, then 0 is the only morphism between them. In that case, either $\text{rank}(\phi_1) \neq \text{rank}(\phi_2)$ or $\text{rank}_A(\Lambda_1) \neq \text{rank}_A(\Lambda_2)$. It is easy to see that under either of these assumptions, $\{u \in \text{Hom}_K(\phi_1, \phi_2) \mid u\Lambda_1 \subseteq \Lambda_2\} = \{0\}$. Hence, from now on we assume that $\text{rank}(\psi_1) = \text{rank}(\psi_2)$.

If $\text{rank}(\phi_1) \neq \text{rank}(\phi_2)$, then $\{u \in \text{Hom}_K(\phi_1, \phi_2) \mid u\Lambda_1 \subseteq \Lambda_2\} = \{0\}$. On the other hand, by Corollary 6.1.12,

$$\text{Hom}_K(\psi_1, \psi_2) \hookrightarrow \text{Hom}_k(\bar{\psi}_1, \bar{\psi}_2) = \text{Hom}_k(\bar{\phi}_1, \bar{\phi}_2).$$

Since ϕ_1 and ϕ_2 have good reduction, we have $\text{rank}(\bar{\phi}_1) \neq \text{rank}(\bar{\phi}_2)$, so $\text{Hom}_k(\bar{\phi}_1, \bar{\phi}_2) = 0$. Therefore, $\text{Hom}_K(\psi_1, \psi_2) = 0$. Hence, from now on, we also assume that ϕ_1 and ϕ_2 have the same rank. Then, since $\text{rank}_A(\Lambda_i) = \text{rank}(\psi_i) - \text{rank}(\phi_i)$, $i = 1, 2$, the lattices have the same rank as well.

Let $0 \neq u \in \text{Hom}_K(\phi_1, \phi_2)$ be such that $u\Lambda_1 \subseteq \Lambda_2$. Consider the entire function $e_2 u$. The kernel

$$\Lambda' = \ker(e_2 u) = \{\alpha \in \mathbb{C}_K \mid u\alpha \in \Lambda_2\}$$

is a discrete $\mathbb{F}_q$-linear subspace of $\mathbb{C}_K$. We obviously have $\Lambda_1 \subset \Lambda'$. It is easy to show using $u\phi_1 = \phi_2 u$ that Λ' is a ϕ_1-module, and $u\Lambda_1$ is a ϕ_2-submodule of Λ_2. Since u has finite kernel, $\text{rank}_A(u\Lambda_1) = \text{rank}_A(\Lambda_2)$. Hence, $u\Lambda_1$ has finite index in Λ_2, and Λ_1 has finite index in Λ'. Applying Lemma 5.1.4 to $\Lambda_1 \subset \Lambda'$, we deduce that there is a unique polynomial $w \in K\{\tau\}$ with the same constant term as u such that $e_2 u = w e_1$. Now

$$
\begin{aligned}
e_2 u = w e_1 \quad &\Longrightarrow \quad e_2 u \phi_1 = w e_1 \phi_1 \\
&\Longrightarrow \quad e_2 \phi_2 u = w \psi_1 e_1 \\
&\Longrightarrow \quad \psi_2 e_2 u = w \psi_1 e_1 \\
&\Longrightarrow \quad \psi_2 w e_1 = w \psi_1 e_1 \\
&\Longrightarrow \quad \psi_2 w = w \psi_1.
\end{aligned}
$$

Hence, $w \in \text{Hom}_K(\psi_1, \psi_2)$. This construction gives an injective homomorphism of A-modules

$$\{u \in \text{Hom}_K(\phi_1, \phi_2) \mid u\Lambda_1 \subseteq \Lambda_2\} \longrightarrow \text{Hom}_K(\psi_1, \psi_2). \tag{6.2.12}$$

Conversely, suppose $w \in \text{Hom}_K(\psi_1, \psi_2)$. Then we have

$$w\psi_1 = \psi_2 w, \quad e_1\phi_1 = \psi_1 e_1, \quad e_2\phi_2 = \psi_2 e_2.$$

We can invert e_1, e_2, w in $K\{\!\{\tau\}\!\}$ by Lemma 2.7.21 (note that $\partial w \neq 0$ because $\mathrm{char}_A(K) = 0$). Then

$$
\begin{aligned}
\phi_1 &= e_1^{-1}\psi_1 e_1 \\
&= e_1^{-1}w^{-1}\psi_2 w e_1 \\
&= e_1^{-1}w^{-1}e_2\phi_2 e_2^{-1}w e_1.
\end{aligned}
$$

Hence, if we put

$$
u = e_2^{-1}w e_1,
$$

then we get $u\phi_1 = \phi_2 u$. This will give us an isogeny $u: \phi_1 \to \phi_2$ if we can show that u is actually a polynomial, and not just a formal power series. By Proposition 6.1.10, $w \in R\{\tau\}$, so $u \in R\{\!\{\tau\}\!\}$ has integral coefficients. Thus, it is meaningful to consider the reduction of u modulo $\mathcal{M}^n$, $n \geq 1$. Since e_1 and e_2 are entire, their coefficients tend to 0. Hence, e_1 and e_2 modulo $\mathcal{M}^n$ are polynomials in $R/\mathcal{M}^n\{\tau\}$. An easy explicit calculation of the coefficients of e_2^{-1} in terms of the coefficients of e_2 (using $e_2^{-1}e_2 = 1$) shows that e_2^{-1} modulo $\mathcal{M}^n$ is also a polynomial for any $n \geq 1$. Thus, the same is true for $u = e_2^{-1}w e_1$. If u is not a polynomial, then the degree of u modulo $\mathcal{M}^n$, as a polynomial in τ, tends to infinity with n. Suppose

$$
\deg_\tau(u \bmod \mathcal{M}^{n-1}) < \deg_\tau(u \bmod \mathcal{M}^n) = s.
$$

Consider $u(\phi_1)_T = (\phi_2)_T u$ modulo $\mathcal{M}^n$. Let f_r and g_r be the leading coefficients of $(\phi_1)_T$ and $(\phi_2)_T$, respectively. Let u_s be the coefficient of τ^s in u. We have

$$
u_s f_r^{q^s} \equiv g_r u_s^{q^r} \pmod{\mathcal{M}^n}.
$$

Since ϕ_1 and ϕ_2 have good reduction, $f_r, g_r \in R^\times$ are units. By assumption, $u_s \in \mathcal{M}^{n-1}$ but $u_s \notin \mathcal{M}^n$, so $u_s^{q^r} \in \mathcal{M}^n$. But then the above congruence, along with the fact that f_r is a unit, implies that $u_s \in \mathcal{M}^n$, a contradiction. Therefore, $u \in \mathrm{Hom}_K(\phi_1, \phi_2)$. Finally, $e_2 u = w e_1$ implies that $u\Lambda_1 \subseteq \Lambda_2$; cf. Corollary 2.7.25. Hence, we obtain an injective homomorphism of A-modules

$$
\mathrm{Hom}_K(\psi_1, \psi_2) \longrightarrow \{u \in \mathrm{Hom}_K(\phi_1, \phi_2) \mid u\Lambda_1 \subseteq \Lambda_2\}. \tag{6.2.13}
$$

It is easy to check that (6.2.12) and (6.2.13) are inverses of each other. This finishes the proof of the theorem.   □

Corollary 6.2.13 *A Drinfeld module over K with complex multiplication has potentially good reduction.*

Proof Let ψ be a Drinfeld module over K with complex multiplication. After passing to a finite extension of K, we may assume that ψ is defined over R and has stable reduction; cf. Lemma 6.1.5. Suppose contrary of the claim that ψ has bad reduction. Let (ϕ, Λ) be the Tate uniformization of ψ. By Theorem 6.2.12, $\mathrm{End}(\psi)$ is an A-submodule of $\mathrm{End}(\phi)$. Hence,

$$\mathrm{rank}_A(\mathrm{End}(\psi)) \leq \mathrm{rank}_A(\mathrm{End}(\phi)).$$

On the other hand, since $\mathrm{char}_A(K) = 0$, Theorem 3.5.3 implies that $\mathrm{rank}_A(\mathrm{End}(\phi)) \leq \mathrm{rank}(\phi)$. Thus,

$$\mathrm{rank}_A(\mathrm{End}(\psi)) \leq \mathrm{rank}_A(\mathrm{End}(\phi)) \leq \mathrm{rank}(\phi) < \mathrm{rank}(\psi).$$

But ψ having complex multiplication means that $\mathrm{rank}_A(\mathrm{End}(\psi)) = \mathrm{rank}(\psi)$, which contradicts the above strict inequality. □

Exercises

6.2.1 This exercise gives an alternative calculation of $v(g_2)$ in Example 6.2.7. The notation and assumptions will be as in that example. Let ω' be a root of $x^q + Tx - \omega$.

(1) Show that the following set is a set of representatives for $C_T^{-1}\Lambda/\Lambda$:

$$\{\alpha\omega' + \beta \mid \alpha \in \mathbb{F}_q, \beta \in C[T]\}.$$

(2) Prove that

$$g_2 = \pm T \left(\prod_{\lambda' \in C_T^{-1}\Lambda/\Lambda}' e_\Lambda(\lambda') \right)^{-1}.$$

(3) Prove that $v(e_\Lambda(\omega')) = v(\omega') < 0$ and $v(e_\Lambda(\beta)) = v(\beta) \geq 0$ for $\beta \in C[T]$.
(4) Deduce that

$$v(g_2) = v(T) - (q^2 - q)v(\omega') - \sum_{\beta \in C[T]}' v(\beta)$$

$$= -(q^2 - q)v(\omega')$$

$$= -(q - 1)v(\omega).$$

6.2.2 Prove the converse of Lemma 6.2.8: If $u = \sum_{i=0}^n u_i \tau^i \in B\{\tau\}^\times$, then u_0 is invertible and $u_1, \ldots, u_n$ are nilpotent.

6.2.3 Let notation and assumptions be as in Theorem 6.2.12. Let $w \in \mathrm{Hom}_K(\psi_1, \psi_2)$ and let u be its corresponding element in $\mathrm{Hom}_K(\phi_1, \phi_2)$.

(a) Prove that $0 \neq \alpha \in \ker(u)$ satisfies $|\alpha| \leq 1$.
(b) Deduce from (a) that $u \colon \Lambda_1 \to \Lambda_2$ is injective.
(c) Prove that there is a short exact sequence of A-modules

$$0 \longrightarrow \ker(u) \longrightarrow \ker(w) \longrightarrow \Lambda_2/u\Lambda_1 \longrightarrow 0.$$

(d) Conclude that

$$\deg_\tau(w) = \log_q[\Lambda_2 : u\Lambda_1] + \deg_\tau(u).$$

6.2.4 Suppose ψ is a Drinfeld module over K of rank 2 and $v(j(\psi)) < 0$. Prove that $\mathrm{End}(\psi) = A$.

6.2.5 Let ψ_1 and ψ_2 be Drinfeld modules over R of rank 2 such that $\bar{\psi}_1 = \bar{\psi}_2 = \bar{C}$. Let (C, Λ_1) and (C, Λ_2) be the Tate uniformizations of ψ_1 and ψ_2, respectively. Let λ_1 and λ_2 be generators of Λ_1 and Λ_2, respectively. Prove that ψ_1 and ψ_2 are isogenous if and only if $C_a(\lambda_1) = C_b(\lambda_2)$ for some nonzero $a, b \in A$.

6.3 Galois Action on Torsion Points

In this section, we consider the action of the absolute Galois group G_K of K on the torsion points of a Drinfeld module ϕ defined over K. It turns out that the action of the inertia subgroup I_K is closely related to the reduction properties of ϕ.

Let S be a set equipped with an action of G_K. We say that S is *unramified* if I_K acts trivially on S.

Theorem 6.3.1 *Let ϕ be a Drinfeld module over K, and let $\mathfrak{l}$ be a prime different from $\mathfrak{p}$. The following properties are equivalent:*

(1) ϕ has good reduction over K.
(2) $\phi[a]$ is unramified for all $a \in A$ coprime to $\mathfrak{p}$.
(3) There are infinitely many $a \in A$, coprime to $\mathfrak{p}$, for which $\phi[a]$ is unramified.
(4) $T_{\mathfrak{l}}(\phi)$ is unramified.

Proof Note that (4) is equivalent to saying that $\phi[\mathfrak{l}^s]$ is unramified for all $s \geq 1$. Hence, (2) $\Rightarrow$ (4) $\Rightarrow$ (3). We will show that (3) $\Rightarrow$ (1) $\Rightarrow$ (2).

(1) $\Rightarrow$ (2). Suppose ϕ has good reduction. By definition, there is $c \in K^\times$ such that $\psi_a = c^{-1}\phi_a c \in R\{\tau\}$ and the leading coefficient of ψ_a is a unit in R for all $a \in A$. Thus, $\bar{\psi}_a(x)$ is a polynomial of the same degree as $\psi_a(x)$. Moreover, if a is coprime to $\mathfrak{p}$, then the polynomial $\bar{\psi}_a(x)$ has distinct roots since $\bar{\psi}_a'(x) = a \pmod{\mathcal{M}} \neq 0$. Hensel's Lemma (see Corollary 2.4.5) implies that the roots of $\psi_a(x)$ are in the maximal unramified extension $\bar{k}K$ of K, so $\psi[a]$ is

unramified. Since $\psi[a] \xrightarrow{\sim} \phi[a], \alpha \mapsto c\alpha$, is an isomorphism of G_K-modules, $\phi[a]$ is unramified.

(3) $\Rightarrow$ (1). Suppose $\phi[a]$ is unramified for infinitely many $a \in A$ such that $\mathfrak{p} \nmid a$. First, we show that ϕ has stable reduction over K. After replacing ϕ by an isomorphic module, we may assume that ϕ has integral coefficients. Let $a \in A$ be an element of positive degree such that $\mathfrak{p} \nmid a$ and $\phi[a]$ is unramified. Since $K(\phi[a])$ is an unramified extension of K, the valuations of the roots of $\phi_a(x)/x$ are integers. Thus, from Theorem 2.5.2, we conclude that the slope of the first line segment of $\mathrm{NP}(\phi_a(x)/x)$ is an integer. By Lemma 6.1.7, ϕ has stable reduction. Thus, after replacing ϕ by an isomorphic module, we may assume that ϕ has integral coefficients and $\bar{\phi}$ is a Drinfeld module of rank $0 < r' \leq r$. We need to show that $r' = r$.

Suppose contrary to the claim that $r' < r$. Let (φ, Λ) be the Tate uniformization of ϕ. Recall that φ has integral coefficients and good reduction, whereas Λ is a φ-lattice in K^{sep} of rank $r - r'$. After passing to a finite extension of K, we can assume that $\Lambda \subset K$. The nonzero elements of Λ have negative valuations (see the proof of Theorem 6.2.11). Let $\omega \in \Lambda$ be a nonzero element such that $v(\lambda) \leq v(\omega)$ for all $0 \neq \lambda \in \Lambda$. Let $a \in A$ be an element which is coprime to $\mathfrak{p}$ and let $d := \deg(a) > 0$. Consider the polynomial $\varphi_a(x) - \omega$. Since the non-constant coefficients of $f(x) := \varphi_a(x) - \omega$ are in R and its leading coefficient is a unit, $\mathrm{NP}(f)$ consists of a single line segment joining $(0, v(\omega))$ with $(q^{r'd}, 0)$. The slope of this line segment is $-v(\omega)/q^{r'd}$. Hence, by Theorem 2.5.2, every root of $f(x)$ has valuation $v(\omega)/q^{r'd}$. Fix a root z of $f(x)$. By (6.2.6),

$$e_\Lambda(z) = z \prod_{\lambda \in \Lambda}{}' \left(1 - \frac{z}{\lambda}\right)$$

$$\in \phi[a]$$

By our choice of ω, we have

$$v(z/\lambda) = \frac{v(\omega)}{q^{r'd}} - v(\lambda) > 0,$$

so $v(1 - z/\lambda) = 0$ for all nonzero $\lambda \in \Lambda$. This implies that

$$v(e_\Lambda(z)) = v(z) = v(\omega)/q^{r'd}.$$

It is clear that for all but finitely many d the number $v(\omega)/q^{r'd}$ is not an integer. When this number is not an integer, $K(e_\Lambda(z))$ is ramified over K. Therefore, for all but finitely many a coprime to $\mathfrak{p}$ the extension $K(\phi[a])$ is ramified over K. This leads to a contradiction. $\square$

Remark 6.3.2 A weaker version of Theorem 6.3.1 was originally proved by Takahashi [Tak82] using a different argument; the advantage of that argument is that it is more elementary as it does not rely on the Tate uniformization. Takahashi's proof is outlined in Exercise 6.3.3.

Let ϕ be a Drinfeld module over K of rank $r \geq 1$. Recall from Chap. 3 that the $\mathfrak{l}$-adic Tate module $T_{\mathfrak{l}}(\phi) \cong A_{\mathfrak{l}}^r$ carries a Galois representation

$$\hat{\rho}_{\phi,\mathfrak{l}} \colon G_K \to \operatorname{Aut}_{A_{\mathfrak{l}}}(T_{\mathfrak{l}}(\phi)) \cong \operatorname{GL}_r(A_{\mathfrak{l}}).$$

By construction, its reduction modulo $\mathfrak{l}$ is the Galois representation on the $\mathfrak{l}$-torsion module:

$$\rho_{\phi,\mathfrak{l}} \colon G_K \to \operatorname{Aut}_{\mathbb{F}_{\mathfrak{l}}}(\phi[\mathfrak{l}]) \cong \operatorname{GL}_r(\mathbb{F}_{\mathfrak{l}}).$$

Corollary 6.3.3 *Let ϕ be a Drinfeld module over K. Let $a \in A$ be an arbitrary element of positive degree such that $\mathfrak{p} \nmid a$.*

(1) *ϕ has potentially good reduction if and only if the image of the inertia group I_K under $\hat{\rho}_{\phi,\mathfrak{l}}$ is finite. Moreover, when this is the case, the kernel of the restriction of $\hat{\rho}_{\phi,\mathfrak{l}}$ to I_K is independent of $\mathfrak{l}$ and contains the wild inertia subgroup I_K^{w}.*
(2) *Suppose ϕ has potentially good reduction. Then $\phi[a]$ is unramified if and only if ϕ has good reduction.*
(3) *Suppose ϕ has potentially stable bad reduction. Then $K(\phi[a])/K$ is wildly ramified for all but finitely many a.*

Proof

(1) Suppose ϕ has good reduction over a finite extension L of K; thanks to Proposition 6.1.6, we may assume that L/K is separable. By Theorem 6.3.1, I_L is in the kernel of $\hat{\rho}_{\phi,\mathfrak{l}}$. On the other hand, $I_L = I_K \cap G_L$, so

$$[I_K : I_L] \leq [G_K : G_L] = [L : K],$$

where the last equality follows from Theorem 1.3.4. Obviously, $\#\hat{\rho}_{\phi,\mathfrak{l}}(I_K) \leq [I_K : I_L]$, so $\hat{\rho}_{\phi,\mathfrak{l}}(I_K)$ is finite.

Conversely, suppose $\hat{\rho}_{\phi,\mathfrak{l}}(I_K)$ is finite. Let $I' := \ker(\hat{\rho}_{\phi,\mathfrak{l}} \colon I_K \to \operatorname{Aut}_{A_{\mathfrak{l}}}(T_{\mathfrak{l}}(\phi)))$. Let σ be an element of G_K mapping to the Frobenius automorphism of $\bar{k}$ under the quotient map $G_K \to G_k$, and let Γ be the closure of the subgroup generated by σ. Then Γ maps isomorphically onto G_k under $G_K \to G_k$. Given $g \in G_K$, we have $gI'g^{-1} \subseteq I_K$ since I_K is normal in G_K. On the other hand, applying $\hat{\rho}_{\phi,\mathfrak{l}}$ to an element in $gI'g^{-1}$, we see that $gI'g^{-1} \subseteq \ker(\hat{\rho}_{\phi,\mathfrak{l}})$. Thus, $gI'g^{-1} = I'$, i.e., $I' \lhd G_K$ is normal. Therefore, the subgroup $H := \langle I', \Gamma \rangle$ of G_K generated by I' and Γ is the set $I'\Gamma = \{i\gamma \mid i \in I', \gamma \in \Gamma\}$, and moreover, $H \cong I' \rtimes \Gamma$ because $I' \cap \Gamma = 1$; see [DF04, p. 175]. Observe that I' is closed in G_K since I' is closed in I_K and I_K is closed in G_K. Thus, H is closed in G_K with finite index, so the fixed field $L := (K^{\mathrm{sep}})^H$ is a finite extension of K by Theorem 1.3.4. From the equality

$H = I'\Gamma$ it is clear that $I_L = H \cap I_K = I'$. We conclude that over L the inertia group acts trivially on $T_{\mathfrak{l}}(\phi)$. Therefore, by Theorem 6.3.1, ϕ has good reduction over L. This proves the first claim of (1).

To prove the second claim of (1), assume ϕ has potentially good reduction. Let I' be the kernel of the restriction of $\hat{\rho}_{\phi,\mathfrak{l}}$ to $I_K = \mathrm{Gal}(K^{\mathrm{sep}}/K^{\mathrm{nr}})$. To find I' we may pass to the maximal unramified extension K^{nr} of K. Let n be the denominator of $e(\phi)$. Combining Lemma 6.1.5 and Proposition 2.6.7, one deduces that n divides $q^r - 1$, and ϕ acquires good reduction over L/K^{nr} if and only if L contains $K^{\mathrm{nr}}(\pi^{1/n})$ as a subfield. Thus, $I' = \mathrm{Gal}(K^{\mathrm{sep}}/K^{\mathrm{nr}}(\pi^{1/n}))$. This group clearly does not depend on $\mathfrak{l}$, and $I_K^{\mathrm{w}} \subset I'$ since $K^{\mathrm{nr}}(\pi^{1/n})/K^{\mathrm{nr}}$ is tamely ramified.

(2) If ϕ has good reduction, then $\phi[a]$ is unramified by Theorem 6.3.1. Now assume $L = K(\phi[a])$ is unramified over K. Then $I_L = I_K$ as subgroups of G_K. By Corollary 6.1.8, ϕ has stable, and therefore good, reduction over L. Hence $I_K = I_L$ acts trivially on $T_{\mathfrak{l}}(\phi)$, so $T_{\mathfrak{l}}(\phi)$ is unramified over K. Now Theorem 6.3.1 implies that ϕ has good reduction over K.

(3) This follows from the proof of Theorem 6.3.1 since the denominator of $v(\omega)/q^{r'\deg(a)}$ is divisible by p for all but finitely many $a \in A$ coprime to $\mathfrak{p}$.

□

Proposition 6.3.4 *Let ϕ be a Drinfeld module over K. Let $\mathfrak{l} \lhd A$ be a prime different from $\mathfrak{p}$. If the inertia group I_K acts on $\phi[\mathfrak{l}]$ by unipotent matrices, then ϕ has stable reduction.*

Proof Because I_K acts on $\phi[\mathfrak{l}]$ by unipotent matrices, the inertia group of the Galois extension $K(\phi[\mathfrak{l}])/K$ is a p-group by Lemma 3.5.11. Hence, the ramification index of $K(\phi[\mathfrak{l}])/K$ is a power of p; cf. Corollary 2.6.11. On the other hand, if ϕ does not have stable reduction over K, then by Lemma 6.1.5 the ramification index of any extension L/K over which ϕ acquires stable reduction is necessarily divisible by some prime not equal to p. This implies that if ϕ does not have stable reduction over K, then ϕ does not have stable reduction over $K(\phi[\mathfrak{l}])$ either. But ϕ does have stable reduction over $K(\phi[\mathfrak{l}])$ by Corollary 6.1.8, so ϕ has stable reduction over K.

□

Remark 6.3.5 It is a theorem of Grothendieck that an abelian variety B over a local field K has semi-stable reduction if and only if the inertia group I_K acts on $T_\ell(B)$ by unipotent matrices, where ℓ is any prime different from the characteristic of the residue field of K. (In case of elliptic curves, semi-stable reduction means that the reduction is either good or multiplicative.) Proposition 6.3.4 proves one direction of the analogue of Grothendieck's theorem for Drinfeld modules. The other direction is generally false, i.e., ϕ might have stable reduction without I_K acting by unipotent matrices on $T_{\mathfrak{l}}(\phi)$; see Example 6.3.6.

Example 6.3.6 This example was shown to me by Chien-Hua Chen. Let $\mathfrak{p} \lhd A$ be a prime not equal to T. Let ϕ be defined by $\phi_T = T + \tau + \mathfrak{p}^{q-1}\tau^2$ over $A_{\mathfrak{p}}$. It is clear that ϕ has bad stable reduction since $\bar{\phi}_T = \bar{T} + \tau$ is a Drinfeld module of rank 1. From the Tate uniformization of ϕ, cf. (6.2.5), we deduce that there is a basis of

$\phi[T]$ over $A/(T)$ such that

$$\rho_{\phi,T}(g) = \begin{pmatrix} 1 & * \\ 0 & \varepsilon(g) \end{pmatrix} \qquad \text{for all } g \in I_K,$$

where ε is the character by which I_K acts on the T-torsion of $\psi_T = T - \mathfrak{p}^{q-1}\tau$; cf. Theorem 3.7.1. Since $\mathfrak{p}\psi_T\mathfrak{p}^{-1} = T - \tau$, we see that $\psi[T]$ is unramified, so $\varepsilon(g) = 1$. Therefore, I_K acts on $\phi[T]$ by unipotent matrices.

Now let ϕ be defined by $\phi_T = T + \tau + \mathfrak{p}\tau^2$ over $A_\mathfrak{p}$. Then again ϕ has bad stable reduction, but ε is the character by which I_K acts on the T-torsion of $\psi_T = T - \mathfrak{p}\tau$. In this case, $\psi[T]$ is ramified, so I_K does not act by unipotent matrices on $\phi[T]$.

For the rest of this section we consider the action of G_K on the $\mathfrak{p}$-primary torsion of a Drinfeld module ϕ. This action is much more complicated than the action on the $\mathfrak{l}$-primary torsion since $\phi[\mathfrak{p}^n]$ is generally ramified, even when ϕ has good reduction. We will only give some partial results.

Assume ϕ is a Drinfeld module over R of rank r and $\bar{\phi}$ is a Drinfeld module. Let H be the height of $\bar{\phi}$. Given $n \geq 1$, let

$$\phi[\mathfrak{p}^n]^0 = \{\alpha \in \phi[\mathfrak{p}^n] : |\alpha| < 1\}.$$

Note that $\phi[\mathfrak{p}^n]^0$ is an A-submodule of $\phi[\mathfrak{p}^n]$; indeed, $\phi[\mathfrak{p}^n]^0$ is obviously closed under addition, and if $|\alpha| < 1$, then $|\phi_a(\alpha)| < 1$ for any $a \in A$ since $\phi_a(x) \in R[x]$. Moreover, since the Galois group preserves the absolute value on K^{sep}, the submodule $\phi[\mathfrak{p}^n]^0$ is G_K-invariant.

Lemma 6.3.7 *With above notation and assumptions, we have*

(1) $\phi[\mathfrak{p}^n]^0 \cong (A/\mathfrak{p}^n)^H$.
(2) *The homomorphism* $\phi[\mathfrak{p}^{n+1}]^0 \longrightarrow \phi[\mathfrak{p}^n]^0$, $\alpha \longmapsto \phi_\mathfrak{p}(\alpha)$, *is surjective.*
(3) *If ϕ has good reduction, then* $\phi[\mathfrak{p}^n]/\phi[\mathfrak{p}^n]^0$ *is unramified.*

Proof Let

$$\phi_{\mathfrak{p}^n}(x) = \mathfrak{p}^n x + \sum_{i=1}^{r \cdot d \cdot n} g_i(\mathfrak{p}^n) x^{q^i}, \qquad \text{where } d = \deg(\mathfrak{p}).$$

Since $\bar{\phi}$ has height H, the coefficient $g_{H \cdot d \cdot n}(\mathfrak{p}^n)$ is a unit in R. This implies that the roots of $\phi_{\mathfrak{p}^n}(x)/x$ with positive valuation correspond to the part of the Newton polygon of $\phi_{\mathfrak{p}^n}(x)/x$ between the vertices $(0, v(\mathfrak{p}^n))$ and $(q^{H \cdot d \cdot n} - 1, 0)$. Hence, $\#\phi[\mathfrak{p}^n]^0 = \#(A/\mathfrak{p}^n)^H$. For $n = 1$, this is sufficient to conclude that $\phi[\mathfrak{p}]^0 \cong (A/\mathfrak{p})^H$ since $\phi[\mathfrak{p}]^0$ can be considered as an $\mathbb{F}_\mathfrak{p}$-vector subspace of $\phi[\mathfrak{p}]$.

It is clear that the image of $\phi[\mathfrak{p}^{n+1}]^0 \to \phi[\mathfrak{p}^n]$, $\alpha \mapsto \phi_\mathfrak{p}(\alpha)$, lies in $\phi[\mathfrak{p}^n]^0$. Let $0 \neq \beta \in \phi[\mathfrak{p}^n]^0$. Note that some of the roots of the polynomial $\phi_\mathfrak{p}(x) - \beta$ have positive valuations since two of the vertices of its Newton polygon are $(0, v(\beta))$ and $(q^{H \cdot d}, 0)$, and $v(\beta) > 0$. It is also clear that such a root α lies in $\phi[\mathfrak{p}^{n+1}]^0$. Thus,

$\phi_{\mathfrak{p}}(\alpha) = \beta$. This implies that there is a short exact sequence

$$0 \longrightarrow \phi[\mathfrak{p}]^0 \longrightarrow \phi[\mathfrak{p}^{n+1}]^0 \xrightarrow{\phi_{\mathfrak{p}}} \phi[\mathfrak{p}^n]^0 \longrightarrow 0.$$

This proves (2), and (1) for $n > 1$ follows from the argument that we have used to prove Theorem 3.5.2.

If ϕ has good reduction, then the leading coefficient of $\phi_{\mathfrak{p}^n}(x)/x$ is a unit. It is easy to see, e.g., by considering the Newton polygon of $\phi_{\mathfrak{p}^n}(x)/x$, that in this case any $\alpha \in \phi[\mathfrak{p}^n]$ satisfies $|\alpha| \leq 1$. Let $\sigma \in I_K$ be an element of the inertia group of K. By definition, $|\sigma(\alpha) - \alpha| < 1$. Since $\sigma(\alpha) - \alpha \in \phi[\mathfrak{p}^n]$, we conclude that $\sigma(\alpha) - \alpha \in \phi[\mathfrak{p}^n]^0$. This implies that I_K acts trivially on $\phi[\mathfrak{p}^n]/\phi[\mathfrak{p}^n]^0$. $\qquad\square$

Remark 6.3.8 With notation of Lemma 6.3.7, suppose ϕ has good reduction and $v(\mathfrak{p}) = 1$. One can show that the inertia group I_K acts on $\phi[\mathfrak{p}^n]^0$ by a character which uniquely extends the $\mathbb{F}_{\mathfrak{p}}$ vector space structure of $\phi[\mathfrak{p}^n]^0$ to a 1-dimensional $\mathbb{F}_{\mathfrak{p}^H}$ vector space structure.

Assume ϕ is defined over R and has stable reduction $\bar{\phi}$. In that case, we have the following exact sequence of A modules:

$$0 \longrightarrow {}^\phi \mathcal{M}_{\mathbb{C}_K} \longrightarrow {}^\phi R_{\mathbb{C}_K} \longrightarrow {}^{\bar{\phi}}\bar{k} \longrightarrow 0.$$

If furthermore we assume that ϕ has good reduction, then every torsion point of ${}^\phi\mathbb{C}_K$ is integral over R, in particular, every torsion point of ${}^\phi\mathbb{C}_K$ lies in $R_{\mathbb{C}_K}$. Hence, from the above exact sequence we get the exact sequence

$$0 \longrightarrow \phi[\mathfrak{p}^n]^0 \longrightarrow \phi[\mathfrak{p}^n] \longrightarrow \bar{\phi}[\mathfrak{p}^n] \longrightarrow 0. \qquad (6.3.1)$$

The exactness is clear except perhaps for the surjectivity $\phi[\mathfrak{p}^n] \longrightarrow \bar{\phi}[\mathfrak{p}^n]$. But we know that $\phi[\mathfrak{p}^n]^0 \cong (A/\mathfrak{p}^n)^H$, $\phi[\mathfrak{p}^n] \cong (A/\mathfrak{p}^n)^r$, and $\bar{\phi}[\mathfrak{p}^n] \cong (A/\mathfrak{p}^n)^{\bar{r}-H}$, where $\bar{r}$ is the rank of $\bar{\phi}$. Since $\bar{r} = r$ by the good reduction assumption, the exactness of (6.3.1) follows.

Taking the inverse limit over the maps $\phi[\mathfrak{p}^{n+1}]^0 \longrightarrow \phi[\mathfrak{p}^n]^0$, we obtain a Galois submodule

$$T_{\mathfrak{p}}(\phi)^0 = \varprojlim \phi[\mathfrak{p}^n]^0 \cong A_{\mathfrak{p}}^H$$

of the Tate module $T_{\mathfrak{p}}(\phi) \cong A_{\mathfrak{p}}^r$ such that $T_{\mathfrak{p}}(\phi)/T_{\mathfrak{p}}(\phi)^0$ is unramified. Moreover, if ϕ has good reduction, then we can canonically identify $T_{\mathfrak{p}}(\phi)/T_{\mathfrak{p}}(\phi)^0$ with $T_{\mathfrak{p}}(\bar{\phi})$. Thus, we have proved the following:

Proposition 6.3.9 *Let ϕ be a Drinfeld module over R with good reduction. Then there is an exact sequence of G_K-modules*

$$0 \longrightarrow T_{\mathfrak{p}}(\phi)^0 \longrightarrow T_{\mathfrak{p}}(\phi) \longrightarrow T_{\mathfrak{p}}(\bar{\phi}) \longrightarrow 0.$$

Lemma 6.3.10 *Let ϕ be a Drinfeld module over R with good reduction. Assume $v(\mathfrak{p}) = 1$ (equiv. $K/F_\mathfrak{p}$ is unramified). Then the ramification index of $K(\phi[\mathfrak{p}^n]^0)/K$ is divisible by $q^{N(n-1)}(q^N - 1)$, where $N = H \cdot \deg(\mathfrak{p})$.*

Proof The coefficients of $\phi_\mathfrak{p}(x) = \sum_{i=0}^{r \cdot \deg(\mathfrak{p})} g_i(\mathfrak{p}) x^{q^i}$ have positive valuations for $0 \le i \le N - 1$, where r is the rank of ϕ. Moreover, by assumption $v(\mathfrak{p}) = 1$ and $v(g_N(\mathfrak{p})) = 0$. Hence, the leftmost line segment of $NP(\phi_\mathfrak{p}(x)/x)$ joins $(0, 1)$ with $(q^N - 1, 0)$. This implies that all nonzero elements of $\phi[\mathfrak{p}]^0$ have valuation $1/(q^N - 1)$. Choose some $0 \ne \pi_1 \in \phi[\mathfrak{p}]^0$ and consider the polynomial $\phi_\mathfrak{p}(x) - \pi_1$ over $K_1 = K(\pi_1)$. Since $v(g_i(\mathfrak{p})) \ge 1$ for $0 \le i \le N - 1$ (cf. Lemma 3.2.14) and

$$v(\mathfrak{p} - \pi_1) = v(\pi_1) = 1/(q^N - 1),$$

the leftmost line segment of $NP(\phi_\mathfrak{p}(x) - \pi_1)$ joins $(0, v(\pi_1))$ with $(q^N, 0)$. Hence, there is $\pi_2 \in \phi[\mathfrak{p}^2]^0$ with $v(\pi_2) = v(\pi_1)/q^N$. Assume inductively that for $m = 2, \dots, n-1$ we have found $\pi_m \in \phi[\mathfrak{p}^m]^0$ such that $v(\pi_m) = 1/(q^N-1)q^{N(m-1)}$ and $\phi_\mathfrak{p}(\pi_m) = \pi_{m-1}$. Put $K_m = K_{m-1}(\pi_m)$. Consider the polynomial $\phi_\mathfrak{p}(x) - \pi_{n-1}$ over $K_{n-1} = K_{n-2}(\pi_{n-1})$. The same argument as in the case of $m = 2$ shows that the leftmost line segment of $NP(\phi_\mathfrak{p}(x) - \pi_{n-1})$ joins $(0, v(\pi_{n-1}))$ with $(q^N, 0)$. Hence, there is $\pi_n \in \phi[\mathfrak{p}^n]^0$ such that $v(\pi_n) = 1/(q^N - 1)q^{N(n-1)}$ and $\phi_\mathfrak{p}(\pi_n) = \pi_{n-1}$.

From the previous construction it is clear that the ramification index of K_n/K is divisible by $(q^N - 1)q^{N(n-1)}$, and since $K_n \subset K(\phi[\mathfrak{p}^n]^0)$, the lemma follows. $\square$

Corollary 6.3.11 *Let ϕ be a Drinfeld module over R with good reduction. Assume $v(\mathfrak{p}) = 1$ and $H = 1$. Then $K(\phi[\mathfrak{p}^n]^0)/K$ is a totally ramified extension of degree*

$$q^{\deg(\mathfrak{p})(n-1)}(q^{\deg(\mathfrak{p})} - 1).$$

Moreover,

$$\mathrm{Gal}(K(\phi[\mathfrak{p}^n]^0)/K) \cong (A/\mathfrak{p}^n)^\times.$$

Proof Denote $L = K(\phi[\mathfrak{p}^n]^0)$ and $d = \deg(\mathfrak{p})$. The action of $\mathrm{Gal}(L/K)$ on $\phi[\mathfrak{p}^n]^0 \cong A/\mathfrak{p}^n$ commutes with the action of A, so $\mathrm{Gal}(L/K)$ is a subgroup of $(A/\mathfrak{p}^n)^\times$. The order of $(A/\mathfrak{p}^n)^\times$ is $q^{dn} - q^{d(n-1)}$. Hence,

$$q^{d(n-1)}(q^d - 1) \ge [L : K].$$

On the other hand, by Lemma 6.3.10, $q^{d(n-1)}(q^d - 1)$ divides the ramification index of L/K. Hence, L/K is totally ramified of degree $q^{d(n-1)}(q^d - 1)$ and $\mathrm{Gal}(L/K) \cong (A/\mathfrak{p}^n)^\times$. $\square$

Example 6.3.12 In this example we point out a few subtleties which might not be apparent from the previous results. The first is that the sequence of Galois modules

$$0 \to \phi[\mathfrak{p}]^0 \longrightarrow \phi[\mathfrak{p}] \longrightarrow \phi[\mathfrak{p}]/\phi[\mathfrak{p}]^0 \to 0 \qquad (6.3.2)$$

usually does not split. The other is that, even though $\phi[\mathfrak{p}]/\phi[\mathfrak{p}]^0$ is unramified, the extension $K(\phi[\mathfrak{p}])/K(\phi[\mathfrak{p}]^0)$ can be ramified, and even wildly ramified.

To make this more explicit, let ϕ be a Drinfeld module of rank 2 over R with good reduction, and assume $H(\bar{\phi}) = 1$. Then the image of the representation

$$\rho_{\phi,\mathfrak{p}} \colon G_K \to \mathrm{Aut}_{\mathbb{F}_{\mathfrak{p}}}(\phi[\mathfrak{p}]) \cong \mathrm{GL}_2(\mathbb{F}_{\mathfrak{p}}),$$

up to conjugation, lies in the subgroup of upper-triangular matrices. By Lemma 6.3.7 and Corollary 6.3.11,

$$\rho_{\phi,\mathfrak{p}}(I_K) = \begin{pmatrix} \mathbb{F}_{\mathfrak{p}}^\times & \rho_{\phi,\mathfrak{p}}(I_K^{\mathrm{w}}) \\ 0 & 1 \end{pmatrix}, \tag{6.3.3}$$

where we have identified $\rho_{\phi,\mathfrak{p}}(I_K^{\mathrm{w}}) = \begin{pmatrix} 1 & * \\ 0 & 1 \end{pmatrix}$ with its image under $\begin{pmatrix} 1 & * \\ 0 & 1 \end{pmatrix} \to \mathbb{F}_{\mathfrak{p}}$, $\begin{pmatrix} 1 & b \\ 0 & 1 \end{pmatrix} \mapsto b$. The above remarks can be summarized as saying that $\rho_{\phi,\mathfrak{p}}(I_K^{\mathrm{w}})$ in the upper right corner in (6.3.3) is not necessarily 0.

For an explicit example of this phenomenon, consider $\phi[T]$ of the Drinfeld module over $K = \mathbb{F}_q(\!(T)\!)$ defined by $\phi_T = T + \tau - \tau^2$. This Drinfeld module has good reduction $\bar{\phi}_T = \tau - \tau^2$. Over $\mathbb{F}_T = \mathbb{F}_q$ the polynomial $\bar{\phi}_T(x) = (x - x^q)^q$ splits completely since the elements of $\mathbb{F}_q$ are the roots of $x - x^q$. Therefore, G_K acts trivially on $\phi[T]/\phi[T]^0 \cong \bar{\phi}[T]$. Thus, $\rho_{\phi,\mathfrak{p}}(G_K) \subseteq \begin{pmatrix} \mathbb{F}_q^\times & \mathbb{F}_q \\ 0 & 1 \end{pmatrix}$, and $[K(\phi[T]) : K] \leq q(q - 1)$. We claim that $K(\phi[T])/K$ is totally ramified of degree $q(q - 1)$, so that $\rho_{\phi,\mathfrak{p}}(G_K) = \rho_{\phi,\mathfrak{p}}(I_K) = \begin{pmatrix} \mathbb{F}_q^\times & \mathbb{F}_q \\ 0 & 1 \end{pmatrix}$.

To prove the previous claim, first note that the Newton polygon of $f(x) := \phi_T(x)/x$ has three vertices, $(0, 1)$, $(q - 1, 0)$, and $(q^2 - 1, 0)$, so $q - 1$ nonzero elements of $\phi[T]$ have valuation $1/(q - 1)$, while the other $q^2 - q$ nonzero elements have valuation 0. This implies that $K(\phi[T]^0)/K$ is totally ramified of degree $q - 1$; in particular, $q - 1$ divides the ramification index of $K(\phi[T])/K$. Next, we observe that $f(x) = g(x^{q-1})$, where $g(x) = T + x - x^{q+1} = T + x(1 - x)^q$. The Newton polygon of $g(x + 1) = T + (x + 1)x^q$ has three vertices, $(0, 1)$, $(q, 0)$, and $(q + 1, 0)$, so one of its roots has valuation 0 and the other q roots have valuation $1/q$. Thus, there is $\alpha \in \phi[T]$ such that $v(\alpha^{q-1} - 1) = 1/q$. This implies that the ramification index of $K(\phi[T])/K$ is also divisible by q. Since q and $q - 1$ are relatively prime, we conclude that $q(q - 1)$ divides the ramification index of $K(\phi[T])/K$. In particular, $q(q - 1)$ divides $[K(\phi[T]) : K]$. Combined with our earlier bound $[K(\phi[T]) : K] \leq q(q - 1)$, we obtain the desired claim.

We conclude this section with a result, which, although stated in terms of Drinfeld modules, properly belongs to local class field theory, or more precisely to the theory of Lubin-Tate formal groups.

Proposition 6.3.13 *Let* $\phi_T = T + g_1\tau + \cdots + g_r\tau^r \in R\{\tau\}$ *be such that* $g_r \equiv 1 \pmod{\pi}$. *Assume* K *is the unramified extension of* $F_{\mathfrak{p}}$ *of degree* r, *and the reduction* $\bar{\phi}$ *of* ϕ *is supersingular, i.e.,* $H(\bar{\phi}) = r$. *Then for any integer* $n \geq 1$, *the extension* $K(\phi[\mathfrak{p}^n])/K$ *is a totally ramified abelian extension of degree* $q^{N(n-1)}(q^N - 1)$, *where* $N = r \cdot \deg(\mathfrak{p})$. *Moreover,*

$$\mathrm{Gal}(K(\phi[\mathfrak{p}^n])/K) \cong (R/\mathfrak{p}^n)^{\times}.$$

Proof For now assume that K is an arbitrary local field, possibly of characteristic 0. Let $\mathcal{F}_{\pi}$ be the set of power series $f(x) \in R[\![x]\!]$ which satisfy the following two conditions:

(i) $f(x) = \pi x +$ terms of degree ≥ 2;
(ii) $f(x) \equiv x^{\#k} \pmod{\pi}$.

Let

$$f^{(n)}(x) = f \circ f \circ \cdots \circ f = f(f(\cdots f(x)\cdots))$$

denote the n-th iterate of $f(x)$. Let $\Lambda_{f,n}$ be the set of elements z of $\mathcal{M}_{\mathbb{C}_K}$ such that $f^{(n)}(z) = 0$. We know from Sect. 2.7 that $f^{(n)}(x)$ has finitely many zeros in $\mathcal{M}_{\mathbb{C}_K}$ and that these zeros are algebraic over K. Lubin and Tate proved in [LT65] that $K(\Lambda_{f,n})/K$ is a totally ramified extension of K with $\mathrm{Gal}(K(\Lambda_{f,n})/K) \cong (R/\pi^n)^{\times}$.

Now back in our setting, we have $k = \mathbb{F}_{\mathfrak{p}^r}$, the degree-$r$ extension of $\mathbb{F}_{\mathfrak{p}}$, and $\mathfrak{p}$ is a uniformizer of K. Therefore, by our assumptions, $\phi_{\mathfrak{p}}(x)$ is in $\mathcal{F}_{\mathfrak{p}}$, and we can apply the Lubin-Tate result. $\square$

Remark 6.3.14 Taking n to infinity in the previous proof one obtains a totally ramified extension K_{π} of K with $\mathrm{Gal}(K_{\pi}/K) \cong R^{\times}$ such that $K_{\pi} \cdot K^{\mathrm{nr}} = K^{\mathrm{ab}}$ is the maximal abelian extension of K. Proving this requires facts from the theory of formal R-modules and local class field theory; cf. [LT65, Mil13]. A formal R-module is an object similar to a formal Drinfeld module that we will discuss in Sect. 6.5.

Exercises

6.3.1 Assume $v(T) = 0$. Give an explicit example of a Drinfeld module over K with stable bad reduction and such that $\phi[T] \subset K$. (Thus, Corollary 6.3.3 (2) is false without the assumption that ϕ has potentially good reduction.)

6.3.2 Let ϕ and ϕ' be Drinfeld modules over K. Assume that there is an isogeny $u : \phi \to \phi'$ defined over K. Show that if ϕ has good (respectively, stable) reduction, then ϕ' has good (respectively, stable) reduction. Conclude that two K-isogenous Drinfeld modules either both have or both do not have good (respectively, stable) reduction.

6.3.3 Let ϕ be a Drinfeld module over R with stable reduction. Let $\mathfrak{l}$ be a prime different from $\mathfrak{p}$. Let

$$\phi_{\mathfrak{l}}(x) = \mathfrak{l}x + g_1(\mathfrak{l})x^q + \cdots + g_n(\mathfrak{l})x^{q^n}$$

be the $\mathfrak{l}$-division polynomial of ϕ, where $n = r \cdot \deg(\mathfrak{l})$. Suppose $c := g_n(\mathfrak{l}) \in M$ and $g_m(\mathfrak{l}) \in R^\times$ for some $1 \le m < n$.

(a) Show that $\phi_{\mathfrak{l}}(x)$ has a root α_1 with $v(\alpha_1) < 0$.
(b) Decompose the polynomial $\phi_{\mathfrak{l}}(x) - \alpha_1$ into linear factors over $\overline{K}$:

$$\phi_{\mathfrak{l}}(x) - \alpha_1 = c \prod_{j=1}^{q^n} (x - \beta_j).$$

Show that

$$\alpha_1^{-1} \phi_{\mathfrak{l}}(\alpha_1 x) = \prod_{j=1}^{q^n} \left(\frac{\alpha_1}{\beta_j} x - 1 \right) + 1.$$

(c) Show that there is at least one β_j such that $v(\alpha_1) < v(\beta_j)$. Denote one of these β_j by α_2.
(d) Show that $v(\alpha_2) < 0$.
(e) Show that there is a sequence of elements $\alpha_s \in \phi[\mathfrak{l}^s]$, $s \ge 1$, such that $\phi_{\mathfrak{l}}(\alpha_{s+1}) = \alpha_s$ and $v(\alpha_s) < v(\alpha_{s+1}) < 0$ for all $s \ge 1$.
(f) Conclude that $\phi[\mathfrak{l}^s]$ can be unramified only for finitely many $s \ge 1$. Therefore, a Drinfeld module ϕ over K has good reduction if and only if $\phi[\mathfrak{l}^s]$ is unramified for all $s \ge 1$.

6.3.4 Let ϕ be a Drinfeld module over K. Let $\mathfrak{l}$ be a prime different from $\mathfrak{p}$. Prove the following:

(a) If ϕ has potentially good reduction, then for any $\sigma \in I_K$ the characteristic polynomial of $\hat{\rho}_{\phi,\mathfrak{l}}(\sigma)$ has coefficients in $\mathbb{F}_q$.
(b) For a suitable finite extension L of K, we have

$$\left(\hat{\rho}_{\phi,\mathfrak{l}}(\sigma) - 1 \right)^2 = 0 \qquad \text{for all } \sigma \in I_L.$$

6.3.5 Let ϕ be a Drinfeld module over R of rank r with good reduction. Assume $v(\mathfrak{p}) = 1$. Let $N = H(\bar{\phi}) \cdot \deg(\mathfrak{p})$. Prove that $K(\phi[\mathfrak{p}]^0)/K$ is a totally ramified cyclic extension of degree $q^N - 1$.

6.3.6 Let $K = \mathbb{F}_q(\!(T)\!)$ and $\mathfrak{p} = T$. Let ϕ be the Drinfeld module of rank 2 over K defined by $\phi_T = T + \tau + \tau^2$. Determine the ramification index and the Galois group of the extension $K(\phi[T])/K$.

6.3.7 Let $K = \mathbb{F}_q(\!(T)\!)$ and $\mathfrak{p} = T$. Give an explicit example of a Drinfeld module over K such that the sequence (6.3.2) splits, and the action of G_K on $\phi[T]/\phi[T]^0$ is nontrivial.

6.4   Rational Torsion Submodule

In this section we prove that for a Drinfeld module ϕ over K the torsion submodule $({}^\phi K)_{\text{tor}}$ of ${}^\phi K$ is finite. There are different proofs of this result in the literature; see, for example, [Den92, Poo97, BK95, Ros03]. We will take the most elementary approach, which demonstrates yet again the utility of Newton polygons. This approach is more or less the one in [BK95], except that in [BK95] the authors do not use Newton polygons, and derive everything directly from estimates on the polynomials $\phi_a(x)$ (this might be the source of their unnecessary assumption that the rank of ϕ is 2).

Proposition 6.4.1 *Let ϕ be a Drinfeld module over R with stable reduction defined by*

$$\phi_T = T + g_1\tau + \cdots + g_r\tau^r.$$

Let $a \in A$ be a nonzero element coprime to $\mathfrak{p}$, and let $0 \neq z \in \phi[a\mathfrak{p}^n]$. If $n \geq 1$, then assume that $z \notin \phi[a\mathfrak{p}^{n-1}]$. We have:

$$-\frac{q}{q-1}\frac{v(g_r)}{q^r-1} \leq v(z) \leq \begin{cases} 0, & \text{if } n = 0, \\ \dfrac{v(\mathfrak{p})}{q^{n\deg(\mathfrak{p})} - q^{(n-1)\deg(\mathfrak{p})}}, & \text{otherwise.} \end{cases}$$

Proof Denote $b = a\mathfrak{p}^n$. For $h \in A$, let $c(h)$ be the leading coefficient of ϕ_h, so $c(T) = g_r$. One easily computes that

$$c(T^m) = g_r^{1+q^r+\cdots+q^{(m-1)r}}.$$

If $b = b_d T^d + \cdots + b_0$ with $b_d, \ldots, b_0 \in \mathbb{F}_q$ and $b_d \neq 0$, then

$$\phi_b = b_d\phi_{T^d} + \cdots + b_1\phi_T + b_0,$$

so $c(b) = b_d \cdot c(T^d)$. This implies

$$v(c(b)) = \frac{q^{dr} - 1}{q^r - 1} v(g_r).$$

By assumption, the coefficients of the polynomial $\phi_b(x)$ are in R. Moreover, because $\bar{\phi}$ is a Drinfeld module, at least one of these coefficients is a unit. Hence, the leftmost line segment of the Newton polygon of $\phi_b(x)/x$ with positive slope starts at $(q^m - 1, 0)$ for some $0 < m \le dr$, whereas the rightmost line segment ends at $(q^{dr} - 1, v(c(b)))$. Since the monomials in $\phi_b(x)/x$ have degrees $q^i - 1$, $0 \le i \le dr$, it is not hard to see that the largest slope of $\mathrm{NP}(\phi_b(x)/x)$ is not larger than the slope of the line segment joining $(q^{dr-1} - 1, 0)$ to $(q^{dr} - 1, v(c(b)))$, which is

$$\frac{v(c(b))}{q^{dr} - q^{dr-1}} = \frac{1}{q^{dr} - q^{dr-1}} \frac{q^{dr} - 1}{q^r - 1} v(g_r)$$

$$\le \frac{q}{q - 1} \frac{1}{q^r - 1} v(g_r).$$

Therefore, if $z \in \overline{K}$ is a root of $\phi_b(x)/x$, then by Theorem 2.5.2,

$$-\frac{q}{q - 1} \frac{v(g_r)}{q^r - 1} \le v(z).$$

Moreover, if $n = 0$, then $\mathrm{NP}(\phi_b(x)/x)$ starts at $(0, 0)$, so all its slopes are non-negative, which implies $v(z) \le 0$.

Now consider $\phi_\mathfrak{p} = \mathfrak{p} + \sum_{i=1}^{dr} g_i(\mathfrak{p})\tau^i$, where now $d = \deg(\mathfrak{p})$. By Lemma 3.2.14, the coefficients $g_i(\mathfrak{p})$, $0 \le i \le d - 1$, are divisible by $\mathfrak{p}$. Therefore, the smallest slope of $\mathrm{NP}(\phi_\mathfrak{p}(x)/x)$ is at least as large as the slope of the line segment joining $(0, v(\mathfrak{p}))$ to $(q^d - 1, 0)$. By Theorem 2.5.2, for any $0 \ne z \in \phi[\mathfrak{p}]$ we have

$$v(z) \le \frac{v(\mathfrak{p})}{q^d - 1}.$$

We claim that more generally, if $z \in \phi[\mathfrak{p}^{n+1}]$ but $z \notin \phi[\mathfrak{p}^n]$, then

$$v(z) \le \frac{v(\mathfrak{p})}{q^{d(n+1)} - q^{dn}}. \tag{6.4.1}$$

We use induction on n, the case $n = 0$ having been settled above. We can assume that $v(z) > 0$. Since $\phi_\mathfrak{p}(z)$ has exact order $\mathfrak{p}^n$, by the induction hypothesis

$$v(\phi_\mathfrak{p}(z)) \le \frac{v(\mathfrak{p})}{q^{dn} - q^{d(n-1)}}.$$

Now, using the fact that $\mathfrak{p}$ divides $g_i(\mathfrak{p})$, $0 \le i \le d-1$, we can write

$$\phi_{\mathfrak{p}}(z) = \mathfrak{p}z + \sum_{i=1}^{q^{dr}} g_i(\mathfrak{p})z^{q^i} = \mathfrak{p}zf(z) + z^{q^d} g(z)$$

for some $f(x), g(x) \in R[x]$. Hence,

$$v(\phi_{\mathfrak{p}}(z)) \ge \min\{v(\mathfrak{p}z), v(z^{q^d})\}.$$

On the other hand, since

$$v(\phi_{\mathfrak{p}}(z)) \le \frac{v(\mathfrak{p})}{q^{dn} - q^{d(n-1)}} < v(\mathfrak{p}) < v(\mathfrak{p}z),$$

we must have

$$q^d \cdot v(z) = v(z^{q^d}) \le v(\phi_{\mathfrak{p}}(z)) \le \frac{v(\mathfrak{p})}{q^{dn} - q^{d(n-1)}}.$$

From $q^d \cdot v(z) \le \frac{v(\mathfrak{p})}{q^{dn} - q^{d(n-1)}}$ we get (6.4.1).

Finally, assume $z \in \phi[a\mathfrak{p}^n]$, $\mathfrak{p} \nmid a$, $n \ge 1$, and $z \notin \phi[a\mathfrak{p}^{n-1}]$. We want to show that the bound (6.4.1) holds for z. We can assume $v(z) > 0$, as otherwise the claim is trivial. Since $\phi_a(z) \in \phi[\mathfrak{p}^n]$, we have

$$v(\phi_a(z)) \le \frac{v(\mathfrak{p})}{q^{dn} - q^{d(n-1)}}. \tag{6.4.2}$$

On the other hand, $\phi_a(z) = az + c_1 z^q + \cdots + c_m z^{q^m}$, where $m = r\deg(a)$, and $c_1, \ldots, c_m \in R$. Since $v(a) = 0$ and $v(z) > 0$, we have a strict inequality $v(az) < v(c_i z^{q^i})$ for all $1 \le i \le m$. Hence, by the strong triangle inequality, $v(\phi_a(z)) = v(z)$. Combining this with (6.4.2) proves the claim.   □

Corollary 6.4.2 *Let ϕ be a Drinfeld module over R with stable reduction defined by*

$$\phi_T = T + g_1\tau + \cdots + g_r\tau^r.$$

If $z \in (\phi\overline{K})_{\mathrm{tor}}$, then

$$-2 \cdot v(g_r) \le v(z) < v(\mathfrak{p}).$$

Proof This immediately follows from Proposition 6.4.1. In fact, the lower bound can be simplified to $-v(g_r) \le v(z)$, assuming $2 < q$ or $1 < r$, with a strict inequality if $v(g_r) \ne 0$.   □

Theorem 6.4.3 *Let ϕ be a Drinfeld module over K. Then $(^{\phi}K)_{\text{tor}}$ is finite.*

Proof Since it is enough to show that $(^{\phi}L)_{\text{tor}}$ is finite for a finite extension L of K, we may assume that ϕ has stable reduction. Moreover, after possibly replacing ϕ by a K-isomorphic module, we may assume that ϕ has integral coefficients. This puts us in the setting of Corollary 6.4.2. We claim that in that case

$$\#(^{\phi}K)_{\text{tor}} \leq (\#k)^{v(\mathfrak{p})+2\cdot v(g_r)}. \tag{6.4.3}$$

Denote $n = v(\mathfrak{p})$ and $m = -2 \cdot v(g_r)$. By Corollary 6.4.2, for nonzero $z \in (^{\phi}K)_{\text{tor}}$ we have $m \leq v(z) < n$. We can uniquely expand $z = \sum_{i \geq m} z_i \pi^i$, where $z_i \in k$. We claim that the set of elements $\{z_j \mid m \leq j < n\}$ uniquely determines z. This implies (6.4.3) since the number of choices of $(z_m, z_{m+1}, \ldots, z_{n-1})$ is $(\#k)^{n-m}$.

Suppose there is $z' \in (^{\phi}K)_{\text{tor}}$ such that the coefficients of its π-adic expansion satisfy $z'_j = z_j$ for all $m \leq j < n$. Then the expansion of $z - z'$ starts with $(z_n - z'_n)\pi^n$, so $v(z - z') \geq n$. On the other hand, since $z - z'$ is again a K-rational torsion element of ϕ, we must have $v(z - z') < n$, unless $z - z' = 0$. $\qquad\square$

Remark 6.4.4 Note that $v(\mathfrak{p})$ is equal to the ramification index $e(K/F_{\mathfrak{p}})$. Hence, (6.4.3) can be restated as

$$\#(^{\phi}K)_{\text{tor}} \leq (\#k)^{e(K/F_{\mathfrak{p}})+2\cdot v(g_r)}. \tag{6.4.4}$$

Exercises

6.4.1 Assume K is fixed. Show that for any given positive integer r there is a Drinfeld module of rank r over K such that $(^{\phi}K)_{\text{tor}} \geq q^r$. Why doesn't this contradict (6.4.3)?

6.4.2 Let $K = \mathbb{F}_q((T))$. Determine $(^{\phi}K)_{\text{tor}}$ for ϕ defined by

(a) $\phi_T = T - (T+1)\tau + \tau^2$.
(b) $\phi_T = T + (T^q - T)\tau + \tau^2$.
(c) $\phi_T = T + \tau + T\tau^2$.

6.4.3 Let ϕ be a Drinfeld module over R with good reduction. Prove the following:

(a) If $\alpha \in R$ with $v(\alpha) > v(\mathfrak{p})/(q-1)$, then $v(\phi_{\mathfrak{p}^n}(\alpha)) = nv(\mathfrak{p}) + v(\alpha)$ for all $n \geq 0$. Conclude that α cannot be a torsion point.

(b) $\#(^{\phi}K)_{\text{tor}} \leq (\#k)^{\lfloor \frac{v(\mathfrak{p})}{q-1} \rfloor + 1}$, where $\left\lfloor \frac{v(\mathfrak{p})}{q-1} \right\rfloor$ is the largest integer less than or equal to $\frac{v(\mathfrak{p})}{q-1}$.

(c) $\#(^\phi K)_{\text{tor}} \le q^{\left(1+\frac{1}{q-1}\right)[K:F_\mathfrak{p}]}$.

6.4.4 Let ϕ be a rank 2 Drinfeld module over R such that $v(j(\phi)) > 0$. Show that if $q - 1$ divides $v(\mathfrak{p})$, then

$$\#(^\phi K)_{\text{tor}} \le q^{[K:F_\mathfrak{p}]/(q-1)}.$$

6.4.5 Show that for each $r \ge 2$ and $a \in A$ there exists a rank r Drinfeld module ϕ over K with a rational torsion point of order a. Hence, $\#(^\phi K)_{\text{tor}}$ cannot be uniformly bounded as ϕ varies over the Drinfeld modules of rank r.

6.5　Formal Drinfeld Modules

Let $\phi \colon A \to R\{\tau\}$ be a Drinfeld module of rank r over R. We will describe a certain formal completion of ϕ which allows to extend ϕ to a homomorphism $\widehat{\phi} \colon A_\mathfrak{p} \to R\{\!\{\tau\}\!\}$. This construction is useful for the study of the kernel of the reduction map $\phi \to \bar{\phi}$, as well as the study of the $\mathfrak{p}$-primary torsion points of ϕ. In particular, it provides a different interpretation of $\phi[\mathfrak{p}^n]^0$ and a different proof of the finiteness of $(^\phi K)_{\text{tor}}$.

In a naive manner, $\widehat{\phi}$ can be described as follows. Denote $d = \deg(\mathfrak{p})$, so that the set

$$A_{<d} = \{a \in A \mid \deg(a) < d\}$$

is a set of representatives for $A/\mathfrak{p}$. Given $a \in A$, write

$$\phi_a = a + g_1(a)\tau + g_2(a)\tau^2 + \cdots,$$

with $g_n(a) = 0$ for $n > r \cdot \deg(a)$. Consider $\mathfrak{p}$ as a uniformizer of $A_\mathfrak{p}$. Since $A_\mathfrak{p}/\mathfrak{p} = A/\mathfrak{p}$, any $\alpha \in A_\mathfrak{p}$ is uniquely represented by a series $\alpha = \sum_{i \ge 0} \alpha_i \mathfrak{p}^i$ with $\alpha_i \in A_{<d}$; cf. Sect. 2.2. Let

$$\widehat{\phi}_\alpha := \sum_{i \ge 0} \phi_{\alpha_i} \phi_{\mathfrak{p}^i} = \sum_{n,m \ge 0} \left(\sum_{i \ge 0} g_m(\alpha_i) \cdot g_n(\mathfrak{p}^i)^{q^m} \right) \tau^{m+n}.$$

By Lemma 3.2.14, $\mathfrak{p}^i$ divides $g_n(\mathfrak{p}^i)$ for $i > n/d$. On the other hand, $|g_m(\alpha_i)|$ remains bounded as $i \to \infty$, since $\{g_m(\alpha_i) \mid i \ge 0\}$ is a finite set. This implies that

$$\left| g_m(\alpha_i) g_n(\mathfrak{p}^i)^{q^m} \right| \to 0 \quad \text{as} \quad i \to \infty,$$

so the series $\sum_{i \geq 0} g_m(\alpha_i) \cdot g_n(\mathfrak{P}^i)^{q^m}$ converges to an element $g_{m,n}(\alpha) \in R$. Thus,

$$\widehat{\phi}_\alpha = \sum_{s \geq 0} \tilde{g}_s(\alpha)\tau^s \in B\{\tau\}, \quad \text{where} \quad \tilde{g}_s(\alpha) = \sum_{\substack{m,n \geq 0 \\ m+n=s}} g_{m,n}(\alpha).$$

It is not hard to check that

  (i) $\tilde{g}_0(\alpha) = \alpha$;
 (ii) if $\alpha \in A$, then $\widehat{\phi}_\alpha = \phi_\alpha$;
(iii) $\widehat{\phi} \colon A_{\mathfrak{P}} \to B\{\tau\}$ is a homomorphism.

One can also check by tedious calculations that $\widehat{\phi}$ does not depend on the choice of $\mathfrak{P}$ as a uniformizer of $A_{\mathfrak{P}}$ or the choice of $A_{<d}$ as a set of representatives of $A/\mathfrak{P}$. A more conceptual construction of of $\widehat{\phi}$ results from considering $B\{\tau\}$ as a topological ring. By Lemma 5.1.1, $B\{\tau\}$ is a local ring with (unique) maximal ideal

$$\mathfrak{M} = \{f \in B\{\tau\} \mid \partial(f) \in M\}.$$

The elements of $\mathfrak{M}^n$ are the power series $f = \sum_{i \geq 0} f_i \tau^i$ such that $f_i \in M^{n-i}$ for $0 \leq i \leq n$. From this it follows that $\bigcap_{n \geq 1} \mathfrak{M}^n = (0)$. We define a valuation on $B\{\tau\}$ by

$$\text{ord}_{\mathfrak{M}}(f) = \max\{n \in \mathbb{Z}_{\geq 0} \mid f \in \mathfrak{M}^n\}.$$

As in Sect. 2.2, one can use this valuation to define an absolute value, and thus a topology, on $B\{\tau\}$, called $\mathfrak{M}$-*adic topology*[1]; cf. [Mat89, p. 57]. It is easy to check that $B\{\tau\}$ is complete in this topology.

Now composing $\phi \colon A \to R\{\tau\}$ with the natural embedding $R\{\tau\} \to R\{\{\tau\}\}$, we get an injective homomorphism $A \to R\{\{\tau\}\}$ which we again denote by ϕ. Note that $\text{ord}_{\mathfrak{M}}(\phi_a) > 0$ if and only if $\mathfrak{p} \mid a$. Hence the valuation $\text{ord}_{\mathfrak{M}}$ induces a valuation on A equivalent to $\text{ord}_{\mathfrak{p}}$. Since $R\{\{\tau\}\}$ is complete with respect to $\text{ord}_{\mathfrak{M}}$ and $\phi \colon A \to R\{\{\tau\}\}$ is continuous, ϕ extends to a homomorphism $\widehat{\phi} \colon A_{\mathfrak{p}} \to R\{\{\tau\}\}$ such that $\partial(\widehat{\phi}_\alpha) = \alpha$ for all $\alpha \in A_{\mathfrak{p}}$. Also note that $\widehat{\phi}$ coincides with ϕ on A. The map $\widehat{\phi}$ is the *formal completion* of ϕ.

At this point we could continue with the study of the formal completion of ϕ, but it is not more difficult to analyze more general objects, called formal Drinfeld modules. In fact the abstraction clarifies which properties of ϕ are important for $\widehat{\phi}$ and which are not. Formal Drinfeld modules were introduced by Rosen [Ros03], although a less explicit but more general version of this can be found in Drinfeld's original paper [Dri74]. For the rest of this section, we largely follow Rosen's article [Ros03].

---

[1] But $B\{\tau\}$ is not a DVR.

Definition 6.5.1 Let O be the ring of integers of a local field containing $\mathbb{F}_q$ as a subfield. Let B be a commutative O-algebra, and let $\iota\colon O \to B$ be the structure homomorphism. A *formal Drinfeld O-module* over B is a ring homomorphism $\widehat{\phi}\colon O \to B\{\{\tau\}\}, \alpha \mapsto \widehat{\phi}_\alpha$, with three properties:

(i) $\partial(\widehat{\phi}_\alpha) = \iota(\alpha)$;
(ii) $\widehat{\phi}(O) \not\subseteq B$;
(iii) $\widehat{\phi}_\varpi \neq 0$ for one (and thus all) uniformizers ϖ of O.

We know from Sect. 2.2 that $O \cong \mathbb{F}_{q^d}[\![\varpi]\!]$, where ϖ is a uniformizer of O and $\mathbb{F}_{q^d}$ is the residue field of O. A formal Drinfeld O-module $\widehat{\phi}\colon O \to B\{\{\tau\}\}$ is an injection since $\widehat{\phi}_\varpi \neq 0$.

The formal completion of a Drinfeld module is clearly a formal Drinfeld A_{p}-module over R. On the other hand, if $\widehat{\phi}$ is the completion of a Drinfeld module, then $\widehat{\phi}_a$ is a polynomial for $a \in A$, whereas this is not necessarily the case for a general formal Drinfeld A_{p}-module over R.

A formal Drinfeld module is merely a ring homomorphism with no actual underlying module. However, if $B = R$ and $\iota(O) \subseteq R$ is an embedding, then the additive power series $\widehat{\phi}_\alpha(x)$ associated with $\widehat{\phi}_\alpha$ converges on $\mathcal{M}$. Hence, through $\widehat{\phi}$, we obtain a new O-module structure $^{\widehat{\phi}}\mathcal{M}$ on $\mathcal{M}$. More generally, given $z \in \mathcal{M}_{\mathbb{C}_K}$, the series $\widehat{\phi}_\alpha(z)$ converges to an element of $\mathcal{M}_{\mathbb{C}_K}$. Thus, we can make $\mathcal{M}_{\mathbb{C}_K}$ into an O module via $\widehat{\phi}$. We analyze this module using some constructions from Chap. 5. More precisely, we construct a logarithm and an exponential function associated with $\widehat{\phi}$, and analyze their radii of convergence. This will allow us to relate $^{\widehat{\phi}}\mathcal{M}_{\mathbb{C}_K}$ to $\mathcal{M}_{\mathbb{C}_K}$ with its O-module structure given by ι, and deduce results about the torsion elements of $\widehat{\phi}$.

To simplify the notation, from now on we write α for $\iota(\alpha) \in R$. Note that $\varpi \in \mathcal{M}$, so $v(\varpi) > 0$ (one can see this using the homomorphism $O \to R \to R/\mathcal{M}$, whose kernel is nonzero, so contains a power of ϖ).

Proposition 6.5.2 *Let $\widehat{\phi}$ be a formal Drinfeld O-module over R. There is a uniquely defined*

$$\log_{\widehat{\phi}} = \sum_{i \geq 0} l_i \tau^i \in K\{\{\tau\}\}$$

such that $l_0 = 1$ and

$$\log_{\widehat{\phi}} \cdot \widehat{\phi}_\alpha = \alpha \cdot \log_{\widehat{\phi}}, \quad \text{for all } \alpha \in O.$$

Moreover, $\log_{\widehat{\phi}}(x) = \sum_{i \geq 0} l_i x^{q^i}$ converges on $\mathcal{M}$ and yields an O-module homomorphism from $^{\widehat{\phi}}\mathcal{M}$ to K. The same claims are valid if R, $\mathcal{M}$, and K are replaced by $R_{\mathbb{C}_K}$, $\mathcal{M}_{\mathbb{C}_K}$, and $\mathbb{C}_K$, respectively.

Proof Denote $\widehat{\phi}_\varpi = \sum_{i \geq 0} g_i \tau^i$. First, we claim that there is a uniquely defined $l = \sum_{i \geq 0} l_i \tau^i \in K\{\{\tau\}\}$ such that $l_0 = 1$ and $l \cdot \widehat{\phi}_\varpi = \varpi \cdot l$. For $l \cdot \widehat{\phi}_\varpi = \varpi \cdot l$ to hold, it is necessary and sufficient that

$$(\varpi - \varpi^{q^n}) \cdot l_n = l_0 \cdot g_n + l_1 \cdot g_{n-1}^q + l_2 \cdot g_{n-2}^{q^2} + \cdots + l_{n-1} \cdot g_1^{q^{n-1}} \qquad \text{for all } n \geq 0.$$
(6.5.1)

Since $\varpi \neq \varpi^{q^n}$ for $n \geq 1$, if we put $l_0 = 1$, then the coefficient l_n is uniquely determined by $l_0, \ldots, l_{n-1}$. Thus, l indeed exists and is unique. Also, note that l has multiplicative inverse l^{-1} in the ring $K\{\{\tau\}\}$ because $l_0 = 1$; cf. Lemma 5.1.1.

Now we claim that $l \cdot \widehat{\phi}_\alpha = \alpha \cdot l$ for all $\alpha \in O$. Indeed, multiplying $l \cdot \widehat{\phi}_\varpi = \varpi \cdot l$ by $\widehat{\phi}_\alpha$ from the right, we get

$$
\begin{aligned}
\varpi \cdot l \cdot \widehat{\phi}_\alpha &= l \cdot \widehat{\phi}_\varpi \cdot \widehat{\phi}_\alpha \\
&= l \cdot \widehat{\phi}_\alpha \cdot \widehat{\phi}_\varpi \\
&= l \cdot \widehat{\phi}_\alpha \cdot l^{-1} \cdot \varpi \cdot l.
\end{aligned}
$$

Denoting $u = l \cdot \widehat{\phi}_\alpha \cdot l^{-1} = \sum_{i \geq 0} u_i \tau^i \in K\{\{\tau\}\}$, we can rewrite the above equation as $\varpi u = u \varpi$. It is trivial to check that only the constants K commute with ϖ in $K\{\{\tau\}\}$. Therefore, $u = \partial u \in K$. On the other hand, from $u = l \cdot \widehat{\phi}_\alpha \cdot l^{-1}$ it is clear that $\partial(u) = \alpha$. Thus, we get $l \cdot \widehat{\phi}_\alpha = \alpha \cdot l$.

Next, we claim that $|l_n| \leq |\varpi|^{-n}$. This is obviously true for $n = 0$. Assume by induction that $|l_i| \leq |\varpi|^{-i}$ holds for $0 \leq i \leq n-1$. Since $|\varpi| < 1$ and $|g_i| \leq 1$ for all $i \geq 0$, from (6.5.1) we obtain

$$
\begin{aligned}
\left|\varpi - \varpi^{q^n}\right| \cdot |l_n| &\leq \max\left(|l_0 \cdot g_n|, |l_1 \cdot g_{n-1}^q|, \ldots, \left|l_{n-1} \cdot g_1^{q^{n-1}}\right|\right) \\
&\leq \max(|l_0|, |l_1|, \ldots, |l_{n-1}|) \\
&\leq \max\left(1, |\varpi|^{-1}, \ldots, |\varpi|^{-(n-1)}\right) \\
&\leq |\varpi|^{-(n-1)}.
\end{aligned}
$$

On the other hand, $\left|\varpi - \varpi^{q^n}\right| \cdot |l_n| = |\varpi| \cdot |l_n|$. Therefore, $|l_n| \leq |\varpi|^{-n}$, as was claimed. Using this inequality, we can estimate the radius of convergence of $l(x) = \sum_{i \geq 0} l_i x^{q^i}$ by

$$\rho(l)^{-1} = \varlimsup_{n \to \infty} |l_n|^{1/q^n} \leq \varlimsup_{n \to \infty} |\varpi|^{-n/q^n} = 1.$$

By Lemma 2.7.6, $l(x)$ converges on $\mathcal{M}$.

The claim that $l(x)$ yields an O-module homomorphism $\widehat{\phi}\mathcal{M} \to K$ follows from the additivity of $l(x)$ and $(l\widehat{\phi}_\alpha)(x)$, and the functional equation $l(\widehat{\phi}_\alpha(x)) =$

$\alpha l(x)$, assuming $(l\widehat{\phi}_\alpha)(z) = l(\widehat{\phi}_\alpha(z))$ for all $z \in \mathcal{M}$. To verify the validity of this numerical evaluation, note that for $\widehat{\phi}_\alpha(x) = \sum_{n\geq 0} \alpha_n x^{q^n} \in R\langle\!\langle x \rangle\!\rangle$ we have $\max\left\{\left|\alpha_n z^{q^n}\right| : n \geq 0\right\} < 1 = \rho(l)$ so we can apply Theorem 2.7.24. $\square$

Let

$$e_{\widehat{\phi}} = \sum_{i\geq 0} e_i \tau^i$$

be the multiplicative inverse of $\log_{\widehat{\phi}}$ in the ring $K\{\!\{\tau\}\!\}$. Multiplying both sides of the equality $1 = e_{\widehat{\phi}} \cdot \log_{\widehat{\phi}}$ by $\widehat{\phi}_\alpha \cdot e_{\widehat{\phi}}$ from the right, one obtains

$$\widehat{\phi}_\alpha \cdot e_{\widehat{\phi}} = e_{\widehat{\phi}} \cdot \log_{\widehat{\phi}} \cdot \widehat{\phi}_\alpha \cdot e_{\widehat{\phi}}$$

$$= e_{\widehat{\phi}} \cdot \alpha \cdot \log_{\widehat{\phi}} \cdot e_{\widehat{\phi}}$$

$$= e_{\widehat{\phi}} \cdot \alpha.$$

Thus, for all $\alpha \in O$ we have the functional equation

$$\widehat{\phi}_\alpha \cdot e_{\widehat{\phi}} = e_{\widehat{\phi}} \cdot \alpha,$$

similar to (5.2.5) from the case of Drinfeld modules over $\mathbb{C}_\infty$. From this equation, for $\alpha = \varpi$, we get the following relations between the coefficients of $e_{\widehat{\phi}}$ and $\widehat{\phi}_\varpi = \sum_{i\geq 0} g_i \tau^i$:

$$e_0 = 1,$$

$$(\varpi^{q^n} - \varpi) \cdot e_n = \left(g_1 e_{n-1}^q + g_2 e_{n-2}^{q^2} + \cdots + g_n e_0^{q^n}\right) \quad \text{for } n \geq 1. \qquad (6.5.2)$$

Proposition 6.5.3 *Let $\widehat{\phi}$ be a formal Drinfeld O-module over R. There is a uniquely defined*

$$e_{\widehat{\phi}} = \sum_{i\geq 0} e_i \tau^i \in K\{\!\{\tau\}\!\}$$

such that $e_0 = 1$ and $\widehat{\phi}_\alpha \cdot e_{\widehat{\phi}} = e_{\widehat{\phi}} \cdot \alpha$ for all $\alpha \in O$. The power series $e_{\widehat{\phi}}(x) = \sum_{i\geq 0} e_i x^{q^i}$ converges on

$$\mathcal{M}^\dagger := \{z \in \mathcal{M} \mid v(z) > v(\varpi)/(q-1)\}$$

and yields an O-module homomorphism from $\mathcal{M}^\dagger$ to $\,{}^{\widehat{\phi}}\mathcal{M}$. The same claims are valid if R, $\mathcal{M}$, and K are replaced by $R_{\mathbb{C}_K}$, $\mathcal{M}_{\mathbb{C}_K}$, and $\mathbb{C}_K$, respectively.

Proof We already proved the existence of $e_{\widehat{\phi}}$ such that $\widehat{\phi}_\alpha \cdot e_{\widehat{\phi}} = e_{\widehat{\phi}} \cdot \alpha$ for all $\alpha \in O$. On the other hand, this functional equation leads to the recursive formula (6.5.2) from which it is clear that $e_0, \ldots, e_{n-1}$ uniquely determine e_n for $n \geq 1$. Thus, the functional equation $\widehat{\phi}_\alpha \cdot e_{\widehat{\phi}} = e_{\widehat{\phi}} \cdot \alpha$ uniquely determines $e_{\widehat{\phi}}$.

We claim that $|e_n| \leq |\varpi|^{-(1+q+\cdots+q^{n-1})}$ for all $n \geq 1$. Since the coefficients of $\widehat{\phi}_\varpi$ are in R, we have $|g_i| \leq 1$ for all $i \geq 0$. In particular, using (6.5.2), we have $|\varpi| \cdot |e_1| = |g_1| \leq 1$, which proves the claim for $n = 1$. Assume by induction that the inequality is true for $|e_1|, \ldots, |e_{n-1}|$. From (6.5.2), we get

$$|\varpi| \cdot |e_n| \leq \max_{1 \leq i \leq n} \left(|g_i| \cdot |e_{n-i}|^{q^i} \right)$$

$$\leq \max_{1 \leq i \leq n} \left(|\varpi|^{-(q^i + q^{i+1} + \cdots + q^{n-1})} \right)$$

$$= |\varpi|^{-(q + q^2 + \cdots + q^{n-1})} .$$

Hence, $|e_n| \leq |\varpi|^{-(1+q+\cdots+q^{n-1})} = |\varpi|^{(1-q^n)/(q-1)}$, as was claimed.

Now

$$\rho(e_{\widehat{\phi}}(x))^{-1} = \varlimsup_{n \to \infty} |e_n|^{1/q^n}$$

$$\leq \varlimsup_{n \to \infty} |\varpi|^{(1-q^n)/(q-1)q^n}$$

$$\leq |\varpi|^{-1/(q-1)} .$$

Therefore, $\rho(e_{\widehat{\phi}}(x)) \geq |\varpi|^{1/(q-1)}$. By Lemma 2.7.6, $e_{\widehat{\phi}}(x)$ converges on $\mathcal{M}^\dagger$. For $z \in \mathcal{M}^\dagger$ we have the estimate

$$\left| e_n z^{q^n} \right| \leq |\varpi|^{-(q^n-1)/(q-1)} |\varpi|^{q^n/(q-1)} < 1,$$

which implies that $e_{\widehat{\phi}}(z) \in \mathcal{M}$. The final claim that $e_{\widehat{\phi}}(x)$ yields an O-module homomorphism $\mathcal{M}^\dagger \to {}^{\widehat{\phi}}\mathcal{M}$ is an immediate consequence of the additivity of $e_{\widehat{\phi}}(x)$ and the functional equation $\widehat{\phi}_\alpha(e_{\widehat{\phi}}(x)) = e_{\widehat{\phi}}(\alpha x)$. (Note that the numerical evaluation is justified in this case by Lemma 2.7.23.) $\square$

Denote

$$({}^{\widehat{\phi}}\mathcal{M})_{\text{tor}} = \left\{ z \in \mathcal{M} \mid \widehat{\phi}_\alpha(z) = 0 \text{ for some } \alpha \in O \right\}.$$

It is easy to see that $({}^{\widehat{\phi}}\mathcal{M})_{\text{tor}}$ is an O-submodule of ${}^{\widehat{\phi}}\mathcal{M}$. Similarly, let $({}^{\widehat{\phi}}\mathcal{M}_{\mathbb{C}_K})_{\text{tor}}$ be the torsion submodule of ${}^{\widehat{\phi}}\mathcal{M}_{\mathbb{C}_K}$.

Suppose $z \in (^{\widehat{\phi}}\mathcal{M}_{\mathbb{C}_K})_{\text{tor}}$ is annihilated by $\widehat{\phi}_\alpha = \sum_{n \geq 0} g_n(\alpha) \tau^n$. If $v(\alpha) = 0$, then α is invertible in O, so $x = \widehat{\phi}_{\alpha^{-1}}(\widehat{\phi}_\alpha(x))$. On the other hand, since $\max\{|g_n(\alpha)z^{q^n}|\} < 1 = \rho(\widehat{\phi}_{\alpha^{-1}}(x))$, we can apply Theorem 2.7.24 to evaluate

$$z = \widehat{\phi}_{\alpha^{-1}}(\widehat{\phi}_\alpha(z)) = \widehat{\phi}_{\alpha^{-1}}(0) = 0.$$

Therefore, a nonzero element of $(^{\widehat{\phi}}\mathcal{M}_{\mathbb{C}_K})_{\text{tor}}$ is necessarily annihilated by $\widehat{\phi}_{\varpi^n}$ for some $n \geq 1$.

We put

$$\widehat{\phi}[\varpi^n] = \{z \in \mathcal{M}_{\mathbb{C}_K} \mid \widehat{\phi}_{\varpi^n}(z) = 0\}.$$

By the results of Sect. 2.7, the zeros of an additive power series in $K\{\!\{\tau\}\!\}$ are algebraic and separable over K. Therefore, $\widehat{\phi}[\varpi^n] \subset K^{\text{sep}}$.

Remark 6.5.4 It is easy to see from the definitions that if ϕ is a Drinfeld module over R with stable reduction, and $\widehat{\phi}$ is its formal completion, then

$$\widehat{\phi}[\mathfrak{p}^n] = \phi[\mathfrak{p}^n]^0.$$

Therefore, by Lemma 6.3.7, if H is the height of the reduction of ϕ, then

$$\widehat{\phi}[\mathfrak{p}^n] \cong (A/\mathfrak{p}^n)^H.$$

A similar statement is true for general formal Drinfeld O-modules, but it requires a different proof; see Exercise 6.5.4.

We analyze $(^{\widehat{\phi}}\mathcal{M})_{\text{tor}}$ by using $e_{\widehat{\phi}}$ and $\log_{\widehat{\phi}}$. For the statement of the next theorem note that if $s > 0$ is a positive integer, then for any $\alpha \in O$ we have $\widehat{\phi}_\alpha(\mathcal{M}^s) \subseteq \mathcal{M}^s$. Therefore, $^{\widehat{\phi}}\mathcal{M}^s$ is an O-submodule of $^{\widehat{\phi}}\mathcal{M}$.

Proposition 6.5.5 *Let $\widehat{\phi}$ be a formal Drinfeld O-module over R. Let $s > v(\varpi)/(q-1)$ be an integer. Then $\log_{\widehat{\phi}}$ and $e_{\widehat{\phi}}$ induce isomorphisms*

$$\log_{\widehat{\phi}} \colon {}^{\widehat{\phi}}\mathcal{M}^s \overset{\sim}{\longrightarrow} \mathcal{M}^s,$$

$$e_{\widehat{\phi}} \colon \mathcal{M}^s \overset{\sim}{\longrightarrow} {}^{\widehat{\phi}}\mathcal{M}^s$$

of O-modules.

Proof Let $\log_{\widehat{\phi}}(x) = \sum_{n \geq 0} l_n x^{q^n}$. By Proposition 6.5.2, $\log_{\widehat{\phi}}$ gives a homomorphism $^{\widehat{\phi}}\mathcal{M} \to K$. In the course of proving that proposition, we established that

$$v(l_n) \geq -n \cdot v(\varpi) \quad \text{for all } n \geq 0.$$

Hence, for $z \in M$ with $v(z) \geq s > v(\varpi)/(q-1)$, we have

$$v(l_n z^{q^n}) \geq -n \cdot v(\varpi) + q^n v(z) \tag{6.5.3}$$

$$\geq v(z)(q^n - n(q-1))$$

$$\geq v(z),$$

with an equality $v(l_n z^{q^n}) = v(z)$ if and only if $n = 0$. This implies that $v(\log_{\widehat{\phi}}(z)) = v(z)$, for $z \in M^s$. In particular, $\log_{\widehat{\phi}}$ is a homomorphism from $\widehat{\phi}M^s$ to M^s.

Now let $e_{\widehat{\phi}}(x) = \sum_{n \geq 0} e_n x^{q^n}$. By Proposition 6.5.3, $e_{\widehat{\phi}}$ gives a homomorphism $M \to \widehat{\phi}M$. In the course of proving that proposition, we established that

$$v(e_n) \geq -\frac{q^n - 1}{q - 1} \cdot v(\varpi) \quad \text{for all } n \geq 0.$$

Hence, for $z \in M$ with $v(z) \geq s > v(\varpi)/(q-1)$, we have

$$v(e_n z^{q^n}) \geq q^n \cdot v(z) - \frac{q^n - 1}{q - 1} \cdot v(\varpi) \tag{6.5.4}$$

$$\geq v(z),$$

with an equality $v(e_n z^{q^n}) = v(z)$ if and only if $n = 0$. This again implies that $v(e_{\widehat{\phi}}(z)) = v(z)$, for $z \in M^s$. In particular, $e_{\widehat{\phi}}$ is a homomorphism from M^s to $\widehat{\phi}M^s$.

Since $(\log_{\widehat{\phi}} \circ e_{\widehat{\phi}})(x) = x$ and $(e_{\widehat{\phi}} \circ \log_{\widehat{\phi}})(x) = x$, to show that $\log_{\widehat{\phi}}$ and $e_{\widehat{\phi}}$ are inverse O-module isomorphisms on M^s, it is enough to show that the numerical evaluation of the composites $\log_{\widehat{\phi}} \circ e_{\widehat{\phi}}$ and $e_{\widehat{\phi}} \circ \log_{\widehat{\phi}}$ at $z \in M^s$ can be made according to $(\log_{\widehat{\phi}} \circ e_{\widehat{\phi}})(z) = \log_{\widehat{\phi}}(e_{\widehat{\phi}}(z))$ and $(e_{\widehat{\phi}} \circ \log_{\widehat{\phi}})(z) = e_{\widehat{\phi}}(\log_{\widehat{\phi}}(z))$. Since the inequity (6.5.3) implies that $\max\{|l_n z^{q^n}|\} < \rho(e_{\widehat{\phi}})$, we can apply Theorem 2.7.24 to conclude that $z = e_{\widehat{\phi}}(\log_{\widehat{\phi}}(z))$. One similarly deduces from (6.5.4) that $z = \log_{\widehat{\phi}}(e_{\widehat{\phi}}(z))$. □

Corollary 6.5.6 *Let $\widehat{\phi}$ be a formal Drinfeld O-module over R. Let $s > v(\varpi)/(q-1)$ be an integer. Then $(\widehat{\phi}M^s)_{\text{tor}} = 0$. In particular, if $v(\varpi) < (q-1)$, then $\widehat{\phi}M$ is torsion-free.*

Proof Obviously, if $O \subseteq R$ acts by multiplication, then M^s has no nonzero torsion elements as an O-module. By Proposition 6.5.5, we have an isomorphism $\widehat{\phi}M^s \cong M^s$ of O-modules, so $\widehat{\phi}M^s$ is also torsion-free. If $v(\varpi) < (q-1)$, then we can take $s = 1$. □

Let L be the field of fractions of O. The injection $\iota: O \to R$ uniquely extends to an embedding $L \hookrightarrow K$, and we consider L as a subfield of K. Let κ be the residue field of O. Note that ι induces an injection $\kappa \hookrightarrow k$. Also note that $e(K/L) = v(\varpi)$, where $e(K/L)$ is the ramification index of K/L.

Proposition 6.5.7 *With previous notation,*

$$\#(\widehat{\phi}M)_{\mathrm{tor}} \leq (\#\kappa)^{[K:L]/(q-1)}.$$

Proof Let $\lfloor e(K/L)/(q-1) \rfloor$ be the largest integer less than or equal to $e(K/L)/(q-1)$, and let $s = \lfloor e(K/L)/(q-1) \rfloor + 1$. By Corollary 6.5.6, $\widehat{\phi}M^s$ is torsion-free. Thus, $(\widehat{\phi}M)_{\mathrm{tor}}$ injects into $\widehat{\phi}M/\widehat{\phi}M^s$. On the other hand, as an abelian group, $\widehat{\phi}M/\widehat{\phi}M^s$ is isomorphic to M/M^s. Thus, $\#(\widehat{\phi}M)_{\mathrm{tor}} \leq \#M/M^s$. The set $\left\{ \sum_{i=1}^{s-1} \alpha_i \pi^i \mid \alpha_1, \ldots, \alpha_{s-1} \in k \right\}$ consists of coset representatives of M^s in M, so $\#(M/M^s) = (\#k)^{s-1}$. On the other hand,

$$\#k = (\#\kappa)^{f(K/L)},$$

where $f(K/F_{\mathfrak{p}}) = [k : \kappa]$ is the residue degree of the extension K/L. Since

$$[K : L] = f(K/L) \cdot e(K/L),$$

we get

$$\begin{aligned}
\#(\widehat{\phi}M)_{\mathrm{tor}} &\leq (\#k)^{s-1} \\
&= (\#\kappa)^{f(K/L) \cdot \lfloor e(K/L)/(q-1) \rfloor} \\
&\leq (\#\kappa)^{[K:L]/(q-1)}.
\end{aligned}$$

$\square$

Now we return to our Drinfeld module $\phi \colon A \to R\{\tau\}$. We proved that $(^{\phi}K)_{\mathrm{tor}}$ is finite. In fact, we gave an explicit bound on its cardinality when ϕ has stable reduction; see (6.4.4). One can use Proposition 6.5.7 to strengthen this bound in the case when ϕ has good reduction.

Theorem 6.5.8 *Let ϕ be a Drinfeld module over R with good reduction. Then*

$$\#(^{\phi}K)_{\mathrm{tor}} \leq (\#k)^{\lfloor e(K/F_{\mathfrak{p}})/(q-1) \rfloor + 1}.$$

Proof We will prove a somewhat stronger statement, which clarifies where the assumption that ϕ has good reduction is used. Suppose ϕ is defined over R and has stable reduction $\bar{\phi}$. Let $\widehat{\phi}$ denote the formal completion of ϕ. Since the action of

A on M via ϕ coincides with its action via $\widehat{\phi}$, we have the following exact sequence of A modules:

$$0 \longrightarrow {}^{\widehat{\phi}}M \longrightarrow {}^{\phi}R \longrightarrow {}^{\bar{\phi}}k \longrightarrow 0.$$

From this exact sequence we derive

$$0 \longrightarrow ({}^{\widehat{\phi}}M)_{\mathrm{tor}} \longrightarrow ({}^{\phi}R)_{\mathrm{tor}} \longrightarrow {}^{\bar{\phi}}k. \qquad (6.5.5)$$

Now, from Proposition 6.5.7 and its proof, we have $\#({}^{\widehat{\phi}}M)_{\mathrm{tor}} \le (\#k)^{\lfloor e(K/F_{\mathfrak{p}})/(q-1)\rfloor}$. Thus,

$$\#({}^{\phi}R)_{\mathrm{tor}} \le (\#k)^{\lfloor e(K/F_{\mathfrak{p}})/(q-1)\rfloor+1}.$$

This bound is valid without assuming that ϕ has good reduction. On the other hand, if we assume that ϕ has good reduction, then the leading coefficients of $\phi_a(x)$, $a \in A$, are units in R. Since the torsion points of ϕ are roots of the division polynomials $\phi_a(x)$, any K-rational torsion point of ϕ will be integral, i.e., will lie in R. Thus, $({}^{\phi}R)_{\mathrm{tor}} = ({}^{\phi}K)_{\mathrm{tor}}$, and the above bound implies the claim of the theorem. □

Remark 6.5.9 In case when ϕ has good reduction, (6.4.4) gives the bound $\#({}^{\phi}K)_{\mathrm{tor}} \le (\#k)^{e(K/F_{\mathfrak{p}})}$ because the good reduction assumption is equivalent to $v(g_r) = 0$. But note that Exercise 6.4.3 outlines a different proof of the bound of Theorem 6.5.8, which does not use formal Drinfeld modules.

Theorem 6.5.10 *Let ϕ be a Drinfeld module over R with good reduction $\bar{\phi}$. If $e(K/F_{\mathfrak{p}}) < (q-1)$, then the reduction modulo M induces an injection*

$$({}^{\psi}K)_{\mathrm{tor}} \lhook\joinrel\longrightarrow {}^{\bar{\psi}}k.$$

Proof This follows from (6.5.5), the fact that $({}^{\phi}R)_{\mathrm{tor}} = ({}^{\phi}K)_{\mathrm{tor}}$ used in the proof of Theorem 6.5.8, and Corollary 6.5.6. □

Exercises

6.5.1 Let $\widehat{\phi}$ be a formal Drinfeld O-module over R, and let $z \in M$. Prove that $\log_{\widehat{\phi}}(z) = 0$ if and only if $z \in ({}^{\widehat{\phi}}M)_{\mathrm{tor}}$.

6.5.2 Give an explicit example which shows that the injection $({}^{\phi}K)_{\mathrm{tor}} \lhook\joinrel\longrightarrow {}^{\bar{\phi}}k$ of Theorem 6.5.10 is not necessarily an isomorphism.

6.5.3 Let $\widehat{\phi} \colon O \to k\{\!\{\tau\}\!\}$ be a formal Drinfeld module. For $f = \sum_{n \geq 0} c_n \tau^n \in k\{\!\{\tau\}\!\}$, define its height $\mathrm{ht}(f)$ to be the smallest subscript n such that $c_n \neq 0$. Prove that there is a natural number $H(\widehat{\phi}) \geq 1$, called the *height* of $\widehat{\phi}$, such that for any $0 \neq \alpha \in O$,

$$\mathrm{ht}(\widehat{\phi}_\alpha) = H(\widehat{\phi}) \cdot \deg(\mathfrak{p}) \cdot \mathrm{ord}_{\mathfrak{p}}(\alpha).$$

6.5.4 Let $\widehat{\phi} \colon O \to R\{\!\{\tau\}\!\}$ be a formal Drinfeld module. Assume the reduction

$$\overline{\widehat{\phi}} \colon O \xrightarrow{\ \widehat{\phi}\ } R\{\!\{\tau\}\!\} \xrightarrow{\ \mathrm{mod}\, \mathcal{M}\ } k\{\!\{\tau\}\!\}$$

of $\widehat{\phi}$ modulo $\mathcal{M}$ is a formal Drinfeld module over k. Let H be the height of $\overline{\widehat{\phi}}$. Prove that

$$\widehat{\phi}[\varpi^n] \cong (A/\varpi^n)^H.$$

Chapter 7
Drinfeld Modules Over Global Fields

Let K be a finite extension of F, considered as an A-field via the natural embeddings

$$\gamma : A \longhookrightarrow F \longhookrightarrow K.$$

To simplify the notation, we will denote $\gamma(a) \in K$ by a.

Let B be the integral closure of A in K, and let ϕ be a Drinfeld module over K. In this chapter we are interested in the following two basic questions:

(1) What can one say about the division fields $K(\phi[\mathfrak{n}])$ of ϕ? For example: When is $\phi[\mathfrak{n}]$ a subset of K? What is the Galois group of $K(\phi[\mathfrak{n}])/K$? Which primes of B split completely in $K(\phi[\mathfrak{n}])$?

(2) What can one say about the A-module structure of ϕK?

The first four sections are devoted to the first question. The precursor is the case of the Carlitz module over F, which we study in Sect. 7.1. The theory of division fields of the Carlitz module was developed by Carlitz [Car38] and Hayes [Hay74]. This theory is strikingly similar to the classical theory of cyclotomic extensions of $\mathbb{Q}$, and its generalization eventually led to an explicit construction of class fields of general function fields, thus providing a solution to the analogue of Hilbert's 12th problem.

In Sect. 7.2, we study the rational torsion submodules of Drinfeld modules over K. Our earlier study of the same question over local fields has direct applications here. We also state some deep conjectures concerning the uniformity and classification of rational torsion submodules.

The division fields of Drinfeld modules of rank ≥ 2 are generally non-abelian, so naturally their theory is richer and more complicated than the theory of Carlitz cyclotomic extensions. To demonstrate the richness of this theory, as well as to give the reader some intuition for these fields, we start the study of division fields of Drinfeld modules in Sect. 7.3 with examples of T-division fields of rank 2 Drinfeld modules over F. We also prove a theorem of Boston and Ose, which states that

© The Author(s), under exclusive license to Springer Nature Switzerland AG 2023   391
M. Papikian, *Drinfeld Modules*, Graduate Texts in Mathematics 296,
https://doi.org/10.1007/978-3-031-19707-9_7

any Galois extension of K with Galois group isomorphic to a subgroup of $\mathrm{GL}_r(\mathbb{F}_q)$ arises as the T-division field of some Drinfeld module of rank r.

In Sect. 7.4, we give congruence conditions under which a given prime of A splits completely in a division field of a Drinfeld module over F. This is a generalization to higher ranks of a key result of the Carlitz cyclotomic theory. The proof heavily relies on the theory of Drinfeld modules over finite fields that we developed in Chap. 4. We also discuss some computational problems related to the explicit description of the action of Frobenius elements on the torsion points of Drinfeld modules.

In Sect. 7.5, we study Drinfeld modules over $\mathbb{C}_\infty$ whose endomorphism rings are as large as possible. More precisely, if ϕ is a Drinfeld module of rank r over $\mathbb{C}_\infty$ then, as we proved in Chap. 3, its endomorphism ring $\mathrm{End}(\phi)$ is an A-order in a totally imaginary field extension of F of degree dividing r. We say that ϕ has complex multiplication by B if B is the integral closure of A in a totally imaginary extension of F of degree r and $\mathrm{End}(\phi) \cong B$. We will show that, up to isomorphism, there are only finitely many Drinfeld modules with complex multiplication by B, and their number is equal to the class number of B. This implies that such Drinfeld modules can be defined over a finite extension H of F. We then identify H with the Hilbert class field of B. The theory of division fields of Drinfeld modules with complex multiplication has many similarities with the theory of Carlitz cyclotomic extensions, e.g., these division fields are abelian extensions of H. In fact, a Drinfeld module with complex multiplication by B can be regarded as a Drinfeld B-module of rank 1, so the Carlitz cyclotomic theory is a special case of the theory of complex multiplication.

In the final section, Sect. 7.6, we study the A-module $^\phi B$ for a given Drinfeld module ϕ over K. We prove an interesting result due to Poonen, which says that $^\phi B$ is isomorphic to the direct sum of its finite torsion submodule and a free A-module of countable rank. This is somewhat counterintuitive since B with its usual A-module structure is a free module of finite rank. To prove this result, we develop the basics of the theory of height functions on Drinfeld modules due to Denis. In the second half of this section, we discuss a construction due to Taelman of a finitely generated submodule $U(\phi/B)$ of $^\phi B$, which plays the role of the group of units in the ring of integers of a number field. Then we discuss Taelman's class number formula and explain why it should be considered as an analogue of the classical class number formula for the residue at 1 of the Dedekind zeta function of a number field. Finally, we discuss a construction of explicit cyclotomic units in $U(C/B)$ due to Anderson.

7.1 Carlitz Cyclotomic Extensions

We start by recalling some basic facts from the theory of cyclotomic extensions of $\mathbb{Q}$; the proofs can be found in Chapter 2 of [Was82] or Sections 13.6 and 14.6 of [DF04]. This is supposed to serve as a motivation for the results that we prove in this section for the division fields of the Carlitz module.

The roots of the polynomial $x^n - 1 \in \mathbb{Q}[x]$ form a multiplicative group μ_n isomorphic to $\mathbb{Z}/n\mathbb{Z}$. A *primitive n-th root* is a generator of μ_n. After fixing a primitive root ζ_n, the isomorphism $\mathbb{Z}/n\mathbb{Z} \to \mu_n$ is given by $m \mapsto \zeta_n^m$. The field $\mathbb{Q}(\zeta_n)$, called the *n-th cyclotomic extension of* $\mathbb{Q}$, is the splitting field of $x^n - 1$, so it does not depend on the choice of ζ_n.

(1) *The minimal polynomial:* Let $\Phi_n(x)$ be the minimal polynomial of ζ_n over $\mathbb{Q}$. We have

$$\Phi_n(x) = \prod_{\substack{\zeta \in \mu_n \\ \zeta \text{ primitive}}} (x - \zeta) = \prod_{\substack{0 \le i < n \\ (i,n)=1}} (x - \zeta_n^i) \in \mathbb{Z}[x].$$

The polynomial $\Phi_n(x)$ does not depend on the choice of the primitive root and has degree

$$\varphi(n) := \#(\mathbb{Z}/n\mathbb{Z})^\times = n \prod_{p \mid n} \left(1 - \frac{1}{p}\right).$$

The polynomial $\Phi_n(x)$ is called the *n-th cyclotomic polynomial*.

(2) *The Galois group:* The map

$$(\mathbb{Z}/n\mathbb{Z})^\times \longrightarrow \text{Gal}(\mathbb{Q}(\zeta_n)/\mathbb{Q})$$

$$m \ (\text{mod } n) \longmapsto (\zeta_n \mapsto \zeta_n^m)$$

is an isomorphism. This isomorphism does not depend on the choice of ζ_n.

(3) *The maximal real subfield:* The subfield of $\mathbb{Q}(\zeta_n)$ fixed by the complex conjugation is the field $\mathbb{Q}(\zeta_n)^+ := \mathbb{Q}(\zeta_n + \zeta_n^{-1})$.

(4) *Ramification:* An odd prime p ramifies in $\mathbb{Q}(\zeta_n)$ if and only if $p \mid n$. The prime 2 ramifies in $\mathbb{Q}(\zeta_n)$ if and only if $4 \mid n$.

(5) *The ring of integers:* $\mathbb{Z}[\zeta_n]$ is the ring of algebraic integers of $\mathbb{Q}(\zeta_n)$.

(6) *Cyclotomic reciprocity law:* Given a prime $p \nmid n$, let f be the smallest positive integer such that $p^f \equiv 1 \ (\text{mod } n)$. Then p splits into $\varphi(n)/f$ distinct primes in $\mathbb{Q}(\zeta_n)$, each of which has residue degree f. In particular, p splits completely in $\mathbb{Q}(\zeta_n)$ if and only if $p \equiv 1 \ (\text{mod } n)$.

(7) *The Kronecker–Weber Theorem:* Every abelian extension of $\mathbb{Q}$ is contained in some $\mathbb{Q}(\zeta_n)$.

It is a consequence of the Chebotarev density theorem that the set of all but finitely many primes that split completely in a given Galois extension of a global field K uniquely determines that extension. To state this more precisely, let us adopt the following notation: Given two sets S and S', write $S \preceq S'$ if $S \subset S' \cup \mathcal{F}$ for some finite set $\mathcal{F}$, and, given a finite Galois extension $K \subset L$, let

$$\text{Spl}(L/K) = \{v \text{ place of } K \mid v \text{ splits completely in } L\}.$$

Theorem 7.1.1 *Let L and M be Galois extensions of K. Then*

$$L \subset M \iff \mathrm{Spl}(M/K) \preceq \mathrm{Spl}(L/K).$$

Proof See [Cox89, Thm. 8.19] or [Mil13, Thm. V.3.25].   □

Remarks 7.1.2

(1) The previous theorem makes the problem of characterizing the primes that split completely in extensions of global fields a central problem in number theory. Results that solve this problem using congruence conditions for a fixed modulus are called "reciprocity laws." For example, the cyclotomic reciprocity mentioned above implies that

$$\mathrm{Spl}(\mathbb{Q}(\zeta_n)/\mathbb{Q}) = \{p \mid p \equiv 1 \ (\mathrm{mod}\ n)\}.$$

(2) The terminology "reciprocity" comes from Gauss' quadratic reciprocity, which can be restated as a congruence condition for splitting of rational primes in the quadratic extension $\mathbb{Q}(\sqrt{p})$ for a given prime p, as we now explain. The *Quadratic Reciprocity Law* is usually stated as follows: Let p and q be distinct odd prime numbers, and define the Legendre symbol as

$$\left(\frac{p}{q}\right) = \begin{cases} 1 & p \equiv n^2 \ (\mathrm{mod}\ q) \text{ for some integer } n; \\ -1 & \text{otherwise.} \end{cases}$$

Then

$$\left(\frac{p}{q}\right)\left(\frac{q}{p}\right) = (-1)^{\frac{p-1}{2}\frac{q-1}{2}}.$$

Now, the prime q splits in $\mathbb{Q}(\sqrt{p})$ if and only if $x^2 - p$ splits into linear factors modulo q, or equivalently $\left(\frac{p}{q}\right) = 1$. Assume for simplicity that $p \equiv 1 \ (\mathrm{mod}\ 4)$. Then the quadratic reciprocity implies that $q \in \mathrm{Spl}(\mathbb{Q}(\sqrt{p})/\mathbb{Q})$ if and only if $\left(\frac{q}{p}\right) = 1$. This does not seem much different from the $\left(\frac{p}{q}\right) = 1$ condition, but the important difference is that $\left(\frac{q}{p}\right) = 1$ is equivalent to congruence conditions modulo p, which is a *fixed* modulus for $\mathbb{Q}(\sqrt{p})/\mathbb{Q}$. For example, this argument gives

$$\mathrm{Spl}(\mathbb{Q}(\sqrt{13})/\mathbb{Q}) = \{q \mid q \equiv 1, 3, 4, 9, 10, 12 \ (\mathrm{mod}\ 13)\},$$

where the listed numbers are the squares in $\mathbb{F}_{13}^{\times}$. If q is a very large prime, then checking whether its residue modulo 13 is one of the numbers above is much easier than checking whether 13 is a square modulo q.

(3) Class field theory is a vast generalization of quadratic and cyclotomic reciprocity laws; it gives congruence conditions for the splitting of primes in abelian

extensions of global fields. There are reciprocity laws also for non-abelian extensions, although this theory is far less complete. We will give one example of such non-abelian reciprocity law in Sect. 7.4. The interested reader can find a nice exposition of the history and generalizations of abelian reciprocity laws in Wyman's article [Wym72].

Returning to the setting of function fields, the obvious analogue of $\mathbb{Q}(\zeta_n)$ for F is the splitting field $F(\zeta_n)$ of $x^n - 1$ over F. If the characteristic p of F divides n, then $x^n - 1 = (x^{n/p} - 1)^p$, so $F(\zeta_n) = F(\zeta_{n/p})$. Thus, we can assume $p \nmid n$. Let s be a positive integer such that $n \mid (q^s - 1)$; such an integer exists because $q \in (\mathbb{Z}/n\mathbb{Z})^\times$. Then ζ_n is a root of $x^{q^s-1} - 1$, so $F(\zeta_n) \subseteq F(\zeta_{q^s-1})$. On the other hand, by Sect. 1.6, the splitting field of $x^{q^s-1} - 1$ is $\mathbb{F}_{q^s} F = \mathbb{F}_{q^s}(T)$. Therefore, by adjoining ζ_n to F we only obtain an extension of the constant field $\mathbb{F}_q$. The splitting behavior of primes in such extensions is not hard to describe; see Exercise 7.1.3. Of course, F has many other extensions; for example, there are infinitely many non-isomorphic quadratic extensions of F, only one of which is $\mathbb{F}_{q^2}(T)$.

To obtain a good analogue of cyclotomic extensions for F, the extensions $\mathbb{Q}(\zeta_n)$ have to be viewed from a different perspective. Consider $\overline{\mathbb{Q}}^\times$ as a $\mathbb{Z}$-module, where $n \in \mathbb{Z}$ acts on $\alpha \in \overline{\mathbb{Q}}^\times$ by $n \circ \alpha = \alpha^n$. Then μ_n is the group of n-torsion elements of $\overline{\mathbb{Q}}^\times$, so cyclotomic extensions are the division fields of $\overline{\mathbb{Q}}^\times$. Now let $\phi \colon A \to F\{\tau\}$ be a Drinfeld module of rank $r \geq 1$ over F. From ϕ we obtain an A-module $^\phi \overline{F}$ whose torsion elements are $\phi[a]$, $0 \neq a \in A$. The division field $F(\phi[a])$ is a Galois extension of F whose Galois group is a subgroup of $\mathrm{Aut}_A(\phi[a]) \cong \mathrm{GL}_r(A/aA)$. In particular, if $r = 1$, then by adjoining the torsion elements of the A-module $^\phi \overline{F}$ to F one obtains abelian extensions. This turns out to be the right idea for constructing a cyclotomic theory over F.

The Drinfeld module of rank 1 that we work with will be the Carlitz module $C_T = T + \tau$ over F.

Definition 7.1.3 Let $\mathfrak{n} \in A_+$. The $\mathfrak{n}$-th *Carlitz cyclotomic extension* of F, denoted $K_\mathfrak{n}$, is the splitting field of $C_\mathfrak{n}(x)$, where $C_\mathfrak{n}(x)$ is the $\mathfrak{n}$-division polynomial of C. A generator of the A-module $C[\mathfrak{n}] \cong A/\mathfrak{n}$ is called a *primitive $\mathfrak{n}$-th root of C*; here, by our usual abuse of notation, we denote the ideal generated by $\mathfrak{n}$ by the same symbol. Thus, if $\zeta_\mathfrak{n}$ is a primitive $\mathfrak{n}$-th root of C, then

$$C[\mathfrak{n}] = \{C_a(\zeta_\mathfrak{n}) \mid a \in A, \deg(a) < \deg(\mathfrak{n})\}.$$

Let $\zeta_\mathfrak{n}$ be a primitive $\mathfrak{n}$-th root of C. Given $a \in A$, the A-submodule of $C[\mathfrak{n}]$ generated by $C_a(\zeta_\mathfrak{n})$ is isomorphic to $A/(\mathfrak{n}/\gcd(a, \mathfrak{n}))$. In particular, $C_a(\zeta_\mathfrak{n})$ is a primitive $\mathfrak{n}$-th root itself if and only if a is coprime to $\mathfrak{n}$.

Definition 7.1.4 Let $\mathfrak{n} \in A_+$. The $\mathfrak{n}$-*th Carlitz cyclotomic polynomial*, denoted $\Phi_{\mathfrak{n}}(x)$, is the polynomial whose roots are the primitive $\mathfrak{n}$-th roots of C:

$$\Phi_{\mathfrak{n}}(x) = \prod_{\substack{\zeta \in C[\mathfrak{n}] \\ \zeta \text{ primitive}}} (x - \zeta)$$

$$= \prod_{\substack{\deg(a) < \deg(\mathfrak{n}) \\ \gcd(a, \mathfrak{n}) = 1}} (x - C_a(\zeta_{\mathfrak{n}})).$$

Note that $\Phi_1(x) = x$ because $C[1] = 0$.

Proposition 7.1.5 *Given* $\mathfrak{n} \in A_+$, *denote* $|\mathfrak{n}| = q^{\deg(\mathfrak{n})}$.

(1) The degree of $\Phi_{\mathfrak{n}}(x)$ *is equal to*

$$|\mathfrak{n}| \prod_{\substack{\mathfrak{p} \in A_+ \text{ prime} \\ \mathfrak{p} | \mathfrak{n}}} \left(1 - \frac{1}{|\mathfrak{p}|}\right).$$

(2) The coefficients of $\Phi_{\mathfrak{n}}(x)$ *are in* A.
(3) Similar to (1.6.2), define the Möbius function on the set of monic nonzero polynomials A_+ *as follows:* $\mu(1) = 1$, *and*

$$\mu(\mathfrak{n}) = -\sum_{\substack{\mathfrak{d} \in A_+ \\ \mathfrak{d} | \mathfrak{n} \\ \deg(\mathfrak{d}) < \deg(\mathfrak{n})}} \mu(\mathfrak{d}), \quad \text{if } \deg(\mathfrak{n}) > 0.$$

   Then

$$\Phi_{\mathfrak{n}}(x) = \prod_{\mathfrak{d} | \mathfrak{n}} C_{\mathfrak{d}}(x)^{\mu(\mathfrak{n}/\mathfrak{d})}. \tag{7.1.1}$$

Proof

(1) We have

$$\deg \Phi_{\mathfrak{n}}(x) = \#\{a \in A \mid \deg(a) < \deg(\mathfrak{n}), \gcd(a, \mathfrak{n}) = 1\} \quad \text{(Definition 7.1.4)}$$

$$= \#(A/\mathfrak{n})^{\times}.$$

Define $\varphi(\mathfrak{n}) = \#(A/\mathfrak{n})^{\times}$ as the analogue of the Euler totient function for A. An explicit formula for $\varphi(\mathfrak{n})$ can be deduced as follows. Let $\mathfrak{n} = \mathfrak{p}_1^{s_1} \cdots \mathfrak{p}_k^{s_k}$ be the prime decomposition of $\mathfrak{n}$. First, from the Chinese Remainder Theorem, one obtains

$$(A/\mathfrak{n})^{\times} \cong \prod_{i=1}^{k} (A/\mathfrak{p}_i^{s_i})^{\times},$$

so $\varphi(\mathfrak{n}) = \prod_{i=1}^{k} \varphi(\mathfrak{p}_i^{s_i})$ is multiplicative. Next, for a prime $\mathfrak{p}$ we have $\#(A/\mathfrak{p})^{\times} = \#\mathbb{F}_{\mathfrak{p}}^{\times} = |\mathfrak{p}| - 1$. To compute $\varphi(\mathfrak{p}^s)$ for $s \geq 2$, observe that reducing the units in $A/\mathfrak{p}^s$ modulo $\mathfrak{p}$ gives the units in $A/\mathfrak{p}$ and produces a short-exact sequence of groups

$$0 \longrightarrow G \longrightarrow (A/\mathfrak{p}^s)^{\times} \longrightarrow (A/\mathfrak{p})^{\times} \longrightarrow 0,$$

where $G = \{1 + \mathfrak{p}a \mid \deg(a) < \deg(\mathfrak{p}^{s-1})\}$. Since $\#G = |\mathfrak{p}|^{s-1}$, we get $\varphi(\mathfrak{p}^s) = |\mathfrak{p}|^{s-1}(|\mathfrak{p}| - 1)$. This leads to a formula similar to the well-known formula for the classical Euler function:

$$\varphi(\mathfrak{n}) = |\mathfrak{n}| \prod_{\substack{\mathfrak{p}|\mathfrak{n} \\ \mathfrak{p}\ \text{prime}}} \left(1 - \frac{1}{|\mathfrak{p}|}\right). \tag{7.1.2}$$

(2) If $\mathfrak{d}$ is an ideal dividing $\mathfrak{n}$, then $C[\mathfrak{d}] \subset C[\mathfrak{n}]$ is the unique submodule of $C[\mathfrak{n}] \cong A/\mathfrak{n}$ isomorphic to $A/\mathfrak{d}$. Moreover, if $\gcd(a, \mathfrak{n}) = \mathfrak{n}/\mathfrak{d}$, then $C_a(\zeta_{\mathfrak{n}})$ generates $C[\mathfrak{d}]$, so $C_a(\zeta_{\mathfrak{n}})$ is a primitive $\mathfrak{d}$-th root of C. Collecting together those roots of $C_{\mathfrak{n}}(x)$ which are primitive $\mathfrak{d}$-th roots, we get the decomposition

$$C_{\mathfrak{n}}(x) = \prod_{\mathfrak{d}|\mathfrak{n}} \prod_{\substack{\zeta \in C[\mathfrak{d}] \\ \zeta\ \text{primitive}}} (x - \zeta) \tag{7.1.3}$$

$$= \prod_{\mathfrak{d}|\mathfrak{n}} \Phi_{\mathfrak{d}}(x).$$

To prove that $\Phi_{\mathfrak{n}}(x)$ has coefficients in A, we use induction on the degree of $\mathfrak{n}$. The claim is obviously true for $\Phi_1(x) = x$. Now suppose that $\deg(\mathfrak{n}) = N$ and $\Phi_{\mathfrak{d}}(x) \in A[x]$ for all $\mathfrak{d} \in A_+$ with $\deg(\mathfrak{d}) < N$. By (7.1.3), we have $C_{\mathfrak{n}}(x) = \Phi_{\mathfrak{n}}(x) \cdot f(x)$, where

$$f(x) = \prod_{\substack{\mathfrak{d}|\mathfrak{n} \\ \mathfrak{d}\neq\mathfrak{n}}} \Phi_{\mathfrak{d}}(x).$$

By the induction hypothesis, $f(x)$ is a monic polynomial in $A[x]$. Since the same is true for $C_{\mathfrak{n}}(x)$, Gauss' Lemma (Corollary 1.1.7) implies that $\Phi_{\mathfrak{n}}(x)$ has coefficients in A.

(3) This follows by applying the multiplicative version of the Möbius inversion formula to (7.1.3); see Exercise 1.6.8.

$\square$

Remark 7.1.6 Recall that Proposition 5.4.3 gives an explicit formula for the coefficients of $C_\mathfrak{n}(x) = \sum_{i \geq 0} g_i(\mathfrak{n}) x^{q^i}$,

$$g_i(\mathfrak{n}) = \frac{\prod_{a \in A_{<i}}(\mathfrak{n} - a)}{D_i}.$$

Therefore, $\Phi_\mathfrak{n}(x)$ can be computed recursively using either (7.1.1) or (7.1.3). For prime powers this is especially simple, as we now explain. Let $\mathfrak{p} \in A$ be a prime. Then

$$\Phi_\mathfrak{p}(x) = C_\mathfrak{p}(x)/x.$$

Before applying (7.1.1) to $\Phi_{\mathfrak{p}^s}(x)$ for $s \geq 2$, we note that

$$\mu(\mathfrak{n}) = \begin{cases} 1 & \text{if } \mathfrak{n} = 1, \\ (-1)^k & \text{if } \mathfrak{n} \text{ is a product of } k \text{ distinct primes,} \\ 0 & \text{if } \mathfrak{n} \text{ is divisible by a square of a prime.} \end{cases}$$

Thus,

$$\Phi_{\mathfrak{p}^s}(x) = \frac{C_{\mathfrak{p}^s}(x)}{C_{\mathfrak{p}^{s-1}}(x)} = \frac{C_\mathfrak{p}(C_{\mathfrak{p}^{s-1}}(x))}{C_{\mathfrak{p}^{s-1}}(x)} = \Phi_\mathfrak{p}(C_{\mathfrak{p}^{s-1}}(x)). \tag{7.1.4}$$

Example 7.1.7 Let $q = 3$ and $\mathfrak{p} = T^3 - T + 1$. Then

$$\Phi_\mathfrak{p}(x) = x^{26} + T(T+1)(T+2)(T^3+2T+2)\mathfrak{p} \cdot x^8 + (T^3+2T+2)\mathfrak{p} \cdot x^2 + \mathfrak{p}.$$

Example 7.1.8 Let $\mathfrak{p} = T$. Then $C_T(x) = Tx + x^q$, and

$$\Phi_T(x) = T + x^{q-1},$$

$$\Phi_{T^2}(x) = T + (Tx + x^q)^{q-1},$$

$$\Phi_{T^3}(x) = T + (T^2 x + (T + T^q)x^q + x^{q^2})^{q-1}.$$

Remark 7.1.9 For a prime number p, the classical cyclotomic polynomial

$$\Phi_p(x) = (x^p - 1)/(x - 1) = x^{p-1} + \cdots + x + 1$$

has a nice uniform shape. On the other hand, the coefficients of $\Phi_p(x) = C_p(x)/x$ depend on p and, as we mentioned in Remark 5.4.5, should be considered as analogues of the binomial coefficients. From this perspective, $\Phi_p(x)$ is more similar to

$$\Phi_p(x+1) = \frac{(x+1)^p - 1}{x} = \sum_{i=1}^{p} \binom{p}{i} x^{i-1}.$$

Proposition 7.1.10 *The polynomial $\Phi_n(x)$ is irreducible over F.*

Proof Let ζ_n be a primitive n-th root of C and let $f(x)$ be the minimal polynomial of ζ_n over F. If we show that any other primitive n-th root of C is also a root of $f(x)$, then from the definitions we get $\Phi_n(x) = f(x)$, and therefore $\Phi_n(x)$ is irreducible.

First, we show that $C_p(\zeta_n)$ is a root of $f(x)$ for any prime p not dividing n. Since ζ_n is a root of Φ_n, its minimal polynomial $f(x)$ divides $\Phi_n(x)$, so we have a decomposition

$$\Phi_n(x) = f(x) \cdot g(x).$$

Note that $f(x)$ is monic and has coefficients in A (as ζ_n is integral over A). Therefore, by Gauss' Lemma and Proposition 7.1.5, all three polynomials Φ_n, f, g are monic and have coefficients in A. Since p and n are coprime, $C_p(\zeta_n)$ is a primitive n-th root of C, so $\Phi_n(C_p(\zeta_n)) = 0$. Now, if $C_p(\zeta_n)$ is not a root of $f(x)$, then $g(C_p(\zeta_n)) = 0$. In that case, $f(x)$ divides $g(C_p(x))$. For a polynomial $h(x) \in A[x]$, denote by $\bar{h}(x) \in \mathbb{F}_p[x]$ the polynomial obtained by reducing its coefficients modulo p. With this notation, $\bar{f}(x)$ divides $\bar{g}(\bar{C}_p(x))$. Since the height $H(C)$ is 1, which is the rank of C, we have $\bar{C}_p(x) = x^{|p|}$. Therefore, $\bar{f}(x)$ divides

$$\bar{g}(\bar{C}_p(x)) = \bar{g}(x^{|p|}) = \bar{g}(x)^{|p|}.$$

This implies that $\bar{f}(x)$ and $\bar{g}(x)$ have a common factor. But then $\Phi_n(x)$, and hence also $C_n(x)$, has a multiple root modulo p. On the other hand, the derivative $\bar{C}_n(x)' = \bar{n} \neq 0$ is a nonzero constant, hence is coprime to $\bar{C}_n(x)$. This implies that $\bar{C}_n(x)$ does not have multiple roots, and we get a contradiction.

Now let ζ' be a primitive n-th root of C. Then $\zeta' = C_a(\zeta_n)$ for some $a \in A$ coprime to n. We can assume that a is monic by adding to a an appropriate multiple of n. Decompose $a = p_1 p_2 \cdots p_s$ into primes. By the previous paragraph, $C_{p_1}(\zeta_n)$ is a root of f; repeating this argument with $C_{p_1}(\zeta_n)$ playing the role of ζ_n, we get that $C_{p_2}(C_{p_1}(\zeta_n)) = C_{p_1 p_2}(\zeta_n)$ is also a root of f; continuing in this manner, we eventually conclude that $C_a(\zeta)$ is a root of f. $\square$

Remark 7.1.11 Recall that for a prime p we have $\Phi_p(x) = C_p(x)/x$. Since $C_p(x)/x$ is an Eisenstein polynomial with respect to p, i.e., its coefficients, except the leading coefficient, are divisible by p and the constant term is not divisible by p^2, the same is true for $\Phi_p(x)$. More generally $\Phi_{p^s}(x)$ for any $s \geq 1$ is Eisenstein (see the proof of Theorem 7.1.16). This observation, combined with Corollary 2.5.6, implies that $\Phi_{p^s}(x)$ is irreducible. (Note that this argument is similar to the proof of the irreducibility of $\Phi_p(x) = 1 + x + \cdots + x^{p-1}$ using the fact that $\Phi_p(x + 1)$ is Eisenstein.) But $\Phi_n(x)$ is not Eisenstein if n is not a prime power (see Exercise 7.1.5), so this argument does not give an alternative proof of the previous proposition for general n.

Since $K_n = F(\zeta_n)$, any automorphism σ of K_n is uniquely determined by its action on ζ_n. Moreover, $\sigma(\zeta_n)$ must be another primitive n-th root of C since the Galois action on $C[n]$ commutes with the action of A. Hence, $\sigma(\zeta_n) = C_a(\zeta_n)$ for a uniquely determined $a \in A$, such that $\deg(a) < \deg(n)$ and $\gcd(a, n) = 1$. It is easy to see that a does not depend on a particular choice of a primitive n-th root: for any other primitive root ζ_n' we have $\zeta_n' = C_b(\zeta_n)$ for some $b \in A$, and therefore

$$\sigma(\zeta_n') = \sigma(C_b(\zeta_n)) = C_b(\sigma(\zeta_n)) = C_b(C_a(\zeta_n)) = C_a(C_b(\zeta_n)) = C_a(\zeta_n').$$

We denote by σ_a the element of $\mathrm{Gal}(K_n/K)$ which maps ζ_n to $C_a(\zeta_n)$.

Theorem 7.1.12 *We have an isomorphism*

$$\mathrm{Gal}(K_n/F) \xrightarrow{\sim} (A/n)^{\times}$$

$$\sigma_a \longmapsto a \pmod{n}.$$

Proof One readily verifies that the indicated map is a homomorphism. It is injective by construction. Proposition 7.1.10 implies that $[K_n : F] = \#(A/n)^{\times}$, so the map is also surjective. $\qquad\qquad\square$

Note that $\mathbb{Q}(\zeta_n)^+$ is the subfield of $\mathbb{Q}(\zeta_n)$ fixed by the image of $\mathbb{Z}^{\times} = \{\pm 1\}$ in $(\mathbb{Z}/n\mathbb{Z})^{\times} \cong \mathrm{Gal}(\mathbb{Q}(\zeta_n)/\mathbb{Q})$. The unique Archimedean place of $\mathbb{Q}$ splits completely in $\mathbb{Q}(\zeta_n)^+$, and all the Archimedean places of $\mathbb{Q}(\zeta_n)^+$ ramify in $\mathbb{Q}(\zeta_n)$ if $n > 2$.[1] The analogue of $\mathbb{Q}(\zeta_n)^+$ for K_n is the following:

Theorem 7.1.13 *The maximal totally real subfield of K_n is the subfield K_n^+ fixed by the image of $A^{\times} = \mathbb{F}_q^{\times}$ in $(A/n)^{\times} \cong \mathrm{Gal}(K_n/F)$, i.e., K_n^+ is the maximal subfield of K_n such that the place ∞ of F splits completely in K_n^+. Every place of K_n^+ over ∞ totally ramifies in K_n, i.e., has ramification index $q - 1$.*

Proof First, we show that $K_n^+ = F(\zeta_n^{q-1})$. Since $\alpha \in \mathbb{F}_q^{\times} \subset \mathrm{Gal}(K_n/F)$ maps ζ_n to $C_\alpha(\zeta_n) = \alpha\zeta_n$, we see that $\sigma_\alpha(\zeta_n^{q-1}) = (\alpha\zeta_n)^{q-1} = \zeta_n^{q-1}$. Hence $\zeta_n^{q-1} \in K_n^+$, and

$$q - 1 = [K_n : K_n^+] \le [F(\zeta_n) : F(\zeta_n^{q-1})] \le q - 1,$$

---

[1] Recall that in a finite extension L/K of number fields a real Archimedean place v of K (i.e., an embedding of K into $\mathbb{C}$ with image contained in $\mathbb{R}$) is said to *ramify* if it extends to an embedding $w: L \to \mathbb{C}$ with non-real image. If all extensions of v to places of L are real, then v is said to *split completely* in L.

which implies that $K_\mathfrak{n}^+ = F(\zeta_\mathfrak{n}^{q-1})$.

The extensions $K_\mathfrak{n}/K_\mathfrak{n}^+$ and $K_\mathfrak{n}^+/F$ are Galois, so the places over ∞ in these respective extensions behave the same way, in the sense that they have the same ramification indices and residue degrees. Fix a place $\widetilde{\infty}$ of $K_\mathfrak{n}^+$ over ∞, and a place $\widetilde{\infty}'$ of $K_\mathfrak{n}$ over $\widetilde{\infty}$. The completion of $K_\mathfrak{n}$ with respect to the absolute value associated with $\widetilde{\infty}'$ is isomorphic to $F_\infty(\zeta_\mathfrak{n})$, and the completion of $K_\mathfrak{n}^+$ with respect to the absolute value associated with $\widetilde{\infty}$ is isomorphic to $F_\infty(\zeta_\mathfrak{n}^{q-1})$. To prove the theorem, it is enough to show that $F_\infty(\zeta_\mathfrak{n}^{q-1}) = F_\infty$, whereas $F_\infty(\zeta_\mathfrak{n})/F_\infty$ is totally ramified of degree $q - 1$ (cf. Proposition 2.8.8).

By Proposition 5.3.1, we have $C[\mathfrak{n}] = \{e_C(a\pi_C/\mathfrak{n}) \mid a \in A_{<d}\}$, where $d = \deg(\mathfrak{n})$ and $e_C(x) = \sum_{i\geq 0} x^{q^i}/D_i$. The functional equation $C_a(e_C(\pi_C/\mathfrak{n})) = e_C(a\pi_C/\mathfrak{n})$ implies that $e_C(\pi_C/\mathfrak{n})$ is a primitive $\mathfrak{n}$-th root of C, so we can take $\zeta_\mathfrak{n} = e_C(\pi_C/\mathfrak{n})$.

By definition (5.4.12), we have $\pi_C = \varpi \mathbf{i}$, where $\varpi \in F_\infty$ and $\mathbf{i} = (-[1])^{1/(q-1)}$. In particular, $F_\infty(\pi_C) = F_\infty(\mathbf{i})$ is a totally ramified extension of F_∞ of degree $q - 1$ and $\pi_C^{q-1} \in F_\infty$. On the other hand,

$$\zeta_\mathfrak{n} = e_C(\pi_C/\mathfrak{n}) = \sum_{i\geq 0} \frac{\pi_C^{q^i}}{\mathfrak{n}^{q^i} D_i} = \pi_C \sum_{i\geq 0} \frac{(\pi_C^{q-1})^{q^{i-1}+q^{i-2}+\cdots+1}}{\mathfrak{n}^{q^i} D_i}.$$

Because $\pi_C^{q-1} \in F_\infty$, we see that $\zeta_\mathfrak{n} = \pi_C \alpha$ for some $\alpha \in F_\infty$. In particular, $F_\infty(\zeta_\mathfrak{n}) = F_\infty(\pi_C)$ is a totally ramified extension of F_∞ of degree $q - 1$ and $F_\infty(\zeta_\mathfrak{n}^{q-1}) = F_\infty(\pi_C^{q-1}) = F_\infty$. $\square$

Because $F_\infty(\zeta_\mathfrak{n})/F_\infty$ is totally ramified, the residue degree of this extension is 1. Hence, the residue field of $F_\infty(\zeta_\mathfrak{n})$ is $\mathbb{F}_q$. This implies that the algebraic closure of $\mathbb{F}_q$ in $F_\infty(\zeta_\mathfrak{n})$ is $\mathbb{F}_q$ itself, i.e., the subfield of $F_\infty(\zeta_\mathfrak{n})$ consisting of elements which are algebraic over $\mathbb{F}_q$ is just $\mathbb{F}_q$ (this is true because the algebraic closure of $\mathbb{F}_q$ in $F_\infty(\zeta_\mathfrak{n})$ naturally injects into the residue field). The same must be true for $K_\mathfrak{n} \subset F_\infty(\zeta_\mathfrak{n})$, i.e., $\mathbb{F}_q$ is its own algebraic closure in $K_\mathfrak{n}$.

Definition 7.1.14 Let K be a finite extension of F with field of constants k. An extension L/K is called *geometric* if k is algebraically closed in L.

We can restate our previous observation as follows:

Corollary 7.1.15 *For any* $\mathfrak{n} \neq 0$, *the extension* $K_\mathfrak{n}/F$ *is geometric.*

Next, we study the ramification and splitting of primes in $K_\mathfrak{n}/F$.

Theorem 7.1.16 *Let* $\mathfrak{n} = \mathfrak{p}^s$, $s \geq 1$, *be a power of a prime* $\mathfrak{p}$. *Let* B *be the integral closure of* A *in* $K_\mathfrak{n}$. *The ideal* $\mathfrak{p}$ *totally ramifies in* $K_\mathfrak{n}/F$, *and* $\mathfrak{P} = \zeta_\mathfrak{n} B$ *is the unique prime ideal of* B *over* $\mathfrak{p}$.

Proof Because the height $H(C)$ of C is equal to its rank (both being 1), all the coefficients of $C_{\mathfrak{p}^s}(x)$, except the leading coefficient, are divisible by $\mathfrak{p}$; see

Lemma 3.2.11. Also, the constant term of $\Phi_{\mathfrak{p}}(x) = C_{\mathfrak{p}}(x)/x$ is equal to $\mathfrak{p}$. By (7.1.4), we have $\Phi_{\mathfrak{p}^s}(x) = \Phi_{\mathfrak{p}}(C_{\mathfrak{p}^{s-1}}(x))$. This implies that $\Phi_{\mathfrak{n}}(x)$ is an Eisenstein polynomial with respect to $\mathfrak{p}$. Hence, by Proposition 2.6.7, $F(\zeta_{\mathfrak{n}})/F$ is totally ramified at $\mathfrak{p}$. In particular, there is a unique prime ideal $\mathfrak{P} \lhd B$ dividing $\mathfrak{p}$. Then $\mathfrak{p}B = \mathfrak{P}^{[K_{\mathfrak{n}}:F]} = \mathfrak{P}^{\varphi(\mathfrak{n})}$ (this follows from Proposition 2.8.8 and Theorem 2.6.3). We need to show that $\mathfrak{P}$ is the principal ideal generated by $\zeta_{\mathfrak{n}}$ in B. Observe that $C_a(x)/x = a + a_1 x^{q-1} + \cdots + x^{q^{\deg(a)}-1}$ has coefficients in A. Therefore, $C_a(\beta)/\beta$ lies in B for any $\beta \in B$. In particular, $C_a(\zeta_{\mathfrak{n}})/\zeta_{\mathfrak{n}} \in B$. Now assume a is coprime to $\mathfrak{n}$. For $b \in A$ such that $ba \equiv 1 \pmod{\mathfrak{n}}$, we have $C_{ba}(\zeta_{\mathfrak{n}}) = \zeta_{\mathfrak{n}}$ and therefore

$$\frac{\zeta_{\mathfrak{n}}}{C_a(\zeta_{\mathfrak{n}})} = \frac{C_{ba}(\zeta_{\mathfrak{n}})}{C_a(\zeta_{\mathfrak{n}})} = \frac{C_b(C_a(\zeta_{\mathfrak{n}}))}{C_a(\zeta_{\mathfrak{n}})} \in B.$$

Thus, $C_a(\zeta_{\mathfrak{n}})/\zeta_{\mathfrak{n}}$ and its inverse $\zeta_{\mathfrak{n}}/C_a(\zeta_{\mathfrak{n}})$ are in B, which implies that $C_a(\zeta_{\mathfrak{n}})/\zeta_{\mathfrak{n}}$ is a unit of B. Now, from Definition 7.1.4, we get

$$\mathfrak{p} = \Phi_{\mathfrak{n}}(0) = \prod_{\substack{\deg(a)<\deg(\mathfrak{n}) \\ (a,\mathfrak{n})=1}} C_a(\zeta_{\mathfrak{n}})$$

$$= \zeta_{\mathfrak{n}}^{\varphi(\mathfrak{n})} \prod_{\substack{\deg(a)<\deg(\mathfrak{n}) \\ (a,\mathfrak{n})=1}} C_a(\zeta_{\mathfrak{n}})/\zeta_{\mathfrak{n}}$$

$$= \zeta_{\mathfrak{n}}^{\varphi(\mathfrak{n})} \cdot u, \qquad u \in B^{\times}.$$

Therefore, $\mathfrak{P}^{\varphi(\mathfrak{n})} = \mathfrak{p}B = (\zeta_{\mathfrak{n}})^{\varphi(\mathfrak{n})}$. Because B is a Dedekind domain, the decomposition of ideals into products of prime ideals is unique, so the previous equality implies $\mathfrak{P} = (\zeta_{\mathfrak{n}})$. $\qquad\square$

Remark 7.1.17 Similar to our observation that the fact that $F_{\infty}(\zeta_{\mathfrak{n}})/F_{\infty}$ is totally ramified implies that $K_{\mathfrak{n}}/F$ is a geometric extension, the fact that $K_{\mathfrak{p}^s}/F$ is totally ramified at $\mathfrak{p}$ implies that this extension is geometric. For general $\mathfrak{n}$ with prime decomposition $\mathfrak{n} = \mathfrak{p}_1^{s_1} \cdots \mathfrak{p}_k^{s_k}$, the cyclotomic extension $K_{\mathfrak{n}}$ is the composite of its subfields $K_{\mathfrak{p}_1^{s_1}}, \ldots, K_{\mathfrak{p}_k^{s_k}}$; see Exercise 7.1.8. But, by just knowing that each $K_{\mathfrak{p}_i^{s_i}}/F$ is a geometric extension, we cannot immediately conclude that $K_{\mathfrak{n}}/F$ is geometric. The problem is that the composite of geometric extensions is not necessarily geometric. For example, $F(\sqrt{T})$ and $F(\sqrt{-T})$ are quadratic geometric extensions of F, although their composite contains $F(\sqrt{-1})$, which is not geometric if -1 is not a square in $\mathbb{F}_q$.

Theorem 7.1.16 implies that if a prime $\mathfrak{p}$ divides $\mathfrak{n}$, then $\mathfrak{p}$ ramifies in $K_{\mathfrak{n}}/F$ since $\mathfrak{p}$ ramifies in the subextension $K_{\mathfrak{p}}/F$. The converse is also true: if $\mathfrak{p}$ does not divide $\mathfrak{n}$, then $\mathfrak{p}$ does not ramify in $K_{\mathfrak{n}}/F$. This follows from the fact that $C_{\mathfrak{n}}(x)$ reduced modulo $\mathfrak{p}$ is a separable polynomial over $\mathbb{F}_{\mathfrak{p}}$ (see the proof of Proposition 7.1.10). It is possible to give a precise description of the splitting of $\mathfrak{p}$ in $K_{\mathfrak{n}}$. Before doing so, we make an observation about general Drinfeld modules over F that will be useful for this and some other problems.

Remark 7.1.18 Let ϕ be a Drinfeld module over F of rank $r \geq 1$. Similar to the terminology of Chap. 6, we say that ϕ has *good reduction* at a prime $\mathfrak{p}$ if the leading coefficient of $\phi_T(x)$ has $\mathfrak{p}$-adic valuation 0 and all other coefficients of $\phi_T(x)$ have non-negative valuations. Note that ϕ has good reduction at all but finitely many primes of A. Let $\mathfrak{p}$ be a prime of good reduction of ϕ and denote by $\bar{\phi}$ the reduction of ϕ modulo $\mathfrak{p}$; explicitly, $\bar{\phi}_T \in \mathbb{F}_\mathfrak{p}\{\tau\}$ is obtained by reducing the coefficients of ϕ_T modulo $\mathfrak{p}$. Then $\bar{\phi}$ is a Drinfeld module over $\mathbb{F}_\mathfrak{p}$ of rank r. Let $\mathfrak{n} \lhd A$ be an ideal coprime to $\mathfrak{p}$. Extend the notation $K_\mathfrak{n} := F(\phi[\mathfrak{n}])$ in Definition 7.1.3 to general Drinfeld modules. The Galois group $\mathrm{Gal}(K_\mathfrak{n}/F)$ is a subgroup of $\mathrm{Aut}_A(\phi[\mathfrak{n}]) \cong \mathrm{GL}_r(A/\mathfrak{n})$; cf. (3.5.5). Fix a prime $\mathfrak{P}$ of $K_\mathfrak{n}$ lying over $\mathfrak{p}$. By Theorem 6.3.1, the prime $\mathfrak{p}$ is unramified in $K_\mathfrak{n}$, so we have a well-defined Frobenius element $\mathrm{Frob}_\mathfrak{P} \in \mathrm{Gal}(K_\mathfrak{n}/F)$, which is uniquely characterized by the property that $\mathrm{Frob}_\mathfrak{P}(\alpha) \equiv \alpha^{|\mathfrak{p}|} \pmod{\mathfrak{P}}$ for all $\alpha \in B$; see Sect. 2.8. We can consider the characteristic polynomial of $\mathrm{Frob}_\mathfrak{P}$ as an element of $\mathrm{GL}_r(A/\mathfrak{n})$. Denote this characteristic polynomial by $P_{\phi,\mathfrak{p},\mathfrak{n}}(x)$. Then $P_{\phi,\mathfrak{p},\mathfrak{n}}(x)$ is a monic polynomial in $(A/\mathfrak{n})[x]$ of degree r, which does not depend on the choice of the prime $\mathfrak{P}$ over $\mathfrak{p}$ since for any other prime $\mathfrak{P}'$ over $\mathfrak{p}$ the elements $\mathrm{Frob}_\mathfrak{P}$ and $\mathrm{Frob}_{\mathfrak{P}'}$ are conjugate in $\mathrm{Gal}(K_\mathfrak{n}/F)$. The reduction modulo $\mathfrak{P}$ induces a canonical isomorphism of A-modules

$$\iota \colon \phi[\mathfrak{n}] \xrightarrow{\sim} \bar{\phi}[\mathfrak{n}].$$

This isomorphism is compatible with the action of $\mathrm{Frob}_\mathfrak{P}$ on $\phi[\mathfrak{n}]$, and the action of $\mathrm{Frob}_{\mathbb{F}_\mathfrak{p}}$ on $\bar{\phi}[\mathfrak{n}]$ since

$$\overline{\mathrm{Frob}_\mathfrak{P}(\alpha)} = \bar{\alpha}^{|\mathfrak{p}|} = \mathrm{Frob}_{\mathbb{F}_\mathfrak{p}}(\bar{\alpha}),$$

where $\bar{*}$ denotes the reduction modulo $\mathfrak{P}$. This implies that $P_{\phi,\mathfrak{p},\mathfrak{n}}(x)$ is equal to the characteristic polynomial of $\mathrm{Frob}_{\mathbb{F}_\mathfrak{p}}$ acting on $\bar{\phi}[\mathfrak{n}]$. On the other hand, by Exercise 4.2.5, the characteristic polynomial of $\mathrm{Frob}_{\mathbb{F}_\mathfrak{p}}$ acting on $\bar{\phi}[\mathfrak{n}]$ is equal to the reduction modulo $\mathfrak{n}$ of the characteristic polynomial of $\mathrm{Frob}_{\mathbb{F}_\mathfrak{p}}$ acting on $T_\mathfrak{l}(\bar{\phi}) \cong A_\mathfrak{l}^{\oplus r}$ for any prime $\mathfrak{l} \neq \mathfrak{p}$. Denote this latter characteristic polynomial by $P_{\bar{\phi}}(x)$. Recall that $P_{\bar{\phi}}(x)$ was studied in Sect. 4.2. In particular, $P_{\bar{\phi}}(x)$ has coefficients in A which do not depend on the choice of $\mathfrak{l}$. Thus,

$$P_{\phi,\mathfrak{p},\mathfrak{n}}(x) \equiv P_{\bar{\phi}}(x) \pmod{\mathfrak{n}}. \tag{7.1.5}$$

The usefulness of this congruence lies in the fact that computing $P_{\bar{\phi}}(x) \in A[x]$ is relatively easy, and, once this polynomial is computed, every $P_{\phi,\mathfrak{p},\mathfrak{n}}(x)$ can be computed by simply reducing the coefficients of $P_{\bar{\phi}}(x)$ modulo $\mathfrak{n}$. Note that this gives the characteristic polynomial of the conjugacy class of the Frobenius at $\mathfrak{p}$ in *distinct* division fields $F(\phi[\mathfrak{n}])$.

Next, we apply Remark 7.1.18 to study the splitting of primes in Carlitz cyclotomic extensions. Slightly generalizing our initial setup, let $C^{(g)}$ be the Drinfeld module of rank 1 over F defined by

$$C_T^{(g)} = T + g\tau, \qquad g \in F^\times. \tag{7.1.6}$$

Theorem 7.1.19 (Carlitz Cyclotomic Reciprocity Law) *Let $\mathfrak{n} \lhd A$ be a nonzero ideal. Let $\mathfrak{p}$ be a prime of A such that $\mathrm{ord}_\mathfrak{p}(\mathfrak{n}) = \mathrm{ord}_\mathfrak{p}(g) = 0$. Denote by $\bar{g} \in \mathbb{F}_\mathfrak{p}$ the reduction of g modulo $\mathfrak{p}$. Let B be the integral closure of A in $F\left(C^{(g)}[\mathfrak{n}]\right)$. Let $\mathfrak{p}B = \prod_{i=1}^{m_\mathfrak{p}} \mathfrak{P}_i$ be the decomposition of $\mathfrak{p}$ into a product of distinct prime ideals of B. The residue degrees $f_\mathfrak{p} := [B/\mathfrak{P}_i : A/\mathfrak{p}]$, $1 \le i \le m_\mathfrak{p}$, are all equal to the order of $\mathfrak{p}/\mathrm{Nr}_{\mathbb{F}_\mathfrak{p}/\mathbb{F}_q}(\bar{g})$ in the multiplicative group $(A/\mathfrak{n})^\times$. In particular, $\mathfrak{p}$ splits completely in $F\left(C^{(g)}[\mathfrak{n}]\right)$ if and only if $\mathfrak{p} \equiv \mathrm{Nr}_{\mathbb{F}_\mathfrak{p}/\mathbb{F}_q}(\bar{g}) \pmod{\mathfrak{n}}$. Finally, if $g = 1$, then $m_\mathfrak{p} = \varphi(\mathfrak{n})/f_\mathfrak{p}$.*

Before giving the proof, we remind the reader that we always implicitly assume that the prime $\mathfrak{p}$ is monic; this is important for the congruence condition in the theorem.

Proof Denote $K = F\left(C^{(g)}[\mathfrak{n}]\right)$. Because the extension K/F is abelian and unramified at $\mathfrak{p}$, there is a well-defined $\mathrm{Frob}_\mathfrak{p} \in \mathrm{Gal}(K/F)$ which does not depend on the choice of a place above $\mathfrak{p}$. By Theorem 4.2.7,

$$P_{\overline{C^{(g)}}}(x) = x - \frac{\mathfrak{p}}{\mathrm{Nr}_{\mathbb{F}_\mathfrak{p}/\mathbb{F}_q}(\bar{g})}.$$

Therefore, by (7.1.5), $\mathrm{Frob}_\mathfrak{p}$ acts on $C^{(g)}[\mathfrak{n}] \cong A/\mathfrak{n}$ by multiplication by $\mathfrak{p}/\mathrm{Nr}_{\mathbb{F}_\mathfrak{p}/\mathbb{F}_q}(\bar{g})$. As the residue degree $f_\mathfrak{p}$ is equal to the order of $\mathrm{Frob}_\mathfrak{p}$ in $\mathrm{Gal}(K/F)$, we conclude that $f_\mathfrak{p}$ is equal to the order of $\mathfrak{p}/\mathrm{Nr}_{\mathbb{F}_\mathfrak{p}/\mathbb{F}_q}(\bar{g})$ in $(A/\mathfrak{n})^\times$. The final claim is a consequence of the formula

$$\varphi(\mathfrak{n}) = [K_\mathfrak{n} : F] = f_\mathfrak{p} \cdot m_\mathfrak{p},$$

cf. Proposition 2.8.8.   □

Remark 7.1.20 The last statement of the theorem can be extended to $C^{(g)}$ as

$$m_\mathfrak{p} = [F(C^{(g)}[\mathfrak{n}]) : F]/f_\mathfrak{p},$$

but generally $[F(C^{(g)}[\mathfrak{n}]) : F]$ can be slightly smaller than $\varphi(\mathfrak{n})$. In fact, it is not hard to show that $\varphi(\mathfrak{n})/[F(C^{(g)}[\mathfrak{n}]) : F]$ is an integer dividing $q - 1$; see the proof of Theorem 7.2.3. On the other hand, determining this integer exactly is a subtle problem. The complete solution is given in [Gek16], which we now describe. First of all, as the isomorphism type of $C^{(g)}$ depends only on the class of $g \in F^\times$ in $F^\times/(F^\times)^{q-1}$, we may assume that g is integral, and not divisible by $(q-1)$-th powers. Then we may write

$$g = \varsigma^{k_0} \mathfrak{p}_1^{k_1} \cdots \mathfrak{p}_s^{k_s},$$

where ς is a generator of $\mathbb{F}_q^\times$, $\mathfrak{p}_1, \ldots, \mathfrak{p}_s$ are $s \geq 0$ different monic primes of A, $0 \leq k_0 < q - 1$, and $0 < k_i < q - 1$ for $1 \leq i \leq s$. Note that $s = 0$, i.e., $g \in \mathbb{F}_q^\times$, is allowed. Let $d_i = \deg(\mathfrak{p}_i)$ for $1 \leq i \leq s$ and $d = \deg(g) = \sum_{i=1}^{s} d_i k_i$. Let

$$k_0^* = \begin{cases} k_0, & \text{if } q \text{ or } d \text{ is even,} \\ \text{the unique } k \equiv k_0 + (q-1)/2 \pmod{q-1} \\ \quad \text{with } 0 \leq k < q - 1, & \text{otherwise.} \end{cases}$$

Theorem 7.1.21 *Let* $\mathfrak{n} \in A$ *be a polynomial of positive degree. With previous notation, assume* $\mathfrak{p}_1, \ldots, \mathfrak{p}_l$ *divide* $\mathfrak{n}$ *and* $\mathfrak{p}_{l+1}, \ldots, \mathfrak{p}_s$ *are coprime with* $\mathfrak{n}$*, where* $1 \leq l \leq s$*. Then*

$$\varphi(\mathfrak{n})/[F(C^{(g)}[\mathfrak{n}]) : F] = \gcd(q - 1, d - 1, k_0^*, k_{l+1}, \ldots, k_s).$$

Example 7.1.22 Applying Theorem 7.1.19 to the Carlitz module C, $\mathfrak{n} = T$, and a prime $\mathfrak{p} \neq T$, we conclude that $x^{q-1} \equiv -T \pmod{\mathfrak{p}}$ is solvable if and only if $\mathfrak{p}(0) = 1$, i.e., the constant term of $\mathfrak{p}$ is 1. As we explain now, this is also a special case of another reciprocity law for A, which goes back to Dedekind, Schmidt, and Carlitz; see [Ros02, Ch. 3].

Let d be a divisor of $q - 1$ and $\mathfrak{p} \in A$ be a prime. For any $a \in A$ not divisible by $\mathfrak{p}$ the image of $a^{(|\mathfrak{p}|-1)/d}$ in $(A/\mathfrak{p})^\times$ has order dividing $q - 1$. Since $\mathbb{F}_\mathfrak{p}^\times$ is cyclic, and its unique subgroup of order $q - 1$ is $\mathbb{F}_q^\times$, there is a unique element in $\mathbb{F}_q^\times$, denoted $\left(\frac{a}{\mathfrak{p}}\right)_d$, such that

$$a^{(|\mathfrak{p}|-1)/d} \equiv \left(\frac{a}{\mathfrak{p}}\right)_d \pmod{\mathfrak{p}}.$$

The symbol $\left(\frac{a}{\mathfrak{p}}\right)_d$ is called the *d-th power residue symbol*. The following properties of this symbol are not hard to prove (cf. [Ros02, p. 24]):

(1) $\left(\frac{ab}{\mathfrak{p}}\right)_d = \left(\frac{a}{\mathfrak{p}}\right)_d \left(\frac{b}{\mathfrak{p}}\right)_d$.

(2) If $\alpha \in \mathbb{F}_q^\times$, then $\left(\frac{\alpha}{\mathfrak{p}}\right)_d = \alpha^{(q-1)\deg(\mathfrak{p})/d}$.

(3) $\left(\frac{a}{\mathfrak{p}}\right)_d = 1$ if and only if $x^d \equiv a \pmod{\mathfrak{p}}$ is solvable.

Now let $\mathfrak{p}$ and $\mathfrak{q}$ be two distinct monic primes in A. Then the *d-th power reciprocity law* states that (see [Ros02, Thm. 3.3])

$$\left(\frac{\mathfrak{q}}{\mathfrak{p}}\right)_d = (-1)^{\frac{(q-1)}{d} \deg(\mathfrak{p}) \deg(\mathfrak{q})} \left(\frac{\mathfrak{p}}{\mathfrak{q}}\right)_d. \tag{7.1.7}$$

When q is odd and $d = 2$, this is an analogue of the quadratic reciprocity since $\left(\frac{a}{\mathfrak{p}}\right)_2 = \pm 1$, depending on whether or not a is a square modulo $\mathfrak{p}$. If $d = q - 1$, then this law was already mentioned in Exercise 4.2.11.

Finally, we compute

$$\left(\frac{-T}{\mathfrak{p}}\right)_{q-1} = \left(\frac{-1}{\mathfrak{p}}\right)_{q-1} \left(\frac{T}{\mathfrak{p}}\right)_{q-1} \qquad \text{(by property (1))}$$

$$= (-1)^{\deg(\mathfrak{p})} \left(\frac{T}{\mathfrak{p}}\right)_{q-1} \qquad \text{(by property (2))}$$

$$= \left(\frac{\mathfrak{p}}{T}\right)_{q-1} \qquad \text{(by (7.1.7))}.$$

By property (3), this implies that $x^{q-1} \equiv -T \pmod{\mathfrak{p}}$ is solvable if and only if $x^{q-1} \equiv \mathfrak{p} \pmod{T}$ is solvable. But $\mathfrak{p}$ modulo T is $\mathfrak{p}(0)$, and $x^{q-1} = \alpha$ is solvable in $\mathbb{F}_T^\times \cong \mathbb{F}_q^\times$ if and only if $\alpha = 1$ (since $\beta^{q-1} = 1$ for any $\beta \in \mathbb{F}_q^\times$). Therefore, $x^{q-1} \equiv -T \pmod{\mathfrak{p}}$ is solvable if and only if $\mathfrak{p}(0) = 1$, which was our statement at the beginning of this example.

By now we have proved the Carlitz module analogues of all the properties of cyclotomic extensions mentioned at the beginning of this section, except for the Kronecker–Weber theorem. On the other hand, it is not hard to see that the obvious restatement of the Kronecker–Weber theorem over F cannot be true, i.e., not every abelian extension of F is contained in some $K_\mathfrak{n}$. For one thing, every $K_\mathfrak{n}/F$ is a geometric extension, but $\mathbb{F}_{q^n} F/F$ is a cyclic Galois extension of order n. The other thing is that ∞ almost completely splits in $K_\mathfrak{n}$ with very small tame ramification, although there are abelian extensions of F where the ramification indices at ∞ are arbitrarily large.

There are two ways to overcome these problems to arrive at a statement which can be reasonably called a Kronecker–Weber theorem over F. The first, which is due to Drinfeld [Dri74], is to put a restriction on the abelian extensions that we consider:

Theorem 7.1.23 *Let K/F be a totally real abelian extension. Then K is contained in $K_\mathfrak{n}^+$ for some nonzero ideal $\mathfrak{n}$ of A.*

The second approach, which is due to Hayes [Hay74], is to expand our supply of "cyclotomic" extensions. Let K' be the composite of all $K_\mathfrak{n}$. Let $K'' := \overline{\mathbb{F}}_q F$

be the maximal constant field extension; this is the composite of the splitting fields of all $x^n - 1$, $n \geq 1$. Finally, consider F as the fraction field of $A' = \mathbb{F}_q[U]$, where $U = 1/T$, and let $C' : A' \to F\{\tau\}$ be the Carlitz module defined by $C'_U = U + \tau$. Let $K_{T^{-n}} := F(C'[U^n])$ be the splitting field of $C'_{U^n}(x)$. This is an abelian extension of F, which is totally ramified at ∞ with ramification index $q^{n-1}(q - 1)$. (The behavior of ∞ in this extension is the same as the behavior of the place corresponding to T in K_{T^n}.) Let $K^+_{T^{-n}}$ be the subfield of $K_{T^{-n}}$ fixed by $\mathbb{F}_q^\times \subset (A'/(U^n))^\times \cong \mathrm{Gal}(K_{T^{-n}}/F)$, and let K''' be the composite of all $K^+_{T^{-n}}$, $n \geq 1$. Hayes proved the following:

Theorem 7.1.24 *The fields K', K'', and K''' are linearly disjoint over F. Every abelian extension of F is contained in the composite of these three fields.*

The proofs of Theorems 7.1.23 and 7.1.24 rely heavily on class field theory and will not be discussed here.

Exercises

7.1.1 By comparing the degrees of both sides of (7.1.3), it follows immediately that

$$|\mathfrak{n}| = \sum_{\mathfrak{d}|\mathfrak{n}} \varphi(\mathfrak{d}).$$

Prove this formula using (7.1.2).

7.1.2 Let d be a divisor of $q - 1$ and let $\zeta \in \mathbb{F}_q^\times$ be an element of order dividing d. Prove that there is $a \in A$ such that $\left(\frac{a}{\mathfrak{p}}\right)_d = \zeta$.

7.1.3 This is essentially Exercise 1.6.4. Let n be a positive integer not divisible by p. Let K be the splitting field of $x^n - 1$ over F. Prove the following:

(a) If $s \geq 1$ is the smallest positive integer such that $q^s \equiv 1 \pmod{n}$, then $K \cong \mathbb{F}_{q^s} F$.
(b) A prime $\mathfrak{p} \lhd A$ of degree d splits in K into $\gcd(d, s)$ distinct primes. In particular, $\mathfrak{p}$ splits completely in K if and only if s divides d.

7.1.4 By a well-known theorem of Gauss, for any prime $p \neq 2$ the groups $(\mathbb{Z}/p^s\mathbb{Z})^\times$ are cyclic for all $s \geq 1$. Now let $\mathfrak{p}$ be a prime of A and $s \geq 1$ be an integer. Prove that $(A/\mathfrak{p}^s)^\times$ is cyclic if and only if one of the following holds:

(1) $s = 1$.
(2) $s = 2$ and $\deg(\mathfrak{p}) = 1$.
(3) $s = 3$, $\deg(\mathfrak{p}) = 1$, and the characteristic of A is 2.

Also, prove that the minimal number of generators of $(A/\mathfrak{p}^s)^\times$ tends to infinity as $s \to \infty$.

7.1.5 Prove the following identities:

(a) $\Phi_{\mathfrak{p}^s\mathfrak{m}}(x) = \Phi_\mathfrak{m}(C_{\mathfrak{p}^s}(x))/\Phi_\mathfrak{m}(C_{\mathfrak{p}^{s-1}}(x))$.

(b) $\Phi_{\mathfrak{p}_1^{s_1}\cdots\mathfrak{p}_k^{s_k}}(x) = \Phi_{\mathfrak{p}_1\cdots\mathfrak{p}_k}(C_{\mathfrak{p}_1^{s_1-1}\cdots\mathfrak{p}_k^{s_k-1}}(x))$.

(c)

$$\Phi_\mathfrak{n}(0) = \begin{cases} 0 & \text{if } \mathfrak{n} = 1, \\ \mathfrak{p} & \text{if } \mathfrak{n} \text{ is a prime power } \mathfrak{p}^s, \\ 1 & \text{if } \mathfrak{n} \text{ is not a prime power.} \end{cases}$$

7.1.6 Prove the following statements:

(a) If $\mathfrak{n} = \mathfrak{p}^s$ is a power of a prime $\mathfrak{p}$, then $K_\mathfrak{n}/K_\mathfrak{n}^+$ is ramified at the place over $\mathfrak{p}$ and at the places over ∞.

(b) If $\mathfrak{n}$ is not a prime power, then $K_\mathfrak{n}/K_\mathfrak{n}^+$ is unramified except at the places over ∞.

7.1.7 Let $\mathfrak{n} \lhd A$ be a nonzero ideal. Let $B_{K_\mathfrak{n}}$ (resp. $B_{K_\mathfrak{n}^+}$) be the integral closure of A in $K_\mathfrak{n}$ (resp. $K_\mathfrak{n}^+$). Prove the following statements:

(a) If $\mathfrak{n} = \mathfrak{p}^s$ is a power of a prime $\mathfrak{p}$, then $B_{K_\mathfrak{n}}^\times = B_{K_\mathfrak{n}^+}^\times$, i.e., every unit of $B_{K_\mathfrak{n}}$ lies in $K_\mathfrak{n}^+$.

(b) Suppose $\mathfrak{n}$ is divisible by two or more primes. Prove that $\zeta_\mathfrak{n} \in B_{K_\mathfrak{n}}^\times$. In particular, $B_{K_\mathfrak{n}^+}^\times \subsetneq B_{K_\mathfrak{n}}^\times$.

7.1.8 Let $\mathfrak{n} \lhd A$ be a nonzero ideal and let $\mathfrak{n} = \prod_{i=1}^m \mathfrak{p}_i^{s_i}$ be its prime decomposition. Let $K_i := F(\zeta_{\mathfrak{p}_i^{s_i}})$, $1 \leq i \leq m$. Prove the following:

(a) $K_i \cap K_j = F$ for $i \neq j$.

(b) $K_i \subset K_\mathfrak{n}$ for $1 \leq i \leq m$.

(c) $K_\mathfrak{n} = K_1 K_2 \cdots K_m$, i.e., $K_\mathfrak{n}$ is the composite of $K_1, \ldots, K_m$.

7.1.9 Let $\mathfrak{n} \lhd A$ be a nonzero ideal. Let $C^{(g)}$ be the Drinfeld module defined by (7.1.6).

(a) Prove that the ramification index and the residue degree of any place of $F(C^{(g)}[\mathfrak{n}])$ over ∞ divides $q - 1$.

(b) For any divisor d of $q - 1$ find g such that the ramification index $e(F_\infty(C^{(g)}[T])/F_\infty)$ is equal to d.

(c) For any divisor d of $q - 1$ find g such that the residue degree $f(F_\infty(C^{(g)}[T])/F_\infty)$ is equal to d.

7.1.10 This exercise outlines an alternative Proof of Theorem 7.1.13 following [Ros02, p. 212] (see also [Hay74]). Assume $\mathfrak{n} \in A_+$ has degree $d > 0$.

(a) Show that exactly $q^{i+1} - q^i$, $0 \le i \le d$, roots of $C_\mathfrak{n}(x)$ have ∞-adic valuation $d - i - \frac{q}{q-1}$. In particular, exactly $q - 1$ roots of $C_\mathfrak{n}(x)$ have ∞-adic valuation $d - \frac{q}{q-1}$.

(b) Let $\zeta_\mathfrak{n}$ be a root of $C_\mathfrak{n}$ with valuation $d - \frac{q}{q-1}$. Show that for any nonzero $a \in A$ of degree $< d$ we have

$$\mathrm{ord}_\infty(C_a(\zeta_\mathfrak{n})) = d - \deg(a) - \frac{q}{q-1}.$$

Conclude that $\zeta_\mathfrak{n}$ is a generator of $C[\mathfrak{n}]$.

(c) Using (b), show that $F_\infty(C[\mathfrak{n}]) = F_\infty(\zeta_\mathfrak{n})$ is a totally tamely ramified extension of F_∞ of degree $q - 1$.

(d) Show that $\zeta_\mathfrak{n}^{q-1} \in F_\infty$.

7.2 Rational Torsion Submodule

Let K be a finite extension of F and let ϕ be a rank r Drinfeld module over K. In this section we discuss some results and conjectures about the rational torsion submodule $(\phi K)_{\mathrm{tor}}$ of ϕ.

Let $\mathfrak{P}$ be a place of K over a prime $\mathfrak{p} \lhd A$, and let $K_\mathfrak{P}$ be the completion of K at $\mathfrak{P}$. By Theorem 6.4.3, $(\phi K_\mathfrak{P})_{\mathrm{tor}}$ is finite. Therefore, $(\phi K)_{\mathrm{tor}}$ is also finite, being a submodule of $(\phi K_\mathfrak{P})_{\mathrm{tor}}$. (For a completely different proof of the finiteness of $(\phi K)_{\mathrm{tor}}$, see Corollary 7.6.13.) One could ask whether there is a number N, depending on K but not on ϕ, such that $\#(\phi K)_{\mathrm{tor}} \le N$ for all Drinfeld modules over K. The next example demonstrates that such a number cannot exist if r is allowed to vary.

Example 7.2.1 Let W be an $\mathbb{F}_q$-vector subspace of K of dimension r. (Since K is infinite dimensional as an $\mathbb{F}_q$-vector space, such W exists for any $r \ge 1$.) By Lemma 3.1.5, the polynomial

$$f(x) = x \prod_{0 \neq w \in W}' \left(1 - \frac{x}{w}\right) = x + \sum_{i=1}^{r} c_i x^{q^i}$$

is $\mathbb{F}_q$-linear. Hence we can define a Drinfeld module of rank r over K by putting $\phi_T(x) = T \cdot f(x)$. By construction, all the roots of $\phi_T(x)$ are in K, so $\#(\phi K)_{\mathrm{tor}} \ge q^r$.

Thus, to have a universal bound on $(\phi K)_{\mathrm{tor}}$ we need to fix the rank of the Drinfeld modules. In case of elliptic curves, the results of Mazur, Merel, and Kamienny imply the existence of a number $N(d)$ depending only on d such that the order of the rational torsion subgroup of an elliptic curve over a number field K of degree d

over $\mathbb{Q}$ is at most $N(d)$; see [Mer96]. This result served as a motivation for the following conjecture formulated by Poonen in [Poo97]:

Conjecture 7.2.2 (Uniform Boundedness Conjecture) *For fixed $r \geq 1$ and $d \geq 1$, there is a number $N(d, r, q)$ depending only on d, r, and q such that $\#(^\phi K)_{\mathrm{tor}} \leq N(d, r, q)$ as K ranges over the extensions of F of degree $\leq d$, and ϕ ranges over the Drinfeld modules of rank r over K.*

This conjecture is known for $r = 1$ (see Theorem 7.2.3), but it is wide open for $r \geq 2$, except for some partial results in the $r = 2$ case; see [Poo97, Pál10, Arm12, Sch03]. Geometrically, the conjecture is equivalent to the statement that the modular variety classifying Drinfeld modules of rank r with an $\mathfrak{n}$-torsion point has no K-rational points for any finite extension K/F of degree d, once $\deg(\mathfrak{n})$ exceeds some constant depending only on r, d, and q. In fact, in the case when $r = 2$, geometric techniques have been used to attack this second version of the conjecture.

Theorem 7.2.3 *We have the following bounds:*

(1) If K/F is a finite extension of degree d and ϕ a Drinfeld module over K of rank 1, then

$$\#(^\phi K)_{\mathrm{tor}} \leq q^{2d}.$$

(2) There is a constant c depending only on q such that if ϕ is a rank 1 Drinfeld module over a finite extension K of F of degree d, then

$$\#(^\phi K)_{\mathrm{tor}} \leq c \cdot d \cdot \log \log d.$$

This bound is the best possible up to a constant factor.

(3) If ϕ is a Drinfeld module over F of rank 1, then $(^\phi F)_{\mathrm{tor}} \cong A/\mathfrak{n}$, where either $\deg(\mathfrak{n}) \leq 1$ or $q = 2$ and $\mathfrak{n} = T(T + 1)$. In particular,

$$\#(^\phi F)_{\mathrm{tor}} \leq \begin{cases} q & \text{if } q \neq 2; \\ q^2 & \text{if } q = 2. \end{cases}$$

Proof

(1) We may assume that $\phi = C^{(g)}$ for some $g \in K^\times$, where $C^{(g)}$ is defined by $C_T^{(g)}(x) = Tx + gx^q$. Let $\mathfrak{T}$ be a place of K lying over T. We consider ϕ as a Drinfeld module over the completion $K_\mathfrak{T}$ of K at $\mathfrak{T}$. Let α be a root of the polynomial $x^{q-1} - g$. Then $\alpha \phi \alpha^{-1} = C$, so ϕ becomes isomorphic to the Carlitz module C over $L := K_\mathfrak{T}(\alpha)$. In particular, ϕ acquires good reduction over L. Let $\mathbb{F}$ be the residue field of L. Let $e(L/F_T)$ and $f(L/F_T)$ be the ramification index and the residue degree of the extension L/F_T, respectively. By Theorem 6.5.8 we have

$$\#(^\phi L)_{\mathrm{tor}} \leq (\#\mathbb{F})^{2 \cdot \frac{e(L/F_T)}{q-1}}.$$

On the other hand, $\#\mathbb{F} = q^{f(L/F_T)}$ and $e(L/F_T) \cdot f(L/F_T) \leq d(q-1)$. Hence, we get

$$\#(^\phi L)_{\mathrm{tor}} \leq q^{2d}.$$

Since $(^\phi K)_{\mathrm{tor}}$ is naturally a submodule of $(^\phi L)_{\mathrm{tor}}$, this gives the desired bound $\#(^\phi K)_{\mathrm{tor}} \leq q^{2d}$.

(2) The following argument is due to Poonen [Poo97, Thm. 8]. Again let α be a root of $x^{q-1} - g$, assuming $\phi = C^{(g)}$. Let $L = K(\alpha)$. Then $[L : F] \leq (q-1)d$. Over L, ϕ is isomorphic to C. Suppose $(^\phi K)_{\mathrm{tor}} \cong A/\mathfrak{n}$. Then $F(C[\mathfrak{n}]) \subseteq L$. On the other hand, by Theorem 7.1.12, $[F(C[\mathfrak{n}]) : F] = \varphi(\mathfrak{n})$. Thus,

$$\varphi(\mathfrak{n}) \leq (q-1)d. \tag{7.2.1}$$

To proceed, we need the following fact related to the classical Mertens' Theorem over function fields: there is a constant c' such that

$$|\mathfrak{n}| \leq c' \cdot \varphi(\mathfrak{n}) \cdot \log\log\varphi(\mathfrak{n}) \quad \text{for all } \mathfrak{n} \in A. \tag{7.2.2}$$

(The proof can be found in [Poo97, p. 581].)

Combining (7.2.1) and (7.2.2), we get

$$\#(^\phi K)_{\mathrm{tor}} = |\mathfrak{n}| \leq c'((q-1)d) \cdot \log\log((q-1)d)$$

$$\leq c \cdot d \log\log d$$

for some constant c depending only on A. This is best possible because one can also show that there is a sequence of $\mathfrak{n}$'s such that $c' \cdot \varphi(\mathfrak{n}) \cdot \log\log\varphi(\mathfrak{n}) \approx |\mathfrak{n}|$, so the bound is the best possible for the Carlitz cyclotomic extensions.

(3) To prove this bound, we use an argument due to Gekeler [Gek16]. (A different proof can be found in [Poo97, Thm. 9].) Let

$$K = F(C[\mathfrak{n}], \alpha) = F(\phi[\mathfrak{n}], \alpha).$$

By Galois theory (cf. [DF04, p. 591]), $\mathrm{Gal}(K/F(C[\mathfrak{n}]))$ is isomorphic to a subgroup of $\mathrm{Gal}(F(\alpha)/F)$, which itself is a subgroup of $\mathbb{F}_q^\times$. Hence, using Theorem 7.1.12, we get

$$\#\,\mathrm{Gal}(K/F) = \#\,\mathrm{Gal}(F(C[\mathfrak{n}])/F) \cdot \#\,\mathrm{Gal}(K/F(C[\mathfrak{n}]))$$

$$= \#(A/\mathfrak{n})^\times \cdot d_1,$$

where $d_1 \mid (q - 1)$. Applying the same argument to the subfield $F(\phi[\mathfrak{n}])$ of K, we get

$$\# \operatorname{Gal}(K/F) = \# \operatorname{Gal}(F(\phi[\mathfrak{n}])/F) \cdot \# \operatorname{Gal}(K/F(\phi[\mathfrak{n}]))$$
$$= \# \operatorname{Gal}(F(\phi[\mathfrak{n}])/F) \cdot d_2,$$

where $d_2 \mid (q - 1)$. On the other hand, $\operatorname{Gal}(F(\phi[\mathfrak{n}])/F)$ is naturally a subgroup of $(A/\mathfrak{n})^\times$. From the above equalities, we obtain

$$[(A/\mathfrak{n})^\times : \operatorname{Gal}(F(\phi[\mathfrak{n}])/F)] = d_2/d_1.$$

Thus, $d_1 \mid d_2 \mid (q-1)$. If $\operatorname{Gal}(F(\phi[\mathfrak{n}])/F) = 1$, then $\#(A/\mathfrak{n})^\times = d_2/d_1 \leq q-1$. Hence, using the formula (7.1.2) for $\#(A/\mathfrak{n})^\times$, we conclude that if $\phi[\mathfrak{n}] \subset F$, then either $\deg(\mathfrak{n}) \leq 1$ or $q = 2$ and $\mathfrak{n} = T(T + 1)$. On the other hand, since $r = 1$, we have $(^\phi F)_{\text{tor}} = \phi[\mathfrak{n}] \subset F$ for some $\mathfrak{n} \lhd A$. It remains to show that the claimed possibilities for $(^\phi F)_{\text{tor}}$ do actually occur.

Example 7.2.1 shows that the T-torsion of rank 1 Drinfeld module can be rational over F. This example is easy to modify to show that the $\mathfrak{n}$-torsion of ϕ for arbitrary $\mathfrak{n}$ of degree 1 can be rational over F. (Simply define ϕ by $\phi_\mathfrak{n}(x) = \mathfrak{n} \cdot f(x)$, where $f(x)$ is the polynomial from Example 7.2.1.) Now assume $q = 2$. Let C be the Carlitz module. Then

$$C_{T(T+1)}(x) = T(T + 1)x + (T^2 + T + 1)x^2 + x^4.$$

Note that $C_{T(T+1)}(1) = 0$, so $C[T(T + 1)]$ is rational over $\mathbb{F}_2(T)$. $\square$

Recall that for a Drinfeld module ϕ over K given by $\phi_T = t + g_1\tau + \cdots + g_r\tau^r$ we defined its basic j-invariants as

$$j_n(\phi) = \frac{g_n^{(q^r-1)/(q^{\gcd(n,r)}-1)}}{g_r^{(q^n-1)/(q^{\gcd(n,r)}-1)}}, \quad 1 \leq n \leq r - 1.$$

Theorem 7.2.4 *Let S be a finite set of primes of A. For $N > 0$, we denote by $D(r, K, S, N)$ the set of rank r Drinfeld modules, ϕ, over K, having the property that for some prime $\mathfrak{P}$ of K, lying over a prime in S, we have*

$$\min\{\operatorname{ord}_\mathfrak{P}(j_n(\phi)) \mid 1 \leq n \leq r - 1\} \geq -N. \tag{7.2.3}$$

Then there is a constant $C(d, S, N)$, depending on $d = [K : F]$, S, and N, such that

$$\#(^\phi K)_{\text{tor}} \leq C(d, S, N)$$

for all $\phi \in D(r, K, S, N)$.

Proof It is clear that it suffices to prove the result when $S = \{\mathfrak{p}\}$ consists of a single prime. We assume this from now on. We can also assume that the prime $\mathfrak{P}$ of K over $\mathfrak{p}$ for which the inequality (7.2.3) holds is fixed. Given $\phi \in D(r, K, S, N)$, we consider ϕ as a Drinfeld module over $K_{\mathfrak{P}}$. By Lemma 6.1.5, there is a separable totally ramified extension L of $K_{\mathfrak{P}}$ such that $[L : K_{\mathfrak{P}}] \leq q^r - 1$ and all $\phi \in D(r, K, S, N)$ have stable reduction over L. Obviously $(^{\phi}K)_{\text{tor}} \subseteq (^{\phi}K_{\mathfrak{P}})_{\text{tor}} \subseteq (^{\phi}L)_{\text{tor}}$, so it is enough to bound $\#(^{\phi}L)_{\text{tor}}$ in terms of d, N, and $\deg(\mathfrak{p})$.

Let v be the normalized valuation on L and ℓ be its residue field. Let $e(L/F_{\mathfrak{p}})$ and $f(L/F_{\mathfrak{p}})$ be the ramification index and the residue degree of $L/F_{\mathfrak{p}}$, respectively. By (6.4.3), for $\phi \in D(r, K, S, N)$ defined by $\phi_T = T + g_1\tau + \cdots + g_r\tau^r$ we have

$$\#(^{\phi}L)_{\text{tor}} \leq (\#\ell)^{v(\mathfrak{p})+2\cdot v(g_r)}. \tag{7.2.4}$$

Suppose the reduction of ϕ over ℓ has rank $n \leq r$. Then, after possibly replacing ϕ by an isomorphic module ϕ' over L, we may assume that ϕ has integral coefficients and $v(g_n) = 0$ (the torsion submodules $(^{\phi}L)_{\text{tor}} \cong (^{\phi'}L)_{\text{tor}}$ are isomorphic). The basic j-invariants do not change under isomorphisms, so the assumption of the theorem implies that

$$-\frac{q^n - 1}{q^{\gcd(n,r)} - 1}v(g_r) = v(j_n(\phi))$$

$$= e(L/F_{\mathfrak{p}}) \cdot \text{ord}_{\mathfrak{p}}(j_n(\phi))$$

$$\geq -e(L/F_{\mathfrak{p}}) \cdot N.$$

Hence

$$v(g_r) \leq \frac{q^{\gcd(n,r)} - 1}{q^n - 1}e(L/F_{\mathfrak{p}}) \cdot N \leq e(L/F_{\mathfrak{p}}) \cdot N.$$

Note that $v(\mathfrak{p}) = e(L/F_{\mathfrak{p}})$ and $\#\ell = q^{\deg(\mathfrak{p}) \cdot f(L/F_{\mathfrak{p}})}$. Therefore, (7.2.4) gives

$$\#(^{\phi}L)_{\text{tor}} \leq q^{\deg(\mathfrak{p})f(L/F_{\mathfrak{p}})e(L/F_{\mathfrak{p}})(1+2N)}.$$

Since

$$f(L/F_{\mathfrak{p}})e(L/F_{\mathfrak{p}}) = [L : F_{\mathfrak{p}}] \leq (q^r - 1)[K_{\mathfrak{P}} : F_{\mathfrak{p}}] \leq (q^r - 1)[K : F],$$

from the previous inequality we get

$$\#(^{\phi}L)_{\text{tor}} \leq q^{\deg(\mathfrak{p}) \cdot d \cdot (q^r-1) \cdot (1+2N)}.$$

$\square$

Remarks 7.2.5

(1) It is easy to see that $D(r, K, S, N)$ is always an infinite set. The integer N is a measure of the "badness" of the bad reduction allowed for ϕ at primes above those in S.
(2) Theorem 7.2.4, in the special case when $r = 2$, is proved in [Ros03, Thm. 5.4] using a different argument.

There is also the following result toward the Uniform Boundedness Conjecture proved in [Poo97] and [Sch03] by applying algebro-geometric techniques to Drinfeld modular curves.

Theorem 7.2.6 *Fix q, d, and a prime $\mathfrak{p} \lhd A$. As K ranges over all extensions of F with $[K : F] = d$ and ϕ ranges over all Drinfeld modules of rank 2 over K, there is a uniform bound on the size of the $\mathfrak{p}$-primary part of $(^\phi K)_{\text{tor}}$.*

In other words, if we write $(^\phi K)_{\text{tor}} \cong A/\mathfrak{m} \oplus A/\mathfrak{n}$ with $\mathfrak{m} \mid \mathfrak{n}$, then in $\mathfrak{n} = \prod \mathfrak{p}_i^{e_i}$ each e_i is bounded by a number dependent only on q, $\mathfrak{p}_i$, and d.

Remarks 7.2.7

(1) The main idea of the Proof of Theorem 7.2.6 goes back to Manin, who proved a similar theorem for elliptic curves in [Man69].
(2) What remains to be shown in order to completely prove Conjecture 7.2.2 in the $r = 2$ case is that for given q and d there are only finitely many primes $\mathfrak{p}$ for which there can exist a K-rational $\mathfrak{p}$-torsion point for a rank 2 Drinfeld module over K with $[K : F] \le d$.

Over F one can try to classify all possible finite A-modules $(^\phi F)_{\text{tor}}$ as ϕ varies. If successfully carried out, such a classification will be a much stronger result than just a uniform bound on $(^\phi F)_{\text{tor}}$. Similar result for elliptic curves was obtained by Mazur in his seminal paper [Maz77]: If E is an elliptic curve over $\mathbb{Q}$, then the torsion subgroup $E(\mathbb{Q})_{\text{tor}}$ is one of the following fifteen groups

$$\mathbb{Z}/N\mathbb{Z}, \qquad 1 \le N \le 10 \quad \text{or} \quad N = 12,$$

$$\mathbb{Z}/2\mathbb{Z} \times \mathbb{Z}/2N\mathbb{Z}, \qquad 1 \le N \le 4.$$

Furthermore, each of these groups does occur as an $E(\mathbb{Q})_{\text{tor}}$. The analogue of Mazur's result for Drinfeld modules of rank 2 is the following conjecture stated by Schweizer in [Sch03]:

Conjecture 7.2.8 *Let ϕ be a Drinfeld module over F of rank 2. Then $(^\phi F)_{\text{tor}}$ is isomorphic to*

$$A/\mathfrak{m} \times A/\mathfrak{n}, \qquad \text{where} \quad \mathfrak{m} \mid \mathfrak{n} \quad \text{and} \quad \deg(\mathfrak{n}) + \deg(\mathfrak{m}) \le 2.$$

Remarks 7.2.9

(1) The Drinfeld modular curve most closely related to the classification of $\mathfrak{n}$-rational torsion points of rank 2 Drinfeld modules is the curve $\Gamma_1(\mathfrak{n}) \setminus \Omega^2$, where

$$\Gamma_1(\mathfrak{n}) := \left\{ \begin{pmatrix} a & b \\ c & d \end{pmatrix} \in \mathrm{GL}_2(A) \mid a \equiv 1 \pmod{\mathfrak{n}} \text{ and } c \equiv 0 \pmod{\mathfrak{n}} \right\}.$$

In literature, this curve is usually denoted by $Y_1(\mathfrak{n})$. It is an algebraic affine curve defined over F. For an extension K of F, the K-rational points of $Y_1(\mathfrak{n})$ are in bijection with the $\overline{K}$-isomorphism classes of rank 2 Drinfeld modules over K with a K-rational $\mathfrak{n}$-torsion point. Thus, one can exclude certain torsion structures by showing that the corresponding Drinfeld modular curve has no K-rational points. We also mention the curve $Y(\mathfrak{n}) = \Gamma(\mathfrak{n}) \setminus \Omega^2$ which parametrizes isomorphism classes of Drinfeld modules with two independent $\mathfrak{n}$-torsion points.

(2) The projective completions of $Y(T)$ and $Y_1(\mathfrak{n})$ with $\deg(\mathfrak{n}) \leq 2$ are projective lines, so have infinitely many F-rational points. The possibilities listed in Conjecture 7.2.8 correspond to these modular curves. Hence, there are infinitely many non-isomorphic Drinfeld modules over F such that $({}^{\phi}F)_{\mathrm{tor}}$ contains an A-submodule isomorphic to $A/TA \oplus A/TA$ or $A/\mathfrak{n}$ with $\deg(\mathfrak{n}) \leq 2$. We will show by explicit examples that all the possibilities for $({}^{\phi}F)_{\mathrm{tor}}$ listed in Conjecture 7.2.8 do occur for Drinfeld modules of rank 2 over F.

(3) The philosophy behind Conjecture 7.2.8, and its counterpart over $\mathbb{Q}$, is that the only rational points on a modular curve of positive genus should be those which have a good reason to exist, namely certain cusps and CM-points. Over F this means that if the completion of $Y_1(\mathfrak{n})$ has genus ≥ 1, then it has no F-rational points at all.

(4) Conjecture 7.2.8 remains largely open. It has been fully proved only when $q = 2$ by Pál [Pál10] by adapting to the function field setting the deep geometric methods developed by Mazur [Maz77, Maz78].

Example 7.2.10 Let ϕ be a Drinfeld module over F defined by $\phi_T = T + (T^q - T)\tau + \tau^2$. Since ϕ has integral coefficients and good reduction at T, we have $\#({}^{\phi}F)_{\mathrm{tor}} \leq q$ by Theorem 6.5.8. Hence, an $\mathfrak{n}$-torsion point of ϕ can be rational only if $\deg(\mathfrak{n}) = 1$. Consider $\mathfrak{n} = T - \alpha$ with $\alpha \in \mathbb{F}_q$. Note that $\phi_{(T-\alpha)}(x)/x = (T - \alpha) + (T^q - T)x^{q-1} + x^{q^2-1}$ is Eisenstein because $T - \alpha$ divides $T^q - T = \prod_{\beta \in \mathbb{F}_q}(T - \beta)$, so this polynomial is irreducible. We conclude that $({}^{\phi}F)_{\mathrm{tor}} = 0$.

Example 7.2.11 Let ϕ be a Drinfeld module over F defined by $\phi_T = T - T\tau + \tau^2$. Since ϕ has integral coefficients and good reduction at T, we have $\#({}^{\phi}F)_{\mathrm{tor}} \leq q$ by Theorem 6.5.8. On the other hand, $\phi_{T-1}(1) = 0$. Thus, $({}^{\phi}F)_{\mathrm{tor}} \cong A/(T - 1)A$.

Example 7.2.12 Let ϕ be a Drinfeld module over F defined by $\phi_T = T + g_1\tau + g_2\tau^2$. Let $\mathfrak{n} = T^2 + \alpha T + \beta$, $\alpha, \beta \in \mathbb{F}_q$, be a monic quadratic polynomial in A. We want to choose $g_1, g_2 \in F$ so that $({}^\phi F)_{\text{tor}} \cong A/\mathfrak{n}A$, generated by 1 over A. Let $z = \phi_T(1) = T + g_1 + g_2$. Now compute

$$\phi_{T+\alpha}(z) = (T+\alpha)z + g_1 z^q + g_2 z^{q^2}$$

$$= (T+\alpha)(T + g_1 + g_2) + g_1(T + g_1 + g_2)^q + g_2(T + g_1 + g_2)^{q^2}.$$

Hence,

$$\phi_\mathfrak{n}(1) = \phi_{T+\alpha}(z) + \beta$$

$$= (T+\alpha)(T + g_1 + g_2) + \beta + g_1(T + g_1 + g_2)^q + g_2(T + g_1 + g_2)^{q^2}.$$

Let $g_2 = -g_1$ and $g := g_1$. Then $z = T \neq 0$ and

$$\phi_\mathfrak{n}(1) = \mathfrak{n} + gT^q - gT^{q^2}.$$

For this expression to be zero, it is necessary and sufficient that

$$g = \frac{\mathfrak{n}}{(T^q - T)^q}.$$

Since $T^q - T$ is the product of degree 1 primes of A, the module $\phi_T = T + g\tau - g\tau^2$ has integral coefficients and good reduction at every prime of degree 2, except possibly $\mathfrak{n}$ if $\mathfrak{n}$ is prime. If $q \neq 2$, then there are at least two monic irreducible quadratics in A, so by Theorem 6.5.8 we have $\#({}^\phi F)_{\text{tor}} \leq q^2$. We conclude that $({}^\phi F)_{\text{tor}} \cong A/\mathfrak{n}A$, assuming $q \neq 2$ or $\mathfrak{n}$ is not prime.

If $q = 2$ and $\mathfrak{n} = T^2 + T + 1$, then ϕ has integral coefficients and good reduction at every prime of degree 3, so $\#({}^\phi F)_{\text{tor}} \leq q^3$. One can then check directly that the cubics $\phi_T(x)$ and $\phi_{T+1}(x)$ are irreducible over F, so ϕ has no F-rational T or $T+1$ torsion points. This again implies that $({}^\phi F)_{\text{tor}} \cong A/\mathfrak{n}A$.

Example 7.2.13 Let ϕ be the Drinfeld module over F defined by

$$\phi_T(x) = Tx \prod_{\alpha,\beta \in \mathbb{F}_q}' \left(1 - \frac{x}{\alpha + \beta T}\right).$$

By construction, $\phi[T]$ is rational over F. On the other hand, observe that in notation of (5.4.5) we have $\phi_T(x) = T \cdot e_{A_{<2}}(x)$. Hence, by (5.4.7),

$$\phi_T(x) = Tx - \frac{T \cdot L_2}{D_1 \cdot L_1^q} x^q + \frac{T \cdot L_2}{D_2} x^{q^2}$$

$$= Tx - \frac{T(T^{q^2} - T)}{(T^q - T)^q} x^q + \frac{T}{(T^q - T)^{q-1}} x^{q^2}.$$

As in the previous example, the Drinfeld module ϕ has integral coefficients and good reduction at every prime of degree 2. Hence, by Theorem 6.5.8, we have $\#(^\phi F)_{\text{tor}} \leq q^2$. We conclude that $(^\phi F)_{\text{tor}} \cong A/TA \oplus A/TA$.

Exercises

7.2.1 Let $C^{(g)}$ be the Drinfeld module over F defined by $C_T^{(g)} = Tx + gx^q$. Let α be a root of $x^{q-1} = g$. Prove that if $F(C[\mathfrak{n}])$ and $F(\alpha)$ are linearly disjoint, then $\text{Gal}(F(C^{(g)}[\mathfrak{n}])/F) \cong (A/\mathfrak{n})^\times$.

7.2.2 Let ϕ be a Drinfeld module of rank 2 over F. Prove that if ϕ has complex multiplication, then $(^\phi F)_{\text{tor}} \leq q$.

7.2.3 Let ϕ be a Drinfeld module of rank $r \geq 1$ over F. Suppose $\phi[\mathfrak{n}] \subset F$, i.e., all $\mathfrak{n}$-torsion points of ϕ are rational over F. Prove that if $q > 2$, then $\deg(\mathfrak{n}) \leq 1$, and if $q = 2$, then $\deg(\mathfrak{n}) \leq 2$.

7.2.4 Let K be a finite extension of F, and let ϕ be a Drinfeld module over K. Show that at most finitely many $\psi \in \text{Twist}(\phi/K)$ have $(^\psi K)_{\text{tor}} \neq 0$. In particular, there are at most finitely many rank 1 Drinfeld modules ϕ over K (up to isomorphism over K) for which $(^\phi K)_{\text{tor}} \neq 0$.

7.3 Division Fields: Examples

Let K be a finite extension of F and let ϕ be a rank r Drinfeld module over K. Recall that from the action of the absolute Galois group G_K on $\phi[\mathfrak{n}]$ we obtain an injective homomorphism (cf. (3.5.5))

$$\rho_{\phi,\mathfrak{n}} \colon \text{Gal}(K(\phi[\mathfrak{n}])/K) \longrightarrow \text{Aut}(\phi[\mathfrak{n}]) \cong \text{GL}_r(A/\mathfrak{n}).$$

The study of rational torsion submodules of Drinfeld modules in Sect. 7.2 can be rephrased as the study of those $\rho_{\phi,\mathfrak{n}}$ which contain the trivial subrepresentation.

But generally $\rho_{\phi,\mathfrak{n}}$ is not only irreducible, but in fact its image is close to being as large as possible. This is a consequence of a series of deep results of Richard Pink and his collaborators, which culminated with the following theorem of Pink and Rütche [PR09]:

Theorem 7.3.1 (Open Image Theorem) *Let ϕ be a Drinfeld module of rank r over a finite extension K of F. Assume $\mathrm{End}(\phi) = A$. Then there is a constant $N(\phi, K)$ depending only on ϕ and K such that*

$$[\mathrm{GL}_r(A/\mathfrak{n}) : \rho_{\phi,\mathfrak{n}}(G_K)] \leq N(\phi, K) \qquad \text{for all nonzero } \mathfrak{n} \lhd A.$$

Remarks 7.3.2

(1) The analogue of Theorem 7.3.1 for elliptic curves over number fields goes back to Serre [Ser72]. In fact, the methods used by Pink and Rütche are modeled on the methods developed by Serre.

(2) Whether the bound in Theorem 7.3.1 can be made uniform, i.e., independent of ϕ, is a major open question. Even in the setting of elliptic curves over $\mathbb{Q}$ the corresponding question is a long-standing open problem known as "Serre's Uniform Boundedness Question": for a non-CM elliptic curve E over $\mathbb{Q}$ it is expected that $\mathrm{Gal}(\mathbb{Q}(E[p])/\mathbb{Q}) \cong \mathrm{GL}_2(\mathbb{F}_p)$ for any prime $p > 37$. Note that in the proof of Theorem 7.2.3 we showed that for $r = 1$ the index of $\mathrm{Gal}(F(\phi[\mathfrak{n}])/F)$ in $(A/\mathfrak{n})^\times$ always divides $q - 1$. See also Exercise 7.3.1.

(3) In [Zyw11], Zywina showed that for the rank-2 Drinfeld module over F defined by $\phi_T = T + \tau - T^{q-1}\tau^2$ the Galois representations $\rho_{\phi,\mathfrak{n}}$ are surjective for all nonzero $\mathfrak{n}$, assuming $q \geq 5$. In [Che22a], Chien-Hua Chen constructed similar examples for $q = 3, 4$, but he also showed that such examples do not exist for $q = 2$. In [Che22b], Chen proved that for the Drinfeld module of rank $r \geq 3$ over F defined by $\phi_T = T + \tau^{r-1} + T^{q-1}\tau^r$ the representations $\rho_{\phi,\mathfrak{n}}$ are surjective for all nonzero $\mathfrak{n}$, assuming r is prime, $q \equiv 1 \pmod{r}$, and the characteristic of F is sufficiently large compared to r.

Nevertheless, for small degree $\mathfrak{n}$ the possibilities for $\rho_{\phi,\mathfrak{n}}(G_K)$ are quite varied. In this section we construct rank 2 Drinfeld modules over F such that $\mathrm{Gal}(F(\phi[T])/F)$ is isomorphic to one of the following subgroups of $\mathrm{GL}_2(\mathbb{F}_q)$:

- $\mathrm{GL}_2(\mathbb{F}_q)$,
- $\mathrm{SL}_2(\mathbb{F}_q)$,
- Borel subgroup,
- Cartan subgroup,
- Normalizer of a Cartan subgroup.

The motivation for singling out these subgroups comes from Serre's celebrated paper on Galois representations arising from the torsion points of elliptic curves [Ser72], where these subgroups play a central role. As is explained in [Ser72], if $G \subseteq \mathrm{GL}_2(\mathbb{F}_p)$ is a subgroup, then (ignoring a few exceptional cases) G is either large in the sense that it contains $\mathrm{SL}_2(\mathbb{F}_p)$ or G is contained in one of the subgroups listed above.[2] Although our constructions are ad hoc, they do employ a few useful tricks, hence can be instructive.

Before proceeding to the examples, we make a small preliminary calculation. Let ϕ be a Drinfeld module of rank 2 defined by

$$\phi_T(x) = Tx + g_1(T)x^q + g_2(T)x^{q^2}, \qquad g_1(T), g_2(T) \in A. \tag{7.3.1}$$

Assume $g_2(c) \neq 0$ for a given $c \in \mathbb{F}_q^\times$. Then ϕ has good reduction at $\mathfrak{p} = T - c$ and its reduction is defined by $\bar{\phi}_T(x) = cx + g_1(c)x^q + g_2(c)x^{q^2}$. Using the method in Remark 4.2.16, we compute that

$$P_{\bar{\phi}}(x) = x^2 + \frac{g_1(c)}{g_2(c)}x + \frac{c}{g_2(c)}.$$

Let $\mathfrak{P}$ be a place of $F(\phi[T])$ over $\mathfrak{p}$. By (7.1.5), $P_{\phi,\mathfrak{p},T}(x) = P_{\bar{\phi}}(x)$ (mod T), or equivalently

$$\mathrm{Tr}(\rho_{\phi,T}(\mathrm{Frob}_{\mathfrak{P}})) = -g_1(c)/g_2(c), \tag{7.3.2}$$

$$\det(\rho_{\phi,T}(\mathrm{Frob}_{\mathfrak{P}})) = c/g_2(c).$$

To simplify the notation, in the examples below we write $\mathrm{Tr}(\mathrm{Frob}_{\mathfrak{P}})$ instead of $\mathrm{Tr}(\rho_{\phi,T}(\mathrm{Frob}_{\mathfrak{P}}))$, and $\det(\mathrm{Frob}_{\mathfrak{P}})$ instead of $\det(\rho_{\phi,T}(\mathrm{Frob}_{\mathfrak{P}}))$.

7.3.1 General Linear Group

Producing explicit examples of ϕ of rank $r \geq 2$ such that $\mathrm{Gal}(F(\phi[T])/F) \cong \mathrm{GL}_r(\mathbb{F}_q)$ is not easy. This type of questions have been studied by Abhyankar (cf. [Abh94]) in connection with the problem of realizing $\mathrm{GL}_r(\mathbb{F}_q)$ and $\mathrm{PGL}_r(\mathbb{F}_q)$ as Galois groups of extensions of F ramified only at T and ∞. In particular, Abhyankar proved[3,4] that if ϕ over F is defined by

$$\phi_T(x) = Tx + Tx^q + x^{q^r},$$

---

[2] It is important for this statement that p is prime. If q is a higher power of p, then, for example, $\mathrm{GL}_2(\mathbb{F}_p)$ is a proper subgroup of $\mathrm{GL}_2(\mathbb{F}_q)$.

[3] Abhyankar only later realized that his approach based on "nice" trinomial equations is related to the theory of Drinfeld modules; cf. [Abh01].

[4] A related result, conjectured by Abhyankar and proved by Breuer [Bre16], says that if ϕ is defined by

$$\phi_T(x) = Tx + g_1 x^q + \cdots + g_r x^{q^r},$$

where $T, g_1, \ldots, g_r$ are algebraically independent over $\mathbb{F}_q$, then over $K' = K(g_1, \ldots, g_r)$ we have $\mathrm{Gal}(K'(\phi[\mathfrak{n}])/K') \cong \mathrm{GL}_r(A/\mathfrak{n})$.

then $\text{Gal}(\phi[T]/F) \cong \text{GL}_r(\mathbb{F}_q)$.

Assuming $q = p$ is prime, we will construct explicit examples of Drinfeld modules of rank 2 with $\text{Gal}(F(\phi[T])/F)$ isomorphic to $\text{GL}_2(\mathbb{F}_q)$ and $\text{SL}_2(\mathbb{F}_q)$ using the following group-theoretic fact (see Proposition 19 in [Ser72]):

Proposition 7.3.3 *Let V be a vector space of dimension 2 over $\mathbb{F}_p$, where p is a prime number. Let $G \subset \text{GL}(V)$ be a subgroup of linear automorphisms of V. Assume that $p \geq 5$ and that the following hypotheses hold:*

(i) *G contains an element s such that $\text{Tr}(s) \neq 0$ and $\text{Tr}(s)^2 - 4\det(s)$ is a nonzero square in $\mathbb{F}_p$.*
(ii) *G contains an element s' such that $\text{Tr}(s') \neq 0$ and $\text{Tr}(s')^2 - 4\det(s')$ is not a square in $\mathbb{F}_p$.*
(iii) *G contains an element s'' such that $u = \text{Tr}(s'')^2/\det(s'')$ is distinct from $0, 1, 2, 4$ and $u^2 - 3u + 1 \neq 0$.*

Then G contains $\text{SL}(V)$. In particular, if $\det \colon G \to \mathbb{F}_p^\times$ is surjective, then $G = \text{GL}(V)$.

Now let ϕ be the Drinfeld module defined by

$$\phi_T(x) = Tx + x^q + x^{q^2}.$$

By (7.3.2), for any $c \in \mathbb{F}_q^\times$ we have

$$\text{Tr}(\text{Frob}_{\mathfrak{P}})^2 - 4\det(\text{Frob}_{\mathfrak{P}}) = 1 - 4c$$

and

$$\text{Tr}(\text{Frob}_{\mathfrak{P}})^2/\det(\text{Frob}_{\mathfrak{P}}) = 1/c.$$

Since c in this case is an arbitrary element of $\mathbb{F}_q^\times$, the hypotheses of Proposition 7.3.3 are satisfied if $q = p \geq 5$ is a prime. Moreover, $\det \colon \text{Gal}(F(\phi[T])/F) \to \mathbb{F}_p^\times$ is surjective, because $\det(\text{Frob}_{\mathfrak{P}}) = c$. Therefore,

$$\text{Gal}(F(\phi[T])/F) \cong \text{GL}_2(\mathbb{F}_p), \qquad \text{for } p \geq 5.$$

The previous statement can be extended to $p = 2, 3$ using brute force computer calculations. The Magma computational algebra system is able to compute the Galois groups of polynomials over F. The following code computes the Galois group of $\phi_T(x)$ for $p = 3$:

```
> F<T>  :=FunctionField(GF(3));
> R<x>  :=PolynomialRing(F);
> f  :=T*x+x^3+x^9;
> G:=GaloisGroup(f);
> # G;
```

The program outputs 48, which is the order of $\mathrm{GL}_2(\mathbb{F}_3)$. Hence, $\mathrm{Gal}(F(\phi[T])/F)$ $\cong \mathrm{GL}_2(\mathbb{F}_3)$. A similar calculation for $p = 2$ gives 6 as the order of the Galois group of $\phi_T(x)$, which again implies that $\mathrm{Gal}(F(\phi[T])/F) \cong \mathrm{GL}_2(\mathbb{F}_2)$.

7.3.2   Special Linear Group

Assume $g_1 = 1$ and $g_2 = T$ in (7.3.1), i.e., consider the Drinfeld module defined by

$$\phi_T(x) = Tx + x^q + Tx^{q^2}.$$

In this case the formulas (7.3.2) give

$$\mathrm{Tr}(\mathrm{Frob}_{\mathfrak{P}})^2 - 4 \cdot \det(\mathrm{Frob}_{\mathfrak{P}}) = 1/c^2 - 4$$

and

$$\mathrm{Tr}(\mathrm{Frob}_{\mathfrak{P}})^2 / \det(\mathrm{Frob}_{\mathfrak{P}}) = 1/c^2.$$

To check the hypotheses of Proposition 7.3.3 as c runs through $\mathbb{F}_q^\times$, we use the following easy lemma:

Lemma 7.3.4 *Assume q is odd and $a \in \mathbb{F}_q^\times$ is fixed. The number of solutions of $x^2 - y^2 = a$ in $\mathbb{F}_q$ is equal to $q - 1$.*

Proof Denote $u = x - y$ and $v = x + y$. Then solving $x^2 - y^2 = a$ is equivalent to solving $uv = a$ with $u, v \in \mathbb{F}_q$. The solutions of this latter equation are the pairs $(u, a/u)$ as u varies in $\mathbb{F}_q^\times$.   □

Now, given $d \in \mathbb{F}_q$, the number of elements $c \in \mathbb{F}_q$ such that $(1/c)^2 - 4 = d^2$ is either 0 or 2. On the other hand, by the previous lemma,

$$\sum_{d \in \mathbb{F}_q} \#\{c \in \mathbb{F}_q^\times \mid (1/c)^2 - 4 = d^2\} = \begin{cases} q - 1 & \text{if } -1 \text{ is a not square in } \mathbb{F}_q; \\ q - 3 & \text{if } -1 \text{ is a square in } \mathbb{F}_q. \end{cases}$$

Therefore, as c varies in $\mathbb{F}_q^\times$, the expression $1/c^2 - 4$ is a square in $\mathbb{F}_q$ exactly $(q-1)/2$ or $(q-3)/2$ times. This implies that (i) and (ii) are satisfied if $q = p \geq 7$. Next, it is easy to check that (iii) is satisfied if $p \geq 11$. Thus, just by examining the Frobenius elements associated with degree 1 primes of A, we conclude that the hypotheses of Proposition 7.3.3 are satisfied if $p \geq 11$. Thus, according to that proposition,

$$\mathrm{SL}_2(\mathbb{F}_p) \subseteq \mathrm{Gal}(F(\phi[T])/F).$$

We claim that this inclusion is an equality. Let $T \neq \mathfrak{p} \lhd A$ be a prime of degree d. Let $\mathfrak{P}$ be a prime of $F(\phi[T])$ over $\mathfrak{p}$. Let $t \in \mathbb{F}_{\mathfrak{p}}$ be the image of T modulo $\mathfrak{p}$. By (7.1.5) and Theorem 4.2.7,

$$\det(\mathrm{Frob}_{\mathfrak{P}}) = (-1)^d \cdot (t \cdot t^q \cdots t^{q^{d-1}})^{-1} \cdot \mathfrak{p}(0)$$

$$= (-1)^d ((-1)^d \mathfrak{p}(0))^{-1} \mathfrak{p}(0)$$

$$= 1,$$

where the second equality follows from the observation that $t, t^q, \ldots, t^{q^{d-1}}$ are the distinct roots of $\mathfrak{p}(x) \in \mathbb{F}_{\mathfrak{p}}[x]$, so their product is equal to $(-1)^d \mathfrak{p}(0)$; here $\mathfrak{p} = T^d + a_{d-1}T^{d-1} + \cdots + a_0$ and $\mathfrak{p}(x) := x^d + a_{d-1}x^{d-1} + \cdots + a_0$. To conclude from this that the determinant of any element of $\mathrm{Gal}(F(\phi[T])/F)$ is 1, we will use Chebotarev's density theorem. This is a well-known theorem in algebraic number theory, which over function fields is a consequence of the Riemann hypothesis for curves over finite fields (see Theorems 9.13A and 9.13B in [Ros02]).

Theorem 7.3.5 (Chebotarev Density Theorem) *Let K/F be a finite Galois extension with Galois group G. Let S be a conjugacy class in G. Then the set of primes of A that are unramified in K and whose associated Frobenius conjugacy class is contained in S has density $\#S/\#G$.*

We will not formally define what "density" is in the theorem; the reader can take this as a natural measure of the frequency of primes satisfying the condition of the theorem among all primes. For us it is sufficient to know that there are infinitely many such primes.

Now, by the Chebotarev density theorem, any element of $\mathrm{Gal}(F(\phi[T])/F)$ is conjugate to some $\mathrm{Frob}_{\mathfrak{P}}$. Thus, $\det(s) = 1$ for all $s \in \mathrm{Gal}(F(\phi[T])/F)$, which implies $\mathrm{Gal}(F(\phi[T])/F) \subseteq \mathrm{SL}_2(\mathbb{F}_p)$. Combining this with the reverse inclusion proved earlier, we get

$$\mathrm{Gal}(F(\phi[T])/F) \cong \mathrm{SL}_2(\mathbb{F}_p), \qquad p \geq 11.$$

In fact, as in the previous example, using computer calculations it is easy to check that this isomorphism holds also for $p = 2, 3, 5, 7$.

7.3.3 Borel Subgroup

A Borel subgroup of $\mathrm{GL}_2(\mathbb{F}_q)$ is a subgroup conjugate to the group of the upper-triangular matrices $\begin{pmatrix} * & * \\ 0 & * \end{pmatrix}$ in $\mathrm{GL}_2(\mathbb{F}_q)$, which we denote by B.

Let a be a generator of $\mathbb{F}_q^\times$. Let α be a root of $x^{q-1} - a$; the other roots of this polynomial are the $\mathbb{F}_q^\times$-multiples of α. The extension $F(\alpha)/F$ is cyclic and everywhere unramified. For a prime $\mathfrak{p}$ of A of degree d we have

$$\text{Frob}_\mathfrak{p}(\alpha) = \alpha^{|\mathfrak{p}|} = b \cdot \alpha \quad \text{for some } b \in \mathbb{F}_q^\times.$$

Clearly

$$b = \alpha^{|\mathfrak{p}|-1} = a^{(|\mathfrak{p}|-1)/(q-1)} = a^{1+q+q^2+\cdots+q^{d-1}} = a^d.$$

Thus, b does not depend on the choice of α. Also, G_F transitively permutes the roots of $x^{q-1} - a$, so this polynomial is irreducible over F.

Consider the Drinfeld module ϕ over F defined by

$$\phi_T(x) = Tx + x^q - \frac{T+a}{a^2}x^{q^2}. \tag{7.3.3}$$

Note that

$$\phi_T(\alpha) = \alpha\left(T + \alpha^{q-1} - \frac{T+a}{a^2}(\alpha^{q-1})^{q+1}\right)$$

$$= \alpha\left(T + a - \frac{T+a}{a^2}a^{q+1}\right) = 0.$$

Therefore, $\mathbb{F}_q\alpha \subseteq \psi[T]$ is a 1-dimensional subspace which is Galois invariant. Choosing α as one of the basis elements of the 2-dimensional $\mathbb{F}_q$-vector space $\phi[T]$, we see that $G := \text{Gal}(F(\phi[T])/F)$ is isomorphic to a subgroup of B.

To prove that $G = B$ it is now enough to show that $\#B = q(q-1)^2$ divides $\#G$. Let $\mathfrak{p} = T + a$. Let $\mathfrak{P}$ be a prime of $F(\phi[T])$ above $\mathfrak{p}$. By Theorem 2.8.11, the decomposition subgroup $D(\mathfrak{P}/\mathfrak{p})$ of G is isomorphic to $\text{Gal}(F_\mathfrak{p}(\phi[T])/F_\mathfrak{p})$. The Newton polygon of $\phi_T(x)/x$ over $F_\mathfrak{p}$ has two segments, with slopes 0 and $1/q(q-1)$, respectively. Therefore, the ramification index $e(F_\mathfrak{p}(\phi[T])/F_\mathfrak{p})$ is divisible by $q(q-1)$. On the other hand, by construction, the extension of $\mathbb{F}_q$ of degree $q-1$ is contained in $F_\mathfrak{p}(\phi[T])$. Hence the residue degree $f(F_\mathfrak{p}(\phi[T])/F_\mathfrak{p})$ is divisible by $q-1$. Since $[F_\mathfrak{p}(\phi[T]) : F_\mathfrak{p}] = e(F_\mathfrak{p}(\phi[T])/F_\mathfrak{p}) \cdot f(F_\mathfrak{p}(\phi[T])/F_\mathfrak{p})$, we conclude that $q(q-1)^2$ divides $\#D(\mathfrak{P}/\mathfrak{p})$, which itself divides $\#G$.

7.3.4 Split Cartan Subgroup

A split Cartan subgroup of $\text{GL}_2(\mathbb{F}_q)$ is a subgroup that is conjugate to the group of diagonal matrices $\begin{pmatrix} * & 0 \\ 0 & * \end{pmatrix}$ in $\text{GL}_2(\mathbb{F}_q)$; such a group is isomorphic to $\mathbb{F}_q^\times \times \mathbb{F}_q^\times$.

Let a be a root of the polynomial $x^{q-1} - T$ and b be a root of $x^{q-1} - (T+1)$. The polynomial $x^{q-1} - T$ is irreducible over F since, for example, it is irreducible over F_T by the Eisenstein criterion. By Proposition 1.3.12, $F(a)$ is the splitting field of $x^{q-1} - T$ and $\mathrm{Gal}(F(a)/F) \cong \mathbb{F}_q^\times$. Similarly, $\mathrm{Gal}(F(b)/F) \cong \mathbb{F}_q^\times$.

Let W be the 2-dimensional $\mathbb{F}_q$-vector subspace of F^{sep} spanned by a and b

$$W = \mathbb{F}_q a + \mathbb{F}_q b.$$

Given $\sigma \in G_F$, we have $\sigma(a) = \alpha a$ and $\sigma(b) = \beta b$ for some $\alpha, \beta \in \mathbb{F}_q^\times$ depending on σ, so the subspace W is stable under the action of G_F. Therefore, the $\mathbb{F}_q$-linear polynomial

$$f(x) = x \prod_{0 \neq w \in W}' \left(1 - \frac{x}{w}\right)$$

has coefficients in F. Define $\phi_T(x) = Tf(x)$. Then ϕ is a Drinfeld module of rank 2 over F and $\phi[T] = W$. We have $F(\phi[T]) = F(W) = F(a, b)$.

Now note that $F(a) \cap F(b) = F$ since, for example, T totally ramifies in $F(a)$ but does not ramify in $F(b)$. Therefore, by Galois theory (cf. [DF04, p. 593]),

$$\mathrm{Gal}(F(\phi[T])/F) \cong \mathrm{Gal}(F(a)/F) \times \mathrm{Gal}(F(b)/F) \cong \mathbb{F}_q^\times \times \mathbb{F}_q^\times.$$

In fact, if we choose a and b as a basis of W and use this basis to define an isomorphism $\mathrm{Aut}(W) \cong \mathrm{GL}_2(\mathbb{F}_q)$, then $\mathrm{Gal}(F(\phi[T])/F) \subset \mathrm{Aut}(W)$ is the subgroup of diagonal matrices.

7.3.5 Non-split Cartan Subgroup

A non-split Cartan subgroup of $\mathrm{GL}_2(\mathbb{F}_q)$ is a subgroup isomorphic to $\mathbb{F}_{q^2}^\times$.

Let $\mathfrak{p} \in A$ be a monic irreducible polynomial of degree 2. Define

$$\phi_T(x) = \frac{T}{\mathfrak{p}} C_{\mathfrak{p}}(x),$$

where $C_T = Tx + x^q$ is the Carlitz module. Obviously $F(\phi[T]) = F(C[\mathfrak{p}])$. Hence, by Theorem 7.1.12,

$$\mathrm{Gal}(F(\phi[T])/F) \cong (A/\mathfrak{p}A)^\times \cong \mathbb{F}_{q^2}^\times.$$

7.3.6   *Normalizer of a Split Cartan Subgroup*

Assume q is odd. Let $D = \begin{pmatrix} \alpha & 0 \\ 0 & \beta \end{pmatrix} \in \mathrm{GL}_2(\mathbb{F}_q)$ be any diagonal matrix with $\alpha \neq \beta$.

Given $S = \begin{pmatrix} a & b \\ c & d \end{pmatrix} \in \mathrm{GL}_2(\mathbb{F}_q)$, an easy calculation shows that the matrix SDS^{-1} is diagonal if and only if either $b = c = 0$ or $a = d = 0$. Hence, taking the split Cartan subgroup C to be the group of diagonal matrices in $\mathrm{GL}_2(\mathbb{F}_q)$, we see that its normalizer, denoted $\mathcal{N}$, is the group generated by C and $\begin{pmatrix} 0 & 1 \\ 1 & 0 \end{pmatrix}$. We conclude that $[\mathcal{N} : C] = 2$ and $\mathcal{N}$ is non-abelian.

We need to construct $\phi_T(x)$ whose splitting field L over F has a non-abelian Galois group and contains a quadratic subfield K such that $\mathrm{Gal}(L/K)$ is isomorphic to the Cartan subgroup.

Let $K = F(\sqrt{T})$, where $\sqrt{T}$ denotes a fixed root of $x^2 - T$. Let $a = \sqrt[q-1]{1 - \sqrt{T}}$ be a fixed root of $x^{q-1} - (1 - \sqrt{T})$; let $b = \sqrt[q-1]{1 + \sqrt{T}}$ be a fixed root of $x^{q-1} - (1 + \sqrt{T})$. Note that $K(a)/K$ is a Galois extension with $\mathrm{Gal}(K(a)/K) \cong \mathbb{F}_q^\times$ but $K(a)/F$ is not Galois. Similarly, $\mathrm{Gal}(K(b)/K) \cong \mathbb{F}_q^\times$. Next, $K(a) \cap K(b) = K$. To see this, note that the prime $T - 1$ of A splits in K into a product of two distinct primes $T - 1 = (\sqrt{T} - 1)(\sqrt{T} + 1)$, and $\sqrt{T} - 1$ totally ramifies in $K(a)$ but does not ramify in $K(b)$. By Galois theory,

$$\mathrm{Gal}(K(a,b)/K) \cong \mathrm{Gal}(K(a)/K) \times \mathrm{Gal}(K(b)/K) \cong \mathbb{F}_q^\times \times \mathbb{F}_q^\times.$$

Let σ be the nontrivial element of $\mathrm{Gal}(K/F)$. Then $\sigma(\sqrt{T}) = -\sqrt{T}$ and

$$(\sigma(a))^{q-1} = \sigma\left(a^{q-1}\right) = \sigma(1 - \sqrt{T}) = 1 + \sqrt{T}.$$

Hence, $\sigma(a)$ is a root of $x^{q-1} - (1 + \sqrt{T})$, so must be an $\mathbb{F}_q$-multiple of b. This implies that the $\mathbb{F}_q$-vector space

$$W = \mathbb{F}_q a + \mathbb{F}_q b$$

is stable under the action of $\mathrm{Gal}(K(a,b)/F)$. Define

$$\phi_T(x) = Tx \prod_{0 \neq w \in W}' \left(1 - \frac{x}{w}\right).$$

Then $\phi_T(x)$ has coefficients in F, and the Galois group of its splitting field is isomorphic to the normalizer of a split Cartan subgroup.

As a more explicit example of this, when $q = 3$, the Drinfeld module constructed above is defined by

$$\phi_T(x) = Tx + \frac{T+1}{T-1}x^3 - \frac{1}{T-1}x^9.$$

We leave it to the reader to check that in this case $\mathcal{N}$ is isomorphic to the dihedral group D_8 with eight elements.

7.3.7   Normalizer of a Non-split Cartan Subgroup

Assume q is odd. Fix a non-square $\varepsilon \in \mathbb{F}_q^\times$. A non-split Cartan subgroup in $\mathrm{GL}_2(\mathbb{F}_q)$ is conjugate to the group

$$C' = \left\{ \begin{pmatrix} a & \varepsilon b \\ b & a \end{pmatrix} \mid a, b \in \mathbb{F}_q \text{ are not both zero} \right\}.$$

As in the split case, a simple calculation shows that the normalizer of C' in $\mathrm{GL}_2(\mathbb{F}_q)$, denoted $\mathcal{N}'$, is the subgroup generated by C' and $\begin{pmatrix} 1 & 0 \\ 0 & -1 \end{pmatrix}$. We conclude again that $[\mathcal{N}' : C'] = 2$ and $\mathcal{N}'$ is non-abelian.

Let

$$\phi_T(x) = Tx - x^{q^2}.$$

By Galois theory (cf. Proposition 1.3.12), the splitting field of $\phi_T(x)$ is the composite of the quadratic extension $\mathbb{F}_{q^2}(T)/F$ with the non-Galois extension $F(\sqrt[q^2-1]{T})/F$ of degree $q^2 - 1$, where $\sqrt[q^2-1]{T}$ denotes a fixed nonzero root of $\phi_T(x)$. The extension $F(\phi[T])/\mathbb{F}_{q^2}(T)$ is Galois with cyclic Galois group of order $q^2 - 1$. Thus, $F(\phi[T])/F$ is a non-abelian extension with Galois group isomorphic to $\mathcal{N}'$.

When $q = 3$, we leave it to the reader to check that $\mathcal{N}' \cong \mathbb{Z}/8\mathbb{Z} \rtimes \mathbb{Z}/2\mathbb{Z}$, but $\mathcal{N}'$ is not isomorphic to D_{16}.

7.3.8   Boston-Ose Theorem

The previous examples, as well as the exercises of this section, indicate that the possibilities for $\mathrm{Gal}(F(\phi[T])/F)$ are quite varied. Still, the next theorem is striking in its scope. Not only it implies that every subgroup of $\mathrm{GL}_r(\mathbb{F}_q)$ arises as the Galois group of the T-division field of some Drinfeld module of rank r, but in fact every

Galois extension whose Galois group is isomorphic to a subgroup of $GL_r(\mathbb{F}_q)$ arises in this manner!

Theorem 7.3.6 (Boston and Ose [BO00]) *Let K be an infinite A-field with* $char_A(K) \neq T$ *and let* $\gamma : A \to K$ *be the structure morphism. Let L/K be a finite Galois extension such that* $Gal(L/K)$ *is isomorphic to a subgroup of* $GL_r(\mathbb{F}_q)$. *Then there exists a Drinfeld module ϕ of rank r over K such that $L = K(\phi[T])$.*

Proof Let $G = Gal(L/K)$ and $\sigma_1, \ldots, \sigma_n$ be the elements of G. According to the Normal Basis Theorem (see Theorem 1.3.13), there exists an element $\alpha \in L$ such that $\sigma_1 \alpha, \ldots, \sigma_n \alpha$ form a basis of L over K. This means that L is a free $K[G]$-module of rank 1. Since K is assumed to be infinite, this implies that the $\mathbb{F}_q$-vector space L contains free $\mathbb{F}_q[G]$-modules of arbitrarily high finite rank.

Let $V \cong \mathbb{F}_q^r$ be the $\mathbb{F}_q[G]$-module corresponding to the embedding of G in $GL_r(\mathbb{F}_q)$. Let $M \subset L$ be a free $\mathbb{F}_q[G]$-module. By duality for group rings, $M^* := Hom_{\mathbb{F}_q[G]}(M, \mathbb{F}_q)$ is also a free $\mathbb{F}_q[G]$-module. We choose M of large enough rank so that M^* surjects onto $V^* := Hom_{\mathbb{F}_q[G]}(V, \mathbb{F}_q)$. Then, taking the dual of the surjection $M^* \to V^*$, we get an injection $V \to M$. Therefore, we get an embedding $V \to L$ of $\mathbb{F}_q[G]$-modules.

Similar to Example 7.2.1, set

$$\phi_T(x) = tx \prod_{0 \neq v \in V}{}' \left(1 - \frac{x}{v}\right), \qquad t = \gamma(T).$$

This defines a Drinfeld module of rank r over L. Since G permutes the elements of V, it fixes the coefficients of $\phi_T(x)$, so ϕ is actually defined over K. If an element $\sigma \in G$ fixes V, then $\sigma = 1$ since by construction G injects into $Aut_{\mathbb{F}_q}(V)$. Therefore, by Galois theory, the inclusion $K(V) \subseteq L$ is an equality. On the other hand, $K(V) = K(\phi[T])$, so $L = K(\phi[T])$. $\square$

Remarks 7.3.7

(1) The assumption that $char_A(K) \neq T$ is not restrictive. If $char_A(K) = T$, then simply define

$$\phi_{T+1}(x) = x \prod_{0 \neq v \in V}{}' \left(1 - \frac{x}{v}\right).$$

This uniquely determines ϕ and $L = K(\phi[T + 1])$.

(2) By Cayley's Theorem, every finite group of order r can be embedded as a subgroup of permutation matrices in $GL_r(\mathbb{F}_q)$. Then, Theorem 7.3.6 implies that every finite Galois extension of an infinite A-field arises as the splitting field of some $\phi_T(x)$. But this last statement is easy to prove directly. Indeed, suppose K is an A-field with $char_A(K) \neq T$ and L/K is the splitting field of a separable polynomial $f(x) \in K[x]$. Let $\alpha_1, \ldots, \alpha_n$ be the roots of $f(x)$

in K^{sep}. Put $V = \mathbb{F}_q \alpha_1 + \cdots + \mathbb{F}_q \alpha_n$ and define $\phi_T(x)$ as in the proof of Theorem 7.3.6.

(3) The proof of the theorem is short, but not easy. The theory of $K[G]$-modules properly belongs to the theory of linear representations of finite groups. This topic is covered in the last part of the textbook by Dummit and Foote, but not to the extent required in the proof. The reader might either take that part of the proof as a "black box" or consult Serre's book [Ser77], especially Exercises 14.5–14.7.

(4) In the case when $r = 1$, one can give a much more elementary proof of Theorem 7.3.6. Let L/K be a Galois extension with $\text{Gal}(L/K) \subset \mathbb{F}_q^\times$. Since $\mathbb{F}_q^\times \subset K$, this is a Kummer extension, i.e., $L = K(\alpha)$, where $\alpha^n = c \in K$ and $n \mid q - 1$. Raising both sides of this equation to $(q - 1)/n$-th power, we may assume that $n = q - 1$. Now define $\phi_T = Tx - \frac{T}{c}x^q$.

(5) Suppose we are given an explicit polynomial $f(x) \in F[x]$ whose Galois group is isomorphic to a subgroup of $\text{GL}_2(\mathbb{F}_q)$. The proof of Theorem 7.3.6 does not give an efficient method for finding an $\mathbb{F}_q$-linear polynomial $\phi_T(x) = Tx + ax^q + bx^{q^2}$ with the same splitting field as $f(x)$. Developing an efficient algorithm for finding $\phi_T(x)$ seems like an interesting question.

(6) There is a classical approach due to Felix Klein to "solving" nonsolvable quintic polynomials using elliptic functions. The idea of this approach is that, given a degree 5 polynomial $f(x) \in \mathbb{Q}[x]$ satisfying certain assumptions, one is able to find an explicit elliptic curve E over $\mathbb{Q}$ such that $\mathbb{Q}(E[5])$ is the splitting field K of $f(x)$; cf. [Ser80, DT02]. Then one can study K using the theory of elliptic curves; for example, one can characterize the primes of $\mathbb{Q}$ that split completely in K using the invariants of E. This leads to reciprocity laws for some non-abelian extensions on $\mathbb{Q}$. In the next section we will do something similar for the division fields of Drinfeld modules.

Exercises

7.3.1 One of the preliminary results in the proof of Theorem 7.3.1 is the following fact:

Theorem 7.3.8 *Let ϕ be a Drinfeld module of rank r over a finite extension K of F. Assume* $\text{End}(\phi) = A$. *Then there is a constant $N(\phi, K)$ depending only on ϕ and K such that for all primes* $\mathfrak{p} \in A$ *of degree* $\geq N(\phi, K)$ *we have $\rho_{\phi, \mathfrak{p}}(G_K) = \text{GL}_r(\mathbb{F}_\mathfrak{p})$.*

Show that the bound $N(\phi, K)$ in the previous theorem cannot be made uniform by showing that $\rho_{\phi, \mathfrak{p}}$ is not surjective for $\phi_T(x) = Tx + x^q - \mathfrak{p}x^{q^2}$ if $q \equiv 1 \pmod 4$ and $\deg(\mathfrak{p})$ is odd.

7.3.2 Let ϕ be the Drinfeld module defined by (7.3.3). Let $\mathfrak{p}$ be a prime not equal to T or $T + a$. Then $\mathfrak{p}$ is unramified in $F(\phi[T])$. Let $\mathfrak{P}$ be a prime in $F(\phi[T])$ over $\mathfrak{p}$. Show that $\rho_{\phi, T}(\mathrm{Frob}_{\mathfrak{P}})$ is conjugate to

$$\begin{pmatrix} a^{\deg(\mathfrak{p})} & * \\ 0 & (-a)^{\deg(\mathfrak{p})} \frac{\mathfrak{p}(0)}{\mathfrak{p}(-a)} \end{pmatrix}.$$

7.3.3 Determine the Galois group of $F(\phi[T])/F$ for the Drinfeld module ϕ defined by (7.3.3), assuming $a \in \mathbb{F}_q^{\times}$ is an arbitrary element, not necessarily a generator.

7.3.4 Let $a, b, \alpha, \beta \in \mathbb{F}_q$ be such that $a \neq b$ and $\alpha \neq 0$, $\beta \neq 0$. Show that there are infinitely many primes $\mathfrak{p} \lhd A$ such that $\mathfrak{p}(a) = \alpha$ and $\mathfrak{p}(b) = \beta$.

7.3.5

(a) Prove that any subgroup of $\mathrm{GL}_2(\mathbb{F}_3)$ is isomorphic to one of the following:

$$1, \quad \mathbb{Z}/2\mathbb{Z}, \quad \mathbb{Z}/3\mathbb{Z}, \quad \mathbb{Z}/4\mathbb{Z}, \quad \mathbb{Z}/6\mathbb{Z}, \quad \mathbb{Z}/8\mathbb{Z}, \quad \mathbb{Z}/2\mathbb{Z} \times \mathbb{Z}/2\mathbb{Z},$$

$$S_3, \quad D_8, \quad Q_8, \quad B \cong D_{12}, \quad \mathcal{N}', \quad \mathrm{SL}_2(\mathbb{F}_3), \quad \mathrm{GL}_2(\mathbb{F}_3),$$

where D_n denotes the dihedral group with n elements, Q_8 is the quaternion group, and $\mathcal{N}'$ is the normalizer of a non-split Cartan described in Sect. 7.3.7.
(b) For each group G in the previous list, construct an explicit Drinfeld module of rank 2 over $F = \mathbb{F}_3(T)$ with $\mathrm{Gal}(F(\phi[T])/F) \cong G$.

7.3.6 Explain why the assumption $\mathrm{End}(\phi) = A$ in Theorem 7.3.1 is necessary.

7.3.7 Assume $q = 2$. Let K be an A-field with $\mathrm{char}_A(K) = 0$. Let L/K be a Galois extension whose Galois group is isomorphic to a subgroup of $\mathrm{GL}_2(\mathbb{F}_q)$. Using only basic Galois theory (as discussed, for example, in [DF04, Ch. 14]), prove that $K = F(\phi[T])$ for some Drinfeld module of rank 2 over K.

7.4 Division Fields: A Reciprocity Theorem

In this section we will extend Theorem 7.1.19 to Drinfeld modules of arbitrary rank. More precisely, we will give congruence conditions modulo $\mathfrak{n}$ which characterize primes splitting completely in $F(\phi[\mathfrak{n}])$ for a given Drinfeld module ϕ over F.

As a first approach to this problem, one can try to use the same argument as in the proof of Theorem 7.1.19. Let ϕ be a Drinfeld module of rank r over F defined by

$$\phi_T(x) = Tx + g_1 x^q + \cdots + g_r x^{q^r}.$$

Let $\mathfrak{p}$ be a prime of good reduction of ϕ and let $\bar\phi$ denote the reduction of ϕ at $\mathfrak{p}$. By (7.1.5), we have $P_{\phi,\mathfrak{p},\mathfrak{n}}(x) \equiv P_{\bar\phi}(x) \pmod{\mathfrak{n}}$. The prime $\mathfrak{p}$ splits completely in $F(\phi[\mathfrak{n}])$ if and only if $\mathrm{Frob}_\mathfrak{p} = 1$, where $\mathrm{Frob}_\mathfrak{p}$ denotes the conjugacy class of the Frobenius at $\mathfrak{p}$ in $\mathrm{Gal}(F(\phi[\mathfrak{n}])/F)$. If $\mathrm{Frob}_\mathfrak{p} = 1$, then

$$P_{\phi,\mathfrak{p},\mathfrak{n}}(x) = (x-1)^r.$$

This leads to congruence conditions for the coefficients of $P_{\bar\phi}$ modulo $\mathfrak{n}$ which are necessary for $\mathfrak{p}$ to split completely in $F(\phi[\mathfrak{n}])$:

$$P_{\bar\phi}(x) \equiv (x-1)^r \pmod{\mathfrak{n}}. \tag{7.4.1}$$

One of these congruences arises from the constant terms of the polynomials above:

$$P_{\bar\phi}(0) = (-1)^{rd-r-d}\, \mathrm{Nr}_{\mathbb{F}_\mathfrak{p}/\mathbb{F}_q}(\bar g_r)^{-1}\mathfrak{p} \qquad (d = \deg(\mathfrak{p}))$$

$$\equiv (-1)^r \pmod{\mathfrak{n}}.$$

Thus:

Lemma 7.4.1 *Assume $\mathfrak{p}$ is a prime of good reduction for ϕ and $\mathfrak{p} \nmid \mathfrak{n}$. If $\mathfrak{p}$ splits completely in $F(\phi[\mathfrak{n}])$, then*

$$\mathfrak{p} \equiv (-1)^{(r-1)\deg(\mathfrak{p})}\, \mathrm{Nr}_{\mathbb{F}_\mathfrak{p}/\mathbb{F}_q}(\bar g_r) \pmod{\mathfrak{n}}.$$

When $r = 1$, the congruence of Lemma 7.4.1 is also sufficient for $\mathfrak{p}$ to split completely in $F(\phi[\mathfrak{n}])$ because $\mathrm{Frob}_\mathfrak{p}$ acts by scalar multiplication on $(A/\mathfrak{n})^\times$; this is essentially the statement of Theorem 7.1.19. On the other hand, when $r \geq 2$, the congruences that arise from (7.4.1) are not sufficient for $\mathrm{Frob}_\mathfrak{p}$ to act as the identity on $\phi[\mathfrak{n}]$. The problem is that $\mathrm{Frob}_\mathfrak{p}$ might not be semi-simple, i.e., the matrix by which $\mathrm{Frob}_\mathfrak{p}$ acts on $\phi[\mathfrak{n}]$ might be conjugate to a matrix in $\mathrm{GL}_r(A/\mathfrak{n})$ with 1's on the main diagonal without actually being the identity matrix. The next example demonstrates this problem.

Example 7.4.2 Let ϕ be the Drinfeld module of rank 2 over $\mathbb{F}_3(T)$ defined by $\phi_T(x) = Tx + x^q + x^{q^2}$. We know from Sect. 7.3.1 that

$$\mathrm{Gal}(F(\phi[T])/F) \cong \mathrm{GL}_2(\mathbb{F}_3).$$

This module has good reduction at all primes of A. Let $\mathfrak{p} = T - 1$ and let $\bar\phi$ be the reduction of ϕ at $\mathfrak{p}$. Then

$$\bar\phi_T(x) = x + x^3 + x^9 = x(x-1)(x+1)(x^3 - x + 1)(x^3 - x - 1),$$

where the last expression is the decomposition of $\bar{\phi}_T(x)$ into a product of irreducible polynomials over $\mathbb{F}_3$. Let α be a root of $x^3 - x + 1$. It is easy to check that

$$\bar{\phi}[T] = \mathbb{F}_3 \oplus \mathbb{F}_3\alpha,$$

and $\mathrm{Frob}_{\mathbb{F}_p}(\alpha) = \alpha^3 = \alpha - 1$. Thus, $\begin{pmatrix} 1 & -1 \\ 0 & 1 \end{pmatrix}$ represents the conjugacy class of Frob_p in $\mathrm{GL}_2(\mathbb{F}_3)$. This matrix is obviously not the identity matrix, so $\mathfrak{p}$ does not split completely in $\phi[T]$. In fact, since Frob_p has order 3, Proposition 2.8.8 implies that $\mathfrak{p}$ splits in $F(\phi[T])$ into a product of 16 primes each having residue degree 3. Finally, using the algorithm described in Sect. 4.2, it is easy to compute that

$$P_{\bar{\phi}}(x) = x^2 + x - \mathfrak{p},$$

which is congruent to $(x - 1)^2$ modulo T. This agrees with (7.4.1).

Thus, besides $P_{\bar{\phi}}(x)$, we need some other invariants associated with ϕ and $\mathfrak{p}$ to distinguish when Frob_p acts as identity on $\phi[\mathfrak{n}]$. It turns out that such an invariant is encoded in the index of the order $A[\pi_p]$ generated by the Frobenius endomorphism π_p of $\bar{\phi}$ in the endomorphism ring $\mathcal{E}_p := \mathrm{End}_{\mathbb{F}_p}(\bar{\phi})$. The starting point is the following fact, which is essentially Exercise 4.2.3.

Proposition 7.4.3 *We have:*

(1) $P_{\bar{\phi}}(x)$ is the minimal polynomial of π_p over A.
(2) $A[\pi_p]$ and $\mathcal{E}_p$ are A-orders in a totally imaginary field extension of F of degree r.

Proof (1) Let $m(x)$ be the minimal polynomial of π_p over A. By Theorem 4.2.2, $P_{\bar{\phi}}(x) = m(x)^s$, where $s = r/[F(\pi_p) : F]$. On the other hand, by Theorem 4.2.7, we have $(P_{\bar{\phi}}(0)) = \mathfrak{p}$. Thus, if $c = m(0) \in A$ is the constant term of $m(x)$, then c^s is equal to the irreducible polynomial $\mathfrak{p}$, up to an $\mathbb{F}_q^\times$ multiple. This implies that $s = 1$, or equivalently $r = [F(\pi_p) : F]$.

(2) Let $K = F(\pi_p)$. By Definition 3.4.22, an A-order in K is an A-subalgebra of K which is finitely generated as an A-module and contains an F-basis of K. Since π_p is integral over A, it is clear that $A[\pi_p]$ is an A-order in K. On the other hand, by Theorem 4.1.3, $F(\pi_p)/F$ is totally imaginary and $\mathcal{E}_p \otimes_A F$ is a central division algebra over $F(\pi_p)$ of dimension $s^2 = 1$. Thus, $\mathcal{E}_p \otimes_A F = F(\pi_p)$, which implies that $\mathcal{E}_p$ is an A-order in $F(\pi_p)$; cf. Remark 3.4.23. $\square$

The previous proposition implies that $A[\pi_p]$ and $\mathcal{E}_p$ are free A-modules of the same rank r. Obviously $A[\pi_p] \subseteq \mathcal{E}_p$ and both contain 1. By Proposition 1.2.3, we can choose an A-basis $\{e_0, e_1, \ldots, e_{r-1}\}$ of $\mathcal{E}_p$ such that $\{a_0 e_0, a_1 e_1, \ldots, a_{r-1} e_{r-1}\}$ is a basis of $A[\pi_p]$ for some monic polynomials $a_0, \ldots, a_{r-1}$ satisfying the divisibility conditions $a_0 \mid a_1 \mid \cdots \mid a_{r-1}$.

Lemma 7.4.4 *With previous notation, $a_0 = 1$ and we can choose $e_0 = 1$.*

Proof This can be proved by adapting the proof of Proposition 1.2.3. We leave the details as an exercise.   □

Lemma 7.4.4 implies that the number of invariant factors of $\mathcal{E}_\mathfrak{p}/A[\pi_\mathfrak{p}]$ of positive degree is at most $r - 1$. Thus,

$$\mathcal{E}_\mathfrak{p}/A[\pi_\mathfrak{p}] \cong A/(b_{\mathfrak{p},1}) \times A/(b_{\mathfrak{p},2}) \times \cdots \times A/(b_{\mathfrak{p},r-1}),$$

for uniquely determined monic polynomials $b_{\mathfrak{p},1}, \ldots, b_{\mathfrak{p},r-1} \in A$ such that

$$b_{\mathfrak{p},1} \mid b_{\mathfrak{p},2} \mid \cdots \mid b_{\mathfrak{p},r-1}.$$

Let

$$P_{\bar{\phi}}(x) = x^r + a_{\mathfrak{p},r-1}x^{r-1} + \cdots + a_{\mathfrak{p},0}.$$

The following theorem was proved by Garai and the author [GP20].

Theorem 7.4.5 *Assume r is coprime to the characteristic of F. Let $\mathfrak{p}$ be a prime of good reduction for ϕ such that $\mathfrak{p} \nmid \mathfrak{n}$. Then $\mathfrak{p}$ splits completely in $F(\phi[\mathfrak{n}])$ if and only if*

$$a_{\mathfrak{p},r-1} \equiv -r \pmod{\mathfrak{n}} \quad and \quad b_{\mathfrak{p},1} \equiv 0 \pmod{\mathfrak{n}}.$$

Proof As we have explained (cf. Remark 7.1.18), $\mathfrak{p}$ splits completely in $F(\phi[\mathfrak{n}])$ if and only if the conjugacy class of $\mathrm{Frob}_\mathfrak{p}$ in $\mathrm{Gal}(F(\phi[\mathfrak{n}])/F)$ is the identity. On the other hand, this latter condition is equivalent to $\mathrm{Frob}_{\mathbb{F}_\mathfrak{p}}$ acting trivially on $\bar{\phi}[\mathfrak{n}]$. Now observe that the action of $\mathrm{Frob}_{\mathbb{F}_\mathfrak{p}}$ on $\bar{\phi}[\mathfrak{n}]$ coincides with the action of the Frobenius endomorphism $\pi_\mathfrak{p}$. Thus, we need to show that the given congruence conditions are equivalent to $\pi_\mathfrak{p}$ acting as 1 on $\bar{\phi}[\mathfrak{n}]$.

First, we prove that $w \in \mathcal{E}_\mathfrak{p}$ acts as 0 on $\bar{\phi}[\mathfrak{n}]$ if and only if $w \in \mathfrak{n}\mathcal{E}_\mathfrak{p}$. If $w \in \mathfrak{n}\mathcal{E}_\mathfrak{p}$, then $w = v\mathfrak{n}$, i.e., $w = v\bar{\phi}_\mathfrak{n}$ for some $v \in \mathcal{E}_\mathfrak{p}$. Obviously $w = v\bar{\phi}_\mathfrak{n}$ acts as 0 on $\bar{\phi}[\mathfrak{n}]$. Conversely, suppose w acts as 0 on $\bar{\phi}[\mathfrak{n}]$. Because $\mathfrak{p} \nmid \mathfrak{n}$, the polynomial $\bar{\phi}_\mathfrak{n}(x)$ is separable. We are assuming that every root of $\bar{\phi}_\mathfrak{n}(x)$ is a root of the $\mathbb{F}_q$-linear polynomial $w(x)$ corresponding to w. By Corollary 3.1.16, $w = v\bar{\phi}_\mathfrak{n}$ for some $v \in \mathbb{F}_\mathfrak{p}\{\tau\}$. We need to show that $v \in \mathcal{E}_\mathfrak{p}$, i.e., v commutes with $\bar{\phi}_T$. Now

$$w\bar{\phi}_T = v\bar{\phi}_\mathfrak{n}\bar{\phi}_T = v\bar{\phi}_T\bar{\phi}_\mathfrak{n}$$

and

$$w\bar{\phi}_T = \bar{\phi}_T w = \bar{\phi}_T v\bar{\phi}_\mathfrak{n}.$$

Thus, $(v\bar{\phi}_T - \bar{\phi}_T v)\bar{\phi}_\mathfrak{n} = 0$. Since $\mathbb{F}_\mathfrak{p}\{\tau\}$ has no zero-divisors and $\bar{\phi}_\mathfrak{n} \neq 0$, we must have $v\bar{\phi}_T = \bar{\phi}_T v$, so $v \in \mathcal{E}_\mathfrak{p}$.

Next, we claim that $\pi_\mathfrak{p}$ acts as a scalar on $\bar{\phi}[\mathfrak{n}]$ if and only if $\mathfrak{n}$ divides $b_{\mathfrak{p},1}$. To see this, choose an A-basis $\{1, e_1, \ldots, e_{r-1}\}$ of $\mathcal{E}_\mathfrak{p}$ such that $\{1, b_{\mathfrak{p},1}e_1, \ldots, b_{\mathfrak{p},r-1}e_{r-1}\}$ is an A-basis of $A[\pi_\mathfrak{p}]$. We can write

$$\pi_\mathfrak{p} = c + a_1 b_{\mathfrak{p},1} e_1 + \cdots + a_{r-1} b_{\mathfrak{p},r-1} e_{r-1}, \qquad \text{for some} \quad c, a_1, \ldots, a_{r-1} \in A.$$

Suppose $\mathfrak{n} \mid b_{\mathfrak{p},1}$. Since $b_{\mathfrak{p},1}$ divides all $b_{\mathfrak{p},i}$, $i \geq 2$, this implies that

$$\pi_\mathfrak{p} = c + \mathfrak{n}w \qquad \text{for some } w \in \mathcal{E}_\mathfrak{p}.$$

Thus, $\pi_\mathfrak{p}$ acts by multiplication by c on $\bar{\phi}[\mathfrak{n}]$. Conversely, suppose $\pi_\mathfrak{p} - c$ annihilates $\bar{\phi}[\mathfrak{n}]$. By the previous paragraph, this implies that $\pi_\mathfrak{p} - c \in \mathfrak{n}\mathcal{E}_\mathfrak{p}$. Thus, $\pi_\mathfrak{p} \in A + \mathfrak{n}\mathcal{E}_\mathfrak{p}$, and therefore $A[\pi_\mathfrak{p}] \subseteq A + \mathfrak{n}\mathcal{E}_\mathfrak{p}$. This inclusion implies that there is a natural surjective quotient homomorphism of A-modules

$$\mathcal{E}_\mathfrak{p}/A[\pi_\mathfrak{p}] \longrightarrow \mathcal{E}_\mathfrak{p}/(A + \mathfrak{n}\mathcal{E}_\mathfrak{p}).$$

Since $\{1, \mathfrak{n}e_1, \ldots, \mathfrak{n}e_{r-1}\}$ is an A-basis of $A + \mathfrak{n}\mathcal{E}_\mathfrak{p}$, we have $\mathcal{E}_\mathfrak{p}/(A + \mathfrak{n}\mathcal{E}_\mathfrak{p}) \cong (A/\mathfrak{n})^{r-1}$. Therefore, there is a surjective quotient homomorphism of A-modules

$$f: A/(b_{\mathfrak{p},1}) \times \cdots \times A/(b_{\mathfrak{p},r-1}) \longrightarrow A/\mathfrak{n} \times \cdots \times A/\mathfrak{n}.$$

To show that this implies $\mathfrak{n} \mid b_{\mathfrak{p},1}$, we can restrict to the primary components of the modules above. Thus, we assume that $\mathfrak{n} = \mathfrak{l}^s$ and $b_{\mathfrak{p},i} = \mathfrak{l}^{s_i}$, $1 \leq i \leq r - 1$, are powers of a prime $\mathfrak{l}$, where $s_1 \leq \cdots \leq s_{r-1}$. We need to show that $s \leq s_1$. For any $m > 0$, the image of $A/\mathfrak{l}^m$ under any A-module homomorphism is isomorphic to $A/\mathfrak{l}^{m'}$ for some $0 \leq m' \leq m$ since the image is cyclic and annihilated by $\mathfrak{l}^m$. Therefore,

$$f(A/\mathfrak{l}^{s_1} \times \cdots \times A/\mathfrak{l}^{s_{r-1}}) \cong A/\mathfrak{l}^{s'_1} \times \cdots \times A/\mathfrak{l}^{s'_{r-1}}.$$

The elementary divisors of this module are $\mathfrak{l}^{s'_1}, \ldots, \mathfrak{l}^{s'_{r-1}}$, and these uniquely determine the module, so we must have $s'_1 = \cdots = s'_{r-1} = s$. Therefore, $s = s'_1 \leq s_1$, as was required to be shown.

Finally, to prove the theorem, note that $\pi_\mathfrak{p}$ acts as the identity on $\bar{\phi}[\mathfrak{n}]$ if and only if

* $\pi_\mathfrak{p}$ acts as a scalar c on $\bar{\phi}[\mathfrak{n}]$, and
* $c \equiv 1 \pmod{\mathfrak{n}}$.

If $\pi_\mathfrak{p}$ acts as a scalar, then its characteristic polynomial satisfies $P_{\bar\phi}(x) \equiv (x - c)^r \pmod{\mathfrak{n}}$. We have

$$(x - c)^r = x^r - rcx^{r-1} + \cdots .$$

Since r is assumed to be coprime to the (usual) characteristic of F, we see that $c \equiv 1 \pmod{\mathfrak{n}}$ if and only if $rc \equiv r \pmod{\mathfrak{n}}$. Hence, $c \equiv 1 \pmod{\mathfrak{n}}$ if and only if $a_{\mathfrak{p},r-1} \equiv -rc \equiv -r \pmod{\mathfrak{n}}$. $\square$

Example 7.4.6 For the Drinfeld module considered in Example 7.4.2 and $\mathfrak{p} = T - 1$ one can show that $b_{\mathfrak{p},1} = 1$. (How to compute $b_{\mathfrak{p},i}$'s will be discussed later in this section.) Thus, because $b_{\mathfrak{p},1} = 1$ is not congruent to 0 modulo T, Theorem 7.4.5 implies that $T - 1$ does not split completely in $F(\phi[T])$. This was established in Example 7.4.2 by a direct calculation. A prime of smallest degree that splits completely in $F(\phi[T])$ is $\mathfrak{p} = T^5 - T^4 - T^2 - 1$. For this prime we have

$$P_{\bar\phi}(x) = x^2 - (T^2 - 1)x - \mathfrak{p} \quad \text{and} \quad b_{\mathfrak{p},1} = T.$$

Thus, $a_{\mathfrak{p},r-1} = -(T^2 - 1) \equiv -2 \pmod{T}$ and $b_{\mathfrak{p},1} \equiv 0 \pmod{T}$.

Remarks 7.4.7

(1) If $\mathfrak{n}$ itself is prime, then in Theorem 7.4.5 one can dispose with the assumption that r is coprime to $p = \operatorname{char}(F)$ as follows: Decompose $r = p^s r'$, $s \geq 0$ with $p \nmid r'$. Then $\mathfrak{p} \nmid \mathfrak{n}$ splits completely in $F(\phi[\mathfrak{n}])$ if and only if

$$a_{\mathfrak{p},p^s} \equiv -r' \pmod{\mathfrak{n}} \quad \text{and} \quad b_{\mathfrak{p},1} \equiv 0 \pmod{\mathfrak{n}}.$$

The proof is essentially the same except at the end we have

$$(x - c)^r = (x^{p^s} - c^{p^s})^{r'} = x^r - r'c^{p^s} x^{p^s(r'-1)} + \cdots .$$

If $\mathfrak{n}$ is prime, then the p-th power map is an automorphism of $(A/\mathfrak{n})^\times$. Thus, $c \equiv 1 \pmod{\mathfrak{n}}$ if and only if $c^{p^s} \equiv 1 \pmod{\mathfrak{n}}$; thus, $c \equiv 1 \pmod{\mathfrak{n}}$ if and only if $a_{\mathfrak{p},p^s} \equiv -r'c^{p^s} \equiv -r' \pmod{\mathfrak{n}}$.

(2) The initial motivation for Theorem 7.4.5 came from a result of Duke and Tóth [DT02] for elliptic curves, although the idea of the proof presented above is different from the ideas in [DT02]. Also, this theorem for Drinfeld modules of rank $r \geq 3$ has no proper analogue for elliptic curves or abelian varieties.

(3) One can show that there is a stronger divisibility property,

$$b_{\mathfrak{p},i} \cdot b_{\mathfrak{p},j} \mid b_{\mathfrak{p},i+j} \quad \text{for all} \quad i + j < r,$$

for the invariant factors of $\mathcal{E}_\mathfrak{p}/A[\pi_\mathfrak{p}]$. In particular, this implies that $b_{\mathfrak{p},1}^i \mid b_{\mathfrak{p},i}$ for all $1 \leq i \leq r - 1$. The proof uses the existence of a special type A-basis of $\mathcal{E}_\mathfrak{p}$; see [GP22, Prop. 2.3].

Corollary 7.4.8 *Let ϕ be a Drinfeld A-module over F of rank r. Given $\mathfrak{n} \in A$, there are infinitely many primes $\mathfrak{p} \lhd A$ such that $\mathfrak{n}$ divides $\mathfrak{b}_{\mathfrak{p},1}$. In particular, for infinitely many primes the inclusion $A[\pi_{\mathfrak{p}}] \subset \mathcal{E}_{\mathfrak{p}}$ is strict.*

Proof From the proof of Theorem 7.4.5, we see that for any prime $\mathfrak{p}$ for which $\pi_{\mathfrak{p}}$ acts as a scalar on $\bar{\phi}[\mathfrak{n}]$ we have $\mathfrak{n} \mid \mathfrak{b}_{\mathfrak{p},1}$ (this part does not use the assumption that r is coprime to the characteristic of F). The primes that split completely in $F(\phi[\mathfrak{n}])$ have this property, and there are infinitely many such primes by the Chebotarev Density Theorem (see Theorem 7.3.5). $\square$

Theorem 7.4.5 is a "reciprocity law" for the division fields $F(\phi[\mathfrak{n}])$ of our Drinfeld module in the sense that it provides necessary and sufficient conditions, expressed as congruences modulo $\mathfrak{n}$, for a given prime $\mathfrak{p}$ to split completely in $F(\phi[\mathfrak{n}])$. On the other hand, when these congruence conditions are not satisfied, one cannot deduce from the theorem how $\mathfrak{p}$ splits in $F(\phi[\mathfrak{n}])$. To have an analogue of Theorem 7.1.12 for $r \geq 2$, we need to know the order of elements in the conjugacy class $\text{Frob}_{\mathfrak{p}}$. The next theorem, proved by Cojocaru and the author [CP15], accomplishes this task for $r = 2$ (this is an analogue of the result of Duke and Tóth mentioned in Remark 7.4.7).

Theorem 7.4.9 *Assume q is odd and $r = 2$. Let ϕ be a Drinfeld module over F with good reduction at $\mathfrak{p}$. Denote by*

$$\Delta_{\mathfrak{p}} = a_{\mathfrak{p},1}^2 - 4a_{\mathfrak{p},0}$$

the discriminant of $P_{\bar{\phi}}(x) = x^2 + a_{\mathfrak{p},1}x + a_{\mathfrak{p},0}$. The matrix

$$\mathcal{F}_{\mathfrak{p}} := \begin{pmatrix} -\frac{a_{\mathfrak{p},1}}{2} & \frac{\Delta_{\mathfrak{p}}}{2\mathfrak{b}_{\mathfrak{p},1}} \\ \frac{\mathfrak{b}_{\mathfrak{p},1}}{2} & -\frac{a_{\mathfrak{p},1}}{2} \end{pmatrix}$$

has entries in A and, for any $\mathfrak{n} \lhd A$ not divisible by $\mathfrak{p}$, the reduction of $\mathcal{F}_{\mathfrak{p}}$ modulo $\mathfrak{n}$ represents in $\text{GL}_2(A/\mathfrak{n})$ the conjugacy class $\text{Frob}_{\mathfrak{p}} \subset \text{Gal}(F(\phi[\mathfrak{n}])/F) \subseteq \text{GL}_2(A/\mathfrak{n})$.

Remarks 7.4.10

(1) We will show that $\Delta_{\mathfrak{p}}$ is divisible by $\mathfrak{b}_{\mathfrak{p},1}^2$. Therefore, if $a_{\mathfrak{p},1} \equiv -2 \pmod{\mathfrak{n}}$ and $\mathfrak{b}_{\mathfrak{p},1} \equiv 0 \pmod{\mathfrak{n}}$, then $\mathcal{F}_{\mathfrak{p}} \equiv 1 \pmod{\mathfrak{n}}$. This means that Theorem 7.4.9 implies Theorem 7.4.5 when $r = 2$.

(2) The matrix $\mathcal{F}_{\mathfrak{p}} \in \text{Mat}_2(A)$ is a "universal" matrix of the Frobenius automorphism at $\mathfrak{p}$ in the division fields of ϕ in the sense that to get a matrix in the conjugacy class of the Frobenius in the Galois groups of different division fields $F(\phi[\mathfrak{n}])$ we just need to reduce $\mathcal{F}_{\mathfrak{p}}$ modulo the corresponding $\mathfrak{n}$.

Example 7.4.11   Let us reexamine Example 7.4.2 using Theorem 7.4.9. We have $q = 3$ and $\mathfrak{p} = T - 1$. As was already mentioned, $P_{\bar{\phi}}(x) = x^2 + x - \mathfrak{p}$, so

$$\Delta_{\mathfrak{p}} = 1 + \mathfrak{p} = T.$$

Moreover, $b_{\mathfrak{p},1} = 1$. Thus,

$$\mathcal{F}_{\mathfrak{p}} = \begin{pmatrix} 1 & -T \\ -1 & 1 \end{pmatrix}.$$

This matrix modulo T is $\begin{pmatrix} 1 & 0 \\ -1 & 1 \end{pmatrix}$. According to Theorem 7.4.9 this latter matrix is in the conjugacy class $\mathrm{Frob}_{\mathfrak{p}} \subset \mathrm{Gal}(F(\phi[T])/F)$. Note that $\begin{pmatrix} 1 & 0 \\ -1 & 1 \end{pmatrix}$ is conjugate to $\begin{pmatrix} 1 & -1 \\ 0 & 1 \end{pmatrix} \in \mathrm{GL}_2(\mathbb{F}_3)$, which we computed directly in Example 7.4.2 to be in the conjugacy class $\mathrm{Frob}_{\mathfrak{p}}$.

Now, as in Example 7.4.6, let $\mathfrak{p} = T^5 - T^4 - T^2 - 1$. For this prime we have $P_{\bar{\phi}}(x) = x^2 - (T^2 - 1)x - \mathfrak{p}$ and $b_{\mathfrak{p},1} = T$. Hence, $\Delta_{\mathfrak{p}} = (T^2 - 1)^2 + \mathfrak{p} = T^5$ and

$$\mathcal{F}_{\mathfrak{p}} := \begin{pmatrix} -(T^2 - 1) & -T^4 \\ -T & -(T^2 - 1) \end{pmatrix}.$$

Since $\mathcal{F}_{\mathfrak{p}} \pmod{T} = 1$, the prime $\mathfrak{p}$ splits completely in $F(\phi[T])/F$.

For the Proof of Theorem 7.4.9 we need two preliminary lemmas about $A[\pi_{\mathfrak{p}}]$ and $\mathcal{E}_{\mathfrak{p}}$.

Lemma 7.4.12   *Assume q is odd and $r = 2$.*

(1) One can uniquely decompose

$$\Delta_{\mathfrak{p}} = c_\pi^2 \cdot \delta_{\mathfrak{p}},$$

so that c_π is monic and $\delta_{\mathfrak{p}}$ is square-free, i.e., $\delta_{\mathfrak{p}}$ is not divisible by a square of any prime.

(2) Let B be the integral closure of A in $F(\pi_{\mathfrak{p}})$. Then

$$B = A[\sqrt{\delta_{\mathfrak{p}}}] = A + A\sqrt{\delta_{\mathfrak{p}}},$$

where $\sqrt{\delta_{\mathfrak{p}}}$ denotes a fixed root of $x^2 - \delta_{\mathfrak{p}}$.

(3) If $O \subseteq B$ is an A-order, then

$$O = A + Ac\sqrt{\delta_{\mathfrak{p}}}, \quad \text{where} \quad B/O \cong A/cA.$$

Proof

(1) This is clear from the decomposition of Δ_p into irreducibles in A.

(2) By Proposition 7.4.3, $F(\pi_p)/F$ is a quadratic extension. Since π_p is integral over A, we have $A[\sqrt{\delta_p}] \subseteq B$. We need to prove the reverse inclusion. Since the characteristic of F is assumed to be odd, the quadratic formula implies that $K := F(\pi_p) = F(\sqrt{\delta_p})$. Now suppose $\alpha = a + b\sqrt{\delta_p} \in K$ is integral over F. Then $\mathrm{Tr}_{K/A}(\alpha) = 2a \in A$. Because 2 is a unit in A, we get $a \in A$. Similarly, $\mathrm{Nr}_{K/F}(\alpha) = a^2 + \delta_p b^2 \in A$, which implies that $\delta_p b^2 \in A$. Since δ_p is square-free, we must have $b \in A$. Thus, $\alpha \in A[\sqrt{\delta_p}]$.

(3) Suppose $O \subseteq B$ is an A-order. Choose $\{1, \sqrt{\delta_p}\}$ as an A-basis of B. Next, choose an A-basis of O of the form $\{1, e\}$. After expanding $e = a + c\sqrt{\delta_p}$ we see that $\{1, c\sqrt{\delta_p}\}$ is also an A-basis of O since $e = a \cdot 1 + c\sqrt{\delta_p}$. With these choices of bases of O and B, it is clear that $B/O \cong A/cA$.

$\square$

By Lemma 7.4.12, we have

$$A[\pi_p] = A + Ac_\pi\sqrt{\delta_p} \quad \text{and} \quad \mathcal{E}_p = A + Ac_\phi\sqrt{\delta_p}$$

for some $c_\pi, c_\phi \in A$. Moreover,

$$\mathcal{E}_p/A[\pi_p] \cong A/(c_\pi/c_\phi)A.$$

Hence

$$b_{p,1} = c_\pi/c_\phi.$$

Lemma 7.4.13 *Assume $r = 2$ and $p \nmid n$. Then $\bar{\phi}[n]$ is a free $\mathcal{E}_p/n\mathcal{E}_p$-module of rank 1.*

Proof Let $n = \mathfrak{l}_1^{s_1} \cdots \mathfrak{l}_m^{s_m}$ be the prime decomposition of n. Since $\bar{\phi}[n] \cong \bar{\phi}[\mathfrak{l}_1^{s_1}] \times \cdots \times \bar{\phi}[\mathfrak{l}_m^{s_m}]$, $\mathcal{E}_p/n\mathcal{E}_p \cong \mathcal{E}_p/\mathfrak{l}_1^{s_1}\mathcal{E}_p \times \cdots \times \mathcal{E}_p/\mathfrak{l}_m^{s_m}\mathcal{E}_p$, and the action of $\mathcal{E}_p$ on $\bar{\phi}[n]$ is compatible with these decompositions, it is enough to prove the claim assuming $n = \mathfrak{l}^s$ is a prime power.

Let $\{1, e\}$ be an A-basis of $\mathcal{E}_p$. It is a simple exercise in linear algebra to show that given a vector space V and a linear transformation S of V, either there is a vector $v \in V$ such that Sv is not a scalar multiple of v or $Sv = cv$ for all $v \in V$ and a fixed constant c. Applying this to the 2-dimensional vector space $\bar{\phi}[\mathfrak{l}]$ over $\mathbb{F}_\mathfrak{l}$, we conclude that either there is $v \in \bar{\phi}[\mathfrak{l}]$ such that $\{v, ev\}$ is an $\mathbb{F}_\mathfrak{l}$-basis of $\bar{\phi}[\mathfrak{l}]$ or $e - c$ acts as 0 on $\bar{\phi}[\mathfrak{l}]$ for some $c \in A$. In the second case, because $\{1, e - c\}$ is still an A-basis of $\mathcal{E}_p$, the map $\mathcal{E}_p \otimes_A A/\mathfrak{l} \to \mathrm{End}_{\mathbb{F}_\mathfrak{l}}(\bar{\phi}[\mathfrak{l}])$ is not injective, which contradicts Lemma 3.4.10. Hence, the second possibility does not occur.

Let $v' \in \bar{\phi}[\mathfrak{l}^s]$ be an element mapping to v under the reduction modulo $\mathfrak{l}$ map

$$\bar{\phi}[\mathfrak{l}^s] \longrightarrow \bar{\phi}[\mathfrak{l}^s] \otimes A/\mathfrak{l} = \bar{\phi}[\mathfrak{l}].$$

By Nakayama's lemma (see Exercise 2.4.5), v' and ev' generate $\bar{\phi}[\mathfrak{l}^s]$ over $A/\mathfrak{l}^s$. As we have observed in the proof of Theorem 7.4.5, an element $y \in \mathcal{E}_\mathfrak{p}$ annihilates $\bar{\phi}[\mathfrak{l}^s]$ if and only if $y \in \mathfrak{l}^s \mathcal{E}_\mathfrak{p}$. Therefore $(\mathcal{E}_\mathfrak{p}/\mathfrak{l}^s \mathcal{E}_\mathfrak{p})v' = \bar{\phi}[\mathfrak{l}^s]$, which implies that $\bar{\phi}[\mathfrak{l}^s]$ is free of rank 1 over $\mathcal{E}_\mathfrak{p}/\mathfrak{l}^s \mathcal{E}_\mathfrak{p}$.   □

Proof of Theorem 7.4.9 Because $\text{Tr}_{F(\sqrt{\delta_\mathfrak{p}})/F}(\pi_\mathfrak{p}) = -a_{\mathfrak{p},1}$ and $\text{Nr}_{F(\sqrt{\delta_\mathfrak{p}})/F}(\pi_\mathfrak{p}) = a_{\mathfrak{p},0}$, we have

$$\pi_\mathfrak{p} = \frac{-a_{\mathfrak{p},1}}{2} + \frac{c_\pi}{2}\sqrt{\delta_\mathfrak{p}} = \frac{-a_{\mathfrak{p},1}}{2} + \frac{b_{\mathfrak{p},1}}{2}c_\phi\sqrt{\delta_\mathfrak{p}}.$$

If we fix $\{1, c_\phi\sqrt{\delta_\mathfrak{p}}\}$ as an A-basis of $\mathcal{E}_\mathfrak{p}$, then a straightforward calculation shows that, with respect to this basis, the action of $\pi_\mathfrak{p}$ on $\mathcal{E}_\mathfrak{p}$ by multiplication is given by the matrix

$$\mathcal{F}_\mathfrak{p} = \begin{pmatrix} -a_{\mathfrak{p},1}/2 & \Delta_\mathfrak{p}/2b_{\mathfrak{p},1} \\ b_{\mathfrak{p},1}/2 & -a_{\mathfrak{p},1}/2 \end{pmatrix}.$$

On the other hand, $\bar{\phi}[\mathfrak{n}]$ is a free $\mathcal{E}_\mathfrak{p}/\mathfrak{n}\mathcal{E}_\mathfrak{p}$-module of rank 1. Therefore, the action of $\pi_\mathfrak{p}$ on $\bar{\phi}[\mathfrak{n}]$ is given by $\mathcal{F}_\mathfrak{p}$ modulo $\mathfrak{n}$. The claim of the theorem now follows from the observation that, up to conjugation, the action of $\pi_\mathfrak{p}$ on $\bar{\phi}[\mathfrak{n}]$ agrees with action of $\text{Frob}_\mathfrak{p}$ on $\phi[\mathfrak{n}]$, as in the Proof of Theorem 7.4.5.   □

Remark 7.4.14 Theorem 7.4.9 has been extended to arbitrary ranks in [GP22], under a mild technical assumption. In that paper, we describe an $r \times r$ matrix $\mathcal{F}_\mathfrak{p}$ with entries in A, whose reduction modulo $\mathfrak{n}$ represents the matrix by which $\text{Frob}_\mathfrak{p}$ acts on $\phi[\mathfrak{n}]$ with respect to a certain basis. The entries of the matrix $\mathcal{F}_\mathfrak{p}$ depend on $b_{\mathfrak{p},1}, \ldots, b_{\mathfrak{p},r-1}$ and the coefficients of certain polynomials $f_1(x), \ldots, f_{r-1}(x)$, which generalize the minimal polynomial of the Frobenius $\pi_\mathfrak{p}$.

Theorem 7.4.5, and its generalization in [GP22], can be used to give an algorithm for computing the invariants $b_{\mathfrak{p},1}, \ldots, b_{\mathfrak{p},r-1}$. We describe this algorithm in the simplest case when $r = 2$ and q is odd.

Computing $b_{\mathfrak{p},1}$

Start by computing $P_{\bar{\phi}}(x)$ using the algorithm described in Sect. 4.2. Then compute the discriminant $\Delta_\mathfrak{p}$ of $P_{\bar{\phi}}(x)$ and c_π. Because $b_{\mathfrak{p},1}$ divides c_π, we get a finite list of possible $b_{\mathfrak{p},1}$'s. Arrange this list according to the degrees of polynomials, from the highest to zero. Then for each b in this list check if there is $c \in A$ such that $\pi_\mathfrak{p} - c \in b\mathcal{E}_\mathfrak{p}$. Obviously we can assume that $\deg(c) < \deg(b)$, so there are only finitely many possibilities for c. Because we have arranged the list of possible $b_{\mathfrak{p},1}$'s by decreasing degrees, the first b that we find satisfying this condition is $b_{\mathfrak{p},1}$.

Given specific b and c, checking whether $\pi_\mathfrak{p} - c \in b\mathcal{E}_\mathfrak{p}$ can be done as follows. First, compute the residue of $\tau^{\deg(\mathfrak{p})} - \bar{\phi}_c$ modulo $\bar{\phi}_b$ using the right division algorithm in $\mathbb{F}_\mathfrak{p}\{\tau\}$. If the residue is nonzero, then $\pi_\mathfrak{p} - c \notin b\mathcal{E}_\mathfrak{p}$. If the residue is 0, then $\tau^{\deg(\mathfrak{p})} - \bar{\phi}_c = u\bar{\phi}_b$ for an explicit $u \in \mathbb{F}_\mathfrak{p}\{\tau\}$ produced by the division algorithm; see Theorem 3.1.13. Now check if the commutation relation $u\bar{\phi}_T = \bar{\phi}_T u$ holds in $\mathbb{F}_p\{\tau\}$; this relation holds if and only if $u \in \mathcal{E}_p$.

Finally, having computed $b_{p,1}$ and c such that $\pi_p - c \in b_{p,1}\mathcal{E}_p$, it is not hard to show that

$$1 \quad \text{and} \quad (\pi_p - c)/b_{p,1} \tag{7.4.2}$$

form an A-basis of $\mathcal{E}_p$; see Exercise 7.4.5. Hence, this algorithm also computes an A-basis of the endomorphism ring $\mathcal{E}_p$.

Example 7.4.15 As we mentioned, the algorithm above can be generalized to work for any $r \geq 2$. The following example is obtained from this generalized algorithm coded in Magma.

Let $q = 5, r = 3$, and

$$\phi_T = T + T\tau + T\tau^2 + \tau^3.$$

Let $p = T^6 + 3T^5 + T^2 + 3T + 3$. In this case,

$$P_{\bar{\phi}}(x) = x^3 + 2T^2x^2 + (3T^4 + T^2 + 3T + 1)x - p,$$

from which one computes that

$$\text{disc}(A[\pi_p]) = (T - 1)^6(T^4 + 2T^3 + 4T^2 + 3T + 4),$$

where the discriminant of an order is defined as in Exercise 7.4.2. Let B be the integral closure of A in $F(\pi_p)$. Then

$$\text{disc}(B) = (T^4 + 2T^3 + 4T^2 + 3T + 4).$$

This implies that (see Exercise 7.4.2)

$$\chi(B/\mathcal{E}_p) \cdot \chi(\mathcal{E}_p/A[\pi]) = \chi(B/A[\pi]) = (T - 1)^3.$$

We conclude that either $b_{p,1} = T - 1$ and $b_{p,2} = (T - 1)^2$ (see Remark 7.4.7 (3)) or $b_{p,1} = 1$ and $b_{p,2} = (T - 1)^n$ for some $0 \leq n \leq 3$. The algorithm confirms that in fact $b_{p,1} = T - 1$ and $b_{p,2} = (T - 1)^2$. In particular, $\mathcal{E}_p = B$. Moreover, an A-basis of $\mathcal{E}_p$ is given by

$$1, \quad e := \frac{\pi_p - 1}{T - 1}, \quad e^2.$$

The element in $\mathbb{F}_p\{\tau\}$ corresponding to e is

$$e = \tau^3 + (2t^5 + 3t^4 + t + 1)\tau^2 + (4t^3 + 2t + 3)\tau + t^5 + 4t^4 + 4t^3 + 4t^2 + 3,$$

where t denotes the image of T under the reduction map $A \to A/\mathfrak{p} = \mathbb{F}_{\mathfrak{p}}$.

Finally, one computes that

$$
\mathcal{F}_{\mathfrak{p}} = \begin{pmatrix} 1 & 0 & T^4 + T^2 + 2T + 1 \\ T - 1 & 1 & 2T^3 + 2T^2 + 2T + 4 \\ 0 & T - 1 & 3(T^2 + 1) \end{pmatrix}.
$$

Note that this matrix is congruent to 1 modulo $\mathfrak{n} \lhd A$, if and only if $\mathfrak{n} = T - 1$, which is the reflection of the fact that $\mathfrak{p}$ splits completely in $F(\phi[\mathfrak{n}])$ if and only if $\mathfrak{n} = T - 1$.

Exercises

7.4.1 Prove Lemma 7.4.4.

7.4.2 Let K/F be a field extension of degree n. Let O be an A-order in K. Let $\{e_1, \ldots, e_n\}$ be an A-basis of O. Define the *discriminant of O* to be the ideal of A generated by $\mathrm{disc}_{K/F}(e_1, \ldots, e_n)$:

$$
\mathrm{disc}(O) = (\mathrm{disc}_{K/F}(e_1, \ldots, e_n)).
$$

(a) Show that $\mathrm{disc}(O)$ does not depend on the choice of the basis $\{e_1, \ldots, e_n\}$. Moreover, $\mathrm{disc}(O) = 0$ if and only if K/F is not separable.

(b) Let $O \subseteq O'$ be an inclusion of orders. Show that

$$
\mathrm{disc}(O) = \chi(O'/O)^2 \cdot \mathrm{disc}(O').
$$

7.4.3 Let ϕ be a Drinfeld module of rank $r \geq 2$ over F and $\mathfrak{p}$ be a prime of good reduction for ϕ. Prove the following statements:

(a) $\left| \mathrm{disc}(A[\pi_{\mathfrak{p}}]) \right| \leq |\mathfrak{p}|^{r-1}$.

(b) $b_{\mathfrak{p},1}^{2(r-1)}$ divides $\mathrm{disc}(A[\pi_{\mathfrak{p}}])$. (In fact, using Remark 7.4.7 (3), one can show that $b_{\mathfrak{p},1}^{r(r-1)}$ divides $\mathrm{disc}(A[\pi_{\mathfrak{p}}])$, but for this exercise one does not need that.)

(c) If r is coprime to the characteristic of F, then $\mathfrak{p}$ does not divide $b_{\mathfrak{p},1}$.

7.4.4 Let ϕ be a Drinfeld module over F of rank $r \geq 2$. Assume r is coprime to the characteristic of F. Let $\mathfrak{p}$ be a prime of good reduction for ϕ. Let

$$
\bar{\phi}\mathbb{F}_{\mathfrak{p}} \cong A/(d_{\mathfrak{p},1}) \times \cdots \times A/(d_{\mathfrak{p},r}), \qquad d_{\mathfrak{p},1} \mid d_{\mathfrak{p},2} \mid \cdots \mid d_{\mathfrak{p},r},
$$

i.e., $d_{\mathfrak{p},1}, \ldots, d_{\mathfrak{p},r}$ are the invariant factors of the A-module $\bar{\phi}\mathbb{F}_{\mathfrak{p}}$. Prove that

$$
d_{\mathfrak{p},1} = \gcd(b_{\mathfrak{p},1}, a_{\mathfrak{p},r-1} + r).
$$

7.4.5 Let $e := (\pi_p - c)/b_{p,1}$ be as in (7.4.2). Let $O = A + Ae \subseteq \mathcal{E}_p$. Show that $\text{disc}(O) = \text{disc}(\mathcal{E}_p)$. Conclude that $O = \mathcal{E}_p$, so $\{1, e\}$ is an A-basis of $\mathcal{E}_p$.

7.5 Complex Multiplication

In this section we study Drinfeld modules with complex multiplication and their division fields. One of the key constructions used to study Drinfeld modules with complex multiplication works in a more general setting, so we start with that construction for arbitrary Drinfeld modules.

Let L be an A-field with a structure morphism $\gamma \colon A \to L$. Let $\phi \colon A \to L\{\tau\}$ be a Drinfeld module of rank r. Let K be a finite extension of F and let $O \subset K$ be an A-order. Assume that there is an embedding $O \hookrightarrow \text{End}_L(\phi) \subset L\{\tau\}$ whose restriction to A is ϕ. We denote this embedding $O \hookrightarrow L\{\tau\}$ also by ϕ to distinguish it from other possible embeddings of O into $L\{\tau\}$.

Let $\mathfrak{A}$ be a nonzero ideal of O. Denote by $I_{\phi,\mathfrak{A}}$ the left ideal of $L\{\tau\}$ generated by all ϕ_α, $\alpha \in \mathfrak{A}$. Since $L\{\tau\}$ has a right division algorithm, $I_{\phi,\mathfrak{A}}$ is a principal ideal (a nonzero element of $I_{\phi,\mathfrak{A}}$ with minimal $\deg_\tau$ will generate this ideal); cf. Corollary 3.1.15. Denote by $\phi_{\mathfrak{A}}$ the *monic* generator of $I_{\phi,\mathfrak{A}}$. We can expand

$$\phi_{\mathfrak{A}} = f_1 \phi_{\alpha_1} + \cdots + f_n \phi_{\alpha_n}, \tag{7.5.1}$$

$$\text{for suitable } f_1, \ldots, f_n \in L\{\tau\} \text{ and } \alpha_1, \ldots, \alpha_n \in \mathfrak{A}.$$

Then, for $\alpha \in O$, we have

$$\phi_{\mathfrak{A}}\phi_\alpha = f_1 \phi_{\alpha_1}\phi_\alpha + \cdots + f_n \phi_{\alpha_n}\phi_\alpha$$

$$= (f_1\phi_\alpha)\phi_{\alpha_1} + \cdots + (f_n\phi_\alpha)\phi_{\alpha_n} \in I_{\phi,\mathfrak{A}}.$$

Therefore, for every $\alpha \in O$, there is a unique $\psi_\alpha \in L\{\tau\}$ such that

$$\phi_{\mathfrak{A}}\phi_\alpha = \psi_\alpha \phi_{\mathfrak{A}}. \tag{7.5.2}$$

Lemma 7.5.1 *With previous notation, the map* $\alpha \mapsto \psi_\alpha$ *gives an embedding* $\psi \colon O \to L\{\tau\}$. *The restriction of* ψ *to* A *is a Drinfeld module of rank* r, *and* $\phi_{\mathfrak{A}}$ *is an isogeny from* ϕ *to* ψ.

Proof From (7.5.2) it is clear that $\deg_\tau(\phi_\alpha) = \deg_\tau(\psi_\alpha)$ for all $\alpha \in O$. Next,

$$\psi_{\alpha+\beta}\phi_{\mathfrak{A}} = \phi_{\mathfrak{A}}\phi_{\alpha+\beta} = \phi_{\mathfrak{A}}(\phi_\alpha + \phi_\beta) = \phi_{\mathfrak{A}}\phi_\alpha + \phi_{\mathfrak{A}}\phi_\beta$$

$$= \psi_\alpha\phi_{\mathfrak{A}} + \psi_\beta\phi_{\mathfrak{A}} = (\psi_\alpha + \psi_\beta)\phi_{\mathfrak{A}}.$$

Hence, $\psi_{\alpha+\beta} = \psi_\alpha + \psi_\beta$. One similarly verifies that $\psi_{\alpha\beta} = \psi_\alpha \psi_\beta$. Therefore, ψ is an injective homomorphism.

To verify that ψ is a Drinfeld module, it remains to show that $\partial(\psi_a) = \gamma(a)$ for all $a \in A$. First, we claim that

$$\mathrm{ht}(\phi_\mathfrak{A}) = \min_{0 \neq \alpha \in \mathfrak{A}} \mathrm{ht}(\phi_\alpha). \tag{7.5.3}$$

Indeed, for any $0 \neq \alpha \in \mathfrak{A}$, we have $\phi_\alpha = f\phi_\mathfrak{A}$ for some $0 \neq f \in L\{\tau\}$, so $\mathrm{ht}(\phi_\mathfrak{A}) \leq \mathrm{ht}(\phi_\alpha)$. On the other hand, (7.5.1) and Lemma 3.1.11 imply that $\mathrm{ht}(\phi_\mathfrak{A}) \geq \min_{0 \neq \alpha \in \mathfrak{A}} \mathrm{ht}(\phi_\alpha)$.

If $\alpha \in \mathfrak{A}$, then $\phi_\alpha : \phi \to \phi$ is an endomorphism, so $\mathrm{ht}(\phi_\alpha) = 0$ if $\mathrm{char}_A(L) = 0$, and $\mathrm{ht}(\phi_\alpha) \equiv 0 \pmod{\deg \mathfrak{p}}$ if $\mathrm{char}_A(L) = \mathfrak{p} \neq 0$; cf. Sect. 3.4. Thus, by (7.5.3), the same holds for $\mathrm{ht}(\phi_\mathfrak{A})$. Now, for an arbitrary $a \in A$, comparing the coefficients of $\tau^{\mathrm{ht}(\phi_\mathfrak{A})}$ on both sides of $\phi_\mathfrak{A}\phi_a = \psi_a\phi_\mathfrak{A}$, one obtains

$$\partial(\psi_a) = \gamma(a)^{q^{\mathrm{ht}(\phi_\mathfrak{A})}} = \gamma(a).$$

$\square$

Denote the Drinfeld module ψ from Lemma 7.5.1 by $\mathfrak{A} * \phi$. Then $\phi_\mathfrak{A}$ gives an isogeny

$$\phi_\mathfrak{A} : \phi \to \mathfrak{A} * \phi \tag{7.5.4}$$

defined over L.

Remark 7.5.2 The Drinfeld module $\mathfrak{A} * \phi$ depends on the choice of the generator $\phi_\mathfrak{A}$ of $I_{\phi,\mathfrak{A}}$. Since any other generator of $I_{\phi,\mathfrak{A}}$ is equal to $c \cdot \phi_\mathfrak{A}$ for some $c \in L^\times$, if instead of the monic generator we choose $c \cdot \phi_\mathfrak{A}$ as a generator of $I_{\phi,\mathfrak{A}}$ and repeat the same construction, then we obtain

$$c\phi_\mathfrak{A}\phi_\alpha = c\psi_\alpha c^{-1} c\phi_\mathfrak{A}.$$

Thus, $\mathfrak{A} * \phi$ gets replaced by an isomorphic module $c(\mathfrak{A} * \phi)c^{-1}$, so the isomorphism class of $\mathfrak{A} * \phi$ is independent of the choice of a generator of $I_{\phi,\mathfrak{A}}$. In particular, if $\mathfrak{A} = aB$ for some nonzero $a \in A$ and c is the leading coefficient of ϕ_a, then $\phi_\mathfrak{A} = c^{-1}\phi_a$, so $(a) * \phi = c^{-1}\phi c$. Thus, $(a) * \phi$ is not necessarily equal to ϕ.

The next proposition relates the endomorphism rings of ϕ and $\mathfrak{A} * \phi$; this proposition is essentially Proposition 3.2 in [Hay79] (see also Proposition 4.7.19 in [Gos96] for a somewhat different proof, but note that the statement of this proposition in Goss's book contains a typo). Define

$$\mathrm{End}(\mathfrak{A}) := \{c \in K \mid c\mathfrak{A} \subset \mathfrak{A}\}.$$

Obviously $O \subseteq \mathrm{End}(\mathfrak{A})$. By Exercise 3.4.9, $\mathrm{End}(\mathfrak{A})$ is an A-order in K; if $\mathrm{End}(\mathfrak{A}) = O$, then the ideal $\mathfrak{A}$ is said to be a *proper* ideal of O.

Proposition 7.5.3 *There is a unique injective ring homomorphism* $u: \mathrm{End}(\mathfrak{A}) \to L\{\tau\}$ *whose restriction to O is* $\mathfrak{A} * \phi$. *This implies that* $\mathrm{End}_L(\mathfrak{A} * \phi)$ *contains an A-subalgebra isomorphic to* $\mathrm{End}(\mathfrak{A})$.

Proof Note that, by construction, the Drinfeld module $\mathfrak{A} * \phi$ is the restriction to A of an embedding $O \to L\{\tau\}$. If $\mathfrak{A}$ is a proper ideal, this observation is sufficient for the proof of the proposition. On the other hand, if $\mathfrak{A}$ is not proper, then $\mathrm{End}(\mathfrak{A})$ is strictly larger than O, so the first claim of the proposition says that $\mathfrak{A} * \phi: O \to L\{\tau\}$ extends to this larger ring. The second claim of the proposition immediately follows from the observation that $\mathrm{End}(\mathfrak{A})$ is commutative, so $u(\mathrm{End}(\mathfrak{A}))$ is in the centralizer of $u(A) = (\mathfrak{A} * \phi)(A)$.

Let $c \in \mathrm{End}(\mathfrak{A})$. Choose some $0 \neq \beta \in \mathfrak{A}$. Since $c\beta \in \mathfrak{A}$, using (7.5.1), we can write

$$\phi_{\mathfrak{A}} \phi_{c\beta} = \left(\sum_{i=1}^{n} f_i \phi_{\alpha_i} \right) \phi_{c\beta} = \sum_{i=1}^{n} f_i \phi_{\alpha_i c\beta}$$

$$= \sum_{i=1}^{n} f_i \phi_{\alpha_i' \beta} = \sum_{i=1}^{n} f_i \phi_{\alpha_i'} \phi_{\beta} = \left(\sum_{i=1}^{n} f_i \phi_{\alpha_i'} \right) \phi_{\beta},$$

where $\alpha_i' := \alpha_i c \in \mathfrak{A}$ for all $1 \leq i \leq n$. On the other hand, $\sum_{i=1}^{n} f_i \phi_{\alpha_i'} \in I_{\phi, \mathfrak{A}}$, so we can write $\sum_{i=1}^{n} f_i \phi_{\alpha_i'} = u \phi_{\mathfrak{A}}$ for some $u \in L\{\tau\}$. Thus,

$$\phi_{\mathfrak{A}} \cdot \phi_{c\beta} = u \cdot \phi_{\mathfrak{A}} \cdot \phi_{\beta}. \tag{7.5.5}$$

Note that $u \neq 0$ if $c \neq 0$. By the definition of $\mathfrak{A} * \phi$, equation (7.5.5) can be rewritten as

$$(\mathfrak{A} * \phi)_{c\beta} \cdot \phi_{\mathfrak{A}} = u \cdot (\mathfrak{A} * \phi)_{\beta} \cdot \phi_{\mathfrak{A}},$$

which implies that

$$(\mathfrak{A} * \phi)_{c\beta} = u \cdot (\mathfrak{A} * \phi)_{\beta}. \tag{7.5.6}$$

A priori, u depends on both c and β, so we write $u = u_{c,\beta}$. On the other hand, for any $\beta' \in \mathfrak{A}$ we have $\beta' c \in \mathfrak{A} \subseteq O$, so $(\mathfrak{A} * \phi)_{\beta\beta'c} = (\mathfrak{A} * \phi)_{\beta'c}(\mathfrak{A} * \phi)_{\beta}$. Using (7.5.6), we can rewrite this equality as

$$u_{c,\beta\beta'} \cdot (\mathfrak{A} * \phi)_{\beta\beta'} = u_{c,\beta'} \cdot (\mathfrak{A} * \phi)_{\beta'}(\mathfrak{A} * \phi)_{\beta}.$$

Therefore, $u_{c,\beta\beta'} = u_{c,\beta'} = u_{c,\beta}$, so u does not depend on β, and we will omit β from notation. Having this fact, and again using (7.5.6), we get

$$u_{cc'} \cdot (\mathfrak{A} * \phi)_\beta = (\mathfrak{A} * \phi)_{cc'\beta} = u_c \cdot (\mathfrak{A} * \phi)_{c'\beta} = u_c u_{c'} \cdot (\mathfrak{A} * \phi)_\beta.$$

Thus, $u_{cc'} = u_c u_{c'}$. One similarly verifies that $u_{c+c'} = u_c + u_{c'}$. We conclude that $\mathrm{End}(\mathfrak{A}) \to L\{\tau\}$, $c \mapsto u_c$, is an injective homomorphism which coincides with $\mathfrak{A} * \phi$ on O. □

Corollary 7.5.4 *Assume that* $\mathrm{End}_L(\phi) \cong O$. *Let B be the integral closure of A in K and let $\mathfrak{C}$ be the conductor of O in B. Then* $\mathrm{End}_L(\mathfrak{C} * \phi) \cong B$. *In particular, ϕ is isogenous over L to a Drinfeld module whose endomorphism ring is the maximal A-order in K.*

Proof By Exercise 3.4.3, because ϕ and $\mathfrak{C} * \phi$ are isogenous over L, we have

$$\mathrm{End}_L^\circ(\mathfrak{C} * \phi) \cong \mathrm{End}_L^\circ(\phi) \cong K.$$

Hence, $\mathrm{End}_L(\mathfrak{C} * \phi)$ is isomorphic to an A-order in K. On the other hand, by Proposition 7.5.3, $B = \mathrm{End}(\mathfrak{C}) \subseteq \mathrm{End}_L(\mathfrak{C} * \phi)$. Since B is the maximal A-order in K, the inclusion $B \subseteq \mathrm{End}_L(\mathfrak{C} * \phi)$ must be an equality. □

Remark 7.5.5 Let $\mathfrak{p} \lhd A$ be a prime. Suppose $\gamma : A \to A/\mathfrak{p} = \mathbb{F}_\mathfrak{p}$ is the quotient map and L is a finite extension of $\mathbb{F}_\mathfrak{p}$. In this case, Proposition 7.5.3 can be generalized, and its proof simplified, by working in the division algebra $L(\tau)$. Let ϕ be a Drinfeld module over L. Denote $R = \mathrm{End}_L(\phi) \subseteq L\{\tau\}$; we do *not* assume that R is commutative. Let $\mathfrak{A}$ be a left ideal of R, that is, $\mathfrak{A} \subseteq R$ is an additive subgroup such that $R\mathfrak{A} \subseteq \mathfrak{A}$. The left ideal $I_{\mathfrak{A},\phi} = L\{\tau\} \cdot \mathfrak{A}$ is still principal, with monic generator $\phi_\mathfrak{A}$. Note that

$$I_{\mathfrak{A},\phi} = L\{\tau\}(\phi_\mathfrak{A}\phi(A)) = L\{\tau\}\phi_\mathfrak{A}.$$

Hence $\phi_\mathfrak{A}\phi(A)\phi_\mathfrak{A}^{-1} \subset L\{\tau\}$, where $\phi_\mathfrak{A}^{-1}$ is the inverse of $\phi_\mathfrak{A}$ in $L(\tau)$. The map

$$\psi : A \longrightarrow L\{\tau\}$$
$$a \longmapsto \phi_\mathfrak{A}\phi_a\phi_\mathfrak{A}^{-1}$$

is the Drinfeld module $\mathfrak{A} * \phi$. The *right order of* $\mathfrak{A}$ in $R \otimes_A F$ is

$$O_r(\mathfrak{A}) = \{\alpha \in R \otimes_A F \mid \mathfrak{A}\alpha \subseteq \mathfrak{A}\}. \tag{7.5.7}$$

It is not hard to show that $O_r(\mathfrak{A})$ is an A-order in $R \otimes_A F$; see Exercise 7.5.3. Let $\alpha \in O_r(\mathfrak{A})$. Since $\mathfrak{A}\alpha \subseteq \mathfrak{A}$, there is $\beta \in L\{\tau\}$ such that $\phi_\mathfrak{A}\alpha = \beta\phi_\mathfrak{A}$. Therefore,

$\phi_{\mathfrak{A}}\alpha\phi_{\mathfrak{A}}^{-1} \in L\{\tau\}$. On the other hand, because $\phi(A)$ is in the center of $O_r(\mathfrak{A})$, for any $a \in A$ we have

$$(\phi_{\mathfrak{A}}\alpha\phi_{\mathfrak{A}}^{-1})(\phi_{\mathfrak{A}}\phi_a\phi_{\mathfrak{A}}^{-1}) = (\phi_{\mathfrak{A}}\phi_a\phi_{\mathfrak{A}}^{-1})(\phi_{\mathfrak{A}}\alpha\phi_{\mathfrak{A}}^{-1}).$$

Thus, $\phi_{\mathfrak{A}}O_r(\mathfrak{A})\phi_{\mathfrak{A}}^{-1} \subseteq \mathrm{End}_L(\mathfrak{A} * \phi)$.

Now assume that K/F is a totally imaginary field extension of degree r. Let B be the integral closure of A in K. We fix an embedding

$$\iota: K \hookrightarrow \mathbb{C}_\infty \tag{7.5.8}$$

Restricted to B, the embedding ι gives $\mathbb{C}_\infty$ a specific B-module structure.

Let ϕ be a Drinfeld module of rank r over $\mathbb{C}_\infty$ such that $\mathrm{End}(\phi) \cong B$, i.e., ϕ has complex multiplication by B; cf. Definition 3.4.20.

Definition 7.5.6 We say that ϕ is *normalizable* if there is an isomorphism $B \xrightarrow{\sim} \mathrm{End}(\phi)$ such that $\partial(\phi_\alpha) = \iota(\alpha)$ for all $\alpha \in B$. Note that when K/F is a normal extension, every Drinfeld module over $\mathbb{C}_\infty$ with CM by B is normalizable since the image of K in $\mathbb{C}_\infty$ is the same under any embedding.

We start our study of Drinfeld modules with CM by B from the perspective of analytic uniformization. In Sect. 5.3, we have used analytic uniformization of Drinfeld modules to construct Drinfeld modules over $\mathbb{C}_\infty$ with an endomorphism ring isomorphic to an order in K. We recall this construction and then develop that idea further.

Remark 7.5.7 All the results about Drinfeld modules with CM that we discuss in this section generalize, with similar proofs, to the case when $\mathrm{End}_L(\phi) \cong O$ is just an A-order in K. But when O is not the maximal order, one has to distinguish between invertible, proper, and improper ideals of O, which creates some technical problems. In some problems where CM Drinfeld modules are used one can reduce the general case $\mathrm{End}_L(\phi) \cong O$ to the case when $O = B$ by replacing ϕ by the isogenous Drinfeld module $\mathfrak{C} * \phi$ of Corollary 7.5.4.

Let $\mathfrak{A}$ be a nonzero ideal of B. As an A-module, $\mathfrak{A}$ is free of rank r. Moreover, because K/F is totally imaginary, the image of $\mathfrak{A}$ under the embedding ι is discrete in $\mathbb{C}_\infty$. Therefore, $\iota(\mathfrak{A}) \subset \mathbb{C}_\infty$ is an A-lattice of rank r. If $c \in \mathbb{C}_\infty^\times$, then $c \cdot \iota(\mathfrak{A})$ is still a lattice. The Drinfeld module $\phi^{c\mathfrak{A}}$ corresponding to this lattice via the analytic uniformization has endomorphism ring

$$\mathrm{End}(\phi^{c\mathfrak{A}}) \cong \{\alpha \in \mathbb{C}_\infty \mid \alpha c\iota(\mathfrak{A}) \subseteq c\iota(\mathfrak{A})\}$$

$$= \{\alpha \in \mathbb{C}_\infty \mid \alpha\iota(\mathfrak{A}) \subseteq \iota(\mathfrak{A})\}$$

$$= \{\alpha \in \iota(K) \mid \alpha\iota(\mathfrak{A}) \subseteq \iota(\mathfrak{A})\} \qquad \text{(since } \mathfrak{A} \subset K\text{)}$$

$$= \iota(B) \qquad \text{(since every ideal of } B \text{ is proper).}$$

Note that by Theorem 5.2.11 we have $\partial(\phi_\alpha^{c\mathfrak{A}}) = \iota(\alpha)$ for all $\alpha \in B$, so $\phi^{c\mathfrak{A}}$ is normalizable. From now on, to simplify the notation, we will generally omit ι from notation.

Conversely, suppose ϕ has CM by B and is normalizable. Let Λ be the lattice corresponding to ϕ. Replacing ϕ by an isomorphic Drinfeld module corresponds to replacing Λ by a homothetic lattice. Thus, after possibly replacing ϕ by an isomorphic Drinfeld module, we may assume that 1 is part of an A-basis of Λ. We know that

$$\mathrm{End}(\phi) \cong \{\alpha \in \mathbb{C}_\infty \mid \alpha\Lambda \subseteq \Lambda\}.$$

Hence, under the assumption that $1 \in \Lambda$, we get an inclusion $B \subseteq \Lambda$ of free A-modules. Since both B and Λ have the same rank r, there is $a \in A$ such that $a\Lambda \subset B$; for example, one can take $a = \chi(\Lambda/B)$. It is clear that $a\Lambda$ is an ideal of B. We conclude that there is an ideal $\mathfrak{A}$ of B and some $c \in \mathbb{C}_\infty^\times$ such that the lattice of ϕ is $c\mathfrak{A}$. This proves the following:

Theorem 7.5.8 *The map of Theorem 5.2.8 induces a bijection between the set of lattices*

$$\{c \cdot \iota(\mathfrak{A}) \mid c \in \mathbb{C}_\infty^\times, \ 0 \neq \mathfrak{A} \ ideal \ of \ B\}$$

and the set of normalizable Drinfeld modules over $\mathbb{C}_\infty$ with CM by B.

Since homothetic lattices give rise to isomorphic Drinfeld modules, the isomorphism classes of normalizable Drinfeld modules with CM by B are in bijection with the homothety classes of lattices of the form $c\mathfrak{A}$. The reader will recognize the similarity of the notion of homothety of lattices of this special form to the usual equivalence of ideals of B from algebraic number theory, where $\mathfrak{A} \sim \mathfrak{B}$ if $\mathfrak{A} = c\mathfrak{B}$ for some $c \in K$. This suggests that the class group of B might play a role in the theory of Drinfeld modules with CM by B. Before exploring this direction, we recall some relevant definitions from algebraic number theory.

A *fractional ideal* of B is a set of the form $\mathfrak{A} = cI$ for some $c \in K$ and an ideal $I \lhd B$, or equivalently, a fractional ideal is a nonzero finitely generated B-submodule of K. Two fractional ideals can be multiplied by the formula $(cI)(c'I') = cc'II'$ and this is independent of the representation of the factors; in fact, if $\mathfrak{A}$ and $\mathfrak{B}$ are fractional ideals, then $\mathfrak{A}\mathfrak{B}$ is the B-submodule of K generated by all ab, where $a \in \mathfrak{A}$ and $b \in \mathfrak{B}$. Given a nonzero fractional ideal $\mathfrak{A}$, put

$$\mathfrak{A}^{-1} = \{\alpha \in K \mid \alpha\mathfrak{A} \subset B\}.$$

By basic algebraic number theory, $\mathfrak{A}^{-1}$ is a fractional ideal of B and $\mathfrak{A}\mathfrak{A}^{-1} = B$. Therefore, nonzero fractional ideals of B form a group under multiplication. The nonzero principal ideals cB, $c \in K^\times$, form a subgroup of this group. The quotient group of the group of nonzero fractional ideals of B by its subgroup of principal ideals is the *class group* of B, denoted $\mathrm{Cl}(B)$. One of the fundamental facts of algebraic number theory is that $\mathrm{Cl}(B)$ is a finite abelian group. Two fractional ideals $\mathfrak{A}$ and $\mathfrak{B}$ of B are said to be *equivalent* if their images are the same in $\mathrm{Cl}(B)$, i.e., if $\mathfrak{A} = c\mathfrak{B}$ for some $c \in K$.

Remarks 7.5.9

(1) From the definition of the class group it is clear that $\mathrm{Cl}(B) = 0$ if and only if B is a PID. On the other hand, because B is a Dedekind domain, B is a PID if and only if it is a UFD. Thus, $\mathrm{Cl}(B)$ measures the failure of unique factorization in B.

(2) The concept of fractional ideal makes sense also for an order O, but in that case not every nonzero fractional ideal is invertible; invertible in this case means $\mathfrak{A}\mathfrak{A}^{-1} = O$. It is easy to show that invertible ideals are proper, but the converse is true if and only if O is a Gorenstein ring; cf. [Mar22, Prop. 3.4].

(3) There is an important geometric interpretation of the class group $\mathrm{Cl}(B)$. This topic lies outside the scope of this book, so we only mention the relevant result, without defining the objects involved; the full proof can be found, for example, in [Lor96]. There is a unique smooth projective geometrically connected curve X over $\mathbb{F}_q$, whose function field is K. Let J_X be the Jacobian variety of X. Let $J_X(\mathbb{F}_q)$ be the group of $\mathbb{F}_q$-rational points on J_X. Then there is a short exact sequence

$$0 \longrightarrow J_X(\mathbb{F}_q) \longrightarrow \mathrm{Cl}(B) \longrightarrow \mathbb{Z}/f_\infty\mathbb{Z} \longrightarrow 0, \qquad (7.5.9)$$

where f_∞ is the residue degree of the unique place $\tilde{\infty}$ of K over ∞.

From our earlier discussion, a lattice of a Drinfeld module with CM by B is essentially a fractional ideal of B, except that we consider B as a subring of the larger field $\mathbb{C}_\infty$. Two such lattices can be multiplied by the same formula as fractional ideals. The next proposition relates the action of ideals on Drinfeld modules with CM by B to the multiplication of fractional ideals.

Proposition 7.5.10 *Let ϕ be a normalizable Drinfeld module over $\mathbb{C}_\infty$ with CM by B. Let $\mathfrak{A} \lhd B$ be a nonzero ideal. Let Λ_ϕ be the lattice of ϕ and let $\Lambda_{\mathfrak{A}*\phi}$ be the lattice of $\mathfrak{A} * \phi$. Then*

$$\Lambda_{\mathfrak{A}*\phi} = \partial(\phi_\mathfrak{A})\mathfrak{A}^{-1}\Lambda_\phi.$$

Proof Let e_ϕ be the exponential function of ϕ and let $e_{\mathfrak{A}*\phi}$ be the exponential function of $\mathfrak{A} * \phi$. Then $\Lambda_{\mathfrak{A}*\phi}$ is the set of zeros of $e_{\mathfrak{A}*\phi}$. By (5.2.7),

$$\phi_\mathfrak{A} \cdot e_\phi = e_{\mathfrak{A}*\phi} \cdot \partial(\phi_\mathfrak{A}).$$

Hence $\partial(\phi_\mathfrak{A})^{-1}\Lambda_{\mathfrak{A}*\phi}$ is the set of zeros of $\phi_\mathfrak{A} \cdot e_\phi$.

Now note that $\phi_{\mathfrak{A}}(z) = 0$ if and only if $\phi_\alpha(z) = 0$ for all $\alpha \in \mathfrak{A}$. Indeed, by definition, $\phi_{\mathfrak{A}}$ is a right divisor of any ϕ_α, so obviously $\phi_{\mathfrak{A}}(z) = 0$ implies $\phi_\alpha(z) = 0$ for all $\alpha \in \mathfrak{A}$. The reverse implication follows from (7.5.1). Thus, using Corollary 2.7.25, we have

$$\phi_{\mathfrak{A}}(e_\phi(z)) = 0 \quad \Longleftrightarrow \quad \phi_\alpha(e_\phi(z)) = 0 \text{ for all } \alpha \in \mathfrak{A}.$$

On the other hand, using (5.2.7) again, we have

$$\phi_\alpha(e_\phi(z)) = e_\phi((\partial\phi_\alpha)z) = e_\phi(\alpha z).$$

Hence, $\phi_\alpha(e_\phi(z)) = 0$ if and only if $\alpha z \in \Lambda_\phi$. This implies

$$\begin{aligned}
\partial(\phi_{\mathfrak{A}})^{-1}\Lambda_{\mathfrak{A}*\phi} &= \{z \in \mathbb{C}_\infty \mid z\alpha \in \Lambda_\phi \text{ for all } \alpha \in \mathfrak{A}\} \\
&= \{z \in \mathbb{C}_\infty \mid z\mathfrak{A} \subseteq \Lambda_\phi\} \\
&= \mathfrak{A}^{-1}\Lambda_\phi.
\end{aligned}$$

$\square$

Definition 7.5.11 Let $CM(B, \iota)$ be the set of isomorphism classes of normalizable Drinfeld modules over $\mathbb{C}_\infty$ with CM by B. We denote the isomorphism class of ϕ by $[\phi]$. For a nonzero fractional ideal $\mathfrak{A}$ of B and $[\phi] \in CM(B, \iota)$, we define $\mathfrak{A}*[\phi]$ to be the isomorphism class of the Drinfeld module corresponding to the lattice $\mathfrak{A}^{-1}\Lambda_\phi$. This is well-defined because two Drinfeld modules over $\mathbb{C}_\infty$ are isomorphic if and only if their lattices are homothetic.

Remark 7.5.12 The previous definition is motivated by Proposition 7.5.10, where $\mathfrak{A}$ was an integral ideal of B. Note that when $\mathfrak{A}$ is an integral ideal Proposition 7.5.10 implies that $\mathfrak{A} * [\phi] = [\mathfrak{A} * \phi]$. Algebraically the action of fraction ideals on ϕ can be defined as follows. Let $a \in A$ be a nonzero element. Let c be the leading coefficient of ϕ_a. Then, from Remark 7.5.2, we have $(a) * \phi = c^{-1}\phi c$. Define $(a^{-1}) * \phi = c\phi c^{-1}$. Now, a fractional ideal $\mathfrak{A}$ is $(a^{-1})\mathfrak{B}$ for some nonzero $a \in A$ and some integral ideal $\mathfrak{B}$, so one can define $\mathfrak{A} * \phi = c(\mathfrak{B} * \phi)c^{-1}$. It is not hard to check that this is well-defined and the induced action on the isomorphism classes $CM(B, \iota)$ agrees with the action in Definition 7.5.11.

Lemma 7.5.13 *Let $\mathfrak{A}$ and $\mathfrak{B}$ be nonzero fractional ideals of B. Then*

(1) $\mathfrak{A} * [\phi] \in CM(B, \iota)$.
(2) $(\mathfrak{A}\mathfrak{B}) * [\phi] = \mathfrak{A} * (\mathfrak{B} * [\phi])$.
(3) $\mathfrak{A} * [\phi] = \mathfrak{B} * [\phi]$ *if and only if $\mathfrak{A}$ and $\mathfrak{B}$ are equivalent.*

Proof

(1) We have

$$\text{End}(\phi^{\mathfrak{A}^{-1}\Lambda_\phi}) \cong \{c \in \mathbb{C}_\infty \mid c\mathfrak{A}^{-1}\Lambda_\phi \subseteq \mathfrak{A}^{-1}\Lambda_\phi\}$$
$$= \{c \in \mathbb{C}_\infty \mid c\Lambda_\phi \subseteq \Lambda_\phi\}$$
$$= B.$$

(2) This is a consequence of the observation that

$$(\mathfrak{A}\mathfrak{B})^{-1}\Lambda_\phi = (\mathfrak{A}^{-1}\mathfrak{B}^{-1})\Lambda_\phi = \mathfrak{A}^{-1}(\mathfrak{B}^{-1}\Lambda_\phi).$$

(3) We can assume that $\Lambda_\phi = \mathfrak{C}$ for some ideal $\mathfrak{C}$ of B. Then

$$\mathfrak{A} * [\phi] = \mathfrak{B} * [\phi] \quad \Longleftrightarrow \quad \mathfrak{A}^{-1}\mathfrak{C} = c\mathfrak{B}^{-1}\mathfrak{C} \text{ for some } c \in \mathbb{C}_\infty^\times$$
$$\Longleftrightarrow \quad \mathfrak{B} = c\mathfrak{A}$$
$$\Longleftrightarrow \quad \mathfrak{A} \text{ and } \mathfrak{B} \text{ have the same image in } \text{Cl}(B),$$

where in the last equivalence we use the observation that if $\mathfrak{B} = c\mathfrak{A}$, then $c \in K$, since $\mathfrak{A}$ and $\mathfrak{B}$ are in K. $\qquad\square$

Note that the previous lemma implies that the principal fractional ideals of B act trivially on $\text{CM}(B, \iota)$. Therefore, the action of the group of fractional ideals on $\text{CM}(B, \iota)$ factors through $\text{Cl}(B)$.

Theorem 7.5.14 $\text{Cl}(B)$ *acts simply transitively on* $\text{CM}(B, \iota)$. *In particular,* $\text{CM}(B, \iota)$ *is a finite set of order* $\# \text{Cl}(B)$.

Proof Let $[\phi]$, $[\psi] \in \text{CM}(B, \iota)$. To show that the class group $\text{Cl}(B)$ acts transitively on $\text{CM}(B, \iota)$ we must find a fractional ideal $\mathfrak{A}$ such that $\mathfrak{A} * [\phi] = [\psi]$. By Theorem 7.5.8, we may assume that $\Lambda_\phi = \mathfrak{C}$ and $\Lambda_\psi = \mathfrak{C}'$ for some ideals $\mathfrak{C}, \mathfrak{C}'$ of B. Let $\mathfrak{A} = \mathfrak{C}'\mathfrak{C}^{-1}$. Then from the definitions it follows that $\mathfrak{A} * [\phi] = [\psi]$. To show that the action is simply transitive, we must show that $\mathfrak{A} * [\phi] = [\phi]$ if and only if $\mathfrak{A}$ is principal. But this we already proved as part (3) of Lemma 7.5.13. $\qquad\square$

The kernel of $\phi_\mathfrak{A}$ is the set of zeros of the $\mathbb{F}_q$-linear polynomial $\phi_\mathfrak{A}(x)$. Consistent with our notation $\phi[a] := \ker(\phi_a)$ for $a \in A$, we denote

$$\phi[\mathfrak{A}] = \ker(\phi_\mathfrak{A}).$$

Note that, when $\mathfrak{A} = aB$ for a nonzero $a \in A$, we have $\phi[\mathfrak{A}] = \phi[a]$.

Proposition 7.5.15 *Assume* $[\phi] \in \text{CM}(B, \iota)$. *Let* $\mathfrak{A}$ *be a nonzero ideal of* B. *Then:*

(1) The kernel $\phi[\mathfrak{A}]$ *is a B-module, where B acts on $\phi[\mathfrak{A}]$ via ϕ.*

(2) *We have an isomorphism of B-modules*

$$\phi[\mathfrak{A}] \cong B/\mathfrak{A}.$$

Proof

(1) To prove that $\phi[\mathfrak{A}]$ is a B-module, one can repeat the already familiar argument: Equation (7.5.2) implies

$$\phi_\mathfrak{A}\phi_\alpha = (\mathfrak{A} * \phi)_\alpha \phi_\mathfrak{A} \quad \text{for all} \quad \alpha \in B.$$

Therefore, if $z \in \phi[\mathfrak{A}]$ and $\alpha \in B$, then

$$\phi_\mathfrak{A}(\phi_\alpha(z)) = (\mathfrak{A} * \phi)_\alpha (\phi_\mathfrak{A}(z)) = (\mathfrak{A} * \phi)_\alpha (0) = 0,$$

so $\phi_\alpha(z) \in \phi[\mathfrak{A}]$.

(2) We have

$$\phi[\mathfrak{A}] \cong \Lambda_{\mathfrak{A}*\phi}/(\partial\phi_\mathfrak{A})\Lambda_\phi \qquad \text{(by Theorem 5.2.11)}$$

$$= \mathfrak{A}^{-1}\Lambda_\phi/\Lambda_\phi \qquad \text{(by Proposition 7.5.10)}.$$

We may assume that $\Lambda_\phi = c\mathfrak{B}$ for some $c \in \mathbb{C}_\infty^\times$ and an integral ideal $\mathfrak{B}$ of B. Since multiplication by c^{-1} induces an isomorphism

$$c\mathfrak{A}^{-1}\mathfrak{B}/c\mathfrak{B} \cong \mathfrak{A}^{-1}\mathfrak{B}/\mathfrak{B}$$

of B-modules, we are reduced to showing that

$$\mathfrak{A}^{-1}\mathfrak{B}/\mathfrak{B} \cong B/\mathfrak{A}. \tag{7.5.10}$$

Since $\mathfrak{A}^{-1}\mathfrak{B}/\mathfrak{B}$ is a $B/\mathfrak{A}$-module, we need to show that it is free of rank 1. Let $\mathfrak{A} = \mathfrak{P}_1^{s_1} \ldots \mathfrak{P}_n^{s_n}$ be the decomposition of $\mathfrak{A}$ into powers of prime ideals. Let $\mathfrak{P}^s$ be one of these prime powers. Then

$$\mathfrak{A}^{-1}\mathfrak{B}/\mathfrak{B} \otimes_B B/\mathfrak{P}^s \cong \mathfrak{A}^{-1}\mathfrak{B}/(\mathfrak{B} + \mathfrak{P}^s\mathfrak{A}^{-1}\mathfrak{B}) = \mathfrak{A}^{-1}\mathfrak{B}/\mathfrak{P}^s\mathfrak{A}^{-1}\mathfrak{B}.$$

By the Chinese remainder theorem, we have $B/\mathfrak{A} \cong \prod_{i=1}^n B/\mathfrak{P}_i^{s_i}$. Hence, it is enough to show that $\mathfrak{A}^{-1}\mathfrak{B}/\mathfrak{P}^s\mathfrak{A}^{-1}\mathfrak{B} \cong B/\mathfrak{P}^s$.

Let $\mathfrak{C}$ be any fractional ideal of B. We will show that $M = \mathfrak{C}/\mathfrak{P}^s\mathfrak{C}$ is a free $B/\mathfrak{P}^s$-module of rank 1. Note that $R = B/\mathfrak{P}^s$ is a local ring since $\mathfrak{P}/\mathfrak{P}^s$ is its only prime ideal. By Nakayama's lemma, it is enough to show that $M/\mathfrak{P}M$ is a one dimensional vector space over $R/\mathfrak{P} = B/\mathfrak{P}$ (see Exercise 2.4.5). Now note that any two elements of $\mathfrak{C}$ are linearly dependent over B, so $M/\mathfrak{P}M$ is at most one dimensional over $B/\mathfrak{P}$. On the other hand, if $M/\mathfrak{P}M = 0$, then by Nakayama's lemma $M = 0$, which implies that $\mathfrak{C} = \mathfrak{P}^s\mathfrak{C}$. This last equality is obviously false, so $M/\mathfrak{P}M$ is one dimensional. $\square$

Remark 7.5.16 It is rather enlightening and useful for some problems to observe that by a natural extension of the definition of Drinfeld modules, one can consider Drinfeld modules with CM by B as Drinfeld B-modules of rank 1, with ι playing the role of γ; cf. Appendix A. Then the isomorphism $\phi[\mathfrak{A}] \cong B/\mathfrak{A}$ is a natural generalization of the isomorphism $C[\mathfrak{n}] \cong A/\mathfrak{n}$ in the case of the Carlitz module. From this perspective the Galois theory of Drinfeld modules with CM can be seen as a generalization of the Carlitz cyclotomic theory.

Assume $[\phi] \in \mathrm{CM}(B, \iota)$ is defined by $\phi_T = T + g_1\tau + \cdots + g_r\tau^r$. By Theorem 3.8.11, there is a finite collection of rational functions $j_{k_1,\ldots,k_l}^{s_1,\ldots,s_l}(\phi)$ in the coefficients $g_1, \ldots, g_r$, which uniquely determine the isomorphism class of ϕ. Let

$$K([\phi]) := \iota(K)\left(\left\{j_{k_1,\ldots,k_l}^{s_1,\ldots,s_l}\right\}\right)$$

be the extension of $\iota(K)$ in $\mathbb{C}_\infty$ obtained by adjoining all the j-invariants of ϕ. The field $K([\phi])$ depends only on the isomorphism class of $[\phi]$ and, by Theorem 3.8.14, $K([\phi])$ is the smallest extension of K which is a field of definition of ϕ.

Theorem 7.5.17 *Let* $[\phi] \in \mathrm{CM}(B, \iota)$. *Denote* $L = K([\phi])$ *and assume that* ϕ *is defined over* L. *Then*

(1) $\mathrm{End}_L(\phi) \cong B$, *i.e., all endomorphisms of* ϕ *are defined over* L.
(2) L *is a finite Galois extension of* K, *with Galois group isomorphic to a subgroup of* $\mathrm{Cl}(B)$.
(3) *The unique place* $\tilde{\infty}$ *of* K *over* ∞ *splits completely in* L/K.
(4) $L(\phi[\mathfrak{A}])/L$ *is a finite abelian extension for any nonzero ideal* $\mathfrak{A} \triangleleft B$.
(5) *The* j-*invariants of* ϕ *are integral over* A, *i.e., the minimal polynomial of a* j-*invariant of* ϕ *over* F *has coefficients in* A.

Proof

(1) For $\sigma \in \mathrm{Aut}(\mathbb{C}_\infty/\iota(K))$, denote by $\sigma(\phi)$ be the Drinfeld module defined by

$$\sigma(\phi_T) = T + \sigma(g_1)\tau + \cdots + \sigma(g_r)\tau^r,$$

where $\phi_T = T + g_1\tau + \cdots + g_r\tau^r$.

Let $u \in \mathrm{End}(\phi)$. We claim that $\sigma(u) = u$ for all $\sigma \in \mathrm{Aut}(\mathbb{C}_\infty/L)$. Because ϕ is defined over L, we have $\sigma(\phi) = \phi$. Hence $\sigma(u) \in \mathrm{End}(\phi)$. On the other hand, by Proposition 3.3.4, the map $\partial \colon \mathrm{End}(\phi) \to \mathbb{C}_\infty$ is injective. Obviously $\partial(\sigma(u)) = \sigma(\partial(u))$. But $\partial(u) \in \iota(K) \subseteq L$, so $\partial(\sigma(u)) = \partial(u)$, which implies that $\sigma(u) = u$. Therefore, $u \in \mathrm{End}_L(\phi)$.

(2) and (3) Let $\sigma \in \mathrm{Aut}(\mathbb{C}_\infty/\iota(K))$. It is clear that if $u \in \mathrm{End}(\phi)$, then $\sigma(u) \in \mathrm{End}(\sigma(\phi))$. Thus,

$$\mathrm{End}(\sigma(\phi)) = \sigma(\mathrm{End}(\phi)) \cong B.$$

On the other hand, because σ fixes $\iota(K)$, we have $\partial(\sigma(u)) = \sigma(\partial(u)) = \partial(u)$. Therefore, $\sigma(\phi)$ is normalizable, so $[\sigma(\phi)] \in \mathrm{CM}(B, \iota)$.

Denote by $j(\phi) = j_{k_1,\ldots,k_l}^{s_1,\ldots,s_l}(\phi)$ one the j-invariants of ϕ. Because $j(\phi)$ is a rational function in the coefficients of ϕ_T, we have $j(\sigma(\phi)) = \sigma(j(\phi))$. On the other hand, since $\mathrm{Cl}(B)$ acts transitively on $\mathrm{CM}(B, \iota)$ and every equivalence class of fractional ideals has an integral representative, there is an ideal $\mathfrak{A} \lhd B$ such that

$$\sigma(j(\phi)) = j(\sigma(\phi)) = j(\mathfrak{A} * \phi).$$

By the construction of $\phi_\mathfrak{A}$, cf. (7.5.4), the isogeny $\phi \to \mathfrak{A} * \phi$ is defined over L, so $j(\mathfrak{A}*\phi) \in L$. This implies that $\sigma(j(\phi)) \in L$, so L is invariant under the action of $\mathrm{Aut}(\mathbb{C}_\infty/\iota(K))$. Thus, L/K is a normal extension. Moreover, because $\mathrm{Cl}(B)$ is finite, $\sigma(j(\phi))$ takes only finitely many values as σ ranges over $\mathrm{Aut}(\mathbb{C}_\infty/\iota(K))$. Therefore, L/K is a finite extension. We need to show that L/K is separable.

Let $K_{\widetilde{\infty}}$ be the completion of $\iota(K)$ at $\widetilde{\infty}$. Because $j(\phi)$ does not change if we replace ϕ by an isomorphic Drinfeld module over $\mathbb{C}_\infty$, we can assume that the lattice Λ of ϕ is an ideal of B. (In that case, ϕ might not be defined over L but this is not important for proving that L/K is separable.) In particular, $\Lambda \subset K_{\widetilde{\infty}}$. From the formulas defining the coefficients of ϕ_T in terms of Λ (see Proposition 5.2.3), we see that the coefficients of ϕ_T are in $K_{\widetilde{\infty}}$. Therefore, $j(\phi) \in K_{\widetilde{\infty}}$, which implies that L is isomorphic to a subfield of $K_{\widetilde{\infty}}$. Now, according to Proposition 2.8.7, the elements of $K_{\widetilde{\infty}}$ which are algebraic over K are necessarily separable over K. Therefore, L/K is a normal and separable extension, so it is Galois. If $\sigma \in \mathrm{Gal}(L/K)$, then by the previous paragraph $\sigma(j(\phi)) = j(\mathfrak{A} * \phi)$ for some nonzero ideal $\mathfrak{A}$ of B which does not depend on the choice of j. The map $\sigma \mapsto \mathfrak{A}$ defines an injective homomorphism $\mathrm{Gal}(L/K) \to \mathrm{Cl}(B)$.

The minimal polynomial $f(x) \in K[x]$ of $j(\phi)$ over K splits completely over L. Since $L \subset K_{\widetilde{\infty}}$, the polynomial $f(x)$ splits completely over $K_{\widetilde{\infty}}$. Since this is true for all the j-invariants of ϕ, the place $\widetilde{\infty}$ splits completely in L.

(4) Let $\mathfrak{A}$ be a nonzero ideal of B. By Proposition 7.5.15, $\phi[\mathfrak{A}]$ is isomorphic to $B/\mathfrak{A}$ as a B-module. On the other hand, because all endomorphisms of ϕ are defined over L, the action of B on $\phi[\mathfrak{A}]$ commutes with the action of $G = \mathrm{Gal}(L(\phi[\mathfrak{A}])/L)$. This implies that G is isomorphic to a subgroup of the group of B-module automorphism of $\phi[\mathfrak{A}]$:

$$\mathrm{Aut}_B(\phi[\mathfrak{A}]) = \mathrm{Aut}_B(B/\mathfrak{A}) \cong (B/\mathfrak{A})^\times.$$

Since $(B/\mathfrak{A})^\times$ is abelian, G is also abelian.

(5) Let B' be the integral closure of A in L. We want to prove that $j(\phi) \in B'$. Note that $j(\phi) \in B'$ if and only if the decomposition $j(\phi)B' = \mathfrak{P}_1^{s_1} \cdots \mathfrak{P}_n^{s_n}$ of the fractional ideal $j(\phi)B'$ into a product of prime ideals of B' contains only positive powers. Thus, $j(\phi)$ is integral over A if and only if $\mathrm{ord}_\mathfrak{P}(j(\phi)) \geq 0$ for every nonzero prime ideal $\mathfrak{P}$ of B'. Let $L_\mathfrak{P}$ be the completion of L with respect to the valuation $\mathrm{ord}_\mathfrak{P}$. By Corollary 6.2.13, a Drinfeld module with CM over $L_\mathfrak{P}$ has potentially good reduction. Let $\mathcal{L}$ be a finite extension of $L_\mathfrak{P}$ over which ϕ acquires good reduction, and let v be the extension of $\mathrm{ord}_\mathfrak{P}$ to $\mathcal{L}$. Since $v(j(\phi))$ is a positive multiple of $\mathrm{ord}_\mathfrak{P}(j(\phi))$, it is enough to show that $v(j(\phi)) \geq 0$. Thus, we may assume that ϕ has good reduction over $L_\mathfrak{P}$. In that case, after possibly replacing ϕ by an isomorphic module, we may assume that $g_1, \ldots, g_{r-1}$ are in the ring of integers R of $L_\mathfrak{P}$ and g_r is a unit in R. Now from the formulas for the j-invariants in Definition 3.8.7 it is clear that $j(\phi) \in R$. Thus, $\mathrm{ord}_\mathfrak{P}(j(\phi)) \geq 0$.

$\square$

It turns out that the field $K([\phi])$ is intimately related to Class Field Theory of K.

Definition 7.5.18 Let $\widetilde{\infty}$ be the unique place of K over the place ∞ of F. The *Hilbert class field* of K with respect to B, denoted H_B, is the maximal unramified abelian extension of $\iota(K)$ in $\mathbb{C}_\infty$ in which $\widetilde{\infty}$ splits completely.

"Unramified" in the previous definition means that no prime of B ramifies in H_B, and "maximal" means that if $L \subset \mathbb{C}_\infty$ is an unramified abelian extension of $\iota(K)$ in which $\widetilde{\infty}$ splits completely, then $L \subseteq H_B$. By Galois theory, the composite $L_1 L_2$ of two abelian extensions L_1 and L_2 of K is again abelian over K; see [DF04, p. 592]. Moreover, from basic ramification and splitting properties of places in extensions of global fields (cf. Theorem 31 in [Mar18]), it follows that if both L_1 and L_2 are unramified over K, then $L_1 L_2/K$ is unramified, and if $\widetilde{\infty}$ splits completely in both L_1 and L_2, then $\widetilde{\infty}$ splits completely in $L_1 L_2/K$. With these facts in hand, and with an appeal to Zorn's lemma, the reader can easily establish that the Hilbert class field exists and is unique, but it is not so clear whether H_B/K will be a finite or infinite extension. The next theorem is a central result in Class Field Theory, which not only gives the finiteness of H_B/K but also specifies its Galois group; cf. [Ros87, Thm. 1.3].

Theorem 7.5.19

(1) The extension H_B/K is a finite Galois extension such that

$$\mathrm{Cl}(B) \cong \mathrm{Gal}(H_B/K)$$

$$\mathfrak{P} \mapsto \mathrm{Frob}_\mathfrak{P},$$

where $\mathfrak{P}$ is a nonzero prime ideal of B and Frob$_\mathfrak{P}$ *is the Frobenius automorphism associated with* $\mathfrak{P}$.

(2) *The extension H_B is uniquely characterized by the following property: except for at most a finite number of exceptions, a nonzero prime ideal $\mathfrak{P}$ of B splits completely in H_B if and only if $\mathfrak{P}$ is principal.*

Example 7.5.20 Assume $q = 3$ and consider the quadratic extension K of F obtained by adjoining a root of the polynomial $y^2 - T(T^2 - T - 1) \in F[y]$. Because the degree of $T(T^2 - T - 1)$ is odd, ∞ ramifies in K/F, so K is imaginary (see Example 3.4.24). The class group $\text{Cl}(B)$ is easy to compute using (7.5.9). Indeed, the curve corresponding to K is the elliptic curve E defined by the Weierstrass equation $y^2 = T(T^2 - T - 1)$. Since the Jacobian variety of an elliptic curve is the curve itself, we just need to compute the number of solutions of $y^2 = T(T^2 - T - 1)$ as y and T vary over $\mathbb{F}_3$. It is trivial to check that $(T = 0, y = 0)$ is the only solution, which combined with the point at "infinity" on E, gives $E(\mathbb{F}_q) \cong \mathbb{Z}/2\mathbb{Z}$. Since $f_\infty = 1$, the short exact sequence (7.5.9) implies that $\text{Cl}(B) \cong \mathbb{Z}/2\mathbb{Z}$. Hence, H_B is a quadratic extension of K. To find this extension, consider

$$L = F\left(\sqrt{T}, \sqrt{T^2 - T - 1}\right),$$

which is a quadratic extension of K. In the extension $F(\sqrt{T})/F$ the place ∞ ramifies, but ∞ splits in $F(\sqrt{T^2 - T - 1})/F$. Hence, $\widetilde{\infty}$ splits in L/K. Similarly, T ramifies in $F(\sqrt{T})/F$, but it is unramified in $F(\sqrt{T^2 - T - 1})/F$, so the ramification index of a prime of L over T is 2. Since T ramifies in K/F, this implies that the prime of B over T does not ramify in L/K. A similar argument implies that the prime of B over $T^2 - T - 1$ does not ramify in L/K. All other primes of A are unramified in both $F(\sqrt{T})$ and $F(\sqrt{T^2 - T - 1})$, so they do not ramify in L/F, and therefore do not ramify in L/K. Thus, no prime of B ramifies in L/K and $\widetilde{\infty}$ splits in L/K. Because these properties uniquely characterize H_B, we conclude that $L = H_B$.

Theorem 7.5.21 (Drinfeld, Hayes) *With above notation, we have* $K([\phi]) = H_B$.

Drinfeld proved this result in [Dri74] using algebro-geometric methods. Hayes proved this result in [Hay79] using methods inspired by the classical theory of elliptic curves with complex multiplication; his methods avoid using geometry at all. We shall follow Hayes' method. (Theorem 7.5.21 in the $r = 2$ case is also proved in [Gek83].) For proving Theorem 7.5.21 the following proposition is essential:

Proposition 7.5.22 *Let $j(\phi)$ denote one of the j-invariants of ϕ. For all but finitely many primes $\mathfrak{P} \lhd B$, we have*

$$\text{Frob}_\mathfrak{P}(j(\phi)) = j(\mathfrak{P} * \phi),$$

where Frob$_\mathfrak{P} \in \text{Gal}(K([\phi])/K)$ *is the Frobenius automorphism associated with* $\mathfrak{P}$.

Proof First, we specify the finite set of primes of B that we exclude from consideration. We choose representatives $\phi_1 = \phi, \ldots, \phi_h$ of the isomorphism classes in $CM(B, \iota)$ defined over $K([\phi])$. Let B' be the integral closure of B in $K([\phi])$. We say that a nonzero prime ideal $\widetilde{\mathfrak{P}}$ of B' is a *good prime* if for all $1 \leq i \leq h$ the coefficients of $(\phi_i)_T$ have non-negative valuations at $\widetilde{\mathfrak{P}}$ and the leading coefficient of $(\phi_i)_T$ has valuation 0 (or equivalently, in the terminology of Chap. 6, each ϕ_i is defined over the completion of B' at $\widetilde{\mathfrak{P}}$ and has good reduction). It is clear that all but finitely many primes of B' are good primes. We consider only those primes of B which are unramified in $K([\phi])$ and for which the primes lying over them in B' are all good; we call these primes of B "good."[5]

Now let $\mathfrak{P}$ be a good prime of B. Let $\widetilde{\mathfrak{P}}$ be a prime of $K([\phi])$ over $\mathfrak{P}$. Denote $|\mathfrak{P}| = \#B/\mathfrak{P}$ and $\deg(\mathfrak{P}) = \log_q(|\mathfrak{P}|)$. By its definition, $\mathrm{Frob}_{\mathfrak{P}}$ is characterized as the unique element of $\mathrm{Gal}(K([\phi])/K)$ such that

$$\mathrm{Frob}_{\mathfrak{P}}(j(\phi)) \equiv j(\phi)^{|\mathfrak{P}|} \pmod{\widetilde{\mathfrak{P}}}$$

for all the j-invariants of ϕ. Hence, we need to prove that

$$j(\mathfrak{P} * \phi) \equiv j(\phi)^{|\mathfrak{P}|} \pmod{\widetilde{\mathfrak{P}}}. \tag{7.5.11}$$

(Note that $j(\mathfrak{P} * \phi)$ and $j(\phi)$ are integral, so the congruence is meaningful.)

Claim Denote by R the ring of integers of the completion of $K([\phi])$ at $\widetilde{\mathfrak{P}}$. For a nonzero ideal $\mathfrak{A}$ of B, the monic polynomial $\phi_{\mathfrak{A}}$ has coefficients in R. Moreover, $\mathfrak{A} * \phi$ is defined over R and its reduction modulo $\widetilde{\mathfrak{P}}$ has rank r.

Proof The Drinfeld module $\mathfrak{A} * \phi$ is isomorphic to one of ϕ_i's that we have fixed earlier, i.e., $c(\mathfrak{A} * \phi)c^{-1} = \phi_i$ for some $1 \leq i \leq h$ and some $c \in K([\phi])^{\times}$. This means that $c\phi_{\mathfrak{A}}$ is an isogeny $\phi \to \phi_i$. By our assumption, ϕ_T and $(\phi_i)_T$ are in $R\{\tau\}$, and the leading coefficients of ϕ_T and $(\phi_i)_T$ are in $R^{\times}$. By Exercise 6.1.5, which itself is an easy consequence of Lemma 6.1.9, $c\phi_{\mathfrak{A}}$ is in $R\{\tau\}$ and its leading coefficient is a unit in R. Since $\phi_{\mathfrak{A}}$ is a monic polynomial, we get $c \in R^{\times}$. But this implies that $\phi_{\mathfrak{A}} \in R\{\tau\}$. Moreover, $\mathfrak{A} * \phi$ is defined over R and its reduction modulo $\widetilde{\mathfrak{P}}$ has rank r (because ϕ_i has these properties). □

Therefore, it is meaningful to reduce $\phi_{\mathfrak{A}}$ modulo $\widetilde{\mathfrak{P}}$. Let $k := B'/\widetilde{\mathfrak{P}}$. We denote the image in $k\{\tau\}$ of a polynomial $f \in R\{\tau\}$ modulo $\widetilde{\mathfrak{P}}$ by $\bar{f}$.

Claim We have $\bar{\phi}_{\mathfrak{P}} = \tau^{\deg(\mathfrak{P})}$.

Proof Let $c\tau^n$ with $c \in k^{\times}$, $n \geq 0$, be the lowest degree nonzero term in $\bar{\phi}_{\mathfrak{P}}$. First, we show that $n \geq 1$, i.e., $\bar{\phi}_{\mathfrak{P}}$ is not separable. For $0 \neq \alpha \in B$, applying Proposition 6.1.10 to $\phi_T \phi_\alpha = \phi_\alpha \phi_T$, we conclude that $\phi_\alpha \in R\{\tau\}$. If $\alpha \in \mathfrak{P}$, then

---

[5] Note that the set of good primes depends on the choice of representatives of the isomorphism classes $CM(B, \iota)$.

we have $\phi_\alpha = f\phi_\mathfrak{P}$ for some $f \in K([\phi])\{\tau\}$. Since $\phi_\mathfrak{P}$ is monic, we can apply the division algorithm (Theorem 3.1.13) in $R\{\tau\}$ to conclude that in fact $f \in R\{\tau\}$. Now from $\alpha = \partial(\phi_\alpha) = \partial(f)\partial(\phi_\mathfrak{P})$ we get that $\mathrm{ord}_{\widetilde{\mathfrak{F}}}(\alpha) \geq \mathrm{ord}_{\widetilde{\mathfrak{F}}}(\partial(\phi_\mathfrak{P}))$. Since $\mathfrak{P}$ is unramified in $K([\phi])$, we have $\mathrm{ord}_\mathfrak{P}(\alpha) = \mathrm{ord}_{\widetilde{\mathfrak{F}}}(\alpha)$. We can choose $\alpha \in \mathfrak{P}$ to have valuation 1, which implies that $0 \leq \mathrm{ord}_{\widetilde{\mathfrak{F}}}(\partial(\phi_\mathfrak{P})) \leq 1$. The same argument applies to every $\mathfrak{A} * \phi$. Now observe that for $m \geq 1$ we have

$$
\begin{aligned}
\phi_{\mathfrak{P}^m} &= (\mathfrak{P} * \phi)_{\mathfrak{P}^{m-1}} \cdot \phi_\mathfrak{P} \\
&= (\mathfrak{P}^2 * \phi)_{\mathfrak{P}^{m-2}} \cdot (\mathfrak{P} * \phi)_\mathfrak{P} \cdot \phi_\mathfrak{P} \\
&= \cdots = (\mathfrak{P}^{m-1} * \phi)_\mathfrak{P} \cdot (\mathfrak{P}^{m-2} * \phi)_\mathfrak{P} \cdots (\mathfrak{P} * \phi)_\mathfrak{P} \cdot \phi_\mathfrak{P}.
\end{aligned}
$$

Therefore,

$$
\mathrm{ord}_{\widetilde{\mathfrak{F}}}(\partial(\phi_{\mathfrak{P}^m})) = \sum_{i=1}^m \mathrm{ord}_{\widetilde{\mathfrak{F}}}(\partial((\mathfrak{P}^{m-i} * \phi)_\mathfrak{P})) \leq m.
$$

We can choose m such that $\mathfrak{P}^m = \alpha B$ is principal. In that case, $\mathrm{ord}_{\widetilde{\mathfrak{F}}}(\alpha) = m$ and $\partial(\phi_{\mathfrak{P}^m}) = \partial(\phi_\alpha) = \alpha$. Thus, the above inequality is an equality, which forces $\mathrm{ord}_{\widetilde{\mathfrak{F}}}(\partial((\mathfrak{P}^{m-i} * \phi)_\mathfrak{P})) = 1$ for all $1 \leq i \leq m$. In particular, $\mathrm{ord}_{\widetilde{\mathfrak{F}}}(\partial(\phi_\mathfrak{P})) = 1$, so $\bar{\phi}_\mathfrak{P}$ indeed has zero constant term.

For any $\alpha \in B$, using (7.5.2), we get $\bar{\phi}_\mathfrak{P} \cdot \bar{\phi}_\alpha = \overline{(\mathfrak{P} * \phi)_\alpha} \cdot \bar{\phi}_\mathfrak{P}$. Considering the lowest degree nonzero terms of both sides, we get $\bar{\alpha}^{q^n} = \bar{\alpha}$. The image of B in k is $B/\mathfrak{P} \cong \mathbb{F}_{q^{\deg(\mathfrak{P})}}$. Hence, n must be divisible by $\deg(\mathfrak{P})$. In particular, $n \geq \deg(\mathfrak{P})$. On the other hand, by Proposition 7.5.15,

$$
\deg(\bar{\phi}_\mathfrak{P}) = \deg(\phi_\mathfrak{P}) = \log_q(B/\mathfrak{P}) = \deg(\mathfrak{P}).
$$

Since $n \leq \deg(\bar{\phi}_\mathfrak{P})$, we get $n \leq \deg(\mathfrak{P})$. Therefore, $n = \deg(\bar{\phi}_\mathfrak{P})$. Finally, because $\phi_\mathfrak{P}$ is monic, we get $\bar{\phi}_\mathfrak{P} = \tau^{\deg(\mathfrak{P})}$. $\square$

Now we return to the proof of the proposition. The congruence (7.5.11) is equivalent to the equality

$$
j(\overline{\mathfrak{P} * \phi}) = j(\bar{\phi})^{|\mathfrak{P}|}
$$

in k. Reducing the equation $\phi_\mathfrak{P} \cdot \phi_T = (\mathfrak{P} * \phi)_T \cdot \phi_\mathfrak{P}$ modulo $\widetilde{\mathfrak{P}}$, one obtains

$$
\begin{aligned}
\overline{\phi_\mathfrak{P} \cdot \phi_T} &= \bar{\phi}_\mathfrak{P} \cdot \bar{\phi}_T \\
&= \tau^{\deg(\mathfrak{P})} \cdot \bar{\phi}_T \\
&= (\overline{\mathfrak{P} * \phi})_T \cdot \tau^{\deg(\mathfrak{P})}.
\end{aligned}
$$

This shows that the coefficients of $(\overline{\mathfrak{P} * \phi})_T$ are the coefficients of $\bar{\phi}_T$ raised to $|\mathfrak{P}|$ power. Hence $j(\overline{\mathfrak{P} * \phi}) = j(\bar{\phi})^{|\mathfrak{P}|}$, as was required to be shown. $\square$

Proof of Theorem 7.5.21 According to Proposition 7.5.22, for any j-invariant $j(\phi)$ of ϕ and for all but finitely many primes $\mathfrak{P} \lhd B$, we have $\mathrm{Frob}_{\mathfrak{P}}(j(\phi)) = j(\mathfrak{P} * \phi)$. Therefore,

$$\mathrm{Frob}_{\mathfrak{P}} = 1 \text{ in } \mathrm{Gal}(K([\phi])/K) \iff j(\mathfrak{P} * \phi) = j(\phi) \text{ for all the } j\text{-invariants of } \phi$$

$$\iff \mathfrak{P} * \phi \cong \phi \quad \text{(Theorem 3.8.11).}$$

On the other hand, we know that $\mathrm{Cl}(B)$ acts simply transitively on $\mathrm{CM}(B, \iota)$. Thus, $\mathfrak{P} * \phi \cong \phi$ if and only if $\mathfrak{P}$ is a principal ideal. We conclude that $\mathrm{Frob}_{\mathfrak{P}} = 1$ if and only if $\mathfrak{P}$ is principal. Since a prime of B splits completely in $K([\phi])$ if and only if $\mathrm{Frob}_{\mathfrak{P}} = 1$, we can apply Theorem 7.5.19 to deduce that $K([\phi]) = H_B$ is the Hilbert class field of K with respect to B. $\square$

Remarks 7.5.23

(1) The previous proof is similar to the proof of Lemma 3.2.11. In fact, we just proved a generalization of that lemma. In the theory of Drinfeld B-modules mentioned earlier (see Appendix A), our module ϕ is a Drinfeld B-module of rank 1. The reduction of a Drinfeld B-module has positive height, so the height of $\bar{\phi}$ is equal to its rank. This implies that $\bar{\phi}_{\mathfrak{P}}$ is purely inseparable.

(2) If $\mathrm{End}(\phi) \cong O$ is isomorphic to an order in B, then $K([\phi])$ is an abelian extension of K, which is ramified at the primes dividing the conductor of O. Hayes proved in [Hay79] that by adjoining the j-invariants of all Drinfeld modules with complex multiplication by an order in B, one obtains the maximal abelian extension of K in which $\tilde{\infty}$ splits completely. Alternatively, he also proved that the same extension can be obtained by fixing one ϕ with $\mathrm{End}(\phi) \cong B$ and adjoining all the torsion points of ϕ to $K([\phi])$, which is analogous to the construction of the Carlitz cyclotomic extensions.

(3) To put Theorem 7.5.21 in a broader context one should mention Hilbert's 12th problem, which asks for explicit constructions of abelian extensions of number fields. The Kronecker–Weber theorem solves this problem for abelian extensions of $\mathbb{Q}$, since it says that every such extension is contained in a cyclotomic extension. Similarly, the classical theory of complex multiplication solves this problem for the imaginary quadratic extensions of $\mathbb{Q}$: the abelian extensions of imaginary quadratic fields are obtained by adjoining the special values of certain elliptic modular functions; see [Sil94, p. 135]. This has been generalized in the 1960s to CM number fields by Shimura and Taniyama using the theory of abelian varieties with complex multiplication, and to totally real fields by Dasgupta and Kakde using p-adic methods in 2021. Nevertheless, the general case of Hilbert's 12th problem remains open. In comparison, the theory of Drinfeld modules completely solves the analogue of Hilbert's 12th problem over function fields; see [Dri74] and [Hay79].

Example 7.5.24 Let $K = F(U)$ with $U^2 = T$. In this case $B = \mathbb{F}_q[U]$. Define a Drinfeld module of rank 2 over K by

$$\phi_T = (U + \tau)(U + \tau) = T + (U + U^q)\tau + \tau^2.$$

It is clear that $u := U + \tau \in K\{\tau\}$ is in $\text{End}_K(\phi)$. Moreover, since $u^2 = \phi_T$, we see that $B \subseteq \text{End}_K(\phi)$. Therefore, $\text{End}_K(\phi) \cong \mathbb{F}_q[U]$, so ϕ has CM by B. Its (unique) j-invariant is

$$j(\phi) = (U + U^q)^{q+1} = T^{\frac{q+1}{2}}(1 + T^{\frac{q-1}{2}})^{q+1}.$$

Note that when q is odd, $j(\phi) \in F$, so ϕ can be defined over F by $T + \tau + j(\phi)^{-1}\tau^2$. On the other hand, when q is even, $j(\phi)$ is inseparable, so $K = F(j(\phi))$. In either case, $H_B = K = K([\phi])$.

Example 7.5.25 Let $q = 3$ and let K be the splitting field of the quadratic equation $y^2 = T(T^2 - T - 1)$ that we already considered in Example 7.5.20. Let B be the integral closure of A in K. Let ϕ be a Drinfeld module of rank 2 over $\mathbb{C}_\infty$ with CM by B. We would like to describe this ϕ explicitly by writing down ϕ_T over H_B. The following calculation is due to Hayes [Hay91].

Let $\phi_T = T + a\tau + \tau^2$, where a is to be determined. Assume the complex multiplication of ϕ is given by

$$\phi_y = y + c_1\tau + c_2\tau^2 + \tau^3.$$

(Note that $\deg_\tau(\phi_y) = 3$ because $\phi_{y^2} = \phi_{T(T^2-T-1)}$ has degree 6 in τ.) By comparing the coefficients of the powers of τ on both sides of the identity $\phi_T\phi_y = \phi_y\phi_T$, one arrives at the system of equations

$$c_1 = a \cdot \frac{y^q - y}{T^q - T},$$

$$c_2 = \frac{y^{q^2} - y + ac_1^q - c_1a^q}{T^{q^2} - T},$$

$$c_1^{q^2} - c_1 = c_2a^{q^2} - ac_2^q + T^{q^3} - T,$$

$$c_2^{q^2} - c_2 = a^{q^3} - a.$$

Substituting the first two equations into the others produces two polynomial identities for a with coefficients in B. Let $f(a) \in K[a]$ be the monic greatest common divisor of these two polynomials. In our case,

$$f(a) = a^2 - (T^2 - 1)(T - 1)ya + T(T + 1)(T^2 - 1)^2.$$

Note that $f(a) \in B[a]$ is quadratic, so $[H_B : K] = 2$. This implies that $\mathrm{Cl}(B) = \mathbb{Z}/2\mathbb{Z}$, in agreement with Example 7.5.20. Solving the quadratic equation for a, we obtain

$$a = (T^2 - 1)\left(z(T^2 + 1) - y(T - 1)\right), \quad \text{where} \quad z^2 = T.$$

From this one computes

$$j(\phi) = a^4 = T^2(T^2 - 1)^4[T(T + 2)(T^2 + 1)(T^3 - T + 2)w$$
$$- (T^8 + T^6 + T^5 + T^4 - T^3 - T^2 - T + 1)],$$

where $w^2 = T^2 - T - 1$.

As a side remark, note that $T, T - 1$, and $T + 1$ divide $j(\phi)$. Let $\mathfrak{p}$ be one of these primes, and $\mathfrak{P}$ be a prime of H_B lying over $\mathfrak{p}$. The reduction $\bar{\phi}$ of ϕ modulo $\mathfrak{P}$ has j-invariant 0. Since $\deg(\mathfrak{p}) = 1$ is odd, the Drinfeld module over $\overline{\mathbb{F}}_\mathfrak{p}$ with j-invariant 0 is supersingular. Therefore, the reduction of ϕ is supersingular. On the other hand, one can show that the reduction of ϕ at $\mathfrak{P}$ is supersingular if and only if $\mathfrak{p}$ does not split in K; see Exercise 7.5.4. It is easy to check that $T, T - 1, T + 1$ indeed do not split in K (T ramifies and $T \pm 1$ remain inert). For a general study of primes dividing the j-invariants of CM Drinfeld modules of rank 2 see [Dor91].

We conclude this section by considering the smallest field of definition of $[\phi] \in \mathrm{CM}(B, \iota)$ over F. The field in question is $F([\phi])$ obtained by adjoining to F all the j-invariants of ϕ. Denote

$$h = \# \mathrm{Cl}(A) = [K([\phi]) : K].$$

Then $[K([\phi]) : F] = h \cdot r$. Obviously

$$[K([\phi]) : F([\phi])] \leq [K : F] = r.$$

Hence

$$h \cdot r \geq [F([\phi]) : F] = \frac{[K([\phi]) : F]}{[K([\phi]) : F([\phi])]} = \frac{h \cdot r}{[K([\phi]) : F([\phi])]} \geq h.$$

Lemma 7.5.26 *With previous notation, if K/F is a Galois extension, then $[F([\phi]) : F] = h$.*

Proof We know that $K([\phi])/K$ is Galois. If K/F is Galois, then $K([\phi])/F$ is separable. Let L be the Galois closure of $K([\phi])/F$ in $\mathbb{C}_\infty$. Let $\sigma \in \mathrm{Gal}(L/F)$. We argue as in the proof of Theorem 7.5.17. Note that $\mathrm{End}(\sigma(\phi)) \cong B$. On the other hand, because K/F is Galois, σ induces an automorphism of $\iota(K)$. Therefore, $\sigma(\phi)$

Table 7.1 F-rational CM j-invariants

q	$j(\phi)$	Equation for ω
Any	0	$\omega = \delta$
Odd	$(T - \beta)^{\frac{q+1}{2}}\left((T - \beta)^{\frac{q-1}{2}} + 1\right)^{q+1}$	$\omega^2 = T - \beta$
Odd	$-(T - \beta)^{\frac{q+1}{2}}\left((T - \beta)^{\frac{q-1}{2}} - 1\right)^{q+1}$	$\omega^2 = \varepsilon(T - \beta)$
2^n	$\alpha^{-1}\left(\ell^{2^{n-1}} + \ell^{2^{n-2}} + \cdots + \ell + 1\right)^{q+1}$	$\omega^2 + \omega = \alpha T + \beta = \ell$
2	$T^3(T + 1)^3$	$\omega^2 + \omega = T^3 + T + 1$
2	$T^6(T + 1)^6$	$\omega^2 + \omega = T^5 + T^3 + 1$
3	$T^4(T + 1)^4(T - 1)^4(T^3 - T - 1)^2$	$\omega^2 = T^3 - T - 1$
3	$-T^4(T + 1)^4(T - 1)^4(T^3 - T + 1)^2$	$\omega^2 = -T^3 + T - 1$
4	$(T^4 + T)^{10}$	$\omega^2 + \omega = T^3 + \delta'$

is normalizable, so $\sigma(\phi) \in \mathrm{CM}(B, \iota)$. This implies that for any j-invariant $j(\phi)$ of ϕ we have $\sigma(j(\phi)) = j(\sigma(\phi)) = j(\mathfrak{A} * \phi)$ for some nonzero $\mathfrak{A} \lhd B$ that depends only on σ. Every embedding $F([\phi]) \hookrightarrow L$ extends to an automorphism of L/F; see [DF04, Thm. 13.27]. Thus, the number of distinct embeddings $F([\phi]) \hookrightarrow L$ is bounded by the class of B. Since $F([\phi])/F$ is separable, we get $[F([\phi]) : F] \le h$. But we already observed that $[F([\phi]) : F] \ge h$, so $[F([\phi]) : F] = h$ as was claimed. $\qquad\square$

Remarks 7.5.27

(1) The claim of Lemma 7.5.26 can be false if K/F is inseparable. Indeed, suppose K/F is quadratic and inseparable. Then q must be even and $K = \mathbb{F}_q(\sqrt{T})$; see Lemma 1.5.16. Therefore, $B = \mathbb{F}_q[\sqrt{T}]$. In this case, $h = 1$ but, by Example 7.5.24, $[F([\phi]) : F] = 2$.

(2) Lemma 7.5.26 and the previous remark imply that a rank 2 Drinfeld module with CM can be defined over F if and only if K is separable over F and $\mathrm{Cl}(B)$ is the trivial group. Table 7.1, taken from [Sch97, p. 324], lists all separable imaginary quadratic extensions K of F with $\mathrm{Cl}(B) = 0$. This is based on the classification of MacRye [Mac71] of such fields (see also [SS15] for a correction of that classification).[6] The table also gives the j-invariant of the unique $[\phi] \in \mathrm{CM}(B, \iota)$; these j-invariants are calculated by the method of Hayes [Hay91]. The notation in the table is the following: $K = F(\omega)$, $\alpha \in \mathbb{F}_q^\times$, $\beta \in \mathbb{F}_q$, $\mathbb{F}_{q^2} = \mathbb{F}_q(\delta)$, $\mathbb{F}_4 = \mathbb{F}_2(\delta')$.

---

[6] This is the function field analogue of a famous result of Heegner and Stark that the nine examples of Gauss are the only imaginary quadratic number fields whose rings of integers are principal ideal domains. It is not hard to check, as in Example 7.5.20, that for the fields listed in the table we have $\mathrm{Cl}(B) = 0$. The hard part is to prove that these are the only fields with this property.

Exercises

7.5.1 Let $q = 3$ and $B = A[\sqrt{T(T^2 - T - 1)}]$. Exhibit an explicit non-principal ideal I of B, thus confirming the claim of Example 7.5.20 that $\mathrm{Cl}(B)$ is nontrivial. Show that I^2 is principal, which agrees with the claim that $\mathrm{Cl}(B) \cong \mathbb{Z}/2\mathbb{Z}$.

7.5.2 With notation of Definition 7.5.18, let $\mathbb{F}_{\tilde{\infty}}$ be the residue field at $\tilde{\infty}$. Prove that the constant field of H_B is $\mathbb{F}_{\tilde{\infty}}$.

7.5.3 Let D be a finite dimensional division F-algebra. Let O be an A-order of D and let $\mathfrak{A}$ be a left ideal of O. Let $O_r(\mathfrak{A})$ be the right order of $\mathfrak{A}$ as defined in (7.5.7). Check the following claims:

(a) $O_r(\mathfrak{A})$ is an A-subalgebra of D.
(b) $\mathfrak{A} \otimes_A F = D$.
(c) $O_r(\mathfrak{A}) \otimes_A F = D$, so $O_r(\mathfrak{A})$ contains an F-basis of D.
(d) There are $a_1, a_2 \in A$ and $m \in \mathfrak{A}$ such that $\frac{a_1}{a_2}m = 1$. Moreover, $a_2 O_r(\mathfrak{A}) \subseteq \mathfrak{A}$.
(e) $O_r(\mathfrak{A})$ is a finitely generated A-module.

Thus, $O_r(\mathfrak{A})$ is an A-order of D.

7.5.4 Let K/F be an imaginary quadratic extension and let ϕ be a Drinfeld module of rank 2 over $\mathbb{C}_\infty$ with $\mathrm{End}^\circ(\phi) \cong K$.

(a) Let ϕ' be another Drinfeld module of rank 2 over $\mathbb{C}_\infty$. Prove that ϕ' is isogenous to ϕ if and only if $\mathrm{End}^\circ(\phi') \cong K$.
(b) Suppose ϕ is defined over a finite extension L of F. Let $\mathfrak{P}$ be a prime of L lying over the prime $\mathfrak{p}$ of A. Assume ϕ has good reduction $\bar{\phi}$ at $\mathfrak{P}$. Prove that $\bar{\phi}$ is an ordinary Drinfeld module of rank 2 over $\mathbb{F}_\mathfrak{p}$ if and only if $\mathfrak{p}$ splits in K.

7.5.5 This exercise generalizes part of Remark 7.5.27. Let $n \geq 1$ be an integer. For an imaginary quadratic extension K/F, let B_K denote the integral closure of A in K. Prove that there exist only finitely many imaginary quadratic extensions K/F such that $\mathrm{CM}(B_K, \iota)$ contains the isomorphism class of a Drinfeld module ϕ which can be defined over an extension L/F of degree $\leq n$.

7.5.6 Let K/F be a totally imaginary field extension of F of degree r and let ϕ be a Drinfeld module of rank r defined over a finite extension L of K with $\mathrm{End}^\circ(\phi) \cong K$.

(a) Let $\theta \colon \mathrm{End}^\circ_L(\phi) \otimes_F F_\mathfrak{p} \hookrightarrow \mathrm{End}_{F_\mathfrak{p}}(V_\mathfrak{p}(\phi))$ be the injection obtained from the action of endomorphisms of ϕ on the Tate module $T_\mathfrak{p}(\phi)$ (see Theorem 3.5.4). Show that the image of θ contains a matrix M whose eigenvalues are all distinct.
(b) Prove that the centralizer of M from (a), that is

$$\{S \in \mathrm{GL}_r(F_\mathfrak{p}) \mid SM = MS\},$$

is an abelian group.

(c) Deduce from (b) that the action of $\mathrm{Gal}(L^{\mathrm{sep}}/L)$ on $T_{\mathfrak{p}}(\phi)$ is abelian, i.e., the image of the representation

$$\mathrm{Gal}(L^{\mathrm{sep}}/L) \to \mathrm{Aut}_{A_{\mathfrak{p}}}(T_{\mathfrak{p}}(\phi))$$

is an abelian group. (Note that the same conclusion can be reached from Theorem 7.5.17.)

7.5.7 Let K/F be an imaginary quadratic extension and let B be the integral closure of A in K. Let ϕ be a Drinfeld module of rank 2 defined over a finite extension L of K with $\mathrm{End}^{\circ}(\phi) \cong K$. Let $\mathfrak{P}$ be a prime of L over a prime $\mathfrak{p}$ of A, where ϕ has good ordinary reduction $\bar{\phi}$. Let k be the residue field at $\mathfrak{P}$. By Corollary 6.1.12, there is a natural injective homomorphism

$$\mathrm{End}_L(\phi) \hookrightarrow \mathrm{End}_k(\bar{\phi}).$$

(a) Show that $\mathrm{End}_k(\bar{\phi})$ is an A-order in B whose conductor is not divisible by $\mathfrak{p}$.
(b) Conclude that if $\mathfrak{p}$ divides the conductor of $\mathrm{End}_L(\phi)$ in B, then the above homomorphism $\mathrm{End}_L(\phi) \to \mathrm{End}_k(\bar{\phi})$ is not surjective.
(c) Let $\mathfrak{c}$ be the conductor of $\mathrm{End}_L(\phi)$ in B, so $\mathrm{End}_L(\phi) \cong A + \mathfrak{c}B$. Decompose $\mathfrak{c} = \mathfrak{p}^m \mathfrak{c}_0$ with $m \geq 0$ and $\mathfrak{p} \nmid \mathfrak{c}_0$. Prove that

$$\mathrm{End}_k(\bar{\phi}) \cong A + \mathfrak{c}_0 B.$$

7.5.8 Assume q is odd. Let $f(T) \in A$ be a monic polynomial of odd degree $n \geq 3$ and let K be the splitting field of $y^2 = f(T)$. Let B be the integral closure of A in K. Let ϕ be a Drinfeld module of rank 2 over H_B with CM by B. Suppose ϕ is defined by $\phi_T = T + a\tau + \tau^2$, where a is integral over A. Let $\mathfrak{P}$ be a prime of H_B dividing a and let $\mathfrak{p} = A \cap \mathfrak{P}$. Prove that $d := \deg_T(\mathfrak{p})$ is odd and $d \leq n$.

7.5.9 Assume $q = 2$. Let K be the splitting field of $f(y) = y^2 + Ty + (T^3 + T^2 + 1) \in F[y]$ and let ϕ be a Drinfeld module over $\mathbb{C}_{\infty}$ of rank 2 with CM by the integral closure of A in K. Compute $j(\phi)$.

7.5.10 Assume K/F is a separable imaginary quadratic extension. Let B be the integral closure of A in K. Let ϕ be a Drinfeld module of rank 2 over $\mathbb{C}_{\infty}$ with CM by B. Prove that $F(j(\phi))/F$ is a Galois extension if and only if ∞ splits completely in $F(j(\phi))$.

7.6   Mordell-Weil Theorem and Class Number Formula

A fundamental theorem in the arithmetic of elliptic curves is the Mordell-Weil theorem, which states that the group of rational points $E(K)$ on an elliptic curve

E over a number field K is finitely generated. In particular, by the structure theorem of finitely generated abelian groups, $E(K)$ is isomorphic to the direct sum of the finite abelian group $E(K)_{\text{tor}}$ and a free abelian group $\mathbb{Z}^n$, where $n \geq 0$ is the *rank* of E over K. Both $\#E(K)_{\text{tor}}$ and the rank are important invariants of E. There are many deep theorems and conjectures about these numbers. First of all, one of the most important open problems about elliptic curves, the Birch and Swinnerton-Dyer conjecture, states that the rank of E is equal to the order of vanishing of the L-function $L(E, s)$ of E at $s = 1$. Also, it is not known[7] whether the rank is uniformly bounded as E varies over all elliptic curves defined over $\mathbb{Q}$, although by a theorem of Mazur $\#\mathbb{E}(\mathbb{Q})_{\text{tor}} \leq 16$.

Now let K be a finite extension of F and let B be the integral closure of A in K. Let ϕ be a Drinfeld module over K, with γ being the natural injection $A \hookrightarrow F \hookrightarrow K$. Then K becomes an A-module via ϕ, and one can ask for a description of this A-module, similar to the description of the group of rational points of an elliptic curve. We already know that $({}^\phi K)_{\text{tor}}$ is finite (in fact, $({}^\phi K_{\mathfrak{P}})_{\text{tor}}$ is finite for the completion of K at an appropriate prime of B). On the other hand, ${}^\phi K$ is not finitely generated, so the analogy with elliptic curves breaks here.[8] The fact that ${}^\phi K$ is not finitely generated is not surprising since K is not finitely generated as an A-module. What is surprising[9] though is that ${}^\phi K$ is isomorphic to a direct sum of its torsion submodule and a free A-module of countable rank; this was proved by Poonen [Poo95].

One might propose that perhaps ${}^\phi B$ would give us a finitely generated A-module, thus some version of the rank of an elliptic curve, if we assume that $\phi: A \to B\{\tau\}$ is defined over B.[10] After all, B is a free finitely generated A-module of rank $[K : F]$. Surprisingly, this is not the case, i.e., ${}^\phi B$ is not finitely generated.

Lemma 7.6.1 *Let $\phi: A \to B\{\tau\}$ be a Drinfeld module. Then ${}^\phi B$ is not finitely generated.*

Proof We show this assuming for simplicity that $K = F$, and leave the general case as an exercise. Suppose ${}^\phi A$ is finitely generated. Then, by Theorem 1.2.4, $({}^\phi A)/T \cdot ({}^\phi A)$ is finite. Let $a_1, \ldots, a_m$ be a set of coset representatives of $T \cdot ({}^\phi A)$

---

[7] Based on heuristics arguments, it was suggested by Park, Poonen, Voight, and Wood [PPVW19] that there are at most finitely many isomorphism classes of elliptic curves over $\mathbb{Q}$ with rank ≥ 22. On the other hand, there is an elliptic curve over $\mathbb{Q}$ whose rank is at least 28; this curve was constructed by Elkies.

[8] For example, when $K = F$ this can be proved as follows: By replacing ϕ by an isomorphic module, we may assume that $\phi: A \to A\{\tau\}$ has integral coefficients. If ${}^\phi F$ is generated by $u_1, \ldots, u_n$, then the primes that divide the denominators of $\phi(A)u_1 + \cdots \phi(A)u_n$ are the primes that divide the denominator of one of the u_i's. Since there are only finitely many such primes, we reach a contradiction.

[9] This is surprising because K with its natural A-module structure is not free, so the twisted action of A on K via ϕ plays a key role in the freeness property.

[10] One can always choose $c \in K^\times$ such that $c\phi_T c^{-1} \in B\{\tau\}$. Thus, by replacing ϕ by the isomorphic module $\psi = c\phi c^{-1}$, we can assume that ϕ is defined over B. Note also that ${}^\phi K \cong {}^\psi K$.

in $^\phi A$. We can write any $a \in A$ as

$$a = \phi_T(b) + a_i \quad \text{for some } b \in A \text{ and } 1 \leq i \leq m.$$

When b has large enough degree, the degree of $\phi_T(b) + a_i$ is $q^r \cdot \deg_T(b) + c$, where c is the degree of the leading coefficient of $\phi_T(x)$ and r is the rank of ϕ ("large enough" here depends on the degrees of the coefficients of ϕ_T and the degrees of a_i's). Since for any natural number there is a polynomial in A of that degree, the previous sentence implies that all but finitely many natural numbers are congruent to c modulo q^r, which is absurd. $\qquad\qquad\qquad\qquad\qquad\qquad\qquad\qquad\qquad\qquad$ $\square$

For a while, it seemed that there was no good analogue of the rank of elliptic curves for Drinfeld modules. This changed around 2010 when in a series of paper Taelman constructed a canonical finitely generated submodule $U(\phi/B)$ of $^\phi B$ and used it to prove analogues of the class number formula and the Birch and Swinnerton-Dyer conjecture in the setting of Drinfeld modules; cf. [Tae10, Tae11, Tae12b, Tae12a].

In this section we prove Poonen's theorem about the structure of $^\phi K$, discuss Taelman's units $U(\phi/B)$, and a construction, due to Anderson, of special explicit elements in $U(C/B)$ analogous to cyclotomic units.

7.6.1 Poonen's Theorem

This subsection is based on [Poo95] and [Den92]. First, we need to introduce some module-theoretic terminology that captures a crucial property related to the freeness that we will be trying to exhibit for $^\phi K$.

Definition 7.6.2 The *rank* of an A-module M is the dimension of the F-vector space $M \otimes_A F$, which is some cardinal number. (Note that this extends our earlier definition of rank in the case when M is finitely generated; cf. Theorem 1.2.4.) An A-module is *tame* if every submodule of finite rank is finitely generated.

Example 7.6.3 Consider F as an A-module via the usual multiplication in F. Since $F \otimes_A F \cong F$, the rank of F is 1. This module is not tame, since F is not finitely generated.

Recall that the torsion submodule M_{tor} of an A-module M is defined by

$$M_{\text{tor}} = \{m \in M \mid am = 0 \text{ for some } 0 \neq a \in A\};$$

it is a submodule by Exercise 1.2.1. It is not hard to show that M_{tor} is the kernel of the natural A-module homomorphism $M \to M \otimes_A K$; see Exercise 9 on page 376 of [DF04].

Lemma 7.6.4 *Let M be an A-module of finite rank, and suppose M_{tor} is finite. Then M/TM has finite cardinality.*

Proof If M' is the image of M in $M \otimes_A F$, then we have the exact sequence

$$0 \to M_{\text{tor}} \to M \to M' \to 0.$$

Tensoring with A/TA yields a right exact sequence

$$M_{\text{tor}}/TM_{\text{tor}} \to M/TM \to M'/TM' \to 0;$$

see [DF04, p. 399]. Since $M_{\text{tor}}/TM_{\text{tor}}$ is finite, it suffices to show that M'/TM' is finite. Hence, without loss of generality, we may assume that $M_{\text{tor}} = 0$ and identify M with its image in $M \otimes_A F$.

The quotient module M/TM is a vector space over $\mathbb{F}_T = A/TA$, and we need to show that it is finite dimensional. Suppose $\dim_{\mathbb{F}_T}(M/TM) \geq d$ for some natural number $d > 0$. Then we can find $m_1, \ldots, m_d \in M$ whose images in M/TM are linearly independent over $\mathbb{F}_T$. We claim that in that case $m_1, \ldots, m_d$ are also linearly independent over F. Indeed, otherwise

$$a_1 m_1 + \cdots + a_d m_d = 0 \tag{7.6.1}$$

for some $a_1, \ldots, a_d \in F$, not all zero. After appropriate scaling, we can assume $a_1, \ldots, a_d \in A$ and $\gcd(a_1, \ldots, a_d) = 1$. But now, reducing (7.6.1) modulo T, one obtains a nontrivial linear relation between the images of $m_1, \ldots, m_d$ in M/TM, in contradiction with an earlier assumption. Thus, $\dim_F(M \otimes_A F) \geq d$. We conclude that

$$\dim_{\mathbb{F}_T}(M/TM) \geq d \quad \Longrightarrow \quad \dim_F(M \otimes_A F) \geq d.$$

Since $M \otimes_A F$ is finite dimensional over F, M/TM must be finite dimensional over $\mathbb{F}_T$. $\square$

Lemma 7.6.5 *Every tame A-module M of countable rank is isomorphic to the direct sum of its torsion submodule M_{tor} with a free A-module of countable rank.*

Proof The quotient module $M' := M/M_{\text{tor}}$ is torsion-free and has countable rank since $M \otimes_A F \cong M' \otimes_A F$.

Suppose we proved the claim of the lemma for M'. Then M' is a free A-module. Let X' be an A-basis of M' and $X \subset M$ be a set of elements that maps bijectively onto X' under the quotient map $M \to M'$. It is clear that the elements of X are linearly independent over A since any linear relation $a_1 x_1 + \cdots + a_n x_n$ in M implies a linear relation $a_1 x_1' + \cdots + a_n x_n'$ in M' between the images of $x_1, \ldots, x_n$. Therefore, the submodule M'' of M generated by X is naturally isomorphic to M'. Note that $M_{\text{tor}} + M'' = M$. Since M'' is free, we also have $M_{\text{tor}} \cap M'' = 0$. Therefore, $M \cong M_{\text{tor}} \oplus M''$.

The previous discussion implies that it is enough to prove the lemma assuming M is torsion-free, which we do from now on. Denote $V = M \otimes_A F$. Because by assumption the vector space V has countable dimension, we can find an increasing sequence of vector subspaces

$$V_1 \subset V_2 \subset \cdots$$

of V such that

$$\dim V_i = i \quad \text{and} \quad \bigcup_{i \geq 1} V_i = V.$$

Let $M_i := M \cup V_i$. Then

$$M_i \subseteq M_{i+1} \quad \text{and} \quad M = \bigcup_{i \geq 1} M_i.$$

Each M_i has finite rank since $M_i \otimes F = V_i$. Hence, by the tameness assumption, M_i is free of finite rank. There is an injection

$$M_i / M_{i-1} \hookrightarrow V_i / V_{i-1} \cong F,$$

so M_i / M_{i-1} embeds as a nonzero finitely generated A-submodule into F. Such a submodule is necessarily free of rank $= 1$. Taking a preimage of a generator of M_i / M_{i-1} in M_i, we can decompose M_i as a direct sum of M_{i-1} and a free A-module $N_i \cong M_i / M_{i-1}$ of rank 1. Thus,

$$M = \bigcup M_i \cong N_1 \oplus N_2 \oplus N_3 \oplus \cdots$$

is isomorphic to a direct sum of countably many free A-modules of rank 1. A direct sum of free modules is itself free, so M is free of countable rank.   □

Let K be a finite extension of F and let B be the integral closure of A in K. Let k be the algebraic closure of $\mathbb{F}_q$ in K; note that $[k : \mathbb{F}_q] \leq [K : F]$. Let $\phi : A \to K\{\tau\}$ be a Drinfeld module. We want to prove that $^\phi K$ is tame. To do so, we will need an auxiliary tool introduced by Denis [Den92] – the height function on $^\phi K$. We will develop only a small part of the theory of height functions with a concrete application in mind.

Definition 7.6.6 Let $\alpha \in K$. The *height of* α, denoted $h(\alpha)$, is defined by

$$h(\alpha) = -\sum_v [k_v : k] \cdot \min\{0, v(\alpha)\},$$

where the sum is over all places of K, k_v denotes the residue field at v, and the valuation v on K is normalized so that $v(K^\times) = \mathbb{Z}$.

Example 7.6.7 If $K = F$ and $\alpha = a/b \in F$ is written in lowest terms with $a, b \in A$, then it is not hard to show that

$$h(\alpha) = \max\{\deg(a), \deg(b)\}.$$

Lemma 7.6.8

(1) For all $\alpha \in K$ and $n \in \mathbb{Z}$, we have

$$h(\alpha^n) = n \cdot h(\alpha).$$

(2) For all $\alpha, \beta \in K$, we have

$$h(\alpha + \beta) \leq h(\alpha) + h(\beta).$$

(3) For a given $N \geq 0$, there are only finitely many $\alpha \in K$ such that $h(\alpha) \leq N$.

Proof

(1) This is clear because for any place v of K we have $v(\alpha^n) = n \cdot v(\alpha)$, so

$$\min\{0, v(\alpha^n)\} = n \cdot \min\{0, v(\alpha)\}.$$

(2) It is enough to show that

$$\min\{0, v(\alpha + \beta)\} \geq \min\{0, v(\alpha)\} + \min\{0, v(\beta)\}.$$

Assume $v(\alpha) < v(\beta)$. In that case, since $\min\{0, v(\beta)\} \leq 0$ and $v(\alpha + \beta) = v(\alpha)$, the desired inequality is obvious. A similar argument applies if $v(\alpha) > v(\beta)$, with the roles of α and β interchanged.

Now assume $v(\alpha) = v(\beta)$. If $v(\alpha) \geq 0$, then $v(\alpha + \beta) \geq 0$, so both sides of the inequality are equal to zero. On the other hand, if $v(\alpha) < 0$, then $v(\alpha+\beta) \geq v(\alpha)$, so

$$\min\{0, v(\alpha + \beta)\} \geq v(\alpha) > v(\alpha) + v(\beta) = \min\{0, v(\alpha)\} + \min\{0, v(\beta)\}.$$

(3) Given $N \geq 0$, it is enough to show that the following two sets are finite:

(i) $S_N = \{v \text{ is a place of } K \mid [k_v : k] \leq N\}$.
(ii) $\mathcal{L}_{N,S} = \{\alpha \in K \mid v(\alpha) \geq -N \text{ for all } v \in S, \text{ and } v(\alpha) \geq 0 \text{ for all } v \notin S\}$, where S is a finite set of places of K.

Proving that S_N is finite is easy. Indeed, let S'_N be the set of places of F lying under the places in S_N. If $w \in S'_N$ and $v \in S_N$ lies over w, then

$$[\mathbb{F}_w : \mathbb{F}_q] \leq [k_v : k] \cdot [k : \mathbb{F}_q] \leq N[K : F].$$

There are only finitely many polynomials in A of degree $\leq N[K : F]$, so S'_N is finite. On the other hand, $\#S_N \leq [K : F] \cdot \#S'_N$, so S_N is also finite.

The proof of the finiteness of $\mathcal{L}_{N,S}$ is more involved. This is one of the preliminary results that one establishes while proving the Riemann-Roch theorem (cf. [Lor96, Lem. IX.3.7]). Here we will only prove this for $K = F$. Write $0 \neq \alpha = a/b \in \mathcal{L}_{N,S}$ with $a, b \in A$ relatively prime. Observe that if a prime $\mathfrak{p} \lhd A$ divides b, then $\mathfrak{p}$ is necessarily in S and $\mathrm{ord}_\mathfrak{p}(b) \leq N$. Thus, $\deg(b)$ is bounded. On the other hand, considering the place ∞, we get $\deg(b) + N \geq \deg(a)$. This implies that there are only finitely many possibilities for a and b. $\square$

Lemma 7.6.9 *Let $f(x) \in K[x]$ be a polynomial of degree n. Then there is a constant c_f, depending only on f, such that for all $\alpha \in K$, we have*

$$n \cdot h(\alpha) - c_f \leq h(f(\alpha)) \leq n \cdot h(\alpha) + c_f.$$

Proof Let $f(x) = a_n x^n + \cdots + a_1 x + a_0$ and let v be a nontrivial normalized valuation on K. There is a positive constant c depending only on f such that:

- If $v(\alpha) < -c$, then $v(f(\alpha)) = n \cdot v(\alpha) + v(a_n)$.
- If $v(\alpha) > c$ and $a_0 \neq 0$, then $v(f(\alpha)) = v(a_0)$.
- If $v(\alpha) > c$ and $a_0 = 0$, then $v(f(\alpha)) \geq 0$.

Therefore, there is a constant $c_{v,f}$ depending only on f and v such that if $v(\alpha) > c$ or $v(\alpha) < -c$, then

$$n \cdot \min\{0, v(\alpha)\} - c_{v,f} \leq \min\{0, v(f(\alpha))\} \leq n \cdot \min\{0, v(\alpha)\} + c_{v,f}. \qquad (7.6.2)$$

On the other hand, the range of values of $\min\{0, v(f(\alpha))\}$ is bounded if $-c \leq v(\alpha) \leq c$. Therefore, we can choose $c_{v,f}$ so that (7.6.2) holds for all $\alpha \in K$.

Let S be the set of places of K where the coefficients $a_0, \ldots, a_n$ have nonzero valuations. Observe that if $v \notin S$, then we can take $c_{v,f} = 0$ in (7.6.2). Indeed, if $v(\alpha) \geq 0$, then $v(f(\alpha)) \geq 0$, so $\min\{0, v(f(\alpha))\} = \{0, v(\alpha)\} = 0$. On the other hand, if $v(\alpha) < 0$, then $v(f(\alpha)) = n \cdot v(\alpha) < 0$, so $\min\{0, v(f(\alpha))\} = n \cdot \{0, v(\alpha)\} = n \cdot v(\alpha)$.

Looking at the definition of the height, we see that

$$n \cdot h(\alpha) - \sum_v [k_v : k] c_{v,f} \leq h(f(\alpha)) \leq n \cdot h(\alpha) + \sum_v [k_v : k] c_{v,f}.$$

Since all but finitely many $c_{v,f}$ are equal to 0, we obtain the claim of the lemma by putting

$$c_f = \sum_v [k_v : k] c_{v,f}.$$

$\square$

Lemma 7.6.10 *Let $\phi: A \to K\{\tau\}$ be a Drinfeld module of rank r and let $\alpha \in K$. Given $b \in A$ of positive degree, the limit*

$$\lim_{n\to\infty} \frac{h(\phi_{b^n}(\alpha))}{q^{nr\,\deg(b)}}$$

exists and is independent of b.

Proof We only prove that the limit exists for $b = T$ and leave the proof of the other claims as an exercise (see Exercise 7.6.4). According to Lemma 7.6.9, for any $n \geq 1$ we have

$$h(\phi_{T^n}(\alpha)) = q^{rn}h(\alpha) + c_{T^n}(\alpha), \tag{7.6.3}$$

where $c_{T^n}: K \to \mathbb{R}$ is a *bounded* function depending only on ϕ_{T^n}. Hence

$$
\begin{aligned}
h(\phi_{T^n}(\alpha)) &= h(\phi_T(\phi_{T^{n-1}}(\alpha)) \\
&= q^r h(\phi_{T^{n-1}}(\alpha)) + c_T(\phi_{T^{n-1}}(\alpha)) \\
&= q^{nr}h(\alpha) + q^r c_{T^{n-1}}(\alpha) + c_T(\phi_{T^{n-1}}(\alpha)),
\end{aligned}
$$

which implies

$$c_{T^n}(\alpha) = q^r c_{T^{n-1}}(\alpha) + c_T(\phi_{T^{n-1}}(\alpha)).$$

Let N be a constant such that $|c_T(\beta)| \leq N$ for all $\beta \subset K$. Then, after dividing the previous equality by q^{nr}, we get

$$\left| \frac{c_{T^n}(\alpha)}{q^{rn}} - \frac{c_{T^{n-1}}(\alpha)}{q^{r(n-1)}} \right| \leq \frac{N}{q^{nr}}. \tag{7.6.4}$$

Now (7.6.4) and (7.6.3) imply that

$$\left| \frac{h(\phi_{T^n}(\alpha))}{q^{rn}} - \frac{h(\phi_{T^{n-1}}(\alpha))}{q^{r(n-1)}} \right| \leq \frac{N}{q^{nr}}.$$

Thus, the sequence $\{h(\phi_{T^n}(\alpha))/q^{rn}\}_{n\geq 1}$ is Cauchy, so it converges to a non-negative real number.   □

The following definition is due to Denis [Den92]. It is modeled on Tate's definition of the canonical height function for elliptic curves; cf. [Sil09, Sec. VIII.9].

Definition 7.6.11 Let $\phi: A \to K\{\tau\}$ be a Drinfeld module of rank r. The *height on ϕK*, denoted $\hat{h}_\phi$, is the function $\hat{h}_\phi: K \to \mathbb{R}_{\geq 0}$ defined by

$$\hat{h}_\phi(\alpha) = \lim_{n\to\infty} \frac{h(\phi_{T^n}(\alpha))}{q^{n\cdot r}}. \tag{7.6.5}$$

Proposition 7.6.12 *Let $\phi \colon A \to K\{\tau\}$ be a Drinfeld module of rank r and let $\hat{h}$ be the height on $^{\phi}K$. Then we have:*

(1) $\hat{h}_{\phi}(\phi_b(\alpha)) = \deg(\phi_b(x)) \cdot \hat{h}_{\phi}(\alpha)$ for all nonzero $b \in A$.
(2) There is a constant $N(\phi) > 0$ depending only on ϕ such that

$$|\hat{h}_{\phi}(\alpha) - h(\alpha)| \leq N(\phi) \quad \text{for all } \alpha \in K.$$

(3) For a given real number N, there are only finitely many $\alpha \in K$ with $\hat{h}_{\phi}(\alpha) \leq N$.
(4) $\hat{h}_{\phi}(\alpha) = 0$ if and only if $\alpha \in (^{\phi}K)_{\text{tor}}$.
(5) $\hat{h}_{\phi}(\alpha + \beta) \leq \hat{h}_{\phi}(\alpha) + \hat{h}_{\phi}(\beta)$.

Proof

(1) Using Lemma 7.6.9, one computes

$$\hat{h}_{\phi}(\phi_b(\alpha)) = \lim_{n \to \infty} \frac{h(\phi_{T^n}(\phi_b(\alpha)))}{q^{rn}}$$

$$= \lim_{n \to \infty} \frac{h(\phi_b(\phi_{T^n}(\alpha)))}{q^{rn}}$$

$$= \lim_{n \to \infty} \frac{\deg(\phi_b) h(\phi_{T^n}(\alpha)) + c_b(\phi_{T^n}(\alpha))}{q^{rn}}$$

$$= \deg(\phi_b) \lim_{n \to \infty} \frac{h(\phi_{T^n}(\alpha))}{q^{rn}} \qquad \text{(since c_b is a bounded}$$

$$\text{function on } K)$$

$$= \deg(\phi_b) \cdot \hat{h}_{\phi}(\alpha).$$

(2) Let N be a constant such that $|c_T(\beta)| \leq N$ for all $\beta \in K$, where c_{T^n} is the function from the proof of Lemma 7.6.10. Then

$$\left| \frac{h(\phi_{T^n}(\alpha))}{q^{rn}} - h(\alpha) \right| = \left| \frac{c_{T^n}(\alpha)}{q^{rn}} \right| \qquad \text{(by (7.6.3))}$$

$$\leq \left| \frac{c_{T^{n-1}}(\alpha)}{q^{r(n-1)}} \right| + \frac{N}{q^{nr}} \qquad \text{(by (7.6.4))}$$

$$\leq \left| \frac{c_{T^{n-2}}(\alpha)}{q^{r(n-2)}} \right| + \frac{N}{q^{(n-1)r}} + \frac{N}{q^{nr}}$$

$$\leq N(1 + q^{-r} + \cdots + q^{-nr})$$

$$\leq N \frac{q^r}{q^r - 1} =: N(\phi).$$

Taking the limit $n \to \infty$, we get $|\hat{h}_\phi(\alpha) - h(\alpha)| \le N(\phi)$, as is claimed.

(3) By Lemma 7.6.8, this property is true for the height h on K, thus also true for $\hat{h}_\phi$ by the preceding property.

(4) Suppose $\alpha \in (^\phi K)_{\text{tor}}$. Then $\phi_b(\alpha) = 0$ for some $b \in A$ with positive degree. Using (1), we get

$$\deg(\phi_b) \cdot \hat{h}_\phi(\alpha) = \hat{h}_\phi(\phi_b(\alpha)) = 0. \qquad (7.6.6)$$

Hence $\hat{h}_\phi(\alpha) = 0$. Conversely, suppose $\hat{h}_\phi(\alpha) = 0$. Then (7.6.6) gives $\hat{h}_\phi(\phi_b(\alpha)) = 0$ for all $b \in A$. Since by (3) there are only finitely many $\beta \in K$ with $\hat{h}_\phi(\beta) = 0$, there must be $b, b' \in A$ such that $b \ne b'$ but $\phi_b(\alpha) = \phi_{b'}(\alpha)$. This implies $\phi_{b-b'}(\alpha) = 0$, so $\alpha \in (^\phi K)_{\text{tor}}$.

(5) We have

$$
\begin{aligned}
\hat{h}_\phi(\alpha + \beta) &= \lim_{n \to \infty} \frac{h(\phi_{T^n}(\alpha + \beta))}{q^{nr}} \\
&= \lim_{n \to \infty} \frac{h(\phi_{T^n}(\alpha) + \phi_{T^n}(\beta))}{q^{nr}} \\
&\le \lim_{n \to \infty} \left(\frac{h(\phi_{T^n}(\alpha))}{q^{nr}} + \frac{h(\phi_{T^n}(\beta))}{q^{nr}} \right) \quad \text{(by Lemma 7.6.8)} \\
&= \hat{h}_\phi(\alpha) + \hat{h}_\phi(\beta).
\end{aligned}
$$

$\square$

Corollary 7.6.13 *Let ϕ be a Drinfeld module over K. Then $(^\phi K)_{\text{tor}}$ is finite.*

Proof This follows from (3) and (4) of Proposition 7.6.12. $\square$

Proposition 7.6.14 *Given a Drinfeld module ϕ over K, the A-module $^\phi K$ is tame.*

Proof Suppose M is a submodule of $^\phi K$ of finite rank. We must show that M is finitely generated. The torsion submodule of $^\phi K$ is finite by Corollary 7.6.13, so the same is true for M. Thus, we may apply Lemma 7.6.4 to deduce that M/TM is finite. Let $S \subset M$ be a set of representatives of M/TM. Let

$$N = \max_{s \in S} \hat{h}_\phi(s)$$

and

$$S' = \{m \in M \mid \hat{h}_\phi(m) \le N\}.$$

By (3) in Proposition 7.6.12, S' is a finite set. We will show that S' generates M.

Let M' be the submodule of M generated by S'. If $M \ne M'$, then by (3) in Proposition 7.6.12, there is an element $m_0 \in M \setminus M'$ which has the minimal height

$\hat{h}_\phi(m_0)$ among the elements of $M \setminus M'$. We can write $m_0 = s + \phi_T(m)$ for some $s \in S$ and $m \in M$. Note that $m \notin M'$, as otherwise $m_0 \in M'$ also (because $\phi_T(m) \in M'$ and $s \in S \subseteq S' \subset M'$). Now

$$2\hat{h}_\phi(m_0) \le 2\hat{h}_\phi(m) \qquad\qquad \text{(by the minimality of } m_0)$$

$$\le \deg(\phi_T(x)) \cdot \hat{h}_\phi(m) \qquad \text{(since } \deg(\phi_T(x)) \ge q \ge 2)$$

$$= \hat{h}_\phi(\phi_T(m)) \qquad\qquad \text{(by (1) in Proposition 7.6.12)}$$

$$= \hat{h}_\phi(m_0 - s)$$

$$\le \hat{h}_\phi(m_0) + \hat{h}_\phi(s) \qquad\quad \text{(by (5) in Proposition 7.6.12)}$$

$$\le \hat{h}_\phi(m_0) + N.$$

Hence $\hat{h}_\phi(m_0) \le N$. This implies that $m_0 \in S' \subset M'$, which contradicts the assumption that $m_0 \notin M'$. Therefore S' generates M, so M is finitely generated, as desired.   □

Theorem 7.6.15 *The A-module $^\phi K$ is isomorphic to the direct sum of its torsion submodule, which is finite, with a free A-module of infinite countable rank. If ϕ is defined over B, then $^\phi B$ is isomorphic to the direct sum of its torsion submodule, which is finite, with a free A-module of infinite countable rank.*

Proof The rank of $^\phi K$ is countable since K is a countable set. It follows from Lemma 7.6.5 and Proposition 7.6.14 that $^\phi K$ is isomorphic to the direct sum of its torsion submodule with a free A-module of countable rank. If the rank is finite, then $^\phi K$ is a finitely generated A-module, which is not the case as we mentioned at the beginning of this section. Thus, the rank of $^\phi K$ is infinite.

The claim about $^\phi B$ follows from the same argument and the observation that a submodule of a tame A-module is itself tame.   □

7.6.2 Taelman's Theorem

To motivate Taelman's construction for Drinfeld modules, it is helpful to recall the classical Dirichlet Unit Theorem and the role that the exponential function plays in it.

Let K be a number field and let O_K be its ring of integers. Consider the map

$$\mathbb{R} \otimes_\mathbb{Q} K = \prod_{w|\infty} K_w \xrightarrow{\exp} \prod_{w|\infty} K_w^\times = (\mathbb{R} \otimes_\mathbb{Q} K)^\times, \qquad (7.6.7)$$

where the products are over the places of K lying over the Archimedean place of $\mathbb{Q}$. The number of places of K over ∞ of $\mathbb{Q}$ is $r_1 + r_2$, where r_1 is the number of real embeddings $K \to \mathbb{C}$ (i.e., those embeddings whose image is in $\mathbb{R}$) and $2r_2$ is the number of non-real embeddings, which come in complex conjugate pairs inducing the same absolute value on K. The exponential sequence (7.6.7) allows us to define the group

$$\exp^{-1}(O_K^\times) = \left\{ \alpha \in \mathbb{R} \otimes_{\mathbb{Q}} K \mid \exp(\alpha) \in O_K^\times \hookrightarrow \prod_{w \mid \infty} K_w^\times \right\}.$$

Note that the map $\exp^{-1}(O_K^\times) \xrightarrow{\exp} O_K^\times$ is not necessarily surjective (its image is formed by the totally positive units at the real places of K), but at least its image has finite index in $O_K^\times$. The map $\exp^{-1}(O_K^\times) \xrightarrow{\exp} O_K^\times$ is neither necessarily injective; in fact,

$$\ker(\exp) = \prod_{w \text{ complex}} (2\pi i \mathbb{Z}).$$

Thus, the rank of $\exp^{-1}(O_K^\times)$ is the rank of $O_K^\times$ plus r_2. The group $\exp^{-1}(O_K^\times)$ lies in the subspace $\mathbb{R} \otimes_{\mathbb{Q}} K^{\mathrm{Tr}=0}$, where

$$K^{\mathrm{Tr}=0} = \{\alpha \in K \mid \mathrm{Tr}_{K/\mathbb{Q}}(\alpha) = 0\},$$

and the Dirichlet Unit Theorem is equivalent to the statement that

$$\exp^{-1}(O_K^\times) \subset \mathbb{R} \otimes K^{\mathrm{Tr}=0}$$

is discrete and co-compact (i.e., $(\mathbb{R} \otimes K^{\mathrm{Tr}=0})/\exp^{-1}(O_K^\times)$ is compact). Indeed, this implies that the rank of $\exp^{-1}(O_K^\times)$ is equal to $[K : \mathbb{Q}] - 1 = r_1 + 2r_2 - 1$, and so the rank of $O_K^\times$ is $r_1 + r_2 - 1$, which is the usual statement of the Dirichlet Unit Theorem (cf. Theorem 38 in [Mar18]). The group $\exp^{-1}(O_K^\times)$ also plays an important role in the statement of the class number formula for the Dedekind zeta function of K, as we will discuss later in this section; cf. [Lor96, p. 275].

In [Tae10], Taelman proved an analogue of the above formulation of Dirichlet's Unit Theorem, with the Carlitz–Drinfeld exponential playing the role of exp, and in [Tae12b] he proved a version of the class number formula for the zeta-value of a Drinfeld module.

From now on K is a finite extension of F and B is the integral closure of A in K. Denote

$$K_\infty = F_\infty \otimes_F K \cong \prod_{w \mid \infty} K_w,$$

where the product is over the places of K lying over the place ∞ of F; see Theorem 2.8.5. Note that K_∞ is not a field if ∞ splits in K/F. Using the inclusions $B \hookrightarrow K \hookrightarrow K_\infty$, we consider B as an A-submodule of K_∞.

For a place w of K over ∞, denote by $|\cdot|_w$ the unique extension of the normalized absolute value on F_∞ to K_w. Define

$$\|\cdot\| : K_\infty \longrightarrow \mathbb{R}_{\geq 0}$$

$$(\alpha_w) \longmapsto \max_{w|\infty} \{|\alpha_w|_w\}.$$

It is trivial to check that $\|\cdot\|$ is a norm on K_∞, when K_∞ is considered as a finite dimensional vector space over F_∞. Moreover, by Proposition 2.3.4, K_∞ is complete with respect to this norm. Recall that we say that a subset $\Lambda \subset K_\infty$ is discrete if the set $\{\lambda \in \Lambda : \|\lambda\| \leq N\}$ is finite for any $N \geq 0$.

Lemma 7.6.16 *We have:*

(1) K_∞ *is locally compact.*
(2) B *is discrete in* K_∞.
(3) K_∞/B *is compact.*

Proof Each K_w, $w \mid \infty$, is a local field, so is locally compact. This implies that $K_\infty = \prod_{w|\infty} K_w$ is also locally compact. If B is not discrete in K_∞, then there is a sequence $\{b_i\}_{i\geq 1}$ of elements of B such that $\|b_i\| \to 0$ as $i \to \infty$. This means that $\max_{w|\infty} \{|b_i|_w\} \to 0$. On the other hand, for each finite place v of K, i.e., a place different from the places over ∞, we have $|b_i|_v \leq 1$. Hence $\prod_v |b_i|_v \to 0$, where the product is over all places of K. This contradicts the product formula in Exercise 2.8.6. Finally, because the rank of B as an A-module is equal to the dimension of K_∞ as an F_∞-vector space, K_∞/B is compact by Lemma 3.4.9. $\square$

Let $\phi: A \to B\{\tau\}$ be a Drinfeld module defined over B. Let e_ϕ be the exponential function associated with ϕ. Note that e_ϕ has coefficients in F_∞, because ϕ_T has coefficients in F_∞, so e_ϕ induces a homomorphism of A-modules

$$K_\infty = \prod_{w|\infty} K_w \xrightarrow{\prod_{w|\infty} e_\phi} \prod_{w|\infty} {}^\phi K_w = {}^\phi K_\infty,$$

which by abuse of notation we denote by e_ϕ. When A acts on K_∞ via ϕ it maps $B \subset K_\infty$ to itself because ϕ has coefficients in B. Thus, we have ${}^\phi B \hookrightarrow {}^\phi K_\infty$. Define

$$e_\phi^{-1}({}^\phi B) = \{\alpha \in K_\infty \mid e_\phi(\alpha) \in {}^\phi B\}.$$

Note that the A-module structure of ${}^\phi B$ plays no role in the definition of $e_\phi^{-1}({}^\phi B)$; the superscript ϕ is included to make it easier to distinguish B in the domain and codomain of the function e_ϕ. The same also applies to the next theorem and its proof.

Theorem 7.6.17 (Taelman [Tae10])

(1) The quotient A-module

$$\mathrm{Cl}(\phi/B) := \frac{{}^\phi K_\infty}{e_\phi(K_\infty) + {}^\phi B}$$

is finite.

(2) The module $e_\phi^{-1}({}^\phi B)$ is a discrete and co-compact A-submodule of K_∞.

Proof

(1) The function e_ϕ is entire on each component of K_∞ and $e_\phi(0) = 0$. Therefore, using Exercise 2.7.3, we can find an open compact subgroup $W \subset K_\infty$, such that $e_\phi(W)$ is open and compact in K_∞. This implies that $e_\phi(K_\infty)$ contains an open compact neighborhood of 0 in K_∞. Therefore, the image of $e_\phi(K_\infty)$ in K_∞/B is open and hence has finite index, because K_∞/B is compact by Lemma 7.6.16. This implies that $B + e_\phi(K_\infty)$, considered as an $\mathbb{F}_q$-vector subspace of K_∞, has finite index in K_∞.

(2) To prove that $e_\phi^{-1}({}^\phi B)$ is discrete in K_∞, suppose this is not the case, i.e., there is an infinite sequence $\{\lambda_i\}_{i\geq 1}$ of elements in $e_\phi^{-1}({}^\phi B)$ all of which satisfy $\|\lambda_i\| \leq N$ for some fixed N. Because K_∞ is locally compact, we can assume that $\{\lambda_i\}_{i\geq 1}$ is Cauchy, after possibly replacing this sequence by a subsequence. Since $\|\lambda_i - \lambda_{i+1}\| \to 0$ and $e_\phi^{-1}({}^\phi B)$ is a submodule of K_∞, the sequence of elements $\lambda_i' := \lambda_i - \lambda_{i+1}$ is in $e_\phi^{-1}({}^\phi B)$ and $\lambda_i' \to 0$. Because e_ϕ is continuous and $e_\phi(0) = 0$, the sequence $\{e_\phi(\lambda_i')\} \subset {}^\phi B$ converges to 0. On the other hand, ${}^\phi B$ is discrete in ${}^\phi K_\infty$, so $e_\phi(\lambda_i') = 0$ for all but finitely many i. Thus, $\lambda_i' \in \ker(e_\phi)$ for all but finitely many i. By Proposition 5.1.3, the kernel of e_ϕ on each K_w is discrete, so $\ker(e_\phi)$ is discrete in K_∞. This implies that $\lambda_i' = 0$ for all sufficiently large i. Thus, $\lambda_i = \lambda_{i+1} = \lambda_{i+2} = \cdots$ for sufficiently large i, which contradicts our initial assumption.

As in the proof of (1), let $W \subset K_\infty$ be an open compact subgroup such that $e_\phi(W)$ is also open and compact in K_∞. Since the kernel of e_ϕ is discrete, we can choose W small enough so that e_ϕ maps W isomorphically onto $e_\phi(W)$. From the proof of (1), ${}^\phi B + e_\phi(W)$ has finite index in ${}^\phi K_\infty$. Because $e_\phi(K_\infty)$ is an $\mathbb{F}_q$-vector subspace of ${}^\phi K_\infty$, the space $({}^\phi B \cap e_\phi(K_\infty)) + e_\phi(W)$ has finite index in $e_\phi(K_\infty)$. It is clear that there is a finite dimensional $\mathbb{F}_q$-vector space V' such that $e_\phi(K_\infty) = {}^\phi B \cap e_\phi(K_\infty) + e_\phi(W) + V'$ (for example, one can take V' to be the $\mathbb{F}_q$-span of coset representatives of ${}^\phi B \cap e_\phi(K_\infty) + e_\phi(W)$ in $e_\phi(K_\infty)$). Let V be an $\mathbb{F}_q$-vector subspace of K_∞ that maps isomorphically

onto V' under e_ϕ, e.g., V could be the $\mathbb{F}_q$-span of a subset of K_∞ that maps
bijectively onto a basis of V' under e_ϕ. Then

$$K_\infty = W + V + e_\phi^{-1}(\phi B).$$

Since V is finite, $W + V$ is compact. Therefore, $K_\infty/e_\phi^{-1}(\phi B)$ is compact.

$\square$

Corollary 7.6.18 $e_\phi^{-1}(\phi B)$ *is a free A-module of rank* $[K : F]$.

Proof Since $e_\phi^{-1}(\phi B) \subset K_\infty$ is discrete and co-compact, $e_\phi^{-1}(\phi B)$ is a free A-
module of rank $\dim_{F_\infty} K_\infty = [K : F]$ by Lemma 3.4.9. $\square$

Let $n = [K : F]$. Let $\{e_1, \ldots, e_n\}$ be an A-basis of $B \subset K_\infty$. Then $\{e_1, \ldots, e_n\}$
is also an F_∞-basis of K_∞. On the other hand, because $e_\phi^{-1}(\phi B) \subset K_\infty$ is discrete
and co-compact, if $\{e_1', \ldots, e_n'\}$ is an A-basis of $e_\phi^{-1}(\phi B)$, then it is also an F_∞-
basis of K_∞. For each $1 \leq i \leq n$, we can write $e_i' = \sum_{j=1}^n c_{i,j} e_j, c_{i,j} \in F_\infty$. The
determinant of the $n \times n$-matrix $(c_{i,j})_{1 \leq i,j \leq n}$ is well-defined, up to an $\mathbb{F}_q^\times$-multiple,
since a different choice of bases of B or $e_\phi^{-1}(\phi B)$ corresponds to multiplying the
matrix $(c_{i,j})_{1 \leq i,j \leq n}$ on the right or on the left by matrices in $\mathrm{GL}_n(A)$.

Definition 7.6.19 The *regulator of* ϕ *over* B, denoted $\mathrm{Reg}(\phi/B)$, is the unique
monic representative in $F_\infty = \mathbb{F}_q((1/T))$ of $\det(c_{i,j})$, where we say that $\alpha = \sum_{i \geq m} \alpha_i (1/T)^i \in \mathbb{F}_q((1/T))$ is monic if $\mathrm{ord}_\infty(\alpha) = m$ and $\alpha_m = 1$.

Remark 7.6.20 Given a finite dimensional real vector space $V \cong \mathbb{R}^n$, a positive-
definite quadratic form $\langle \cdot, \cdot \rangle$ on V, and a lattice $\Lambda \subset V$, an extremely important
invariant of Λ is its *discriminant with respect to* $\langle \cdot, \cdot \rangle$ defined as

$$\mathrm{disc}(\Lambda) = \det\left((\langle \lambda_i, \lambda_j \rangle)\right)_{1 \leq i,j \leq n},$$

where $\{\lambda_1, \lambda_2, \ldots, \lambda_n\}$ is a $\mathbb{Z}$-basis of Λ; cf. Exercise 1.2.10.

For example, when the positive-definite quadratic form on $\mathbb{R}^n$ is the usual inner
product $\langle \vec{x}, \vec{y} \rangle = x_1 y_1 + \cdots + x_n y_n$, and we expand each $\lambda_i = \sum_{j=1}^n c_{i,j} e_i$ in terms
of the standard basis of $\mathbb{R}^n$, then

$$\mathrm{disc}(\Lambda) = (\det(c_{i,j})_{1 \leq i,j \leq n})^2$$

is the square of the volume of V/Λ.

Some examples with number theoretic importance are the following:

- The *discriminant of a number field* K, denoted $\mathrm{disc}(K)$, is the discriminant of
  its ring of integers $O_K \subset \mathbb{R} \otimes_\mathbb{Q} K$ with respect to the trace form $\langle \alpha, \beta \rangle = \mathrm{Tr}_{K/\mathbb{Q}}(\alpha\beta)$.

- The *regulator of a number field* K, denoted $\mathrm{Reg}(K)$, is the positive square root of the discriminant of $\log(O_K^\times/(O_K^\times)_{\mathrm{tor}}) \subset \mathbb{R}^{r+s-1}$ with respect to the usual inner product.
- The *regulator of an elliptic curve* E over a number field K, denoted $\mathrm{Reg}(E/K)$, is the discriminant of $E(K)/E(K)_{\mathrm{tor}}$ in $E(K) \otimes_{\mathbb{Z}} \mathbb{R}$ with respect to the Néron-Tate pairing; cf. [Sil09].

The definition of the regulator of a Drinfeld module ϕ over B clearly fits into this context, and most closely resembles the regulator of a number field.

To motivate Taelman's Class Number Formula, we first recall the classical Class Number Formula. Let K be a number field and let O_K be its ring of integers. For a nonzero ideal $I \lhd O_K$, denote $|I| = \#O_K/I$. The Dedekind zeta function of K is defined as

$$\zeta_K(s) = \sideset{}{'}\sum_{I \lhd O_K} |I|^{-s}$$

$$= \sideset{}{'}\prod_{\mathfrak{P} \lhd O_K} \left(1 - \frac{1}{|\mathfrak{P}|^s}\right)^{-1}, \quad \mathrm{Re}(s) > 1,$$

where the sum is over all nonzero ideals of O_K, while the product is over all nonzero prime ideals. It is known that $\zeta_K(s)$ converges absolutely for $\mathrm{Re}(s) > 1$ and extends to a meromorphic function defined for all complex s with only one simple pole at $s = 1$. The Class Number Formula is a formula for the residue of $\zeta_K(s)$ at $s = 1$ in terms of the arithmetic invariants of K:

$$\lim_{s \to 1}(s - 1)\zeta_K(s) = \frac{2^{r_1} \cdot (2\pi)^{r_2} \cdot \mathrm{Reg}(K) \cdot \#\mathrm{Cl}(K)}{\#(O_K^\times)_{\mathrm{tor}} \cdot \sqrt{|\mathrm{disc}(K)|}}, \tag{7.6.8}$$

where $\mathrm{Cl}(K)$ is the ideal class group of O_K.

Returning to the setting of function fields, let K be a finite extension of $F = \mathbb{F}_q(T)$ and let B be the integral closure of A in K. As in the number field case, given a nonzero ideal $I \lhd B$, we can define $|I| = \#B/I$ and consider

$$\zeta_B(s) = = \sideset{}{'}\sum_{I \lhd B} |I|^{-s}$$

$$= \sideset{}{'}\prod_{\mathfrak{P} \lhd B} \left(1 - \frac{1}{|\mathfrak{P}|^s}\right)^{-1}, \quad \mathrm{Re}(s) > 1.$$

For function fields, the primes that belong to B and the places that lie over ∞ should be treated on equal footing since the choice of the place $\infty = 1/T$ of F plays no special role for K, at least when there are no Drinfeld modules involved. Therefore, it is natural to extend the above product to a product over all places of K

and consider

$$\zeta_K(s) = \prod_v \left(1 - \frac{1}{q_v^s}\right)^{-1}, \qquad \mathrm{Re}(s) > 1,$$

where q_v is the order of the residue field at v (note that $|\mathfrak{P}| = q_{\mathfrak{P}}$). Assume $\mathbb{F}_q$ is the field of constants of K, and let g be the genus of K (or rather, the genus of the projective smooth curve X corresponding to K). Then it turns out that there is a polynomial $L_K(x) \in \mathbb{Z}[x]$ of degree $2g$ such that

$$\zeta_K(s) = \frac{L_K(q^{-s})}{(1 - q^{-s})(1 - q^{1-s})}.$$

Moreover, $L_K(1/q) = \#J_X(\mathbb{F}_q)/q^g$. Thus,

$$\lim_{s \to 1}(s - 1)\zeta_K(s) = \frac{\#J_X(\mathbb{F}_q)}{q^{g-1}(q - 1)\log(q)}.$$

Since $\#J_X(\mathbb{F}_q)$ is closely related to the class number of B (see (7.5.9) for the special case when K/F is imaginary), g is closely related to the degree of the discriminant of K/F (via the Riemann-Hurwitz genus formula), and $q - 1$ is the order of the group of roots of unity in K, we see that the residue of $\zeta_K(s)$ at $s = 1$ is similar to the residue of its number field counterpart. The facts that we have listed above about $\zeta_K(s)$ for a function field K were proved by André Weil in the 1940s, along with the analogue of the Riemann Hypothesis,[11] which says that all the roots of $\zeta_K(s)$ lie on the line $\mathrm{Re}(s) = 1/2$, or equivalently, the inverse roots of $L_K(x)$ all have absolute value $\sqrt{q}$. To prove this, Weil first revised the foundations of algebraic geometry to make certain geometric techniques available over fields of positive characteristic.

Now $\#(B/I)$ is not the only way to measure the "size" of B/I. Assume $I \neq B$. Since B/I is a finite A-module, we can decompose it as $B/I \cong A/a_1 A \oplus \cdots \oplus A/a_n A$ for uniquely determined monic polynomials $a_1, \ldots, a_n \in A$ of positive degrees satisfying the divisibility condition $a_1 \mid a_2 \mid \cdots \mid a_n$. Recall that we defined (cf. Definition 1.2.5)

$$\chi(B/I) = a_1 \cdots a_n,$$

as a substitute for the cardinality of a finite abelian group. In addition, put $\chi(B/B) = 1$. We define

$$\zeta_C(B, 1) = \sideset{}{'}\sum_{I \lhd B} \frac{1}{\chi(B/I)},$$

---

[11] For hyperelliptic curves all of this was conjectured by E. Artin and the important case when X is an elliptic curve was proved by H. Hasse; cf. Remark 4.2.1.

where I ranges over the nonzero ideals of B. In contrast with the Dedekind zeta function, this sum converges in F_∞ since $\deg(\chi(B/I)) \to \infty$. Note that

$$\zeta_C(A, 1) = \sum_{a \in A_+} \frac{1}{a}$$

is the Carlitz zeta-value $\zeta_C(1)$ that we already encountered in Sect. 5.4. Next, observe that the Chinese remainder theorem implies that if I and I' are relatively prime ideals of B, then $B/II' \cong B/I \oplus B/I'$. Therefore,

$$\chi(B/II') = \chi(B/I) \cdot \chi(B/I').$$

Using the unique factorization of ideals in B into prime ideals, we get

$$\zeta_C(B, 1) = \prod_{\mathfrak{P} \lhd B}' \left(1 - \frac{1}{\chi(B/\mathfrak{P})}\right)^{-1}$$

$$= \prod_{\mathfrak{P} \lhd B}' \frac{\chi(B/\mathfrak{P})}{\chi(B/\mathfrak{P}) - 1},$$

where $\mathfrak{P}$ ranges over all nonzero prime ideals of B.

Lemma 7.6.21 *Let $\mathfrak{p}$ be the prime of A lying under the prime $\mathfrak{P} \lhd B$. Denote $\mathbb{F}_{\mathfrak{P}} = B/\mathfrak{P}$. Then*

$$\chi(B/\mathfrak{P}) = \mathfrak{p}^{[\mathbb{F}_{\mathfrak{P}}:\mathbb{F}_{\mathfrak{p}}]}.$$

Proof Note that the finite field $\mathbb{F}_{\mathfrak{P}}$ is a field extension of $\mathbb{F}_{\mathfrak{p}}$. Thus, as an A-module, $\mathbb{F}_{\mathfrak{P}}$ is an $\mathbb{F}_{\mathfrak{p}}$-vector space of dimension $[\mathbb{F}_{\mathfrak{P}} : \mathbb{F}_{\mathfrak{p}}]$. This implies that $\mathfrak{p}$ annihilates $\mathbb{F}_{\mathfrak{P}}$. Therefore, if we write $B/\mathfrak{P} \cong A/a_1 A \oplus \cdots \oplus A/a_n A$ in terms of its invariant factors, all the a_i's divide $\mathfrak{p}$. Since $\mathfrak{p}$ is irreducible, we get $a_i = \mathfrak{p}$, $1 \le i \le n$. $\qquad \square$

Now recall that in Example 4.2.12 we have computed that

$$\chi(^C(B/\mathfrak{P})) = \mathfrak{p}^{[\mathbb{F}_{\mathfrak{P}}:\mathbb{F}_{\mathfrak{p}}]} - 1,$$

where C denotes the Carlitz module. Therefore, we can rewrite $\zeta_C(B, 1)$ as

$$\zeta_C(B, 1) = \prod_{\mathfrak{P} \lhd B}' \frac{\chi(B/\mathfrak{P})}{\chi(^C(B/\mathfrak{P}))}.$$

This last expression makes it obvious that we can extend the definition of $\zeta_C(B, 1)$ to an arbitrary Drinfeld module ϕ over B by defining

$$\zeta_\phi(B, 1) = \prod_{\mathfrak{P} \lhd B}{}' \frac{\chi(B/\mathfrak{P})}{\chi(^\phi(B/\mathfrak{P}))}. \tag{7.6.9}$$

We leave it as an exercise for the reader to show that the product on the right-hand side of the above definition converges in F_∞. We point out that ϕ is allowed to have bad reductions and even reductions of "rank 0" at some of the primes of B (in other words, the non-constant coefficients of $\phi_T = \gamma(T) + g_1\tau + \cdots + g_r\tau^r$ might all belong to the same $\mathfrak{P}$). In this last case, we get an isomorphism of A-modules $^\phi(B/\mathfrak{P}) \cong B/\mathfrak{P}$, and in particular $\chi(B/\mathfrak{P}) = \chi(^\phi(B/\mathfrak{P}))$. This can happen only for finitely many $\mathfrak{P}$. Another point to emphasize is that $\zeta_\phi(B, 1)$ depends on the chosen model of ϕ over B, i.e., if ψ is a Drinfeld module over B isomorphic to ϕ over K, then $\zeta_\phi(B, 1)$ and $\zeta_\psi(B, 1)$ will in general differ by an $F^\times$-multiple.

Remark 7.6.22 Besides being similar to the residue of the Dedekind zeta function of a number field at 1, the product (7.6.9) is also similar to a product whose investigation lead Birch and Swinnerton-Dyer to their famous conjecture. Let E be an elliptic curve over $\mathbb{Q}$. Given a prime p, denote by $\#E(\mathbb{F}_p)$ the number of $\mathbb{F}_p$-rational points on the reduction of E at p. In the early 1960s, using one of the first computers available, Birch and Swinnerton-Dyer estimated

$$\prod_{p \leq N} \frac{p}{\#E(\mathbb{F}_p)}$$

for certain elliptic curves as N grows. This product formally approaches the value $L(E, 1)$ of the L-function of E at 1. Birch and Swinnerton-Dyer observed that

$$\prod_{p \leq N} \frac{p}{\#E(\mathbb{F}_p)} \sim \frac{C}{(\log N)^r}, \tag{7.6.10}$$

for some constant C and $r = \operatorname{rank} E(\mathbb{Q})$, and this lead them to conjecture that the order of vanishing of $L(E, s)$ at $s = 1$ is equal to the rank of the group $E(\mathbb{Q})$.[12] Although there have been many important breakthroughs, this conjecture remains open and is generally considered as the most important open problem about elliptic curves; see [Wil06].

Finally, we are ready to state Taelman's Class Number Formula, which is the following remarkable relationship:

---

[12] Goldfeld proved in [Gol82] that if the asymptotic of the partial products $\prod_{p \leq N} p/\#E(\mathbb{F}_p)$ is indeed as indicated in (7.6.10), then $\operatorname{rank} E(\mathbb{Q}) = \operatorname{ord}_{s=1} L(E, s)$ and moreover the Riemann hypothesis holds for $L(E, s)$, i.e., $L(E, s) \neq 0$ for $\operatorname{Re}(s) > 1$.

Theorem 7.6.23 (Taelman [Tae12b]) *For every Drinfeld module ϕ over B, we have*

$$\zeta_\phi(B, 1) = \mathrm{Reg}(\phi/B) \cdot \chi(\mathrm{Cl}(\phi/B)).$$

We already explained why $\mathrm{Reg}(\phi/B)$ is analogous to the regulator of a number field, although in the previous formula, compared with (7.6.8), $\mathrm{Reg}(\phi/B)$ also absorbs the discriminant of a number field and the order of the group of roots of unity in such a field. On the other hand, it is less clear why $\mathrm{Cl}(\phi/B)$ should be considered as an analogue of the class group of a number field or the Tate-Shafarevich group of an elliptic curve. In [Tae11], Taelman gives a more algebraic construction of $e_\phi^{-1}(^\phi B)$ and $\mathrm{Cl}(\phi/B)$ and shows that they are related to Ext modules in the category of shtukas. The usual class groups are similarly related to Ext groups, so this gives some justification for calling $\mathrm{Cl}(\phi/B)$ the class module of a Drinfeld module. But more importantly, Taelman [Tae12a] and then Anglés and Taelman [AT13, AT15] proved theorems about $\mathrm{Cl}(\phi/B)$ for the Carlitz module which are very similar to well-known classical theorems about the class groups of cyclotomic extensions, such as the Herbrand-Ribet theorem and the Mazur-Wiles theorem.

The proof of Theorem 7.6.23 lies outside the scope of this book, although Taelman's proof of this theorem in [Tae12b] is essentially self-contained and mostly uses methods of non-Archimedean analysis. We will only prove the theorem in the special case when $B = A$ and $\phi = C$.

Lemma 7.6.24 *Let C be the Carlitz module considered as being defined over A, i.e.,*

$$C : A \to A\{\tau\}, \quad C_T = T + \tau.$$

Then

(1) $\mathrm{Cl}(C/A) = 0$.
(2) $e_\phi^{-1}(^C A) = \log_C(1) \cdot A$.
(3) $\mathrm{Reg}(C/A) = \log_C(1)$.

Proof

(1) By Proposition 5.4.1, the exponential $e_C(x)$ and the logarithm $\log_C(x)$ have coefficients in the ring of integers A_∞ of F_∞. Moreover, e_C is entire and, by Exercise 5.4.4, the radius of convergence of $\log_C$ is $q^{q/(q-1)}$. Thus, using Theorem 2.7.24, we have $e_C(\log_C(\alpha)) = \alpha$ for all $\alpha \in A_\infty$, so $e_C : A_\infty \to A_\infty$ is surjective. In particular, $A_\infty \subset e_\phi(F_\infty)$. On the other hand, $F_\infty = A_\infty + A$ as $\mathbb{F}_q$-vector spaces, so $^\phi F_\infty = e_\phi(F_\infty) + {}^\phi A$. Thus,

$$\mathrm{Cl}(C/A) = \frac{^\phi F_\infty}{e_\phi(F_\infty) + {}^\phi A} = 0.$$

(2) By Corollary 7.6.18, $e_\phi^{-1}(^C A)$ is a free A-module of rank 1. Note that $\log_C(1) \in$
     $e_\phi^{-1}(^C A)$, since $\log_C$ converges at 1 and $e_C(\log_C(1)) = 1 \in A$. We claim
     that $\log_C(1)$ generates $e_\phi^{-1}(^C A)$. Let $u \in F_\infty$ be a generator of the A-module
     $e_\phi^{-1}(^C A)$. Then $b = e_C(u) \in A$ and $\log_C(1) = a \cdot u$ for some nonzero $a \in A$.
     Applying e_C to $\log_C(1) = a \cdot u$, we get

$$1 = e_C(au) = C_a(e_C(u)) = C_a(b).$$

We leave to the reader the verification of the following formula:

$$\deg_T(C_a(b)) = \begin{cases} -\infty & \text{if } b = 0; \\ 0 & \text{if } \deg_T(b) = \deg_T(a) = 0; \\ q^{\deg_T(a)-1} & \text{if } \deg_T(b) = 0, \deg_T(a) > 0; \\ q^{\deg_T(a)} \deg_T(b) & \text{if } \deg_T(b) > 0. \end{cases}$$

This formula and the equality $1 = C_a(b)$ imply that $\deg_T(b) = \deg_T(a) = 0$. In
particular, $\log_C(1)$ is an $\mathbb{F}_q^\times$-multiple of u, so $\log_C(1)$ is a generator of $e_\phi^{-1}(^C A)$.

(3) Observe from Proposition 5.4.1 that $\log_C(1) = 1+m$, with $|m| < 1$, so $\log_C(1)$
     is monic. Therefore, the claim $\text{Reg}(C/A) = \log_C(1)$ immediately follows from
     (2) and the definition of the regulator.   □

Theorem 7.6.25 *We have*

$$\zeta_C(A, 1) = \text{Reg}(C/A) \cdot \chi(\text{Cl}(C/A)).$$

Proof Lemma 7.6.24, combined with our earlier observation that $\zeta_C(A, 1) =$
$\zeta_C(1)$, implies that Theorem 7.6.23, in the case when $B = A$ and $\phi = C$, is
equivalent to the equality $\zeta_C(1) = \log_C(1)$. This we already proved as part of
Theorem 5.4.14.   □

7.6.3   Anderson's Theorem

Let ϕ be a Drinfeld module over B. We know that the A-module $e_\phi^{-1}(^\phi B)$ is free
and finitely generated. Let

$$U(\phi/B) = \left\{ e_\phi(\alpha) \mid \alpha \in e_\phi^{-1}(^\phi B) \right\}.$$

This is a finitely generated A-submodule of $^\phi B$, which, from the discussion at the
beginning of Sect. 7.6.2, can be thought of as the analogue of the group of totally

positive units of a number ring. It is a natural question to ask whether one can construct an explicit set of generators of $U(\phi/B)$. In this section we discuss an approach to this question due to Anderson in the case when K is a Carlitz cyclotomic extension and $\phi = C$. To motivate what is to follow, we again recall some classical facts from number theory.

Let p be an odd prime and let $\zeta_p = \exp(2\pi i/p)$. The ring of integers of the cyclotomic extension $\mathbb{Q}(\zeta_p)$ is $\mathbb{Z}[\zeta_p]$. Since $\mathbb{Q}(\zeta_p)/\mathbb{Q}$ is totally imaginary, by Dirichlet's Unit Theorem the group of units $\mathbb{Z}[\zeta_p]^\times$ has rank $(p - 3)/2$. It is not hard to show that the numbers

$$\frac{1 - \zeta_p^a}{1 - \zeta_p} \quad \text{for} \quad a = 1, \ldots, p - 1$$

are in $\mathbb{Z}[\zeta_p]^\times$; see [Was82, p. 2]. The subgroup $\mathscr{C}_p \subseteq \mathbb{Z}[\zeta_p]^\times$ generated by these numbers and the roots of unity $\pm\zeta_p^a$ is the group of *cyclotomic units* of $\mathbb{Q}(\zeta_p)$. It is a central result of the theory of cyclotomic fields, due to Dirichlet and Kummer, that the index $[\mathbb{Z}[\zeta_p]^\times : \mathscr{C}_p]$ is finite and in fact is equal to the class number of $\mathbb{Q}(\zeta_p)^+$; see [Was82, Ch. 8]. Furthermore, the famous (still open) Kummer-Vandiver conjecture predicts that p does not divide $[\mathbb{Z}[\zeta_p]^\times : \mathscr{C}_p]$.

Now let $\mathfrak{p} \lhd A$ be a prime, let $\zeta_\mathfrak{p} = e_C(\pi_C/\mathfrak{p})$, let $K = F(\zeta_\mathfrak{p})$ be the $\mathfrak{p}$-th Carlitz cyclotomic extension of F, and let B be the integral closure of A in K. Motivated by the above classical results, Anderson [And96] constructed an analogue of cyclotomic units for $F(\zeta_\mathfrak{p})$ and proved a Dirichlet-Kummer type theorem for them. He also proved formulas relating the values at 1 of the Goss analogues of Dirichlet's L-functions to Carlitz logarithms of algebraic numbers in $\mathbb{C}_\infty$ similar to the classical formulas that can be found, for example, in [Was82, p. 37].

To describe Anderson's results, we first need to introduce certain formal power series. Let x and z be independent indeterminates over $\mathbb{C}_\infty$. Given $m \in \mathbb{Z}_{\geq 0}$, define

$$\ell_m(x, z) = \sum_{a \in A_+} \frac{C_a(x)^m}{a} z^{q^{\deg(a)}},$$

and put

$$S_m(x, z) := e_C(\ell_m(x, z))$$

$$= \sum_{i=0}^{\infty} \sum_{a \in A_+} \frac{1}{D_i} \left(\frac{C_a(x)^m}{a}\right)^{q^i} z^{q^{i+\deg(a)}}.$$

Theorem 7.6.26 (Anderson) *The power series $S_m(x, z)$ is a polynomial in x and z with coefficients in A.*

Proof Anderson's proof of this theorem in [And96] is indirect and quite ingenious. First, he defines a function $\|\cdot\| : F[x] \to \mathbb{R}_{\geq 0}$ by[13]

$$\|f\| := \sup_{\alpha \in F_\infty/A} |f(e_C(\pi_C \alpha))| \tag{7.6.11}$$

and shows that for all $f, g \in F[x]$ we have

$$\|f + g\| \leq \max(\|f\|, \|g\|), \tag{7.6.12}$$

$$\|fg\| \leq \|f\| \cdot \|g\|,$$

$$\|f\| = 0 \text{ if and only if } f = 0.$$

Next, writing $S_m(x, z) = \sum_{n \geq 0} w_n(x) z^n$, he shows that $\|w_n\| \to 0$ as $n \to \infty$ using explicit estimates. Finally, he proves that $w_n(x) \in A[x]$ and observes that for $w(x) \in A[x]$ the inequality $\|w\| \leq 1$ implies that $w(x) \in \mathbb{F}_q$. Hence, $w_n \to 0$ is possible if and only if $w_n = 0$ for all sufficiently large n. $\qquad\square$

Example 7.6.27 Anderson deduced from his estimates that $S_m(x, z) = x^m z$, provided $0 \leq m < q$. He also computed several explicit examples when $m \geq q$, one of which is the following for $q = 3$:

$$S_6(x, z) = x^6 z + (2T(1 + T)(2 + T)x^6 + x^{12})z^3 + x^{18}z^9.$$

These polynomials become increasingly complicated as m grows; cf. [And96, p. 190].

Remarks 7.6.28

(1) Because of Theorem 7.6.26, one says that the power series $\ell_m(x, z)$ is *log-algebraic*:[14] it is formally the logarithm of an algebraic object over A, a polynomial. Note that because $S_1(x, 1) = x$, we have

$$\log_C(x) = \sum_{a \in A_+} \frac{C_a(x)}{a},$$

which is similar to the power series expansion of the classical logarithm

$$\log(1 - x) = -\sum_{n \geq 1} \frac{x^n}{n}.$$

---

[13] This is well-defined because F_∞/A is compact.

[14] As in the title of [And96].

(2) In the case when m is a sum of less than q q-powers, Thakur gave a different (elementary) proof of Theorem 7.6.26 in [Tha04, §8.10] based on explicit calculations with the coefficients of $S_m(x, z)$. This has the advantage that it provides explicit formulas for the polynomials $S_m(x, z)$, gives bounds on the degrees, and shows that q occurs as a parameter. Another proof of Thakur's theorem was given by Papanikolas in [Pap22] using hyperderivatives of polynomials over finite fields.

Following Thakur's argument in [Tha04, §8.10], we give a direct proof of a very special case of Theorem 7.6.26:

Proposition 7.6.29 *We have*

$$S_1(x, z) = xz.$$

Proof For $i \in \mathbb{Z}_{\geq 0}$, denote

$$A_{i+} = \{a \in A_+ \mid \deg(a) = i\}.$$

Write

$$S_1(x, z) = \sum_{n=0}^{\infty} \sum_{a \in A_+} \frac{1}{D_n} \left(\frac{C_a(x)}{a} \right)^{q^n} z^{q^{n+\deg(a)}}$$

$$= \sum_{n=0}^{\infty} w_n z^{q^n},$$

where

$$w_n = \sum_{i=0}^{n} \frac{1}{D_{n-i}} \left(\sum_{a \in A_{i+}} \frac{C_a(x)}{a} \right)^{q^{n-i}}. \tag{7.6.13}$$

Next, we separately examine the sum $\sum_{a \in A_{i+}} C_a(x)/a$. We have

$$C_a(x) = \sum_{n=0}^{\deg(a)} \left\{ \begin{matrix} a \\ n \end{matrix} \right\} x^{q^n},$$

where, by Lemma 5.4.2 and Proposition 5.4.3, the coefficient $\left\{ \begin{smallmatrix} a \\ n \end{smallmatrix} \right\}$ is equal to

$$\left\{ \begin{matrix} a \\ n \end{matrix} \right\} = \sum_{i=0}^{n} (-1)^{n-i} \frac{a^{q^i}}{D_i L_{n-i}^{q^i}}.$$

Thus,

$$C_a(x) = \sum_{n=0}^{\deg(a)} \sum_{i=0}^{n} (-1)^{n-i} \frac{a^{q^i} x^{q^n}}{D_i L_{n-i}^{q^i}}$$

$$= \sum_{i=0}^{\deg(a)} \frac{a^{q^i}}{D_i} \sum_{n=i}^{\deg(a)} (-1)^{n-i} \frac{x^{q^n}}{L_{n-i}^{q^i}}.$$

Using this, we get

$$\sum_{a \in A_{i+}} \frac{C_a(x)}{a} = \sum_{a \in A_{i+}} \frac{1}{a} \sum_{j=0}^{i} \frac{a^{q^j}}{D_j} \sum_{k=j}^{i} (-1)^{k-j} \frac{x^{q^k}}{L_{k-j}^{q^j}}$$

$$= \sum_{j=0}^{i} \frac{1}{D_j} \sum_{k=j}^{i} (-1)^{k-j} \frac{x^{q^k}}{L_{k-j}^{q^j}} S_i(q^j - 1),$$

where

$$S_i(q^j - 1) = \sum_{a \in A_{i+}} a^{q^j - 1}.$$

On the other hand, by Proposition 5.4.12,

$$S_i(q^j - 1) = \begin{cases} 0 & \text{if } 0 \le j < i, \\ (-1)^i D_i / L_i & \text{if } j = i. \end{cases}$$

Therefore,

$$\sum_{a \in A_{i+}} \frac{C_a(x)}{a} = (-1)^i \frac{x^{q^i}}{L_i}.$$

This is a nice identity in its own right; substituting this identity into (7.6.13), we get

$$w_n = \sum_{i=0}^{n} \frac{1}{D_{n-i}} \left((-1)^i \frac{x^{q^i}}{L_i} \right)^{q^{n-i}} = x^{q^n} \sum_{i=0}^{n} (-1)^i \frac{1}{D_{n-i} L_i^{q^{n-i}}}.$$

Finally, by Exercise 5.4.1, the second sum is 0 for $n \ge 1$, so

$$w_n = \begin{cases} x & n = 0, \\ 0 & n \ge 1. \end{cases}$$

(Note that Exercise 5.4.1 is just a special case of (5.1.9).) This implies that $S_1(x, z) = xz$.   □

Now returning to the Carlitz cyclotomic extension $K = F(\zeta_\mathfrak{p})$, we define the module $\mathscr{C}_\mathfrak{p}$ of *Anderson-Carlitz cyclotomic units* as the A-span under the Carlitz module of the B-valued points

$$\{S_m(e_C(\pi_C a/\mathfrak{p}), 1) \mid m \in \mathbb{Z}_{\geq 0}, 0 \neq a \in A, \deg(a) < \deg(\mathfrak{p})\}.$$

(Note that $e_C(\pi_C a/\mathfrak{p}) = C_a(\zeta_\mathfrak{p}) \in B$, so $S_m(e_C(\pi_C a/\mathfrak{p}), 1) \in B$ because $S_m(x, z)$ is a polynomial with coefficients in A.) From this definition it is not clear whether $\mathscr{C}_\mathfrak{p}$ is finitely generated, although this follows from the next proposition:

Proposition 7.6.30 *The module $\mathscr{C}_\mathfrak{p}$ is an A-submodule of $U(C/B)$. In particular, $\mathscr{C}_\mathfrak{p}$ is finitely generated.*

Proof As we have observed, $S_m(e_C(\pi_C b/\mathfrak{p}), 1) \in B$ for all $0 \neq b \in A$ and $m \in \mathbb{Z}_{\geq 0}$. On the other hand, by definition, we have

$$S_m(e_C(\pi_C b/\mathfrak{p}), 1) = S_m(C_b(\zeta_\mathfrak{p}), 1) = e_C\left(\sum_{a \in A_+} \frac{C_{ab}(\zeta_\mathfrak{p})^m}{a}\right).$$

It is easy to see that the inner sum converges in K_∞, so $S_m(e_C(\pi_C b/\mathfrak{p}), 1)$ is the exponential of an element in K_∞. Combining these two observations and comparing them to the definition of $U(C/B)$, we conclude that $S_m(e_C(\pi_C b/\mathfrak{p}), 1) \in U(C/B)$.   □

Of course, Anderson's paper [And96] predates [Tae10] by about fifteen years. In [And96], Anderson proves that $\mathscr{C}_\mathfrak{p}$ is finitely generated by a completely different argument and computes its rank. This is quite deep and is accomplished by relating $\mathscr{C}_\mathfrak{p}$ to special values of Goss L-functions.

Theorem 7.6.31 (Anderson) *The A-module $\mathscr{C}_\mathfrak{p}$ is finitely generated and has rank*

$$(|\mathfrak{p}| - 1)(q - 2)/(q - 1).$$

Next, we show that $U(C/B)$ has the same rank as its submodule $\mathscr{C}_\mathfrak{p}$.

Proposition 7.6.32 *The rank of $U(C/B)$ is*

$$(|\mathfrak{p}| - 1)(q - 2)/(q - 1).$$

Proof By Corollary 7.6.18, the rank of $e_C^{-1}(^C B)$ is

$$[F(\zeta_\mathfrak{p}) : F] = |\mathfrak{p}| - 1.$$

The place ∞ splits completely in $F(\zeta_{\mathfrak{p}})^+$ and then every place of $F(\zeta_{\mathfrak{p}})^+$ over ∞ totally ramifies in $K = F(\zeta_{\mathfrak{p}})$ with ramification index $q - 1$; cf. Theorem 7.1.13. Thus, the number of places of K over ∞ is $(|\mathfrak{p}| - 1)/(q - 1)$. In the proof of Theorem 7.1.13 we showed that $\pi_C \in K_{\infty'}$ for any place $\infty' \mid \infty$. Therefore, the kernel of

$$e_C \colon e_C^{-1}(^C B) \longrightarrow U(C/B)$$

is

$$\prod_{\infty' \mid \infty} \pi_C A.$$

This latter A-module has rank $(|\mathfrak{p}| - 1)/(q - 1)$, so the rank of $U(C/B)$ is

$$(|\mathfrak{p}| - 1) - \frac{|\mathfrak{p}| - 1}{q - 1} = \frac{(|\mathfrak{p}| - 1)(q - 2)}{q - 1}.$$

$\square$

Definition 7.6.33 Let R be a Dedekind domain and let M be an R-module. Let $N \subseteq M$ be an R-submodule. The *divisible closure* of N in M, denoted $\sqrt{N}$, is the set of all $m \in M$ such that $r \cdot m \in N$ for some nonzero $r \in R$.

Remarks 7.6.34

(1) Theorem 7.6.31 and Proposition 7.6.32 imply that, without a priori having the module $U(C/B)$ of units, one can *define* $U(C/B)$ as the divisible closure of $\mathscr{C}_{\mathfrak{p}}$ in $^C B$; see Exercise 7.6.13.

(2) In [AT15], Anglès and Taelman proved that

$$\chi(\mathrm{Cl}(C/B^+)) = \chi(U(C/B)/\mathscr{C}_{\mathfrak{p}}),$$

where B^+ is the integral closure of A in $F(\zeta_{\mathfrak{p}})^+$, in analogy with the Dirichlet-Kummer theorem. On the other hand, in [AT13] they gave examples where $\chi(\mathrm{Cl}(C/B^+))$ is divisible by $\mathfrak{p}$, so the analogue of the Kummer-Vandiver conjecture fails in this context.

(3) Recall that $\zeta_{\mathfrak{p}} B$ is the unique prime ideal of B over $\mathfrak{p}$; see Theorem 7.1.16. Thus, $C_a(\zeta_{\mathfrak{p}})/\zeta_{\mathfrak{p}}$ lies in $B^\times = (B^+)^\times$ for any $0 \neq a \in A_{<\deg(\mathfrak{p})}$, where the equality $B^\times = (B^+)^\times$ follows from Exercise 7.1.7. Note that $C_a(\zeta_{\mathfrak{p}})/\zeta_{\mathfrak{p}}$ are the analogues of classical cyclotomic units $(\zeta_p^n - 1)/(\zeta_p - 1), 0 < n < p$. Let $\mathscr{C}_{\mathfrak{p}}$ be the subgroup of $(B^+)^\times$ generated by

$$\left\{ C_a(\zeta_{\mathfrak{p}})/\zeta_{\mathfrak{p}} \mid 0 \neq a \in A_{<\deg(\mathfrak{p})} \right\}.$$

In [GaRo81], Galovich and Rosen proved that

$$\# \mathrm{Cl}(B^+) = [(B^+)^\times : \mathscr{C}_p],$$

where $\mathrm{Cl}(B^+)$ is the usual class group of B^+. This result is a "geometric" analogue of the Dirichlet-Kummer theorem, whereas the result of Anglès and Taelman $\chi(\mathrm{Cl}(C/B^+)) = \chi(U(C/B)/\mathscr{C}_p)$ is an "arithmetic" analogue of that theorem.

Exercises

7.6.1 Determine if the direct product $\prod_{n=1}^{\infty} A$ of countably many copies of A is free and/or tame.

7.6.2 Prove the claim in Example 7.6.7.

7.6.3 Let $\alpha \in K^\times$. Prove that $h(\alpha) = 0$ if and only if α is a root of unity.

7.6.4 Complete the proof of Lemma 7.6.10.

7.6.5 Let ϕ be a Drinfeld module over K. Show that if $\gamma \in (^\phi K)_{\mathrm{tor}}$, then $\hat{h}_\phi(\alpha + \gamma) = \hat{h}_\phi(\alpha)$ for all $\alpha \in K$.

7.6.6 Let ϕ and ψ be Drinfeld modules over K. Let $u: \phi_1 \to \phi_2$ be an isogeny over K. Prove that

$$\deg(u(x)) \cdot \hat{h}_\phi(\alpha) = \hat{h}_\psi(u(\alpha)) \qquad \text{for all } \alpha \in K.$$

Note that this implies (1) in Proposition 7.6.12.

7.6.7 Let v be a place of K and let K_v denote the completion of K at v. For $\alpha \in K_v$, define $\tilde{v} = \min\{0, v(\alpha)\}$. Let $\phi: A \to K_v\{\tau\}$ be a Drinfeld module of rank r. Given $b \in A$ of positive degree, show that the limit

$$\hat{h}_{\phi,v}(\alpha) = \lim_{n \to \infty} \frac{\tilde{v}(\phi_{b^n}(\alpha))}{q^{nr \deg(b)}}$$

exists and is independent of b. This local version of the height was introduced by Poonen in [Poo95]. Prove the following:

(a) The function $\hat{h}_{\phi,v}(\alpha) - \tilde{v}(\alpha)$ is bounded independently of $\alpha \in K_v$.
(b) We have $\hat{h}_{\phi,v}(\alpha \pm \beta) \geq \min\{\hat{h}_{\phi,v}(\alpha), \hat{h}_{\phi,v}(\beta)\}$ for all $\alpha, \beta \in K$.
(c) If ϕ is a Drinfeld module over K, then for all $\alpha \in K$, we have

$$\hat{h}_\phi(\alpha) = -\sum_v [k_v : k] \cdot \hat{h}_{\phi,v}(\alpha),$$

where the sum is over all places of K and k is the algebraic closure of $\mathbb{F}_q$ in K.

7.6.8 Let ϕ be a Drinfeld module defined over F_∞. In Chap. 5, we showed that the exponential e_ϕ induces the following short exact sequence of A-modules:

$$0 \longrightarrow \Lambda_\phi \longrightarrow \mathbb{C}_\infty \xrightarrow{\;e_\phi\;} {}^\phi\mathbb{C}_\infty \longrightarrow 0.$$

Prove that this restricts to a short exact sequence

$$0 \longrightarrow \Lambda_\phi \longrightarrow F_\infty^{\mathrm{sep}} \xrightarrow{\;e_\phi\;} {}^\phi F_\infty^{\mathrm{sep}} \longrightarrow 0.$$

Moreover, the above sequence is equivariant with respect to the action of $G :=$ $\mathrm{Gal}(F_\infty^{\mathrm{sep}}/F_\infty)$, and taking invariants gives an exact sequence

$$0 \longrightarrow \Lambda_\phi^G \longrightarrow F_\infty \xrightarrow{\;e_\phi\;} {}^\phi F_\infty \longrightarrow H^1(G, \Lambda_\phi) \longrightarrow 0.$$

7.6.9 Let K/F be a finite extension and let $\phi \colon A \to K\{\tau\}$ be a Drinfeld module. Show that ${}^\phi K_\infty / e_\phi(K_\infty)$ is not finitely generated as an A-module.

7.6.10 Show that the product (7.6.9) converges in F_∞.

7.6.11 Prove that $S_0(x, z) = z$.

7.6.12 Let $\|\cdot\|$ be the Anderson norm defined in (7.6.11). Prove that $\|\cdot\|$ has the following properties:

(a) The properties listed in (7.6.12).
(b) If $f \in A[x]$ and $\|f\| < 1$, then $f = 0$.
(c) For all $f \in F[x]$ and $a \in A$,

$$\|f(C_a(x))\| = \|f\|.$$

7.6.13 Let $\mathfrak{p} \lhd A$ be a prime, let $\zeta_\mathfrak{p} = e_C(\pi_C/\mathfrak{p})$, let $K = F(\zeta_\mathfrak{p})$, and let B be the integral closure of A in K. Let $\mathscr{C}_\mathfrak{p}$ be the module of Anderson-Carlitz cyclotomic units. Prove the following:

(a) $({}^C B)_{\mathrm{tor}} = U(C/B)_{\mathrm{tor}} = (\mathscr{C}_\mathfrak{p})_{\mathrm{tor}}$.
(b) $U(C/B)$ is the divisible closure of $\mathscr{C}_\mathfrak{p}$ in ${}^C B$.

Appendix A
Drinfeld Modules for General Function Rings

In this appendix, we define Drinfeld modules for general function rings and discuss the differences and similarities of the general theory of Drinfeld modules with the special case of $\mathbb{F}_q[T]$ that was the focus of the previous chapters.

Let F be a finite extension of $\mathbb{F}_q(T)$ and let ∞ be a fixed place of F. We will assume that F is a geometric extension of $\mathbb{F}_q(T)$, so $\mathbb{F}_q$ is the field of constants of F. We define a subring F as

$$A := \{\alpha \in F \mid v(\alpha) \geq 0 \text{ for all places } v \neq \infty \text{ of } F\}.$$

The ring A is a Dedekind domain but generally it is not a PID; we call such rings *function rings*.

Example A.1 Let $F = \mathbb{F}_q(T)$. If ∞ is the place with uniformizer $1/T$, then $A = \mathbb{F}_q[T]$. On the other hand, if we choose ∞ to be the place corresponding to an irreducible polynomial $\mathfrak{p}$ of degree $d \geq 1$, then the ring A consists of fractions $a/\mathfrak{p}^s$ such that $a \in \mathbb{F}_q[T]$, $s \geq 0$, and

$$\deg(a) \leq s \cdot d.$$

This last inequality is needed to satisfy the assumption that $\mathrm{ord}_{1/T}(\alpha) \geq 0$. The exact sequence (7.5.9) implies that $\mathrm{Cl}(A) \cong \mathbb{Z}/d\mathbb{Z}$ since in this case the curve X is the projective line $\mathbb{P}^1_{\mathbb{F}_q}$ over $\mathbb{F}_q$, so its Jacobian variety J_X is trivial. It is also easy to see directly that A is not a PID when $d \geq 2$; indeed, the ideal $(1/\mathfrak{p}, T/\mathfrak{p})$ of A is not principal.

Another explicit example of possible A in our current setting, which is not a PID, is the ring B from Example 7.5.20.

Next, we define a function $|\cdot| : F \to \mathbb{R}_{\geq 0}$ by first putting $|0| = 0$ and $|a| = \#(A/aA)$ for all $0 \neq a \in A$, and then extending it to F by $|a/b| = |a|/|b|$ for all nonzero $a, b \in A$. We leave it to the reader to verify that $|\cdot|$ is a non-Archimedean

© The Author(s), under exclusive license to Springer Nature Switzerland AG 2023
M. Papikian, *Drinfeld Modules*, Graduate Texts in Mathematics 296,
https://doi.org/10.1007/978-3-031-19707-9

absolute value on F. Since $|a| \geq 1$ for all nonzero $a \in A$, this absolute value is
an absolute value at ∞ (which is well-defined up to equivalence). Furthermore, for
$0 \neq \alpha \in F$ we define $\deg(\alpha) = \log_q |\alpha|$; note that $-\deg$ is a non-Archimedean
valuation on F equivalent to a valuation at ∞. These definitions extend to the ideals
of A: for a nonzero ideal $\mathfrak{n} \triangleleft A$ we put $|\mathfrak{n}| = \#(A/\mathfrak{n})$ and $\deg(\mathfrak{n}) = \log_q |\mathfrak{n}|$. Note that
in our usual setting of $A = \mathbb{F}_q[T]$, $|\cdot|$ is the normalized absolute value at $\infty = 1/T$
defined in Chap. 3 and $\deg = \deg_T$. From now on we fix $|\cdot|$ as the absolute value at
∞. Let F_∞ be the completion of F with respect to $|\cdot|$, and let $\mathbb{C}_\infty$ be the completion
of an algebraic closure of F_∞.

Let K be an A-field by which, as before, we mean a field equipped with a ring
homomorphism $\gamma: A \to K$.

Definition A.2 A *Drinfeld A-module* over K is an $\mathbb{F}_q$-algebra homomorphism

$$\phi: A \longrightarrow K\{\tau\}$$

$$a \longmapsto \phi_a$$

such that $\phi(A) \not\subset K$ and $\partial \phi_a = \gamma(a)$.

Remarks A.3

(1) The main difference of Definition A.2 in comparison to Definition 3.2.2 is the
    absence of "rank." In fact, since A is generally not monogenic over $\mathbb{F}_q$, it is
    not immediately clear whether the notion of rank even makes sense (but see
    Corollary A.14).

(2) Unlike the case of $\mathbb{F}_q[T]$, given an A-field K it is not clear whether there are
    Drinfeld A-modules over K. For example, if $K = F$, then there are no Drinfeld
    A-modules defined over F if the class group of A is nontrivial. To see this,
    observe that a Drinfeld $\mathbb{F}_q[T]$-module ϕ with complex multiplication by the
    maximal $\mathbb{F}_q[T]$-order B in a totally imaginary extension F of $\mathbb{F}_q(T)$ can be
    considered as a Drinfeld B-module over $\mathbb{C}_\infty$, with $\iota: B \hookrightarrow \mathbb{C}_\infty$ in (7.5.8)
    playing the role of γ. In Sect. 7.5, we proved that the smallest field of definition
    of ϕ is the Hilbert class field of B. This is a nontrivial extension of F if the class
    group of B is nontrivial.

(3) Given a function ring A, choose $T \in A$ transcendental over $\mathbb{F}_q$. Then a Drinfeld
    A-module ϕ can be considered as a Drinfeld $\mathbb{F}_q[T]$-module ψ defined by $\psi_T =
    \phi_T$. The rank of ψ will be the rank of ϕ multiplied by the rank A as a free
    $\mathbb{F}_q[T]$-module; also note that $A \subseteq \operatorname{End}(\psi)$. This allows one to reduce certain
    problems about general Drinfeld modules to Drinfeld $\mathbb{F}_q[T]$-modules of higher
    rank.

From now on we assume that A is a fixed function ring and by "Drinfeld module"
we mean Drinfeld A-module.

Definition A.4 A *morphism* $u: \phi \to \psi$ of Drinfeld modules over K is an element
$u \in K\{\tau\}$ such that $u\phi_a = \psi_a u$ for all $a \in A$. A nonzero morphism is called an
isogeny.

Lemma A.5 *Let ϕ be a Drinfeld module over K. Then*

(1) ϕ is an embedding.
(2) There exists $r \in \mathbb{Q}_{>0}$ such that

$$\deg_x \phi_a(x) = |a|^r \quad \text{for all } 0 \neq a \in A.$$

(3) Assume $\mathfrak{p} := \text{char}_A(K)$ is nonzero. Let $\text{ord}_\mathfrak{p}$ be the normalized valuation on F associated with the prime $\mathfrak{p}$. There exists $H \in \mathbb{Q}_{>0}$ such that

$$\text{ht}(\phi_a) = H \cdot \text{ord}_\mathfrak{p}(a) \cdot \deg(\mathfrak{p}) \quad \text{for all } 0 \neq a \in A.$$

Proof

(1) Let $\mathfrak{n}$ be the kernel of ϕ. If $\mathfrak{n} \neq 0$, then $\mathfrak{n}$ is a maximal ideal since A is a Dedekind domain and $K\{\tau\}$ has no zero-divisors. In that case, $\phi(A) \cong A/\mathfrak{n}$ is a finite field, so $\phi(A) \subseteq K$ since $K\{\tau\}^\times = K^\times$. This contradicts one of the assumptions of Definition A.2.
(2) Clearly

$$\deg_x \phi_a(x) \geq 0 \text{ for all nonzero } a.$$

$$\deg_x \phi_a(x) > 0 \text{ for some } a \text{ since } \phi(A) \not\subset K.$$

$$\deg_x \phi_{ab}(x) = \deg_x \phi_a(\phi_b(x)) = \deg_x \phi_a(x) \cdot \deg_x \phi_b(x).$$

$$\deg_x \phi_{a+b}(x) = \deg_x(\phi_a(x) + \phi_b(x)) \leq \max(\deg_x \phi_a(x), \deg_x \phi_b(x)).$$

Thus, $A \twoheadrightarrow \mathbb{Z}_{\geq 0}, u \mapsto \deg_x \phi_a(x)$, extends to a nontrivial absolute value on F, which must be equivalent to $|\cdot|$. Hence there exists $r > 0$ such that $\deg_x \phi_a(x) = |a|^r$ for all nonzero $a \in A$.
(3) By Lemma 3.1.11, ht: $A \to \mathbb{Z}_{\geq 0} \cup \{+\infty\}, a \mapsto \text{ht}\, \phi_a$, is a valuation on A, which can be uniquely extended to a valuation on F. If $\text{char}_A(K)$ is nonzero, then ht on F is a nontrivial valuation such that $\text{ht}(a) > 0$ for all $a \in \mathfrak{p}$. Hence ht is equivalent to $\text{ord}_\mathfrak{p}$, so $\text{ht} = H \cdot \deg(\mathfrak{p}) \cdot \text{ord}_\mathfrak{p}$ for some $H \in \mathbb{Q}_{>0}$.

$\square$

Definition A.6 The number r in Lemma A.5 is called the *rank* of ϕ, and the number H is called the *height* of ϕ. We will prove shortly that both of these numbers are integers and $0 < H \leq r$.

As in Sect. 3.3, we denote the A-module of morphism $\phi \to \psi$ over K by $\text{Hom}_K(\phi, \psi)$.

Lemma A.7 *Suppose $u\colon \phi \to \psi$ is an isogeny. We have:*

(1) ϕ and ψ have the same rank and height.

(2) If $\operatorname{char}_A(K) = 0$, then u is separable and the map

$$\operatorname{Hom}_K(\phi, \psi) \longrightarrow K$$

$$u \longmapsto \partial u$$

is an injective homomorphism. In particular, in this case, $\operatorname{End}_K(\phi)$ is a commutative ring.

Proof Essentially the same proof as for Proposition 3.3.4, with ϕ_T replaced by an arbitrary ϕ_a of positive degree. □

Let $\mathfrak{n} \lhd A$ be a nonzero ideal. Let ϕ be a Drinfeld module over K. Let $I_{\phi,\mathfrak{n}} = K\{\tau\}\phi(\mathfrak{n})$ be the left ideal of $K\{\tau\}$ generated by all ϕ_a, $a \in \mathfrak{n}$; cf. Sect. 7.5. Since $K\{\tau\}$ has a right division algorithm, $I_{\phi,\mathfrak{n}}$ is a principal ideal generated by some $\phi_\mathfrak{n} \in K\{\tau\}$, well-defined up to a $K^\times$-multiple. If $\mathfrak{n} = (a)$ is principal in A, then ϕ_a can be taken as $\phi_\mathfrak{n}$; on the other hand, if $\mathfrak{n}$ is not principal, then $\phi_\mathfrak{n}$ is not in $\phi(A)$ and does not necessarily commute with the elements $\phi_a \in K\{\tau\}$, $a \in A$. We define $\phi[\mathfrak{n}]$ as the set of zeros of $\phi_\mathfrak{n}$ (without multiplicities),

$$\phi[\mathfrak{n}] := \{\alpha \in \overline{K} \mid \phi_\mathfrak{n}(\alpha) = 0\}.$$

For nonzero $a \in A$, we define $\phi[a]$ to be the set of zeros of $\phi_a(x)$ in $\overline{K}$. Note that $\phi[a] = \phi[aA]$.

Lemma A.8 *We have*

$$\phi[\mathfrak{n}] = \bigcap_{a \in \mathfrak{n}} \phi[a].$$

Proof Let $z \in \bigcap_{a \in \mathfrak{n}} \phi[a]$. We can write, cf. (7.5.1),

$$\phi_\mathfrak{n} = f_1\phi_{a_1} + \cdots + f_m\phi_{a_m},$$

for suitable $f_1, \ldots, f_m \in K\{\tau\}$ and $a_1, \ldots, a_m \in \mathfrak{n}$. Now $\phi_\mathfrak{n}(z) = \sum f_i(\phi_{a_i}(z)) = 0$, so $z \in \phi[\mathfrak{n}]$. Conversely, suppose $z \in \phi[\mathfrak{n}]$ and $a \in \mathfrak{n}$. Then $\phi_a \in I_{\phi,\mathfrak{n}}$, so $\phi_a = f\phi_\mathfrak{n}$ for some $f \in K\{\tau\}$. Now $\phi_a(z) = f(\phi_\mathfrak{n}(z)) = 0$, so $z \in \phi[a]$. □

Using the previous lemma, it is easy to check that $\phi[\mathfrak{n}]$ is an A-module via $b \circ \alpha = \phi_b(\alpha)$, since each $\phi[a]$ is an A-module. In fact, $\phi[\mathfrak{n}]$ is an $A/\mathfrak{n}$-module since any $a \in \mathfrak{n}$ acts as 0 on $\phi[\mathfrak{n}]$. To deduce the structure of this module, we will use the structure theorem for finitely generated modules over Dedekind domains:

Theorem A.9 *Let R be a Dedekind domain and let M be a finitely generated R-module. Then*

$$M \cong R^{\oplus n} \oplus I \oplus M_{\text{tor}},$$

where $n \geq 0$, $I \lhd R$ is an ideal of R, and

$$M_{\text{tor}} \cong R/P_1^{e_1} \oplus R/P_2^{e_2} \oplus \cdots \oplus R/P_s^{e_s}$$

for some $s \geq 0$ and powers $P_i^{e_i}$, $e_i \geq 1$, of (not necessarily distinct) prime ideals. The number n is uniquely determined by M, the ideals $P_i^{e_i}$ for $i = 1, \ldots, s$ are unique, and the ideal I is unique up to multiplication by a principal ideal.

Proof See Theorem 22 on page 771 of [DF04]. □

Definition A.10 Let L be the fraction field of the Dedekind domain R, and let M be a finitely generated R-module. The number $\dim_L(M \otimes_R L)$ is called the *rank* of M. Note that the rank of the module M in Theorem A.9 is n if $I = 0$ and $n + 1$ if $I \neq 0$.

Remark A.11 Note that $\mathbb{F}_q[T]$ (or more generally a PID) is a Dedekind domain whose ideals are principal. In that case, the ideal I in Theorem A.9 is isomorphic to R as an R-module (assuming it is nonzero), so Theorem A.9 implies the structure theorem of Sect. 1.2.

Theorem A.12 *Let ϕ be a Drinfeld module over K, let $\mathfrak{p}$ be a prime, and let $n \in \mathbb{Z}_{>0}$.*

(1) If $\mathfrak{p} \neq \operatorname{char}_A(K)$, then

$$\phi[\mathfrak{p}^n] \cong \prod_{i=1}^{r} A/\mathfrak{p}^n,$$

where r is the rank of ϕ.
(2) If $\mathfrak{p} = \operatorname{char}_A(K)$, then

$$\phi[\mathfrak{p}^n] \cong \prod_{i=1}^{r-H} A/\mathfrak{p}^n,$$

where H is the height of ϕ.

Proof Suppose $\mathfrak{q}$ is a prime different from $\mathfrak{p}$ and $z \in \phi[\mathfrak{p}^n] \cap \phi[\mathfrak{q}]$. Since $\mathfrak{p}^n + \mathfrak{q} = A$, z is a root of $\phi_1(x) = x$, so $z = 0$. This observation, combined with Theorem A.9, implies that

$$\phi[\mathfrak{p}^n] \cong A/\mathfrak{p}^{e_1} \oplus \cdots \oplus A/\mathfrak{p}^{e_s}. \tag{A.1}$$

Under this isomorphism, the submodule $\phi[\mathfrak{p}] \subseteq \phi[\mathfrak{p}^n]$ corresponds to $(A/\mathfrak{p})^{\oplus s}$ on the right hand side, so s does not depend on n. Next, we claim that $e_1 = \cdots = e_s = n$. Note that this claim will follow if we show that $\phi[\mathfrak{p}^m] \cong (A/\mathfrak{p}^m)^{\oplus s}$ for some $m \geq n$ since under this isomorphism the submodule $\phi[\mathfrak{p}^n] \subseteq \phi[\mathfrak{p}^m]$ corresponds to the submodule of $(A/\mathfrak{p}^m)^{\oplus s}$ annihilated by $\mathfrak{p}^n$, which is $(A/\mathfrak{p}^n)^{\oplus s}$.

The class number of A is finite, so there is $m \geq n$ such that $\mathfrak{p}^m = (a)$ is a principal ideal. Suppose

$$\phi[a^2] = \phi[\mathfrak{p}^{2m}] \cong A/\mathfrak{p}^{i_1} \oplus \cdots \oplus A/\mathfrak{p}^{i_s}.$$

Assume $1 \leq i_1, \ldots, i_\ell < m$ and $m \leq i_{\ell+1}, \ldots, i_s$. Note that $i_1, \ldots, i_s \leq 2m$ since $\phi[\mathfrak{p}^{2m}]$ is annihilated by $\mathfrak{p}^{2m}$. With this notation and assumptions, we have

$$\phi[a] = \phi[\mathfrak{p}^m] \cong A/\mathfrak{p}^{i_1} \oplus \cdots \oplus A/\mathfrak{p}^{i_\ell} \oplus (A/\mathfrak{p}^m)^{\oplus(s-\ell)}.$$

Hence $\#\phi[a] = |\mathfrak{p}|^{i_1 + \cdots + i_\ell + (r-s)m}$. On the other hand, $\#\phi[a] = q^{\deg_\tau \phi_a - \operatorname{ht}\phi_a}$. Since $\deg_\tau \phi_{a^2} = 2 \cdot \deg_\tau \phi_a$ and $\operatorname{ht}\phi_{a^2} = 2 \cdot \operatorname{ht}\phi_a$, we get

$$\#\phi[a^2] = q^{\deg_\tau \phi_{a^2} - \operatorname{ht}\phi_{a^2}} = (\#\phi[a])^2.$$

Thus,

$$i_1 + \cdots + i_s = 2(i_1 + \cdots + i_\ell + (r-s)m).$$

This implies that

$$i_1 + \cdots + i_\ell = (i_{\ell+1} - 2m) + \cdots + (i_s - 2m) \leq 0.$$

Hence $\ell = 0$, i.e., none of $i_1, \ldots, i_s$ is strictly less than m, so we conclude that $\phi[\mathfrak{p}^m] \cong (A/\mathfrak{p}^m)^{\oplus s}$. Finally, note that

$$\deg_\tau \phi_a = r \cdot \deg(a) = r \cdot \deg(\mathfrak{p}) \cdot m,$$

and

$$\operatorname{ht}\phi_a = \begin{cases} 0, & \text{if } \mathfrak{p} \neq \operatorname{char}_A(K); \\ H \cdot \deg(\mathfrak{p}) \cdot m, & \text{if } \mathfrak{p} = \operatorname{char}_A(K). \end{cases}$$

Hence the number of distinct roots of $\phi_a(x)$ is $|a|^r$ (respectively, $|a|^{r-H}$) if $\mathfrak{p} \neq \operatorname{char}_A(K)$ (respectively, $\mathfrak{p} = \operatorname{char}_A(K)$). This implies that $s = r$ (respectively, $s = r - H$) if $\mathfrak{p} \neq \operatorname{char}_A(K)$ (respectively, $\mathfrak{p} = \operatorname{char}_A(K)$). $\qquad\square$

Remarks A.13

(1) The previous proof is similar to the proof of Theorem 3.5.2, but we point out that the proof of Theorem 3.5.2 does not transfer to the present setting since $\phi_{\mathfrak{p}}$ and $\phi_{\mathfrak{p}^n}$ are generally not elements of $\phi(A)$, so a priori one does not know their degrees and heights.

(2) For a similar reason one has to modify the definition of the Tate module $T_{\mathfrak{p}}(\phi)$ in (3.5.2). Let $m \geq 1$ be an integer such that $\mathfrak{p}^m = (a)$ is principal. Then we have a surjective homomorphism of A-modules

$$\phi_{a^n} : \phi[\mathfrak{p}^{m(n+n')}] \to \phi[\mathfrak{p}^{mn'}]$$

for all integers $n, n' > 0$. The Tate module is then defined as

$$T_{\mathfrak{p}}(\phi) = \varprojlim_n \phi[\mathfrak{p}^{mn}],$$

where the transition morphisms are ϕ_{a^s} for all $s > 0$. Alternatively, one can take Exercise 3.5.1,

$$T_{\mathfrak{p}}(\phi) \cong \operatorname{Hom}_{A_{\mathfrak{p}}}(F_{\mathfrak{p}}/A_{\mathfrak{p}}, {}^{\phi}K^{\mathrm{sep}}),$$

as the definition of $T_{\mathfrak{p}}(\phi)$ and this makes sense regardless of whether A is a PID or not.

Corollary A.14 *Let ϕ be a Drinfeld module over K.*

(1) *The rank r of ϕ is a positive integer.*
(2) *If $\operatorname{char}_A(K) \neq 0$, then the height H of ϕ is a positive integer such that $1 \leq H \leq r$.*

Proof This is an immediate consequence of Theorem A.12. □

We leave the proof of the following corollary as an exercise:

Corollary A.15 *Let $\mathfrak{n}$ be a nonzero ideal of A, and let ϕ be a Drinfeld module over K. Then*

$$\deg_x \phi_{\mathfrak{n}}(x) = |\mathfrak{n}|^r,$$

and, assuming $\operatorname{char}_A(K) \nmid \mathfrak{n}$,

$$\phi[\mathfrak{n}] \cong (A/\mathfrak{n})^{\oplus r}.$$

Essentially the same argument as in Sect. 3.4.2 can be used to prove the following:

Theorem A.16 *Let ϕ and ψ be Drinfeld modules of rank r over K. Then $\mathrm{Hom}_K(\phi, \psi)$ is a finitely generated torsion-free A-module of rank $\leq r^2$.*

The more refined statements about $\mathrm{Hom}_K(\phi, \psi)$ and $\mathrm{End}_K(\phi)$ proved earlier in this book for $A = \mathbb{F}_q[T]$ are also true for general A; for example, $\mathrm{End}_K(\phi) \otimes_A F_\infty$ is a division algebra.

Now we briefly discuss the analytic uniformization of Drinfeld modules. Under the natural embeddings $A \hookrightarrow F \hookrightarrow F_\infty \hookrightarrow \mathbb{C}_\infty$ the image of A is discrete in $\mathbb{C}_\infty$ since for any fixed $N \geq 0$ there are only finitely many $a \in A$ with $|a| \leq N$ (this follows from the proof of Lemma 7.6.8).

As in Definition 5.2.1, an *A-lattice* in $\mathbb{C}_\infty$ is a discrete finitely generated A-submodule of $\mathbb{C}_\infty$. Note that such submodules need not be free; cf. Theorem A.9. The *rank* of a lattice Λ is its rank as an A-module.

Example A.17 Choose $v_1, \ldots, v_r \in \mathbb{C}_\infty$ and nonzero ideals $\mathfrak{n}_1, \ldots, \mathfrak{n}_r$ of A. Then one can show following the proof of Proposition 5.3.8 that $\Lambda = \mathfrak{n}_1 v_1 + \cdots + \mathfrak{n}_r v_r$ is a lattice of rank r if and only if $v_1, \ldots, v_r$ are linearly independent over F_∞.

Let Λ be a lattice of rank r. The Carlitz–Drinfeld exponential function e_Λ of Λ is an entire function; for any $a \in A$ we have the functional equation $e_\Lambda(ax) = \phi_a(e_\Lambda(x))$ for some $\phi_a(x) \in \mathbb{C}_\infty\langle x \rangle$; the map $a \mapsto \phi_a$ defines a Drinfeld module over $\mathbb{C}_\infty$ of rank r. The proofs of these statements are exactly the same as the proofs for the $A = \mathbb{F}_q[T]$ case in Chap. 5. (Note that we still have $a^{-1}\Lambda/\Lambda \cong (A/aA)^r$, even if Λ is not free, since for any ideal $\mathfrak{n}$ of A we have $a^{-1}\mathfrak{n}/\mathfrak{n} \cong A/aA$; see (7.5.10).) Conversely, given a Drinfeld module ϕ of rank r over $\mathbb{C}_\infty$ consider a power series

$$e_\phi = e_0 + e_1\tau + e_2\tau^2 + \cdots$$

with unknown coefficients e_i. Choose some $a \in A$ of positive degree. Let $\phi_a = a + g_1\tau + \cdots + g_m\tau^m$. The condition $e_\phi a = \phi_a e_\phi$ leads to the system of equations in e_i's

$$(a^{q^n} - a) \cdot e_n = e_{n-1}^q \cdot g_1 + e_{n-2}^{q^2} \cdot g_2 + \cdots + e_{n-m}^{q^m} \cdot g_m, \qquad n \geq 0,$$

where we put $e_0 = 1$ and $e_i = 0$ for $i < 0$. Clearly this system has a unique solution. We denote by $e_\phi \in \mathbb{C}_\infty\{\{\tau\}\}$ the power series determined by this system.

Lemma A.18 *The functional equation $e_\phi b = \phi_b e_\phi$ holds for all $b \in A$.*

Proof We can assume that $b \neq 0$. Then in $\mathbb{C}_\infty\{\{\tau\}\}$ we have

$$\phi_b e_\phi b^{-1} = \phi_b(\phi_a e_\phi a^{-1})b^{-1} = \phi_a(\phi_b e_\phi b^{-1})a^{-1}.$$

Hence $ea = \phi_a e$, where $e = \phi_b e_\phi b^{-1}$; note also that the constant term of e is 1. But e_ϕ is the unique power series with constant term 1 that satisfies the equation $ea = \phi_a e$, so $e_\phi = \phi_b e_\phi b^{-1}$.   □

One checks as in Sect. 5.2 that $e_\phi(x)$ is entire and that its set of zeros Λ_ϕ is a lattice giving rise to ϕ. Moreover, the same argument as in the proof of Theorem 5.2.11 (with T replaces by an arbitrary non-constant $a \in A$) shows that

$$\mathrm{Hom}_K(\phi, \psi) \cong \mathrm{Hom}(\Lambda_\phi, \Lambda_\psi).$$

Therefore, the category of Drinfeld modules over $\mathbb{C}_\infty$ is equivalent to the category of lattices. In particular, the homothety classes of lattices of rank r are in bijection with the isomorphism classes of rank-r Drinfeld modules over $\mathbb{C}_\infty$.

We conclude this appendix with a few remarks about how (and if) the other results discussed in the book transfer to the setting of general function rings:

Chapter 3:   The Weil pairing discussed in Sect. 3.7 exists for general A and in terms of Anderson motives can be defined as in Sect. 3.7.1; see [vdH04]. The discussion of the Poonen-Weil pairing in Sect. 3.7.3 also does not use the assumption $A = \mathbb{F}_q[T]$; it is defined and studied in [Poo96] for general A. On the other hand, the definition of the Weil pairing by explicit formulas is specific for $\mathbb{F}_q[T]$.

The classification of isomorphism classes of Drinfeld modules in terms of j-invariants in Sect. 3.8 is also specific to $\mathbb{F}_q[T]$. On the other hand, for general A, Drinfeld proved in [Dri74] that there is an affine variety of dimension $r - 1$ which classifies Drinfeld A-modules of rank r up to isomorphisms. This implies that there is a finite set of (non-explicit) parameters that distinguishes the isomorphism classes of Drinfeld modules over the algebraic closure of the base field. These parameters are quite complicated even for $r = 2$: in [Gek86, p. 73], Gekeler gives a formula for the genus of the modular curve of rank-2 Drinfeld A-modules and this genus grows with $\# \mathrm{Cl}(A)$. (When $A = \mathbb{F}_q[T]$, the modular curve is just the projective line over F with $j(\phi)$ being the coordinate.)

Chapter 4:   The main results of this chapter are true for general A, with essentially the same proofs. The reader might consult [Dri77, Gek91, Gek92, Yu95], [Lau96, Ch. 2], [Gos96, §4.12] for the details.

Chapter 5:   Proposition 5.5.1 is true for general A and can be proved using the same argument. The other results in Sect. 5.5 are more specific to $\mathbb{F}_q[T]$, although the key facts of Theorem 5.5.10 that $[K(\Lambda) : K]$ is unbounded but the residue degree of $K(\Lambda)$ over K remains bounded as ϕ varies over the Drinfeld modules of fixed rank should be true for general A.

Chapter 6:   The invariant (6.1.1) for general A is defined in [Dri74] as

$$e(\phi) = \min_a \ \min_{1 \le i \le r} \frac{v(g_i(a))}{q^i - 1},$$

where a runs over a set of (finitely many) generators of the ring A and $g_i(a)$'s are the coefficients of $\phi_a = \sum_{i\geq 0} g_i(a)\tau^i$. Then ϕ defined over K has potentially stable reduction over a finite extension L/K if $e(\phi) \cdot e(L/K) \in \mathbb{Z}$.

The existence of Tate uniformization for general A is proved in [Dri74, §7].

The proof of Theorem 6.3.1 works for general A, so ϕ has good reduction over K if and only if $\phi[a]$ is unramified for infinitely many $a \in A$.

The computations in Sect. 6.4 seem specific to $\mathbb{F}_q[T]$, but they can be easily extended to arbitrary A as follows. Fix a non-constant $T \in A$ and consider ϕ as an $\mathbb{F}_q[T]$-Drinfeld module of higher rank. The set $({}^\phi K)_{\text{tor}}$ is the same whether we consider ϕ as a Drinfeld module for A or $\mathbb{F}_q[T]$ (since the torsion submodule can be characterized as the set of elements which have finite orbits under iteration of ϕ_T). Thus, the estimates in Sect. 6.4 are valid for general A with r being the rank of ϕ over $\mathbb{F}_q[T]$. In particular, $({}^\phi K)_{\text{tor}}$ is finite for a local field K.

The formal Drinfeld modules were defined by Rosen [Ros03] for the ring of integers of an arbitrary local field of positive characteristic and Sect. 6.5 discusses them in that generality.

Chapter 7:   The cyclotomic theory of rank-1 Drinfeld modules (Sect. 7.1) and the theory of complex multiplication (Sect. 7.5) are essentially equivalent to each other if A is allowed to be arbitrary. The reader is encouraged to consult Hayes' beautiful article [Hay79] for the details.

The Uniform Boundedness Conjecture 7.2.2 is expected to hold for general A; in fact, it reduces to the $\mathbb{F}_q[T]$-case if the rank is allowed to increase, using the same trick as in the previous remark. But, as we mentioned, the conjecture is wide open even for $r = 2$ and $A = \mathbb{F}_q[T]$.

The computations and most of the results of Sect. 7.3 and 7.4 are specific to $\mathbb{F}_q[T]$. A notable exception is the Open Image Theorem 7.3.1, which was proved by Pink and his collaborators in the most general sense possible; see [PR09] for fields of A-characteristic 0 and [DP12] for fields of nonzero A-characteristic.

Theorem 7.6.15 was proved by Poonen for general A; the proof is essentially the same as presented in the book. Taelman initially stated and proved his class number formula [Tae12b] only for $\mathbb{F}_q[T]$. Extending his results to arbitrary A turned out to be quite complicated. This has been done by Debry [Deb16], assuming p does not divide $\#\mathrm{Cl}(A)$, by Mornev [Mor18], assuming the Drinfeld module in question has everywhere good reduction, and by Anglès, Ngo Dac and Tavares Ribeiro [ANDTR22] in full generality using their theory of Anderson-Stark units, which generalizes Anderson's theory of cyclotomic units discussed in Sect. 7.6.3. Finally, we mention that Anderson himself, inspired by some calculations of Thakur and ideas from the theory of partial differential equations, established in [And94] a version of log-algebraicity of harmonic series $\exp(-\sum_{n=1}^{\infty} x^n/n) = 1 - x$ for rank-1 Drinfeld modules for general A.

Exercises

A.1 Let $\mathfrak{p} \lhd A$ be a prime and let ϕ be a Drinfeld module over K. Then $\phi_{\mathfrak{p}}$ defines an isogeny $\phi \to \psi$. Without using Theorem A.12, show that for all n there is a short exact sequence

$$0 \longrightarrow \phi[\mathfrak{p}] \longrightarrow \phi[\mathfrak{p}^n] \xrightarrow{\phi_{\mathfrak{p}}} \psi[\mathfrak{p}^{n-1}] \longrightarrow 0.$$

A.2 Prove Corollary A.15.

A.3 Let ϕ be a Drinfeld module over $\mathbb{C}_\infty$ and let Λ be its associated lattice. Show that for any nonzero ideal $\mathfrak{n} \lhd A$ there is an isomorphism of A-modules $\phi[\mathfrak{n}] \cong \Lambda/\mathfrak{n}\Lambda$.

Appendix B
Notes on Exercises

In this appendix we provide references for some exercises that are based on published results, as well as hints and solutions for a few other selected exercises.

Chapter 1

(1.2.6) (a) We use induction on n. If $n = 2$, then by assumption $(a_1, a_2) = A$, so there are $r_1, r_2 \in A$ such that $r_1 a_1 + r_2 a_2 = 1$. In this case, $\begin{pmatrix} a_1 & a_2 \\ -r_2 & r_1 \end{pmatrix}$ is the desired matrix. Assume now that $n \geq 3$ and we have proved the statement for $n - 1$. If $a_1 = \cdots = a_{n-1} = 0$, then $a_n \in A^\times$, and we can take our matrix to be $\begin{pmatrix} 0 & a_n \\ I_{n-1} & 0 \end{pmatrix}$. Now suppose not all $a_1, \ldots, a_{n-1}$ are zero. Let d be their greatest common divisor, so that $a_1 = db_1, \ldots, a_{n-1} = db_{n-1}$ and $\gcd(b_1, \ldots, b_{n-1}) = 1$. By the induction assumption, there exists an $(n-2) \times (n-1)$ matrix B with entries in A such that $\det \begin{pmatrix} b_1 & \cdots & b_{n-1} \\ & B & \end{pmatrix} = 1$. Since a_n must be relatively prime to d, there exist $r_1, r_2 \in A$ such that $r_2 d - r_1 a_n = 1$. The matrix

$$\begin{pmatrix} db_1 & \cdots & db_{n-1} & a_n \\ & B & & 0 \\ r_1 b_1 & \cdots & r_1 b_{n-1} & r_2 \end{pmatrix}$$

has determinant $r_2 d - r_1 a_n = 1$, as can be easily seen by expanding the determinant along the last column.

(1.3.1) See Proposition 21 on page 529 and Exercise 22 on page 531 in [DF04]. We also note that the inequality $[FK : k] \leq [F : k] \cdot [K : k]$ always holds.

© The Author(s), under exclusive license to Springer Nature Switzerland AG 2023
M. Papikian, *Drinfeld Modules*, Graduate Texts in Mathematics 296,
https://doi.org/10.1007/978-3-031-19707-9

(1.3.2) Let r and s be generators of D_{2n} such that $r^n = s^2 = 1$ and $rs = sr^{-1}$. Let
F be the fixed field of the subgroup generated by s. The extension F/K has
degree n but is not Galois since $\langle s \rangle$ is not normal in D_{2n} if $n \geq 3$. Because
L/K is Galois, the primitive element theorem applies to F. Thus, $F = K(\alpha)$. Let $f(x) \in K[x]$ be the minimal polynomial of α. This polynomial
is irreducible in $K[x]$, has degree n, and does not split over F (since F/K
is not normal). On the other hand, $f(x)$ splits in L since L/K is Galois.
Therefore, L is the splitting field of $f(x)$.

(1.3.3) See [DF04, p. 647].

(1.3.4) (b) This is not easy to prove; see [Ser73, p. 77].

   (c) See [DF04, p. 647].

(1.5.2) (b) This might be false if K is finite; see [FR17, Rem. 4.3].

(1.5.5) See [DF04, pp. 650–651].

(1.6.6) (a) Observe that $\mathbb{F}_{q^n} \to \mathbb{F}_{q^n}, \alpha \mapsto \alpha^q - \alpha$, is an $\mathbb{F}_q$-linear map and compute
the dimension of its image.

   (b) Use the fact that given an integer $n \geq 1$ and a cyclic multiplicative
group G of order m generated by g, the subgroup of G generated g^n is
its unique subgroup of order $m/\gcd(m, n)$; cf. [DF04, p. 58].

(1.7.2) See [Rei03, Thm. 30.4].

(1.7.3) Suppose L is maximal and $\alpha \in \mathrm{Cent}_D(L)$. Then $L(\alpha)$ is a subfield of D
containing L, so $\alpha \in L$. Since $L \subseteq \mathrm{Cent}_D(L)$, we get $L = \mathrm{Cent}_D(L)$.
Conversely, suppose $L = \mathrm{Cent}_D(L)$ and L' is a field extension of L in D.
Then $L' \subseteq \mathrm{Cent}_D(L)$, so $L' = L$.

(1.7.4) (a) Use Theorem 1.7.10 and the fact that a finite field has a unique extension
of given degree.

   (b) Every element of D is in some maximal subfield, so this follows from
Theorem 1.7.23.

   (d) The sets $\{\alpha L^\times \alpha^{-1}\}$ are not disjoint since each contains 1.

Chapter 2

(2.1.3) See Lemma 3.3 on page 21 in [Cas86].

(2.2.4) See Lemma 1.1 on page 92 in [Cas86].

(2.2.5) See Lemma 2.1 on page 95 in [Cas86].

(2.2.6) By Exercise 2.1.3, there are $c_1, \ldots, c_n \in K$ such that $|c_i|_i > 1$ and $|c_i|_j < 1$ for $i \neq j$. Consider $\sum_{i=1}^n \frac{c_i^m}{1+c_i^m} b_i$. See the proof of Theorem 3.1 on page
22 in [Cas86] for the details.

(2.2.9) First, use the binomial theorem to show that

$$|x + y|^m \leq c(m+1) \max(|x|, |y|)^m$$

for all $m \geq 1$.

(2.2.12) See page 53 in [Cas86].
(2.4.4) See the proof of Theorem 2.6.3 (1).
(2.6.2) Use the discussion in [DF04, pp. 613–615].
(2.6.6) See Corollary 2 on page 136 in [Cas86].
(2.6.8) (b) Note that for any $\alpha \in R_L$ such that $\alpha \equiv \bar{\alpha} \pmod{\pi_L}$ we have $v_L(f(\alpha)) \geq 1$. Prove that if $v_L(f(\alpha)) \geq 2$, then $v_L(f(\alpha + \pi_L)) = 1$.
    (c) See Proposition 12 on page 57 in [Ser79]
(2.6.6) (b) Show that two Artin–Schreier extensions of K given by the polynomials $x^p - x - a$ and $x^p - x - b$ with $a, b \in K$ are isomorphic if and only if $a = \omega b + c^p - c$ for some $\omega \in \mathbb{F}_p^\times$ and $c \in K$.
(2.7.2) See Lemma 4.1 on page 60 in [Cas86].
(2.8.4) See Chapter 4 in [Mar18].
(2.8.5) Use the Primitive Element Theorem to write $\widehat{K} \cong \mathbb{F}_p[x]/(f(x))$ for some monic irreducible separable $f(x) \in \mathbb{F}_p[x]$. Then apply Exercise 2.3.3 to deduce that one can choose $f(x)$ to be in $F[x]$.
(2.8.6) Use Exercise 2.8.2 to reduce to the case when $K = \mathbb{F}_q(T)$ or $\mathbb{Q}$, where the product formula can be checked by a direct computation.

Chapter 3

(3.1.4) (a) Let $m = p^n$ be the degree of f. By definition,

$$\mathrm{disc}(f) = (-1)^{m(m-1)/2} a_n^{2m-2} \prod_{i \neq j} (\alpha_i - \alpha_j),$$

where the product is over the roots of f. The roots of f form a vector space W over $\mathbb{F}_p$. Fix some $\alpha \in W$. Then

$$\prod_{\substack{\beta \in W \\ \beta \neq \alpha}} (\alpha - \beta) = \prod_{0 \neq \gamma \in W} \gamma.$$

Hence

$$\mathrm{disc}(f) = (-1)^{m(m-1)/2} a_n^{2m-2} \left(\prod_{0 \neq \alpha \in W} \alpha \right)^m.$$

We can decompose

$$f(x) = a_n x \prod_{0 \neq \alpha \in W} (x - \alpha),$$

which shows that

$$\prod_{0\neq\alpha\in W} \alpha = a_0/a_n.$$

Combining with the previous formula for the discriminant, the formula follows.

(b) Using the product rule $(u\cdot v)' = u'v+uv'$ from calculus, observe that the *logarithmic derivative* $g'(x)/g(x)$ of any nonzero polynomial $g(x) \in K[x]$ is given by the formula

$$\frac{g'(x)}{g(x)} = \sum_{i=1}^{d} \frac{m_i}{x - z_i},$$

where $z_1,\ldots,z_d$ are the distinct roots of $g(x)$ in $\overline{K}$, and m_i is the multiplicity of z_i for $1 \leq i \leq d$.

(3.1.6) (a) The claim is equivalent to the equality

$$f_W(x) = f_{W_{n-1}}(x)^q - f_{W_{n-1}}(w_n)^{q-1}f_{W_{n-1}}(x).$$

Note that both sides are monic polynomials of the same degree. Hence, it is enough to show that any $w \in W$ is a root of $f_{W_{n-1}}(x)^q - f_{W_{n-1}}(w_n)^{q-1}f_{W_{n-1}}(x)$. We can write $w = v + aw_n$ for a uniquely determined $v \in W_{n-1}$ and $a \in \mathbb{F}_q$. Since $f_{W_{n-1}}(x)$ is $\mathbb{F}_q$-linear and vanishes at v, we have $f_{W_{n-1}}(w) = f_{W_{n-1}}(v) + af_{W_{n-1}}(w_n) = af_{W_{n-1}}(w_n)$. Since $a^q = a$, we get

$$f_{W_{n-1}}(w)^q = af_{W_{n-1}}(w_n)^q = f_V(w_n)^{q-1}f_V(w).$$

(3.1.7) Use Exercise 3.1.6.
(3.4.2) See [And86, Lem. 1.4.5].
(3.4.5) See [KP22, p. 137].
(3.4.6) See [KP22, Prop. 7.4].
(3.5.2) Show that $\mathrm{Gal}(K'/K)$ acts on $\mathrm{Hom}_{K'}(\phi, \psi)$ through a quotient which is isomorphic to a subgroup of $\Gamma(\mathfrak{p})$, and then apply Proposition 3.5.8; cf. [KP22, Prop. 8.2].
(3.5.3) $V_{\mathfrak{p}}(\phi)$ is a vector space over $F_{\mathfrak{p}}$ of dimension r equipped with an action of $L \otimes_F F_{\mathfrak{p}}$. Let $e_{\mathfrak{P}}$ be the idempotent in $L \otimes_F F_{\mathfrak{p}}$ corresponding to its direct factor $L_{\mathfrak{P}}$. Let $V_{\mathfrak{P}} = e_{\mathfrak{P}} \cdot V_{\mathfrak{p}}(\phi)$. Then

$$V_{\mathfrak{p}}(\phi) = \bigoplus_{\mathfrak{P}|\mathfrak{p}} V_{\mathfrak{P}}.$$

Note that each $V_\mathfrak{P}$ is a vector space over $L_\mathfrak{P}$. Let $d_\mathfrak{P} = \dim_{L_\mathfrak{P}} V_\mathfrak{P}$. Then

$$\sum_{\mathfrak{P}|p} d_\mathfrak{P} \cdot [L_\mathfrak{P} : F_p] = \dim_{F_p} V_p(\phi) = r,$$

and

$$\sum_{\mathfrak{P}|p} [L_\mathfrak{P} : F_p] = [L : F].$$

From these two equalities it is clear that $[L : F] \leq r$.

(3.5.5) Write $g = I + N$, where N is a nilpotent matrix. Then $N^r = 0$, since the minimal polynomial of N is x^m for some $m \leq r$. If $p^s > r$, then $g^{p^s} = (I + N)^{p^s} = I + N^{p^s} = I$. Next, consider g for which N is a single Jordan block, i.e., the entries on the diagonal and right above the diagonal of g are equal to 1, with all other entries being 0. This element has order $p^{\lceil \log_p r \rceil}$. Indeed, the order of this element is a power of p. If $p^s < r$, then $g^{p^s} = I + N^{p^s} \neq I$, since $N^{p^s} \neq 0$ (the order of N is r).

(3.5.6) See Lemma 2.1 on page 46 in [Cas86].

(3.7.2) This is Lemma 5.2 in [vdH04]. First of all, by the properties of tensor products of modules, we have an isomorphism

$$\left(\bigotimes_{\overline{K}[T]}^r \overline{M}_\phi \right) \otimes_A A/aA \cong \bigotimes_{\overline{K}[T]}^r \left(\overline{M}_\phi \otimes_A A/aA \right) \cong \bigotimes_{\overline{K}[T]}^r \overline{M}_\phi / a\overline{M}_\phi$$

of $\overline{K}[T]$-modules. Consequently, we have a canonical isomorphism of $\overline{K}\{\tau\} \otimes_{\mathbb{F}_q} A/aA$-modules

$$\overline{M}_\psi / a\overline{M}_\psi \xrightarrow{\sim} \bigwedge_{K[T]}^r \overline{M}_\phi / a\overline{M}_\phi.$$

Using Lemma 3.6.1, we can rewrite this isomorphism as

$$\left(\overline{M}_\psi / a\overline{M}_\psi \right)^\tau \otimes_{\mathbb{F}_q} \tilde{K} \xrightarrow{\sim} \bigwedge_{K[T]}^r \left(\left(\overline{M}_\phi / a\overline{M}_\phi \right)^\tau \otimes_{\mathbb{F}_q} \tilde{K} \right)$$

$$\cong \left(\bigwedge_{K[T]}^r \left(\overline{M}_\phi / a\overline{M}_\phi \right)^\tau \right) \otimes_{\mathbb{F}_q} \tilde{K}.$$

This implies the desired isomorphism.

(3.7.6) Let $n = \deg f$. Suppose $0 \neq \beta \in \ker f^*$. Then $\tau^n \cdot f^* \in K\{\tau\}$ vanishes
        at β. Extend $\{\beta\}$ to a basis of the $\mathbb{F}_q$-vector space $\ker(\tau^n \cdot f^*)$ and use
        Exercise 3.1.6 to write

$$\tau^n \cdot f^* = h \cdot (\beta^{q-1} - \tau)$$

for some $h \in K\{\tau\}$. Now multiply both sides by τ^{-n} on the left.
(3.7.7) See Theorem 15 in [Ore33] or Theorem 1.7.13 in [Gos96].
(3.7.9) See [Poo96, pp. 804–809].
(3.8.1) See [Gek08, (1.4)].
(3.8.2) See [Ham03, Example 2.2].
(3.8.3) See [Gek08, (1.6)].
(3.8.5) This exercise was suggested by an anonymous referee.

Chapter 4

(4.1.6) See [Ang96, Lem. 3.7.1].
(4.2.6) If $H(\phi) \geq 2$, then $P(x) \equiv x^{H(\phi)}g(x) \pmod{\mathfrak{p}}$ implies that the coefficient
        a_1 of $P(x) = x^r + a_{r-1}x^{r-1} + \cdots + a_1 x + a_0$ must be zero as $\deg(a_1) < d$.
        Now if $\phi[\mathfrak{l}]$ is rational, then $P(x) \equiv (x-1)^r \pmod{\mathfrak{l}}$. This implies that rx
        is 0, a contradiction.
(4.2.8) See [Ang96, Thm. 4.2].
(4.3.1) As in the proof of Corollary 4.2.4, it is enough to show that

$$\mathrm{Hom}_k(\phi, \psi) \otimes_A F_{\mathfrak{l}} \longrightarrow \mathrm{Hom}_{F_{\mathfrak{l}}[\pi]}(V_{\mathfrak{l}}(\phi), V_{\mathfrak{l}}(\psi)) \tag{B.1}$$

is an isomorphism. We proved that $V_{\mathfrak{l}}(\phi)$ is a free $F_{\mathfrak{l}}[\pi]$-module of rank
$s_\phi := r/[\widetilde{F}_\phi : F]$. Moreover, considering π as an indeterminate x, we have
an isomorphism of $F_{\mathfrak{l}}[x]$-modules

$$V_{\mathfrak{l}}(\phi) \cong (F_{\mathfrak{l}}[x]/(m_\phi(x)))^{\oplus s_\phi}.$$

It is clear that there is a nonzero homomorphism $V_{\mathfrak{l}}(\phi) \to V_{\mathfrak{l}}(\psi)$ of
$F_{\mathfrak{l}}[x]$-modules if and only if $m_\phi(x)$ and $m_\psi(x)$ have a common irreducible
factor over $F_{\mathfrak{l}}$. On the other hand, the greatest common divisor of $m_\phi(x)$
and $m_\psi(x)$ in $F[x]$ is the same as the greatest common divisor of these

polynomials in $F_1[x]$. Since $m_\phi(x)$ and $m_\psi(x)$ are irreducible in $F[x]$, we conclude that

$$\dim_{F_1} \mathrm{Hom}_{F_1[x]}(V_1(\phi), V_1(\psi))$$

$$= \begin{cases} 0 & \text{if } m_\phi(x) \neq m_\psi(x); \\ s_\phi^2 \cdot [\widetilde{F}_\phi \otimes_F F_1 : F_1] = \frac{r^2}{[\widetilde{F}_\phi : F]} & \text{if } m_\phi(x) = m_\psi(x). \end{cases}$$

Now consider $\mathrm{Hom}_k(\phi, \psi) \otimes_A F_1$. If ϕ and ψ are not isogenous, then this F_1-vector space is 0. On the other hand, if $u : \phi \to \psi$ is an isogeny defined over k, then $uw \in \mathrm{Hom}_k(\phi, \psi) \otimes_A F_1$ for any $w \in D_\phi$, so

$$\dim_{F_1} \mathrm{Hom}_k(\phi, \psi) \otimes_A F_1 = \dim_F \mathrm{Hom}_k(\phi, \psi) \otimes_A F$$

$$\geq \dim_F D_\phi$$

$$= \frac{r^2}{[\widetilde{F}_\phi : F]} \qquad \text{(Theorem 4.1.3).}$$

By Theorem 4.3.2, ϕ and ψ are isogenous over k if and only if $m_\phi(x) = m_\psi(x)$. Therefore, the previous two paragraphs imply that

$$\dim_{F_1} \mathrm{Hom}_k(\phi, \psi) \otimes_A F_1 \geq \dim_{F_1} \mathrm{Hom}_{F_1[G_k]}(V_1(\phi), V_1(\psi)).$$

Since the map (B.1) is injective by Theorem 3.5.4, the above inequality implies that (B.1) is an isomorphism.

(4.3.6) See [Yu95, Prop. 1].

(4.3.8) See [Lau96, p. 26].

(4.4.3) If $\mathfrak{p} = 1$, then ϕ is supersingular, so assume $\mathfrak{p} \neq T$. Note that $\phi_T = (\sqrt{t} + \tau)^2$, so $\mathbb{F}_q(\sqrt{T}) \subset D := \mathrm{End}(\phi)$. If ϕ is ordinary, then $\mathbb{F}_q(\sqrt{T}) = D$ and $\mathfrak{p}$ splits in $\mathbb{F}_q(\sqrt{T})$. On the other hand, if ϕ is supersingular, then D is a quaternion algebra ramified at $\mathfrak{p}$, so $\mathfrak{p}$ does not split in $\mathbb{F}_q(\sqrt{T})$. Thus, ϕ is supersingular if and only if $\mathfrak{p}$ does not split in $\mathbb{F}_q(\sqrt{T})$, i.e., $\left(\frac{T}{\mathfrak{p}}\right) = -1$. By the quadratic reciprocity, cf. (7.1.7), we have

$$\left(\frac{T}{\mathfrak{p}}\right) = (-1)^{\frac{q-1}{2}d}\left(\frac{\mathfrak{p}}{T}\right).$$

On the other hand,

$$\left(\frac{\mathfrak{p}}{T}\right) = \left(\frac{p_0}{T}\right) = p_0^{\frac{q-1}{2}}.$$

(4.4.4) See [Gek83, Thm. 5.9].

(4.4.6) (a) If $\pi \in A$, then π acts as a scalar on $\phi[\mathfrak{l}]$. If there is a rational $\mathfrak{l}$-torsion point, this scalar is 1 modulo $\mathfrak{l}$, so π acts as 1 on $\phi[\mathfrak{l}]$. This implies that all $\mathfrak{l}$-torsion points are defined over k.

(b) Let $m(x)$ be the minimal polynomial of π over $\phi(A)$. Let r_1 be the degree of $m(x)$. Write $m(x) = m_{\text{sep}}(x^{p^s})$; cf. Lemma 1.1.14. By assumption,

$$P(x) \equiv (x - a)^r \pmod{\mathfrak{l}}$$

for some $a \in A$. Since $P(x) = m(x)^{r/r_1}$, we must have

$$m(x) \equiv (x - a)^{r_1} \pmod{\mathfrak{l}}.$$

Suppose $\mathfrak{l}$ does not divide the discriminant $\text{disc}(m_{\text{sep}}(x))$. Then the reduction $\bar{m}_{\text{sep}}(x)$ of m_{sep} modulo $\mathfrak{l}$ has distinct roots in $\overline{\mathbb{F}}_{\mathfrak{l}}$,

$$\bar{m}_{\text{sep}}(x) \equiv (x - \alpha_1) \cdots (x - \alpha_{r_2}),$$

where $r_2 = r_1/p^s$ Thus,

$$\bar{m}_{\text{sep}}(x^{p^s}) \equiv ((x - \beta_1) \cdots (x - \beta_{r_2}))^{p^s},$$

where $\beta_i^{p^s} = \alpha_i$. Note that $\beta \mapsto \beta^{p^s}$ is an automorphism of $\overline{\mathbb{F}}_{\mathfrak{l}}$, so $\beta_i \neq \beta_j$. This contradicts $m(x) \equiv (x - a)^{r_1} \pmod{\mathfrak{l}}$, unless $r_2 = 1$. Thus, $\mathfrak{l}$ must divide $\text{disc}(m_{\text{sep}})$ or $r_2 = 1$. The absolute value of the roots of m_{sep} in $\mathbb{C}_\infty$ can be bounded from above in terms of $\#k$ and r. Hence, $\left| \text{disc}(m_{\text{sep}}) \right| \leq C(k, r)$ for some constant $C(k, r)$ which depends only on k and r, but not on ϕ. Since $\mathfrak{l} \mid \text{disc}(m_{\text{sep}})$, we must have $|\mathfrak{l}| = q^{\deg(\mathfrak{l})} \leq C(k, r)$. Thus, for $\deg(\mathfrak{l}) > \log_q C(k, r)$, we must have $m_{\text{sep}}(x) = x - b$ for some $b \in A$. Therefore, $m(x) = x^{p^s} - b$ is purely inseparable or $s = 0$. In either case, there is a unique place in $\widetilde{F}$ over $\mathfrak{p}$, so ϕ is supersingular.

Chapter 5

(5.1.1) Follow the proof of Theorem 2.7.10, except that both in the statement of the theorem, and throughout the proof, one must be careful with the order of the factors as we are working with a non-commutative ring. See also [Ros03, Thm. 3.2].

(5.1.5) See [Gos80, p. 408].

(5.3.3) Let $n = r/[L : F]$. Choose $\omega_2, \ldots, \omega_n \in \mathbb{C}_\infty$ so that they are algebraically independent over F_∞. Consider $\Lambda = O + O\omega_2 + \cdots O\omega_n$.

(5.3.6) See [BBP21, Ch. 3].

(5.4.7) See [Gek88, (11.4)].

(5.4.11) See [Tha04, pp. 162–163].

(5.5.2) To show that $|\mu_2|$ uniquely determines $|\lambda_2|$, note that the function

$$x \prod_{\substack{\lambda \in \Lambda \\ |\lambda| \le x}}' \frac{x}{|\lambda|}$$

is continuous on $\mathbb{R}_{\ge 0}$ and strictly increasing. Hence, for a given $|\mu_2|$, Eq. (5.5.5) has a unique solution $|\lambda_2|$.

(5.5.4) This problem was suggested by Florian Breuer.

(5.5.6) See [Gek19b, Cor. 3.6].

(5.5.7) See [Gek19b, Cor. 3.2].

(5.5.11) First, show that the extension $F_\infty(\Lambda)/F_\infty$ is unramified if and only if $\log_q |\lambda_1|$ and $\log_q |\lambda_2|$ are both integers.

(5.5.14) See Proposition 4.4 and Lemma 4.6 in [Tag93], and Lemma 4.1 and Lemma 4.2 in [BPR21].

(5.5.15) This problem was suggested by Florian Breuer. See Lemma 4.2 in [BPR21].

Chapter 6

(6.1.7) See [KP22, Prop. 7.1]. (b) Show that for $\alpha \in \phi[\mathfrak{n}]$ we have $\overline{u(\alpha)} = \bar{u}(\bar\alpha) = 0$, so $v(u(\alpha)) > 0$. Deduce from this that $u(\alpha) = 0$. Hence $\phi[\mathfrak{n}] \subset \ker(u)$. Apply the division algorithm to deduce that $u = w\phi_\mathfrak{n}$.

(6.1.8) This exercise was suggested by Florian Breuer.

(6.2.2) Use the fact that the intersection of all prime ideals of B is the nilradical of B; see [DF04, p. 674].

(6.3.3) See [Tak82].

(6.4.3) (a) Use induction.

(b) Show that the canonical homomorphism $R \to R/M^m$ is injective on $({}^\phi R)_{\mathrm{tor}}$ if $m = \left\lfloor \frac{v(\mathfrak{p})}{q-1} \right\rfloor + 1$.

(6.4.5) Use the Tate uniformization of Drinfeld modules; cf. [Poo97, Prop. 2].

(6.5.1) Note that $\widehat{\phi}_{\varpi^n}(\mathcal{M}) \subseteq \mathcal{M}^{n+1}$ for any $n \ge 1$.

(6.5.3) See [Ros03, Lem. 1.3].

(6.5.4) Use Exercise 5.1.1. See [Ros03, Thm. 3.3] for the details.

Chapter 7

(7.1.4) Note that an element of the form $1 + a\mathfrak{p}^m$, $a \in A$ and $s/2 \le m < s$, has order p in $(A/\mathfrak{p}^s)^\times$.

(7.1.10) (d) Note that the action of the Galois group preserves the absolute values
of the roots of $C_\mathfrak{n}(x)$, so any $\sigma \in \mathrm{Gal}(F_\infty(C[\mathfrak{n}])/F_\infty)$ maps $\zeta_\mathfrak{n}$ to some
$\alpha\zeta_\mathfrak{n}, \alpha \in \mathbb{F}_q^\times$.

(7.2.2) See [Sch03, Prop. 3.4].

(7.2.3) See [Sch03, Prop. 4.4].

(7.2.4) See [Poo97, Thm. 3 and 4].

(7.3.1) This exercise was suggested by Chien-Hua Chen.

(7.3.4) Consider the Carlitz cyclotomic extension $F(C[(T-a)(T-b)])/F)$ and
apply the Chebotarev density theorem.

(7.3.7) Note that $\mathrm{GL}_2(\mathbb{F}_2) \cong S_3$, whose only nontrivial subgroups are
$\mathbb{Z}/2\mathbb{Z}, \mathbb{Z}/3\mathbb{Z}$, and S_3. If $\mathrm{Gal}(K/F) \cong \mathbb{Z}/3\mathbb{Z}$ or S_3, then K is the splitting
field of an irreducible cubic. Since the characteristic is not 3, after an
appropriate change of variables we may assume that the cubic is given
by $x^3 + ax + b$ for some $a, b \in F$. Then the corresponding linearized
polynomial is $f(x) = x^4 + ax^2 + bx$.

(7.5.7) (a) Show that the conductor of $A[\pi_k]$ in B is not divisible by $\mathfrak{p}$, where
$\pi_k = \tau^{[k:\mathbb{F}_q]}$ is the Frobenius endomorphism of $\bar{\phi}$.

(7.5.8) See [Hay91, Prop. 2].

(7.5.9) See [Hay91, Example 2].

(7.6.4) Use (1) and (2) in Proposition 7.6.12.

(7.6.5) See [Den92, Thm. 1].

(7.6.6) See [Den92, Cor. 2].

(7.6.7) See Propositions 1 and 6 in [Poo95].

(7.6.8) See pages 419 and 420 in [Sil09].

(7.6.12) See [And96].

References

[Abh94]   Shreeram S. Abhyankar, *Nice equations for nice groups*, Israel J. Math. **88** (1994), no. 1–3, 1–23.

[Abh01]   ———, *Resolution of singularities and modular Galois theory*, Bull. Amer. Math. Soc. (N.S.) **38** (2001), no. 2, 131–169.

[And86]   Greg W. Anderson, *t-motives*, Duke Math. J. **53** (1986), no. 2, 457–502.

[And94]   ———, *Rank one elliptic A-modules and A-harmonic series*, Duke Math. J. **73** (1994), no. 3, 491–542.

[And96]   ———, *Log-algebraicity of twisted A-harmonic series and special values of L-series in characteristic p*, J. Number Theory **60** (1996), no. 1, 165–209.

[ANDTR20]   Bruno Anglès, Tuan Ngo Dac, and Floric Tavares Ribeiro, *Recent developments in the theory of Anderson modules*, Acta Math. Vietnam. **45** (2020), no. 1, 199–216.

[ANDTR22]   ———, *A class formula for admissible Anderson modules*, Invent. Math. **229** (2022), no. 2, 563–606.

[AnglesMM97] Bruno Anglès, *On some characteristic polynomials attached to finite Drinfeld modules*. Manuscripta Math., **93** (1997), no. 3, 369–379.

[Ang96]   Bruno Anglès, *On some subrings of Ore polynomials connected with finite Drinfeld modules*, J. Algebra **181** (1996), no. 2, 507–522.

[Ang01]   ———, *On Gekeler's conjecture for function fields*, J. Number Theory **87** (2001), no. 2, 242–252.

[Apo76]   Tom M. Apostol, *Introduction to analytic number theory*, Undergraduate Texts in Mathematics, Springer-Verlag, New York-Heidelberg, 1976.

[Arm12]   Cécile Armana, *Torsion des modules de Drinfeld de rang 2 et formes modulaires de Drinfeld*, Algebra Number Theory **6** (2012), no. 6, 1239–1288.

[AT90]   Greg W. Anderson and Dinesh S. Thakur, *Tensor powers of the Carlitz module and zeta values*, Ann. of Math. (2) **132** (1990), no. 1, 159–191.

[AT13]   Bruno Anglès and Lenny Taelman, *On a problem à la Kummer-Vandiver for function fields*, J. Number Theory **133** (2013), no. 3, 830–841.

[AT15]   ———, *Arithmetic of characteristic p special L-values*, Proc. Lond. Math. Soc. (3) **110** (2015), no. 4, 1000–1032, With an appendix by Vincent Bosser.

[Ax70]   James Ax, *Zeros of polynomials over local fields—The Galois action*, J. Algebra **15** (1970), 417–428.

[BB18]   Alp Bassa and Peter Beelen, *On the Deuring polynomial for Drinfeld modules in Legendre form*, Acta Arith. **186** (2018), no. 2, 179–190.

[BBP21]   Dirk Basson, Florian Breuer, and Richard Pink, *Drinfeld modular forms of arbitrary rank*, preprint (2021).

© The Author(s), under exclusive license to Springer Nature Switzerland AG 2023
M. Papikian, *Drinfeld Modules*, Graduate Texts in Mathematics 296,
https://doi.org/10.1007/978-3-031-19707-9

[BK95]   Sunghan Bae and Ja Kyung Koo, *Torsion points of Drinfeld modules*, Canad. Math. Bull. **38** (1995), no. 1, 3–10.

[BO00]   Nigel Boston and David T. Ose, *Characteristic p Galois representations that arise from Drinfeld modules*, Canad. Math. Bull. **43** (2000), no. 3, 282–293.

[Boy99]   P. Boyer, *Mauvaise réduction des variétés de Drinfeld et correspondance de Langlands locale*, Invent. Math. **138** (1999), no. 3, 573–629.

[BPR21]   Florian Breuer, Fabien Pazuki, and Mahefason Heriniaina Razafinjatovo, *Heights and isogenies of Drinfeld modules*, Acta Arith. **197** (2021), no. 2, 111–128.

[BR86]   Mauro Beltrametti and Lorenzo Robbiano, *Introduction to the theory of weighted projective spaces*, Exposition. Math. **4** (1986), no. 2, 111–162.

[BR16]   Florian Breuer and Hans-Georg Rück, *Drinfeld modular polynomials in higher rank II: Kronecker congruences*, J. Number Theory **165** (2016), 1–14.

[Bre16]   Florian Breuer, *Explicit Drinfeld moduli schemes and Abhyankar's generalized iteration conjecture*, J. Number Theory **160** (2016), 432–450.

[Bro92]   M. L. Brown, *Singular moduli and supersingular moduli of Drinfeld modules*, Invent. Math. **110** (1992), no. 2, 419–439.

[BS66]   A. I. Borevich and I. R. Shafarevich, *Number theory*, Pure and Applied Mathematics, Vol. 20, Academic Press, New York-London, 1966, Translated from the Russian by Newcomb Greenleaf.

[Car35]   Leonard Carlitz, *On certain functions connected with polynomials in a Galois field*, Duke Math. J. **1** (1935), no. 2, 137–168.

[Car37]   _____, *An analogue of the von Staudt-Clausen theorem*, Duke Math. J. **3** (1937), no. 3, 503–517.

[Car38]   _____, *A class of polynomials*, Trans. Amer. Math. Soc. **43** (1938), no. 2, 167–182.

[Car40]   L. Carlitz, *An analogue of the Staudt-Clausen theorem*, Duke Math. J. **7** (1940), 62–67.

[Cas86]   J. W. S. Cassels, *Local fields*, London Mathematical Society Student Texts, vol. 3, Cambridge University Press, Cambridge, 1986.

[Che22a]   Chien-Hua Chen, *Exceptional cases of adelic surjectivity for Drinfeld modules of rank 2*, Acta Arith. **202** (2022), no. 4, 361–377.

[Che22b]   _____, *On surjectivity of adelic Galois representation for Drinfeld modules*, Ph.D. thesis, Pennsylvania State University, 2022.

[CL13]   Imin Chen and Yoonjin Lee, *Newton polygons, successive minima, and different bounds for Drinfeld modules of rank 2*, Proc. Amer. Math. Soc. **141** (2013), no. 1, 83–91.

[Coh95]   P. M. Cohn, *Skew fields*, Encyclopedia of Mathematics and its Applications, vol. 57, Cambridge University Press, Cambridge, 1995, Theory of general division rings.

[Con]   Keith Conrad, *Expository papers*, https://kconrad.math.uconn.edu/blurbs/.

[Cox89]   David A. Cox, *Primes of the form $x^2 + ny^2$*, A Wiley-Interscience Publication, John Wiley & Sons, Inc., New York, 1989, Fermat, class field theory and complex multiplication.

[CP12]   Chieh-Yu Chang and Matthew A. Papanikolas, *Algebraic independence of periods and logarithms of Drinfeld modules*, J. Amer. Math. Soc. **25** (2012), no. 1, 123–150, With an appendix by Brian Conrad.

[CP15]   Alina Carmen Cojocaru and Mihran Papikian, *Drinfeld modules, Frobenius endomorphisms, and CM-liftings*, Int. Math. Res. Not. IMRN (2015), no. 17, 7787–7825.

[Deb16]   Christophe Debry, *Towards a class number formula for Drinfeld modules*, Ph.D. thesis, University of Amsterdam, 2016.

[Den92]   Laurent Denis, *Hauteurs canoniques et modules de Drinfeld*, Math. Ann. **294** (1992), no. 2, 213–223.

[DF04]   David S. Dummit and Richard M. Foote, *Abstract algebra*, third ed., John Wiley
            & Sons, Inc., Hoboken, NJ, 2004.

[DH87]   Pierre Deligne and Dale Husemoller, *Survey of Drinfeld modules*, Current trends
            in arithmetical algebraic geometry (Arcata, Calif., 1985), Contemp. Math., vol. 67,
            Amer. Math. Soc., Providence, RI, 1987, pp. 25–91.

[Dor91]   David R. Dorman, *On singular moduli for rank 2 Drinfeld modules*, Compositio
            Math. **80** (1991), no. 3, 235–256.

[DP12]   Anna Devic and Richard Pink, *Adelic openness for Drinfeld modules in special
            characteristic*, J. Number Theory **132** (2012), no. 7, 1583–1625.

[Dri74]   V. G. Drinfeld, *Elliptic modules*, Mat. Sb. (N.S.) **94** (1974), 594–627.

[Dri77]   ———, *Elliptic modules. II*, Mat. Sb. (N.S.) **102(144)** (1977), no. 2, 182–194,
            325.

[Dri80]   ———, *Langlands' conjecture for GL(2) over functional fields*, Proceedings
            of the International Congress of Mathematicians (Helsinki, 1978), Acad. Sci.
            Fennica, Helsinki, 1980, pp. 565–574.

[DT02]   W. Duke and Á. Tóth, *The splitting of primes in division fields of elliptic curves*,
            Experiment. Math. **11** (2002), no. 4, 555–565 (2003).

[Elk99]   Noam D. Elkies, *Linearized algebra and finite groups of Lie type. I. Linear and
            symplectic groups*, Applications of curves over finite fields (Seattle, WA, 1997),
            Contemp. Math., vol. 245, Amer. Math. Soc., Providence, RI, 1999, pp. 77–107.

[Fla53]   Harley Flanders, *The norm function of an algebraic field extension*, Pacific J. Math.
            3 (1953), 103–113.

[Fla55]   ———, *The norm function of an algebraic field extension. II*, Pacific J. Math. **5**
            (1955), 519–528.

[FR17]   Uriya A. First and Zinovy Reichstein, *On the number of generators of an algebra*,
            C. R. Math. Acad. Sci. Paris **355** (2017), no. 1, 5–9.

[FvdP04]   Jean Fresnel and Marius van der Put, *Rigid analytic geometry and its applications*,
            Progress in Mathematics, vol. 218, Birkhäuser Boston, Inc., Boston, MA, 2004.

[GaRo81]   Steven Galovich and Michael Rosen, *The class number of cyclotomic function
            fields*, J. Number Theory **13** (1981), no. 3, 363–375.

[Gar02]   Francis Gardeyn, *Une borne pour l'action de l'inertie sauvage sur la torsion d'un
            module de Drinfeld*, Arch. Math. (Basel) **79** (2002), no. 4, 241–251.

[Gek83]   Ernst-Ulrich Gekeler, *Zur Arithmetik von Drinfeld-Moduln*, Math. Ann. **262**
            (1983), no. 2, 167–182.

[Gek86]   ———, *Drinfeld modular curves*, Lecture Notes in Mathematics, vol. 1231,
            Springer-Verlag, Berlin, 1986.

[Gek88]   ———, *On the coefficients of Drinfeld modular forms*, Invent. Math. **93** (1988),
            no. 3, 667–700.

[Gek91]   ———, *On finite Drinfeld modules*, J. Algebra **141** (1991), no. 1, 187–203.

[Gek92]   ———, *On the arithmetic of some division algebras*, Comment. Math. Helv. **67**
            (1992), no. 2, 316–333.

[Gek08]   ———, *Frobenius distributions of Drinfeld modules over finite fields*, Trans.
            Amer. Math. Soc. **360** (2008), no. 4, 1695–1721.

[Gek16]   ———, *The Galois image of twisted Carlitz modules*, J. Number Theory **163**
            (2016), 316–330.

[Gek17]   ———, *On Drinfeld modular forms of higher rank*, J. Théor. Nombres Bordeaux
            29 (2017), no. 3, 875–902.

[Gek19a]   ———, *On the field generated by the periods of a Drinfeld module*, Arch. Math.
            (Basel) **113** (2019), no. 6, 581–591.

[Gek19b]   ———, *Towers of GL(r)-type of modular curves*, J. Reine Angew. Math. **754**
            (2019), 87–141.

[Gol82]   Dorian Goldfeld, *Sur les produits partiels eulériens attachés aux courbes ellip-
            tiques*, C. R. Acad. Sci. Paris Sér. I Math. **294** (1982), no. 14, 471–474.

[Gos78]   David Goss, *von Staudt for* $\mathbf{F}_q[T]$, Duke Math. J. **45** (1978), no. 4, 885–910.

[Gos79]   _____, *v-adic zeta functions, L-series and measures for function fields*, Invent. Math. **55** (1979), no. 2, 107–119, With an addendum.

[Gos80]   _____, *The algebraist's upper half-plane*, Bull. Amer. Math. Soc. (N.S.) **2** (1980), no. 3, 391–415.

[Gos96]   _____, *Basic structures of function field arithmetic*, Ergebnisse der Mathematik und ihrer Grenzgebiete (3) [Results in Mathematics and Related Areas (3)], vol. 35, Springer-Verlag, Berlin, 1996.

[Gou20]   Fernando Q. Gouvêa, *p-adic numbers*, third ed., Universitext, Springer, Cham, [2020] ©2020, An introduction.

[GP20]   Sumita Garai and Mihran Papikian, *Endomorphism rings of reductions of Drinfeld modules*, J. Number Theory **212** (2020), 18–39.

[GP22]   _____, *Computing endomorphism rings and Frobenius matrices of Drinfeld modules*, J. Number Theory **237** (2022), 145–164.

[GvdPRVG97] E.-U. Gekeler, M. van der Put, M. Reversat, and J. Van Geel (eds.), *Drinfeld modules, modular schemes and applications*, World Scientific Publishing Co., Inc., River Edge, NJ, 1997.

[Ham93]   Yoshinori Hamahata, *Tensor products of Drinfeld modules and v-adic representations*, Manuscripta Math. **79** (1993), no. 3–4, 307–327.

[Ham03]   _____, *The values of J-invariants for Drinfeld modules*, Manuscripta Math. **112** (2003), no. 1, 93–108.

[Hay74]   D. R. Hayes, *Explicit class field theory for rational function fields*, Trans. Amer. Math. Soc. **189** (1974), 77–91.

[Hay79]   David R. Hayes, *Explicit class field theory in global function fields*, Studies in algebra and number theory, Adv. in Math. Suppl. Stud., vol. 6, Academic Press, New York-London, 1979, pp. 173–217.

[Hay91]   _____, *On the reduction of rank-one Drinfeld modules*, Math. Comp. **57** (1991), no. 195, 339–349.

[HK21]   Everett W. Howe and Kiran S. Kedlaya, *Every positive integer is the order of an ordinary abelian variety over* $\mathbb{F}_2$, Res. Number Theory. **7** (2021), no. 4, 59.

[HY00]   Liang-Chung Hsia and Jing Yu, *On characteristic polynomials of geometric Frobenius associated to Drinfeld modules*, Compositio Math. **122** (2000), no. 3, 261–280.

[Kat21]   Jeffrey Katen, *Explicit Weil-pairing for Drinfeld modules*, Int. J. Number Theory **17** (2021), no. 9, 2131–2152.

[Ked17]   Kiran S. Kedlaya, *On the algebraicity of generalized power series*, Beitr. Algebra Geom. **58** (2017), no. 3, 499–527.

[Kob84]   Neal Koblitz, *p-adic numbers, p-adic analysis, and zeta-functions*, second ed., Graduate Texts in Mathematics, vol. 58, Springer-Verlag, New York, 1984.

[KP22]   Nikolas Kuhn and Richard Pink, *Finding endomorphisms of Drinfeld modules*, J. Number Theory **232** (2022), 118–154.

[Laf02]   Laurent Lafforgue, *Chtoucas de Drinfeld et correspondance de Langlands*, Invent. Math. **147** (2002), no. 1, 1–241.

[Laf18]   Vincent Lafforgue, *Chtoucas pour les groupes réductifs et paramétrisation de Langlands globale*, J. Amer. Math. Soc. **31** (2018), no. 3, 719–891.

[Lan56]   Serge Lang, *Algebraic groups over finite fields*, Amer. J. Math. **78** (1956), 555–563.

[Lan02]   _____, *Algebra*, third ed., Graduate Texts in Mathematics, vol. 211, Springer-Verlag, New York, 2002.

[Lau96]   Gérard Laumon, *Cohomology of Drinfeld modular varieties. Part I*, Cambridge Studies in Advanced Mathematics, vol. 41, Cambridge University Press, Cambridge, 1996, Geometry, counting of points and local harmonic analysis.

[Lau97]   _____, *Cohomology of Drinfeld modular varieties. Part II*, Cambridge Studies in Advanced Mathematics, vol. 56, Cambridge University Press, Cambridge, 1997, Automorphic forms, trace formulas and Langlands correspondence, With an appendix by Jean-Loup Waldspurger.

[Leh09]   Thomas Lehmkuhl, *Compactification of the Drinfeld modular surfaces*, Mem. Amer. Math. Soc. **197** (2009), no. 921, xii+94.

[LN94]   Rudolf Lidl and Harald Niederreiter, *Introduction to finite fields and their applications*, first ed., Cambridge University Press, Cambridge, 1994.

[Lor96]   Dino Lorenzini, *An invitation to arithmetic geometry*, Graduate Studies in Mathematics, vol. 9, American Mathematical Society, Providence, RI, 1996.

[LRS93]   G. Laumon, M. Rapoport, and U. Stuhler, *$\mathscr{D}$-elliptic sheaves and the Langlands correspondence*, Invent. Math. **113** (1993), no. 2, 217–338.

[LT65]   Jonathan Lubin and John Tate, *Formal complex multiplication in local fields*, Ann. of Math. (2) **81** (1965), 380–387.

[Mac71]   R. E. MacRae, *On unique factorization in certain rings of algebraic functions*, J. Algebra **17** (1971), 243–261.

[Man69]   Ju. I. Manin, *The p-torsion of elliptic curves is uniformly bounded*, Izv. Akad. Nauk SSSR Ser. Mat. **33** (1969), 459–465.

[Mar18]   Daniel A. Marcus, *Number fields*, Universitext, Springer, Cham, 2018, Second edition of [MR0457396], With a foreword by Barry Mazur.

[Mar22]   Stefano Marseglia, *Cohen-Macaulay type of orders, generators and ideal classes*, arXiv 2206.03758 [math.AC] (2022).

[Mat89]   Hideyuki Matsumura, *Commutative ring theory*, second ed., Cambridge Studies in Advanced Mathematics, vol. 8, Cambridge University Press, Cambridge, 1989, Translated from the Japanese by M. Reid.

[Mau19]   Andreas Maurischat, *On field extensions given by periods of Drinfeld modules*, Arch. Math. (Basel) **113** (2019), no. 3, 247–254.

[Maz77]   B. Mazur, *Modular curves and the Eisenstein ideal*, Inst. Hautes Études Sci. Publ. Math. (1977), no. 47, 33–186 (1978), With an appendix by Mazur and M. Rapoport.

[Maz78]   ———, *Rational isogenies of prime degree (with an appendix by D. Goldfeld)*, Invent. Math. **44** (1978), no. 2, 129–162.

[Mer96]   Loïc Merel, *Bornes pour la torsion des courbes elliptiques sur les corps de nombres*, Invent. Math. **124** (1996), no. 1–3, 437–449.

[Mil86]   J. S. Milne, *Abelian varieties*, Arithmetic geometry (Storrs, Conn., 1984), Springer, New York, 1986, pp. 103–150.

[Mil12a]   James S. Milne, *Fields and Galois theory (v4.30)*, 2012, Available at www.jmilne.org/, p. 124.

[Mil12b]   ———, *Motives – Grothendieck's dream*, available at www.jmilne.org.

[Mil13]   ———, *Class field theory (v4.02)*, 2013, Available at www.jmilne.org, pp. 281+viii.

[Mor18]   Maxim Mornev, *Shtuka cohomology and special values of Goss l-functions*, Ph.D. thesis, University of Amsterdam, 2018.

[Mum70]   David Mumford, *Abelian varieties*, Tata Institute of Fundamental Research Studies in Mathematics, vol. 5, Published for the Tata Institute of Fundamental Research, Bombay; Oxford University Press, London, 1970.

[Oka91]   Shozo Okada, *Kummer's theory for function fields*, J. Number Theory **38** (1991), no. 2, 212–215.

[Ore33]   Oystein Ore, *On a special class of polynomials*, Trans. Amer. Math. Soc. **35** (1933), no. 3, 559–584.

[Ouk20]   Hassan Oukhaba, *On local fields generated by division values of formal Drinfeld modules*, Glasg. Math. J. **62** (2020), no. 2, 459–472.

[Pál10]   Ambrus Pál, *On the torsion of Drinfeld modules of rank two*, J. Reine Angew. Math. **640** (2010), 1–45.

[Pap22]   Matthew A. Papanikolas, *Hyperderivative power sums, Vandermonde matrices, and Carlitz multiplication coefficients*, J. Number Theory **232** (2022), 317–354.

[Pin06]   Richard Pink, *The Galois representations associated to a Drinfeld module in special characteristic. II. Openness*, J. Number Theory **116** (2006), no. 2, 348–372.

[Poo95]   Bjorn Poonen, *Local height functions and the Mordell-Weil theorem for Drinfeld modules*, Compositio Math. **97** (1995), no. 3, 349–368.

[Poo96]   _____, *Fractional power series and pairings on Drinfeld modules*, J. Amer. Math. Soc. **9** (1996), no. 3, 783–812.

[Poo97]   _____, *Torsion in rank 1 Drinfeld modules and the uniform boundedness conjecture*, Math. Ann. **308** (1997), no. 4, 571–586.

[Poo98]   _____, *Drinfeld modules with no supersingular primes*, Internat. Math. Res. Notices (1998), no. 3, 151–159.

[Pot98]   Igor Yu. Potemine, *Minimal terminal Q-factorial models of Drinfeld coarse moduli schemes*, Math. Phys. Anal. Geom. **1** (1998), no. 2, 171–191.

[PPVW19]   Jennifer Park, Bjorn Poonen, John Voight, and Melanie Matchett Wood, *A heuristic for boundedness of ranks of elliptic curves*, J. Eur. Math. Soc. (JEMS) **21** (2019), no. 9, 2859–2903.

[PR03]   Matthew A. Papanikolas and Niranjan Ramachandran, *A Weil-Barsotti formula for Drinfeld modules*, J. Number Theory **98** (2003), no. 2, 407–431.

[PR09]   Richard Pink and Egon Rütsche, *Adelic openness for Drinfeld modules in generic characteristic*, J. Number Theory **129** (2009), no. 4, 882–907.

[Rei03]   I. Reiner, *Maximal orders*, London Mathematical Society Monographs. New Series, vol. 28, The Clarendon Press, Oxford University Press, Oxford, 2003, Corrected reprint of the 1975 original, With a foreword by M. J. Taylor.

[Rob00]   Alain M. Robert, *A course in p-adic analysis*, Graduate Texts in Mathematics, vol. 198, Springer-Verlag, New York, 2000.

[Ros87]   Michael Rosen, *The Hilbert class field in function fields*, Exposition. Math. **5** (1987), no. 4, 365–378.

[Ros02]   _____, *Number theory in function fields*, Graduate Texts in Mathematics, vol. 210, Springer-Verlag, New York, 2002.

[Ros03]   _____, *Formal Drinfeld modules*, J. Number Theory **103** (2003), no. 2, 234–256.

[Sch95]   Andreas Schweizer, *On the Drinfeld modular polynomial $\Phi_T(X, Y)$*, J. Number Theory **52** (1995), no. 1, 53–68.

[Sch97]   _____, *On singular and supersingular invariants of Drinfeld modules*, Ann. Fac. Sci. Toulouse Math. (6) **6** (1997), no. 2, 319–334.

[Sch03]   _____, *On the uniform boundedness conjecture for Drinfeld modules*, Math. Z. **244** (2003), no. 3, 601–614.

[Ser72]   Jean-Pierre Serre, *Propriétés galoisiennes des points d'ordre fini des courbes elliptiques*, Invent. Math. **15** (1972), no. 4, 259–331.

[Ser73]   _____, *A course in arithmetic*, Springer-Verlag, New York-Heidelberg, 1973, Translated from the French, Graduate Texts in Mathematics, No. 7.

[Ser77]   _____, *Linear representations of finite groups*, Springer-Verlag, New York-Heidelberg, 1977, Translated from the second French edition by Leonard L. Scott, Graduate Texts in Mathematics, Vol. 42.

[Ser79]   _____, *Local fields*, Graduate Texts in Mathematics, vol. 67, Springer-Verlag, New York-Berlin, 1979, Translated from the French by Marvin Jay Greenberg.

[Ser80]   _____, *Extensions icosaédriques*, Seminar on Number Theory, 1979–1980 (French), Univ. Bordeaux I, Talence, 1980, pp. Exp. No. 19, 7.

[She98]   Jeffrey T. Sheats, *The Riemann hypothesis for the Goss zeta function for $\mathbf{F}_q[T]$*, J. Number Theory **71** (1998), no. 1, 121–157.

[Shi75]   Goro Shimura, *On the real points of an arithmetic quotient of a bounded symmetric domain*, Math. Ann. **215** (1975), 135–164.

[Sie32]   C. L. Siegel, *Über die Perioden elliptischer Funktionen*, J. Reine Angew. Math. **167** (1932), 62–69.

[Sil86]   Joseph H. Silverman, *The arithmetic of elliptic curves*, Graduate Texts in Mathematics, vol. 106, Springer-Verlag, New York, 1986.

[Sil94]   _____, *Advanced topics in the arithmetic of elliptic curves*, Graduate Texts in Mathematics, vol. 151, Springer-Verlag, New York, 1994.

[Sil09]   ——, *The arithmetic of elliptic curves*, second ed., Graduate Texts in Mathematics, vol. 106, Springer, Dordrecht, 2009.

[SS15]   Qibin Shen and Shuhui Shi, *Function fields of class number one*, J. Number Theory **154** (2015), 375–379.

[Tae10]   Lenny Taelman, *A Dirichlet unit theorem for Drinfeld modules*, Math. Ann. **348** (2010), no. 4, 899–907.

[Tae11]   ——, *The Carlitz shtuka*, J. Number Theory **131** (2011), no. 3, 410–418.

[Tae12a]   ——, *A Herbrand-Ribet theorem for function fields*, Invent. Math. **188** (2012), no. 2, 253–275.

[Tae12b]   ——, *Special L-values of Drinfeld modules*, Ann. of Math. (2) **175** (2012), no. 1, 369–391.

[Tag92]   Yuichiro Taguchi, *Ramifications arising from Drinfeld modules*, The arithmetic of function fields (Columbus, OH, 1991), Ohio State Univ. Math. Res. Inst. Publ., vol. 2, de Gruyter, Berlin, 1992, pp. 171–187.

[Tag93]   ——, *Semi-simplicity of the Galois representations attached to Drinfeld modules over fields of "infinite characteristics"*, J. Number Theory **44** (1993), no. 3, 292–314.

[Tag95a]   ——, *A duality for finite t-modules*, J. Math. Sci. Univ. Tokyo **2** (1995), no. 3, 563–588.

[Tag95b]   ——, *The Tate conjecture for t-motives*, Proc. Amer. Math. Soc. **123** (1995), no. 11, 3285–3287.

[Tak82]   Toyofumi Takahashi, *Good reduction of elliptic modules*, J. Math. Soc. Japan **34** (1982), no. 3, 475–487.

[Tha04]   Dinesh S. Thakur, *Function field arithmetic*, World Scientific Publishing Co., Inc., River Edge, NJ, 2004.

[vBCL$^+$21]   Raymond van Bommel, Edgar Costa, Wanlin Li, Bjorn Poonen, and Alexander Smith, *Abelian varieties of prescribed order over finite fields*, arXiv 2106.13651 [math.NT] (2021).

[vdH04]   Gert-Jan van der Heiden, *Weil pairing for Drinfeld modules*, Monatsh. Math. **143** (2004), no. 2, 115–143.

[Vig80]   Marie-France Vignéras, *Arithmétique des algèbres de quaternions*, Lecture Notes in Mathematics, vol. 800, Springer, Berlin, 1980.

[VS06]   Gabriel Daniel Villa Salvador, *Topics in the theory of algebraic function fields*, Mathematics: Theory & Applications, Birkhäuser Boston, Inc., Boston, MA, 2006.

[Wad41]   L. I. Wade, *Certain quantities transcendental over $GF(p^n, x)$*, Duke Math. J. **8** (1941), 701–720.

[Was82]   Lawrence C. Washington, *Introduction to cyclotomic fields*, Graduate Texts in Mathematics, vol. 83, Springer-Verlag, New York, 1982.

[Wat79]   William C. Waterhouse, *Introduction to affine group schemes*, Graduate Texts in Mathematics, vol. 66, Springer-Verlag, New York-Berlin, 1979.

[Wil06]   Andrew Wiles, *The Birch and Swinnerton-Dyer conjecture*, The millennium prize problems, Clay Math. Inst., Cambridge, MA, 2006, pp. 31–41.

[Wym72]   B. F. Wyman, *What is a reciprocity law?*, Amer. Math. Monthly **79** (1972), 571–586; correction, ibid. 80 (1973), 281.

[Yu86]   Jing Yu, *Transcendence and Drinfeld modules*, Invent. Math. **83** (1986), no. 3, 507–517.

[Yu91]   ——, *Transcendence and special zeta values in characteristic p*, Ann. of Math. (2) **134** (1991), no. 1, 1–23.

[Yu95]   Jiu-Kang Yu, *Isogenies of Drinfeld modules over finite fields*, J. Number Theory **54** (1995), no. 1, 161–171.

[YZ17]   Zhiwei Yun and Wei Zhang, *Shtukas and the Taylor expansion of L-functions*, Ann. of Math. (2) **186** (2017), no. 3, 767–911.

[Zyw11]   David Zywina, *Drinfeld modules with maximal Galois action on their torsion points*, arXiv:1110.4365 [math.NT] (2011).

Index

Printed in the United States
by Baker & Taylor Publisher Services